T0201618

RSMeans
ESTIMATING
HANDBOOK

THIRD EDITION

RSMeans

R.S. Means Company, Inc.
A division of Reed Construction Data
Construction Publishers & Consultants
63 Smiths Lane
Kingston, MA 02364-3008
781-422-5000
www.rsmeans.com

Managing editor: Andrea Sillah. Editor: Jessica deMartin.
Production manager: Michael Kokernak. Production coordinator: Jill Goodman.
Composition: Sheryl Rose. Proofreaders: Jill Goodman and Mary Lou Geary.
Technical support: Sharon Proulx. Cover design: Paul Robertson.

Printed in the United States of America

SKY10047003_050323

Library of Congress Catalog Number pending

ISBN 978-0-87629-273-0

Table of Contents

Foreword

How This Book Is Arranged

This book provides technical data required to prepare quantity takeoffs and complete construction cost estimates for major construction projects. It will help design and construction professionals compare the cost of design alternatives, perform cost analysis and value engineering, and review estimate quotes and change orders prepared by others. The book includes information on sizing, productivity, equipment requirements, design standards, and engineering factors— all organized according to the latest 2004 CSI MasterFormat classification system.

This edition has been completely updated with new and expanded technical information and arranged according to the newest MasterFormat system for ease of use. Added content covers a range of construction specialties.

RSMeans Estimating Handbook, Third Edition, was reviewed and updated by the RSMeans engineering staff and an independent reviewer. (See "About the Editors" and the "Acknowledgments" for more about the contributors to this publication.)

The CSI MasterFormat Divisions:

Division 01 — General Requirements
Division 02 — Existing Conditions
Division 03 — Concrete
Division 04 — Masonry
Division 05 — Metals
Division 06 — Wood, Plastics & Composites
Division 07 — Thermal & Moisture Protection
Division 08 — Openings
Division 09 — Finishes
Division 10 — Specialties
Division 11 — Equipment
Division 12 — Furnishings
Division 13 — Special Construction
Division 14 — Conveying Equipment
Division 21 — Fire Suppression
Division 22 — Plumbing
Division 23 — Heating, Ventilating, & Air Conditioning
Division 25 — Integrated Automation
Division 26 — Electrical
Division 27 — Communications
Division 28 — Electronic Safety & Security
Division 31 — Earthwork

Division 32 — Exterior Improvements
Division 33 — Utilities
Division 34 — Transportation
Division 35 — Waterway and Marine
Division 40 — Process Integration
Division 41 — Material Processing & Handling Equipment
Division 42 — Process Heating, Cooling & Drying Equipment
Division 43 — Process Gas & Liquid Handling, Purification & Storage Equipment
Division 44 — Pollution Control Equipment
Division 45 — Industry-Specific Manufacturing Equipment
Division 48 — Electrical Power Generation

Each division of this book is divided into four sections: an introduction, estimating tips, a checklist, and estimating data section.

- The introduction provides an overview of what is included in each division.
- The tips provide shortcuts, hints on how to avoid pitfalls, and other handy estimating information.
- The checklists will help to make sure that all items are included in an estimate.
- The estimating data, the main section of each division, includes numerous charts and tables with information and assistance for all aspects of construction estimating.

The appendices of the book include the Historical Cost Indexes and City Cost Indexes, in addition to abbreviations, conversion tables, equivalents, formulas, and other general information useful when preparing construction estimates. The comprehensive index can be used to pinpoint any specific topic.

About the Editors

The following individuals were instrumental in shaping the content of this new third edition.

Barbara Balboni is a senior engineer with RSMeans and consults with clients in both the government and corporate sectors. Specializing in custom cost modeling, she has developed a methodology for benchmark studies for major builders involved in multi-year phased projects. Ms. Balboni is lead engineer for the U.S. Department of Energy's electronic cost modeling system and for the Bureau of Labor Statistic's new construction cost indices. She serves as senior editor for *RSMeans Square Foot Costs, RSMeans Assemblies Cost Data,* and *RSMeans Interior Cost Data,* and as technical editor for several RSMeans reference publications. In addition, Ms. Balboni is a training instructor, responsible for teaching Unit Price, Plan Reading and Material Takeoff, and Square Foot Estimating. Her past experiences include project manager for Blanchard Architectural Associates and The McKenna Group. Ms. Balboni earned a degree in engineering, with highest honors from Wentworth Institute of Technology. She is a member of the American Institute of Architects and the Boston Society of Architects.

John Chiang, PE, is a senior electrical engineer with RSMeans with expertise in the residential, commercial, and industrial construction markets with over twenty-five years of cost engineering experience in CSI Divisions 26, 27, and 28, Electrical, Communications, and Safety and Security. Recent consulting projects include value engineering data for Southwire, the TRACES cost estimating database, the Bureau of Labor Statistics cost index, and code research for Security Industrial Association. He implements "time and motion" studies for building product manufacturers, which combines with an RSMeans cost analysis of installation and provides input to pricing models and business strategy.

Mr. Chiang is a registered engineer in Texas and holds an MS in industrial engineering from the University of Texas, and a BS in electronics engineering. He is a senior member of IEEE, and a member of the National Society of Professional Engineers as well as the Massachusetts Society of Professional Engineers.

David G. Drain, PE, is an engineer with RSMeans and has technical responsibility for the research and publishing of cost data for Division 33 – Utilities. He also contributes to Division 44 – Pollution Control Equipment, grey water, and water reuse items. Mr. Drain also works on consulting with clients in the government and corporate sectors including the U.S. Army Corps of Engineers, Hunt Building Company, Dow, and BASF. He has worked at RSMeans since 2006. Before joining RSMeans he held positions with Stone & Webster, Charles T. Main, Sverdrup, Massachusetts Water Resources Authority, Perot Systems Government

Services, and Stenbeck & Taylor. Mr. Drain holds a BS in civil engineering from Lafayette College, Easton, Pennsylvania. He is a licensed professional engineer, certified soil evaluator, and certified system inspector. In addition he has had specialized training in toxics use reduction and risk assessment for water utilities.

Robert Kuchta is an engineer with RSMeans. He holds a BS in civil engineering from Wentworth Institute of Technology. Mr. Kuchta is responsible for the technical editing of Division 8 – Openings in the RSMeans cost database and is senior editor of *RSMeans Light Commercial Cost Data*. He has been a training instructor since 2003. He served in the U.S. Navy as a CONUS Seabees operations chief, and has worked with Ralph M. Parsons, Brown & Root Government Services, and The Shaw Group/Beneco Enterprises. He is a licensed construction supervisor in Massachusetts and a surveyor in training.

Joe Macaluso, a certified cost consultant and member of the Association for the Advancement of Cost Engineering International, currently chairs the Government and Public Works Special Interest Group and has been awarded the Technical Excellence Award for 2008. He also chairs the New York Interagency Engineering Council. Mr. Macaluso is currently the construction cost estimator for the Empire State Development Corporation, an agency that provides financial and technical assistance for projects throughout New York State. Over the past twenty-three years he has prepared and reviewed construction cost estimates, written budgets and schedules for major public works projects, negotiated change orders, and worked in the field of contract administration. In addition, he has taught cost estimating at Long Island University and the City University of New York.

Robert W. Mewis, CCC, is a senior engineer/editor with RSMeans. His background includes hands-on construction experience, professional estimating, and project management. Mr. Mewis graduated from Fitchburg State College with a degree in industrial education, and attained a certificate of professional achievement in construction supervision from Northeastern University. He has worked for construction consulting firms, general contractors, and subcontractors, performing construction estimating from the conceptual stage to hard bidding. Mr. Mewis holds the designation of certified cost consultant from the Association for the Advancement of Cost Engineering and is a member of the American Society of Professional Estimators.

Mel Mossman, PE, is the chief mechanical engineer at RSMeans. He has been senior editor of *RSMeans Mechanical Cost Data, RSMeans Plumbing Cost Data,* and *RSMeans Facilities Construction Cost Data* for more than thirty-four years. He is a contributing editor on all other RSMeans cost products; is responsible for the development, maintenance, and update of the mechanical database; responds to all mechanical technical cost inquiries; and provides technical support on consulting projects. A registered professional engineer, Mr. Mossman has taught the RSMeans Mechanical/Electrical Cost Estimating seminar for many years and has also lectured at ASHRAE events.

Stephen Plotner is a senior engineer with RSMeans and has technical responsibility for the research and publishing of cost data for Division 3 – Concrete and for Division 5 – Metals. He also has technical responsibility as senior editor for RSMeans cost data titles including the *RSMeans Facilities Construction Cost Data, RSMeans Facilities Maintenance and Repair Cost Data,* and *RSMeans Concrete & Masonry Cost Data.* Mr. Plotner also consults with clients in the government and corporate sectors including the U.S. Army Corps of Engineers, Hunt Building Company, Clark Realty Builders, Forest City, the Bureau of Indian Affairs, and the U.S. Department of Energy. Mr. Plotner also teaches RSMeans seminars around the world on such subjects as construction cost estimating and facilities maintenance. Joining RSMeans in 1996, Mr. Plotner formerly held positions with Payless Cashways Building Materials (Northeast regional facilities administrator), Financial Construction Inc. (owner and president), Financial Building Corp (construction project manager and estimator), and Veitas & Veitas Engineers (structural engineer, construction superintendent). Mr. Plotner holds a BS in civil engineering from Northeastern University, Boston.

Eugene Spencer is a senior engineer with RSMeans. He consults with clients in both government and private sectors. He has over thirty years of experience in private and government construction and facilities maintenance on complex projects such as hospitals, medical research facilities, high-rise buildings, and utilities installation. Mr. Spencer was the lead engineer in the development of a facilities maintenance assessment tool for General Motors plants throughout the U.S. He has participated in developing a model to predict flood damage on highways for the U.S. Army Corps of Engineers. Mr. Spencer also served as project manager for Daniel O'Connell Sons Construction, and he has held positions at Bechtel, Morrison Knudsen, Turner Construction and the U.S. Navy's CEC Corp. He earned a BS in Civil Engineering at Arizona State University. Mr. Spencer also teaches RSmeans classes for mechanical/electrical estimating, site work estimating, and facilities maintenance estimating.

Phillip R. Waier, PE, joined RSMeans twenty years ago as a principal engineer. Mr. Waier specializes in advising clients in government and legal sectors. He is currently program manager for the U.S. Army Corps of Engineers' electronic global cost estimating database and lead consultant for Pentagon and General Services Administration engagements. As a commercial cost construction authority, he examines material and local cost factors and recently completed a study to determine national asphalt and cement usage for the Portland Cement Association. He has technical responsibility for RSMeans cost estimating data guides, including *RSMeans Building Construction Cost Data* and *RSMeans Facilities Maintenance & Repair Cost Data.* Mr. Waier serves as an expert witness in class action suits involving commercial and residential building products. He previously held engineering and management positions with Mobil Oil Corporation, Janus Incorporated, and Charles T. Main/Metcalf and Eddy, Inc. Mr. Waier earned an MS degree in Civil Engineering at Villanova University and is a registered professional engineer in Massachusetts and Rhode Island.

Acknowledgments

The editors would like to express our appreciation to the following individuals who reviewed and provided valuable input for this book in its earlier editions.

Robert Cox, PhD, is the associate director of and an associate professor at the M.E. Rinker School of Building Construction at the University of Florida. He currently teaches cost analysis and construction project simulation and has taught construction management courses in planning and scheduling, estimating, and productivity improvement. Dr. Cox's industry experience includes professional estimating, project engineering and management, and business development.

Roy Gilley, AIA, PE, is principal architect at Gilley Design Associates. He was a partner at Gilley-Hinkel Architects for twenty years, after having served as staff architect for firms in Connecticut and the Washington, D.C., area. His architectural and design projects range from single-family residences to multi-million dollar commercial facilities including high-rise housing, schools, shopping centers, hospitals and nursing homes, warehouses, theaters, and research laboratories. Among Mr. Gilley's specialties are master planning, building inspection and evaluation, feasibility investigation and planning, programming, architectural and interior design, historic preservation, contract documents, specifications, project coordination, contract administration, and cost estimating.

Jeff Goldman has more than twenty years of experience in estimating construction costs and bidding projects from both the contractor's and owner's perspectives, on commercial, industrial, and retail projects. He currently manages the estimating department for the construction and real estate divisions of Shaws Supermarkets, and maintains his own cost consulting firm in Duxbury, Massachusetts. Mr. Goldman has worked in cost estimating and consulting as chief estimator for Tishman Construction, engineer/editor for RSMeans project manager/senior estimator for Barr and Barr, senior estimator for Spaulding and Slye/Colliers International, senior estimator in charge of Changes and Claims for Bechtel/Parsons Brinckerhoff on the Central Artery Project, and lead estimator for The Austin Company. His experience has enabled him to develop many cost models for conceptual estimating, a primary part of his own cost consulting business.

Kenneth K. Humphreys, PE, CCE, is the secretary/treasurer of the International Cost Engineering Council. He served as executive director of the Association for the Advancement of Cost Engineering from 1971 to 1992. He currently practices as a consulting engineer. He is a noted authority on the subject of cost engineering, and is the author of several major books in the field. His career includes more than twenty-five years of industrial and academic experience in the steel and coal industries as a research engineer with the United States Steel Corporation, as an Associate Director of the Coal Research Bureau of the State of West Virginia, and as an assistant dean of the College of Mineral and Energy Resources, West Virginia University.

Patricia Jackson, PE, is president and owner of Jackson A&E Associates, Inc., a Texas architectural and engineering consulting firm that offers owners representation and program management for the construction of new facilities, building expansion, and renovation projects nationwide. She is a degreed architect and licensed structural engineer with twenty-five years of experience in the construction industry. Prior to her work with Jackson A&E, Ms. Jackson was vice president of project management for Aguirre Corporation of Dallas, Texas; Director of engineering operations and development for RSMeans and manager of technical programs for the Construction Specifications Institute, among other positions.

Martin F. Joyce is executive vice president of Bond Brothers, a Massachusetts contracting and construction management firm that concentrates in both the building and the civil and utility fields. Mr. Joyce has served in several supervisory, estimating, and management positions during his more than thirty-year career with Bond Brothers, and has managed a wide range of construction projects.

Thomas Sinacore is a manager in Greyhawk North America's project controls group and is responsible for assignments in its scheduling and cost estimating practice. He is experienced in project management, planning, scheduling, estimating, and project delay and disruption claims analysis. His cost estimating work includes quantity surveys, change order reviews, design development budgets, and construction estimates. He is a certified cost engineer and is skilled in economic analysis, financial modeling, forecasting, and statistical and linear programming applications. Greyhawk is a construction management and consulting firm providing services in diverse practice areas and in regions throughout the United States.

Joseph E. Wallwork, also of Greyhawk North America, is an expert in construction scheduling, cost estimating, and construction management. He has more than twenty years of experience in all aspects of design and construction. His past positions include manager of estimating and scheduling for the headquarters office of a major national engineering, design, and construction management firm; project manager and estimator for a national public works contractor; and resident engineer in charge of construction for power and wastewater facilities. Mr. Wallwork, as a contractor, had hands-on experience taking a project from the bidding stage through construction planning, scheduling, management, and completion.

We are also grateful to the following organizations, which have granted permission to reproduce specific charts or text from their publications:

American Institute of Architects

American Institute of Steel Construction

Heating, Refrigerating, and Air Conditioning Engineers

Anvil International (formerly Grinnell Corporation)

Asphalt Roofing Manufacturers Association

The Association for the Advancement of Cost Engineering

Brick Industry Association

Caterpillar Tractor Company

Concrete Reinforcing Steel Institute

Craftsman Book Company

Gypsum Association

National Landscape Association

Occupational Safety & Health Administration (OSHA)

Painting & Decorating Contractors of America

Peckham Industries, Inc.

Rinker Materials Corporation

Steel Door Institute

U.S. Gypsum Corporation

Western Wood Products Association

Introduction
Estimating Review

Estimating Review

The purpose of this book is to provide estimators with reliable information to assist in determining quantities of materials, as well as labor required for construction projects. Construction estimating can be separated into two basic components: how many units are required and reasonable costs for those units. The determination of quantities is commonly called the *quantity takeoff*. It is, simply, the tabulation of the physical units of materials needed for the construction of the project. While this may sound like a straightforward task, any estimator knows that the process can be quite involved. For example, when estimating a concrete footing (excavation and backfill not included), the items necessary include formwork (forms, ties or spreaders, keyway, inserts, form oil, delivery, erecting, stripping, and cleaning); reinforcing (delivery, cutting, bending, ties, splices, chairs, and other accessories); and the concrete itself (placing, compacting, finishing, curing, protecting, and patching). A working knowledge of construction materials and methods is a must for a successful estimator. This knowledge helps to ensure that all items are accounted for and tabulated correctly—a sound basis for a good estimate.

The determination of "reasonable" costs for tabulated quantities is one of the main reasons why estimates vary. What may be reasonable for one job may be unrealistic for another. Labor rates (and productivity) will vary from region to region. Costs for the same materials also vary—from city to city and even from supplier to supplier within the same town. The experienced estimator evaluates each project individually, investigates fluctuations in costs, and uses that knowledge as an advantage.

Estimating for building construction is certainly not as simple a task as many people believe. Before one can obtain a quantity and apply a cost per unit to that quantity, many hours, days, or even weeks of hard work may be put into a detailed project before arriving at that one "magic number."

Types of Estimates

Several different levels of estimates are used to project construction costs. Each has a different purpose. The various types may be referred to by different names, some of which may not be recognized by all as necessary or definitive. Most estimators will agree to several basic levels, each of which has its place in the construction estimating process.

Figure 0-1 indicates various types, or levels, of estimates defined by national industry organizations, including suggested ranges of estimating accuracy.

Figure 0-1   Comparison of Estimate Classification Systems

ANSI Standards Z94.0	AACE Classification Standard	ASPE Classification Standard	Means Estimate Categories
Order of Magnitude Estimate	Class 5	Level 1	Order of Magnitude
Budget Estimate	Class 4	Level 2	Square Foot/Cubic Foot
	Class 3	Level 3	Assembly (System)
Definitive Estimate	Class 2	Level 4	Unit Price
		Level 5	
	Class 1	Level 6	

RSMeans, through its publications and seminars, has traditionally presented the following four levels of estimates. They are described in further detail below.

Order of Magnitude Estimates: The order of magnitude estimate could be loosely described as an educated guess. It can be completed in a matter of minutes. Accuracy is –30% to +50%.

Square Foot and Cubic Foot Estimates: This type of estimate is most often used when only the proposed size and use of a planned building is known. Accuracy is –20% to +30%.

Assemblies (Systems) Estimate: An assemblies estimate is best used as a budgetary tool in the planning stages of a project. Accuracy is expected at –10% to +20%.

Unit Price Estimate: Working plans and full specifications are required to complete a unit price estimate. It is the most accurate of the four types, but is also the most time-consuming. Used primarily for bidding purposes, accuracy is –5% to +10%.

Order of Magnitude Estimates: The order of magnitude estimate can be completed when only minimal information is available. The use and size of the proposed structure should be known. The "units" can be general. For example: "A small office building for a service company in a suburban industrial park will cost about $1,000,000." This type of statement (or estimate) can be made after a few minutes of thought, drawing on experience, and by making comparisons to similar projects. End product units and scaling factor estimating methods may also be used. While this rough figure might be appropriate for a project in one region of the country, an adjustment may be required for a change of location and for cost changes over time (price changes, inflation, etc.).

Square Foot and Cubic Foot Estimates: The use of these types of estimates is most appropriate prior to the preparation of plans or preliminary drawings, when budgetary parameters are being analyzed and established. Costs may be broken down into different components or elements, and then according to the relationship of each component to the project as a whole, in terms of costs per square foot. This breakdown enables the designer, planner, or estimator to adjust certain components according to the unique requirements of the proposed project.

Historical data for square foot costs of new construction are plentiful. However, the best source of square foot costs is the estimator's own cost records for similar projects, adjusted to the parameters of the current project. While helpful for preparing preliminary budgets, square foot and cubic foot estimates can also be useful as checks against other, more detailed estimates. While slightly more time is required than with order of magnitude estimates, a greater accuracy is achieved due to more specific definition of the project.

A variation of the square foot and cubic foot methods is the "end units" method. Instead of expressing costs in dollars per square foot or dollars per cubic foot, costs are expressed in units related to the particular type of project, such as dollars per bed for hospitals, per student for schools, or per car for parking garages.

Exponent Estimating Technique *(courtesy of Martin Kahn)*
The cost of similar facilities or components of different sizes varies with the size raised to a power (often 0.6 relationship is refered to as the six-tenths rule)

$$\frac{C_1}{C_2} = \left(\frac{Q_1}{Q_2}\right)^x$$

where
    C = Cost
    Q = Size or capacity
    X = Exponent

Assemblies (or Systems) Estimates: Ever-increasing design and construction costs make budgeting and cost efficiency increasingly important in the early stages of building projects. Unit price estimating, because of the time and detailed information required, is not suited as a budgetary or planning tool. A faster and more cost-effective method for the planning phase of a building project is the "assemblies," or "systems" estimate.

The assemblies method is a logical, sequential approach that reflects how a building is constructed. Seven "UNIFORMAT II" divisions organize building construction into major components that can be used in assemblies estimates. These UNIFORMAT II divisions are listed below.

A — Substructure
B — Shell
C — Interiors
D — Services
E — Equipment & Furnishings

F — Special Construction
G — Building Site Work

Each division is further broken down into individual assemblies. Each individual assembly incorporates several different items into a system that is commonly used in building construction.

In the assemblies format, a construction component may appear within more than one division. For example, concrete is found in Substructure, as well as in Shell and Equipment. Conversely, each division may incorporate many different areas of construction and the labor of different trades.

A great advantage of the assemblies estimate is that the estimator/designer is able to substitute one system for another during design development and can quickly determine the cost differential. The owner can then anticipate accurate budgetary requirements before final details and dimensions are established.

Unlike unit price estimates, the assemblies method does not require final design details, but the estimators who use it must have a solid background knowledge of construction materials and methods, code requirements, design options, and budgetary restrictions.

The assemblies estimate should not be used as a substitute for the unit price estimate. While the assemblies approach can be an invaluable tool in the planning stages of a project, it should be supported by unit price estimating when greater accuracy is required.

Unit Price Estimates

The unit price estimate is the most accurate and detailed of the four estimate types, and therefore takes the most time to complete. Detailed plans and specifications must be available to the unit price estimator to determine the quantities of materials, equipment, and labor. Current and accurate costs for these items (unit prices) are also necessary. All decisions regarding the building's materials and methods must have been made before this type of estimate can be completed. There are fewer variables, and the estimate can, therefore, be more accurate.

Because of the detail involved and the need for accuracy, unit price estimates require a great deal of time and expense to complete properly. For this reason, unit price estimating is often used for construction bidding. It can also be effective for determining certain detailed costs in conceptual budgets or during design development.

Most construction specification manuals and cost reference books, such as *RSMeans Building Construction Cost Data*, divide all unit price estimating information into the 49* CSI MasterFormat divisions as described and listed in the Foreword of this book.

*Some numbers are reserved for future use.

Before Starting the Estimate

In recent years, plans and specifications have become massive volumes containing a wealth of information. It is of utmost importance that the estimator read all contract documents thoroughly. These documents exist to protect all parties involved in the construction process. The contract documents are prepared so that the estimators will be bidding equally and competitively, ensuring that all items in a project are included. The contract documents protect the designer (the architect or engineer) by ensuring that all work is supplied and installed as specified. The owner also benefits from thorough and complete construction documents, being guaranteed a measure of quality control and a complete job. Finally, the contractor benefits because the scope of work is well defined, eliminating the gray areas of what is implied, but not stated. "Extras" are more readily avoided. Change orders, if required, are accepted with less argument if the original contract documents are complete, well stated, and most important, read by all concerned parties.

During the first review of the specifications, all items to be estimated should be identified and noted. The General Conditions, Supplemental Conditions, and Special Conditions sections of the specifications should be examined carefully. These sections describe the items that have a direct bearing on the proposed project, but may not be part of the actual, physical construction. An office trailer, temporary utilities, and testing are examples.

While analyzing the plans and specifications, the estimator should evaluate the different portions of the project to determine which areas warrant the most attention. For example, if a building is to have a steel framework with a glass and aluminum frame skin, then more time should be spent estimating Division 05–Metals and Division 08–Doors & Windows, than Division 06–Wood & Plastics.

The estimator should always visit the project site to ascertain the overall scope and to address specific conditions that may not be apparent from review of the plans. Items to look for include site access, proximity to resources including utilities, and adequacy of space for storage and equipment.

Figures 0-2 and 0-3 are charts showing the relative percentage of different construction components by MasterFormat Division and Assemblies (UNIFORMAT II) Division, respectively. These charts have been developed to represent the average percentages for new construction as a whole. All commonly used building types are included. The estimator should determine, for a given project, the relative proportions of each component, and estimating time should be allocated accordingly. More time and care should be given to estimating those areas that contribute more to the cost of the project.

Figure 0-2 Cost Distribution by MasterFormat Division

NO.	DIVISION	%
015433	CONTRACTOR EQUIPMENT*	6.6%
3120,3130	Earth Moving & Earthwork	4.1
3160,3170	LB Elements & Tunneling	0.1
3210	Bases, Ballasts & Paving	0.3
3310,3330,3340,3350	Utility Services & Drainage	0.2
3230	Site Improvements	0.1
3290	Planting	0.6
0241,31-34	SITE & INFRASTRUCTURE, DEMO	5.4
0310	Conc. Forming & Accessories	2.8
0320	Concrete Reinforcing	2.3
0330	Cast-in-Place Concrete	5.2
0340	Precast Concrete	2.1
03	CONCRETE	12.4
0405	Basic Masonry Mat. & Methods	0.6
0420	Unit Masonry	6.0
0440	Stone Assemblies	0.4
04	MASONRY	7.0

NO.	DIVISION	%
0510	Structural Metal Framing	3.9%
0520,0530	Metal Joists & Decking	9.2
0550	Metal Fabrications	1.0
05	METALS	14.1
0610	Rough Carpentry	0.7
0620	Finish Carpentry	0.3
06	WOOD & PLASTICS & Composites	1.0
0710	Dampproofing & Waterproofing	0.1
0720, 0780	Thermal, Fire & Smoke Protection	1.4
0740, 0750	Roofing & Siding	1.2
0760	Flashing & Sheet Metal	0.3
07	THERMAL & MOISTURE PROT.	3.0
0810, 0830	Doors & Frames	6.1
0840, 0880	Glazing & Curtain Walls	1.0
08	Openings	7.1

NO.	DIVISION	%
0920	Plaster & Gypsum Board	1.9%
0930, 0966	Tile & Terrazzo	2.3
0950, 0980	Ceilings & Acoustical Treatment	2.9
0960	Flooring	1.9
0970, 0990	Wall Finishes & Painting/Coating	0.9
09	FINISHES	9.9
Covers	Divs 10-14, 25,28,41,43,44	6.0
2210, 2230, 2240, 2320	Piping, Pumps, Plumbing Equipment	13.4
2113	Fire Suppression Sprinkler Systems	3.0
2350	Central Heating Equipment	2.6
2330, 2340, 2360, 2370, 2380	Air Conditioning & Ventilation	3.1
21, 22, 23	FIRE SUPPRESS., PLUMB. & HVAC	22.1
26, 27, 3370	ELECTRICAL, COMMUN. & UTIL.	12.0
	TOTAL (Div. 1-16)	100.0%

* Percentage for contractor equipment is spread among divisions and included above for information only

Figure 0-3 *Cost Distribution by UNIFORMAT II Division (Assemblies)*

Division No.	Building System	Percentage		Division No.	Building System	Percentage
A	Substructure	6.3%		D10	Services: Conveying	3.9%
B10	Shell: Superstructure	19.8		D20-40	Mechanical	22.1
B20	Exterior Closure	11.5		D50	Electrical	12.0
B30	Roofing	2.9		E10	Equipment & Furnishings	2.1
C	Interior Construction	15.5		G	Site Work	3.9
					Total (Div. A-G)	100.0%

The Quantity Takeoff

The quantity takeoff can be thought of as two processes: quantifying and tabulating. In the quantifying process, materials, and work items are scaled, counted, and calculated. In the tabulating process, the resultant quantities are tabulated in a way that costs can be assigned to them.

Quantifying can be performed manually, electronically assisted, or digitally. The manual method is done with pencil and paper, architect's scales, and calculators. The electronically assisted method typically employs the use of a device called a *digitizer*. The digitizer allows the quantity surveyor to save time because the scaling operation is much more efficient. Using a pointing device, the quantity surveyor simply points to the beginning and end points of a line on a plan, and it is automatically scaled. Similarly, square footages of areas can be determined, and items counted. The digital method is used with CAD (digitally produced) plans or BIM (building information modeling) based projects. No paper is required since CAD plans are viewed on a computer monitor, and dimensions can be determined by highlighting items or through an electronically generated report. With the necessary programming, not only quantities, but descriptions, labor-hours, and raw costs, can be linked to items on the plan. This information can be used to start a quantity takeoff and estimate within the software, or exported into a separate estimating software package. The estimating information must be modified utilizing the estimator's knowledge and experience.

For the tabulation process, the resultant calculations from any of the quantification methods are transferred, tabulated, and stored—either manually on paper (columnar sheets or preprinted forms) or electronically (using a spreadsheet or database program). An advantage of spreadsheet and database programs is that quantification and tabulation are performed concurrently. In addition, multiple codes can be assigned to each takeoff item so that they can be easily sorted or filtered by project phase, location, expected order of work, or any user-defined criteria.

When working with the plans during the quantity takeoff, consistency is the most important consideration. If each job is approached in the same manner, a pattern will develop, such as moving from the lower floors to the top, clockwise, or counterclockwise. Consistency is more important than the method used and helps to avoid duplications, omissions, and errors. Preprinted forms and spreadsheet templates create documents that provide an excellent means for developing consistent patterns.

Figure 0-4 is an example of a pre-printed form. Preprinted forms can also be used as guides for creating customized spreadsheet templates. Templates are created ahead of time by the user and include all key labels, formulas, and formatting. Only the variables need to be input when estimates are produced.

Figure 0-4 Quantity Sheet

RSMeans Forms

QUANTITY SHEET

		SHEET NO.		
PROJECT		ESTIMATE NO.		
LOCATION	ARCHITECT	DATE		
TAKE OFF BY	EXTENSIONS BY:	CHECKED BY		

DESCRIPTION	NO.	DIMENSIONS		UNIT	UNIT	UNIT	UNIT

RSMeans

General Rules for the Quantity Takeoff

General rules have been established for improving the speed, ease, and accuracy of the takeoff process. Quantity estimators should adhere to these rules. No estimate will be reliable if a mistake is made in the quantity takeoff, no matter how precise the unit price information may be.

General Rule 1

When taking off quantities, follow the guidelines provided by the person who will be applying unit prices to the quantities (the cost estimator). The takeoff should be clear and informative in order to prevent misinterpretation. Use symbols, sketches, or footnotes to clarify ambiguities in the takeoff. The quantity estimator should be thought of as an assistant to the cost estimator.

General Rule 2

A takeoff list is not just a list of materials, but a list of measurements separated into categories to which unit prices are applied. When working with spreadsheets, entering quantities should be made as simple as possible. (See Figure 0-4.)
The name of the building component is entered in the far left-hand column, labeled "Description." The number of components called out on the plans is listed next, followed by their dimensions (such as length, width, and depth or height). Quantities of items that are taken off for the component are listed in the subsequent columns. When taking off strip footings, for example, the associated items include structural excavation, concrete, formwork, backfill, and disposal. Appropriate units of measure, such as cubic yards of concrete, square feet of forms, and linear feet of pour strips, are applied to each item.

If more than one building component is entered, or if there are several different sizes of the same component, then quantities are listed in the appropriate columns and totaled. In this way, the quantity estimator can calculate, for instance, the total number of cubic yards of concrete needed for strip footings for the entire building and present it in one sum at the bottom of the column labeled "Concrete."

General Rule 3

A quantity estimate is not intended for use in direct purchase of materials. In fact, many items on a quantity estimate have no material value. These are called *work items* and are simply areas that require labor, such as fine grading or finishing concrete surfaces. Work items may not appear on the plans, but are nonetheless required to complete the job. The quantity estimator should pay close attention to areas that may contain labor requirements. Items that do have material value are called *material items*. Both material items and work items are assembled on the same form for eventual pricing out, or cost estimating.

Any item that has a cost value should be assigned a unit of measure, even if it is only in lump sum (LS) form. The term "lump sum" is used for certain work items that cannot be measured or expressed in any other way. "LS" calls the estimator's attention to an item that requires a cost allowance based on judgment.

General Rule 4

A quantity estimator may begin the takeoff with any building component and proceed in the order of his/her choice. A good approach is to follow roughly the order of the actual field construction, such as from the footings upward to the roof. This provides the quantity estimator with the clearest mental picture of the project. If a project consists of more than one building, each structure should be taken off separately, since unit costs may vary from structure to structure.

General Rule 5

Check the plan, drawings, and details carefully for changes in scale, plans reduced to one-half or one-quarter their original size, notes such as "NTS" (Not To Scale), or discrepancies between the specifications and the plans. Be consistent when listing dimensions.

General Rule 6

Where possible, use the dimensions stated in the plans instead of measuring by scale, but make a habit of frequently checking printed dimensions with a scale or with mental arithmetic to spot drafters' errors. Always express dimensions in the same order, such as:

> length × width × height (or depth)

General Rule 7

Use a systematic procedure when working with plans. For instance, take measurements in a clockwise direction around a floor or roof plan, first recording the measurements of items displayed horizontally on the plans, and then recording those shown vertically. This method is most useful when taking off two-dimensional areas.

General Rule 8

Whenever possible, the items in a quantity takeoff should be identified by their location on the plans.

General Rule 9

Use decimals in quantity takeoffs instead of fractions, because they are faster, more precise, and easy to use on a calculator or spreadsheet. Plan dimensions that are given in feet and inches are converted to decimal feet, that is, feet and tenths of a foot.

General Rule 10

Since estimating is not an exact science, the lists of quantities need not be overly precise. An example of unnecessary precision is using dimensions to the 1/8" calculating excavation quantities. However, a reasonable degree of precision is expected. No detailed estimator wants to be accused of ballpark estimating.

In most cases, the use of two decimal places is sufficient for quantity surveying purposes (12' 4-1/2" = 12.38') and easy to enter into a calculator or computer. However, when writing the product of the calculation, decimals are usually meaningless. Develop rules for precision that are consistent with measurement capabilities. An example follows.

Item	Input	Output
Earthwork	Nearest 0.10 feet	Nearest CF or CY
Concrete	Nearest 0.01 feet	Nearest CF or CY
Formwork	Nearest 0.01 feet	Nearest SF
Finishing and Precast	Nearest 0.01 feet	Nearest SF
Lumber	Nearest 0.10 feet	Nearest BF
Finishes	Nearest 0.10 feet	Nearest SF or SY

The quantity estimator must also learn the standards of each industry. For instance, a lumber dimension of 12' 1-1/2" must be rounded up to 14' due to standard sawmill cutting practices.

Finally, do not convert units until all items in a column are totaled. For instance, keep concrete in cubic feet (CF) until all of the quantities listed in the concrete column have been added together. Then convert to cubic yards (CY).

General Rule 11

The quantity estimator should add an allowance for waste to certain quantities. Before the waste allowance is made, the quantities are referred to as *net quantities*. After the allowance for waste is added, the quantities are considered *gross quantities*.

General Rule 12

Ideally, a second quantity takeoff should be performed by a separate individual or team to ensure that no items have been omitted or duplicated. Unfortunately, the personnel to perform a second estimate are usually not available, or the cost for additional resources may be prohibitive. Typically, the quantity estimator systematically must check his or her own work. In fact, if the takeoff is done manually, the dimensions taken from plans should be checked by the quantity estimator, while the extensions should be checked by another qualified person.

General Rule 13

One way to avoid omissions and duplications is to mark the plans as items are taken off, such as making colored pencil or highlighter shadings and check marks directly on the plans. Most quantity estimators have their own methods of marking plans. Usually a combination of methods is used, rather than a single method, as one kind of mark may be effective in taking off one particular category, and different marks effective for other categories. The quantity estimator may assume that any item on a plan that has not been marked has not been taken off yet.

When work is interrupted, for whatever reason, select a natural stopping point and mark it clearly so that when work resumes, no items are missed or duplicated.

Systematic application of these rules will make the quantity estimator's job faster, easier, and more accurate.

Costing the Estimate

When the quantities have been determined, then prices, or unit costs, must be applied to determine total costs. No matter which source of cost information is used, the system and sequence of pricing should be the same as those used for the quantity takeoff. This consistent approach should continue through both accounting and cost control during construction of the project.

Unit price estimates for building construction are typically organized according to the 49 divisions of the CSI MasterFormat. However, other organizational structures can be used. Within each division, the components or individual construction items are identified, listed, and assigned costs. This kind of definition and detail is necessary to complete an accurate estimate. In addition, each "item" can be broken down further into material, labor, and equipment components.

Types of Costs: All costs included in a unit price estimate can be divided into two types: direct and indirect. Direct costs are those directly linked to the physical construction of a project, those costs without which the project could not be completed. The material, labor, and equipment costs mentioned above, as well as subcontract costs, are all direct costs. These may also be referred to as "bare," or "unburdened" costs.

Figure 0-5 Project Cost Organizational Chart

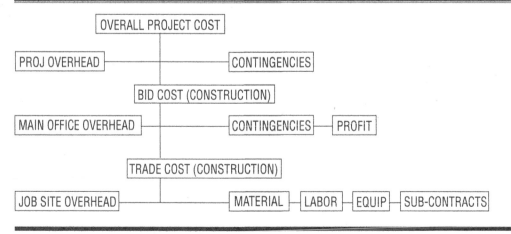

Indirect, or overhead, costs are those costs which are incurred in achieving project completion, but not applicable to any specific task. They may include items such as supervision, temporary facilities, insurance, professional services, and contingencies. Indirect costs are separated into two groups: job site overhead, and main office overhead. Job site overhead costs are those indirect costs associated with a job site. They can be estimated in detail, but are typically calculated as a percentage of direct costs and included in CSI MasterFormat Division 1 of the estimate. Main office overhead costs are costs associated with the operation of the contractor's home (or main) office. Overhead costs are typically calculated as a percentage of the total project cost and added at the end of the estimate. Some costs, such as professional services, may be either project overhead or main office overhead cost, depending on how the resource is used.

Compiling the Data: At the costing stage of the estimate, there is typically a large amount of data that must be assembled, analyzed, and organized. Generally, the information contained in these documents falls into the following major categories, and can exist in paper or digital form.

- Quantity takeoffs for all general contractor items
- Material supplier written quotations and published prices
- Material supplier telephone quotations
- Subcontractor quotations
- Equipment supplier quotations and published prices
- Cost analysis
- Historic cost data from previous projects
- Costs and historic data from independent sources

A system is needed to efficiently handle this mass of data and to ensure that everything will get transferred (and only once) from the quantity takeoff to the cost estimate. Some general rules for this procedure are:

- Code each document with a division number in a consistent place.
- Use telephone quotation forms and templates for uniformity when recording telephone quotes.
- Document the source of every quantity and price.
- Use a logical, consistent directory filing system and file naming convention.
- Back up all important data.

All subcontract costs should be properly noted and listed separately. These costs contain the subcontractors' markups, and will be treated differently from other direct costs when the estimator calculates the project overhead, profit, and contingency allowance.

Additional Tips: When estimating manually, follow these guidelines:

- Write on only one side of a page if possible.
- Keep each type of document in its pile (quantities, material, subcontractors, equipment) filed in order by division number.
- Keep the entire estimate file in one or more compartmented folders.

When using electronic spreadsheets:

- Save commonly used templates in a separate folder.
- Spot-check important formula results with manual calculations to verify results.
- Frequently save spreadsheets to avoid losing work.
- Combine related spreadsheets into workbooks.
- Use a naming convention that indicates whether a spreadsheet is in progress or complete.
- When entering quantities into a spreadsheet cell, include useful numerical information. For example, *if 10% is added to a quantity of 300*, instead of simply entering *330*, enter *=300*1.1*

The Estimate Summary: When the pricing of all direct costs is complete, the estimator has two choices: all further price changes and adjustments can be made on the Cost Analysis or Consolidated Estimate document, or total costs for each subdivision can be transferred to an Estimate Summary document so that all further price changes, until bid time, will be recorded on one document.

Unless the estimate has a limited number of items, it is recommended that costs be transferred to an Estimate Summary document. This step should be double-checked, since an error of transposition may easily occur.

If a company has certain standard listings that are used repeatedly, it would save valuable time to have a customized Estimate Summary document with the items that need to be listed. Appropriate column headings or categories for any estimate summary are:

- Material
- Labor
- Equipment
- Subcontractor
- Total

As items are listed in the proper columns, each category is added, and appropriate markups applied to the total dollar values. Generally, the sum of each column has different percentages added near the end of the estimate for indirect costs:

- Sales tax
- Main office overhead

- Profit
- Contingencies

Since over 50% of the work on a typical building project is performed by subcontractors, two aspects of bid preparation deserve special attention. The first is the subcontractor's scope of work, a clear understanding of which is essential, not only to compare the bids of competing subs, but to ensure that the estimator has included all the items that the subcontractors may have excluded (such as cutting and patching, temporary protection, scaffolding, hoisting, etc.). The second is that subcontractor prices typically do not arrive until bid day, which leaves little time to analyze competing bids, and makes last-minute gaps in coverage difficult to address if the estimator has not done the necessary coordination with these subs beforehand.

Division One
General Requirements

Introduction

The "General Requirements" section of the CSI MasterFormat is a mixture of performance and prescriptive specifications, used as the clearinghouse for items that do not apply directly to construction, the costs of which are customarily spread out over the entire project. These costs are also referred to as project overhead, or job site overhead.

The term "General Conditions" is sometimes interchanged with "General Requirements." Purists would argue that general conditions are the portions of the contract documents in which the rights, responsibilities, and relationships of the involved parties are itemized, whereas general requirements are cost items as described above. For estimating purposes, either term is acceptable. The important thing is to include project overhead costs in your estimate.

This chapter provides tables and charts for assistance in calculating the general requirements items listed below. It also includes tables relating to main office overhead, typically added at the end of the estimate. Instructions are given for the use of each chart, along with some background information on the circumstances under which each might be used. Preceeding the estimating charts, in this and all succeeding chapters, is an estimator's checklist, as well as estimating tips and precautions.

- Overhead and profit
- Construction time requirements
- Architectural fees for repair and/or remodeling projects
- Engineering fees
- Construction management fees
- General contractor's overhead and profit
- Installing contractor's overhead and profit
- Main office expenses
- Insurance
- Bonds
- Permits
- Tools
- Unemployment/social security taxes
- Scheduling costs
- Surveying costs
- Testing services costs
- Temporary utility installation
- Project cleanup

Estimating Data

The following tables present guidelines for project overhead and general conditions items. Please note that these percentages should be used as indicators of what may be expected, but that each project should be evaluated individually.

Table of Contents

Estimating Tips

Salaries

When figuring general requirements, do not include the salaries of the project managers or any other personnel who are not directly site-related. Their wages are included in "Main Office" overhead.

Cleanup

Always allow for cleanup in the estimate. No matter how clean a subcontractor leaves an area, it is almost always necessary to clean the area again.

Weather Considerations

In locations where snow is likely, allow for the expense of snowplowing if the project begins in, ends in, or works through the winter season. Another consideration is melting snow, which will inevitably end up in trenches, pits, or other low areas. Consequently, pumping costs should also be carried.

Site Visit

Always visit a proposed site. Do not rely on someone else's judgment unless statements in the contract require it.

Check for:

- Site access—Can loaded trucks move into and out of the site easily? Is the site in/near a residential area? Are there height/weight restrictions?
- How far away are the utilities that can be hooked into for temporary power?
- Site drainage—Is the area marshy? Will there be water problems when it rains?
- Do any utilities need to be relocated?
- Will any adjacent structures be affected?

If any of these items apply to the project, the associated costs must be estimated and included in the project bid/estimate.

The cost for installing temporary utilities, especially lighting, may be included in the specifications for the respective trades. Check this out to avoid adding an unnecessary cost to the estimate.

Checklist ✓

For an estimate to be reliable, all items must be accounted for. A complete estimate can also limit contingencies. The following checklist can be used to help ensure that all items are included.

☐ **Direct Overhead Costs (Project Overhead)**

☐ **Personnel**
- ☐ Superintendent
- ☐ Project manager (if for that project only)
- ☐ Field engineer (if for that project only)
- ☐ Cost engineer (if for that project only)
- ☐ Scheduler
- ☐ Warehouse personnel (if for that project only)
- ☐ Watchman/guard dogs
- ☐ Tool room keeper (if for that project only)
- ☐ Timekeeper (if for that project only)
- ☐ Foreman (working directly for the contractor)
- ☐ Site safety manager

☐ **Bonds**
- ☐ Surety
  - ☐ Bid
  - ☐ Payment
  - ☐ Performance

☐ **Miscellaneous**
- ☐ Vehicles
- ☐ Permits
- ☐ Licenses
- ☐ Tools and equipment
- ☐ Photographs
- ☐ Surveying

- ☐ Testing
- ☐ Job signs
- ☐ Pumping
- ☐ Dust control
- ☐ Scaffolding
- ☐ Lifting/hoisting
- ☐ Cleanup (periodic)
- ☐ Final cleanup
- ☐ Damage/repair to adjoining buildings and/or public ways

☐ **Temporary Facilities**
- ☐ Field office expense
  - ☐ Set-up and removal
  - ☐ Light
  - ☐ Water
  - ☐ Telephone/internet
  - ☐ Supplies
  - ☐ Equipment
  - ☐ Fax machine
  - ☐ Copy machine
  - ☐ Blueprint machine
  - ☐ Coffee machine
  - ☐ Computers/software
- ☐ Temporary light and power
- ☐ Temporary heat
- ☐ Temporary water
- ☐ Cell phones/radios
- ☐ Toilet facilities
- ☐ Enclosures
- ☐ Storage trailers
- ☐ Fencing
- ☐ Barricades and signals
- ☐ Construction road
- ☐ Project sign

☐ *Indirect Costs (Main Office Overhead)*

☐ *Salaries*
- ☐ President
- ☐ Executives
- ☐ Secretaries/reception
- ☐ Estimators/schedulers
- ☐ Project managers
- ☐ Construction manager
- ☐ Cost engineers
- ☐ Purchasing agent
- ☐ Cost/bookkeeping
- ☐ Engineers
- ☐ Other office personnel
- ☐ Yard personnel
  - ☐ Tool manager
  - ☐ Mechanics/maintenance
  - ☐ Drivers
  - ☐ Equipment operators

☐ *Office*
- ☐ Rent/cost of ownership
- ☐ Electricity
- ☐ Gas
- ☐ Water
- ☐ Sewer
- ☐ Telephone
- ☐ Internet
- ☐ Postage

- ☐ Office equipment
- ☐ Furniture/furnishings
- ☐ Office supplies
- ☐ Advertising
- ☐ Literature
- ☐ Club/association dues

☐ *Professional Services*
- ☐ Legal
- ☐ Accounting
- ☐ Architectural
- ☐ Engineering

☐ *Vehicles*
- ☐ Cars/trucks
- ☐ Cost of operation
- ☐ Mileage expenses

☐ *Insurance*
- ☐ Fire
- ☐ Property damage
- ☐ Vehicles
- ☐ Public liability
- ☐ Windstorm
- ☐ Workers' compensation
- ☐ Unemployment
- ☐ Social security
- ☐ Flood
- ☐ Theft
- ☐ Elevator

Figure 1-1 Architectural Fees

Building Type	Total Project Size in Thousands of Dollars						
	100	250	500	1,000	5,000	10,000	50,000
Factories, garages, warehouses, repetitive housing	9.0%	8.0%	7.0%	6.2%	5.3%	4.9%	4.5%
Apartments, banks, schools, libraries, offices, municipal buildings	12.2	12.3	9.2	8.0	7.0	6.6	6.2
Churches, hospitals, homes, laboratories, museums, research	15.0	13.6	12.7	11.9	9.5	8.8	8.0
Memorials, monumental work, decorative furnishings	—	16.0	14.5	13.1	10.0	9.0	8.3

In this figure, typical percentage fees are tabulated by project size for professional architectural service. Fees may vary from those listed depending on the degree of design difficulty and economic conditions in a particular area.

Rates can be interpolated horizontally and vertically. Various portions of the same project requiring different rates should be adjusted proportionately. For alterations, add 50% to the fee for the first $500,000 of project cost, and add 25% to the fee for the project cost over $500,000.

Architectural fees tabulated above include engineering fees. Civil engineering fees are included in the architectural fee for project sites requiring minimal design, such as city sites. However, separate civil engineering fees must be added when utility connections require design, drainage calculations are needed, stepped foundations are required, or provisions are required to protect adjacent wetlands or to meet special requirements such as LEED or historical preservation.

Figure 1-2 Architectural Fees for Smaller Projects

Architectural Fees for project sizes of:	
Up to $10,000	15%
to $25,000	13%
to $100,000	10%
to $500,000	8%
to $1,000,000	7%

The listed fees are approximate for smaller projects, such as repair work and/or remodeling existing structures.

Division 1

Figure 1-3 Engineering Fees as Percentages of a Project or Subcontract

Engineering Fees for:	Minimum	Maximum
Planning consultant/project	0.5%	2.5%
Structural/per Project	1.0%	2.5%
Percent of Subcontract:		
Electrical	4.1%	10.1%
Elevator/Conveying Systems	2.5%	5.0%
Food Service/Kitchen Equipment	8.0%	12.0%
Landscaping & Site Development	2.5%	6.0%
Mechanical, Plumbing & HVAC	4.1%	10.1%

Figure 1-4 Structural Engineering Fees

Structural engineering fees based on the type and size of a construction project.				
Type of Construction	\$500	\$500–\$1,000	\$1,000–\$5,000	Over \$5,000
Industrial buildings, factories, and warehouses	Technical payroll times 2.0 to 2.5	1.60%	1.25%	1.00%
Hotels, apartments, offices, dormitories, hospitals, public buildings, and food stores		2.00%	1.70%	1.20%
Museums, banks, churches, and cathedrals		2.00%	1.75%	1.25%
Thin shells, prestressed concrete, Earthquake resistive		2.00%	1.75%	1.50%
Parking ramps, auditoriums, stadiums, convention halls, hangars, and boiler houses		2.50%	2.00%	1.75%
Special buildings, major alterations, underpinning, and future expansion		Add to above 0.5%	Add to above 0.5%	Add to above 0.5%

Total project size (in thousands of dollars) spans the four size columns.

For complex reinforced concrete or unusually complicated structures, add 20% to 50% of above percentages. These fees are included in the architectural fees.

Figure 1-5 Mechanical and Electrical Engineering Fees

Mechanical and electrical engineering fees are based on the value of the subcontract.

Type of Construction	Subcontract Size							
	$25,000	$50,000	$100,000	$225,000	$350,000	$500,000	$750,000	$1,000,000
Simple structures	6.4%	5.7%	4.8%	4.5%	4.4%	4.3%	4.2%	4.1%
Intermediate structures	8.0	7.3	6.5	5.6	5.1	5.0	4.9	4.8
Complex structures	10.1	9.0	9.0	8.0	7.5	7.5	7.0	7.0

For renovations, add 15% to 25% to applicable fee.

These fees are for engineering services only, unless otherwise noted. They are included in the architectural fees shown in Figures 1-1 and 1-2. If engineering fees are to be listed separately, they must be deducted from the architectural fees.

Figure 1-6 Engineering Fees for Smaller Projects as a Percentage of Project Costs

Project Type	Minimum	Maximum
Landscape & Site Development	2.5%	6.0%
Elevator/Conveying Systems	2.5%	5.0%
Mechanical (Plumbing & HVAC)	5.7%	9.0%
Structural	1.6%	2.5%

The listed fees are approximate for smaller projects, such as repair work and/or the remodeling of existing structures.

Figure 1-7 Sustainable Design Fees

Design Fees	Conventional Building	Green Building
Architectural	3.1% to 11%	3.1% to 16%
Structural	1.0% to 2.5%	1.0% to 3.5%
Mechanical	0.4% to 1.3%	0.6% to 2.6%
Electrical	0.2% to 0.5%	0.2% to 0.5%
Landscape Architect	0.1% to 0.3%	0.1% to 1.3%
Construction Manager	2.5% to 10%	3.5% to 15%
LEED® Process	0%	0.25% to 0.5%

Figure 1-8 Construction Management Fees as a Percentage of
Project Costs

Project Size	Minimum	Maximum
$1,000,000 and above	4.5%	7.5%
$5,000,000 and above	2.5%	4.0%

These fees reflect the range of costs charged by construction management for normally incurred construction management services for new projects.

Figure 1-9 Construction Management Fees for Smaller Projects

Project Size	Average Fee
For work to $10,000	10%
To $25,000	9%
To $100,000	8%
To $500,000	7%
To $1,000,000	6%

These fees reflect the range of costs charged by construction firms for management of smaller projects, such as the repair and/or remodeling of existing structures.

Figure 1-10 Fees for Progress Schedules

Type of Schedule	Minimum	Maximum
Critical path, as % of architectural fee	.50%	1.00%
Rule of thumb, CPM, per job	.03%	.05%
CPM & cost control, per job	.04%	.08%

These items are usually included in the fees of a construction management firm, if one is used. If an independent firm is contracted to prepare the progress schedules, this table indicates the range of fees one can expect to be charged.

Figure 1-11   Installing Contractor's Overhead and Profit

Below are the average installing contractor's percentage markups applied to base labor rates to arrive at typical billing rates.

Column A: Labor rates are based on union wages averaged for 30 major U.S. cities. Base rates including fringe benefits are listed hourly and daily. These figures are the sum of the wage rate and employer-paid fringe benefits such as vacation pay, employer-paid health and welfare costs, pension costs, plus appropriate training and industry advancement funds costs.

Column B: Workers' compensation rates are the national average of state rates established for each trade

Column C: Column C lists average fixed overhead figures for all trades. Included are federal and state unemployment costs set at 6.2%; social security taxes (FICA) set at 7.65%; builder's risk insurance costs set at 0.44%; and public liability costs set at 2.02%. All the percentages except those for social security taxes vary from state to state as well as from company to company.

Column D and E: Percentages in Columns D and E are based on the presumption that the installing contractor has annual billing of $4,000,000 and up. Overhead percentages may increase with smaller annual billing. The overhead percentages for any given contractor may vary greatly and depend on a number of factors, such as the contractor's annual volume, engineering and logistical support costs, and staff requirements. The figures for overhead and profit will also vary depending on the type of job, the job location, and the prevailing economic conditions. All factors should be examined very carefully for each job.

Column F: Column F lists the total of Columns B, C, D, and E.

Column G: Column G is Column A (hourly base labor rate) multiplied by the percentage in Column F (O&P percentage).

Column H: Column H is the total of Column A (hourly base labor rate) plus Column G (Total O&P).

Column I: Column I is Column H multiplied by 8 hours.

		A		B	C	D	E	F	G	H	I
		Base Rate Incl. Fringes		Work-ers' Comp. Ins.	Average Fixed Over-head	Over-head	Profit	Total Overhead & Profit		Rate with O & P	
Abbr.	Trade	Hourly	Daily					%	Amount	Hourly	Daily
Skwk	Skilled Workers Average (35 trades)	$40.85	$326.80	15.5%	16.3%	13.0%	10%	54.8%	$22.40	$63.25	$506.00
	Helpers Average (5 trades)	30.30	242.40	17.2		11.0		54.5	16.50	46.80	374.40
	Foreman Average, Inside (.50 over trade)	41.35	330.80	15.5		13.0		54.8	22.65	64.00	512.00
	Foreman Average, Outside ($2.00 over trade)	42.85	342.80	15.5		13.0		54.8	23.50	66.35	530.80
Clab	Common Building Laborers	31.60	252.80	17.8		11.0		55.1	17.40	49.00	392.00
Asbe	Asbestos/Insulation Workers/Pipe Coverers	44.10	352.80	14.4		16.0		56.7	25.00	69.10	552.80
Boil	Boilermakers	50.20	401.60	11.4		16.0		53.7	26.95	77.15	617.20
Bric	Bricklayers	40.50	324.00	14.4		11.0		51.7	20.95	61.45	491.60
Brhe	Bricklayer Helpers	32.15	257.20	14.4		11.0		51.7	16.60	48.75	390.00
Carp	Carpenters	39.95	319.60	17.8		11.0		55.1	22.00	61.95	495.60
Cefi	Cement Finishers	38.30	306.40	9.1		11.0		46.4	17.75	56.05	448.40
Elec	Electricians	47.00	376.00	6.5		16.0		48.8	22.95	69.95	559.60
Elev	Elevator Constructors	56.60	452.80	6.6		16.0		48.9	27.70	84.30	674.40
Eqhv	Equipment Operators, Crane or Shovel	42.55	340.40	10.0		14.0		50.3	21.40	63.95	511.60
Eqmd	Equipment Operators, Medium Equipment	41.35	330.80	10.0		14.0		50.3	20.80	62.15	497.20
Eqlt	Equipment Operators, Light Equipment	39.05	312.40	10.0		14.0		50.3	19.65	58.70	469.60
Eqol	Equipment Operators, Oilers	36.80	294.40	10.0		14.0		50.3	18.50	55.30	442.40
Eqmm	Equipment Operators, Master Mechanics	42.70	341.60	10.0		14.0		50.3	21.50	64.20	513.60
Glaz	Glaziers	38.60	308.80	13.9		11.0		51.2	19.75	58.35	466.80
Lath	Lathers	35.55	284.40	10.6		11.0		47.9	17.05	52.60	420.80
Marb	Marble Setters	38.65	309.20	14.4		11.0		51.7	20.00	58.65	469.20
Mill	Millwrights	41.60	332.80	9.5		11.0		46.8	19.45	61.05	488.40
Mstz	Mosaic & Terrazzo Workers	37.70	301.60	9.1		11.0		46.4	17.50	55.20	441.60
Pord	Painters, Ordinary	35.20	281.60	12.5		11.0		49.8	17.55	52.75	422.00
Psst	Painters, Structural Steel	36.00	288.00	44.2		11.0		81.5	29.35	65.35	522.80
Pape	Paper Hangers	35.45	283.60	12.5		11.0		49.8	17.65	53.10	424.80
Pile	Pile Drivers	38.50	308.00	20.1		16.0		62.4	24.00	62.50	500.00
Plas	Plasterers	36.15	289.20	13.6		11.0		50.9	18.40	54.55	436.40
Plah	Plasterer Helpers	32.30	258.40	13.6		11.0		50.9	16.45	48.75	390.00
Plum	Plumbers	48.75	390.00	7.8		16.0		50.1	24.40	73.15	585.20
Rodm	Rodmen (Reinforcing)	44.55	356.40	23.2		14.0		63.5	28.30	72.85	582.80
Rofc	Roofers, Composition	34.25	274.00	31.2		11.0		68.5	23.45	57.70	461.60
Rots	Roofers, Tile & Slate	34.10	272.80	31.2		11.0		68.5	23.35	57.45	459.60
Rohe	Roofers, Helpers (Composition)	25.35	202.80	31.2		11.0		68.5	17.35	42.70	341.60
Shee	Sheet Metal Workers	47.20	377.60	11.6		16.0		53.9	25.45	72.65	581.20
Spri	Sprinkler Installers	48.15	385.20	7.8		16.0		50.1	24.10	72.25	578.00
Stpi	Steamfitters or Pipefitters	49.35	394.80	7.8		16.0		50.1	24.70	74.05	592.40
Ston	Stone Masons	40.80	326.40	14.4		11.0		51.7	21.10	61.90	495.20
Sswk	Structural Steel Workers	44.70	357.60	37.9		14.0		78.2	34.95	79.65	637.20
Tif	Tile Layers	38.10	304.80	9.1		11.0		46.4	17.70	55.80	446.40
Tilh	Tile Layers Helpers	30.05	240.40	9.1		11.0		46.4	13.95	44.00	352.00
Trlt	Truck Drivers, Light	30.95	247.60	16.6		11.0		53.9	16.70	47.65	381.20
Trhv	Truck Drivers, Heavy	31.95	255.60	16.6		11.0		53.9	17.20	49.15	393.20
Sswl	Welders, Structural Steel	44.70	357.60	37.9		14.0		78.2	34.95	79.65	637.20
Wrck	*Wrecking	31.60	252.80	34.1		11.0		71.4	22.55	54.15	433.20

*Not included in averages

2009 averages

Figure 1-12 Workers' Compensation

The table below tabulates the national averages for workers' compensation insurance rates by trade and type of building. The average "Insurance Rate" is multiplied by the "% of Building Cost" for each trade. This produces the "Workers' Compensation" cost by % of total labor cost, to be added for each trade by building type to determine the weighted average workers' compensation rate for the building types analyzed.

Trade	Insurance Rate (% Labor Cost) Range		Average	% of Building Cost Office Bldgs.	Schools & Apts.	Mfg.	Workers' Compensation Office Bldgs.	Schools & Apts.	Mfg.
Excavation, Grading, etc.	4.2 % to	17.5%	10.0%	4.8%	4.9%	4.5%	0.48%	0.49%	0.45%
Piles & Foundations	5.9 to	40.3	20.1	7.1	5.2	8.7	1.43	1.05	1.75
Concrete	5.1 to	26.8	14.6	5.0	14.8	3.7	0.73	2.16	0.54
Masonry	4.8 to	43.3	14.4	6.9	7.5	1.9	0.99	1.08	0.27
Structural Steel	5.9 to	104.1	37.9	10.7	3.9	17.6	4.06	1.48	6.67
Miscellaneous & Ornamental Metals	3.4 to	22.8	10.7	2.8	4.0	3.6	0.30	0.43	0.39
Carpentry & Millwork	5.9 to	53.2	17.8	3.7	4.0	0.5	0.66	0.71	0.09
Metal or Composition Siding	5.9 to	38	16.6	2.3	0.3	4.3	0.38	0.05	0.71
Roofing	5.9 to	77.1	31.2	2.3	2.6	3.1	0.72	0.81	0.97
Doors & Hardware	4.9 to	32	11.6	0.9	1.4	0.4	0.10	0.16	0.05
Sash & Glazing	5.9 to	32.4	13.9	3.5	4.0	1.0	0.49	0.56	0.14
Lath & Plaster	3.3 to	43.9	13.6	3.3	6.9	0.8	0.45	0.94	0.11
Tile, Marble & Floors	3.1 to	17.9	9.1	2.6	3.0	0.5	0.24	0.27	0.05
Acoustical Ceilings	2.6 to	45.6	10.6	2.4	0.2	0.3	0.25	0.02	0.03
Painting	4.7 to	29.6	12.5	1.5	1.6	1.6	0.19	0.20	0.20
Interior Partitions	5.9 to	53.2	17.8	3.9	4.3	4.4	0.69	0.77	0.78
Miscellaneous Items	2.1 to	168.2	16.0	5.2	3.7	9.7	0.83	0.59	1.55
Elevators	2.8 to	11.8	6.6	2.1	1.1	2.2	0.14	0.07	0.15
Sprinklers	2.5 to	14.3	7.8	0.5	—	2.0	0.04	—	0.16
Plumbing	2.9 to	12.4	7.8	4.9	7.2	5.2	0.38	0.56	0.41
Heat., Vent., Air Conditioning	4.3 to	23	11.6	13.5	11.0	12.9	1.57	1.28	1.50
Electrical	2.8 to	11.5	6.5	10.1	8.4	11.1	0.66	0.55	0.72
Total	2.1 % to	168.2%	—	100.0%	100.0%	100.0%	15.78%	14.23%	17.69%
			Overall Weighted Average	15.90%					

Figure 1-13 Workers' Compensation Rates by States

The table below lists the weighted average workers' compensation base rate for each state with a factor comparing this with the national average of 15.5%.

State	Weighted Average	Factor	State	Weighted Average	Factor	State	Weighted Average	Factor
Alabama	24.2%	156	Kentucky	18.4%	119	North Dakota	13.7%	88
Alaska	21.7	140	Louisiana	28.3	183	Ohio	14.8	95
Arizona	9.5	61	Maine	15.4	99	Oklahoma	14.6	94
Arkansas	13.2	85	Maryland	16.7	108	Oregon	13.5	87
California	19.6	126	Massachusetts	12.5	81	Pennsylvania	13.8	89
Colorado	13.2	85	Michigan	17.3	112	Rhode Island	21.2	137
Connecticut	21.0	135	Minnesota	25.7	166	South Carolina	18.3	118
Delaware	15.1	97	Mississippi	19.1	123	South Dakota	19.1	123
District of Columbia	13.7	88	Missouri	17.2	111	Tennessee	15.1	97
Florida	13.8	89	Montana	16.6	107	Texas	12.2	79
Georgia	26.2	169	Nebraska	23.0	148	Utah	12.3	79
Hawaii	17.0	110	Nevada	11.6	75	Vermont	18.0	116
Idaho	10.8	70	New Hampshire	20.0	129	Virginia	11.7	75
Illinois	21.6	139	New Jersey	13.4	86	Washington	9.8	63
Indiana	5.9	38	New Mexico	18.5	119	West Virginia	9.2	59
Iowa	11.5	74	New York	12.6	81	Wisconsin	15.0	97
Kansas	8.9	57	North Carolina	18.9	122	Wyoming	6.5	42
			Weighted Average for U.S. is	15.9% of payroll = 100%				

Figure 1-14   Unemployment Taxes and Social Security Taxes

State unemployment tax rates vary not only from state to state, but also with the experience rating of the contractor. The federal unemployment tax rate is 6.2% of the first $7,000 of wages. This is reduced by a credit of up to 5.4% for timely payment to the state. The minimum federal unemployment tax is 0.8% after all credits. Social security (FICA) for 2009 is estimated at time of publication to be 7.65% of wages up to $102,000.

Figure 1-15a   Labor Rate Breakdown

This is an example of a labor rate worksheet. The burdened labor rate consists of the base rate, supplemental benefits (sometimes called fringe benefits), payroll taxes, and insurance. Labor rates may also include liability insurance, bonding, overhead, and profit; when these items are not included in the labor rate, they are added as separate line items. It is important to remember the labor rate is not what the worker actually receives in his or her paycheck, but the total the employer must pay for that worker. Base rates and benefit rates can be set by unions, project labor agreements, state or federal prevailing wage rate agreements, or as a result of market conditions. Rates can be obtained from RSMeans cost data books, unions, state or federal prevailing wage schedules, ENR magazine, or other sources. Taxes and insurance rates can also be obtained from RSMeans cost data books or other sources. An Excel® spreadsheet can easily be developed to automate the calculations.

LABOR RATE WORKSHEET

Project:	Anywhere Office Building
Date:	9/12/09

Base Rate Time Period	From: 1/1/09	To: 12/31/09		
Trade	Laborer			
Base Rate Classification	C		(a) Base Rate $	37.99

Benefits:		
Vacation (included in base wage rate)	$	
Health & Welfare	$	7.50
Pension	$	7.19
Annuity	$	2.90
Education/Apprentice	$	-
Security Fund	$	-
Supplemental Unemployment	$	2.80
Training	$	0.41
(b) Total Benefits	$	20.80

(c) Base Rate & Benefits (a)+(b)	$	58.79

Payroll Tax & Insurance:	
F.I.C.A./Social Security	7.65%
Federal Unemployment	0.80%
State Unemployment	7.10%
Workmen's Compensation Rate Code # 6217	8.29%
Disability	0.02%
(d) Total Paryoll Tax & Insurance	23.86%

(e)Total Rate with Payroll Tax & Insurance (a) x (d)	$9.06

Total Labor Rate (c)+(e)	$	67.85

Other Mark-ups (if applicable)	
Liability Insurance	5%
Bonding	1%
Overhead & Profit	15%

Figure 1-15b   Labor Hour Cost for Crews

This is an example of how labor rates for crews are calculated. Typically equipment is accounted for as a separate column in estimates; however, it may also be included in the hourly crew rate. An Excel® spreadsheet template can easily be developed to automate the calculations.

	Qty	Description	Hourly Rate	Daily Rate	% of Total	Weighted Cost	Cost/Labor Hr
			Composite Labor Rate				
Labor	2	Laborers	$30.25	$242.00	67%	$20.27	$32.73
	1	Equipment Operator	$37.75	$302.00	33%	$12.46	
Equipment	1	Backhoe Loader	$35.68	$285.40	100%	$285.40	$11.89
Labor & Eq							$44.62

Figure 1-16   Determining Available Workdays

The chart below is a sample format used to determine the number of workdays available per month for exterior construction. The example shown might be typical for the mid-Atlantic region of the U.S. Each contractor's calculation will be unique to his labor practices and climate. The basic steps involved are:

1. Determine total available workdays per month, counting weekdays and Saturdays, if applicable, and deducting holidays.
2. Deduct average time allowed per month for vacations and sick days (based on historical data).
3. Obtain weather data from the local office of the National Weather Service, and subtract anticipated lost days due to poor weather.

		Available Workdays		
Month	No. of Potential Workdays	Lost Days Due to Weather	Lost Days Due to Vaca./Sick Time	Actual Workdays per Worker
January	22	10	2	10
February	19	10	3	6
March	22	8	3	11
April	21	8	2	11
May	22	6	1	15
June	21	4	2	15
July	21	3	3	15
August	23	3	4	16
September	19	4	3	12
October	22	6	2	14
November	21	8	3	10
December	20	10	4	6
	253	80	32	141

Figure 1-17 Insurance Rates

This table represents approximate values relative to total project cost for the most common types of basic insurance coverage.

Type	Minimum	Maximum
Builder's Risk	.24%	.64%
All-risk Type	.25%	.62%
Contractor's Equipment Floater	.50%	1.50%
Public Liability, Average	—	2.02%

Figure 1-18 Builder's Risk Insurance Rates

Builder's risk insurance covers a building during construction. Premiums are paid by the owner or the contractor. Blasting, collapse, and underground insurance would raise total insurance costs above those listed. A floater policy on materials delivered to the job runs $.75 to $1.25 for $100 value. Contractor equipment insurance runs $.50 to $1.50 per $100 value. Insurance for miscellaneous tools up to $1,500 value runs from $3 to $7.50 per $100 value.

Tabulated below are New England builder's risk insurance rates in dollars per $100 value for over $1,000 deductible. For $25,000 deductible, rates can be reduced 13% to 34%. On contracts over $1,000,000, rates may be lower than those tabulated. Policies are written annually for the total completed value in place. For "all risk" insurance (excluding flood, earthquake, and certain other perils), add $.025 to total rates below.

Coverage	Frame Construction (Class 1) Range		Average	Brick Construction (Class 4) Range		Average	Fire Resistive (Class 6) Range		Average
Fire Insurance	$.350 to	$ .850	$.600	$.158 to	$.189	$.174	$.052 to	$.080	$.070
Extended Coverage	.115 to	.200	.158	.080 to	.105	.101	.081 to	.105	.100
Vandalism	.012 to	.016	.014	.008 to	.011	.011	.008 to	.011	.010
Total Annual Rate	$.477 to	$1.066	$.772	$.246 to	$.305	$.286	$.141 to	$.196	$.180

Figure 1-19 Performance Bond Rates

This table shows the performance bond rate for a job scheduled to be completed in 12 months. Add 1% of the premium cost per month for jobs requiring more than 12 months to complete. The rates are standard rates offered to contractors considered by the bonding company to be financially sound and capable of doing the work. The rates quoted are suggested averages. Actual rates vary from contractor to contractor, and from bonding company to bonding company. Contractors should prequalify through a bonding company agency before submitting a bid on a contract that requires a bond.

Contract Amount	Building Construction Class B Projects			Highways & Bridges Class A New Construction			Class A-1 Highway Resurfacing		
First $ 100,000 bid		$25.00 per M			$15.00 per M			$9.40 per M	
Next 400,000 bid	$ 2,500 plus	$15.00	per M	$ 1,500 plus	$10.00	per M	$ 940 plus	$7.20	per M
Next 2,000,000 bid	8,500 plus	10.00	per M	5,500 plus	7.00	per M	3,820 plus	5.00	per M
Next 2,500,000 bid	28,500 plus	7.50	per M	19,500 plus	5.50	per M	15,820 plus	4.50	per M
Next 2,500,000 bid	47,250 plus	7.00	per M	33,250 plus	5.00	per M	28,320 plus	4.50	per M
Over 7,500,000 bid	64,750 plus	6.00	per M	45,750 plus	4.50	per M	39,570 plus	4.00	per M

Figure 1-20 Permit Rates

Permit costs vary greatly depending on many factors, such as type of project, proposed occupancy, location, local codes, need for variances or changes of zoning, etc. This table provides a rule of thumb to use when local conditions cannot be determined.

Project Permits	Minimum	Maximum
Rule of thumb, most cities	.50%	2%

Figure 1-21 Small Tools Allowance

A variety of small tools must be purchased in the course of almost every job, whether to complete particular tasks required for that project or to replace tools that have "mysteriously disappeared." This table provides a rule of thumb that can be used to assign a value to this often overlooked item.

Small Tools Allowance	Minimum	Maximum
As % of contractor's work	.50%	2%

Figure 1-22 Construction Time Requirements

This table shows average construction time in months for different types and sizes of building projects. Design time runs 25% to 40% of construction time.

Building Type	Size S.F.	Project Value	Construction Duration
Industrial/Warehouse	100,000	$8,000,000	14 Months
	500,000	$32,000,000	19 Months
	1,000,000	$75,000,000	21 Months
Offices/Retail	50,000	$7,000,000	15 Months
	250,000	$28,000,000	23 months
	500,000	$58,000,000	34 months
Institutional/Hospitals/Laboratory	200,000	$45,000,000	31 months
	500,000	$110,000,000	52 months
	750,000	$160,000,000	55 months
	1,000,000	$210,000,000	60 months

In order to estimate the general requirements of a construction project, it is necessary to have an approximate project duration time. Duration must be determined because many items, such as supervision and temporary facilities, are directly time variable. The average durations presented in this chart will vary depending on such factors as location, complexity, time of year started, local economic conditions, materials required, or the need for the completed project.

Figure 1-23 Survey Data

This table provides the information needed to compute the time required to survey a parcel or to perform the layout for a planned structure. To estimate the costs of collecting survey data, multiply the number of units on the plans (number of acres, LF, etc.) by the amount listed in the labor-hours column for the appropriate task. Multiply this number by the local labor rate. The result should be the approximate bare cost (without overhead and profit) for surveying.

Survey Data	Crew Makeup	Daily Output	Labor-Hours	Unit
Surveying, conventional, topographical, Minimum	1 Chief of Party 1 Instrument Person 1 Rodman/Chainman	3.30	7.270	Acre
Maximum	1 Chief of Party 1 Instrument Person 2 Rodmen/Chainmen	.60	53.330	Acre
Lot location and lines, for large quantities Minimum	1 Chief of Party 1 Instrument Person 1 Rodman/Chainman	2	12.000	Acre
Average	"	2.5	19.200	Acre
Maximum, for small quantities	1 Chief of Party 1 Instrument Person 2 Rodmen/Chainmen	1	32.000	Acre
Monuments, 3' long	1 Chief of Party 1 Instrument Person 1 Rodman/Chainman	10	2.500	Ea.
Property lines, perimeter, cleared land	"	1.000	.024	LF

Figure 1-24 Testing Services

This table lists approximate values for testing services for different types of structures.

Type of Building	Minimum	Maximum
Concrete, costing $1,000,000	0.5%	5%
Steel	0.5%	1.0%
Building $10,000,000	0.35%	0.5%

Figure 1-25 Final Cleaning

To estimate the cleanup costs for a structure, whether continuous (daily) or the final cleanup, determine the floor area (total), divide by 1,000, and multiply this number by the appropriate number listed in the labor-hour column for your task. The result will be the number of labor-hours needed for that task. Multiply this by your local labor rates to determine the bare costs (without overhead and profit) for your task.

Final Cleaning	Crew Makeup	Daily Output	Labor-Hours	Unit
Cleanup of floor area, continuous, per day	2 Building Laborers .25 Truck Driver (light) .25 Light Truck 1.5 Ton	24	.75	M.S.F.
Final	"	11.50	1.570	M.S.F.
Cleanup after job completion, allow .30% per total job				

Figure 1-26 Overtime

One way to improve the completion date of a project or eliminate negative float from a schedule is to compress activity duration times. This can be achieved by increasing the crew size or working overtime with the proposed crew.

To determine the costs of working overtime to compress activity duration times, consider the following examples. Below is an overtime efficiency and cost chart based on a five-, six-, or seven-day week with an eight- to twelve-hour day. Payroll percentage increases for time and a half and double time are shown for the various working days.

Days per Week	Hours per Day	Production Efficiency					Payroll Cost Factors	
		1 Week	2 Weeks	3 Weeks	4 Weeks	Average 4 Weeks	@1-1/2 Times	@2 Times
	8	100%	100%	100%	100%	100%	100%	100%
	9	100	100	95	90	96.25	105.6	111.1
5	10	100	95	90	85	91.25	110.0	120.0
	11	95	90	75	65	81.25	113.6	127.3
	12	90	85	70	60	76.25	116.7	133.3
	8	100	100	95	90	96.25	108.3	116.7
	9	100	95	90	85	92.50	113.0	125.9
6	10	95	90	85	80	87.50	116.7	133.3
	11	95	85	70	65	78.75	119.7	139.4
	12	90	80	65	60	73.75	122.2	144.4
	8	100	95	85	75	88.75	114.3	128.6
	9	95	90	80	70	83.75	118.3	136.5
7	10	90	85	75	65	78.75	121.4	142.9
	11	85	80	65	60	72.50	124.0	148.1
	12	85	75	60	55	68.75	126.2	152.4

Figure 1-27 General Contractor's Overhead, Project Overhead, and Main Office Overhead

There are two distinct types of overhead on a construction project: project overhead and main office overhead. Project overhead includes those costs at a construction site not directly associated with the installation of construction materials. Examples of project overhead costs include the following:

1. Superintendent
2. Construction office and storage trailers
3. Temporary sanitary facilities
4. Temporary utilities
5. Security fencing
6. Photographs
7. Cleanup
8. Performance and payment bonds

The above project overhead items are also referred to as general requirements and therefore are estimated in Division 1. Division 1 is the first division listed in the CSI MasterFormat, but it is usually the last division estimated. The sum of the costs in Divisions 1 through 49 is referred to as the sum of the direct costs.

All construction projects also include indirect costs. The primary components of indirect costs are the contractor's main office overhead and profit. The amount of the main office overhead expense varies depending on the following:

1. Owner's compensation
2. Project manager's and estimator's wages
3. Clerical support wages
4. Office rent and utilities
5. Corporate legal and accounting costs
6. Advertising
7. Automobile expenses
8. Association dues
9. Travel and entertainment expenses

These costs are usually calculated as a percentage of annual sales volume. This percentage can range from 35% for a small contractor doing less than $500,000 to 5% for a large contractor with sales in excess of $100 million.

Figure 1-28   General Contractor's Main Office Expense

A general contractor's main office expense consists of many items. The percentage of main office expense declines with the increased annual volume of the contractor. Typical main office expenses range from 2% to 20%, with the median about 7.2% of total volume. This equals about 7.7% of direct costs. The following are approximate percentages of total overhead for different items usually included in a general contractor's main office overhead. With different accounting procedures, these percentages may vary.

Item	Typical Range	Average
Manager's clerical and estimators' salaries	40% to 55%	48%
Profit sharing, pension and bonus plans	2 to 20	12
Insurance	5 to 8	6
Estimating and project management (not including salaries)	5 to 9	7
Legal, accounting and data processing	0.5 to 5	3
Automobile and light truck expense	2 to 8	5
Depreciation and overhead capital expensitures	2 to 6	4
Maintenance of office equipment	0.1 to 1.5	1
Office rental	3 to 5	4
Utilities, including phone and light	1 to 3	2
Miscellaneous	5 to 15	8
Total		100%

Figure 1-29   General Contractor's Main Office Expense as a Percentage of Annual Volume

This table represents approximate ranges of the cost of maintaining the main office, including salaries, as a percentage of the total dollar volume a contractor expects to bill in their fiscal year.

Annual Volume	% of Annual Volume
Under $1,000,000	17.5%
Up to $2,500,000	8.0%
Up to $4,000,000	6.8%
Up to $7,000,000	5.6%
Up to $10,000,000	5.1%
Over $10,000,000	3.9%

Figure 1-30 General Contractor's Main Office Expense for Smaller Projects

This table provides main office expenses (as a percentage of annual volume) for contractors specializing in smaller projects, such as repair and/or remodeling of existing structures.

Annual Volume	% of Annual Volume	
	Minimum	Maximum
To $50,000	20%	30%
To $100,000	17%	22%
To $250,000	16%	19%
To $500,000	14%	16%
To $1,000,000	8%	10%

Figure 1-31 Project Overhead as a Percentage of Project Costs

Overhead is defined as costs that are associated with a construction project, but not directly with the actual construction. This table shows percentages that can be used as a rule of thumb for estimating overhead costs.

Overhead as a Percentage of Direct Costs		Overhead and Profit Allowance—Add to Items That Do Not Include Subcontractor's O&P— Average	Allowance to Add to Items That Do Include Subcontractor's O&P		Typical by Size of Project	
Minimum	5%	25%	Minimum	5%	under $100,000	30%
Average	13%		Average	10%	$500,000	25%
Maximum	30%		Maximum	15%	$2,000,000	20%
					over $10,000,000	15%

Figure 1-32a Sales Tax by State

State sales tax on materials is tabulated below (5 states have no sales tax). Many states allow local jurisdictions, such as a county or city, to levy additional sales tax. Some projects may be sales tax exempt, particularly those constructed with public funds.

State	Tax (%)	State	Tax (%)	State	Tax (%)	State	Tax (%)
Alabama	4	Illinois	6.25	Montana	0	Rhode Island	7
Alaska	0	Indiana	6	Nebraska	5.5	South Carolina	6
Arizona	5.6	Iowa	5	Nevada	6.5	South Dakota	4
Arkansas	6	Kansas	5.3	New Hampshire	0	Tennessee	7
California	7.25	Kentucky	6	New Jersey	7	Texas	6.25
Colorado	2.9	Louisiana	4	New Mexico	5	Utah	4.65
Connecticut	6	Maine	5	New York	4	Vermont	6
Delaware	0	Maryland	6	North Carolina	4.25	Virginia	5
District of Columbia	5.75	Massachusetts	6.25	North Dakota	5	Washington	6.5
Florida	6	Michigan	6	Ohio	5.5	West Virginia	6
Georgia	4	Minnesota	6.5	Oklahoma	4.5	Wisconsin	5
Hawaii	4	Mississippi	7	Oregon	0	Wyoming	4
Idaho	6	Missouri	4.225	Pennsylvania	6	Average	4.91 %

Figure 1-32b Sales Tax by Province (Canada)

GST - a value-added tax, which the government imposes on most goods and services provided in or imported into Canada. PST - a retail sales tax, which five of the provinces impose on the price of most goods and some services. QST - a value-added tax, similar to the federal GST, which Quebec imposes. HST - three provinces have combined their retail sales tax with the federal GST into one harmonized tax.

Province	PST (%)	QST (%)	GST(%)	HST(%)
Alberta	0	0	6	0
British Columbia	7	0	6	0
Manitoba	7	0	6	0
New Brunswick	0	0	0	13
Newfoundland	0	0	0	13
Northwest Territories	0	0	6	0
Nova Scotia	0	0	0	13
Ontario	8	0	6	0
Prince Edward Island	10	0	6	0
Quebec	0	7.5	6	0
Saskatchewan	5	0	6	0
Yukon	0	0	6	0

Figure 1-33 General Commissioning Requirements

	Unit	Total Incl. O&P
Commissioning Including documentation of design intent, performance verification, O&M, training, minimum Maximum	Project "	.50% .75%

Notes

Division Two
Existing Conditions

Introduction

Division 2 – Existing Conditions includes site demolition and remediation, building relocation, site surveys, and geotechnical investigations. The scope of work included in demolition can range from the removal of a window to the complete dismantling and removal of existing structures. Estimators must assess the particular project and determine whether the takeoff should be performed by the piece, square foot, cubic foot, or whole unit.

In RSMeans cost data, whole-building demolition covers above-ground portions only, including loading and haul-away. If slabs, footings, foundations, and below-grade walls are being removed, they must be estimated separately, and costs for waste handling and transportation away from the site will have to be added. In Division 2, selective demolition includes site-related items such as underground utilities, site improvements, and roadways, but some selective demo is also covered in other divisions.

A green approach is to reduce the amount of demolition by using as much of the existing structure as possible. However, if that is not feasible, an alternative is deconstruction, which is the process of disassembling building components and recovering them for reuse. Non-salvageable materials are recycled, and as little as possible is hauled to landfills. While disposal costs with deconstruction will be lower than with standard demolition, the added cost of labor for uninstalling and processing the materials needs to be considered, as does the added time to the overall project schedule.

To assist in estimating existing conditions, the following categories of charts and tables have been included:

- Demolition
- Deconstruction
- Hazardous Materials

Estimating Data

The following tables present estimating guidelines for items found in Division 2. Please note that these guidelines are intended as indicators of what may be expected, but that each project must be evaluated individually.

Table of Contents

Division 2

Estimating Tips

Selective Demolition

Historic preservation often requires that the contractor remove materials from the existing structure, rehab them, and replace them. The estimator must be aware of any related measures and precautions such as special handling, storage, or security that must be taken when doing selective demolition and cutting and patching.

Subsurface Investigations

Many companies, eager to get started on their projects, shortchange the site investigation process. For the relatively small amounts of time and money required, it is not a good idea to skimp on this important item. The untimely discovery of even one subsurface "abnormality" can be a painful lesson. An example is finding that a site was unknowingly used at some point in the past as a spoils site for a nearby industrial park, and that there is ten feet of bad soil to excavate from under the stiff clay cap that was to be built on.

Deconstruction Scheduling

Because salvaging and processing materials for deconstruction takes longer than demolition, additional time needs to be scheduled for it. Planning ahead by finding buyers for salvaged materials is one way to save time.

Estimating Mold Remediation

Mold remediators must take special considerations into account when planning the removal of mold-infested materials. Limited access to small areas such as crawl spaces can slow work down and add time and cost to a project. Protective measures, including sealing off contaminated areas to avoid cross-contaminating other parts of a building, must be taken. Moldy debris may have to be packaged in vacuum-sealed plastic bags before it can be removed from the structure. These steps require additional time, materials, and labor.

Checklist ✓

For an estimate to be reliable, all items must be accounted for. A complete estimate can also limit contingencies. The following checklist can be used to help ensure that all items are included.

☐ **Demolition**
- ☐ Demolition of structure
- ☐ Rubbish disposal
- ☐ Remove from site

☐ **Deconstruction**
- ☐ Disassembly
- ☐ Processing
- ☐ Storage
- ☐ Purchaser

☐ **Hazardous Material Remediation**
- ☐ Lead paint removal/ encapsulation
- ☐ Asbestos removal/encapsulation

Figure 2-1   Demolition Defined

Whole Building Demolition - Demolition of the whole building with no concern for any particular building element, component, or material type being demolished. This type of demolition is accomplished with large pieces of construction equipment that break up the structure, load it into trucks and haul it to a disposal site, but disposal or dump fees are not included. Demolition of below-grade foundation elements, such as footings, foundation walls, grade beams, slabs on grade, etc., is not included. Certain mechanical equipment containing flammable liquids or ozone-depleting refrigerants, electric lighting elements, communication equipment components, and other building elements may contain hazardous waste, and must be removed, either selectively or carefully, as hazardous waste before the building can be demolished.

Foundation Demolition - Demolition of below-grade foundation footings, foundation walls, grade beams, and slabs on grade. This type of demolition is accomplished by hand or pneumatic hand tools, and does not include saw cutting, or handling, loading, hauling, or disposal of the debris.

Gutting - Removal of building interior finishes and electrical/mechanical systems down to the load-bearing and subfloor elements of the rough building frame, with no concern for any particular building element, component, or material type being demolished. This type of demolition is accomplished by hand or pneumatic hand tools, and includes loading into trucks (but not hauling), disposal or dump fees, scaffolding, or shoring. Certain mechanical equipment containing flammable liquids or ozone-depleting refrigerants, electric lighting elements, communication equipment components, and other building elements may contain hazardous waste, and must be removed, either selectively or carefully, as hazardous waste, before the building is gutted.

Selective Demolition - Demolition of a selected building element, component, or finish, with some concern for surrounding or adjacent elements, components, or finishes. This type of demolition is accomplished by hand or pneumatic hand tools, and does not include handling, loading, storing, hauling, or disposal of the debris, scaffolding, or shoring. "Gutting" methods may be used in order to save time, but damage that is caused to surrounding or adjacent elements, components, or finishes may have to be repaired at a later time.

Careful Removal - Removal of a piece of service equipment, building element or component, or material type, with great concern for both the removed item and surrounding or adjacent elements, components or finishes. The purpose of careful removal may be to protect the removed item for later re-use, preserve a higher salvage value of the removed item, or replace an item while taking care to protect surrounding or adjacent elements, components, connections, or finishes from cosmetic and/or structural damage. An approximation of the time required to perform this type of removal is 1/3 to 1/2 the time it would take to install a new item of like kind. This type of removal is accomplished by hand or pneumatic hand tools, and does not include loading, hauling, or storing the removed item, scaffolding, shoring, or lifting equipment.

Cutout Demolition - Demolition of a small quantity of floor, wall, roof, or other assembly, with concern for the appearance and structural integrity of the surrounding materials. This type of demolition is accomplished by hand or pneumatic hand tools, and does not include saw cutting, handling, loading, hauling, or disposal of debris, scaffolding, or shoring.

Rubbish Handling - Work activities that involve handling, loading, or hauling of debris. Generally, the cost of rubbish handling must be added to the cost of all types of demolition,

(continued on next page)

Figure 2-1 Demolition Defined *(continued from previous page)*

with the exception of whole building demolition.

Minor Site Demolition - Demolition of site elements outside the footprint of a building. This type of demolition is accomplished by hand or penumatic hand tools, or with larger pieces of construction equipment, and may include loading a removed item into a truck. (Check the crew for equipment used.) It does not include saw cutting, hauling, or disposal of debris, and, sometimes, handling or loading.

Figure 2-2 Dumpsters

Dumpster rental costs on construction sites are presented in two ways.

The cost per week rental includes the delivery of the Dumpster, its pulling or emptying once per week, and its final removal. The assumption is made that the Dumpster contractor could choose to empty a Dumpster by simply bringing in an empty unit and removing the full one. These costs also include the disposal of the materials in the Dumpster.

The alternate pricing can be used when actual planned conditions are not approximated by the weekly numbers. Delivery, rent per day, rent per month, and disposal fee are separate lines that can be used for non-weekly pickup of Dumpsters.

Figure 2-3 Deconstruction – Labor Summary of Tasks Performed

Component	Tasks (hours)				Component Total	Labor-hours/unit
	Disassembly	Processing	Prod. Support	Non-prod.		
Interior						
1. Interior doors, frames, trim	5.75	5.25	----	----	11.0	0.55/each
Baseboards	4.75	5.0			9.75	0.19/lf
2. Kitchen cabinets	2.75	0.5			3.25	0.27/each
Plumbing fixtures	7.75	1.75	----	----	9.5	0.59/each
Radiators	1.5	0.5			2.0	0.13/each
Appliances	0.25	2.75			3.0	0.60/each
3. Bathroom floor tile	2.50	0.50	----	----	3.0	0.038/sf
4. Oak strip flooring	19.25	27.0	0.25		46.50	0.038/sf
5. Plaster - upper level	34.25	10.0	5.50	----	49.75	0.012/sf(plaster area)
6. Plaster - lower level	23.75	10.75	2.0		36.50	0.009/sf(plaster area)
7. Piping and wiring	6.75	3.25	0.50	----	10.50	0.0072/lbs
8. Interior partition walls	6.25	24.75	3.0	----	34.0	0.18/lf
9. Windows and window trim	10.0	2.50	0.50	----	13.0	0.54 each
10. Ceiling joists	1.0	4.75	0.5	----	6.25	0.0075/lf
11. Interior load-bearing walls	2.75	15.5	1.75	----	20.0	0.027/lf
12. Second level sub-floor	16.0	6.0	1.25	----	23.25	0.023/sf
13. Second level joists	7.25	16.25	1.5	----	25.0	0.027/lf
14. First level sub-floor	7.75	8.0	----	----	15.75	0.016/sf
15. First level joists	7.0	10.0	----	----	17.0	0.020/lf
16. Stairs	2.5	0.75	0.75	----	4.0	0.3/riser
Exterior						
17. Gutters, fascias, rakes	2.25	1.0	----	----	3.25	0.014/lf
18. Chimney	33.25	40.5	4.75	----	78.5	0.16/cu.ft.
19. Gable ends	8.0	3.0	0.75	----	11.75	0.053/sf
20. Masonry walls - upper section	14.75	104.5	20.5	----	139.75	0.25/sf(brick area)
21. Masonry walls - lower section	15.75	84	5.25	----	105.0	0.078/sf(brick area)
Roof						
22. Roofing material	17.75	18.25	1.75	----	37.75	2.68/100 sf
23. Roof Sheathing boards	21.25	14.5	1.5	----	37.25	0.028/sf
24. Roof framing	7.25	9.75	7	----	24.0	0.021/lf
25. Shed roof framing at entry	1.25	2.25	----	----	3.5	0.036/lf
Building Subtotal	291.25	433.5	59	----	783.75	
26. Talk shop	----	----	29	29.5	58.5	NA
27. Supervision	----	----	9.5	----	9.5	
28. Meetings, paper work, daily roll-out and roll-in of tools, etc.	----	----	38	43.5	81.5	
29. Research monitoring	----	----	----	89.5	89.5	
30. Lunch, breaks, idle	----	----	----	118.75	118.75	
Business Subtotal	----	----	76.5	280.25	357.75	
Grand Total	291.25	433.5	135.5	280.25	1141.5	

Note: Averages from new deconstruction studies are being compiled for RSMeans 2010 cost data and may differ slightly from numbers printed here.

(courtesy of NAHB Research Center, www.nahbrc.org Deconstruction: Building Disassembly and Material Salvage – The Riverdale Case Study *[1997])*

Figure 2-4   Underground Storage Tank Removal

Underground storage tank removal can be divided into two categories: non-leaking and leaking. Prior to removing an underground storage tank, tests should be made, with the proper authorities present, to determine whether a tank has been leaking or the surrounding soil has been contaminated.

To safely remove liquid underground storage tanks:

1. Excavate to the top of the tank.
2. Disconnect all piping.
3. Open all tank vents and access ports.
4. Remove all liquids and/or sludge.
5. Purge the tank with an inert gas.
6. Provide access to the inside of the tank and clean out the interior using proper personal protective equipment (PPE).
7. Excavate soil surrounding the tank using proper PPE for on-site personnel.
8. Pull and properly dispose of the tank.
9. Clean up the site of all contaminated material.
10. Install new tanks or close the excavation.

Figure 2-5   Steps for Estimating Asbestos Removal

Asbestos removal is accomplished by a specialty contractor who understands the federal and state regulations regarding the handling and disposal of the material. The process of asbestos removal is divided into many individual steps. An accurate estimate can be calculated only after all the steps have been priced.

The steps are generally as follows:

1. Obtain an asbestos abatement plan from an industrial hygienist.
2. Monitor the air quality in and around the removal area and along the path of travel between the removal area and transport area. This establishes the background contamination.
3. Construct a two-part decontamination chamber at entrance to removal area.
4. Install a HEPA filter to create a negative pressure in the removal area.
5. Install wall, floor, and ceiling protection as required by the plan, usually 2 layers of fireproof 6 mil polyethylene.
6. Industrial hygienist visually inspects work area to verify compliance with plan.
7. Provide temporary supports for conduit and piping affected by the removal process.
8. Proceed with asbestos removal and bagging process. Monitor air quality as described in Step #2. Discontinue operations when contaminate levels exceed applicable standards.
9. Document the legal disposal of materials in accordance with EPA standards.
10. Thoroughly clean removal area including all ledges, crevices, and surfaces.
11. Post-abatement inspection by industrial hygienist to verify plan compliance.
12. Provide a certificate from a licensed industrial hygienist attesting that contaminate levels are within acceptable standards before returning area to regular use.

Figure 2-6   Lead Paint Remediation Methods

Lead paint remediation can be accomplished by the following methods:

1. Abrasive blast
2. Chemical stripping
3. Power tool cleaning with vacuum collection system
4. Encapsulation
5. Remove and replace
6. Enclosure

Each of these methods has strengths and weaknesses depending on the specific circumstances of the project. The following is an overview of each method:

1. **Abrasive blasting** is usually accomplished with sand or recyclable metallic blast. Before work can begin, the area must be contained to ensure the blast material with lead does not escape to the atmosphere. The use of vacuum blast greatly reduces the containment requirements. Lead abatement equipment that may be associated with this work includes a negative air machine. In addition, it is necessary to have an industrial hygienist monitor the project on a continual basis. When the work is complete, the spent blast sand with lead must be disposed of as a hazardous material. If metallic shot was used, the lead is separated from the shot and disposed of as hazardous material. Worker protection includes disposable clothing and respiratory protection.

2. **Chemical stripping** requires strong chemicals be applied to the surface to remove the lead paint. Before the work can begin, the area under/adjacent to the work area must be covered to catch the chemical and removed lead. After the chemical is applied to the painted surface, it is usually covered with paper. The chemical is left in place for the specified period, then the paper with lead paint is pulled or scraped off. The process may require several chemical applications.

The paper with chemicals and lead paint adhered to it, plus the containment and loose scrapings collected by a HEPA (High Efficiency Particulate Air Filter) vac, must be disposed of as a hazardous material. The chemical stripping process usually requires a neutralizing agent and several wash downs after the paint is removed. Worker protection includes a neoprene or other compatible protective clothing and respiratory protection with face shield. An industrial hygienist is required intermittently during the process.

3. **Power tool cleaning** is accomplished using shrouded needle blasting guns. The shrouding with different end configurations is held up against the surface to be cleaned. The area is blasted with hardened needles and the shroud captures the lead with a HEPA vac and deposits it in a holding tank. An industrial hygienist monitors the project; protective clothing and a respirator are required until air samples prove otherwise. When the work is complete, the lead must be disposed of as a hazardous material.

4. **Encapsulation** is a method that leaves the well bonded lead paint in place after the peeling paint has been removed. Before the work can begin, the area under/adjacent to the work must be covered to catch the scrapings. The scraped surface is then washed with a detergent and rinsed. The prepared surface is covered with approximately 10 mils of paint. A reinforcing fabric can also be embedded in the paint covering. The scraped paint and containment must be disposed of as a hazardous material. Workers must wear protective clothing and respirators.

5. **Remove and replace** is an effective way to remove lead paint from windows, gypsum walls, and concrete masonry surfaces. The painted materials are removed and new materials are installed. Workers should wear a respirator and Tyvek® suit. The

(continued on next page)

Figure 2-6 Lead Paint Remediation Methods *(continued from previous page)*

demolished materials must be disposed of as a hazardous waste if it fails the TCLP (Toxicity Characteristic Leaching Process) test.

6. **Enclosure** is the process that permanently seals lead painted materials in place. This process has many applications such as covering lead painted drywall with new drywall, covering exterior construction with Tyvek paper, and then re-siding, or

covering lead painted structural members with aluminum or plastic. The seams on all enclosing materials must be securely sealed. An industrial hygienist monitors the project, and protective clothing and a respirator are required until air samples prove otherwise.

All the processes require clearance monitoring and wipe testing as required by the hygienist.

Figure 2-7 Installation Time in Labor-Hours for Lead Paint Removal by Chemicals

Hazardous Material Remediation	Crew	Daily Output	Labor-Hours	Unit
Removal Existing lead paint, by chemicals, per application				
Baseboard, to 6" wide	1 Painter	64	.125	L.F.
To 12" wide		32	.250	"
Balustrades, one side		28	.286	S.F.
Cabinets simple design		32	.250	
Ornate design		25	.320	
Cornice, simple design		60	.133	
Ornate design		20	.400	
Doors, one side, flush		84	.095	
Two panel		80	.100	
Four panel		45	.178	▼
For trim, one side, add		64	.125	L.F.
Fence, picket, one side		30	.267	S.F.
Grilles, one side, simple design		30	.267	
Ornate design		25	.320	▼
Pipes, to 4" diameter		90	.089	L.F.
To 8" diameter		50	.160	
To 12" diameter		36	.222	
To 16" diameter		20	.400	▼

Note: One chemical application will remove approximately 3 layers of paint.

Figure 2-8 Installation Time in Labor-Hours for Lead Paint Encapsulation

Hazardous Material Remediation	Crew	Labor-Hours	Unit
LEAD PAINT ENCAPSULATION, water based polymer coating, 14 mil DFT, interior, brushwork, trim, under 6"	1 Painter	.033	L.F.
6" to 12" wide		.044	
Balustrades		.027	
Pipe to 4" diameter		.016	
To 8" diameter		.021	
To 12" diameter		.032	
To 16" diameter		.047	
Cabinets, ornate design		.040	S.F.
Simple design		.032	"
Doors, 3' x 7', both sides, incl. frame & trim			
Flush		1.333	Ea.
French, 10–15 lite		2.667	
Panel		2.000	
Louvered		2.909	
Windows, per interior side, per 15 S.F.,			
1–6 lite		.571	Ea.
7–10 lite		1.067	
12 lite		1.391	
Radiators		1.000	
Grilles, vents		.029	S.F.
Walls, roller, drywall, or plaster		.008	
With spunbonded reinforcing fabric		.011	
Wood		.010	

Division Three
Concrete

Introduction

Concrete is one of the most versatile and widely used materials in the construction industry. It can be custom-mixed for compressive strength or color application, placed in almost any shape or form, and finished to resemble masonry, stone, and other surfaces. Concrete is durable and virtually maintenance-free. It is also fairly easy to work with and installs relatively quickly.

Concrete estimating is generally divided into five basic areas:

- Concrete materials and placement
- Formwork
- Reinforcing
- Finishing
- Precast concrete

Concrete quantities are usually taken off in measures of cubic yards, as this is how supply companies charge for it. Formwork is taken off in square feet (generally in square feet of area in contact with the concrete as opposed to the actual area of the forms used). Reinforcing is normally taken off in tons of steel. Finishing is taken off in square feet of finished surface. Precast concrete quantities are taken off in either square feet or by the piece, depending on usage and the type of unit being considered.

Although concrete requires a large amount of energy to produce, there have been recent strides by manufacturers to reduce the environmental impacts of processing and using it. By adding blast furnace slag or coal fly ash (industrial waste products), the cement content of concrete is reduced along with the pollution and greenhouse gasses associated with its production. Using recycled aggregate reduces the amount of material needing to be transported to landfills. Concrete can also be formulated to be very porous; in this form it is useful in reducing pollution due to storm water runoff by allowing the water to seep into the ground where it can be absorbed naturally.

Traditional concrete does not have a high insulating value, but it does have a high mass. This can be used to its advantage. In passive solar spaces, concrete floors can soak up heat during hot summer days and release it during the cooler evenings. When additional insulation value is required for underground structures, permanent insulating formwork works well and saves the additional step of adding insulation after the concrete is cured.

Another green approach is finishing the concrete surface to a higher level of quality, possibly adding color or texture in lieu of adding carpeting or tile and their associated costs, both monetary and environmental.

For estimating concrete quantities, this division includes charts and tables in the following areas:

- General information
- General formulas
- Proportionate quantities for concrete forms and reinforcing
- Concrete handling
- Material quantities for concrete
- Concrete volume for various applications
- Formwork
- Reinforcing
- Installation times for concrete in labor-hours

Division 3

Estimating Data

The following tables present guidelines for estimating items found in Division 3 – Concrete. Please note that these guidelines are intended as indicators of what may be expected, but that each project must be evaluated individually.

Table of Contents

Division 3

Estimating Tips

Concrete Block-outs

When estimating quantities of concrete for floor slabs or walls, do not bother to deduct small areas (two square feet or so) unless there are a large number of these areas, as this can take up more estimating time than the areas are worth. Also remember that you will be adding approximately 3% to the total volume for waste, thereby making these small areas even less significant.

Reinforcing Steel Adds

When estimating the amount of reinforcing steel, either bar or mesh, if no lap specifications are given, add 10% to your quantities for lapping, splices, and waste.

Check All Plans

It cannot be stressed enough that all plans for concrete must be checked. Concrete has been known to show up in all sections of construction drawings in various forms and uses (e.g., equipment pads located only on mechanical or electrical drawings, grouting requirements on steel drawings, etc.). Assuming all concrete requirements are indicated on the structural and architectural drawings can be a costly error.

Cold Weather Pouring of Footings

When placing concrete in cold weather, it may not always be necessary to use heating devices to keep concrete warm while it cures. In many cases, insulating blankets and straw are all that are needed. Each situation must be evaluated individually, a worthwhile exercise in view of the considerable expense that can be saved.

Concrete Placement — Direct Chute

When estimating the placement of concrete by direct chute, the forms available generally determine the volume of concrete placed per pour. The more forms you have available, the more concrete can be placed in the course of a day.

Form Uses

When estimating the number of forms and reuses for a job, remember to have enough forms on hand to keep the forming crew(s) busy while the previously poured concrete sets and starts to cure.

Concrete Placement — Bucket, Pump, or Conveyor

When placing concrete by methods involving a bucket and crane, pumping system, conveyor belt system, or other mechanical system, set up enough forms to keep the above systems productive for the entire day. Usually the cost for the use of the equipment for a full day will be charged to you even if you only use it for part of a day.

Sequencing the Pour

During the estimating phase of the project, whatever method or sequence you envision for placing concrete should be documented thoroughly. This will allow those in the field to know how you arrived at the estimated costs and which methods they should use (or try to out-do, in terms of cost effectiveness). Another reason for detailed documentation is that during the documentation process, it may become evident that you cannot place the concrete as you planned or that you can do it differently, more efficiently, or at less cost.

Testing Concrete

When estimating the amount of concrete compression testing that will be necessary for a project, figure on a minimum of one test per pour on smaller pours and a minimum of one test for each fifty yards of concrete placed. Each test should consist of taking a set of three cylinders minimum.

Finishing Concrete

Rule of thumb for finishing concrete: allow 1,000 SF of slab per each cement finisher.

Drilling Anchor Bolts

Consider drilling anchor bolts into the concrete, where design permits, to save time, labor, and materials on layout and templates during the pour.

Checklist ✓

For an estimate to be reliable, all items must be accounted for. A complete estimate can also limit contingencies. The following checklist can be used to help ensure that all items are included.

☐ *Applications*
- ☐ Aprons
- ☐ Architectural concrete
- ☐ Bases
  - ☐ Flagpole
  - ☐ Light
  - ☐ Sign
- ☐ Beams
- ☐ Cant strips
- ☐ Columns
- ☐ Copings
- ☐ Curbs
- ☐ Drives
- ☐ Elevated slabs incl. on metal deck
- ☐ Embedded items
- ☐ Equipment pads
- ☐ Exterior walls
- ☐ Footings
  - ☐ Column
  - ☐ Spread
  - ☐ Strip (wall)
  - ☐ Grade beams
  - ☐ Pile caps
- ☐ Foundation mats
- ☐ Foundation walls
- ☐ Girders
- ☐ Gutters
- ☐ Panels
- ☐ Piers
- ☐ Platforms
- ☐ Ramps
- ☐ Retention walls
- ☐ Sills
- ☐ Slab-on-grade
- ☐ Stair fill
- ☐ Stairs
  - ☐ On grade
  - ☐ Planters
- ☐ Sump pits
- ☐ Toppings
- ☐ Walks

☐ *Concrete*
- ☐ Aggregate
- ☐ Air entraining agent
- ☐ Cement
- ☐ Color
- ☐ Compressive strength
- ☐ Enhancer
- ☐ Hardeners
- ☐ Heated
- ☐ Mix design
- ☐ Other admixtures
- ☐ Retarder
- ☐ Sand
- ☐ Water

☐ *Formwork*
- ☐ Block-outs
- ☐ Bracing
- ☐ Chamfer strips
- ☐ Clamps
- ☐ Clean
- ☐ Erection
- ☐ Expansion joints
- ☐ Inserts
- ☐ Keyways
- ☐ Liners
- ☐ Oil
- ☐ Release agent
- ☐ Removal

- ☐ Repair
- ☐ Scaffolding
- ☐ Shoring
- ☐ Storage
- ☐ Ties
- ☐ Transportation
- ☐ Vapor barrier
- ☐ Beam forms
- ☐ Column forms
- ☐ Curb forms
- ☐ Domes
- ☐ Edge forms
- ☐ Fuel pump forms
- ☐ Gutter forms
- ☐ Pans
- ☐ Ramp forms
- ☐ Slab forms
- ☐ Stair forms
- ☐ Wall forms
- ☐ Joints
  - ☐ Expansion
  - ☐ Construction

☐ *Reinforcing*

- ☐ Bars
  - ☐ Grade of steel
  - ☐ Galvanized
  - ☐ Epoxy coating
- ☐ Wire mesh
  - ☐ Galvanized
  - ☐ Epoxy coating
- ☐ Fiber reinforcing
- ☐ Forming
- ☐ Ties & stirrups
- ☐ Accessories
- ☐ Cutting
- ☐ Chairs
- ☐ Bolsters

☐ *Ready-Mix Concrete*

- ☐ Transporting
  - ☐ Mixer truck

- ☐ Hand or power buggies
- ☐ Conveyer
- ☐ Placing
  - ☐ Direct chute
  - ☐ Crane & bucket
  - ☐ Pump truck
- ☐ Consolidating

☐ *Curing and Protection*

- ☐ Admixtures
- ☐ Sprays
- ☐ Straw
- ☐ Blankets
- ☐ Heating
- ☐ Canvas
- ☐ Plastic sheeting

☐ *Finish*

- ☐ Screed
- ☐ Darby
- ☐ Float
- ☐ Broom
- ☐ Trowel

☐ *Toppings*

- ☐ Abrasive
- ☐ Aggregate
- ☐ Colors
- ☐ Dustproof
- ☐ Epoxy
- ☐ Granolithic
- ☐ Hardeners
- ☐ Integral

☐ *Patch*

☐ *Rub*

☐ *Grind*

☐ *Fill*

☐ *Sandblast*

Division 3

☐ *Caulk*

☐ *Precast Concrete*
    ☐ Architectural
    ☐ Beams
       ☐ Ts
       ☐ Ls
    ☐ Columns
    ☐ Floor slabs
       ☐ Slabs
       ☐ Planks
       ☐ Hollow core

☐ Tees
☐ Double tees
☐ Multiple tees
☐ Joists
☐ Lift slaps
☐ Lintels
☐ Lightweight
☐ Stairs
☐ Wall panels
    ☐ Tilt-ups
    ☐ Precast
    ☐ Architectural

Figure 3-1 Cast-in-Place Formed Concrete

Although the concrete and reinforcing material costs for a grade beam on the ground and exterior spandrel beam on the twentieth floor may be the same, forming, placing, stripping, and rubbing costs are not. This is the reason for performing a detailed breakdown of costs for an estimate.

The cost of formed concrete-in-place is always a function of the area of forms or square feet of contact area required. Per cubic yard of concrete, an 8" foundation wall will require 50% more form material, form labor, stripping, rubbing, and waterproofing than a 12" wall. Still, many contractors continue to estimate formed concrete-in-place by the cubic yard without considering the wall thickness.

Quantities can be compiled and listed in various formats, but EACH of the quantities MUST be computed. It is important that once a format for compiling quantities is established, the same format should be used as a standard operating estimating procedure. This standardization minimizes errors and omissions and presents the various quanitities in an orderly manner for pricing, cost control, and future estimating.

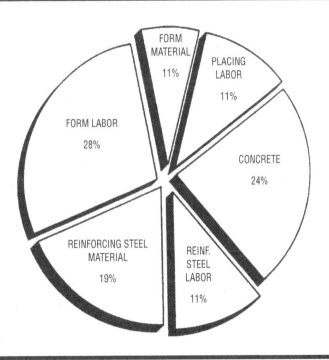

Figure 3-2 Formwork: Ratio of Board Feet to Square Feet of Net Contact Area

Job-built forms for different concrete components have various lumber requirements. This table provides the board feet of lumber required for each square foot of form contact area. In this table, the surface of the form is figured at 1 BF per SF, and the studs and bracing are figured in the usual methods of calculating board feet.

Formwork	Ratio
Wall or spread footings	2.0 B.F./S.F.
Column and pier footings	2.5 B.F./S.F.
Columns and piers	3.0 B.F./S.F.
Walls - to 12' 0" high	2.5 B.F./S.F.
Walls - over 12' 0" high	3.5 B.F./S.F.
Wall pilasters	4.0 B.F./S.F.
Beams (without shores)	4.0 B.F./S.F.
Supported slabs (without shores)	2.0 B.F./S.F.
Metal pan centering (without shores)	1.5 B.F./S.F.
Wood shores	1.8 B.F./S.F.
Stairs, without soffits	2.5 B.F./S.F.
Stairs, with soffits	6.0 B.F./S.F.

Figure 3-3 Formwork Labor-Hours

Labor-hours for formwork vary greatly from one component to another. This table is convenient for planning the number of forms to be fabricated or provided, depending on the number of reuses anticipated.

Item	Unit	Hours Required			Total Hours 1 Use	Multiple Use		
		Fabricate	Erect & Strip	Clean & Move		2 Use	3 Use	4 Use
Beam and Girder, interior beams, 12" wide	100 S.F.	6.4	8.3	1.3	16.0	13.3	12.4	12.0
Hung from steel beams		5.8	7.7	1.3	14.8	12.4	11.6	11.2
Beam sides only, 36" high		5.8	7.2	1.3	14.3	11.9	11.1	10.7
Beam bottoms only, 24" wide		6.6	13.0	1.3	20.9	18.1	17.2	16.7
Box out for openings		9.9	10.0	1.1	21.0	16.6	15.1	14.3
Buttress forms, to 8' high		6.0	6.5	1.2	13.7	11.2	10.4	10.0
Centering, steel, 3/4" rib lath			1.0		1.0			
3/8" rib lath or slab form	↓		0.9		0.9			
Chamfer strip or keyway	100 L.F.		1.5		1.5	1.5	1.5	1.5
Columns, fiber tube 8" diameter			20.6		20.6			
12"			21.3		21.3			
16"			22.9		22.9			
20"			23.7		23.7			
24"	↓		24.6		24.6			
30"			25.6		25.6			
Round Steel, 12" diameter			22.0		22.0	22.0	22.0	22.0
16"			25.6		25.6	25.6	25.6	25.6
20"			30.5		30.5	30.5	30.5	30.5
24"	↓		37.7		37.7	37.7	37.7	37.7
Plywood 8" x 8"	100 S.F.	7.0	11.0	1.2	19.2	16.2	15.2	14.7
12" x 12"		6.0	10.5	1.2	17.7	15.2	14.4	14.0
16" x 16"		5.9	10.0	1.2	17.1	14.7	13.8	13.4
24" x 24"		5.8	9.8	1.2	16.8	14.4	13.6	13.2
Steel framed plywood 8" x 8"			10.0	1.0	11.0	11.0	11.0	11.0
12" x 12"			9.3	1.0	10.3	10.3	10.3	10.3
16" x 16"			8.5	1.0	9.5	9.5	9.5	9.5
24" x 24"			7.8	1.0	8.8	8.8	8.8	8.8
Drop head forms, plywood		9.0	12.5	1.5	23.0	19.0	17.7	17.0
Coping forms		8.5	15.0	1.5	25.0	21.3	20.0	19.4
Culvert, box			14.5	4.3	18.8	18.8	18.8	18.8
Curb forms, 6" to 12" high, on grade		5.0	8.5	1.2	14.7	12.7	12.1	11.7
On elevated slabs	↓	6.0	10.8	1.2	18.0	15.5	14.7	14.3
Edge forms to 6" high, on grade	100 L.F.	2.0	3.5	0.6	6.1	5.6	5.4	5.3
7" to 12" high	100 S.F.	2.5	5.0	1.0	8.5	7.8	7.5	7.4
Equipment foundations	"	10.0	18.0	2.0	30.0	25.5	24.0	23.3
Flat slabs, including drops		3.5	6.0	1.2	10.7	9.5	9.0	8.8
Hung from steel		3.0	5.5	1.2	9.7	8.7	8.4	8.2
Closed deck for domes		3.0	5.8	1.2	10.0	9.0	8.7	8.5
Open deck for pans		2.2	5.3	1.0	8.5	7.9	7.7	7.6
Footings, continuous, 12" high		3.5	3.5	1.5	8.5	7.3	6.8	6.6
Spread, 12" high		4.7	4.2	1.6	10.5	8.7	8.0	7.7
Pile caps, square or rectangular		4.5	5.0	1.5	11.0	9.3	8.7	8.4
Grade beams, 24" deep		2.5	5.3	1.2	9.0	8.3	8.0	7.9
Lintel or Sill forms		8.0	17.0	2.0	27.0	23.5	22.3	21.8
Spandrel beams, 12" wide		9.0	11.2	1.3	21.5	17.5	16.2	15.5
Stairs			25.0	4.0	29.0	29.0	29.0	29.0
Trench forms in floor		4.5	14.0	1.5	20.0	18.3	17.7	17.4
Walls, Plywood, at grade, to 8' high		5.0	6.5	1.5	13.0	11.0	9.7	9.5
8' to 16'		7.5	8.0	1.5	17.0	13.8	12.7	12.1
16' to 20'		9.0	10.0	1.5	20.5	16.5	15.2	14.5
Foundation walls, to 8' high		4.5	6.5	1.0	12.0	10.3	9.7	9.4
8' to 16' high		5.5	7.5	1.0	14.0	11.8	11.0	10.6
Retaining wall to 12' high, battered		6.0	8.5	1.5	16.0	13.5	12.7	12.3
Radial walls to 12' high, smooth		8.0	9.5	2.0	19.5	16.0	14.8	14.3
But in 2' chords		7.0	8.0	1.5	16.5	13.5	12.5	12.0
Prefabricated modular, to 8' high		—	4.3	1.0	5.3	5.3	5.3	5.3
Steel, to 8' high		—	6.8	1.2	8.0	8.0	8.0	8.0
8' to 16' high		—	9.1	1.5	10.6	10.3	10.2	10.2
Steel framed plywood to 8' high		—	6.8	1.2	8.0	7.5	7.3	7.2
8' to 16' high	↓	—	9.3	1.2	10.5	9.5	9.2	9.0

Division 3

Figure 3-4 Typical Metal Pan Erection Procedures

Metal or fiberglass pans are used for joist slab construction. They can be erected either on a solid platform or on centering as described below. Each manufacturer provides information on erection and concrete volumes for its products. The following information (from CECO Corporation) is the type of information provided by pan manufacturers.

1. All beam, girder, and column forms are placed and braced by the general contractor.
2. The shores and stringers are erected.
3. The soffit forms are placed on the stringers, and the shores are adjusted to support the soffit forms at the elevation set by the beam and girder forms.
4. The wood headers for bridging joists, beam tees, and special headers for electrical outlets, etc., are then placed.
5. Lines for joist edges are marked with a chalkline.
6. The end forms are placed first and are nailed to the soffit with barbed nails through the nail holes in the flanges. Note that the soffit form does not require special fillers to accommodate the tapered end form.
7. The 3' intermediates are set and nailed. Overlaps are 1" minimum and 5" maximum.
8. Various joist lengths are accommodated by closing the center gap with 1', 2', or 3' long intermediates. For the special 10" and 15" filler widths, which are furnished in 3' lengths only, joist length variations are accommodated by overlapping.
9. Pans and soffit boards are oiled to facilitate removal.
10. All reinforcing steel and mesh, all hanger wires, plumbing sleeves, anchors, electrical outlet boxes, etc., are placed by the general contractor or his subcontractors.
11. Concrete is placed by the general contractor.
12. Shores and centering are removed when the concrete has gained its proper strength.
13. Steel forms are removed and reconditioned for the next use.

Figure 3-5 Symbols and Abbreviations Used in Reinforcing Steel Drawings

#	Indicates size of deformed bar number
$\varnothing$	Round, used mainly for plain round bars
@	Spacing, center to center, each at
⟵⟶	Direction in which bars extend, span
↔	Limits of area covered by bars

PL	Plain bar	OF	Outside face	₵	Center line
Bt	Bent	NF	Near face	∠	Angle
Str	Straight	FF	Far Face	WWF	Welded wire fabric
Stir	Stirrup	EF	Each face		
Sp	Spiral	Bot	Bottom		
Ct	Column tie	EW	Each way		
IF	Inside face	T	Top		

Figure 3-6 Reinforcing Steel Weights and Measures

Bar Desig-nation No.**	Nominal Weight Lb./Ft.	U.S. Customary Units Nominal Dimensions*			Nominal Weight kg/m	SI Units Nominal Dimensions*		
		Diameter in.	Cross Sectional Area, in.²	Perimeter in.		Diameter mm	Cross Sectional Area, cm²	Perimeter mm
3	0.376	0.375	0.11	1.178	0.560	9.52	0.71	29.9
4	0.668	0.500	0.20	1.571	0.994	12.70	1.29	39.9
5	1.043	0.625	0.31	1.963	1.552	15.88	2.00	49.9
6	1.502	0.750	0.44	2.356	2.235	19.05	2.84	59.8
7	2.044	0.875	0.60	2.749	3.042	22.22	3.87	69.8
8	2.670	1.000	0.79	3.142	3.973	25.40	5.10	79.8
9	3.400	1.128	1.00	3.544	5.059	28.65	6.45	90.0
10	4.303	1.270	1.27	3.990	6.403	32.26	8.19	101.4
11	5.313	1.410	1.56	4.430	7.906	35.81	10.06	112.5
14	7.65	1.693	2.25	5.32	11.384	43.00	14.52	135.1
18	13.60	2.257	4.00	7.09	20.238	57.33	25.81	180.1

*The nominal dimensions of a deformed bar are equivalent to those of a plain round bar having the same weight per foot as the deformed bar.

**Bar numbers are based on the number of eighths of an inch in the nominal diameter of the bars.

Figure 3-7 Weight of Steel Reinforcing per SF in Wall (PSF)

The table below lists the weight per SF for reinforcing steel in walls. Weights will be correct for any grade steel reinforcing. For bars in two directions, add weights for each size and spacing.

C/C Spacing in Inches	Bar Size								
	#3 Wt. (PSF)	#4 Wt. (PSF)	#5 Wt. (PSF)	#6 Wt. (PSF)	#7 Wt. (PSF)	#8 Wt. (PSF)	#9 Wt. (PSF)	#10 Wt. (PSF)	#11 Wt. (PSF)
2"	2.26	4.01	6.26	9.01	12.27				
3"	1.50	2.67	4.17	6.01	8.18	10.68	13.60	17.21	21.25
4"	1.13	2.01	3.13	4.51	6.13	8.10	10.20	12.91	15.94
5"	.90	1.60	2.50	3.60	4.91	6.41	8.16	10.33	12.75
6"	.752	1.34	2.09	3.00	4.09	5.34	6.80	8.61	10.63
8"	.564	1.00	1.57	2.25	3.07	4.01	5.10	6.46	7.97
10"	.451	.802	1.25	1.80	2.45	3.20	4.08	5.16	6.38
12"	.376	.668	1.04	1.50	2.04	2.67	3.40	4.30	5.31
18"	.251	.445	.695	1.00	1.32	1.78	2.27	2.86	3.54
24"	.188	.334	.522	.751	1.02	1.34	1.70	2.15	2.66
30"	.150	.267	.417	.600	.817	1.07	1.36	1.72	2.13
36"	.125	.223	.348	.501	.681	.890	1.13	1.43	1.77
42"	.107	.191	.298	.429	.584	.763	.97	1.17	1.52
48"	.094	.167	.261	.376	.511	.668	.85	1.08	1.33

Figure 3-8   Common Stock Styles of Welded Wire Fabric

The following table and graphic provide the specifications, sizes, and weights of welded wire fabric used for slab reinforcement.

	New Designation	Old Designation	Steel Area per Foot				Approximate Weight per 100 S.F.	
	Spacing - Cross Section Area (in.) - Sq. In. 100)	Spacing Wire Gauge (in.) - (AS & W)	Longitudinal		Transverse			
			in.	cm.	in.	cm	Lbs.	kg
Rolls	6 x 6 - W1.4 x W1.4	6 x 6 - 10 x 10	0.028	0.071	0.028	0.071	21	9.53
	6 x 6 - W2.0 x W2.0	6 x 6 - 8 x 8 (1)	0.040	0.102	0.040	0.102	29	13.15
	6 x 6 - W2.9 x W2.9	6 x 6 - 6 x 6	0.058	0.147	0.053	0.147	42	19.05
	6 x 6 - W4.0 x W4.0	6 x 6 - 4 x 4	0.080	0.203	0.080	0.203	58	26.31
	4 x 4 - W1.4 x W1.4	4 x 4 - 10 x 10	0.042	0.107	0.042	0.107	31	14.06
	4 x 4 - W2.0 x W2.0	4 x 4 - 8 x 8 (1)	0.060	0.152	0.060	0.152	43	19.50
	4 x 4 - W2.9 x W2.9	4 x 4 - 6 x 6	0.087	0.221	0.087	0.221	62	28.12
	4 x 4 - W4.0 x W4.0	4 x 4 - 4 x 4	0.120	0.305	0.120	0.305	85	38.56
Sheets	6 x 6 - W2.9 x W2.9	6 x 6 - 6 x 6	0.058	0.147	0.058	0.147	42	19.05
	6 x 6 - W4.0 x W4.0	6 x 6 - 4 x 4	0.080	0.203	0.080	0.203	58	26.31
	6 x 6 - W5.5 x W5.5	6 x 5 - 2 x 2 (2)	0.110	0.279	0.110	0.279	80	36.29
	6 x 6 - W4.0 x W4.0	4 x 4 - 4 x 4	0.120	0.305	0.120	0.305	85	38.56

Notes:

Exact W-number size for 8 gauge is W2.1

Exact W-number size for 2 gauge is W5.4

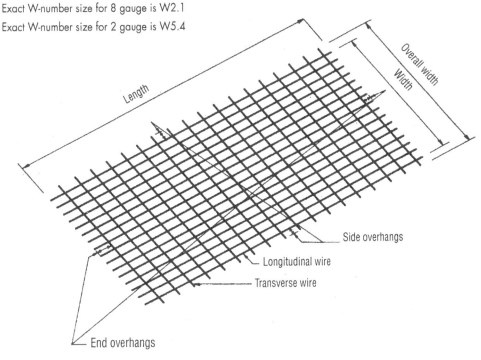

Welded Wire Fabric

Division 3

Figure 3-9a Estimated Weight of Welded Wire Fabric for All Styles Having Uniform Spacings and Gauges of Transverse Members

Steel Wire Gauge Numbers	Approximate Weights in Lbs. per 100 S.F.—Based on 60" Width C. to C. of Outside Longitudinal Wires							
	Weight of Transverse Members*							
	Spacing							
	2"	3"	4"	6"	8"	10"	12"	16"
000	256.95	170.95	128.22	85.48	61.11	51.29	42.74	32.05
000	217.31	144.87	108.66	72.44	54.33	43.46	36.22	27.16
00	181.16	120.78	90.58	60.39	45.29	36.23	30.19	22.65
0	155.37	103.58	77.69	51.79	38.84	31.07	25.90	19.42
1	132.43	88.29	66.22	44.14	33.11	26.49	22.07	16.55
2	113.96	75.97	56.98	37.99	28.49	22.79	18.99	14.24
1/4"	103.33	68.89	51.67	34.44	25.83	20.67	17.22	12.92
3	98.21	65.47	49.10	32.74	24.55	19.64	16.37	12.28
4	83.95	55.97	41.97	27.98	20.99	16.79	13.99	10.49
5	70.87	47.24	35.43	23.62	17.72	14.17	11.81	8.86
6	60.96	40.64	30.48	20.32	15.24	12.19	10.16	7.62
7	51.81	34.54	25.90	17.27	12.95	10.36	8.63	6.48
8	43.40	28.93	21.70	14.47	10.85	8.68	7.23	5.43
9	36.37	24.25	18.18	12.12	9.09	7.27	6.06	4.55
10	30.14	20.09	15.07	10.05	7.53	6.03	5.02	3.77
11	24.01	16.01	12.01	8.00	6.00	4.80	4.00	3.00
12	18.41	12.27	9.20	6.14	4.60	6.38	3.07	2.30
13	13.84	9.23	6.92	4.61	3.46	2.77	2.31	1.73
14	10.58	7.06	5.29	3.53	2.65	2.12	1.76	1.32
15	8.57	5.72	4.29	2.86	2.14	1.71	1.43	1.07
16	6.46	4.31	3.23	2.15	1.62	1.29	1.08	.81

Note: To determine the weight of any type, add the weights of the longitudinal and transverse members, adjust the total to the nearest lb., considering 0.5 lb. or over to be 1 lb., and less then 0.5 lb. to be zero.

*Includes weight of 1" projection beyond longitudinal selvage wires.

(courtesy Concrete Reinforcing Steel Institute)

Figure 3-9b Estimated Weight of Welded Wire Fabric for All Styles Having Uniform Spacings and Gauges of Longitudinal Members

Steel Wire Gauge Numbers	Approximate Weights in Lbs. per 100 S.F.—Based on 60" Width C. to C. of Outside Longitudinal Wires						
	Weight of Longitudinal Members						
	Spacing						
	2"	3"	4"	6"	8"	10"	12"
0000000	397.05	268.97	204.93	140.89	108.87	89.66	76.85
000000	352.22	238.60	181.79	124.98	96.58	79.53	68.17
00000	306.47	207.61	158.18	108.75	84.03	69.20	59.31
0000	256.43	173.71	132.35	90.99	70.31	57.90	49.63
000	217.31	147.21	112.16	77.11	59.59	49.07	42.06
00	181.16	122.72	93.50	64.28	49.67	40.91	35.06
0	155.37	105.25	80.19	55.13	42.60	35.08	30.07
1	132.43	89.71	68.35	46.99	36.31	29.90	25.63
2	113.96	77.20	58.82	40.44	31.25	25.73	22.06
1/4"	103.33	70.00	53.33	36.67	28.33	23.33	20.00
3	98.21	66.53	50.69	34.85	26.93	22.18	19.01
4	83.95	56.87	43.33	29.79	23.02	18.96	16.25
5	70.87	48.01	36.58	25.15	19.43	16.00	13.72
6	60.96	41.29	31.46	21.63	16.71	13.76	11.80
7	51.81	35.10	26.74	18.38	14.21	11.70	10.03
8	43.40	29.40	22.40	15.40	11.90	9.80	8.40
9	36.37	24.64	18.77	12.91	9.97	8.21	7.04
10	30.14	20.42	15.56	10.69	8.26	6.81	5.83
11	24.01	16.27	12.39	8.52	6.58	5.42	4.65
12	18.41	12.47	9.50	6.53	5.05	4.16	3.56
13	13.84	9.38	7.15	4.91	3.80	3.13	2.68
14	10.58	7.17	5.46	3.76	2.90	2.39	2.05
15	8.57	5.81	4.43	3.04	2.35	1.94	1.66
16	6.46	4.38	3.33	2.29	1.77	1.46	1.25

Note: To determine the weight of any type, add the weights of the longitudinal and transverse members, adjust the total to the nearest lb., considering 0.5 lb. or over to be 1 lb., and less than 0.5 lb. to be zero.

(courtesy Concrete Reinforcing Steel Institute)

Figure 3-10 Wire Gauge, Designation, and Dimensions in Welded Wire Fabric for Concrete Reinforcing

American Steel and Wire Gauge Number	W & D Size Numbers Smooth	W & D Size Numbers Deformed	Area in.²	Area cm³	Nominal Diameter in.	Nominal Diameter mm
	W31	D31	0.310	2.000	0.628	15.95
	W30	D30	0.300	1.936	0.618	15.70
	W28	D28	0.280	1.807	0.597	15.16
	W26	D26	0.260	1.679	0.575	14.61
	W24	D24	0.240	1.549	0.553	14.05
	W22	D22	0.220	1.419	0.529	13.44
	W20	D20	0.200	1.290	0.504	12.80
0000000			0.189	1.219	0.490	12.45
	W18	D18	0.180	1.167	0.478	12.14
000000			0.167	1.076	0.4615	11.72
	W16	D16	0.160	1.032	0.451	11.46
00000			0.146	0.942	0.4305	10.94
	W14	D14	0.140	0.903	0.422	10.72
0000			0.122	0.787	0.394	10.01
	W12	D12	0.120	0.774	0.390	9.91
	W11	D11	0.110	0.710	0.374	9.54
	W10.5		0.105	0.678	0.366	9.30
000			0.103	0.665	0.3625	9.21
	W10	D10	0.100	0.645	0.356	9.04
	W9.5		0.095	0.613	0.348	8.84
	W9	D9	0.090	0.581	0.338	8.59
			0.086	0.556	0.331	8.41
00	W8.5		0.085	0.548	0.329	8.36

American Steel and Wire Gauge Number	W & D Size Numbers Smooth	W & D Size Numbers Deformed	Area in.²	Area cm²	Nominal Diameter in.	Nominal Diameter mm
	W8	D8	0.080	0.516	0.319	8.10
	W7.5		0.075	0.484	0.309	7.85
0			0.074	0.478	0.3065	7.79
	W7	D7	0.070	0.452	0.298	7.57
	W6.5		0.065	0.419	0.288	7.32
1			0.063	0.407	0.283	7.19
	W6	D6	0.060	0.387	0.279	7.01
	W5.5		0.055	0.355	0.264	6.71
2			0.054	0.348	0.2625	6.67
	W5	D5	0.050	0.323	0.252	6.40
3			0.047	0.303	0.244	6.20
	W4.5		0.045	0.290	0.240	6.10
4	W4	D4	0.040	0.258	0.225	5.72
	W3.5		0.035	0.226	0.211	5.36
5			0.034	0.219	0.207	5.26
	W3		0.030	0.194	0.195	4.95
6	W2.9		0.029	0.187	0.192	4.88
7	W2.5		0.025	0.161	0.177	4.37
8	W2.1		0.021	0.136	0.162	4.12
	W2		0.020	0.129	0.159	4.04
9			0.017	0.111	0.148	3.76
	W1.5		0.015	0.097	0.138	3.51
10	W1.4		0.014	0.090	0.135	3.43

Figure 3-11 Types of Welded Wire Fabric Most Commonly Used in Construction

Designation	Size of Square Mesh in.	Size of Square Mesh mm	Wire Designation	Sectional Area in.²/Ft.	Sectional Area cm²/m	Weight Lbs./100 Ft.²	Weight kg/m²
6 x 6 – 10/10	6	152.4	W1.4	0.029	.631	21	1.026
6 x 6 – 8/8	6	152.4	W2.1	0.041	.868	30	1.4660
6 x 6 – 6/6	6	152.4	W2.9	0.058	1.227	42	2.053
6 x 6 – 4/4	6	152.4	W4	.080	1.693	58	2.834
4 x 4 – 10/10	4	101.6	W1.4	.043	0.910	31	1.515
4 x 4 – 8/8	4	101.6	W2.1	.062	1.312	44	2.150
4 x 4 – 6/6	4	101.6	W2.9	.087	1.841	61	2.981
4 x 4 – 4/4	4	101.6	W4	.119	2.518	85	4.154

Figure 3-12 Preferred and Acceptable Methods for Transporting Concrete to the Job Site and at the Job Site

Job	Equipment	Screw Spreaders	Tremies	Drop Chutes	Chutes	Barrows, Buggies	Concrete Pumps	Pneumatic Guns	Belt Conveyors	Elevators	Cranes, Buckets	Mobile Continuous Mixers	Nonagitating Trucks	Truck Mixers	Truck Agitators
	Job Site Mix											●		X	
	Short Haul												●	X	●
	Long Haul													●	X
Horizontally	Short				●	●	●	●	●		●	●	●	●	●
Horizontally	Medium						●		X	X	X				
Horizontally	Long						X		●						
Vertically	Short		●	●			●	●	●		X				
Vertically	Medium						X			X	●				
Vertically	Long						X			●	X				
	Footings			●			X				X	●	X	X	X
	Foundations			●			X				X	●	X	X	X
	Walls			●	X		X	●			X	●	X	X	X
	Columns					X	●				●				
	Beams					X	●				●				
	Underwater		●												
	Elevated Slabs					●	●			X	●	●			
	Different Shapes						X	●							
	Mass Concrete						X		●		X		X	X	X
	Slabs on Ground	●										X	●	●	●
	Repairs							●				X			

● Preferred Method   X Acceptable Method

Division 3

Figure 3-13 Data on Handling and Transporting Concrete

On most construction jobs, the method of concrete placement is an important consideration for the estimator and the project manager. This table is a guide for planning time and equipment for concrete placement.

Quantity	Rate
Unloading 6 cubic yard mixer truck, 1 operator	Minimum time 2 minutes, average time 7 minutes
Wheeling, using 4-1/2 cubic feet wheelbarrows,	
1 Laborer	
Up to 100'	Average 1-1/2 cubic yards per hour
up to 200'	Average 1 cubic yard per hour
Wheeling, using 8 cubic feet hand buggies,	
1 laborer	
up to 100'	Average 5 cubic yards per hour
up to 200'	Average 3 cubic yards per hour
Transporting, using 28 cubic feet power buggies,	
1 laborer	
up to 500'	Average 20 cubic feet per hour
up to 1000'	Average 15 cubic feet per hour
Portable conveyor, 16" belt, 30° elevation,	
1 laborer	
100 FPM belt speed	30 C.Y./Hr. max., 15 C.Y./Hr. average
200 FPM belt speed	60 C.Y./Hr. max., 30 C.Y./Hr. average
300 FPM belt speed	90 C.Y./Hr. max., 45 C.Y./Hr. average
400 FPM belt speed	120 C.Y./Hr. max., 60 C.Y./Hr. average
500 FPM belt speed	150 C.Y./Hr. max., 75 C.Y./Hr. average
600 FPM belt speed	180 C.Y./Hr. max., 90 C.Y./Hr. average
Feeder conveyors, up to 600',	
5 laborers	
500 FPM belt speed	150 C.Y./Hr. max., 50 C.Y./Hr. average
600 FPM belt speed	180 C.Y./Hr. max., 60 C.Y./Hr. average
Side discharge conveyor,	
1 laborer	
fed by portable conveyor	40–60 C.Y./Hr. average
fed by feed conveyor	80–100 C.Y./Hr. average
fed by crane & bucket	35–60 C.Y./Hr. average
Mobile crane	
1 operator*	
2 laborers	
1 cubic yard bucket	40 C.Y./Hr. average
2 cubic yard bucket	60 C.Y./Hr. average
Tower crane	
1 operator*	
2 laborers	
1 cubic yard bucket	35–40 C.Y./Hr. average
Small line pumping systems,	
1 foot vertical = 6 feet horizontally	
1 90 degree bend = 40 feet horizontally	
1 45 degree bend = 20 feet horizontally	
1 30 degree bend = 13 feet horizontally	
1 foot rubber hose - 1-1/2 feet of steel tubing	
1 operator*	
average output, 1000' horizontally	40–50 C.Y./Hr. Average

*In some regions an oiler or helper may be needed, according to union rules.

(*courtesy* Concrete Estimating Handbook, *Michael F. Kenny, Van Nostrand Reinhold Company*)

Figure 3-14 Types of Cement and Concrete and Their Major Uses

Type of Portland Cement	Type of Concrete	Major Use
Type I, Normal	Standard-type concrete	General construction
Type IA, air-entraining	Standard with air entraining (more workability, and resistance to freezing and thawing)	General construction
Type I/II	Standard concrete, used for either Type I or II requirements	General construction
Type II, Moderate	Slower setting, fewer heat generation, and smaller volume change than Types I and IA; develops strength in 28 days	General construction and situations involving moderate sulfate action
Type IIA, air-entraining	Same as concrete using Type II moderate cement, but with better workability and resistance to freezing and thawing	General construction and situations involving moderate sulfate action
Type III, High-Early-Strength	Rapid setting, higher heat generation (offsets freezing), some volume change, develops strength in 7 days	Construction requiring rapid development of strength
Type IIIA, air-entraining	Same as concrete using Type III high-early-strength, but with better workability and resistance to freezing and thawing	Construction requiring rapid development of strength
Type IV, Low Heat of Hydration	Slow setting, low heat generation, small volume change, good strength with age	Massive concrete construction
Type V, Sulfate-Resisting	High resistance to sulfate attack, fairly low heat generation, high strength with age	Construction exposed to ground water or soil that contains sulfates

Division 3

Figure 3-15   Fly Ash and Concrete Strength

 Concrete made with fly ash will typically be slightly lower in strength than straight cement concrete up to 28 days, equal strength at 28 days, and substantially higher strength within a year's time.

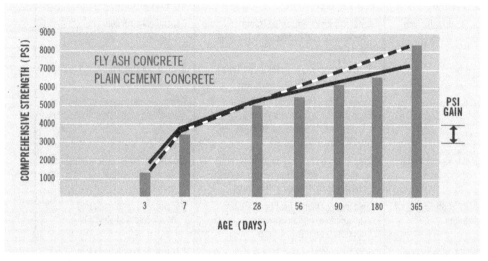

(courtesy of Headwaters Resources, Inc.)

Figure 3-16 Quantities of Cement, Sand, and Stone for One Cubic Yard of Various Concrete Mixes

Use this table to determine the quantities of cement, sand, and stone for small quantities of site-mixed concrete. Cement is listed in sacks or cubic feet, and aggregate in cubic yards.

Concrete (C.Y.)	Mix 1 1:1:1-3/4			Mix 2 1:2:2.25			Mix 3 1:2.25:3			Mix 4 1:3:4		
	Cement (Sacks)	Sand (C.Y.)	Stone (C.Y.)	Cement (Sacks)	Sand (C.Y.)	Stone (C.Y.)	Cement (Sacks)	Sand (C.Y.)	Stone (C.Y.)	Cement (Sacks)	Sand (C.Y.)	Stone (C.Y.)
1	10	.37	.63	7.75	.56	.65	6.25	.52	.70	5.0	.56	.74
2	20	.74	1.26	15.50	1.12	1.30	12.50	1.04	1.40	10.0	1.12	1.48
3	30	1.11	1.89	23.25	1.68	1.95	18.75	1.56	2.10	15.0	1.68	2.22
4	40	1.48	2.52	31.00	2.24	2.60	25.00	2.08	2.80	20.0	2.24	2.96
5	50	1.85	3.15	38.75	2.80	3.25	31.25	2.60	3.50	25.0	2.80	3.70
6	60	2.22	3.78	46.50	3.36	3.90	37.50	3.12	4.20	30.0	3.36	4.44
7	70	2.59	4.41	54.25	3.92	4.55	43.75	3.64	4.90	35.0	3.92	5.18\
8	80	2.96	5.04	62.00	4.48	5.20	50.00	4.16	5.60	40.0	4.48	5.92
9	90	3.33	5.67	69.75	5.04	5.85	56.25	4.68	6.30	45.0	5.04	6.66
10	100	3.70	6.30	77.50	5.60	6.50	62.50	5.20	7.00	50.0	5.60	7.40
11	110	4.07	6.93	85.25	6.16	7.15	68.75	5.72	7.70	55.0	6.16	8.14
12	120	4.44	7.56	93.00	6.72	7.80	75.00	6.24	8.40	60.0	6.72	8.88
13	130	4.82	8.20	100.76	7.28	8.46	81.26	6.76	9.10	65.0	7.28	9.62
14	140	5.18	8.82	108.50	7.84	9.10	87.50	7.28	9.80	70.0	7.84	10.36
15	150	5.56	9.46	116.26	8.40	9.76	93.76	7.80	10.50	75.0	8.40	11.10
16	160	5.92	10.08	124.00	8.96	10.40	100.00	8.32	11.20	80.0	8.96	11.84
17	170	6.30	10.72	131.76	9.52	11.06	106.26	8.84	11.90	85.0	9.52	12.58
18	180	6.66	11.34	139.50	10.08	11.70	112.50	9.36	12.60	90.0	10.08	13.32
19	190	7.04	11.98	147.26	10.64	12.36	118.76	9.84	13.30	95.0	10.64	14.06
20	200	7.40	12.60	155.00	11.20	13.00	125.00	10.40	14.00	100.0	11.20	14.80

(courtesy Estimating Tables for Home Building, *Paul I. Thomas, Craftsman Book Company*)

(continued on next page)

Division 3

Figure 3-16 Quantities of Cement, Sand, and Stone for One Cubic Yard of Various Concrete Mixes *(continued from previous page)*

Concrete (C.Y.)	Mix 1 1:1:1-3/4			Mix 2 1:2:2.25			Mix 3 1:2.25:3			Mix 4 1:3:4		
	Cement (Sacks)	Sand (C.Y.)	Stone (C.Y.)	Cement (Sacks)	Sand (C.Y.)	Stone (C.Y.)	Cement (Sacks)	Sand (C.Y.)	Stone (C.Y.)	Cement (Sacks)	Sand (C.Y.)	Stone (C.Y.)
21	210	7.77	13.23	162.75	11.76	13.65	131.25	10.92	14.70	105.0	11.76	15.54
22	220	8.14	13.86	170.05	12.32	14.30	137.50	11.44	15.40	110.0	12.32	16.28
23	230	8.51	14.49	178.25	12.88	14.95	143.75	11.96	16.10	115.0	12.88	17.02
24	240	8.88	15.12	186.00	13.44	15.60	150.00	12.48	16.80	120.0	13.44	17.76
25	250	9.25	15.75	193.75	14.00	16.25	156.25	13.00	17.50	125.0	14.00	18.50
26	260	9.64	16.40	201.52	14.56	16.92	162.52	13.52	18.20	130.0	14.56	19.24
27	270	10.00	17.00	209.26	15.12	17.56	168.76	14.04	18.90	135.0	15.02	20.00
28	280	10.36	17.64	217.00	15.68	18.20	175.00	14.56	19.60	140.0	15.68	20.72
29	290	10.74	18.28	224.76	16.24	18.86	181.26	15.08	20.30	145.0	16.24	21.46

Figure 3-17 Fly Ash Mix Design

green▸ Concrete mixes are designed by selecting the proportions of the mix components that will develop the required strength, produce a workable consistency concrete that can be handled and placed easily, attain sufficient durability under exposure to in-service environmental conditions, and be economical. Procedures for proportioning fly ash concrete mixes differ slightly from those for conventional concrete mixes.

One mix design approach commonly used in proportioning fly ash concrete mixes is to use a mix design with all Portland cement, remove some of the Portland cement, and then add fly ash to compensate for the cement that is removed.

Figure 3-18 Concrete Quantities for Pile Caps – 3'0" OC

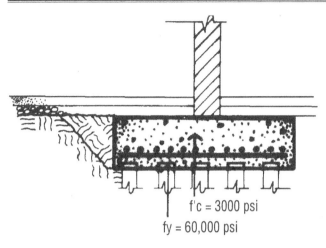

f'c = 3000 psi
fy = 60,000 psi

Section Through Pile Cap

Load	Number of Piles @ 3'-0" O.C. per Footing Cluster									
Work-ing (K)	2 (C.Y.)	4 (C.Y.)	6 (C.Y.)	8 (C.Y.)	10 (C.Y.)	12 (C.Y.)	14 (C.Y.)	16 (C.Y.)	18 (C.Y.)	20 (C.Y.)
50	(.9)	(1.9)	(3.3)	(4.9)	(5.7)	(7.8)	(9.9)	(11.1)	(14.4)	(16.5)
100	(1.0)	(2.2)	(3.3)	(4.9)	(5.7)	(7.8)	(9.9)	(11.1)	(14.4)	(16.5)
200	(1.0)	(2.2)	(4.0)	(4.9)	(5.7)	(7.8)	(9.9)	(11.1)	(14.4)	(16.5)
400	(1.1)	(2.6)	(5.2)	(6.3)	(7.4)	(8.2)	(13.7)	(11.1)	(14.4)	(16.5)
800		(2.9)	(5.8)	(7.5)	(9.2)	(13.6)	(17.6)	(15.9)	(19.7)	(22.1)
1200			(5.8)	(8.3)	(9.7)	(14.2)	(18.3)	(20.4)	(21.2)	(22.7)
1600				(9.8)	(11.4)	(14.5)	(19.5)	(20.4)	(24.6)	(27.2)
2000				(9.8)	(11.4)	(16.6)	(24.1)	(21.7)	(26.0)	(28.8)
3000						(17.5)		(26.5)	(30.3)	(32.9)
4000								(30.2)	(30.7)	(36.5)

Figure 3-19 Concrete Quantities for Pile Caps – 4'6" OC

Load	Number of Piles @ 4'-6" O.C. per Cap					
Working (K)	2 (C.Y.)	3 (C.Y.)	4 (C.Y.)	5 (C.Y.)	6 (C.Y.)	7 (C.Y.)
50	(2.3)	(3.6)	(5.6)	(11.0)	(13.7)	(12.9)
100	(2.3)	(3.6)	(5.6)	(11.0)	(13.7)	(12.9)
200	(2.3)	(3.6)	(5.6)	(11.0)	(13.7)	(12.9)
400	(3.0)	(3.6)	(5.6)	(11.0)	(13.7)	(12.9)
800			(6.2)	(11.5)	(14.0)	(12.9)
1200				(13.0)	(13.7)	(13.4)
1600						(14.0)

Figure 3-20 Concrete Footings/Volume – 6" and 8" Deep

Ground Floor Area	Cubic Yards of Concrete Needed for Size of Footing Shown									
	6" Deep by:					8" Deep by:				
	12"	16"	18"	20"	24"	12"	16"	18"	20"	24"
200	1.11	1.48	1.67	1.85	2.22	1.48	1.97	2.22	2.47	2.96
300	1.37	1.83	2.06	2.28	2.74	1.83	2.44	2.74	3.04	3.65
400	1.57	2.10	2.36	2.62	3.15	2.10	2.80	3.15	3.49	4.19
500	1.76	2.35	2.64	2.93	3.52	2.35	3.13	3.52	3.90	4.68
600	1.92	2.57	2.89	3.20	3.85	2.57	3.43	3.85	4.27	5.13
700	2.07	2.77	3.11	3.45	4.14	2.77	3.69	4.14	4.60	5.52
800	2.22	2.96	3.34	3.70	4.44	2.96	3.96	4.44	4.93	5.92
900	2.35	3.14	3.53	3.91	4.70	3.14	4.19	4.70	5.22	6.26
1000	2.48	3.31	3.73	4.13	4.96	3.31	4.42	4.96	5.51	6.61
1100	2.61	3.48	3.92	4.34	5.22	3.48	4.65	5.22	5.80	6.95
1200	2.72	3.63	4.09	4.53	5.44	3.63	4.85	5.44	6.04	7.25
1300	2.83	3.78	4.25	4.71	5.66	3.78	5.05	5.66	6.29	7.54
1400	2.94	3.93	4.42	4.90	5.88	3.93	5.25	5.88	6.53	7.84
1500	3.03	4.05	4.56	5.05	6.07	4.05	5.41	6.08	6.74	8.09
1600	3.15	4.20	4.73	5.24	6.29	4.20	5.61	6.29	6.99	8.38
1700	3.24	4.32	4.87	5.39	6.48	4.32	5.77	6.48	7.19	8.63
1800	3.33	4.45	5.00	5.54	6.66	4.45	5.94	6.66	7.40	8.87
1900	3.42	4.57	5.14	5.70	6.85	4.57	6.10	6.85	7.60	9.12
2000	3.52	4.69	5.28	5.85	7.03	4.69	6.27	7.03	7.81	9.37
2100	3.59	4.79	5.39	5.98	7.18	4.79	6.40	7.18	7.97	9.56
2200	3.68	4.92	5.53	6.13	7.36	4.92	6.57	7.36	8.18	9.81
2300	3.76	5.01	5.64	6.25	7.51	5.01	6.70	7.51	8.34	10.01
2400	3.85	5.14	5.78	6.41	7.70	5.14	6.86	7.70	8.55	10.25
2500	3.92	5.24	5.89	6.53	7.84	5.24	6.99	7.84	8.71	10.45
2600	4.00	5.34	6.00	6.65	7.99	5.34	7.13	7.99	8.88	10.65
2700	4.07	5.43	6.12	6.78	8.14	5.43	7.26	8.14	9.04	10.85
2800	4.14	5.53	6.23	6.90	8.29	5.53	7.39	8.29	9.21	11.04
2900	4.22	5.63	6.34	7.02	8.44	5.63	7.52	8.44	9.37	11.24
3000	4.31	5.76	6.48	7.18	8.62	5.76	7.69	8.62	9.58	11.49

(*courtesy* Estimating Tables for Home Building, *Paul I. Thomas, Craftsman Book Company*)

Division 3

Figure 3-21 Concrete Footings/Volume – 10" and 12" Deep

Ground Floor Area	Cubic Yards of Concrete Needed for Size of Footing Shown									
	10" Deep by:					12" Deep by:				
	12"	16"	18"	20"	24"	12"	16"	18"	20"	24"
200	1.85	2.47	2.78	3.08	3.70	2.22	2.96	3.33	3.70	4.44
300	2.28	3.04	3.43	3.80	4.57	2.74	3.65	4.11	4.57	5.48
400	2.62	3.49	3.94	4.37	5.24	3.15	4.19	4.72	5.24	6.29
500	2.93	3.90	4.40	4.88	5.86	3.52	4.68	5.27	5.86	7.03
600	3.20	4.27	4.82	5.35	6.42	3.85	5.13	5.77	6.42	7.70
700	3.45	4.60	5.19	5.76	6.91	4.14	5.52	6.22	6.91	8.29
800	3.70	4.93	5.56	6.17	7.40	4.44	5.92	6.66	7.40	8.88
900	3.91	5.22	5.88	6.53	7.84	4.70	6.26	7.05	7.84	9.40
1000	4.13	5.51	6.20	6.89	8.27	4.96	6.61	7.44	8.27	9.92
1100	4.34	5.80	6.53	7.25	8.70	5.22	6.95	7.83	8.70	10.43
1200	4.53	6.04	6.81	7.56	9.07	5.44	7.25	8.16	9.07	10.89
1300	4.71	6.29	7.08	7.86	9.44	5.66	7.54	8.49	9.44	11.32
1400	4.90	6.53	7.36	8.17	9.81	5.88	7.84	8.82	9.81	11.77
1500	5.05	6.74	7.59	8.43	10.12	6.08	8.09	9.10	10.12	12.14
1600	5.24	6.99	7.87	8.74	10.49	6.29	8.38	9.44	10.49	12.58
1700	5.39	7.19	8.10	9.00	10.80	6.48	8.63	9.71	10.80	12.95
1800	5.54	7.40	8.33	9.25	11.11	6.66	8.87	9.99	11.11	13.32
1900	5.70	7.60	8.57	9.51	11.41	6.85	9.12	10.27	11.41	13.69
2000	5.85	7.81	8.80	9.77	11.72	7.03	9.37	10.55	11.72	14.06
2100	5.98	7.97	8.98	9.97	11.97	7.18	9.56	10.77	11.97	14.36
2200	6.13	8.18	9.21	10.23	12.28	7.36	9.81	11.04	12.28	14.73
2300	6.25	8.34	9.40	10.43	12.53	7.51	10.01	11.27	12.53	15.02
2400	6.41	8.55	9.63	10.69	12.83	7.70	10.25	11.54	12.83	15.39
2500	6.53	8.71	9.82	10.90	13.08	7.84	10.45	11.77	13.08	15.69
2600	6.65	8.88	10.00	11.10	13.33	7.99	10.65	11.99	13.33	15.98
2700	6.78	9.04	10.19	11.31	13.57	8.14	10.85	12.21	13.57	16.28
2800	6.90	9.21	10.37	11.51	13.82	8.29	11.04	12.43	13.82	16.58
2900	7.02	9.37	10.56	11.72	14.07	8.44	11.24	12.65	14.07	16.87
3000	7.18	9.58	10.79	11.98	14.38	8.62	11.49	12.93	14.38	17.24

(courtesy Estimating Tables for Home Building, *Paul I. Thomas, Craftsman Book Company*)

Figure 3-22 Cubic Yards of Concrete per Linear Foot of Wall

To use this table, determine the height of the wall, find the factor in the "Thickness" column, and multiply this factor by the length of the wall.

Example: Calculate the volume of concrete in a wall 40 feet long, 10 inches thick, and 8 feet high.

Factor × length = .2480 × 40 = 9.9, say 10 CY.

Wall Height in Feet	Thickness of Wall in Inches						
	6"	7"	8"	9"	10"	11"	12"
1	.0185	.0217	.0248	.0279	.0310	.0341	.0372
2	.0360	.0434	.0496	.0558	.0620	.0682	.0744
3	.0556	.0651	.0744	.0837	.0930	.1023	.1116
4	.0741	.0868	.0992	.1116	.1240	.1364	.1488
5	.0926	.1085	.1240	.1395	.1550	.1705	.1860
6	.1111	.1302	.1488	.1674	.1860	.2046	.2232
7	.1296	.1519	.1736	.1953	.2170	.2387	.2606
8	.1482	.1736	.1984	.2232	.2480	.2728	.2976
9	.1667	.1953	.2232	.2511	.2790	.3069	.3348
10	.1852	.2170	.2480	.2790	.3100	.3410	.3720
11	.2046	.2387	.2728	.3069	.3410	.3751	.4092
12	.2232	.2604	.2976	.3348	.3720	.4092	.4464

Division 3

Figure 3-23 Concrete Walls/Volume

(No deduction has been made for openings.) Interpolate for intermediate heights of wall.

Ground Floor Area	Cubic Yards of Concrete Needed for Size Wall Shown									
	8" Wall by Height of:					9" Wall by Height of:				
	1'	2'	4'	6'	8'	1'	2'	4'	6'	8'
200	1.49	2.98	5.95	8.93	11.90	1.67	3.25	6.70	10.04	13.39
300	1.83	3.67	7.34	11.01	14.68	2.06	4.13	8.26	12.39	16.52
400	2.11	4.22	8.43	12.65	16.86	2.37	4.74	9.49	14.23	18.97
500	2.36	4.72	9.42	14.14	18.85	2.65	5.30	10.60	15.90	21.20
600	2.58	5.16	10.32	15.48	20.63	2.90	5.80	11.61	17.41	23.21
700	2.78	5.56	11.11	16.67	22.22	3.13	6.25	12.50	18.75	25.00
800	2.97	5.95	11.90	17.86	23.81	3.35	6.70	13.39	20.09	26.78
900	3.15	6.30	12.60	18.90	25.20	3.54	7.09	14.17	21.26	28.35
1000	3.33	6.65	13.29	19.94	26.59	3.74	7.48	14.95	22.43	29.91
1100	3.50	6.99	13.99	20.98	27.97	3.93	7.87	15.74	23.60	31.47
1200	3.64	7.29	14.58	21.87	29.16	4.10	8.20	16.41	24.61	32.81
1300	3.79	7.59	15.18	22.77	30.36	4.27	8.54	17.07	25.61	34.15
1400	3.95	7.89	15.77	23.66	31.55	4.44	8.87	17.74	26.62	35.49
1500	4.06	8.13	16.27	24.40	32.54	4.56	9.15	18.30	27.45	36.60
1600	4.22	8.43	16.86	25.30	33.73	4.75	9.49	18.97	28.46	37.94
1700	4.34	8.68	17.39	26.04	34.72	4.89	9.77	19.53	29.30	39.06
1800	4.47	8.93	17.86	26.78	35.71	5.02	10.04	20.09	30.13	40.18
1900	4.59	9.18	18.35	27.53	36.70	5.16	10.32	20.65	30.97	41.29
2000	4.71	9.42	18.85	28.27	37.70	5.30	10.60	21.20	31.81	42.41
2100	4.81	9.62	19.24	28.87	38.49	5.42	10.83	21.65	32.48	43.30
2200	4.93	9.87	19.74	29.61	39.48	5.55	11.10	22.21	33.31	44.42
2300	5.03	10.07	20.14	30.21	40.28	5.67	11.33	22.65	33.98	45.31
2400	5.16	10.32	20.63	30.95	41.27	5.81	11.61	23.21	34.82	46.43
2500	5.26	10.52	21.03	31.55	42.06	5.92	11.83	23.66	35.49	47.32
2600	5.35	10.71	21.43	32.14	42.85	6.02	12.05	24.11	36.16	48.21
2700	5.45	10.91	21.82	32.74	43.65	6.14	12.28	24.55	36.83	49.10
2800	5.56	11.11	22.22	33.33	44.44	6.25	12.50	25.00	37.50	50.00
2900	5.66	11.31	22.62	33.93	45.24	6.36	12.72	25.44	38.17	50.89
3000	5.78	11.56	23.11	34.67	46.23	6.50	13.00	26.00	39.00	52.00

(*courtesy* Estimating Tables for Home Building, *Paul I. Thomas, Craftsman Book Company*)

(continued on next page)

Figure 3-23 Concrete Walls/Volume (continued from previous page)

Ground Floor Area	10" Wall by Height of:					12" Wall by Height of:				
	1'	2'	4'	6'	8'	1'	2'	4'	6'	8'
200	1.86	3.72	7.44	11.16	14.88	2.23	4.46	8.93	13.39	17.86
300	2.30	4.59	9.18	13.76	18.35	2.75	5.51	11.01	16.52	22.08
400	2.63	5.27	10.54	15.81	21.08	3.16	6.32	12.65	18.97	25.30
500	2.94	5.89	11.78	17.67	23.56	3.53	7.07	14.14	21.20	28.27
600	3.23	6.45	12.90	19.34	25.79	3.87	7.74	15.48	23.21	30.95
700	3.47	6.94	13.89	20.83	27.78	4.16	8.33	16.67	25.00	33.33
800	3.72	7.44	14.88	22.32	29.76	4.46	8.93	17.86	26.78	35.71
900	3.93	7.87	15.75	23.62	31.50	4.73	9.45	18.90	28.35	37.80
1000	4.15	8.31	16.62	24.92	33.23	4.99	9.97	19.94	29.91	39.88
1100	4.37	8.74	17.48	26.23	34.97	5.25	10.49	20.98	31.47	41.96
1200	4.56	9.11	18.23	27.34	36.46	5.47	10.94	21.87	32.81	43.75
1300	4.74	9.49	18.97	28.46	37.94	5.69	11.38	22.77	34.15	45.53
1400	4.93	9.86	19.72	29.57	39.43	5.92	11.83	23.66	35.49	47.32
1500	5.09	10.17	20.34	30.50	40.67	6.10	12.20	24.40	36.60	48.81
1600	5.27	10.54	21.08	31.62	42.16	6.33	12.65	25.30	37.94	50.59
1700	5.43	10.85	21.70	32.55	43.40	6.51	13.02	26.04	39.06	52.08
1800	5.53	11.16	22.32	33.48	44.64	6.70	13.39	26.78	40.18	53.57
1900	5.70	11.47	22.94	34.41	45.88	6.88	13.76	27.53	41.29	55.06
2000	5.89	11.78	23.56	35.34	47.12	7.07	14.14	28.27	42.41	56.54
2100	6.01	12.03	24.06	36.08	48.11	7.22	14.43	28.87	43.30	57.73
2200	6.17	12.34	24.68	37.01	49.35	7.40	14.81	29.61	44.42	59.22
2300	6.30	12.59	25.17	37.76	50.34	7.55	15.10	30.21	45.31	60.41
2400	6.45	12.90	25.79	38.69	51.58	7.74	15.48	30.95	46.43	61.90
2500	6.57	13.14	26.29	39.43	52.58	7.89	15.77	31.55	47.32	63.09
2600	6.70	13.39	26.78	40.18	53.57	8.04	16.07	32.14	48.21	64.28
2700	6.82	13.64	27.28	40.92	54.56	8.19	16.37	32.74	49.10	65.47
2800	6.95	13.89	27.78	41.66	55.55	8.33	16.67	33.33	50.00	66.66
2900	7.07	14.14	28.27	42.41	56.54	8.48	16.96	33.93	50.89	67.85
3000	7.22	14.45	28.89	43.34	57.78	8.67	17.34	34.67	52.00	69.34

Division 3

Figure 3-24 Square, Rectangular, and Round Column Forms

Multiply the factors by the length of columns to determine the volume of concrete for forms.

Square and Rectangular Column Forms							
Column Size (in Inches)	Volume-C.F./L.F.						
	12"	14"	16"	18"	20"	22"	24"
12	.994	1.161	1.327	1.494	1.661	1.827	1.994
14	1.161	1.355	1.550	1.744	1.938	2.133	2.327
16	1.327	1.550	1.772	1.994	2.216	2.438	2.661
18	1.494	1.744	1.994	2.244	2.494	2.744	2.994
20	1.661	1.938	2.216	2.494	2.772	3.050	3.327
22	1.827	2.133	2.438	2.744	3.050	3.355	3.661
24	1.994	2.327	2.661	2.994	3.327	3.661	3.994

Round Column Forms			
Column Diameter (in Inches)	Volume (C.F./L.F.)	Column Diameter (in Inches)	Volume (C.F./L.F.)
12	0.785	20	2.182
14	1.069	24	3.142
16	1.396	30	4.909
18	1.767	36	7.069

Figure 3-25 Typical Range of Risers for Various Story Heights

This table is to be used as a quick reference for stair design. Use it to determine concrete finish area, form area, and concrete volume.

General Design: Maximum height between landings is 12'; usual stair angle is 20° to 50°, with 30° to 35° the ideal.

Maximum riser height is 7" for commercial, 7-3/4" for residential.

Minimum rise height is 4".

Minimum tread width is 11" for commercial, 10" for residential.

Story Height (Ft.)	Minimum Risers	Maximum Ht. (inches)	Tread Width (inches)	Maximum Risers	Minimum Riser Ht. (inches)	Tread Width (inches)
8	14	6.86	11.14	24	4	14
8.5	15	6.80	11.20	25.5	4	14
9	16	6.75	11.25	27	4	14
9.5	17	6.71	11.29	28.5	4	14
10	18	6.67	11.33	30	4	14
10.5	18	7.00	11.00	31.5	4	14
11	19	6.95	11.05	33	4	14
11.5	20	6.90	11.10	34.5	4	14
12	21	6.86	11.14	36	4	14
12.5	22	6.82	11.18	37.5	4	14
13	23	6.78	11.22	39	4	14
13.5	24	6.75	11.25	40.5	4	14
14	24	7.00	11.00	42	4	14

Division 3

Figure 3-26 Floor Selection Guide

This table provides general information to assist in the selection of a concrete floor surface to meet specific use requirements. Service conditions are listed in order of increasing severity, and types of floors in order of increasing durability to permit approximate comparisons. Special considerations (move-in abuse, future use, safety factor, etc.) may justify a highly durable floor even for mild day-to-day conditions.

Service Require ments	Type of Floor	Relative Cost Index	Strength of Surface	Relative Abrasion Resistance	Other Charactleristics of Floor Surface
Light Traffic	Floor no. 1: Monolithic Concrete Floor. Adequate thickness, reinforcement and strength; properly proportioned mix 3" to 4" slump and minimum bleeding. Good placing and finishing practice with proper tools and proper timing; thorough curing.	100	3500–4500 psi	100	Susceptible to dusting from (1) laitance cuased by use of high slump concrete and/or over-troweling, (2) improper curing, or (3) fracturing of aggregate at the surface under heavy usage, load or impact.
Light Traffic	Floor No. 2: Liquid Hardener Applied to Floor No. 1 Membrane-type compounds should not be used for curing when liquid hardener is to be used. Apply hardener no sooner than 28 days after floor is installed.	105	3500–4500 psi	100	Liquid hardener temporarily arrests dusting caused by (1) and (2) above. Requires repeated application of hardeners. Does not arrest dusting from (3).
Light and/or Moderate Duty Traffic	Floor No. 3: Dry Shake of Natural Aggregate Applied Over Freshly Floated Surface of Floor No. 1. Well-graded quartz, traprock, emery or granite aggregate is dry mixed with portland cement. Proprietary products contain plasticizing agents and, when desired colorfast pigments. Heavy applications of shake increases thickness of high-strength surface and service life of floor. Proper timing of finishing procedures and throrough curing are essential.	non-colored: 115-120 colored: 125-135	8000–12,000 psi	200	Provides "scuff proofing" for high frequency foot traffic. Built-in colored surface, if desired. Dense surface is more resistant to (1) scaling from cycles of freeze and thaw and use of de-icing salts, and (2) mild corrosive materials. Easy to clean. Aggregate at surface fractures under high frequency, hard wheeled traffic and impact.
Moderate Duty Traffic	Floor No. 4: High Strength Natural Aggregate Topping (two) course) Applied Over Fresh or Set Base Slag. 3/4" to 2" thick topping: properly proportioned mix containing 1/4" – 3/8" aggregate, designed for 1" to 2" slump and 8000-10,000 psi 28 day. Aggregates commonly used are quartz, traprock, emery or granite.	over fresh slab: 130-180 over set slab: 140-185	8000–12,000 psi	200	Same as Floor No. 3 plus ability of surface to withstand heavier loads. Aggregate at surface fractures under high frequency, hard wheeled traffic and impact.

(continued on next page)

Figure 3-26 Floor Selection Guide *(continued from previous page)*

Service Require ments	Type of Floor	Relative Cost	Strength of Surface	Relative Abrasion Resistance	Other Charactieristics of Floor Surface
Moderate and/or Heavy Duty Traffic	Floor No. 5: Dry Shake of Metallic Aggregate Applied Over Freshly Floated Surface of Floor No. 1 or Floor No. 4, Scientifically graded metallic aggregate, free of rust, oil and non-ferrous particles is combined with plasticizing agents and, where desired, colorfast pigments. Mix dry with Portland cement. Heavy application of shake increases thickness of high-strength iron-armoured surface and service life of flor, Proper timing of finishing Procedures and proper curing are essential.	monolithic floor: 130–140 two course floor: 160–170	12,000 psi	800	Ductile metallic surface is non-dusting, dense and resistant to oil and grease. Easy to clean. Provides built-in color and/or durable, slip-resistant finish where desired. Withstands impact and high frequency or heavy load traffic from hard wheeled material handling equipment.
Heavy Duty Traffic and/or Extra Heavy Duty Traffic (Key Area Floors)	Floor No. 6: High Strength Metallic Aggregate Topping Applied Over Fresh or Set Base Slab. Specially formulated all-iron aggregate topping, Anvil Top, requires only addition of water. Apply 1/2"-1" thickness using procedures recommended by field service man.	1/2" over fresh base slab: 380–400 1" over set base slab: 490–550	12,000 psi	800* *Plus greater thickness of iron-armour	Service life under high concentration of heavy industrial traffic is 10 to 15 times greater than high-strength concrete toppings. Withstands heavy impact, heavy abrasion and high point loads. Dense surface is easy to clean.

(courtesy Modern Plant Operation and Maintenance, Winter 1971/2)

Division 3

Figure 3-27 Concrete Quantities for Metal Pan Floor Slabs

Concrete Quantities/20" Widths*				
Depth of Metal Pan	Width of Joist	C.F. of Concrete per S.F. for Various Slab Thicknesses		Additional Concrete for Tapered End Forms, C.F. per L.F. of Bearing Wall or Beam (one side only)
		2-1/2"	3"	4-1/2"	
6"	4"	.303	.345	.470	.13
	5"	.319	.361	.486	.12
	6"	.334	.375	.501	.12
8"	4"	.339	.381	.506	.17
	5"	.361	.402	.527	.16
	6"	.380	.422	.547	.16
10"	4"	.377	.419	.544	.21
	5"	.404	.445	.570	.20
	6"	.428	.470	.595	.19
12"	4"	.418	.459	.584	.25
	5"	.449	.491	.616	.24
	6"	.479	.520	.645	.23
14"	5"	.497	.538	.664	.28
	6"	.531	.573	.698	.27
	7"	.562	.604	.729	.26
16"	6"	.585	.627	.752	.31
	7"	.621	.663	.788	.30
	8"	.654	.695	.820	.29
20"	7"	.744	.786	.911	.37
	8"	.785	.826	.951	.36
	9"	.822	.864	.989	.35

*Apply only for areas over flange type forms and joists between them. Bridging joists, special headers, beams, tees, etc., not included.

(courtesy CECO Corporation)

(continued on next page)

Figure 3-27 Concrete Quantities for Metal Pan Floor Slabs
(continued from previous page)

Concrete Quantities/20" Widths*				
Depth of Metal Pan	Width of Joist	C.F. of Concrete per S.F. for Various Slab Thicknesses		Additional Concrete for Tapered End Forms, C.F. per L.F. of Bearing Wall or Beam (one side only)
		2-1/2"	3"	4-1/2"	
6"	5"	.288	.329	.454	.11
	6"	.299	.341	.466	.10
	7"	.310	.352	.477	.10
8"	5"	.317	.359	.484	.14
	6"	.333	.374	.499	.14
	7"	.347	.389	.514	.14
10"	5"	.348	.390	.515	.18
	6"	.367	.409	.534	.17
	7"	.386	.427	.552	.17
12"	5"	.381	.422	.547	.21
	6"	.404	.445	.570	.21
	7"	.425	.464	.592	.20
14"	5"	.415	.456	.581	.25
	6"	.441	.483	.608	.24
	7"	.467	.508	.633	.24
16"	6"	.481	.522	.647	.28
	7"	.509	.551	.676	.27
	8"	.537	.578	.703	.26
20"	7"	.599	.641	.766	.34
	8"	.633	.675	.800	.33
	9"	.665	.707	.832	.32

Division 3

Figure 3-28 Concrete Quantities for Metal Pan Floor Slabs
Using Adjustable Forms

Concrete Quantities/20" Widths*				
Depth of Metal Pan	Width of Joist	C.F. of Concrete per S.F. for Various Slab Thicknesses		Additional Concrete for Tapered End Forms, C.F. per L.F. of Bearing Wall or Beam (one side only)
		2"	2-1/3"	3"	
6"	4-1/2"	.279	.321	.362	.12
	5-1/2"	.295	.337	.378	.12
	7-1/2"	.322	.364	.405	.11
8"	4-1/2"	.309	.350	.393	.16
	5-1/2"	.331	.372	.414	.16
	7-1/2"	.367	.409	.451	.15
10"	4-1/2"	.340	.381	.424	.20
	5-1/2"	.367	.408	.450	.19
	7-1/2"	.413	.455	.496	.18
12"	4-1/2"	.371	.412	.455	.24
	5-1/2"	.403	.444	.486	.23
	7-1/2"	.458	.500	.542	.22
14"	4-1/2"	.402	.443	.485	.28
	5-1/2"	.438	.480	.522	.27
	7-1/2"	.504	.545	.587	.26

*Apply only for areas over flange type forms and joists between them. Bridging joists, special headers, beams, tees, etc., not included.

(courtesy CECO Corporation)

(continued on next page)

Figure 3-28 Concrete Quantities for Metal Pan Floor Slabs
Using Adjustable Forms *(continued from previous page)*

Concrete Quantities/30" Widths*					
Depth of Metal Pan	Width of Joist	C.F. of Concrete per S.F. for Various Slab Thicknesses			Additional Concrete for Tapered End Forms, C.F. per L.F. of Bearing Wall or Beam (one side only)
		2-1/2"	3"	3-1/2"	
6"	5-1/2"	.300	.341	.383	.11
	6-1/2"	.311	.353	.395	.11
	7-1/2"	.322	.364	.406	.11
8"	5-1/2"	.326	.368	.408	.15
	6-1/2"	.341	.383	.424	.15
	7-1/2"	.355	.397	.438	.14
10"	5-1/2"	.352	.394	.434	.19
	6-1/2"	.371	.413	.454	.18
	7-1/2"	.389	.430	.472	.18
12"	5-1/2"	.378	.420	.461	.22
	6-1/2"	.401	.443	.484	.22
	7-1/2"	.422	.464	.505	.21
14"	5-1/2"	.404	.446	.486	.26
	6-1/2"	.430	.472	.514	.25
	7-1/2"	.456	.497	.539	.24

Figure 3-29 Voids Created by Various Size Metal Pan Forms

Depth of Pan Form	C.F. of Void Created per L.F. (for various widths of metal pan forms)				**C.F. per Tapered End	
	30" Width	20" Width	15" Width	10" Width	30" Width	20" Width
8"	1.628	1.072	.794	.516	4.47	2.88
10"	2.023	1.329	.982	.634	5.55	3.57
12"	2.414	1.581	1.165	.748	6.62	4.24
14"	2.801	1.829	1.343	.857	7.67	4.89
16"	3.183	2.072	1.516	.961	8.72	5.54
20"	3.933	2.544	1.850	1.155	10.76	6.78

**Total void created by standard 3'-0" length tapered end form.

(courtesy CECO Corporation)

Division 3

Figure 3-30 Steel Domes — Two-Way Joist Construction

Two-way joist or waffle slabs are formed with metal or fiberglass domes. These are erected on a flat form in accordance with the manufacturer's specifications. The information in this chart is typical of that found in sheets from manufacturers.

Voids Created with 2'-0" Design Module		
Depth of Dome	Overall Plan Size 24" x 24"	
	Plan Size of Void	C.F. of Void
8"	19" x 19"	1.41
10"	19" x 19"	1.90
12"	19" x 19"	2.14
14"	19" x 19"	2.44

Voids Created with 3'-0' Design Module		
Standard Sizes	Overall Plan Size 36" x 36"	
Depth of Dome	Plan Size of Void	C.F. of Void
8"	30" x 30"	3.85
10"	30" x 30"	4.78
12"	30" x 30"	5.53
14"	30" x 30"	6.54
16"	30" x 30"	7.44
20"	30" x 30"	9.16

Filler Sizes	Overall Plan Size 26" x 36"		Overall Plan Size 26" x 26"	
Depth of Dome	Plan Size of Void	C.F. of Void	Plan Size of Void	C.F. of Void
8"	20" x 30"	2.54	20" x 20"	1.65
10"	20" x 30"	3.13	20" x 20"	2.06
12"	20" x 30"	3.63	20" x 20"	2.41
14"	20" x 30"	4.27	20" x 20"	2.87
16"	20" x 30"	4.85	20" x 20"	3.14
20"	20" x 30"	5.90	20" x 20"	3.81

U.S. Patent No. 2,580,785 applies to all sizes.

(courtesy CECO Corporation)

Figure 3-31 Precast Concrete Wall Panels

This table lists the practical size limits for precast concrete wall panels by thickness. Panels are either solid or insulated with plain, colored, or textured finishes. Transportation is an important cost factor. Engineering data is available from fabricators to assist with construction details. Usual minimum job size for economical use of panels is about 5,000 SF.

Thickness	Maximum Size	Thickness	Maximum Size
3"	50 S.F.	6"	300 S.F.
4"	150 S.F.	7"	300 S.F.
5"	200 S.F.	8"	300 S.F.

Figure 3-32 Daily Erection Rate of Precast Concrete Panels

This table lists the daily erection rate of precast concrete panels for a six-person crew with a crane for high-rise and low-rise structures.

Panel Size L x H	Area per Panel (S.F.)	Low-Rise Daily Production Range		High-Rise Daily Production Range	
		Pieces	Area (S.F.)	Pieces	Area (S.F.)
4' x 4'	16	10 / 20	160 / 320	9 / 18	144 / 288
4' x 8'	32	10 / 19	320 / 608	9 / 17	288 / 544
8' x 8'	64	9 / 18	576 / 1152	8 / 16	512 / 1024
10' x 10'	100	8 / 15	800 / 1500	7 / 14	700 / 1400
15' x 10'	150	7 / 12	1050 / 1800	6 / 11	900 / 1650
20' x 10'	200	6 / 8	1200 / 1600	5 / 7	1000 / 1400
30' x 10'	300	5 / 7	1500 / 2100	4 / 7	1200 / 2100

Division 3

Figure 3-33 Installation Time in Labor-Hours for Strip Footings

Description	Labor-Hours	Unit
Formwork	.066	S.F.C.A.
Formwork Keyway	.015	L.F.
Reinforcing		
#4 to #7	15.238	ton
#8 to #14	8.889	ton
Dowels		
#4	.128	Ea.
#6	.152	Ea.
Place Concrete,		
Direct Chute	.400	C.Y.
Pumped	.640	C.Y.
Crane and Bucket	.711	C.Y.
Concrete in Place		
36" x 12" Reinforced	1.630	C.Y.

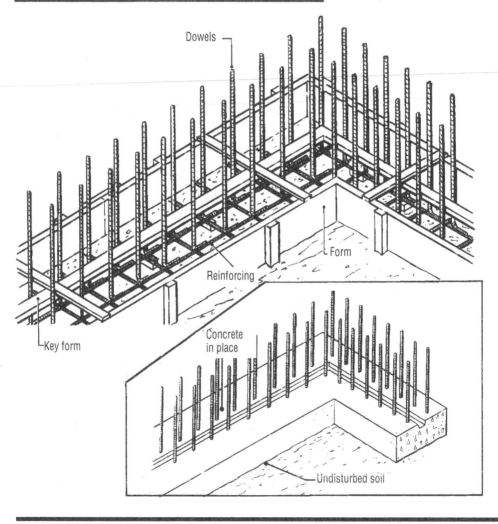

Figure 3-34 Installation Time in Labor-Hours for Insulating Concrete Forming

Description	Labor-Hours	Unit
Insulating Forms, Left in Place		
S.F. is for exterior face, but includes forms for both faces		
4" wall, straight block, 16" x 48" (5.33 S.F.)	.178	Ea.
90 corner block, exterior 16" x 38" x 22" (6.67 S.F.)	.213	
45 corner block, exterior 16" x 34" x 18" (5.78 S.F.)	.213	
6" wall, straight block, 16" x 48" (5.33 S.F.)	.178	
90 corner block, exterior 16" x 32" x 24" (6.22 S.F.)	.213	
45 corner block, exterior 16" x 26" x 18" (4.89 S.F.)	.213	
Bridge ledge block, 16" x 48" (5.33 S.F.)	.200	
Taper top block, 16" x 48" (5.33 S.F.)	.200	
8" wall, straight block, 16" x 48" (5.33 S.F.)	.178	
90 corner block, exterior 16" x 34" x 26" (6.67 S.F.)	.213	
45 corner block, exterior 16" x 28" x 20" (5.33 S.F.)	.213	
Brick ledge block, 16" x 48" (5.33 S.F.)	.200	
Taper top block, 16" x 48" (5.33 S.F.)	.200	

Division 3

Figure 3-35   Installation Time in Labor-Hours for Foundation Walls

Description	Labor-Hours	Unit
Forms		
Job-Built Plywood		
1 Use/Month	.130	S.F.C.A.
4 Uses/Month	.095	S.F.C.A.
Modular Prefabricated Plywood		
1 Use/Month	.053	S.F.C.A.
4 Uses/Month	.049	S.F.C.A.
Steel Framed Plywood		
1 Use/Month	.080	S.F.C.A.
4 Uses/Month	.072	S.F.C.A.
Box Out Openings to 10 S.F.	2.000	Ea.
Brick Shelf	.200	S.F.C.A.
Bulkhead	.181	L.F.
Corbel to 12" Wide	.320	L.F.
Pilasters	.178	S.F.C.A.
Waterstop Dumbbell	.055	L.F.
Reinforcing		
#3 to #7	10.667	ton
#8 to #14	8.000	ton
Place Concrete 12" Walls		
Direct Chute	.480	C.Y.
Pumped	.674	C.Y.
With Crane and Bucket	.711	C.Y.
Finish, Break Ties and Patch Voids	.015	S.F.
Burlap Rub with Grout	.018	S.F.
Concrete in Place 12" Thick	5.926	C.Y.

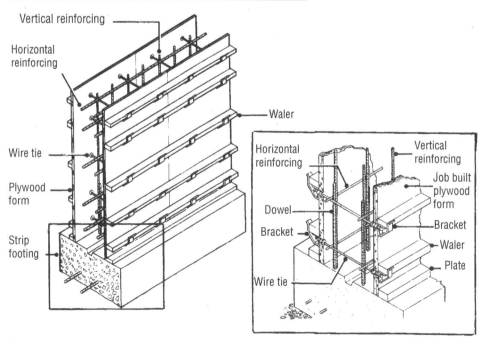

Figure 3-36 Installation Time in Labor-Hours for Grade Beams

Description	Labor-Hours	Unit
Forms	.083	S.F.C.A.
Brick or Slab Shelf	.200	S.F.C.A.
Bulkhead	.181	L.F.
Reinforcing		
#3 to #7	20.000	ton
#8 to #18	11.852	ton
Place Concrete		
Direct Chute	.320	C.Y.
Pumped	.492	C.Y.
With Crane and Bucket	.533	C.Y.
Finish, Break Ties and Patch Voids	.015	S.F.

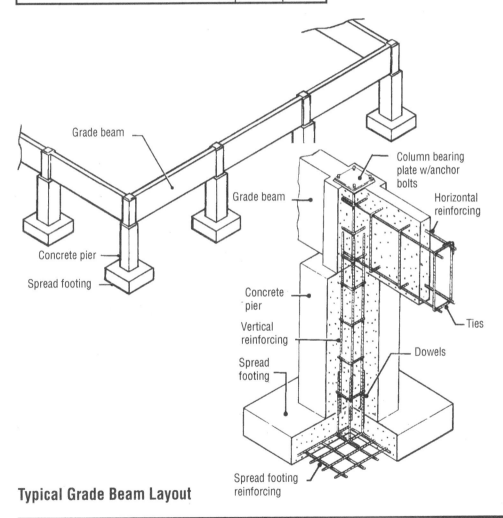

Typical Grade Beam Layout

Figure 3-37 Installation Time in Labor-Hours for Slab-on-Grade

Description	Labor-Hours	Unit
Slab-on-Grade		
Fine Grade	.010	S.Y.
Gravel Under Floor Slab 6" Deep Compacted	.005	S.F.
Polyethylene Vapor Barrier	.216	Sq.
Reinforcing WWF 6 x 6 (W1.4/W1.4)	.457	C.S.F.
Place and Vibrate Concrete 4" Thick Direct Chute	.436	C.Y.
Expansion Joint Premolded Bituminous Fiber		
1/2" x 6"	.021	L.F.
Edge Forms in Place to 6" High 4 Uses on Grade	.053	L.F.
Curing w/Sprayed Membrane Curing Compound	.168	C.S.F.
Finishing Floor		
Monolithic Screed Finish	.009	S.F.
Steel Trowel Finish	.015	S.F.

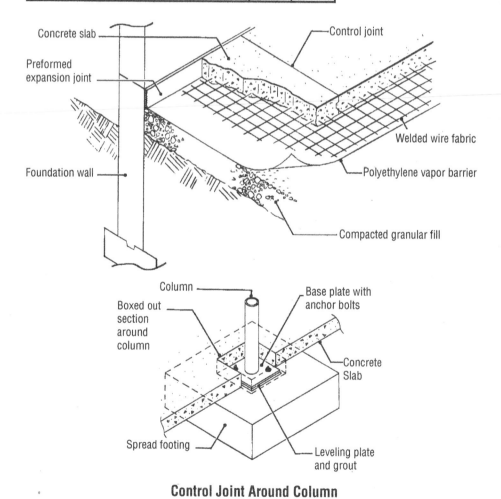

Control Joint Around Column

Figure 3-38 Installation Time in Labor-Hours for Concrete Flat-Plate

Description	Labor-Hours	Unit
Forms in Place		
Columns Square	.136	S.F.C.A.
Round Fiber Tube	.221	L.F.
Round Steel	.256	L.F.
Capitols	2.667	Ea.
Flat Plate	.086	S.F.
Edge Forms	.091	S.F.C.A.
Reinforcing Columns		
#3 to #7	21.333	ton
#8 to #14	13.913	ton
Spirals	14.545	ton
Butt Splice, Clamp Sleeve and Wedge	.373	Ea.
Mechanical Full Tension Splice with Filler Metal	.903	Ea.
Elevated Slab	11.034	ton
Hoisting Reinforcing	.609	ton
Place Concrete		
Pumped	.492	C.Y.
With Crane and Bucket	.582	C.Y.
Steel Trowel Finish	.015	S.F.
Concrete in Place Including		
Forms, Reinforcing and Finish	6.467	C.Y.

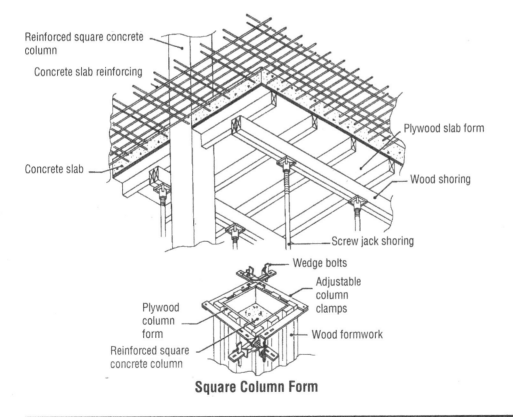

Square Column Form

Figure 3-39 Installation Time in Labor-Hours for Concrete Flat
Slab with Drop Panels

Description	Labor-Hours	Unit
Forms in Place		
Columns Square	.136	S.F.C.A.
Round Fiber Tube	.221	L.F.
Round Steel	.256	L.F.
Capitols	2.667	Ea.
Flat Slab with Drops	.088	S.F.
Edge Forms	.091	S.F.C.A.
Reinforcing Columns		
#3 to #7	21.333	ton
#8 to #14	13.913	ton
Spirals	14.545	ton
Butt Splice, Clamp Sleeve and Wedge	.373	Ea.
Mechanical Full Tension Splice with Filler Metal	.903	Ea.
Elevated Slab	11.034	ton
Hoisting Reinforcing	.609	ton
Place Concrete		
Pumped	.492	C.Y.
With Crane and Bucket	.582	C.Y.
Steel Trowel Finish	.015	S.F.
Concrete in Place Including		
Forms, Reinforcing and Finish	5.403	C.Y.

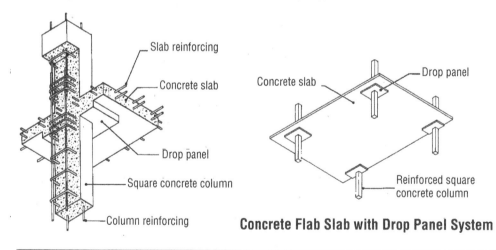

Slab reinforcing

Concrete slab

Drop panel

Square concrete column

Column reinforcing

Concrete slab

Drop panel

Reinforced square concrete column

Concrete Flab Slab with Drop Panel System

Figure 3-40 Installation Time in Labor-Hours for Concrete Waffle Slabs

Description	Labor-Hours	Unit
Forms in Place		
Joists and 19" Domes	.097	S.F.
Joists and 30" Domes	.102	S.F.
Edge Forms	.091	S.F.C.A.
Reinforcing		
Joists	20.000	ton
Slabs 6 x 6 w4/w4	.593	C.S.F.
Placing Concrete		
Pumped	.582	C.Y.
With Crane and Bucket	.674	C.Y.
Steel Trowel Finish	.015	S.F.
Concrete in Place Including		
Forms, Reinforcing and Finish	5.950	C.Y.

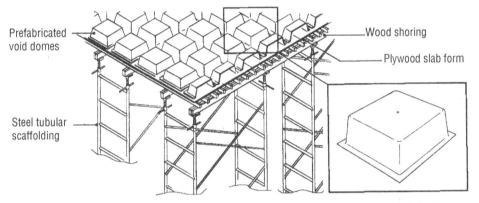

Prefabricated void domes

Wood shoring

Plywood slab form

Steel tubular scaffolding

Prefabricated Void Form

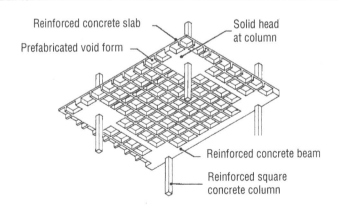

Reinforced concrete slab

Prefabricated void form

Solid head at column

Reinforced concrete beam

Reinforced square concrete column

Concrete Waffle Slab System

Figure 3-41   Installation Time in Labor-Hours for Precast Concrete Planks

2'-0"

3'-4"

Description	Labor-Hours	Unit
Erect and Grout Prestressed Slabs		
6" Deep	1.029	Ea.
8" Deep	1.108	Ea.
10" Deep	1.200	Ea.
12" Deep	1.309	Ea.
Average		
8" Deep	.010	S.F.

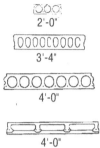

4'-0"

4'-0"

8'-0"

8'-0"

Precast Plank Profiles

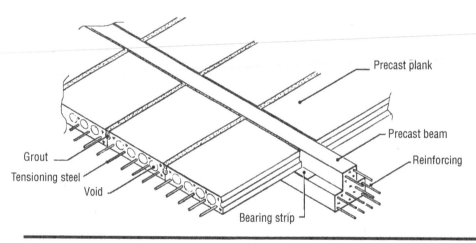

Precast plank

Precast beam

Reinforcing

Grout

Tensioning steel

Void

Bearing strip

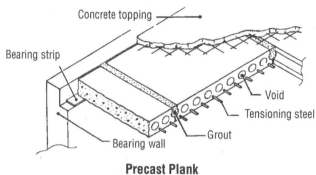

Concrete topping

Bearing strip

Void

Tensioning steel

Bearing wall

Grout

Precast Plank

Figure 3-42 Installation Time in Labor-Hours for Precast Single T Beams

Description	Labor-Hours	Unit
Single T Erection		
Span		
40'	7.200	Ea.
80'	9.000	Ea.
100'	12.000	Ea.
120'	14.400	Ea.
Welded Wire Fabric 6 x 6-10/10	.457	C.S.F.
Place Concrete		
Pumped	.582	C.Y.
With Crane and Bucket	.674	C.Y.
Broom Finish	.012	S.F.
Steel Trowel Finish	.015	S.F.

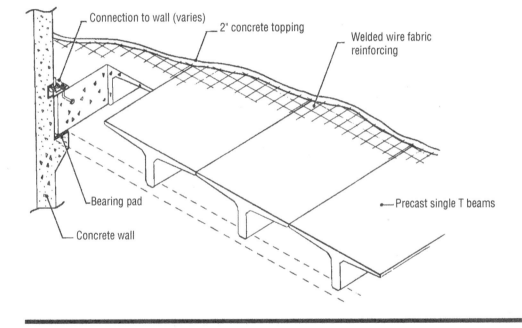

Connection to wall (varies)

2" concrete topping

Welded wire fabric reinforcing

Bearing pad

Concrete wall

Precast single T beams

Division 3

Figure 3-43 Installation Time in Labor-Hours for Precast Columns, Beams, and Double Ts

Description	Labor-Hours	Unit
Columns		
12' High	1.895	Ea.
24' High	2.400	Ea.
Beams		
20' Span		
12" x 20"	2.250	Ea
18" x 36"	3.000	Ea.
24" x 44"	3.273	Ea.
30" Span		
12" x 36"	3.000	Ea.
18" x 44"	3.600	Ea.
24" x 52"	4.500	Ea.
40' Span		
12" x 52"	3.600	Ea.
18" x 52"	4.500	Ea.
24" x 52"	6.000	Ea.
Double T's		
45' Span		
12" Deep x 8' Wide	3.273	Ea.
18" Deep x 8' Wide	3.600	Ea.
50' Span, 24" Deep x 8' Wide	4.500	Ea.
60' Span, 32" Deep x 10' Wide	5.143	Ea.
Welded Wire Fabric 6 x 6-10/10	.457	C.S.F.
Place Concrete		
Pumped	.582	C.Y.
With Crane and Bucket	.674	C.Y.
Broom Finish	.012	S.F.
Steel Trowel Finish	.015	S.F.

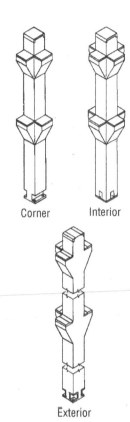

Corner   Interior

Exterior

Precast Concrete Columns

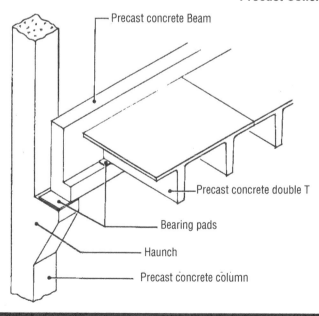

Precast concrete Beam

Precast concrete double T

Bearing pads

Haunch

Precast concrete column

Figure 3-44 Installation Time in Labor-Hours for Lift-Slab Construction

Description	Labor-Hours	Unit
Edge Forms 7" to 12", 4 uses	.074	S.F.C.A.
Box Out for Slab Openings	.120	L.F.
Reinforcing #3 to #7	13.913	ton
Post Tensioning Strand		
100 kip	.021	lb.
200 kip	.019	lb.
Placing Concrete		
Direct Chute	.291	C.Y.
Pumped	.388	C.Y.
Steel Trowel Finish	.015	S.F.

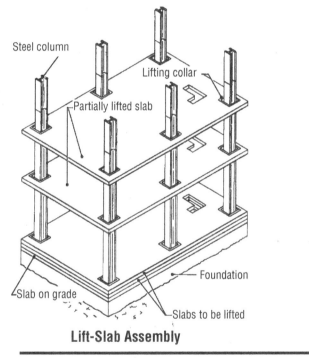

Lift-Slab Assembly

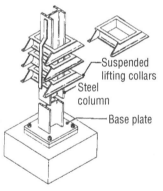

**Method of Suspending
Lifting Collars**

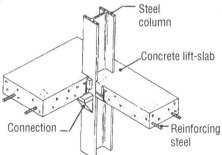

Typical Slab to Column Connection

Figure 3-45 Installation Time in Labor-Hours for Post-Tensioned Slabs and Beams

Description	Labor-Hours	Unit
Prestressing Steel Post-Tensioned in Field		
Grouted Strand 100 kip		
50' Span	.053	lb.
100' Span	.038	lb.
200' Span	.024	lb.
300 kip		
50' Span	.024	lb.
100' Span	.020	lb.
200' Span	.018	lb.
Ungrouted Strand 100 kip		
50' Span	.025	lb.
100' Span or 200' Span	.021	lb.
200 kip		
50' span	.022	lb.
100' Span or 200' Span	.019	lb.
Grouted Bars 42 kip		
50' Span	.025	lb.
75' Span	.020	lb.
143 kip		
50' Span	.020	lb.
75' Span	.015	lb.
Ungrouted Bars 42 kip		
50' Span	.023	lb.
75' Span	.018	lb.
143 kip		
50' Span	.019	lb.
75' Span	.015	lb.
Ungrouted Single Strand, 100' Slab		
25 kip	.027	lb.
35 kip	.022	lb.

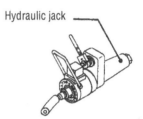

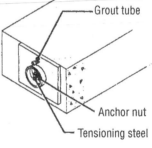

Hydraulic Tensioning Jack

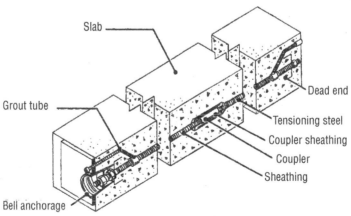

Post-Tensioning Assembly

Division Four
Masonry

Introduction

The word "masonry" may refer to a variety of items, including brick, block, structural tile, glass block, fieldstone, cut stone, and their respective installations. Brick, block, and tile are often called "masonry units," and working with them is commonly referred to as "unit masonry." Masonry construction is popular for a variety of reasons. It is fireproof, capable of withstanding abuse, and requires relatively little maintenance. It can also be an aesthetically pleasing and a relatively economical form of construction.

The two most common units in use are brick and concrete block. Brick is probably the best known type of unit masonry. It is very adaptable and can be installed in a variety of pleasing patterns, textures, and colors.

The major classifications of brick include face brick, common brick, glazed brick, fire brick, and pavers. Face bricks have a better appearance and are used when the bricks are to be in view. Common bricks are used when appearance is not as important, such as for the backup of a face brick veneer wall. Glazed bricks have a ceramic or other glazing material on one or two sides of the faces. They are used in a variety of circumstances, such as in restrooms, cafeterias, and kitchens, where durability and cleanliness are important. They may also be used as decorative wainscoting or for an attractive building exterior. Fire bricks are used in areas where high temperatures are encountered. Brick pavers are used as a wearing surface or flooring.

Concrete masonry units—concrete block—are used primarily for walls, partitions, or backup material for other veneers, such as face brick. The basic concrete block has actual face dimensions of 7-5/8" × 15-5/8", but is referred to as an 8" × 16" unit. Standard block thicknesses include 2", 3", 4", 6", 8", and 12" units. Concrete block is also available in a variety of textures and colors.

Masonry is generally priced by the piece and/or converted into a cost per square foot. The following topics are covered by the estimating charts and tables in this chapter:

- Weather factors
- Mortar and accessories
- Brick masonry
- Concrete unit masonry
- Glass block

green There are many masonry products that can improve the environmental attributes of this division's materials. Adobe brick, for example, is cured in the sun instead of a kiln used for standard bricks, eliminating the heavy energy requirements and environmental impacts caused by operating these huge kilns. Bricks can also be made from cleaned petroleum contaminated soil, which can help defray the costs of cleaning contaminated sites.

Because of their high mass, concrete masonry units or bricks can be used in walls where it is desirable to absorb the radiated heat from the sun during the warm days and release it during the cooler nights. Autoclave aerated concrete can be formed into masonry units that offer additional attributes of reduced weight and increased thermal insulating values compared to standard concrete masonry units. Concrete masonry units with integral insulation are another green choice because the addition of rigid insulation to the cores of the blocks markedly improves the insulating value of the units. Bricks lend themselves very well to being reused; the finished product adds character to what may have been an ordinary surface.

Unlike cast-in-place concrete, which is ordered from plants close to the project site, masonry may come from many miles away. To keep it green, it is recommended that you try to use masonry that is produced close to the project site. This reduces the environmental impacts of transporting the material many miles.

Additional green considerations include the longevity, ease of maintenance, and good insulation attributes. Masonry units also represent some of the most salvageable and reusable building materials.

Division 4

Estimating Data

The following tables present guidelines for estimating masonry items.

Table of Contents

Estimating Tips

Masonry Wall Block-outs

When estimating quantities for masonry walls, do not deduct areas less than two square feet in area. They will more than likely use cut whole block and should be configured as such.

Masonry Accessories

Remember to include miscellaneous items in your masonry estimate; they tend to get overlooked. These items include but are not limited to flashing, reinforcing, anchors, wall ties, inserts, bearing plates, lintels, support angles, and channels, allowances for joist pockets, waterproofing, cleanup, final cleaning and pointing, steam cleaning, acid or power wash, color for mortar, and control joints. For a more detailed list, see the checklist located after these tips.

Site Cleanup

If the plan at the time of the estimate is to erect masonry walls after concrete slabs have been poured, keep in mind that the masonry contractor will be responsible for cleaning the dropped mortar off of the concrete slab. Allow for this cleanup.

Bracing

One commonly overlooked item in masonry estimating is an allowance for bracing walls. Until the structural system is tied into the masonry wall systems, the walls can be blown over relatively easily.

Panelization

Plan ahead. Could your next project, which is not scheduled to start until later, be built with pre-assembled panels? If you have the labor available today, it might be economical to have the contractor pre-build the walls in panelized sections, then deliver and quickly erect them at the site. This could save quite a bit of time on the project.

Special Brick

When a project calls for special brick, such as utility size or glazed, remember that these (especially glazed) more than likely will be special order. The order and manufacture time can be surprisingly long. Paying a premium or extra charge may be the only way to ensure faster, on-time delivery.

Split Face Block

Split face block will take longer than common brick to set. This is because these blocks have a somewhat irregular depth dimension (on account of the splitting process). They do not look "right" if set by lining up the squared corners. Adjustments must be made to have them line up properly.

Division 4

Bricklaying Productivity

The national average productivity for laying brick ranges from 400 bricks per day (considered "low productivity") to 600 bricks per day (considered "high productivity").

Anti-Graffiti Products

If not specified, consider adding as an option the application of one of the various anti-graffiti or vandalism products. These coatings are roller, brush, or spray-applied to close the pores of the brick, thereby preventing permanent damage. Remember that graffiti and vandalism are present in all cities and towns, not just major metropolitan areas. Cleaning of masonry is best accomplished by specialized proprietary masonry cleaners, used per the manufacturer's directions.

Checklist ✓

For an estimate to be reliable, all items must be accounted for. A complete estimate can also limit contingencies. The following checklist can be used to help ensure that all items are included.

☐ **Applications**

☐ **Exterior Walls**
   ☐ Load bearing
   ☐ Non-load bearing

☐ **Interior Walls**
   ☐ Load bearing
   ☐ Non-load bearing

☐ **Solid Walls**

☐ **Cavity Walls**

☐ **Veneer Walls**

☐ **Flooring**

☐ **Brick**
   ☐ Common
   ☐ Face
   ☐ Cement
   ☐ Fire
   ☐ Color
      ☐ Special color
      ☐ Glazed
      ☐ Special finish
   ☐ Size
      ☐ Standard
      ☐ Jumbo
      ☐ Norman
      ☐ Roman
      ☐ Engineer
      ☐ Double
   ☐ Bond pattern

   ☐ Common
   ☐ Running
   ☐ English
   ☐ Flemish
   ☐ Stack
  ☐ Reinforcing
  ☐ Wall ties
  ☐ Grouting/fill

☐ **Concrete Block**
  ☐ Exterior
  ☐ Interior
  ☐ Regular
  ☐ Lightweight
  ☐ Solid
  ☐ Hollow core
  ☐ Finish
    ☐ 1 side
    ☐ 2 side
    ☐ Ribbed
    ☐ Fluted
    ☐ Split face
    ☐ Ground
    ☐ Colored
    ☐ Glazed
  ☐ Bond beams
  ☐ Lintels
  ☐ Pilasters
  ☐ Grouting/fill
  ☐ Reinforcing
    ☐ Bars
    ☐ Wall reinforcing (truss type)
  ☐ Wall ties
  ☐ Anchor bolts

Division 4

☐ *Glass Block*

☐ *Structural Tile*

☐ *Terra Cotta*

☐ *Stone*
- ☐ Ashlar
- ☐ Rubble
- ☐ Cut stone
  - ☐ Bases
  - ☐ Curbs
  - ☐ Facing panels
  - ☐ Flooring
  - ☐ Soffits
  - ☐ Simulated stone
  - ☐ Showers
  - ☐ Stairs
  - ☐ Stair treads
  - ☐ Thresholds
  - ☐ Window sills
  - ☐ Window stools

☐ *Mortar*
- ☐ Type
  - ☐ K
  - ☐ O
  - ☐ N
  - ☐ S
  - ☐ M
- ☐ Color
- ☐ Admixtures

☐ *Joints*
- ☐ Concave
- ☐ Struck
- ☐ Flush
- ☐ Raked
- ☐ Weathered
- ☐ Stripped

☐ *Cleaning Masonry*
- ☐ Sandblasting
- ☐ Steam clean
- ☐ Acid wash
- ☐ Power wash
- ☐ Final clean

☐ *Miscellaneous*
- ☐ Anchors
- ☐ Bolts
- ☐ Control joints
- ☐ Copings
- ☐ Dowels
- ☐ Expansion joints
- ☐ Flashings
- ☐ Insulation
- ☐ Insets
- ☐ Reglets
- ☐ Pointing
- ☐ Waterproofing
- ☐ Weep holes
- ☐ Window sills
- ☐ Window stools
- ☐ Vents

Figure 4-1 Weather Factors Affecting Masonry Productivity

Relative Humidity %	\ Temperature °F 40°	45°	50°	55°	60°	65°	70°	75°	80°	85°	90°	95°
100°		.626	.635	.659	.673	.683	.683	.666	.620			
95°	.605	.637	.650	.664	.683	.689	.705	.680	.635	.630		
90°	.615	.635	.657	.671	.697	.717	.726	.727	.690	.640		
85°	.615	.650	.670	.688	.724	.737	.752	.760	.725	.673	.600	
80°	.636	.650	.675	.710	.745	.765	.790	.795	.750	.760	.670	
75°	.636	.665	.690	.727	.765	.790	.835	.860	.780	.679	.610	
70°	.636	.675	.700	.740	.790	.816	.909	.930	.781	.680	.613	
65°	.636	.670	.700	.747	.795	.855	.950	.948	.781	.680	.610	
60°	.619	.670	.708	.747	.786	.820	.947	.973	.779	.673	.613	
55°	.619	.655	.700	.737	.775	.818	.909	.854	.885	.705	.640	.614
50°	.614	.655	.685	.727	.760	.790	.844	.908	.945	.765	.772	.623
45°	.610	.630	.673	.700	.742	.762	.790	.830	.820	.755	.683	.628
40°	.593	.614	.637	.685	.727	.742	.762	.768	.754	.697	.645	.615
35°		.604	.630	.660	.688	.706	.727	.717	.694	.670	.635	
30°			.612	.628	.648	.667	.685	.670	.648	.648	.620	
25°				.610	.630	.631	.620	.620	.625			
	40°	45°	50°	55°	60°	65°	70°	75°	80°	85°	90°	95°

Division 4

Weather often affects the "workability" of mortar and thus the productivity of masons. This table can be used to factor in the effects of weather on productivity. Apply the following steps:

1. Obtain the average temperature and humidity (local climatological data) for the month or time period needed for the region in question. (This information can be obtained from the U.S. weather services in the area, local airports, or possibly the local newspapers.)

2. Enter the table at the top with the average temperature obtained in Step 1.

3. Follow that column down to the average humidity reading obtained in Step 1.

4. Read the productivity factor at that intersection of humidity and temperature.

5. Multiply your average expected productivity for masonry installation by this factor. The result is your productivity adjusted for temperature conditions.

Figure 4-2 General Mortar Information

General information concerning mortar materials, mixes, and mixing.

Cement

A bag of Portland cement contains 94 lbs. When packed, Portland cement weights 108-1/2 lbs. per cubic foot.

A cubic foot of cement paste requires about 94 lbs. of cement.

Masonry Cement (prepackaged)

Type N – for normal masonry construction

Type S – for masonry construction that requires more than twice the strength of Type N

Type M – for masonry construction that requires more than three times the strength of Type N

Hydrated Lime

One cubic foot weighs about 40 lbs.

One paper package contains about 200 lbs. net.

One 100 lb. package makes about 2-1/4 cubic feet lime putty.

Prepared Mortar

A pre-mixed mixture of cement, lime, and gypsum, in varying proportions, depending on the manufacturer.

One paper package contains 70 lbs. net, and is equal to one cubic foot.

Proportioning of Mortar Materials

1:3 Cement mortar (by volume):

    Cement: 27 cubic feet ÷ 3 = 9 cubic feet = 9 bags

    Sand: 27 cubic feet   = 1 cubic yard

    Mortar yield: = 1 cubic yard

    (The cement fills the voids in the sand and there is little increase in bulk.)

1:2:9 Cement-lime-mortar (by volume):

    Cement: 27 cubic feet ÷ 9 = 3 cubic feet = 3 bags

    Lime: 27 cubic feet ÷ 2   = 13-1/2 cubic feet = 6 bags

    Sand: 27 cubic feet   = 1 cubic yard

    Mortar yield: = 1 cubic yard

    (This is actually a 1:3 cementitious-sand mixture)

1:3 Prepared mortar (by volume):

    Prepared mortar: 27 cubic feet ÷ 3 = 9 cubic feet = 9 bags

    Sand: 27 cubic feet   = 1 cubic yard

    Mortar yield: = 1 cubic yard

Job Mixing

A two-bag mortar mixer, with two bags of cementing agent and 30 number 2 shovels of sand will yield two wheelbarrows of mortar or about seven cubic feet.

A three-bag mixer, with three bags of cementing agent and 45 number 2 shovels of sand will yield three wheelbarrows of mortar, or about 10-1/2 cubic feet.

A two-bag mixer is adequate for jobs using up to 25 masons, and they can be supplied by an experienced mortar maker. Assistance in bringing materials to the machine is necessary.

On larger jobs, a three-bag mixer would be required. Mortar would be deposited on a mortar box and then wheeled to the masons, or the box lifted and deposited by forklift. Additional labor would be required.

(courtesy Masonry Estimating Handbook, *Michael F. Kenny, Construction Publishing Company, Inc.)*

Figure 4-3 Brick Mortar Mix Proportions

This chart shows some common mortar types, the mixing proportions, and their general uses.

Brick Mortar Mixes*					
Type	Portland Cement	Hydrated Lime	Sand (maximum)**	Strength	Use
M	1	1/4	3-3/4	High	General use where high strength is required, especially good compressive strength; work that is below grade and in contact with earth.
S	1	1/2	4-1/2	High	Okay for general use, especially good where high lateral strength is desired.
N	1	1	6	Medium	General use when masonry is exposed above grade; best to use when high compressive and lateral strengths are not required.
0	1	2	9	Low	Do not use when masonry is exposed to severe weathering; acceptable for non-loadbearing walls of solid units and interior non-loadbearing partitions of hollow units.

* The water should be of the quality of drinking water. Use as much as is needed to bring the mix to a suitably plastic and workable state.

** The sand should be damp and loose. A general rule for sand content is that it should be not less than 2-¼ times or more than 3 times the sum of the cement and lime volumes.

Figure 4-4 Compressive Strengths of Mortar

This table lists the expected 28-day compressive strengths for some common types of mortar mixes.

Mortar Type	Average Compressive Strength at 28 Days
M	2500 p.s.i.
S	1800 p.s.i.
N	750 p.s.i.
0	350 p.s.i.
K	75 p.s.i.

(courtesy Masonry & Concrete Construction, *Ken Nolan, Craftsman Book Co.*)

Division 4

Figure 4-5 Mortar Quantities for Brick Masonry

This table can be used to determine the amounts of mortar needed for brick masonry walls for various joint thicknesses and types of common brick.

Brickwork	Joint Thickness (in inches)	Actual Requirement (in C.F. per 1,000)	Requirement with 15% Waste (in C.F. per 1,000)
Standard	3/8	8.6	0.4
Standard	1/2	11.7	0.5
Modular	3/8	7.6	0.3
Modular	1/2	10.4	0.4
Roman	3/8	10.7	0.5
Roman	1/2	14.4	0.6
Norman	3/8	11.2	0.5
Norman	1/2	15.1	0.6
Type	Thickness (in inches)	Actual Requirement (in C.F. per S.F.)	Requirement with 15% Waste (in C.F. per 1,000 S.F.)
Parging or Backplastering	3/8	0.03	1.3
	1/2	0.04	1.7

Note: Quantities are based on 4" thick brickwork. For each additional 4" brickwork, or where masonry units are used in backup, add parging or backplastering.

Modular dimensions are measured from center to center of masonry joints. To fit into the system, the unit or number of units must be the size of the module, less the thickness of the joint. A standard face brick with 1/2" joint would measure 8-1/2" long, whereas a modular brick with the same joint would measure 8".

Figure 4-6 Mortar Quantities for Concrete Block Masonry

This table can be used to determine the amounts of mortar needed for concrete block masonry walls for joint thicknesses for various types of masonry units.

Type	Joint Thickness (in inches)	Actual Requirement (in C.F. per 1,000)	Requirement with 15% Waste (in C.F. per 1,000)
Concrete Block			
Shell Bedding			
All thicknesses	3/8	0.6	0.7
Full Bedding			
12 x 8 x 16, 3-Core	3/8	0.9	1.0
12 x 8 x 16, 2-Core	3/8	0.8	0.9
8 x 8 x 16, 3-Core	3/8	0.7	0.8
8 x 8 x 16, 2-Core	3/8	0.7	0.8
6 x 8 x 16, 3-Core	3/8	0.7	0.8
6 x 8 x 16, 2-Core	3/8	0.7	0.8
4 x 8 x 16, 3-Core	3/8	0.7	0.8
4 x 8 x 16, Solid	3/8	0.8	0.9

Type	Joint Thickness (in inches)	Actual Requirement (in C.F. per 1,000)	Requirement with 15% Waste (in C.F. per 1,000)
4S Glazed Structural Units			
(nominal 2-1/2" x 8")			
4 SA (2" thick)	1/4	2.6	0.1
4 S (4" thick)	1/4	5.6	0.2
4D Glazed Structural Units			
(nominal 5" x 8")			
4 DCA (2" thick)	1/4	3.3	0.1
4 DC (4" thick)	1/3	7.1	0.3
4 DC 60 (6" thick)	1/4	10.9	0.5
4 DC 80 (8" thick)	1/4	13.7	0.6
6T Glazed Structural Units			
(nominal 5" x 12")			
6 TCA (2" thick)	1/4	4.2	0.2
6 TC (4" thick)	1/4	9.3	0.4
6 TC 60 (6" thick)	1/4	14.2	0.6
6 TC 80 (8" thick)	1/4	19.1	0.8
6P Glazed Structural Units -			
1/4" Joints (nominal 4" x 12")			
6 PCA (2" thick)		4.0	0.2
6 PC (4" thick)		8.6	0.4
6 PC 60 (6" thick)		13.1	0.8
6 PC 80 (8" thick)		17.7	0.8
8W Glazed Structural Units			
(nominal 8" x 16")			
8 WCA (2" thick)	1/4	6.0	0.3
8 WCA (2" thick)	3/8	9.1	0.4
8 WC (4" thick)	1/4	12.9	0.6
8 WC (4" thick)	3/8	18.4	0.8

Division 4

(continued on next page)

Figure 4-6 Mortar Quantities for Concrete Block Masonry
(continued from previous page)

Type	Joint Thickness (in inches)	Actual Requirement (in C.F. per 1,000)	Requirement with 15% Waste (in C.F. per 1,000)
Spectra-Glaze® Units –			
3/8" Joints (nominal 4" x 16")			
44S (4" thick)		16.0	0.7
64S (6" thick)		24.5	1.0
84S (8" thick)		33.0	1.4
Spectra-Glaze® Units–			
3/8" Joints (nominal 8" x 16")			
2S (2" thick)		9.0	0.4
4S (4" thick)		19.2	0.8
6S (6" thick)		29.5	1.3
8S (8" thick)		39.7	1.7
10S (10" thick)		50.0	2.1
12S (12" thick)		60.2	2.6
Structural Clay Backup and Wall Tile			
5 x 12 (4" thick)	1/2	20.0	0.9
5 x 12 (6" thick)	1/2	30.4	1.3
5 x 12 (8" thick)	1/2	40.5	1.7
8 x 12 (6" thick)	1/2	35.6	1.5
8 x 12 (8" thick)	1/2	47.5	2.0
8 x 12 (12" thick)	1/2	71.2	3.0
Structural Clay Partition Tile			
12 x 12 (4" thick)	1/2	28.4	1.2
12 x 12 (6" thick)	1/2	42.5	1.8
12 x 12 (8" thick)	1/2	56.7	2.4
12 x 12 (10" thick)	1/2	70.9	3.0
12 x 12 (12" thick)	1/2	85.1	3.6
Gypsum Block (all 12" x 30")			
2" thick	1/4	12.2	0.5
	3/8	18.4	0.8
	1/2	24.6	1.0

(continued on next page)

Figure 4-6 Mortar Quantities for Concrete Block Masonry
(continued from previous page)

Type	Joint Thickness (in inches)	Actual Requirement (in C.F. per 1,000)	Requirement with 15% Waste (in C.F. per 1,000)
Gypsum Block (all 12" x 30") cont.			
3" thick	1/4	18.3	0.8
	3/8	27.6	1.2
	1/2	36.9	1.6
4" thick	1/4	24.5	1.0
	3/8	36.8	1.6
	1/2	49.2	2.1
6" thick	1/4	36.7	1.6
	3/8	55.2	2.4
	1/2	73.8	3.1
Glass Block (all 3-7/8" thick)			
5-3/4 x 5-3/4	1/4		
7-3/4 x 7-3/4	1/4		
11-3/4 x 11-3/4	1/4		
Firebrick			
Thin joints, including waste 300 lbs. fireclay per 1,000.			

Note: Gypsum partition tile cement should be used. 900 lbs. are required per cubic yard, plus 1 cubic yard of sand. 1,000 units equal 2,610 SF.

Division 4

Figure 4-7 Brick Nomenclature

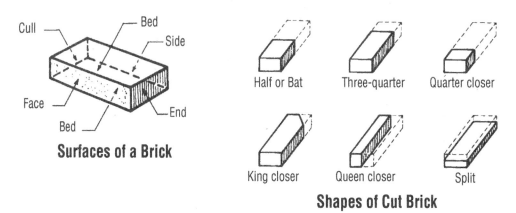

Surfaces of a Brick

Shapes of Cut Brick

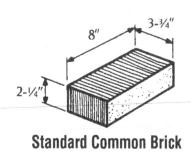

Standard Common Brick

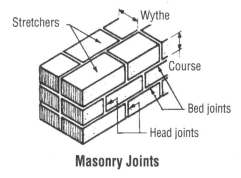

Masonry Joints

(continued on next page)

Figure 4-7 Brick Nomenclature *(continued from previous page)*

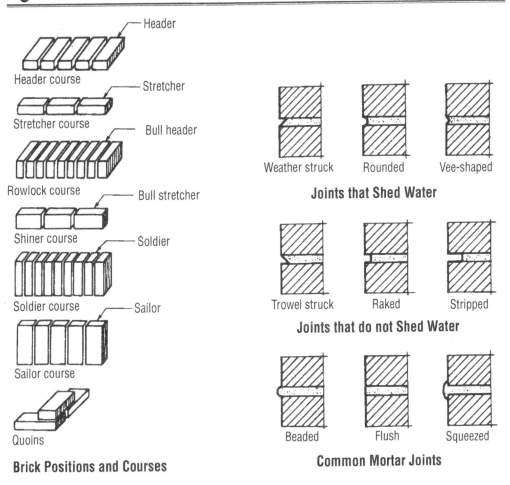

Brick Positions and Courses

Header course — Header
Stretcher course — Stretcher
Rowlock course — Bull header
Shiner course — Bull stretcher
Soldier course — Soldier
Sailor course — Sailor
Quoins

Joints that Shed Water

Weather struck   Rounded   Vee-shaped

Joints that do not Shed Water

Trowel struck   Raked   Stripped

Common Mortar Joints

Beaded   Flush   Squeezed

Figure 4-8 Heights of Brick Masonry Walls by Course

Course No.	Standard 2-1/4" Brick (3/8" Joint)	Course No.	Standard 2-1/4" Brick (3/8" joint)
1	2-5/8"	26	5' 8-1/4"
2	5-1/4"	27	5' 10-7/8"
3	7-7/8"	28	6' 1-1/2"
4	10-1/2"	29	6' 4-1/8"
5	1' 1-1/8"	30	6' 6-3/4"
6	1' 3-3/4"	31	6' 9-3/8"
7	1' 6-3/8"	32	7' 0"
8	1' 9"	33	7' 2-5/8"
9	1' 11-5/8"	34	7' 5-1/4"
10	2' 2-1/4"	35	7' 7-7/8"
11	2' 4-7/8"	36	7' 10-1/2"
12	2' 7-1/2"	37	8' 1-1/8"
13	2' 10-1/8"	38	8' 3-3/4"
14	3' 0-3/4"	39	8' 6-3/8"
15	3' 3-3/8"	40	8' 9"
16	3' 6"	41	8' 11-5/8"
17	3' 8-5/8"	42	9' 2-1/4"
18	3' 11-1/4"	43	9' 4-7/8"
19	4' 1-7/8"	44	9' 7-1/2"
20	4' 4-1/2"	45	9' 10-1/8"
21	4' 7-1/8"	46	10' 0-3/4"
22	4' 9-3/4"	47	10' 3-3/8"
23	5' 0-3/8"	48	10' 6"
24	5' 3"	49	10' 8-5/8"
25	5' 5-5/8"	50	10' 11-1/4"

Note: When standard brick is used in conjunction with concrete block, the brick coursing corresponds to height modules, i.e., 3 courses of brick lay up to 8".

Figure 4-9 Sizes of Modular Brick

Unit Designation	Nominal Dimensions (in inches)			Joint Thickness (in inches)	Manufactured Dimensions (in inches)			Modular Coursing (in inches)
	t	h	l		t	h	l	
Standard Modular	4	2-2/3	8	3/8	3-5/8	2-1/4	7-5/8	3C = 8
				1/2	3-1/2	2-1/4	7-1/2	
Engineer	4	3-1/5	8	3/8	3-5/8	2-13/16	7-5/8	5C = 16
				1/2	3-1/2	2-11/16	7-1/2	
Economy 8 or Jumbo Closure	4	4	8	3/8	3-5/8	3-5/8	7-5/8	1C = 4
				1/2	3-1/2	3-1/2	7-1/2	
Double	4	5-1/3	8	3/8	3-5/8	4-15/16	7-5/8	3C = 16
				1/2	3-1/2	4-13/16	7-1/2	
Roman	4	2	12	3/8	3-5/8	1-5/8	11-5/8	2C = 4
				1/2	3-1/2	1-1/2	11-1/2	
Norman	4	2-2/3	12	3/8	3-5/8	2-1/4	11-5/8	3C = 8
				1/2	3-1/2	2-1/4	11-1/2	
Norwegian	4	3-1/5	12	3/8	3-5/8	2-13/16	11-5/8	5C = 16
				1/2	3-1/2	2-11/16	11-1/2	
Economy 12 or Jumbo Utility	4	4	12	3/8	3-5/8	3-5/8	11-5/8	1C = 4
				1/2	3-1/2	3-1/2	11-1/2	
Triple	4	5-1/3	12	3/8	3-5/8	4-15/16	11-5/8	3C = 16
				1/2	3-1/2	4-13/16	11-1/2	
SCR brick	6	2-2/3	12	3/8	5-5/8	2-1/4	11-5/8	3C = 8
				1/2	5-1/2	2-1/4	11-1/2	
6-in Norwegian	6	3-1/5	12	3/8	5-5/8	2-13/16	11-5/8	5C = 16
				1/2	5-1/2	2-11/16	11-1/2	
6-in Jumbo	6	4	12	3/8	5-5/8	3-5/8	11-5/8	1C = 4
				1/2	5-1/2	3-1/2	11-1/2	
8-in Jumbo	8	4	12	3/8	7-5/8	3-5/8	11-5/8	1C = 4
				1/2	7-1/2	3-1/2	11-1/2	

(courtesy Masonry and Concrete Construction, *Ken Nolan, Craftsman Book Company)*

Division 4

Figure 4-10 Brick Bonding Patterns

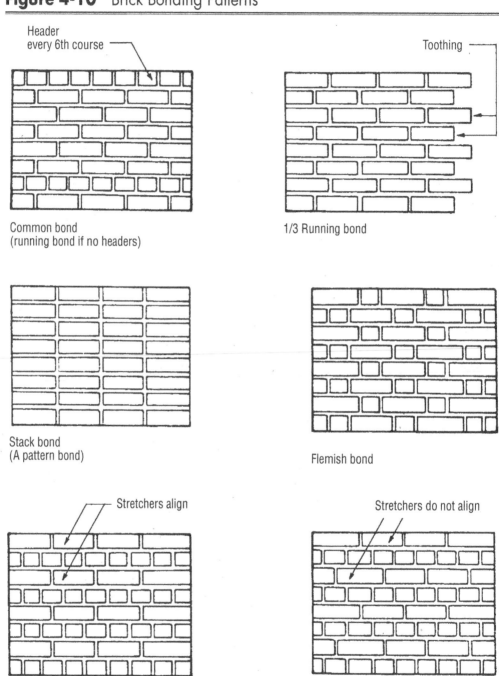

Header
every 6th course

Toothing

Common bond
(running bond if no headers)

1/3 Running bond

Stack bond
(A pattern bond)

Flemish bond

Stretchers align

Stretchers do not align

English bond

Cross bond
(english, flemish, or dutch)

Figure 4-11   Brick Requirements and Waste Factors

This table provides the number of bricks required per SF and waste factors based on brick type and bonding pattern.

Type of Brick	Description	Waste		Bricks/S.F.
Standard	Running Bond	net		6.55
		5% waste		6.88
	Common Bond	net		7.64
	(6th course headers)	5% waste		8.02
	English Bond	net		9.83
	(full headers every other course)	5% waste		10.32
	Flemish Bond	net		9.83
	(alternate headers and stretchers			
	each course)	5% waste		10.32
	8" Wall, Running Bond	net		13.10
		5% waste		13.76
	8" Wall, Common Bond	net		13.10
		5% waste		13.76
	12" Wall, Running Bond	net	(Face Brick)	13.10
			(Common Brick)	6.55
		5% waste	(Face Brick)	13.76
	12" Wall, Common Bond	net	(Face Brick)	15.28
			(Common Brick)	4.37
		5% waste	(Face Brick)	16.04
			(Common Brick)	4.59
Concrete	Running Bond	net		6.86
		5% waste		7.20
Modular	Running Bond	net		7.19
		5% waste		7.55
Modular Roman	Running Bond	net		6.00
		5% waste		6.30
Modular Norman	Running Bond	net		4.57
		5% waste		4.80
Fire	(2-1/4" x 8")	net		6.98
		5% waste		7.33
	(2-1/2" x 9")	net		5.67
		5% waste		5.95

Division 4

Figure 4-12 Brick Quantities per SF

Number of bricks per SF based on brick type and size of joints.

Brick Type & Size	\multicolumn{6}{c}{Size of Joint (in inches)}					
	1/4	1/3	3/8	1/2	5/8	3/4
Standard face brick (8" x 2-1/4")	6.98	6.70	6.55	6.16	5.81	5.49
Standard common brick (8" x 2-1/4")	6.98	6.70	6.55	6.16	5.81	5.49
Concrete brick (7-5/8" x 2-1/4")	7.31	7.00	6.86	6.45	6.07	5.73
Modular brick (7-1/2" x 2-1/6")	7.68	7.35	7.19	6.73	6.35	5.98
Modular Roman brick (11-5/8" x 1-5/8")	6.47	6.15	6.00	5.59	5.22	4.90
Modular Norman brick (11-5/8" x 2-1/4")	4.85	4.66	4.57	4.32	4.09	3.88

Note: Above constants are net, i.e., no waste is included. (See Figure 4-11.)

Figure 4-13 Adjustments to Brick Quantity Factors

This table provides factors for determining the additional quantities of brick needed when specific bonding patterns are used.

\multicolumn{5}{c}{For Other Bonds Standard Size Add to S.F. Quantities}				
Bond Type	Description	Factor	Description	Factor
Common	Full header every fifth course	+20%	Header = W x H exposed	+100%
	Full header every sixth course	+16.7%	Rowlock = H x W exposed	+100%
English	Full header every second course	+50%	Rowlock stretcher = L x W exposed	+33.3%
Flemish	Alternate headers every course	+33.3%	Soldier = H x L exposed	—
	every sixth course	+5.6%	Sailor = W x L exposed	−33.3%

(See "Brick Quantities per SF" table above for basic quantities of brick per SF.)

Figure 4-14 Brick Masonry Systems

The following tables show typical quantities of brick and mortar, and installation time per SF and per thousand bricks for the systems described. The crew used for these examples consists of 3 bricklayers and 2 bricklayer helpers.

Brick in Place			
Item	8" Common Brick Wall 8" x 2-2/3" x 4"	Select Common Face 8" x 2-2/3" x 4"	Red Face Brick 8" x 2-2/3" x 4"
1030 brick delivered Type N mortar Installation	12.5 C.F. 0.556 Days	10.3 C.F. 0.667 Days	10.3 C.F. 0.667 Days
Total per S.F. of wall	13.5 bricks/S.F.	6.75 bricks/S.F.	6.75 bricks/S.F.

This table is for common bond with 3/8" concave joints and includes 3% waste for brick and 25% waste for mortar.

Both the number of bricks per SF and the waste factors are based on the brick type and the type of bonding pattern chosen.

Brick Veneer in Place			
Item	8" Common Brick Wall 8" x 2-2/3" x 4"	Select Common Face 8" x 2-2/3" x 4"	Red Face Brick 8" x 2-2/3" x 4"
1030 brick delivered Type N mortar Installation using 3 Bricklayers and 2 Helpers	12.5 C.F. 0.556 Days	10.3 C.F. 0.667 Days	10.3 C.F. 0.667 Days
Total per S.F. of wall	13.5 bricks/S.F.	6.75 bricks/S.F.	6.75 bricks/S.F.

This table is for common bond with 3/8" concave joints and includes 3% waste for brick and 25% waste for mortar.

Reinforced Brick in Walls		
Item	8" to 9" Thick Wall	16" to 17" Thick Wall
1030 brick (select common) Type PL mortar Reinforcing bars delivered Installation using 3 Bricklayers and 2 Helpers	12.5 C.F. 50 lb. 0.571 Days	13.9 C.F. 50 lb. 0.513 Days
Total per S.F. of wall	8" wall, 13.5 brick/S.F.	16" wall, 27.0 brick/S.F.

This table is for common bond with 3/8" concave joints and includes 3% waste. Standard 8" x 2-2/3" x 4" bricks.

(continued on next page)

Division 4

Figure 4-14 Brick Masonry Systems *(continued from previous page)*

Brick Chimneys				
Quantities	**16" x 16"**	**20" x 20"**	**20" x 24"**	**20" x 32"**
Brick	28 brick	37 brick	42 brick	51 brick
Type M mortar	.5 C.F.	.6 C.F.	1.0 C.F.	1.3 C.F.
Flue tile (square)	8" x 8"	12" x 12"	2 @ 8" x 12"	2@ 12" x 12"
Install tile & brick using 1 Bricklayer and 1 Helper	0.55 day	0.73 day	.083 day	.10 day

Note: Use of 2/3 of the above amounts if the chimney butts against a straight wall. Labor for chimney brick using 1 bricklayer and 1 bricklayer helper is 31 hours per thousand bricks. An 8″ x 12″ flue takes 33 bricks, and two 8″ x 8″ flues take 37 bricks.

Figure 4-15 Typical Concrete Block Systems and Nomenclature

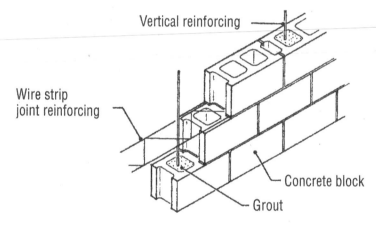

Grouted and Reinforced Block

(continued on next page)

Figure 4-15 Typical Concrete Block Systems and Nomenclature
(continued from previous page)

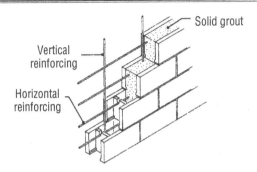

Solid grout

Vertical reinforcing

Horizontal reinforcing

Interlocking Concrete Block

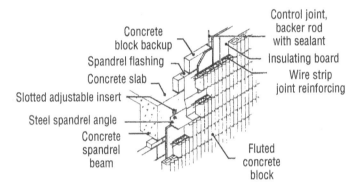

Concrete block backup

Spandrel flashing

Concrete slab

Slotted adjustable insert

Steel spandrel angle

Concrete spandrel beam

Control joint, backer rod with sealant

Insulating board

Wire strip joint reinforcing

Fluted concrete block

Block Face Cavity Wall System

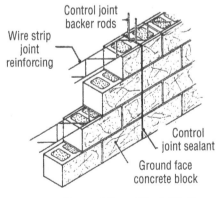

Control joint backer rods

Wire strip joint reinforcing

Control joint sealant

Ground face concrete block

Ground Face Block Wall

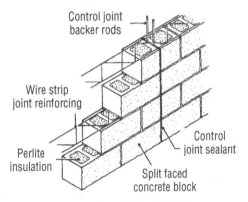

Control joint backer rods

Wire strip joint reinforcing

Perlite insulation

Control joint sealant

Split faced concrete block

Split Faced Block Wall

(continued on next page)

Division 4

Figure 4-15 Typical Concrete Block Systems and Nomenclature
(continued from previous page)

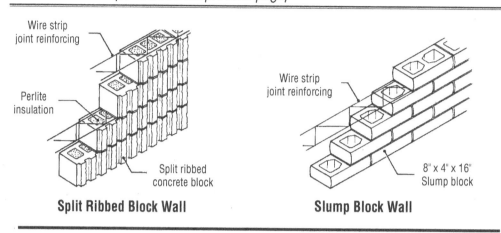

Split Ribbed Block Wall   **Slump Block Wall**

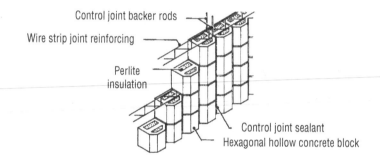

Hexagonal Block

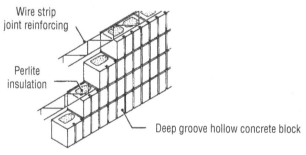

Deep Groove Hollow Block

(continued on next page)

Figure 4-15   Typical Concrete Block Systems and Nomenclature
(continued from previous page)

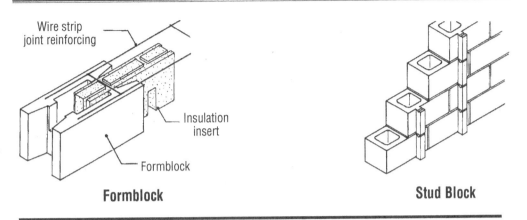

Formblock   **Stud Block**

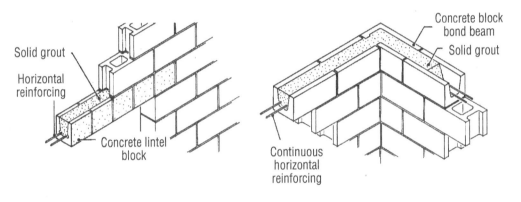

Lintels   **Bond Beam**

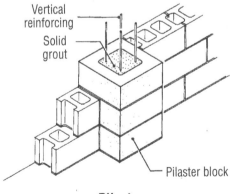

Pilaster

(continued on next page)

Division 4

Figure 4-15 Typical Concrete Block Systems and Nomenclature
(continued from previous page)

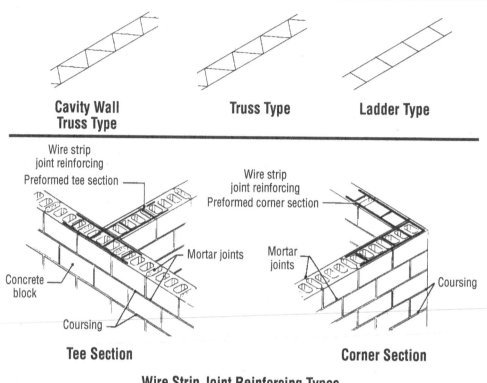

Cavity Wall Truss Type   **Truss Type**   **Ladder Type**

Wire strip joint reinforcing

Preformed tee section

Wire strip joint reinforcing

Preformed corner section

Mortar joints

Mortar joints

Concrete block

Coursing

Coursing

Coursing

Tee Section   **Corner Section**

Wire Strip Joint Reinforcing Types

Figure 4-16   Concrete Block Masonry Conversion Factors

Block Type	Description	Waste	Quantity
8" x 16" Concrete Block	All Thicknesses	net	1.125
		5% waste	1.181
4S Glazed Structural Units		net	6.71
		5% waste	7.05
	Horizontal Bullnose	net	1.49*
		5% waste	1.56*
	Vertical Bullnose	net	4.50*
		5% waste	4.73*
4D Glazed Structural Units		net	3.37
		5% waste	3.54
	Horizontal Bullnose	net	1.50*
		5% waste	1.58*
	Vertical Bullnose	net	2.25*
		5% waste	2.36*
6P Glazed Structural Units		net	3.00
		5% waste	3.15
	Horizontal Bullnose	net	1.00*
		5% waste	1.05*
	Vertical Bullnose	net	3.00
		5% waste	3.15
6T Glazed Structural Units		net	2.25
		5% waste	2.36
	Horizontal Bullnose	net	1.00*
		5% waste	1.05*
	Vertical Bullnose	net	2.25*
		5% waste	2.36*
8W Glazed Structural Units		net	1.125
		5% waste	1.181
	Horizontal Bullnose	net	0.75*
		5% waste	0.79*
	Vertical Bullnose	net	1.50*
		5% waste	1.58*
Spectra-Glaze® Units	8" x 16" (nominal)	net	1.125
		5% waste	1.181
	Horizontal Bullnose	net	0.75*
		5% waste	0.79*
	Vertical Bullnose	net	1.50*
		5% waste	1.58*

Division 4

(continued on next page)

Figure 4-16   Concrete Block Masonry Conversion Factors
(continued from previous page)

Block Type	Description	Waste	Quantity
Structural Clay Tile	5" x 12"	net	2.09
		5% waste	2.20
	8" x 12"	net	1.36
		5% waste	1.42
	12" x 12"	net	0.92
		5% waste	0.97
Gypsum Block	12" x 30"	net	0.38
		5% waste	0.40
Glass Block	5-3/4" x 5-3/4"	net	4.00
	7-3/4" x 7-3/4"	net	2.25
	11-3/4" x 11-3/4"	net	1.00

Note: Factors given in units per square foot except those where * occurs, and these are in units for linear foot. Quantities of block and waste factors per square foot based upon the type of block.

(courtesy Masonry Estimating Handbook, Michael F. Kenny, Construction Publishing Company, Inc.)

Figure 4-17   Weights of Block Partitions

The approximate weight of various sized block walls per square foot area of wall.

Concrete Blocks Nominal Size	Approximate Weight per S.F.	
	Standard	Lightweight
2" x 8" x 16"	20 PSF	15 PSF
4"	30	20
6"	42	30
8"	55	38
10"	70	47
12"	85	55

Figure 4-18 Height of Concrete Block Walls by Course

Course No.	Concrete Block (3/8" joint)	6T Series G.S.U. (1/4" joint)	8W Series G.S.U. (1/4" joint)
1	0' 8"	5-1/3"	0' 8"
2	1' 4"	10-2/3"	1' 4"
3	2' 0"	1' 4"	2' 0"
4	2' 8"	1' 9-1/3"	2' 8"
5	3' 4"	2' 2-2/3"	3' 4"
6	4' 0"	2' 8"	4' 0"
7	4' 8"	3' 1-1/3"	4' 8"
8	5' 4"	3' 6-2/3"	5' 4"
9	6' 0"	4' 0"	6' 0"
10	6' 8"	4' 5-1/3"	6' 8"
11	7' 4"	4' 10-2/3"	7' 4"
12	8' 0"	5' 4"	8' 0"
13	8' 8"	5' 9-1/3"	8' 8"
14	9' 4"	6' 2-2/3"	9' 4"
15	10' 0"	6' 8"	10' 0"
16	10' 8"	7' 1-1/3"	10' 8"
17	11' 4"	7' 6-2/3"	11' 4"
18	12' 0"	8' 0"	12' 0"
19	12' 8"	8' 5-1/3"	12' 8"
20	13' 4"	8' 10-2/3"	13' 4"
21	14' 0"	9' 4"	14' 0"
22	14' 8"	9' 9-1/3"	14' 8"
23	15' 4"	10' 2-2/3"	15' 4"
24	16' 0"	10' 8"	16' 0"
25	16' 8"	11' 1-1/3"	16' 8"
26	17' 4"	11' 6-2/3"	17' 4"
27	18' 0"	12' 0"	18' 0"
28	18' 8"	12' 5-1/3"	18' 8"
29	19' 4"	12' 10-2/3"	19' 4"
30	20' 0"	13' 4"	20' 0"
31	20' 8"	13' 9-1/3"	20' 8"
32	21' 4"	14' 2-2/3"	21' 4"
33	22' 0"	14' 8"	22' 0"
34	22' 8"	15' 1-1/3"	22' 8"
35	23' 4"	15' 6-2/3"	23' 4"
36	24' 0"	16' 0"	24' 0"
37	24' 8"	16' 5-1/3"	24' 8"
38	25' 4"	16' 10-2/3"	25' 4"
39	26' 0"	17' 4"	26' 0"
40	26' 8"	17' 9-1/3"	26' 8"
41	27' 4"	18' 2-2/3"	27' 4"
42	28' 0"	18' 8"	28' 0"
43	28' 8"	19' 1-1/3"	28' 8"
44	29' 4"	19' 6-2/3"	29' 4"
45	30' 0"	20' 0"	30' 0"
46	30' 8"	20' 5-1/2"	30' 8"
47	31' 4"	20' 10-2/3"	31' 4"
48	32' 0"	21' 4"	32' 0"
49	32' 8"	21' 9-1/3"	32' 8"
50	33' 4"	22' 2-2/3"	33' 4"

Note: When standard brick is used in conjunction with concrete block, the brick coursing corresponds to block height modules, i.e., 3 courses of brick lay up to 8".

(*courtesy* Masonry Estimating Handbook, *Michael F. Kenny, Construction Publishing Company, Inc.*)

Division 4

Figure 4-19 Concrete Block Required for Walls

Ground Floor Area	Number of 8" x 16" Blocks Needed for Height of Wall Shown									
	2'-0"	2'-8"	3'-4"	4'-0"	4'-8"	5'-4"	6'-0"	6'-8"	7'-4"	8'-0"
200	135	180	225	270	315	360	405	450	495	540
300	167	222	278	333	389	444	500	555	611	666
400	191	255	319	383	446	510	574	638	701	765
500	214	285	356	428	499	570	641	713	784	855
600	234	312	390	468	546	624	702	780	858	936
700	252	336	420	504	588	672	756	840	924	1008
800	270	360	450	540	630	720	810	900	990	1080
900	286	381	476	572	667	762	857	953	1048	1143
1000	302	402	503	604	704	804	905	1005	1106	1206
1100	317	423	529	634	740	846	952	1058	1163	1269
1200	331	441	551	662	772	882	992	1103	1213	1323
1300	344	459	574	689	803	918	1033	1148	1262	1377
1400	358	477	596	716	835	954	1073	1193	1312	1431
1500	369	492	615	738	861	984	1107	1230	1353	1476
1600	383	510	638	765	893	1020	1148	1275	1403	1530
1700	394	525	636	788	919	1050	1181	1313	1444	1575
1800	405	540	675	810	945	1080	1215	1350	1485	1620
1900	416	555	694	833	971	1110	1249	1388	1526	1665
2000	428	570	713	855	998	1140	1283	1425	1568	1710
2100	437	582	728	873	1019	1164	1310	1455	1601	1746
2200	448	597	746	896	1045	1194	1343	1493	1642	1791
2300	457	609	761	914	1066	1218	1370	1523	1675	1827
2400	468	624	780	936	1092	1248	1404	1560	1716	1872
2500	477	636	795	954	1113	1272	1431	1590	1749	1908
2600	486	648	810	972	1134	1296	1458	1620	1782	1944
2700	495	660	825	990	1155	1320	1485	1650	1815	1980
2800	504	672	840	1008	1176	1344	1512	1680	1848	2016
2900	513	684	855	1026	1197	1368	1539	1710	1881	2052
3000	524	699	874	1049	1223	1398	1573	1748	1922	2097

This table provides the approximate number of 8" x 16" concrete blocks needed for the foundation walls for structures of various sizes. To use this chart:

1. In the left-most column, find the ground floor area for the structure.

2. Read across the page until you reach the column that matches the height of your foundation wall. The number in that box represents the number of blocks needed for that foundation.

Note: No deductions have been made for openings.

(courtesy How to Estimate Building Losses and Construction Costs, Paul I. Thomas, Prentice Hall)

Figure 4-20 Volume of Grout Fill for Concrete Block Walls

Center to Center Spacing	6" C.M.U. Per S.F. Volume in C.F.		8" C.M.U. Per S.F. Volume in C.F.		12" C.M.U. Per S.F. Volume in C.F.	
Grouted Cores	40% Solid	75% Solid	40% Solid	75% Solid	40% Solid	75% Solid
All cores grouted solid	.27	.11	.36	.15	.55	.23
cores grouted 16" O.C.	.14	.06	.18	.08	.28	.12
cores grouted 24" O.C.	.09	.04	.12	.05	.18	.08
cores grouted 32" O.C.	.07	.03	.09	.04	.14	.06
cores grouted 40" O.C.	.05	.02	.07	.03	.11	.05
cores grouted 48% O.C.	.04	.02	.06	.03	.09	.04

Note: Costs are based on high-lift grouting method.

Low-lift grouting is used when the wall is built to a maximum height of 5'. The grout is pumped or poured into the cores of the concrete block. The operation is repeated after each five additional feet of wall height has been completed. High-lift grouting is used when the wall has been built to the full story height. Some of the advantages are: the vertical reinforcing steel can be placed after the wall is completed, and the grout can be supplied by a ready-mix concrete supplier so that it may be pumped in a continuous operation.

Figure 4-21 Quantities for Glass Block Partitions

	Per 100 S.F.		Per 1000 Block				
Size	No. of Block	Mortar 1/4" Joint	Asphalt Emulsion	Caulk	Expansion Joint	Panel Anchors	Wall Mesh
6" x 6"	410 ea.	5.0 C.F.	.17 gal.	1.5 gal.	80 L.F.	20 ea.	500 L.F.
8" x 8"	230	3.6	.33	2.8	140	36	670
12" x 12"	102	2.3	.67	6.0	312	80	1000
Approximate quantity per 100 S.F.			.07 gal.	.6 gal.	32 L.F.	9 ea.	51, 68, 102 L.F.

Division 4

Figure 4-22 Typical Installation Practices for Reinforcing Masonry Walls

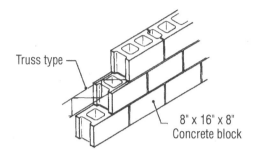

Truss type

8" x 16" x 8"
Concrete block

Wire Strip Joint Reinforcing

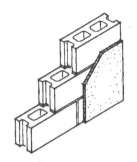

Plaster Direct to Block

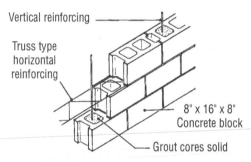

Vertical reinforcing

Truss type
horizontal
reinforcing

8" x 16" x 8"
Concrete block

Grout cores solid

Reinforced Concrete Block Wall

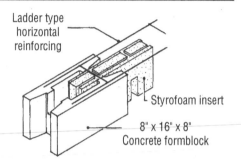

Ladder type
horizontal
reinforcing

Styrofoam insert

8" x 16" x 8"
Concrete formblock

Insulated Concrete Block

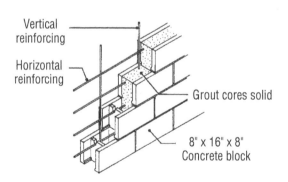

Vertical
reinforcing

Horizontal
reinforcing

Grout cores solid

8" x 16" x 8"
Concrete block

Interlocking Concrete Block

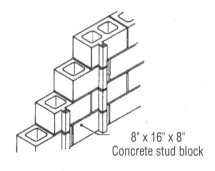

8" x 16" x 8"
Concrete stud block

Self-furring Concrete Block

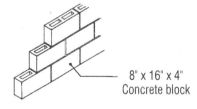

8" x 16" x 4"
Concrete block

Nonbearing Concrete Block Partition

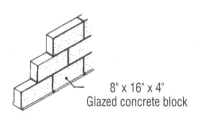

8" x 16" x 4"
Glazed concrete block

Glazed Concrete Block

Figure 4-23 Quantities for Joint Reinforcing and Masonry Ties

Number of LF of masonry joint reinforcement needed per unit.

Joint Reinforcement			
Number of Courses Between Reinforcement	**L.F. of Reinforcement**		
2	0.67 L.F.	per block	(net)
2	0.74 L.F.		(incl. 10% waste)
2	0.75 L.F.	per S.F. block	(net)
2	0.83 L.F.		(incl. 10% waste)
3	0.44 L.F.	per block	(net)
3	0.48 L.F.		(incl. 10% waste)
3	0.50 L.F.	per S.F. block	(net)
3	0.55 L.F.		(incl. 10% waste)

Masonry joint ties can be estimated on a per SF basis as shown in this table.

Masonry Ties		
Ties		**Ties per S.F. of Wall Area**
One tie every 6 courses of brick	1' 0" O.C.	1.5
	2' 0" O.C.	0.75
One Tie 2' 0" O.C., each way		0.25

Division 4

Figure 4-24   Installation Time in Labor-Hours for Brick Masonry

Description	Labor-Hours	Unit
Brick Wall		
Veneer		
4" Thick		
Running Bond		
Standard Brick (6.75/S.F.)	.182	S.F.
Engineer Brick (5.63/S.F.)	.154	S.F.
Economy Brick (4.50/S.F.)	.129	S.F.
Roman Brick (6.00/S.F.)	.160	S.F.
Norman Brick (4.50/S.F.)	.125	S.F.
Norwegian Brick (3.75/S.F.)	.107	S.F.
Utility Brick (3.00/S.F.)	.089	S.F.
Common Bond, Standard Brick (7.88/S.F.)	.216	S.F.
Flemish Bond, Standard Brick (9.00/S.F.)	.267	S.F.
English Bond, Standard Brick (10.13/S.F.)	.286	S.F.
Stack Bond, Standard Brick (6.75/S.F.)	.200	S.F.
6" Thick		
Running Bond		
S.C.R. Brick (4.50/S.F.)	.129	S.F.
Jumbo Brick (3.00/S.F.)	.092	S.F.
Backup		
4" Thick		
Running Bond		
Standard Brick (6.75/S.F.)	.167	S.F.
Solid, Unreinforced		
8" Thick Running Bond (13.50/S.F.)	.296	S.F.
12" Thick, Running Bond (20.25/S.F.)	.421	S.F.
Solid, Rod Reinforced		
8" Thick, Running Bond	.308	S.F.
12" Thick, Running Bond	.444	S.F.
Cavity		
4" Thick		
4" Backup	.242	S.F.
6" Backup	.276	S.F.
Brick Chimney		
16" x 16", Standard Brick w/8" x 8" Flue	.889	V.L.F.
16" x 16", Standard Brick w/8" x 12" Flue	1.000	V.L.F.
20" x 20", Standard Brick w/12" x 12" Flue	1.140	V.L.F.
Brick Column		
8" x 8", Standard Brick 9.0 V.L.F.	.286	V.L.F.
12" x 12", Standard Brick 20.3 V.L.F.	.640	V.L.F.
20" x 20", Standard Brick 56.3 V.L.F.	1.780	V.L.F.
Brick Coping		
Precast, 10" Wide, or Limestone, 4" Wide	.178	L.F.
Precast 14" Wide, or Limestone, 6" Wide	.200	L.F.

(continued on next page)

Figure 4-24 Installation Time in Labor-Hours for Brick Masonry
(continued from previous page)

Description	Labor-Hours	Unit
Brick Fireplace		
30" x 24" Opening, Plain Brickwork	40.000	Ea.
Firebox Only, Fire Brick (110/Ea.)	8.000	Ea.
Brick Prefabricated Wall Panels, 4" Thick		
Minimum	.093	S.F.
Maximum	.144	S.F.
Brick Steps	53.330	M
Window Sill		
Brick on Edge	.200	L.F.
Precast, 6" Wide	.229	L.F.
Needle Brick and Shore, Solid Brick		
8" Thick	6.450	Ea.
12" Thick	8.160	Ea.
Repoint Brick		
Hard Mortar		
Running Bond	.100	S.F.
English Bond	.123	S.F.
Soft Mortar		
Running Bond	.080	S.F.
English Bond	.098	S.F.
Toothing Brick		
Hard Mortar	.267	V.L.F.
Soft Mortar	.200	V.L.F.
Sandblast Brick		
Wet System		
Minimum	.024	S.F.
Maximum	.057	S.F.
Dry System		
Minimum	.013	S.F.
Maximum	.027	S.F.
Sawing Brick, Per Inch of Depth	.027	L.F.
Steam Clean, Face Brick	.033	S.F.
Wash Brick, Smooth	.014	S.F.

Division 4

Figure 4-25   Installation Time in Labor-Hours for Adobe Brick

Description	Labor-Hours	Unit
Abode Brick, Semi-stabilized, with cement mortar		
Brick, 10" x 4" x 14", 2.6/S.F.	.071	S.F.
12" x 4" x 16", 2.3/S.F.	.069	
10" x 4" x 16", 2.3/S.F.	.068	
8" x 4" x 16", 2.3/S.F.	.071	
4" x 4" x 16", 2.3/S.F.	.074	
6" x 4" x 16", 2.3/S.F.	.074	
4" x 4" x 12", 3.0/S.F.	.077	
8" x 4" x 12", 3.0/S.F.	.077	

Figure 4-26 Installation Time in Labor-Hours for Block Walls, Partitions, and Accessories

Description	Labor-Hours	Unit
Foundation Walls, Trowel Cut Joints, Parged 1/2" Thick, 1 Side, 8" x 16" Face		
Hollow		
8" Thick	.093	S.F.
12" Thick	.122	S.F.
Solid		
8" Thick	.096	S.F.
12" Thick	.126	S.F.
Backup Walls, Tooled Joint 1 Side, 8" x 16" Face		
4" Thick	.091	S.F.
8" Thick	.100	S.F.
Partition Walls, Tooled Joint 2 Sides 8" x 16" Face		
Hollow		
4" Thick	.093	S.F.
8" Thick	.107	S.F.
12" Thick	.141	S.F.
Solid		
4" Thick	.096	S.F.
8" Thick	.111	S.F.
12" Thick	.148	S.F.
Stud Block Walls, Tooled Joints 2 Sides 8" x 16" Face		
6" Thick and 2", Plain	.098	S.F.
Embossed	.103	S.F.
10" Thick and 2", Plain	.108	S.F.
Embossed	.114	S.F.
6" Thick and 2" Each Side, Plain	.114	S.F.
Acoustical Slotted Block Walls Tooled 2 Sides		
4" Thick	.127	S.F.
8" Thick	.151	S.F.
Glazed Block Walls, Tooled Joint 2 Sides 8" x 16", Glazed 1 Face		
4" Thick	.116	s.F.
8" Thick	.129	S.F.
12" Thick	.171	S.F.
8" x 16", Glazed 2 Faces		
4" Thick	.129	S.F.
8" Thick	.148	S.F.
8" x 16", Corner		
4" Thick	.140	Ea.

Division 4

(continued on next page)

Figure 4-26 Installation Time in Labor-Hours for Block Walls, Partitions, and Accessories *(continued from previous page)*

Description	Labor-Hours	Unit
Structural Facing Tile, Tooled 2 Sides		
5" x 12", Glazed 1 Face		
4" Thick	.182	S.F.
8" Thick	.222	S.F.
5" x 12", Glazed 2 Faces		
4" Thick	.205	S.F.
8" Thick	.246	S.F.
8" x 16", Glazed 1 Face		
4" Thick	.116	S.F.
8' Thick	.129	S.F.
8" x 16", Glazed 2 Faces		
4" Thick	.123	S.F.
8" Thick	.137	S.F.
Exterior Walls, Tooled Joint 2 Sides, Insulated		
8" x 16" Face, Regular Weight		
8" Thick	.110	S.F.
12" Thick	.145	S.F.
Lightweight		
8" Thick	.104	S.F.
12" Thick	.137	S.F.
Architectural Block Walls, Tooled Joint 2 Sides		
8" x 16" Face		
4" Thick	.116	S.F.
8" Thick	.138	S.F.
12" Thick	.181	S.F.
Interlocking Block Walls, Fully Grouted		
Vertical Reinforcing		
8" Thick	.131	S.F.
12" Thick	.145	S.F.
16" Thick	.173	S.F.
Bond Beam, Grouted, 2 Horizontal Rebars		
8" x 16" Face, Regular Weight		
8" Thick	.133	L.F.
12" Thick	.192	L.F.
Lightweight		
8" Thick	.131	L.F.
12" Thick	.188	L.F.
Lintels, Grouted, 2 Horizontal Rebars		
8" x 16" Face, 8" Thick	.119	L.F.
16" x 16" Face, 8" Thick	.131	L.F.
Control Joint 4" Wall	.013	L.F.
8" Wall	.020	L.F.
Grouting Bond Beams and Lintels		
8" Deep Pumped, 8" Thick	.018	L.F.
12" Thick	.025	L.F.
Concrete Block Cores Solid		
4" Thick By Hand	.035	S.F.
8" Thick Pumped	.038	S.F.
Cavity Walls 2" Space Pumped	.016	S.F.
6" Space	.034	S.F.

(continued on next page)

Figure 4-26 Installation Time in Labor-Hours for Block Walls, Partitions, and Accessories *(continued from previous page)*

Description	Labor-Hours	Unit
Joint Reinforcing		
Wire Strips Regular Truss to 6" Wide	.267	C.L.F.
12" Wide	.400	C.L.F.
Cavity Wall with Drip Section to 6" Wide	.267	C.L.F.
12" Wide	.400	C.L.F.
Lintels Steel Angles Minimum	.008	lb.
Maximum	.016	lb.
Wall Ties	.762	C
Coping for 12" Wall Stock Units, Aluminum	.200	L.F.
Precast Concrete	.188	L.F.
Structural Reinforcing, Placed Horizontal,		
#3 and #4 Bars	.018	lb.
#5 and #6 Bars	.010	lb.
Placed vertical, #3 and #4 Bars	.023	lb.
#5 and #6 Bars	.012	lb.
Acoustical Slotted Block		
4" Thick	.127	S.F.
6" Thick	.138	S.F.
8" Thick	.151	S.F.
12" Thick	.163	s.F.
Lightweight Block		
4" Thick	.090	S.F.
6" Thick	.095	S.F.
8" Thick	.100	S.F.
10" Thick	.103	S.F.
12" Thick	.130	S.F.
Regular Block		
Hollow		
4" Thick	.093	S.F.
6" Thick	.100	S.F.
8" Thick	.107	S.F.
10" Thick	.111	S.F.
12" Thick	.141	S.F.
Solid		
4" Thick	.095	S.F.
6" Thick	.105	S.F.
8" Thick	.113	S.F.
12" Thick	.150	S.F.
Glazed Concrete Block		
Single face 8" x 16"		
2" Thick	.111	S.F.
4" Thick	.116	S.F.
6" Thick	.121	S.F.
8" Thick	.129	S.F.
12" Thick	.171	S.F.
Double Face		
4" Thick	.129	S.F.
6" Thick	.138	S.F.
8" Thick	.148	S.F.

Division 4

(continued on next page)

Figure 4-26   Installation Time in Labor-Hours for Block Walls, Partitions, and Accessories *(continued from previous page)*

Description	Labor-Hours	Unit
Joint Reinforcing Wire Strips		
4" and 6" Wall	.267	C.L.F.
8" Wall	.320	C.L.F.
10" and 12" Wall	.400	C.L.F.
Steel Bars Horizontal		
#3 and #4	.018	lb.
#5 and #6	.010	lb.
Vertical		
#3 and #4	.023	lb.
#5 and #6	.012	lb.
Grout Cores Solid		
By Hand 6" Thick	.035	S.F.
Pumped 8" Thick	.038	S.F.
10" Thick	.039	S.F.
12" Thick	.040	S.F.

Division Five
Metals

Introduction

Steel is among the most versatile materials used in the construction industry. It can be formed, pressed, rolled, cut, bolted, and welded to be used in a myriad of situations. Steel for the construction industry is usually shop-fabricated to conform with shop drawings and quality standards. The field erection process includes delivery, shakeout (sorting by sequence), hoisting into place, temporary connection, plumbing (or truing), and final connection (welding, bolting, etc.).

This process must be planned thoroughly, as the sequence will dictate what type of hoisting equipment can be used, the interferences that will be created or avoided, and what other work can continue as the steel is being erected. The overall erection time, and ultimately the final cost, will be affected.

green Mining iron ore takes a great toll on the environment, but with the emergence of the "mini mill," the steel industry has moved toward reducing the environmental impacts associated with its products. Mini mills have two distinct environmental and economic advantages: first, steel scrap is the primary raw material used in them to produce steel, and second, their furnaces can be easily turned on and off depending on demand, greatly reducing the plant's energy consumption.

The following topics are covered in charts and tables to aid in estimating metals:

- General symbols and nomenclature
- Steel quantity guidelines
- Weights and dimensions of steel shapes and sections
- Weights and depths of open web steel joists
- Weights and dimensions of steel bars, plates, and sheets
- Steel erecting labor-hour tables

Estimating Data

The following tables present effective estimating guidelines for items found in Division 5 – Metals. Please note that while these guidelines can be used as indicators of what may be expected, each project must be evaluated individually.

Table of Contents

Division 5

Estimating Tips

Plates and Connections

When estimating the total tonnage of structural steel, add 10% to the total weight to allow for plates, connections, and waste, as a rule of thumb.

Shop-Applied Finish Paint

When the specifications call for the finish coat of paint to be applied prior to installation, allow considerable time for touch up. Every time you lift, move, weld, bolt, or alter the position of a piece of steel, you will need to touch up the finish.

Joist Spacings

It is possible to lower overall costs for steel joist-type construction by having greater spacing between joists. The economies of needing fewer joists can well offset the costs of the larger or heavier joists that may be required to replace them.

Joist Bridging

It generally costs less to install joist systems that can utilize horizontal bridging (as opposed to cross or other types of bridging).

Checklist ✓

For an estimate to be reliable, all items must be accounted for. A complete estimate can also limit contingencies. The following checklist can be used to help ensure that all items are included.

☐ **Anchor Bolts**

☐ **Base Plates**

☐ **Leveling Plates**

☐ **Structural Steel**
- ☐ Columns
- ☐ Beams
- ☐ Girders
- ☐ Joists
- ☐ Angles
- ☐ Clip angles
- ☐ Channels
- ☐ Connectors
- ☐ Cross bracing
- ☐ Wind bracing
- ☐ Bridging
- ☐ Girts
- ☐ Hanger rods

☐ **Decking**
- ☐ Roof
- ☐ Floor

☐ **Galvanizing/ Galvanized Materials**

☐ **Miscellaneous Framing**
- ☐ Roof frames
- ☐ Load-bearing studs, joists, rafters, trusses
- ☐ Equipment support framing
- ☐ Light gauge framing
- ☐ Miscellaneous

☐ **Stairs**

☐ **Fastening**
- ☐ Welding
- ☐ Bolts
- ☐ Expansion anchors
- ☐ Epoxy anchors
- ☐ Machinery anchors
- ☐ Weld studs
- ☐ Shear connectors
- ☐ Rivets
- ☐ Studs
- ☐ Timber connectors
- ☐ Machine screws

☐ **Painting**
- ☐ Shop
- ☐ Field
- ☐ Finish

☐ **Seismic Considerations**

☐ **Miscellaneous Metals**
- ☐ Bumper posts
- ☐ Corner guards
- ☐ Crane rails
- ☐ Curb angles
- ☐ Decorative iron and steel
- ☐ Expansion joints and covers
- ☐ Floor gratings
- ☐ Ladders
- ☐ Lintels

☐ **Miscellaneous Frames**

☐ **Pipe Supports**

Division 5

- ☐ *Railings*
  - ☐ Handrails
  - ☐ Balcony rails
  - ☐ Stair
  - ☐ Wall
- ☐ *Stair Treads*
- ☐ *Trench Covers*

- ☐ *Walkways*
- ☐ *Window Guards*
- ☐ *Wire Products*
- ☐ *Railroad Trackwork*
- ☐ *Metal Cleaning*

Figure 5-1 Common Steel Sections

The upper section of this table shows the name, shape, common designation, and basic characteristics of commonly used steel. The lower portion explains how to read the designations used for the illustrated sections above.

Shape & Designation	Name & Characteristics	Shape & Designation	Name & Characteristics
W	W Shape / Parallel flange surfaces	MC	Miscellaneous Channel / Infrequently rolled by some producers
S	American Standard Beam (I Beam) / Sloped inner flange	L	Angle / Equal or unequal legs, constant thickness
M	Miscellaneous Beams / Cannot be classified as W, HP or S; infrequently rolled by some producers	T	Structural Tee / Cut from W, M or S on center of web
C	American Standard Channel / Sloped inner flange	HP	Bearing Pile / Parallel flanges and equal flange and web thickness

Common drawing designations follow:

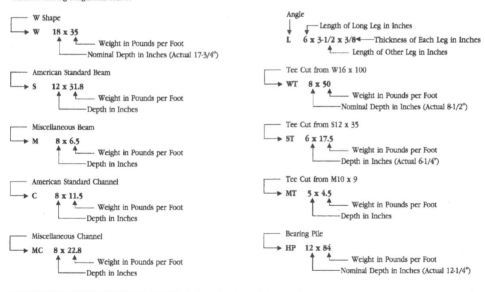

Figure 5-2 Common Structural Steel Specifications

ASTM A992 (formerly A36, then A572 Grade 50) is the all-purpose carbon-grade steel widely used in building and bridge construction.

The other high-strength steels listed below may each have certain advantages over ASTM A992 structural carbon steel, depending on the application. They have proven to be economical choices where, due to lighter members, the reduction of dead load and the associated savings in shipping cost can be significant.

ASTM A588 atmospheric weathering, high-strength low-alloy steels can be used in the bare (uncoated) conditions, where exposure to normal atmosphere causes a tightly adherant oxide to form on the surface, protecting the steel from further oxidation. ASTM A242 corrosion-resistant, high-strength low-alloy steels have enhanced atmospheric corrosion resistance of at least two times that of carbon structural steels with copper, or four times that of carbon structural steels without copper. The reduction or elimination of maintenance resulting from the use of these steels often offsets their higher initial cost.

Steel Type	ASTM Designation	Minimum Yield Stress in KSI	Shapes Available
Carbon	A36	36	All structural shape groups, and plates & bars up thru 8" thick
	A529	50	Structural shape group 1, and plates & bars up thru 2" thick
High-Strength Low-Alloy Quenched & Self-Tempered	A913	50	All structural shape groups
		60	
		65	
		70	
High-Strength Low-Alloy Columbium-Vanadium	A572	42	All structural shape groups, and plates & bars up thru 6" thick
		50	All structural shape groups, and plates & bars up thru 4" thick
		55	Structural shape groups 1 & 2, and plates & bars up thru 2" thick
		60	Structural shape groups 1 & 2, and plates & bars up thru 1-1/4" thick
		65	Structural shape group 1, and plates & bars up thru 1-1/4" thick
High-Strength Low-Alloy Columbium-Vanadium	A992	50	All structural shape groups
Weathering High-Strength Low-Alloy	A242	42	Structural shape groups 4 & 5, and plates & bars over 1-1/2" up thru 4" thick
		46	Structural shape group 3, and plates & bars over 3/4" up thru 1-1/2" thick
		50	Structural shape groups 1 & 2, and plates & bars up thru 3/4" thick
Weathering High-Strength Low-Alloy	A588	42	Plates & bars over 5" up thru 8" thick
		46	Plates & bars over 4" up thru 5" thick
		50	All structural shape groups, and plates & bars up thru 4" thick
Quenched and Tempered Low-Alloy	A852	70	Plates & bars up thru 4" thick
Quenched and Tempered Alloy	A514	90	Plates & bars over 2-1/2" up thru 6" thick
		100	Plates & bars up thru 2-1/2" thick

Figure 5-3 Estimating Steel Quantities

One estimate on erection is that a crane can handle 35 to 60 pieces per day. Say the average is 45. With usual sizes of beams, girders, and columns, this would amount to about 20 tons per day. The type of connection greatly affects the speed of erection. Moment connections for continuous design slow down production and increase erection costs.

Short open web bar joists can be set at the rate of 75 to 80 per day, with 50 being the average for setting long span joists.

After main members are calculated, add the following for usual allowances: base plates 2% to 3%; column splices 4% to 5%; and miscellaneous details 4% to 5%, for a total of 10% to 13% in addition to main members.

Ratio of column to beam tonnage varies depending on the type of steel used, typical spans, story heights, and live loads.

It is more economical to keep the column size constant and to vary the strength of the column by using high strength steels. This also saves floor space. Buildings have recently gone as high as ten stories with 8" high strength columns. For light columns less than W8 x 31, concrete filled steel columns are economical.

High strength steels may be used in columns and beams to save floor space and to meet head room requirements. High strength steel in some sizes sometimes requires long lead times.

Round, square, and rectangular columns, both plain and concrete filled, are readily available and save floor area, but are higher in cost per pound than rolled columns. For high unbraced columns, tube columns may be less expensive.

Below are average minimum figures for the weights of the structural steel frame for different types of buildings using A36 steel, rolled shapes, and simple joints. For economy in domes, rise to span ratio = 0.13. Open web joist framing systems will reduce weights by 10% to 40%. Composite design can reduce steel weight by up to 25%, but additional concrete floor slab thickness may be required. Continuous design can reduce the weights up to 20%. There are many building codes with different live load requirements and different structural requirements, such as hurricane and earthquake loadings, which can alter the figures.

Division 5

Structural Steel Weights per S.F. of Floor Area									
Type of Building	No. of Stories	Avg. Spans	L.L. #/S.F.	Lbs. Per S.F.	Type of Building	No. of Stories	Avg. Spans	L.L. #/S.F.	Lbs. Per S.F.
Steel Frame Mfg.	1	20'x20'	40	8	Apartments	2-8	20'x20'	40	8
		30'x30'		13		9-25			14
		40'x40'		18	Office	to 10			10
Parking garage	4	Various	80	8.5		20	Various	80	18
Domes (Schwedler)*	1	200'	30	10		30			26
		300'		15		over 50			35

Figure 5-4   Dimensions for Various Steel Shapes

The following tables represent the shapes commonly manufactured and available in the steel industry. The tables also give the physical dimensions for the steel shapes.

		Web	Flange	
Designation	Depth (in.)	Thickness (in.)	Width (in.)	Thickness (in.)
W 36x300	36-3/4	15/16	16-5/8	1-11/16
x280	36-1/2	7/8	16-5/8	1-9/16
x260	36-1/4	13/16	16-1/2	1-7/16
x245	36-1/8	13/16	16-1/2	1-3/8
x230	35-7/8	3/4	16-1/2	1-1/4
W 36x210	36-3/4	13/16	12-1/8	1-3/8
x194	36-1/2	3/4	12-1/8	1-1/4
x182	36-3/8	3/4	12-1/8	1-3/16
x170	36-1/8	11/16	12	1-1/8
x160	36	5/8	12	1
x150	35-7/8	5/8	12	15/16
x135	35-1/2	5/8	12	13/16
W 33x241	34-1/8	13/16	15-7/8	1-3/8
x221	33-7/8	3/4	15-3/4	1-1/4
x201	33-5/8	11/16	15-3/4	1-1/8
W 33x152	33-1/2	5/8	11-5/8	1-1/16
x141	33-1/4	5/8	11-1/2	15/16
x130	33-1/8	9/16	11-1/2	7/8
x118	32-7/8	9/16	11-1/2	3/4
W 30x211	31	3/4	15-1/8	1-5/16
x191	30-5/8	11/16	15	1-3/16
x173	30-1/2	5/8	15	1-1/16
W 30x132	30-1/4	5/8	10-1/2	1
x124	30-1/8	9/16	10-1/2	15/16
x116	30	9/16	10-1/2	7/8
x108	29-7/8	9/16	10-1/2	3/4
x 99	29-5/8	1/2	10-1/2	11/16
W 27x178	27-3/4	3/4	14-1/8	1-3/16
x161	27-5/8	11/16	14	1-1/16
x146	27-3/8	5/8	14	1
W 27x114	27-1/4	9/16	10-1/8	15/16
x102	27-1/8	1/2	10	13/16
x 94	26-7/8	1/2	10	3/4
x 84	26-3/4	7/16	10	5/8
W 24x162	25	11/16	13	1-1/4
x146	24-3/4	5/8	12-7/8	1-1/16
x131	24-1/2	5/8	12-7/8	15/16
x117	24-1/4	9/16	12-3/4	7/8
x104	24	1/2	12-3/4	3/4
W 24x 94	24-1/4	1/2	9-1/8	7/8
x 84	24-1/8	1/2	9	3/4
x 76	23-7/8	7/16	9	11/16
x 68	23-3/4	7/16	9	9/16
W 24x 62	23-3/4	7/16	7	9/16
x 55	23-5/8	3/8	7	1/2

(continued on next page)

Figure 5-4 Dimensions for Various Steel Shapes
(continued from previous page)

W Shapes Dimensions				
		Web	**Flange**	
Designation	**Depth (in.)**	**Thickness (in.)**	**Width (in.)**	**Thickness (in.)**
W 21x147	22	3/4	12-1/2	1-1/8
x132	21-7/8	5/8	12-1/2	1-1/16
x122	21-5/8	5/8	12-3/8	15/16
x111	21-1/2	9/16	12-3/8	7/8
x101	21-3/8	1/2	12-1/4	13/16
W 21x 93	21-5/8	9/16	8-3/8	15/16
x 83	21-3/8	1/2	8-3/8	13/16
x 73	21-1/4	7/16	8-1/4	3/4
x 68	21-1/8	7/16	8-1/4	11/16
x 62	21	3/8	8-1/4	5/8
W 21x 57	21	3/8	6-1/2	5/8
x 50	20-7/8	3/8	6-1/2	9/16
x 44	20-5/8	3/8	6-1/2	7/16
W 18x119	19	5/8	11-1/4	1-1/16
x106	18-3/4	9/16	11-1/4	15/16
x 97	18-5/8	9/16	11-1/8	7/8
x 86	18-3/8	1/2	11-1/8	3/4
x 76	18-1/4	7/16	11	11/16
W 18x 71	18-1/2	1/2	7-5/8	13/16
x 65	18-3/8	7/16	7-5/8	3/4
x 60	18-1/4	7/16	7-1/2	11/16
x 55	18-1/8	3/8	7-1/2	5/8
x 50	18	3/8	7-1/2	9/16
W 18x 46	18	3/8	6	5/8
x 40	17-7/8	5/16	6	1/2
x 35	17-3/4	5/16	6	7/16
W 16x100	17	9/16	10-3/8	1
x 89	16-3/4	1/2	10-3/8	7/8
x 77	16-1/2	7/16	10-1/4	3/4
x 67	16-3/8	3/8	10-1/4	11/16
W 16x 57	16-3/8	7/16	7-1/8	11/16
x 50	16-1/4	3/8	7-1/8	5/8
x 45	16-1/8	3/8	7	9/16
x 40	16	5/16	7	1/2
x 36	15-7/8	5/16	7	7/16
W 16x 31	15-7/8	1/4	5-1/2	7/16
x 26	15-3/4	1/4	5-1/2	3/8
W 14x730	22-3/8	3-1/16	17-7/8	4-15/16
x665	21-5/8	2-13/16	17-5/8	4-1/2
x605	20-7/8	2-5/8	17-3/8	4-3/16
x550	20-1/4	2-3/8	17-1/4	3-13/16
x500	19-5/8	2-3/16	17	3-1/2
x455	19	2	16-7/8	3-3/16

Division 5

(continued on next page)

Figure 5-4 Dimensions for Various Steel Shapes
(continued from previous page)

		Web	Flange	
W Shapes Dimensions				
		Web	**Flange**	
Designation	**Depth (in.)**	**Thickness (in.)**	**Width (in.)**	**Thickness (in.)**
W 14x426	18-5/8	1-7/8	16-3/4	3-1/16
x398	18-1/4	1-3/4	16-5/8	2-7/8
x370	17-7/8	1-5/8	16-1/2	2-11/16
x342	17-1/2	1-9/16	16-3/8	2-1/2
x311	17-1/8	1-7/16	16-1/4	2-1/4
x283	16-3/4	1-5/16	16-1/8	2-1/16
x257	16-3/8	1-3/16	16	1-7/8
x233	16	1-1/16	15-7/8	1-3/4
x211	15-3/4	1	15-3/4	1-9/16
x193	15-1/2	7/8	15-3/4	1-7/16
x176	15-1/4	13/16	15-5/8	1-5/16
x159	15	3/4	15-5/8	1-3/16
x145	14-3/4	11/16	15-1/2	1-1/16
W 14x132	14-5/8	5/8	14-3/4	1
x120	14-1/2	9/16	14-5/8	15/16
x109	14-3/8	1/2	14-5/8	7/8
x 99	14-1/8	1/2	14-5/8	3/4
x 90	14	7/16	14-1/2	11/16
W 14x 82	14-1/4	1/2	10-1/8	7/8
x 74	14-1/8	7/16	10-1/8	13/16
x 68	14	7/16	10	3/4
x 61	13-7/8	3/8	10	5/8
W 14x 53	13-7/8	3/8	8	11/16
x 48	13-3/4	5/16	8	5/8
x 43	13-5/8	5/16	8	1/2
W 14x 38	14-1/8	5/16	6-3/4	1/2
x 34	14	5/16	6-3/4	7/16
x 30	13-7/8	1/4	6-3/4	3/8
W 14x 26	13-7/8	1/4	5	7/16
x 22	13-3/4	1/4	5	5/16
W 12x336	16-7/8	1-3/4	13-3/8	2-15/16
x305	16-3/8	1-5/8	13-1/4	2-11/16
x279	15-7/8	1-1/2	13-1/8	2-1/2
x252	15-3/8	1-3/8	13	2-1/4
x230	15	1-5/16	12-7/8	2-1/16
x210	14-3/4	1-3/16	12-3/4	1-7/8
x190	14-3/8	1-1/16	12-5/8	1-3/4
x170	14	15/16	12-5/8	1-9/16
x152	13-3/4	7/8	12-1/2	1-3/8
x136	13-3/8	13/16	12-3/8	1-1/4
x120	13-1/8	11/16	12-3/8	1-1/8
x106	12-7/8	5/8	12-1/4	1
x 96	12-3/4	9/16	12-1/8	7/8
x 87	12-1/2	1/2	12-1/8	13/16
x 79	12-3/8	1/2	12-1/8	3/8
x 72	12-1/4	7/16	12	11/16
x 65	12-1/8	3/8	12	5/8

(continued on next page)

Figure 5-4 Dimensions for Various Steel Shapes
(continued from previous page)

W Shapes Dimensions				
		Web	Flange	
Designation	Depth (in.)	Thickness (in.)	Width (in.)	Thickness (in.)
W 12x 58	12-1/4	3/8	10	5/8
x 53	12	3/8	10	9/16
W 12x 50	12-1/4	3/8	8-1/8	5/8
x 45	12	5/16	8	9/16
x 40	12	5/16	8	1/2
W 12x 35	12-1/2	5/16	6-1/2	1/2
x 30	12-3/8	1/4	6-1/2	7/16
x 26	12-1/4	1/4	6-1/2	3/8
W 12x 22	12-1/4	1/4	4	7/16
x 19	12-1/8	1/4	4	3/8
x 16	12	1/4	4	1/4
x 14	11-7/8	3/16	4	1/4
W 10x112	11-3/8	3/4	10-3/8	1-1/4
x100	11-1/8	11/16	10-3/8	1-1/8
x 88	10-7/8	5/8	10-1/4	1
x 77	10-5/8	1/2	10-1/4	7/8
x 68	10-3/8	1/2	10-1/8	3/4
x 60	10-1/4	7/16	10-1/8	11/16
x 54	10-1/8	3/8	10	5/8
x 49	10	5/16	10	9/16
W 10x 45	10-1/8	3/8	8	5/8
x 39	9-7/8	5/16	8	1/2
x 33	9-3/4	5/16	8	7/16
W 10x 30	10-1/2	5/16	5-3/4	1/2
x 26	10-3/8	1/4	5-3/4	7/16
x 22	10-1/8	1/4	5-3/4	3/8
W 10x 19	10-1/4	1/4	4	3/8
x 17	10-1/8	1/4	4	5/16
x 15	10	1/4	4	1/4
x 12	9-7/8	3/16	4	3/16
W 8x 67	9	9/16	8-1/4	15/16
x 58	8-3/4	1/2	8-1/4	13/16
x 48	8-1/2	3/8	8-1/8	11/16
x 40	8-1/4	3/8	8-1/8	9/16
x 35	8-1/8	5/16	8	1/2
x 31	8	5/16	8	7/16
W 8x 28	8	5/16	6-1/2	7/16
x 24	7-7/8	1/4	6-1/2	3/8
W 8x 21	8-1/4	1/4	5-1/4	3/8
x 18	8-1/8	1/4	5-1/4	5/16
W 8x 15	8-1/8	1/4	4	5/16
x 13	8	1/4	4	1/4
x 10	7-7/8	3/16	4	3/16

(continued on next page)

Division 5

Figure 5-4 Dimensions for Various Steel Shapes
(continued from previous page)

		Web	Flange	
Designation	**Depth (in.)**	**Thickness (in.)**	**Width (in.)**	**Thickness (in.)**
W Shapes Dimensions				
W 6x 25	6-3/8	5/16	6-1/8	7/16
x 20	6-1/4	1/4	6	3/8
x 15	6	1/4	6	1/4
W 6x 16	6-1/4	1/4	4	3/8
x 12	6	1/4	4	1/4
x 9	5-7/8	3/16	4	3/16
W 5x 19	5-1/8	1/4	5	7/16
x 16	5	1/4	5	3/8
W 4x 13	4-1/8	1/4	4	3/8
M Shapes Dimensions				
M 14x 18	14	3/16	4	1/4
M 12x 11.8	12	3/16	3-1/8	1/4
M 10x 9	10	3/16	2-3/4	3/16
M 8x 6.5	8	1/8	2-1/4	3/16
M 6x 20	6	1/4	6	3/8
x 4.4	6	1/8	1-7/8	3/16
M 5x 18.9	5	5/16	5	7/16
M 4x 13	4	1/4	4	3/8
S Shapes Dimensions				
S 24x121	24-1/2	13/16	8	1-1/16
x106	24-1/2	5/8	7-7/8	1-1/16
x100	24	3/4	7-1/4	7/8
x 90	24	5/8	7-1/8	7/8
x 80	24	1/2	7	7/8
S 20x 96	20-1/4	13/16	7-1/4	15/16
x 86	20-1/4	11/16	7	15/16
x 75	20	5/8	6-3/8	13/16
x 66	20	1/2	6-1/4	13/16
S 18x 70	18	11/16	6-1/4	11/16
x 54.7	18	7/16	6	11/16
S 15x 50	15	9/16	5-5/8	5/8
x 42.9	15	7/16	5-1/2	5/8
S 12x 50	12	11/16	5-1/2	11/16
x 40.8	12	7/16	5-1/4	11/16
x 35	12	7/16	5-1/8	9/16
x 31.8	12	3/8	5	9/16
S 10x 35	10	5/8	5	1/2
x 25.4	10	5/16	4-5/8	1/2
S 8x 23	8	7/16	4-1/8	7/16
x 18.4	8	1/4	4	7/16
S 7x 20	7	7/16	3-7/8	3/8
x 15.3	7	1/4	3-5/8	3/8

(continued on next page)

Figure 5-4 Dimensions for Various Steel Shapes
(continued from previous page)

		Web	Flange	
Designation	**Depth (in.)**	**Thickness (in.)**	**Width (in.)**	**Thickness (in.)**
S Shapes Dimensions				
S 6x 17.25	6	7/16	3-5/8	3/8
x 12.5	6	1/4	3-3/8	3/8
S 5x 14.75	5	1/2	3-1/4	5/16
x 10	5	3/16	3	5/16
S 4x 9.5	4	5/16	2-3/4	5/16
x 7.7	4	3/16	2-5/8	5/16
S 3x 7.5	3	3/8	2-1/2	1/4
x 5.7	3	3/16	2-3/8	1/4
H P Shapes Dimensions				
HP 14x117	14-1/4	13/16	14-7/8	13/16
x102	14	11/16	14-3/4	11/16
x 89	13-7/8	5/8	14-3/4	5/8
x 73	13-5/8	1/2	14-5/8	1/2
HP 13x100	13-1/8	3/4	13-1/4	3/4
x 87	13	11/16	13-1/8	11/16
x 73	12-3/4	9/16	13	9/16
x 60	12-1/2	7/16	12-7/8	7/16
HP 12x 84	12-1/4	11/16	12-1/4	11/16
x 74	12-1/8	5/8	12-1/4	5/8
x 63	12	1/2	12-1/8	1/2
x 53	11-3/4	7/16	12	7/16
HP 10x 57	10	9/16	10-1/4	9/16
x 42	9-3/4	7/16	10-1/8	7/16
HP 8x 36	8	7/16	8-1/8	7/16
Channels American Standard Dimensions				
MC18x 58	18	11/16	4-1/4	5/8
x 51.9	18	5/8	4-1/8	5/8
x 45.8	18	1/2	4	5/8
x 42.7	18	7/16	4	5/8
MC13x 50	13	13/16	4-3/8	5/8
x 40	13	9/16	4-1/8	5/8
x 35	13	7/16	4-1/8	5/8
x 31.8	13	3/8	4	5/8
MC12x 50	12	13/16	4-1/8	11/16
x 45	12	11/16	4	11/16
x 40	12	9/16	3-7/8	11/16
x 35	12	7/16	3-3/4	11/16
x 37	12	5/8	3-5/8	5/8
x 32.9	12	1/2	3-1/2	5/8
x 30.9	12	7/16	3-1/2	5/8
x 10.6	12	3/16	1-1/2	5/16

Division 5

(continued on next page)

Figure 5-4 Dimensions for Various Steel Shapes

(continued from previous page)

Channels American Standard Dimensions				
		Web	**Flange**	
Designation	**Depth (in.)**	**Thickness (in.)**	**Width (in.)**	**Thickness (in.)**
MC10x 41.1	10	13/16	4-3/8	9/16
x 33.6	10	9/16	4-1/8	9/16
x 28.5	10	7/16	4	9/16
x 28.3	10	1/2	3-1/2	9/16
x 25.3	10	7/16	3-1/2	1/2
x 24.9	10	3/8	3-3/8	9/16
x 21.9	10	5/16	3-1/2	1/2
x 8.4	10	3/16	1-1/2	1/4
x 6.5	10	1/8	1-1/8	3/16
MC 9x 25.4	9	7/15	3-1/2	9/16
x 23.9	9	3/8	3-1/2	9/16
MC 8x 22.8	8	7/16	3-1/2	1/2
x 21.4	8	3/8	3-1/2	1/2
x 20	8	3/8	3	1/2
x 18.7	8	3/8	3	1/2
x 8.5	8	3/16	1-7/8	5/16
MC 7x 22.7	7	1/2	3-5/8	1/2
x 19.1	7	3/8	3-1/2	1/2
x 17.6	7	3/8	3	1/2
MC 6x 18	6	3/8	3-1/2	1/2
x 16.3	6	3/8	3	1/2
x 15.3	6	5/16	3-1/2	3/8
x 15.1	6	5/16	3	1/2
x 12	6	5/16	2-1/2	3/8
C 15x 50	15	11/16	3-3/4	5/8
x 40	15	1/2	3-1/2	5/8
x 33.9	15	3/8	3-3/8	5/8
C 12x 30	12	1/2	3-1/8	1/2
x 25	12	3/8	3	1/2
x 20.7	12	5/16	3	1/2
C 10x 30	10	11/16	3	7/16
x 25	10	1/2	2-7/8	7/16
x 20	10	3/8	2-3/4	7/16
x 15.3	10	1/4	2-5/8	7/16
C 9x 20	9	7/16	2-5/8	7/16
x 15	9	5/16	2-1/2	7/16
x 13.4	9	1/4	2-3/8	7/16
C 8x 18.75	8	1/2	2-1/2	3/8
x 13.75	8	5/16	2-3/8	3/8
x 11.5	8	1/4	2-1/4	3/8
C 7x 14.75	7	7/16	2-1/4	3/8
x 12.25	7	5/16	2-1/4	3/8
x 9.8	7	3/16	2-1/8	3/8
C 6x 13	6	7/16	2-1/8	5/16
x 10.5	6	5/16	2	5/16
x 8.2	6	3/16	1-7/8	5/16

(continued on next page)

Figure 5-4 Dimensions for Various Steel Shapes
(continued from previous page)

Channels American Standard Dimensions				
		Web		Flange	
Designation	Depth (in.)	Thickness (in.)	Width (in.)	Thickness (in.)
C 5x 9	5	5/16	1-7/8	5/16
x 6.7	5	3/16	1-3/4	5/16
C 4x 7.25	4	5/16	1-3/4	5/16
x 5.4	4	3/16	1-5/8	5/16
C 3x 6	3	3/8	1-5/8	1/4
x 5	3	1/4	1-1/2	1/4
x 4.1	3	3/16	1-3/8	1/4

Angles Equal Legs and Unequal Legs Properties for Designing			
Size and Thickness (in.)	Weight per Foot (lb.)	Size and Thickness (in.)	Weight per Foot (lb.)
L 9 x4 x 5/8	26.3	L 4 x4 x 3/4	18.5
9/16	23.8	5/8	15.7
1/2	21.3	1/2	12.8
		7/16	11.3
L 8 x8 x1-1/8	56.9	3/8	9.8
1	51.0	5/16	8.2
7/8	45.0	1/4	6.6
3/4	38.9		
5/8	32.7	L 4 x3-1/2x 5/8	14.7
9/16	29.6	1/2	11.9
1/2	26.4	7/16	10.6
		3/8	9.1
L 8 x6 x1	44.2	5/16	7.7
7/8	39.1	1/4	6.2
3/4	33.8		
5/8	28.5	L 3 x2-1/2x 1/2	8.5
9/16	25.7	7/16	7.6
1/2	23.0	3/8	6.6
7/16	20.2	5/16	5.6
		1/4	4.5
L 8 x4 x1	37.4	3/16	3.39
3/4	28.7		
9/16	21.9	L 3 x2 x 1/2	7.7
1/2	19.6	7/16	6.8
		3/8	5.9
L 7 x4 x 3/4	26.2	5/16	5.0
5/8	22.1	1/4	4.1
1/2	17.9	3/16	3.07
3/8	13.6		
		L 2-1/2x2-1/2x 1/2	7.7
L 5 x3-1/2x 3/4	19.8	3/8	5.9
5/8	16.8	5/16	5.0
1/2	13.6	1/4	4.1
7/16	12.0	3/16	3.07
3/8	10.4		
5/16	8.7	L 2-1/2x2 x 3/8	5.3
1/4	7.0	5/16	4.5
		1/4	3.62
L 5 x3 x 5/8	15.7	3/16	2.75
1/2	12.8		
7/16	11.3	L 2 x2 x 3/8	4.7
3/8	9.8	5/16	3.92
5/16	8.2	1/4	3.19
1/4	6.6	3/16	2.44
		1/8	1.65

Division 5

(continued on next page)

Figure 5-4 Dimensions for Various Steel Shapes
(continued from previous page)

Designation	Depth (in.)	Stem Thickness (in.)	Flange Width (in.)	Flange Thickness (in.)
Structural Tees Cut from W Shapes Dimensions				
WT 18 x150	18-3/8	15/16	16-5/8	1-11/16
x140	18-1/4	7/8	16-5/8	1-9/16
x130	18-1/8	13/16	16-1/2	1-7/16
x122.5	18	13-16	16-1/2	1-3/8
x115	18	3/4	16-1/2	1-1/4
x105	18-3/8	13/16	12-1/8	1-3/8
x 97	18-1/4	3/4	12-1/8	1-1/4
x 91	18-1/8	3/4	12-1/8	1-3/16
x 85	18-1/8	11/16	12	1-1/8
x 80	18	5/8	12	1
x 75	17-7/8	5/8	12	15/16
x 67.5	17-3/4	5/8	12	13/16
WT 16.5x120.5	17-1/8	13/16	15-7/8	1-3/8
x110.5	17	3/4	15-3/4	1-1/4
x100.5	16-7/8	11/16	15-3/4	1-1/8
x 76	16-3/4	5/8	11-5/8	1-1/16
x 70.5	16-5/8	5/8	11-1/2	15/16
x 65	16-1/2	9/16	11-1/2	7/8
x 59	16-3/8	9/16	11-1/2	3/4
WT 15 x105.5	15-1/2	3/4	15-1/8	1-5/16
x 95.5	15-3/8	11/16	15	1-3/16
x 86.5	15-1/4	5/8	15	1-1/16
x 66	15-1/8	5/8	10-1/2	1
x 62	15-1/8	9/16	10-1/2	15/16
x 58	15	9/16	10-1/2	7/8
x 54	14-7/8	9/16	10-1/2	3/4
x 49.5	14-7/8	1/2	10-1/2	11/16
WT 13.5x 89	13-7/8	3/4	14-1/8	1-3/16
x 80.5	13-3/4	11/16	14	1-1/16
x 73	13-3/4	5/8	14	1
x 57	13-5/8	9/16	10-1/8	15/16
x 51	13-1/2	1/2	10	13/16
x 47	13-1/2	1/2	10	3/4
x 42	13-3/8	7/16	10	5/8
WT 12 x 81	12-1/2	11/16	13	1-1/4
x 73	12-3/8	5/8	12-7/8	1-1/16
x 65.5	12-1/4	5/8	12-7/8	15/16
x 58.5	12-1/8	9/16	12-3/4	7/8
x 52	12	1/2	12-3/4	3/4
x 47	12-1/8	1/2	9-1/8	7/8
x 42	12	1/2	9	3/4
x 38	12	7/16	9	11/16
x 34	11-7/8	7/16	9	9/16
x 31	11-7/8	7/16	7	9/16
x 27.5	11-3/4	3/8	7	1/2

(continued on next page)

Figure 5-4 Dimensions for Various Steel Shapes
(continued from previous page)

Designation	Depth (in.)	Stem Thickness (in.)	Flange Width (in.)	Flange Thickness (in.)
Structural Tees Cut from W Shapes Dimensions				
		Stem	**Flange**	
	Depth (in.)	**Thickness (in.)**	**Width (in.)**	**Thickness (in.)**
WT 10.5x 73.5	11	3/4	12-1/2	1-1/8
x 66	10-7/8	5/8	12-1/2	1-1/16
x 61	10-7/8	5/8	12-3/8	15/16
x 55.5	10-3/4	9/16	12-3/8	7/8
x 50.5	10-5/8	1/2	12-1/4	13/16
x 46.5	10-3/4	9/16	8-3/8	15/16
x 41.5	10-3/4	1/2	8-3/8	13/16
x 36.5	10-5/8	7/16	8-1/4	3/4
x 34	10-5/8	7/16	8-1/4	11/16
x 31	10-1/2	3/8	8-1/4	5/8
x 28.5	10-1/2	3/8	6-1/2	5/8
x 25	10-3/8	3/8	6-1/2	9/16
x 22	10-3/8	3/8	6-1/2	7/16
WT 9x 59.5	9-1/2	5/8	11-1/4	1-1/16
x 53	9-3/8	9/16	11-1/4	15/16
x 48.5	9-1/4	9/16	11-1/8	7/8
x 43	9-1/4	1/2	11-1/8	3/4
x 38	9-1/8	7/16	11	11/16
x 35.5	9-1/4	1/2	7-5/8	13/16
x 32.5	9-1/8	7/16	7-5/8	3/4
x 30	9-1/8	7/16	7-1/2	11/16
x 27.5	9	3/8	7-1/2	5/8
x 25	9	3/8	7-1/2	9/16
x 23	9	3/8	6	5/8
x 20	9	5/16	6	1/2
x 17.5	8-7/8	5/16	6	7/16
WT 8x 50	8-1/2	9/16	10-3/8	1
x 44.5	8-3/8	1/2	10-3/8	7/8
x 38.5	8-1/4	7/16	10-1/4	3/4
x 33.5	8-1/8	3/8	10-1/4	11/16
x 28.5	8-1/4	7/16	7-1/8	11/16
x 25	8-1/8	3/8	7-1/8	5/8
x 22.5	8-1/8	3/8	7	9/16
x 20	8	5/16	7	1/2
x 18	7-7/8	5/16	7	7/16
x 15.5	8	1/4	5-1/2	7/16
x 13	7-7/8	1/4	5-1/2	3/8

Division 5

(continued on next page)

Figure 5-4 Dimensions for Various Steel Shapes
(continued from previous page)

Designation	Depth (in.)	Stem Thickness (in.)	Flange Width (in.)	Flange Thickness (in.)
WT 7x365	11-1/4	3-1/16	17-7/8	4-15/16
x332.5	10-7/8	2-13/16	17-5/8	4-1/2
x302.5	10-1/2	2-5/8	17-3/8	4-3/16
x275	10-1/8	2-3/8	17-1/4	3-13/16
x250	9-3/4	2-3/16	17	3-1/2
x227.5	9-1/2	2	16-7/8	3-3/16
x213	9-3/8	1-7/8	16-3/4	3-1/16
x199	9-1/8	1-3/4	16-5/8	2-7/8
x185	9	1-5/8	16-1/2	2-11/16
x171	8-3/4	1-9/16	16-3/8	2-1/2
x155.5	8-1/2	1-7/16	16-1/4	2-1/4
x141.5	8-3/8	1-5/16	16-1/8	2-1/16
x128.5	8-1/4	1-3/16	16	1-7/8
x116.5	8	1-1/16	15-7/8	1-3/4
x105.5	7-7/8	1	15-3/4	1-9/16
x 96.5	7-3/4	7/8	15-3/4	1-7/16
x 88	7-5/8	13/16	15-5/8	1-5/16
x 79.5	7-1/2	3/4	15-5/8	1-3/16
x 72.5	7-3/8	11/16	15-1/2	1-1/16
x 66	7-3/8	5/8	14-3/4	1
x 60	7-1/4	9/16	14-5/8	15/16
x 54.5	7-1/8	1/2	14-5/8	7/8
x 49.5	7-1/8	1/2	14-5/8	3/4
x 45	7	7/16	14-1/2	11/16
x 41	7-1/8	1/2	10-1/8	7/8
x 37	7-1/8	7/16	10-1/8	13/16
x 34	7	7/16	10	3/4
x 30.5	7	3/8	10	5/8
x 26.5	7	3/8	8	11/16
x 24	6-7/8	5/16	8	5/8
x 21.5	6-7/8	5/16	8	1/2
x 19	7	5/16	6-3/4	1/2
x 17	7	5/16	6-3/4	7/16
x 15	6-7/8	1/4	6-3/4	3/8
x 13	7	1/4	5	7/16
x 11	6-7/8	1/4	5	5/16

The table is titled: **Structural Tees Cut from W Shapes Dimensions**

(continued on next page)

Figure 5-4 Dimensions for Various Steel Shapes
(continued from previous page)

		Stem	Flange	
Designation	Depth (in.)	Thickness (in.)	Width (in.)	Thickness (in.)
WT 6 x168	8-3/8	1-3/4	13-3/8	2-15/16
x152.5	8-1/8	1-5/8	13-1/4	2-11/16
x139.5	7-7/8	1-1/2	13-1/8	2-1/2
x126	7-3/4	1-3/8	13	2-1/4
x115	7-1/2	1-5/16	12-7/8	2-1/16
x105	7-3/8	1-3/16	12-3/4	1-7/8
x 95	7-1/4	1-1/16	12-5/8	1-3/4
x 85	7	15/16	12-5/8	1-9/16
x 76	6-7/8	7/8	12-1/2	1-3/8
x 68	6-3/4	13/16	12-3/8	1-1/4
x 60	6-1/2	11/16	12-3/8	1-1/8
x 53	6-1/2	5/8	12-1/4	1
x 48	6-3/8	9/16	12-1/8	7/8
x 43.5	6-1/4	1/2	12-1/8	13/16
x 39.5	6-1/4	1/2	12-1/8	3/4
x 36	6-1/8	7/16	12	11/16
x 32.5	6	3/8	12	5/8
x 29	6-1/8	3/8	10	5/8
x 26.5	6	3/8	10	9/16
x 25	6-1/8	3/8	8-1/8	5/8
x 22.5	6	5/16	8	9/16
x 20	6	5/16	8	1/2
x 17.5	6-1/4	5/16	6-1/2	1/2
x 15	6-1/8	1/4	6-1/2	7/16
x 13	6-1/8	1/4	6-1/2	3/8
x 11	6-1/8	1/4	4	7/16
x 9.5	6-1/8	1/4	4	3/8
x 8	6	1/4	4	1/4
x 7	6	3/16	4	1/4

Table title: Structural Tees Cut from W Shapes Dimensions

Division 5

(continued on next page)

Figure 5-4 Dimensions for Various Steel Shapes
(continued from previous page)

		Stem	Flange	
Designation	Depth (in.)	Thickness (in.)	Width (in.)	Thickness (in.)
Structural Tees Cut from W Shapes Dimensions				
WT 5 x 56	5-5/8	3/4	10-3/8	1-1/4
x 50	5-1/2	11/16	10-3/8	1-1/8
x 44	5-3/8	5/8	10-1/4	1
x 38.5	5-1/4	1/2	10-1/4	7/8
x 34	5-1/4	1/2	10-1/8	3/4
x 30	5-1/8	7/16	10-1/8	11/16
x 27	5	3/8	10	5/8
x 24.5	5	5/16	10	9/16
x 22.5	5	3/8	8	5/8
x 19.5	5	5/16	8	1/2
x 16.5	4-7/8	5/16	8	7/16
x 15	5-1/4	5/16	5-3/4	1/2
x 13	5-1/8	1/4	5-3/4	7/16
x 11	5-1/8	1/4	5-3/4	3/8
x 9.5	5-1/8	1/4	4	3/8
x 8.5	5	1/4	4	5/16
x 7.5	5	1/4	4	1/4
6	4-7/8	3/16	4	3/16
WT 4 x 33.5	4-1/2	9/16	8-1/4	15/16
x 29	4-3/8	1/2	8-1/4	13/16
x 24	4-1/4	3/8	8-1/8	11/16
x 20	4-1/8	3/8	8-1/8	9/16
x 17.5	4	5/16	8	1/2
x 15.5	4	5/16	8	7/16
x 14	4	5/16	6-1/2	7/16
x 12	4	1/4	6-1/2	3/8
x 10.5	4-1/8	1/4	5-1/4	3/8
x 9	4-1/8	1/4	5-1/4	5/16
x 7.5	4	1/4	4	5/16
x 6.5	4	1/4	4	1/4
x 5	4	3/16	4	3/16
WT 3 x 12.5	3-1/4	5/16	6-1/8	7/16
x 10	3-1/8	1/4	6	3/8
x 7.5	3	1/4	6	1/4
x 8	3-1/8	1/4	4	3/8
x 6	3	1/4	4	1/4
x 4.5	3	3/16	4	3/16
WT 2.5x 9.5	2-5/8	1/4	5	7/16
x 8	2-1/2	1/4	5	3/8
WT 2 x 6.5	2-1/8	1/4	4	3/8

(continued on next page)

Figure 5-4 Dimensions for Various Steel Shapes
(continued from previous page)

Structural Tees Cut from **M** Shapes Dimensions				
		Stem	Flange	
Designation	Depth (in.)	Thickness (in.)	Width (in.)	Thickness (in.)
MT 7 x 9	7	3/16	4	1/4
MT 6 x 5.9	6	3/16	3-1/8	1/4
MT 5 x 4.5	5	3/16	2-3/4	3/16
MT 4 x 3.25	4	1/8	2-1/4	3/16
MT 3 x 10	3	1/4	6	3/8
x 2.2	3	1/8	1-7/8	3/16
MT 2.5x 9.45	2-1/2	5/16	5	7/16
MT 2 x 6.5	2	1/4	4	3/8
Structural Tees Cut from **S** Shapes Dimensions				
ST 12 x 60.5	12-1/4	13/16	8	1-1/16
x 53	12-1/4	5/8	7-7/8	1-1/16
x 50	12	3/4	7-1/4	7/8
x 45	12	5/8	7-1/8	7/8
x 40	12	1/2	7	7/8
ST 10 x 48	10-1/8	13/16	7-1/4	15/16
x 43	10-1/8	11/16	7	15/16
x 37.5	10	5/8	6-3/8	13/16
x 33	10	1/2	6-1/4	13/16
ST 9 x 35	9	11/16	6-1/4	11/16
x 27.35	9	7/16	6	11/16
ST 7.5x 25	7-1/2	9/16	5-5/8	5/8
x 21.45	7-1/2	7/16	5-1/2	5/8
ST 6 x 25	6	11/16	5-1/2	11/16
x 20.4	6	7/16	5-1/4	11/16
x 17.5	6	7/16	5-1/8	9/16
x 15.9	6	3/8	5	9/16
ST 5 x 17.5	5	5/8	5	1/2
x 12.7	5	5/16	4-5/8	1/2
ST 4 x 11.5	4	7/16	4-1/8	7/16
x 9.2	4	1/4	4	7/16
ST 3.5x 10	3-1/2	7/16	3-7/8	3/8
x 7.65	3-1/2	1/4	3-5/8	3/8
ST 3 x 8.625	3	7/16	3-5/8	3/8
x 6.25	3	1/4	3-3/8	3/8
ST 2.5x 7.375	2-1/2	1/2	3-1/4	5/16
x 5	2-1/2	3/16	3	5/16
ST 2 x 4.75	2	5/16	2-3/4	5/16
x 3.85	2	3/16	2-5/8	5/16
ST 1.5x 3.75	1-1/2	3/8	2-1/2	1/4
x 2.85	1-1/2	3/16	2-3/8	1/4

Division 5

Figure 5-5   External Dimensions for Various Steel Shapes

This figure relates the surface and box-out areas for various structural steel shapes. This information is especially helpful when estimating fireproofing or boxing out a column with drywall or other materials.

Surface Areas and Box Areas—W Shapes—Square Feet per Foot of Length				
	Case A	Case B	Case C	Case D
Designation				
W 36x300	9.99	11.40	7.51	8.90
x280	9.95	11.30	7.47	8.85
x260	9.90	11.30	7.42	8.80
x245	9.87	11.20	7.39	8.77
x230	9.84	11.20	7.36	8.73
x210	8.91	9.93	7.13	8.15
x194	8.88	9.89	7.09	8.10
x182	8.85	9.85	7.06	8.07
x170	8.82	9.82	7.03	8.03
x160	8.79	9.79	7.00	8.00
x150	8.76	9.76	6.97	7.97
x135	8.71	9.70	6.92	7.92
W 33x241	9.42	10.70	7.02	8.34
x221	9.38	10.70	6.97	8.29
x201	9.33	10.60	6.93	8.24
x152	8.27	9.23	6.55	7.51
x141	8.23	9.19	6.51	7.47
x130	8.20	9.15	6.47	7.43
x118	8.15	9 11	6.43	7.39
W 30x211	8.71	9 97	6.42	7.67
x191	8.66	9.92	6.37	7.62
x173	8.62	9.87	6.32	7.57
x132	7.49	8.37	5.93	6.81
x124	7.47	8.34	5.90	6.78
x116	7.44	8.31	5.88	6.75
x108	7.41	8.28	5.84	6.72
x 99	7 37	8.25	5.81	6.68
W 27x178	7.95	9.12	5.81	6.98
x161	7.91	9.08	5.77	6.94
x146	7.87	9.03	5.73	6.89
x114	6.88	7.72	5.39	6.23
x102	6.85	7.68	5.35	6.18
x 94	6.82	7.65	5.32	6.15
x 84	6.78	7.61	5.28	6.11
W 24x162	7.22	8.30	5.25	6.33
x146	7.17	8.24	5.20	6.27
x131	7.12	8.19	5.15	6.22
x117	7 08	8.15	5.11	6.18
x104	7.04	8.11	5 07	6.14
x 94	6.16	6.92	4.81	5.56
x 84	6.12	6.87	4.77	5.52
x 76	6.09	6.84	4.74	5.49
x 68	6.06	6.80	4.70	5.45
x 62	5.57	6.16	4.54	5.13
x 55	5.54	6.13	4.51	5.10

Case A: Shape perimeter, minus one flange surface
Case B: Shape perimeter
Case C: Box perimeter, equal to one flange surface plus twice the depth
Case D: Box perimeter, equal to two flange surfaces plus twice the depth

(continued on next page)

Figure 5-5 External Dimensions for Various Steel Shapes
(continued from previous page)

Surface Areas and Box Areas—W Shapes—Square Feet per Foot of Length				
	Case A	Case B	Case C	Case D
Designation				
W 21x147	6.61	7.66	4.72	5.76
x132	6.57	7.61	4.68	5.71
x122	6.54	7.57	4.65	5.68
x111	6.51	7.54	4.61	5.64
x101	6.48	7.50	4.58	5.61
x 93	5.54	6.24	4.31	5.01
x 83	5.50	6.20	4.27	4.96
x 73	5.47	6.16	4.23	4.92
x 68	5.45	6.14	4.21	4.90
x 62	5.42	6.11	4.19	4.87
x 57	5.01	5.56	4.06	4.60
x 50	4.97	5.51	4.02	4.56
x 44	4.94	5.48	3.99	4.53
W 18x119	5.81	6.75	4.10	5.04
x106	5.77	6.70	4.06	4.99
x 97	5.74	6.67	4.03	4.96
x 86	5.70	6.62	3.99	4.91
x 76	5.67	6.59	3.95	4.87
x 71	4.85	5.48	3.71	4.35
x 65	4.82	5.46	3.69	4.32
x 60	4.80	5.43	3.67	4.30
x 55	4.78	5.41	3.65	4.27
x 50	4.76	5.38	3.62	4.25
x 46	4.41	4.91	3.52	4.02
x 40	4.38	4.88	3.48	3.99
x 35	4.34	4.84	3.45	3.95
W 16x100	5.28	6.15	3.70	4.57
x 89	5.24	6.10	3.66	4.52
x 77	5.19	6.05	3.61	4.47
x 67	5.16	6.01	3.57	4.43
x 57	4.39	4.98	3.33	3.93
x 50	4.36	4.95	3.30	3.89
x 45	4.33	4.92	3.27	3.86
x 40	4.31	4.89	3.25	3.83
x 36	4.28	4.87	3.23	3.81
x 31	3.92	4.39	3.11	3.57
x 26	3.89	4.35	3.07	3.53

Case A: Shape perimeter, minus one flange surface
Case B: Shape perimeter
Case C: Box perimeter, equal to one flange surface plus twice the depth
Case D: Box perimeter, equal to two flange surfaces plus twice the depth

Division 5

(continued on next page)

Figure 5-5 External Dimensions for Various Steel Shapes
(continued from previous page)

Surface Areas and Box Areas—W Shapes—Square Feet per Foot of Length				
	Case A	Case B	Case C	Case D
Designation				
W 14x730	7.61	9.10	5.23	6.72
x665	7.46	8.93	5.08	6.55
x605	7.32	8.77	4.94	6.39
x550	7.19	8.62	4.81	6.24
x500	7.07	8.49	4.68	6.10
x455	6.96	8.36	4.57	5.98
x426	6.89	8.28	4.50	5.89
x398	6.81	8.20	4.43	5.81
x370	6.74	8.12	4.36	5.73
x342	6.67	8.03	4.29	5.65
x311	6.59	7.94	4.21	5.56
x283	6.52	7.86	4.13	5.48
x257	6.45	7.78	4.06	5.40
x233	6.38	7.71	4.00	5.32
x211	6.32	7.64	3.94	5.25
x193	6.27	7.58	3.89	5.20
x176	6.22	7.53	3.84	5.15
x159	6.18	7.47	3.79	5.09
x145	6.14	7.43	3.76	5.05
x132	5.93	7.16	3.67	4.90
x120	5.90	7.12	3.64	4.86
x109	5.86	7.08	3.60	4.82
x 99	5.83	7.05	3.57	4.79
x 90	5.81	7.02	3.55	4.76
x 82	4.75	5.59	3.23	4.07
x 74	4.72	5.56	3.20	4.04
x 68	4.69	5.53	3.18	4.01
x 61	4.67	5.50	3.15	3.98
x 53	4.19	4.86	2.99	3.66
x 48	4.16	4.83	2.97	3.64
x 43	4.14	4.80	2.94	3.61
x 38	3.93	4.50	2.91	3.48
x 34	3.91	4.47	2.89	3.45
x 30	3.89	4.45	2.87	3.43
x 26	3.47	3.89	2.74	3.16
x 22	3.44	3.86	2.71	3.12

Case A: Shape perimeter, minus one flange surface
Case B: Shape perimeter
Case C: Box perimeter, equal to one flange surface plus twice the depth
Case D: Box perimeter, equal to two flange surfaces plus twice the depth

(continued on next page)

Figure 5-5 External Dimensions for Various Steel Shapes
(continued from previous page)

Surface Areas and Box Areas—W Shapes—Square Feet per Foot of Length				
Designation	Case A	Case B	Case C	Case D
W 12x336	5.77	6.88	3.92	5.03
x305	5.67	6.77	3.82	4.93
x279	5.59	6.68	3.74	4.83
x252	5.50	6.58	3.65	4.74
x230	5.43	6.51	3.58	4.66
x210	5.37	6.43	3.52	4.58
x190	5.30	6.36	3.45	4.51
x170	5.23	6.28	3.39	4.43
x152	5.17	6.21	3.33	4.37
x136	5.12	6.15	3.27	4.30
x120	5.06	6.09	3.21	4.24
x106	5.02	6.03	3.17	4.19
x 96	4.98	5.99	3.13	4.15
x 87	4.95	5.96	3.10	4.11
x 79	4.92	5.93	3.07	4.08
x 72	4.89	5.90	3.05	4.05
x 65	4.87	5.87	3.02	4.02
x 58	4.39	5.22	2.87	3.70
x 53	4.37	5.20	2.84	3.68
x 50	3.90	4.58	2.71	3.38
x 45	3.88	4.55	2.68	3.35
x 40	3.86	4.52	2.66	3.32
x 35	3.63	4.18	2.63	3.18
x 30	3.60	4.14	2.60	3.14
x 26	3.58	4.12	2.58	3.12
x 22	2.97	3.31	2.39	2.72
x 19	2.95	3.28	2.36	2.69
x 16	2.92	3.25	2.33	2.66
x 14	2.90	3.23	2.32	2.65

Case A: Shape perimeter, minus one flange surface
Case B: Shape perimeter
Case C: Box perimeter, equal to one flange surface plus twice the depth
Case D: Box perimeter, equal to two flange surfaces plus twice the depth

Division 5

(continued on next page)

Figure 5-5 External Dimensions for Various Steel Shapes
(continued from previous page)

Surface Areas and Box Areas—W Shapes—Square Feet per Foot of Length				
	Case A	Case B	Case C	Case D
Designation				
W 10x112	4.30	5.17	2.76	3.63
x100	4.25	5.11	2.71	3.57
x 88	4.20	5.06	2.66	3.52
x 77	4.15	5.00	2.62	3.47
x 68	4.12	4.96	2.58	3.42
x 60	4.08	4.92	2.54	3.38
x 54	4.06	4.89	2.52	3.35
x 49	4.04	4.87	2.50	3.33
x 45	3.56	4.23	2.35	3.02
x 39	3.53	4.19	2.32	2.98
x 33	3.49	4.16	2.29	2.95
x 30	3.10	3.59	2.23	2.71
x 26	3.08	3.56	2.20	2.68
x 22	3.05	3.53	2.17	2.65
x 19	2.63	2.96	2.04	2.38
x 17	2.60	2.94	2.02	2.35
x 15	2.58	2.92	2.00	2.33
x 12	2.56	2.89	1.98	2.31
W 8x 67	3.42	4.11	2.19	2.88
x 58	3.37	4.06	2.14	2.83
x 48	3.32	4.00	2.09	2.77
x 40	3.28	3.95	2.05	2.72
x 35	3.25	3.92	2.02	2.69
x 31	3.23	3.89	2.00	2.67
x 28	2.87	3.42	1.89	2.43
x 24	2.85	3.39	1.86	2.40
x 21	2.61	3.05	1.82	2.26
x 18	2.59	3.03	1.79	2.23
x 15	2.27	2.61	1.69	2.02
x 13	2.25	2.58	1.67	2.00
x 10	2.23	2.56	1.64	1.97
W 6x 25	2.49	3.00	1.57	2.08
x 20	2.46	2.96	1.54	2.04
x 15	2.42	2.92	1.50	2.00
x 16	1.98	2.31	1.38	1.72
x 12	1.93	2.26	1.34	1.67
x 9	1.90	2.23	1.31	1.64
W 5x 19	2.04	2.45	1.28	1.70
x 16	2.01	2.43	1.25	1.67
W 4x 13	1.63	1.96	1.03	1.37

Case A: Shape perimeter, minus one flange surface
Case B: Shape perimeter
Case C: Box perimeter, equal to one flange surface plus twice the depth
Case D: Box perimeter, equal to two flange surfaces plus twice the depth

Figure 5-6 Dimensions and Weights for Rectangular and Square Structural Steel Tubing

Structural Tubing Rectangular Dimensions		
Nominal Size (in.)	Wall Thickness (in.)	Weight per Foot (Lbs.)
20 x 12	1/2 3/8 5/16	103.30 78.52 65.87
20 x 8	1/2 3/8 5/16	89.68 68.31 57.36
20 x 4	1/2 3/8 5/16	76.07 58.10 48.86
18 x 6	1/2 3/8 5/16	76.07 58.10 48.86
16 x 12	1/2 3/8 5/16	89.68 68.31 57.36
16 x 8	1/2 3/8 5/16	76.07 58.10 48.86
16 x 4	1/2 3/8 5/16	62.46 47.90 40.35
14 x 10	1/2 3/8 5/16	76.07 58.10 48.86
14 x 6	1/2 3/8 5/16 1/4	62.46 47.90 40.35 32.63
14 x 4	1/2 3/8 5/16 1/4	55.66 42.79 36.10 29.23
12 x 8	5/8 1/2 3/8 5/16 1/4	76.33 62.46 47.90 40.35 32.63
12 x 6	1/2 3/8 5/16 1/4 3/16	55.66 42.79 36.10 29.23 22.18
12 x 4	1/2 3/8 5/16 1/4 3/16	48.85 37.69 31.84 25.82 19.63
12 x 2	1/4 3/16	22.42 17.08

(continued on next page)

Division 5

Figure 5-6 Dimensions and Weights for Rectangular and Square
Structural Steel Tubing *(continued from previous page)*

Structural Tubing Rectangular Dimensions		
Nominal Size (in.)	Wall Thickness (in.)	Weight per Foot (Lbs.)
10 x 6	1/2	48.85
	3/8	37.69
	5/16	31.84
	1/4	25.82
	3/16	19.63
10 x 4	1/2	42.05
	3/8	32.58
	5/16	27.59
	1/4	22.42
	3/16	17.08
10 x 2	3/8	27.48
	5/16	23.34
	1/4	19.02
	3/16	14.53
8 x 6	1/2	42.05
	3/8	32.58
	5/16	27.59
	1/4	22.42
	3/16	17.08
8 x 4	1/2	35.24
	3/8	27.48
	5/16	23.34
	1/4	19.02
	3/16	14.53
8 x 3	3/8	24.93
	5/16	21.21
	1/4	17.32
	3/16	13.25
8 x 2	3/8	22.37
	5/16	19.08
	1/4	15.62
	3/16	11.97
7 x 5	1/2	35.24
	3/8	27.48
	5/16	23.34
	1/4	19.02
	3/16	14.53
7 x 4	3/8	24.93
	5/16	21.21
	1/4	17.32
	3/16	13.25
7 x 3	3/8	22.37
	5/16	19.08
	1/4	15.62
	3/16	11.97
6 x 4	1/2	28.43
	3/8	22.37
	5/16	19.08
	1/4	15.62
	3/16	11.97

(continued on next page)

Figure 5-6   Dimensions and Weights for Rectangular and Square
              Structural Steel Tubing *(continued from previous page)*

Structural Tubing Rectangular Dimensions		
Nominal Size	Wall Thickness	Weight per Foot
(in.)	(in.)	(Lbs.)
6 x 3	3/8 5/16 1/4 3/16	19.82 16.96 13.91 10.70
6 x 2	3/8 5/16 1/4 3/16	17.27 14.83 12.21 9.42
5 x 4	3/8 5/16 1/4 3/16	19.82 16.96 13.91 10.70
5 x 3	1/2 3/8 5/16 1/4 3/16	21.63 17.27 14.83 12.21 9.42
5 x 2	5/16 1/4 3/16	12.70 10.51 8.15
4 x 3	5/16 1/4 3/16	12.70 10.51 8.15
4 x 2	5/16 1/4 3/16	10.58 8.81 6.87
3 x 2	1/4 3/16	7.11 5.59

Division 5

(continued on next page)

Figure 5-6 Dimensions and Weights for Rectangular and Square Structural Steel Tubing *(continued from previous page)*

Structural Tubing Square Dimensions		
Nominal Size (in.)	Wall Thickness (in.)	Weight per Foot (Lbs.)
16 x 16	1/2 3/8 5/16	103.30 78.52 65.87
14 x 14	1/2 3/8 5/16	89.68 68.31 57.36
12 x 12	1/2 3/8 5/16 1/4	76.07 58.10 48.86 39.43
10 x 10	5/8 1/2 3/8 5/16 1/4	76.33 62.46 47.90 40.35 32.63
8 x 8	5/8 1/2 3/8 5/16 1/4 3/16	59.32 48.85 37.69 31.84 25.82 19.63
7 x 7	1/2 3/8 5/16 1/4 3/16	42.05 32.58 27.59 22.42 17.08
6 x 6	1/2 3/8 5/16 1/4 3/16	35.24 27.48 23.34 19.02 14.53
5 x 5	1/2 3/8 5/16 1/4 3/16	28.43 22.37 19.08 15.62 11.97
4 x 4	1/2 3/8 5/16 1/4 3/16	21.63 17.27 14.83 12.21 9.42
3.5 x 3.5	5/16 1/4 3/16	12.70 10.51 8.15
3 x 3	5/16 1/4 3/16	10.58 8.81 6.87
2.5 x 2.5	1/4 3/16	7.11 5.59
2 x 2	1/4 3/16	5.41 4.32

Figure 5-7 Weights of Square and Round Bars

Size in Inches	Weight in Lbs. per Foot	
	Square	Round
0		
1/16	0.013	0.010
1/8	0.053	0.042
3/16	0.120	0.094
1/4	0.213	0.167
5/16	0.332	0.261
3/8	0.479	0.376
7/16	0.651	0.512
1/2	0.851	0.668
9/16	1.077	0.846
5/8	1.329	1.044
11/16	1.608	1.263
3/4	1.914	1.503
13/16	2.246	1.764
7/8	2.605	2.046
15/16	2.991	2.349
1	3.403	2.673
1/16	3.841	3.017
1/8	4.307	3.382
3/16	4.798	3.769
1/4	5.317	4.176
5/16	5.862	4.604
3/8	6.433	5.053
7/16	7.032	5.523
1/2	7.656	6.013
9/16	8.308	6.525
5/8	8.985	7.057
11/16	9.690	7.610
3/4	10.421	8.185
13/16	11.179	8.780
7/8	11.963	9.396
15/16	12.774	10.032
2	13.611	10.690
1/16	14.475	11.369
1/8	15.366	12.068
3/16	16.283	12.788
1/4	17.227	13.530
5/16	18.197	14.292
3/8	19.194	15.075
7/16	20.217	15.879
1/2	21.267	16.703
9/16	22.344	17.549
5/8	23.447	18.415
11/16	24.577	19.303
3/4	25.734	20.211
13/16	26.917	21.140
7/8	28.126	22.090
15/16	29.362	23.061

Division 5

(continued on next page)

Figure 5-7 Weights of Square and Round Bars
(continued from previous page)

| Size in Inches | Weight in Lbs. per Foot | |
	Square	Round
3	30.63	24.05
1/16	31.91	25.07
1/8	33.23	26.10
3/16	34.57	27.15
1/4	35.94	28.23
5/16	37.34	29.32
3/8	38.76	30.44
7/16	40.21	31.58
1/2	41.68	32.74
9/16	43.19	33.92
5/8	44.71	35.12
11/16	46.27	36.34
3/4	47.85	37.58
13/16	49.46	38.85
7/8	51.09	40.13
15/16	52.76	41.43
4	54.44	42.76
1/16	56.16	44.11
1/8	57.90	45.47
3/16	59.67	46.86
1/4	61.46	48.27
5/16	63.28	49.70
3/8	65.13	51.15
7/16	67.01	52.63
1/2	68.91	54.12
9/16	70.83	55.63
5/8	72.79	57.17
11/16	74.77	58.72
3/4	76.78	60.30
13/16	78.81	61.90
7/8	80.87	63.51
15/16	82.96	65.15
5	85.07	66.81
1/16	87.21	68.49
1/8	89.38	70.20
3/16	91.57	71.92
1/4	93.79	73.66
5/16	96.04	75.43
3/8	98.31	77.21
7/16	100.61	79.02
1/2	102.93	80.84
9/16	105.29	82.69
5/8	107.67	84.56
11/16	110.07	86.45
3/4	112.50	88.36
13/16	114.96	90.29
7/8	117.45	92.24
15/16	119.96	94.22

(continued on next page)

Figure 5-7 Weights of Square and Round Bars
(continued from previous page)

Size in Inches	Weight in Lbs. per Foot	
	Square	Round
6	122.50	96.21
1/16	125.07	98.23
1/8	127.66	100.26
3/16	130.28	102.32
1/4	132.92	104.40
5/16	135.59	106.49
3/8	138.29	108.61
7/16	141.02	110.75
1/2	143.77	112.91
9/16	146.55	115.10
5/8	149.35	117.30
11/16	152.18	119.52
3/4	155.04	121.77
13/16	157.92	124.03
7/8	160.83	126.32
15/16	163.77	128.63
7	166.74	130.95
1/16	169.73	133.30
1/8	172.74	135.67
3/16	175.79	138.06
1/4	178.86	140.48
5/16	181.96	142.91
3/8	185.08	145.36
7/16	188.23	147.84
1/2	191.41	150.33
9/16	194.61	152.85
5/8	197.84	155.38
11/16	201.10	157.94
3/4	204.38	160.52
13/16	207.69	163.12
7/8	211.03	165.74
15/16	214.39	168.38
8	217.78	171.04
1/16	221.19	173.73
1/8	224.64	176.43
3/16	228.11	179.15
1/4	231.60	181.90
5/16	235.12	184.67
3/8	238.67	187.45
7/16	242.25	190.26
1/2	245.85	193.09
9/16	249.48	195.94
5/8	253.13	198.81
11/16	256.82	201.70
3/4	260.53	204.62
13/16	264.26	207.55
7/8	268.02	210.50
15/16	271.81	213.48

Division 5

(continued on next page)

Figure 5-7 Weights of Square and Round Bars
(continued from previous page)

Size in Inches	Weight in Lbs. per Foot	
	Square	Round
9	275.63	216.48
1/16	279.47	219.49
1/8	283.33	222.53
3/16	287.23	225.59
1/4	291.15	228.67
5/16	295.10	231.77
3/8	299.07	234.89
7/16	303.07	238.03
1/2	307.10	241.20
9/16	311.15	244.38
5/8	315.24	247.59
11/16	319.34	250.81
3/4	323.48	254.06
13/16	327.64	257.33
7/8	331.82	260.61
15/16	336.04	263.92
10	340.28	267.25
1/16	344.54	270.60
1/8	348.84	273.98
3/16	353.16	277.37
1/4	357.50	280.78
5/16	361.88	284.22
3/8	366.28	287.67
7/16	370.70	291.15
1/2	375.16	294.65
9/16	379.64	298.17
5/8	384.14	301.70
11/16	388.67	305.26
3/4	393.23	308.84
13/16	397.82	312.45
7/8	402.43	316.07
15/16	407.07	319.71
11	411.74	323.38
1/16	416.43	327.06
1/8	421.15	330.77
3/16	425.89	334.49
1/4	430.66	338.24
5/16	435.46	342.01
3/8	440.29	345.80
7/16	445.14	349.61
1/2	450.02	353.44
9/16	454.92	357.30
5/8	459.85	361.17
11/16	464.81	365.06
3/4	469.80	368.98
13/16	474.81	372.91
7/8	479.84	376.87
15/16	484.91	380.85
12	490.00	384.85

Figure 5-8 Weights of Rectangular Sections

Weight of Rectangular Sections in Pounds per Linear Foot														
Width in Inches	Thickness in Inches													
	3/16	1/4	5/16	3/8	7/16	1/2	9/16	5/8	11/16	3/4	13/16	7/8	15/16	1
1/4	0.16	0.21	0.27	0.32	0.37	0.43	0.48	0.53	0.58	0.64	0.69	0.74	0.80	0.85
1/2	0.32	0.43	0.53	0.64	0.74	0.85	0.96	1.06	1.17	1.28	1.38	1.49	1.60	1.70
3/4	0.48	0.64	0.80	0.96	1.12	1.28	1.44	1.60	1.75	1.91	2.07	2.23	2.39	2.55
1	0.64	0.85	1.06	1.28	1.49	1.70	1.91	2.13	2.34	2.55	2.76	2.98	3.19	3.40
1-1/4	0.80	1.06	1.33	1.60	1.86	2.13	2.39	2.66	2.92	3.19	3.46	3.72	3.99	4.25
1-1/2	0.96	1.28	1.60	1.91	2.23	2.55	2.87	3.19	3.51	3.83	4.15	4.47	4.79	5.10
1-3/4	1.12	1.49	1.86	2.23	2.61	2.98	3.35	3.72	4.09	4.47	4.84	5.21	5.58	5.95
2	1.28	1.70	2.13	2.55	2.98	3.40	3.83	4.25	4.68	5.10	5.53	5.95	6.38	6.81
2-1/4	1.44	1.91	2.39	2.87	3.35	3.83	4.31	4.79	5.26	5.74	6.22	6.70	7.18	7.66
2-1/2	1.60	2.13	2.66	3.19	3.72	4.25	4.79	5.32	5.85	6.38	6.91	7.44	7.98	8.51
2-3/4	1.75	2.34	2.92	3.51	4.09	4.68	5.26	5.85	6.43	7.02	7.60	8.19	8.77	9.36
3	1.91	2.55	3.19	3.83	4.47	5.10	5.74	6.38	7.02	7.66	8.29	8.93	9.57	10.2
3-1/4	2.07	2.76	3.46	4.15	4.84	5.53	6.22	6.91	7.60	8.29	8.99	9.68	10.4	11.1
3-1/2	2.23	2.98	3.72	4.47	5.21	5.95	6.70	7.44	8.19	8.93	9.68	10.4	11.2	11.9
3-3/4	2.39	3.19	3.99	4.79	5.58	6.38	7.18	7.98	8.77	9.57	10.4	11.2	12.0	12.8
4	2.55	3.40	4.25	5.10	5.95	6.81	7.66	8.51	9.36	10.2	11.1	11.9	12.8	13.6
4-1/4	2.71	3.62	4.52	5.42	6.33	7.23	8.13	9.04	9.94	10.8	11.8	12.7	13.6	14.5
4-1/2	2.87	3.83	4.79	5.74	6.70	7.66	8.61	9.57	10.5	11.5	12.4	13.4	14.4	15.3
4-3/4	3.03	4.04	5.05	6.06	7.07	8.08	9.09	10.1	11.1	12.1	13.1	14.1	15.2	16.2
5	3.19	4.25	5.32	6.38	7.44	8.51	9.57	10.6	11.7	12.8	13.8	14.9	16.0	17.0
5-1/4	3.35	4.47	5.58	6.70	7.82	8.93	10.0	11.2	12.3	13.4	14.5	15.6	16.7	17.9
5-1/2	3.51	4.68	5.85	7.02	8.19	9.36	10.5	11.7	12.9	14.0	15.2	16.4	17.5	18.7
5-3/4	3.67	4.89	6.11	7.34	8.56	9.78	11.0	12.2	13.5	14.7	15.9	17.1	18.3	19.6
6	3.83	5.10	6.38	7.66	8.93	10.2	11.5	12.8	14.0	15.3	16.6	17.9	19.1	20.4
6-1/4	3.99	5.32	6.65	7.98	9.30	10.6	12.0	13.3	14.6	16.0	17.3	18.6	19.9	21.3
6-1/2	4.15	5.53	6.91	8.29	9.68	11.1	12.4	13.8	15.2	16.6	18.0	19.4	20.7	22.1
6-3/4	4.31	5.74	7.18	8.61	10.0	11.5	12.9	14.4	15.8	17.2	18.7	20.1	21.5	23.0
7	4.47	5.95	7.44	8.93	10.4	11.9	13.4	14.9	16.4	17.9	19.4	20.8	22.3	23.8
7-1/4	4.63	6.17	7.71	9.25	10.8	12.3	13.9	15.4	17.0	18.5	20.0	21.6	23.1	24.7
7-1/2	4.79	6.38	7.98	9.57	11.2	12.8	14.4	16.0	17.5	19.1	20.7	22.3	23.9	25.5
7-3/4	4.94	6.59	8.24	9.89	11.5	13.2	14.8	16.5	18.1	19.8	21.4	23.1	24.7	26.4
8	5.10	6.81	8.51	10.2	11.9	13.6	15.3	17.0	18.7	20.4	22.1	23.8	25.5	27.2
8-1/2	5.42	7.23	9.04	10.8	12.7	14.5	16.3	18.1	19.9	21.7	23.5	25.3	27.1	28.9
9	5.74	7.66	9.57	11.5	13.4	15.3	17.2	19.1	21.1	23.0	24.9	26.8	28.7	30.6
9-1/2	6.06	8.08	10.1	12.1	14.1	16.2	18.2	20.2	22.2	24.2	26.3	28.3	30.3	32.3
10	6.38	8.51	10.6	12.8	14.9	17.0	19.1	21.3	23.4	25.5	27.6	29.8	31.9	34.0
10-1/2	6.70	8.93	11.2	13.4	15.6	17.9	20.1	22.3	24.6	26.8	29.0	31.3	33.5	35.7
11	7.02	9.36	11.7	14.0	16.4	18.7	21.1	23.4	25.7	28.1	30.4	32.8	35.1	37.4
11-1/2	7.34	9.78	12.2	14.7	17.1	19.6	22.0	24.5	26.9	29.3	31.8	34.2	36.7	39.1
12	7.66	10.2	12.8	15.3	17.9	20.4	23.0	25.5	28.1	30.6	33.2	35.7	38.3	40.8

Division 5

Figure 5-9 Weights of Flat Rolled Steel

Thick-ness (in inches)	Weight			Thick-ness (in inches)	Weight		
	Lbs./S.F.	Lbs./ Sq. Meter	Kgs./ Sq. Meter		Lbs./S.F.	Lbs./ Sq. Meter	Kgs./ Sq. Meter
.01	.408	4.392	1.992	.51	20.808	223.975	101.593
.02	.816	8.783	3.984	.52	21.216	228.366	103.585
.03	1.224	13.175	5.976	.53	21.624	232.758	105.577
.04	1.632	17.567	7.968	.54	22.032	237.150	107.569
.05	2.040	21.958	9.960	.55	22.440	241.541	109.561
.06	2.448	26.350	11.952	.56	22.848	245.933	111.553
.07	2.856	30.742	13.944	.57	23.256	250.324	113.545
.08	3.264	35.133	15.936	.58	23.664	254.716	115.537
.09	3.672	39.525	17.928	.59	24.072	259.108	117.529
.10	4.080	43.917	19.920	.60	24.480	263.499	119.521
.11	4.488	48.308	21.912	.61	24.888	267.891	121.513
.12	4.896	52.700	23.904	.62	25.296	272.283	123.505
.13	5.304	57.092	25.896	.63	25.704	276.674	125.497
.14	5.712	61.483	27.888	.64	26.112	281.066	127.489
.15	6.120	65.875	29.880	.65	26.520	285.458	129.481
.16	6.528	70.267	31.872	.66	26.928	289.849	131.474
.17	6.936	74.658	33.864	.67	27.336	294.241	133.466
.18	7.344	79.050	35.856	.68	27.744	298.633	135.458
.19	7.752	83.441	37.848	.69	28.152	303.024	137.450
.20	8.160	87.833	39.840	.70	28.560	307.416	139.442
.21	8.568	92.225	41.832	.71	28.968	311.808	141.434
.22	8.976	96.616	43.825	.72	29.376	316.199	143.426
.23	9.384	101.008	45.817	.73	29.784	320.591	145.418
.24	9.792	105.400	47.809	.74	30.192	324.983	147.410
.25	10.200	109.791	49.801	.75	30.600	329.374	149.402
.26	10.608	114.183	51.793	.76	31.008	333.766	151.394
.27	11.016	118.575	53.785	.77	31.416	338.158	153.386
.28	11.424	122.966	55.777	.78	31.824	342.549	155.378
.29	11.832	127.358	57.769	.79	32.232	346.941	157.370
.30	12.240	131.750	59.761	.80	32.640	351.333	159.362
.31	12.648	136.141	61.753	.81	33.048	355.724	161.354
.32	13.056	140.533	63.745	.82	33.456	360.116	163.346
.33	13.464	144.925	65.737	.83	33.864	364.508	165.338
.34	13.872	149.316	67.729	.84	34.272	368.899	167.330
.35	14.280	153.708	69.721	.85	34.680	373.291	169.322
.36	14.688	158.100	71.713	.86	35.088	377.683	171.314
.37	15.096	162.491	73.705	.87	35.496	382.074	173.306
.38	15.504	166.883	75.697	.88	35.904	386.466	175.298
.39	15.912	171.275	77.689	.89	36.312	390.858	177.290
.40	16.320	175.666	79.681	.90	36.720	395.249	179.282
.41	16.728	180.058	81.673	.91	37.128	399.641	181.274
.42	17.136	184.450	83.665	.92	37.536	404.033	183.266
.43	17.544	188.841	85.657	.93	37.944	408.424	185.258
.44	17.952	193.233	87.649	.94	38.352	412.816	187.250
.45	18.360	197.625	89.641	.95	38.760	417.207	189.242
.46	18.768	202.016	91.633	.96	39.168	421.599	191.234
.47	19.176	206.408	93.625	.97	39.576	425.991	193.226
.48	19.584	210.800	95.617	.98	39.984	430.382	195.218
.49	19.992	215.191	97.609	.99	40.392	434.774	197.210
.50	20.400	219.583	99.601	1.00	40.800	439.166	199.202

Figure 5-10 Weights of Rolled Floor Plates

This table lists weights for floor plates. Note that various mills offer various raised patterns and surface projections in a variety of widths. A maximum width of 96" and a maximum thickness of 1" are available.

Gauge	Rolled Floor Plates: Standard Thicknesses and Weights				
	Theoretical Weight per S.F. (in Lbs.)	Nominal Thickness (in inches)	Theoretical Weight per S.F. (in Lbs.)	Nominal Thickness (in inches)	Theoretical Weight per S.F. (in Lbs.)
18	2.40	1/8"	6.16	1/2"	21.47
16	3.00	3/16"	8.71	9/16"	24.02
14	3.75	1/4"	11.26	5/8"	26.58
13	4.50	5/16"	13.81	3/4"	31.68
12	5.25	3/8"	16.37	7/8"	36.78
		7/16"	18.92	1"	41.89

Thickness of rolled floor plates is measured near the edge and does not take the raised pattern into account.

Figure 5-11 Dimensions and Weights of Sheet Steel

Gauge No.	Approximate Thickness			Weight		
	Inches (in decimal parts)	Millimeters				
	Steel	Steel		per S.F. in Ounces	per S.F. in Lbs.	per Square Meter in Kg.
0000000	0.4782	12.146		320	20	97.65
000000	.4484	11.389		300	18.75	91.55
00000	.4185	10.630		280	17.50	85.44
0000	.3886	9.870		260	16.25	79.33
000	.3587	9.111		240	15	73.24
00	.3288	8.352		220	13.75	67.13
0	.2989	7.592		200	12.50	61.03
1	.2690	6.833		180	11.25	54.93
2	.2541	6.454		170	10.625	51.88
3	.2391	6.073		160	10	48.82
4	.2242	5.695		150	9.375	45.77
5	.2092	5.314		140	8.75	42.72
6	.1943	4.935		130	8.125	39.67
7	.1793	4.554		120	7.5	36.32
8	.1644	4.176		110	6.875	33.57
9	.1495	3.797		100	6.25	30.52
10	.1345	3.416		90	5.625	27.46
11	.1196	3.038		80	5	24.41
12	.1046	2.657		70	4.375	21.36
13	.0897	2.278		60	3.75	18.31
14	.0747	1.897		50	3.125	15.26
15	.0673	1.709		45	2.8125	13.73
16	.0598	1.519		40	2.5	12.21
17	.0538	1.367		36	2.25	10.99
18	.0478	1.214		32	2	9.765
19	.0418	1.062		28	1.75	8.544
20	.0359	.912		24	1.50	7.324
21	.0329	.836		22	1.375	6.713
22	.0299	.759		20	1.25	6.103
23	.0269	.683		18	1.125	5.49
24	.0239	.607		16	1	4.882
25	.0209	.531		14	0.875	4.272
26	.0179	.455		12	.75	3.662
27	.0164	.417		11	.6875	3.357
28	.0149	.378		10	.625	3.052

Figure 5-12 Standard Welding Symbols

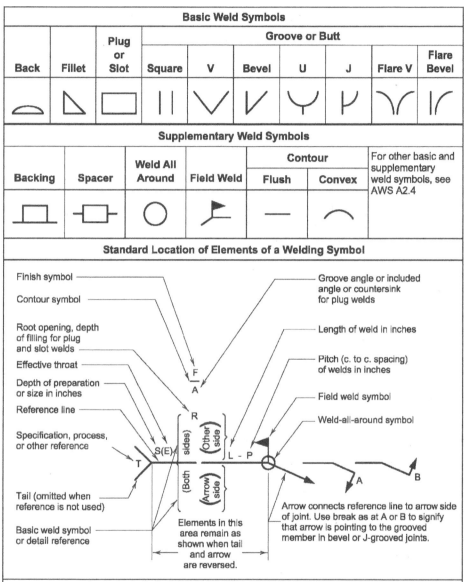

Basic Weld Symbols									
		Plug or Slot	Groove or Butt						
Back	Fillet		Square	V	Bevel	U	J	Flare V	Flare Bevel

Supplementary Weld Symbols						For other basic and supplementary weld symbols, see AWS A2.4
		Weld All Around	Field Weld	Contour		
Backing	Spacer			Flush	Convex	

Standard Location of Elements of a Welding Symbol

Finish symbol

Contour symbol

Root opening, depth of filling for plug and slot welds

Effective throat

Depth of preparation or size in inches

Reference line

Specification, process, or other reference

Tail (omitted when reference is not used)

Basic weld symbol or detail reference

Groove angle or included angle or countersink for plug welds

Length of weld in inches

Pitch (c. to c. spacing) of welds in inches

Field weld symbol

Weld-all-around symbol

Arrow connects reference line to arrow side of joint. Use break as at A or B to signify that arrow is pointing to the grooved member in bevel or J-grooved joints.

F
A
R
S(E)
T
(Both sides)
(Arrow side)
(Other side)
L - P
A
B

Elements in this area remain as shown when tail and arrow are reversed.

Note:
Size, weld symbol, length of weld, and spacing must read in that order, from left to right, along the reference line. Neither orientation of reference nor location of the arrow alters this rule.

The perpendicular leg of △, V, P, Γ, weld symbols must be at left.

Arrow and other side welds are of the same size unless otherwise shown. Dimensions of fillet welds must be shown on both the arrow side and the other side symbol.

The point of the field weld symbol must point toward the tail.

Symbols apply between abrupt changes in direction of welding unless governed by the "all around" symbol or otherwise dimensioned.

These symbols do not explicitly provide for the case that frequently occurs in structural work, where duplicate material (such as stiffeners) occurs on the far side of a web or gusset plate. The fabricating industry has adopted this convention: that when the billing of the detail material discloses the existence of a member on the far side as well as on the near side, the welding shown for the near side shall be duplicated on the far side.

(courtesy of American Institute of Steel Construction)

Division 5

Figure 5-13 Coating Structural Steel

This figure shows the expected average coverage of coating structural steel with red oxide rust inhibitive paint on an aluminum paint. On field-welded jobs, shop coat is necessarily omitted. All painting must be done in the field and usually consists of two coats. The table below shows paint coverage and daily production for field painting.

Type Construction	Surface Area per Ton	Coat	One Gallon Covers		In 8 Hrs. Painter Covers		Average per Ton Spray	
			Brush	Spray	Brush	Spray	Gallons	Labor-hours
Light Structural	300 S.F. to 500 S.F.	1st	500 S.F.	455 S.F.	640 S.F.	2000 S.F.	0.9 gals.	1.6 L.H.
		2nd	450	410	800	2400	1.0	1.3
		3rd	450	410	960	3200	1.0	1.0
Medium	150 S.F. to 300 S.F.	All	400	365	1600	3200	0.6	0.6
Heavy Structural	50 S.f. to 100 S.F.	1st	400	365	1920	4000	0.2	0.2
		2nd	400	365	2000	4000	0.2	0.2
		3rd	400	365	2000	4000	0.2	0.2
Weighted Average	225 S.F.	All	400	365	1350	3000	0.6	0.6

Figure 5-14 Standard Joist Details

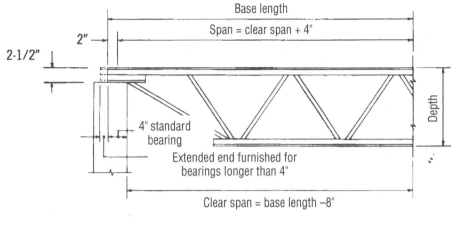

K Series

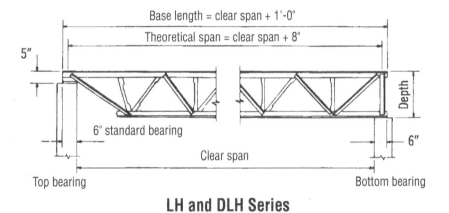

LH and DLH Series

(continued on next page)

Division 5

Figure 5-14 Standard Joist Details *(continued from previous page)*

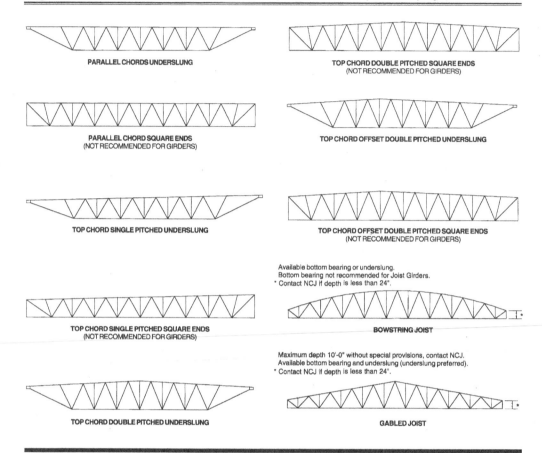

PARALLEL CHORDS UNDERSLUNG

TOP CHORD DOUBLE PITCHED SQUARE ENDS
(NOT RECOMMENDED FOR GIRDERS)

PARALLEL CHORD SQUARE ENDS
(NOT RECOMMENDED FOR GIRDERS)

TOP CHORD OFFSET DOUBLE PITCHED UNDERSLUNG

TOP CHORD SINGLE PITCHED UNDERSLUNG

TOP CHORD OFFSET DOUBLE PITCHED SQUARE ENDS
(NOT RECOMMENDED FOR GIRDERS)

Available bottom bearing or underslung.
Bottom bearing not recommended for Joist Girders.
* Contact NCJ if depth is less than 24".

TOP CHORD SINGLE PITCHED SQUARE ENDS
(NOT RECOMMENDED FOR GIRDERS)

BOWSTRING JOIST

Maximum depth 10'-0" without special provisions, contact NCJ.
Available bottom bearing and underslung (underslung preferred).
* Contact NCJ if depth is less than 24".

TOP CHORD DOUBLE PITCHED UNDERSLUNG

GABLED JOIST

Figure 5-15   Weights of Open Web Steel Joists (K Series)

The approximate weights per linear foot as shown in the following table do not include any accessories or connection plates.

Joist Designation	8K1	10K1	12K1	12K3	12K5	14K1	14K3	14K4	14K6	16K2	16K3
Depth (in.)	8	10	12	12	12	14	14	14	14	16	16
Approx. Wt. (Lbs./Ft.)	5.1	5.0	5.0	5.7	7.1	5.2	6.0	6.7	7.7	5.5	6.3
Joist Designation	16K4	16K5	16K6	16K7	16K9	18K3	18K4	18K5	18K6	18K7	18K9
Depth (in.)	16	16	16	16	16	18	18	18	18	18	18
Approx. Wt. (Lbs./Ft.)	7.0	7.5	8.1	8.6	10.0	6.6	7.2	7.7	8.5	9.0	10.2
Joist Designation	18K10	20K3	20K4	20K5	20K6	20K7	20K9	20K10	22K4	22K5	22K6
Depth (in.)	18	20	20	20	20	20	20	20	22	22	22
Approx. Wt. (Lbs./Ft.)	11.7	6.7	7.6	8.2	8.9	9.3	10.8	12.2	8.0	8.8	9.2
Joist Designation	22K7	22K9	22K10	22K11	24K4	24K5	24K6	24K7	24K8	24K9	24K10
Depth (in.)	22	22	22	22	24	24	24	24	24	24	24
Approx. Wt. (Lbs./Ft.)	9.7	11.3	12.6	13.8	8.4	9.3	9.7	10.1	11.5	12.0	13.1
Joist Designation	24K12	26K5	26K6	26K7	26K8	26K9	26K10	26K12	28K6	28K7	28K8
Depth (in.)	24	26	26	26	26	26	26	26	28	28	28
Approx. Wt. (Lbs./Ft.)	16.0	9.8	10.6	10.9	12.1	12.2	13.8	16.6	11.4	11.8	12.7
Joist Designation	28K9	28K10	28K12	30K7	30K8	30K9	30K10	30K11	30K12		
Depth (in.)	28	28	28	30	30	30	30	30	30		
Approx. Wt. (Lbs./Ft.)	13.0	14.3	17.1	12.3	13.2	13.4	15.0	16.4	17.6		

Division 5

Figure 5-16 Weights of Open Web Steel Joists (LH Series)

The approximate weights per linear foot as shown in the following table do not include any accessories or connection plates.

Joist Designation	18LH02	18LH03	18LH04	18LH05	18LH06	18LH07	18LH08	18LH09		
Depth (in.)	18	18	18	18	18	18	18	18		
Approx. Wt. (Lbs./Ft.)	10	11	12	15	15	17	19	21		
Joist Designation	20LH02	20LH03	20LH04	20LH05	20LH06	20LH07	20LH08	20LH09	20LH10	
Depth (in.)	20	20	20	20	20	20	20	20	20	
Approx. Wt. (Lbs./Ft.)	10	11	12	14	15	17	19	21	23	
Joist Designation	24LH03	24LH04	24LH05	24LH06	24LH07	24LH08	24LH09	24LH10	24LH11	
Depth (in.)	24	24	24	24	24	24	24	24	24	
Approx. Wt. (Lbs./Ft.)	11	12	13	16	17	18	21	23	25	
Joist Designation	28LH05	28LH06	28LH07	28LH08	28LH09	28LH10	28LH11	28LH12	28LH13	
Depth (in.)	28	28	28	28	28	28	28	28	28	
Approx. Wt. (Lbs./Ft.)	13	16	17	18	21	23	25	27	30	
Joist Designation	32LH06	32LH07	32LH08	32LH09	32LH10	32LH11	32LH12	32LH13	32LH14	32LH15
Depth (in.)	32	32	32	32	32	32	32	32	32	32
Approx. Wt. (Lbs./Ft.)	14	16	17	21	21	24	27	30	33	35
Joist Designation	36LH07	36LH08	36LH09	36LH10	36LH11	36LH12	36LH13	36LH14	36LH15	
Depth (in.)	36	36	36	36	36	36	36	36	36	
Approx. Wt. (Lbs./Ft.)	16	18	21	21	23	25	30	36	36	
Joist Designation	40LH08	40LH09	40LH10	40LH11	40LH12	40LH13	40LH14	40LH15	40LH16	
Depth (in.)	40	40	40	40	40	40	40	40	40	
Approx. Wt. (Lbs./Ft.)	16	21	21	22	25	30	35	36	42	
Joist Designation	44LH09	44LH10	44LH11	44LH12	44LH13	44LH14	44LH15	44LH16	44LH17	
Depth (in.)	44	44	44	44	44	44	44	44	44	
Approx. Wt. (Lbs./Ft.)	19	21	22	25	30	31	36	42	47	
Joist Designation	48LH10	48LH11	48LH12	48LH13	48LH14	48LH15	48LH16	48LH17		
Depth (in.)	48	48	48	48	48	48	48	48		
Approx. Wt. (Lbs./Ft.)	21	22	25	29	32	36	42	47		

Figure 5-17 Weights of Open Web Steel Joists (DLH Series)

The approximate weights per linear foot as shown in the following table do not include any accessories or connection plates.

Joist Designation	52DLH10	52DLH11	52DLH12	52DLH13	52DLH14	52DLH15	52DLH16	52DLH17
Depth (in.)	52	52	52	52	52	52	52	52
Approx. Wt. (Lbs./Ft.)	25	26	29	34	39	42	45	52
Joist Designation	56DLH11	56DLH12	56DLH13	56DLH14	56DLH15	56DLH16	56DLH17	
Depth (in.)	56	56	56	56	56	56	56	
Approx. Wt. (Lbs./Ft.)	26	30	34	39	42	46	51	
Joist Designation	60DLH12	60DLH13	60DLH14	60DLH15	60DLH16	60DLH17	60DLH18	
Depth (in.)	60	60	60	60	60	60	60	
Approx. Wt. (Lbs./Ft.)	29	35	40	43	46	52	59	
Joist Designation	64DLH12	64DLH13	64DLH14	64DLH15	64DLH16	64DLH17	64DLH18	
Depth (in.)	64	64	64	64	64	64	64	
Approx. Wt. (Lbs./Ft.)	31	34	40	43	46	52	59	
Joist Designation	68DLH13	68DLH14	68DLH15	68DLH16	68DLH17	68DLH18	68DLH19	
Depth (in.)	68	68	68	68	68	68	68	
Approx. Wt. (Lbs./Ft.)	37	40	44	49	55	61	67	
Joist Designation	72DLH14	72DLH15	72DLH16	72DLH17	72DLH18	72DLH19		
Depth (in.)	72	72	72	72	72	72		
Approx. Wt. (Lbs./Ft.)	41	44	50	56	59	70		

Division 5

Figure 5-18 Steel Floor Grating and Treads

This table lists the materials weight per SF for welded steel grating.

Bearing Bar Size (in.)	Steel Bearing Bars, 1-3/16" O.C. Cross Bars 4" O.C. Weight	Steel Bearing Bars, 15/16" O.C. Cross Bars 4" O.C. Weight	Steel Bearing Bars, 1-3/16" O.C. Cross Bars 2" O.C. Weight	Steel Bearing Bars, 15/16" O.C. Cross Bars 2" O.C. Weight
3/4 x 1/8"	4.1#	5.0#	4.8#	5.7#
1 x 1/8	5.2	6.4	5.9	7.1
1-1/4 x 1/8	6.3	7.9	7.0	8.6
1-1/4 x 3/16	9.1	11.5	9.8	12.2
1-1/2 x 1/8	7.4	9.3	8.1	10.0
1-3/4 x 3/16	12.5	15.8	13.2	16.5
2 x 3/16	14.1	18.0	14.8	18.7
2-1/4 x 3/16	15.7	20.0	16.4	20.7

Figure 5-19 Aluminum Floor Grating

This table lists the material weight per SF for aluminum grating, alloy 6063.

Bearing Bar Size in Inches	Bearing Bars 1-3/16" O.C.		For Close Mesh Add to 4" O.C. Chart	
	Cross Bars 4" O.C. Weight	Cross Bars 2" O.C. Weight	Total Weight	Add to Cost
1 x 1/8	2.0#	2.1#	3.1#	75%
1-1/4 x 1/8	2.4	2.5	3.8	75%
1-1/4 x 3/16	3.3	3.5	5.1	60%
1-1/2 x 1/8	2.9	3.0	4.6	75%
1-3/4 x 3/16	4.6	4.8	7.2	60%
2 x 3/16	5.3	5.5	8.1	60%
2-1/4 x 3/16	5.9	6.1	9.1	60%

Figure 5-20 Installation Time in Labor-Hours for Various Structural Steel Building Components

The following tables show the expected average installation times for various structural steel shapes. Table A presents installation times for column shapes, Table B for beams, Table C for light framing and bolts, and Table D for various projects.

Table A		
Description	**Labor-Hours**	**Unit**
Columns		
Steel Concrete Filled		
3-1/2" Diameter	9.33	Ea.
6-5/8" Diameter	1.120	Ea.
Steel Pipe		
3" Diameter	.933	Ea.
8" Diameter	1.120	Ea.
12" Diameter	1.244	Ea.
Structural Tubing		
4" x 4"	.966	Ea.
8" x 8"	1.120	Ea.
12" x 8"	1.167	Ea.
W Shape 2 Tier		
W8 x 31	.052	L.F.
W8 x 67	.057	L.F.
W10 x 45	.054	L.F.
W10 x 112	.058	L.F.
W12 x 50	.054	L.F.
W12 x 190	.061	L.F.
W14 x 74	.057	L.F.
W14 x 176	.061	L.F.

Table B				
Description	**Labor-Hours**	**Unit**	**Labor-Hours**	**Unit**
Beams, W Shape				
W6 x 9	.949	Ea.	.093	L.F.
W10 x 22	1.037	Ea.	.085	L.F.
W12 x 26	1.037	Ea.	.064	L.F.
W14 x 34	1.333	Ea.	.069	L.F.
W16 x 31	1.333	Ea.	.062	L.F.
W18 x 50	2.162	Ea.	.088	L.F.
W21 x 62	2.222	Ea.	.077	L.F.
W24 x 76	2.353	Ea.	.072	L.F.
W27 x 94	2.581	Ea.	.067	L.F.
W30 x 108	2.857	Ea.	.067	L.F.
W33 x 130	3.200	Ea.	.071	L.F.
W26 x 300	3.810	Ea.	.077	L.F.

Division 5

(continued on next page)

Figure 5-20 Installation Time in Labor-Hours for Various Structural
Steel Building Components *(continued from previous page)*

Table C		
Description	**Labor-Hours**	**Unit**
Light Framing		
Angles 4" and Larger	.055	Lbs.
Less than 4"	.091	Lbs.
Channels 8" and Larger	.048	Lbs.
Less than 8"	.072	Lbs.
Cross Bracing Angles	.055	Lbs.
Rods	.034	Lbs.
Hanging Lintels	.069	Lbs.
High-Strength Bolts in Place		
3/4" Bolts	.070	Ea.
7/8" Bolts	.076	

Table D				
Description	**Labor-Hours**	**Unit**	**Labor-Hours**	**Unit**
Apartments, Nursing Homes, etc.				
1–2 Stories	4.211	Piece	7.767	ton
3–6 Stories	4.444	Piece	7.921	ton
7–15 Stories	4.923	Piece	9.014	ton
Over 15 Stories	5.333	Piece	9.209	ton
Offices, Hospitals, etc.				
1–2 Stories	4.211	Piece	7.767	ton
3–6 Stories	4.741	Piece	8.889	ton
7–15 Stories	4.923	Piece	9.014	ton
Over 15 Stories	5.120	Piece	9.209	ton
Industrial Buildings				
1 Story	3.478	Piece	6.202	ton

Figure 5-21 Installation Time in Labor-Hours for the Erection of
Steel Superstructure Systems

Description	Labor-Hours	Unit
Steel Columns Concrete Filled		
4" Diameter	.072	L.F.
5" Diameter	.055	L.F.
6-5/8" Diameter	.047	L.F.
Steel Pipe		
3" to 5" Diameter	.008	Lb.
6" to 12" Diameter	.004	Lb.
Structural Tubing		
6" x 6"	.007	Lb.
10" x 10"	.004	Lb.
W Shapes		
W8 x 31	.052	L.F.
W10 x 45	.054	L.F.
W12 x 50	.054	L.F.
W14 x 74	.057	L.F.
Beams, W Shapes, Average	.080	L.F.
Steel Joists K Series Horizontal Bridging		
To 30' Span	6.667	ton
30' to 50' Span	4.706	ton
(Includes One Row of Bolted Cross Bridging for Spans Over 40' Where Required)		
LH Series Bolted Cross Bridging		
Spans to 96'	6.154	ton
DLH Series Bolted Cross Bridging		
Spans to 144' Shipped in 2 Pieces	6.154	ton
Joist Girders	6.154	ton
Trusses, Factory Fabricated, with Chords	7.273	ton
Metal Decking Open Type		
1-1/2" Deep		
22 Gauge	.007	S.F.
18 and 20 Gauge	.008	S.F.
3" Deep		
20 and 22 Gauge	.009	S.F.
18 Gauge	.010	S.F.
16 Gauge	.011	S.F.
4-1/2" Deep		
20 Gauge	.012	S.F.
18 Gauge	.013	S.F.
16 Gauge	.014	S.F.
7-1/2" Deep		
18 Gauge	.019	S.F.
16 Gauge	.020	S.F.

Division 5

(continued on next page)

Figure 5-21   Installation Time in Labor-Hours for the Erection of
Steel Superstructure Systems *(continued from previous page)*

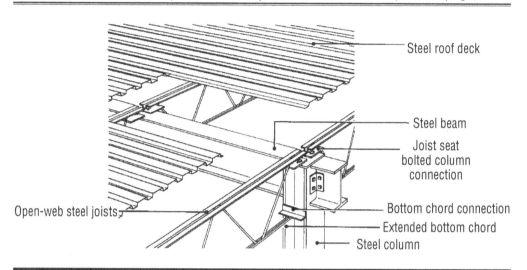

Figure 5-22 Installation Time in Labor-Hours for Steel Roof Decks

Description	Labor-Hours	Unit
Roof Deck Open Type		
1-1/2" Deep		
22 Gauge	.007	S.F.
18 and 20 Gauge	.008	S.F.
3" Deep		
22 and 20 Gauge	.009	S.F.
18 Gauge	.010	S.F.
16 Gauge	.011	S.F.
4-1/2" Deep		
20 Gauge	.012	S.F.
6" Deep		
18 Gauge	.016	S.F.
7-1/2" Deep		
18 Gauge	.019	S.F.
Roof Deck Cellular Units Galv.		
3" Deep 20-20 Gauge	.023	S.F.
4-1/2" Deep 20-18 Gauge	.029	S.F.

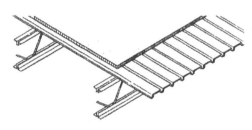

Roof Deck System with Insulation

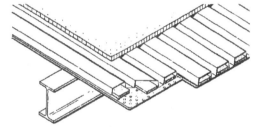

Acoustic Deck System

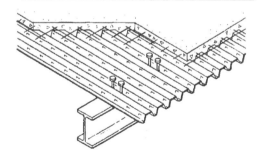

Composite Beam, Deck and Slab

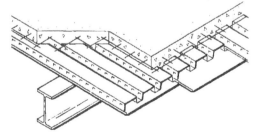

Cellular Deck System

Division 5

Figure 5-23 Installation Time in Labor-Hours for Platforms
and Walkways

Description	Labor-Hours	Unit
Light Framing, Steel		
L's 4" and Larger	0.55	Lb.
Less than 4"	.091	Lb.
C's 8" and Larger	.048	Lb.
Less than 8"	.072	Lb.
Aluminum Shapes		
Grating, Aluminum		
1-1/4" Deep	.032	S.F.
2-1/4" Deep	.032	S.F.
Fiberglass	.040	L.F.
Steel	.093	S.F.
Steel Pipe Railing w/Toe Plate	.160	L.F.

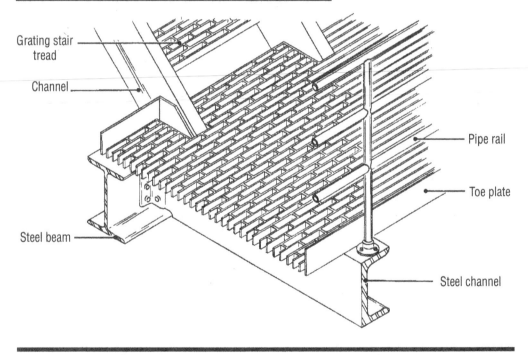

Grating stair tread

Channel

Steel beam

Pipe rail

Toe plate

Steel channel

Division Six

Wood, Plastics & Composites

Introduction

Wood, unlike most processed building products, is an organic material that can be used in its natural state. Wood can be easily shaped or cut to size on the job site or in the shop. Its strength is influenced by density, moisture content, and natural defects such as knots, grain, etc. Structurally, wood may be used for joists, posts, columns, beams, and trusses. It may also be laminated and formed into structural shapes, such as columns, arches, and vaults, with fiberboard, particleboard, plywood, or wood decking to form composite panels. Wood may be left exposed and finished for interior or exterior use and may be purchased pressure-treated to provide resistance to moisture, rot, or exposure.

Structural lumber is stress-graded for bending, tension parallel to the grain, horizontal shear, compression perpendicular to the grain, compression parallel to the grain, and modulus of elasticity.

Structural lumber is usually dressed on four sides and, therefore, reduced in size by approximately 1/2" in each direction. For example, the actual dimensions of a 6" × 10" beam dressed is 5-1/2" × 9-1/2".

Wood structures are fastened by nails, pins, dowels, screws, bolts, and adhesives, or by fabricated metal connectors, tailored to perform specific connection functions.

This division includes tables for the following:

- Framing systems
- Board foot measure of jobs
- Stud requirements
- Interior/exterior framing
- Roof framing systems
- Installation times
- Nail quantities
- Millwork

Unlike concrete and steel, wood does not take as much energy to process into forms usable for construction, so it is a fairly environmentally friendly product *if* it is harvested in a responsible, sustainable method. Obtaining wood that comes with a certification from an independent third party attesting to its sustainability carries a price premium of about 5% on lumber mill material costs. Engineered wood products are considered environmentally friendly because they require less wood content compared to traditional wood building products. It is important that binders used in wood products such as engineered wood, plywood, and cabinetry be low in volatile organic compounds (VOCs). Vapors that VOCs release (off-gas) are suspected to be harmful, especially in enclosed spaces.

Estimating Data

The following tables present estimating guidelines for items found in Division 6 – Wood, Plastics, and Composites. Please note that these guidelines are intended as indicators of what may generally be expected, but that each project must be evaluated individually.

Table of Contents

Division 6

Estimating Tips

Items Listed
Note that in construction documents, not all rough carpentry items are listed or noted. Protective treads, inserts, and rails are examples of necessary but not listed items that the contractor must provide.

Item Search
The search for carpentry items must include roofing sections, wall sections, and all detail drawings.

Temporary Construction
Temporary construction should be included in Division 1 – General Requirements, though it may sometimes be listed in Division 6. Double check to ensure that items are not double-priced.

Treated Lumber
Treated lumber should always be used when the lumber will come in direct contact with concrete, masonry, or earth.

Bridging
Even when bridging is not shown on joist drawings, always include it, as it helps to distribute concentrated loads to the adjacent joists.

Blocking
One of the most overlooked areas of wood blocking is at roof edges. Almost always, a built-up roof or membrane roof system will require some sort of blocking and/or cant strip system.

Millwork
When budgeting millwork, a rule of thumb is that the total cost of millwork items will be two to three times the cost of the materials required. Do not overlook adding protection to all millwork, especially if the work is done by others.

Pricing
Do not rely on yesterday's material quotes. Due to fluctuations in the lumber market, caution is warranted.

Estimating Help
Short of time? Many lumber yards retain competent estimators who will provide material lists/estimates from your plans for little or no cost. This is also a good method for checking your own takeoff.

Division 6

Checklist ✔

For an estimate to be reliable, all terms must be accounted for. A complete estimate can also limit contingencies. The following checklist can be used to help ensure that all terms are included.

☐ **Structural**
- ☐ Columns
- ☐ Beams
  - ☐ Box
  - ☐ Laminated
- ☐ Joists
- ☐ Posts and girts
- ☐ Purlins
- ☐ Rafters
- ☐ Trusses

☐ **Blocking**

☐ **Bracing**

☐ **Bridging**

☐ **Cabinets**

☐ **Casings**

☐ **Ceiling Beams, Decorative**

☐ **Chair Rails**

☐ **Countertops**

☐ **Door Bucks**

☐ **Finish Carpentry**
- ☐ Casings
- ☐ Closets
- ☐ Cornices
- ☐ Cupolas
- ☐ Molding and trim
- ☐ Paneling
- ☐ Railings

- ☐ Shelving
- ☐ Stairs
- ☐ Thresholds
- ☐ Wainscoting

☐ **Fire Stopping**

☐ **Floor Planking**

☐ **Nailer Plates**

☐ **Siding**
- ☐ Sheathing
  - ☐ Hardboard
  - ☐ Particleboard
  - ☐ Plywood
- ☐ Cedar
- ☐ Redwood
- ☐ Fir
- ☐ White pine

☐ **Roof Carpentry**
- ☐ Cants
- ☐ Decking
- ☐ Sheathing

☐ **Sills**

☐ **Soffits**

☐ **Subflooring**

☐ **Underlayment**

☐ **Wall Framing**
- ☐ Studs
- ☐ Plates
- ☐ Sills

Figure 6-1 Wood Framing Nomenclature

These graphics illustrate the common terms for the various components of wood framing. Whether for residential (shown here) or commercial, the terminology is the same.

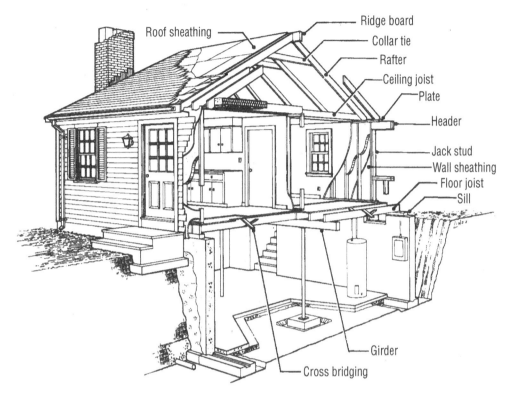

Wood Framing System

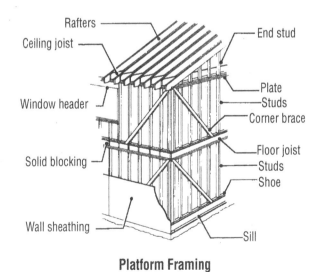

Platform Framing

(continued on next page)

Division 6

Figure 6-1   Wood Framing Nomenclature *(continued from previous page)*

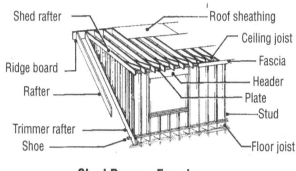

Shed Dormer Framing

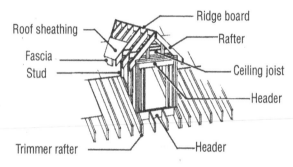

Gable Dormer Framing

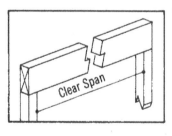

Single beam

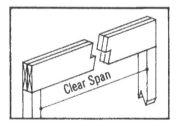

Double beam

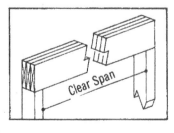

Triple beam

Wood Beams and Columns

(continued on next page)

Figure 6-1   Wood Framing Nomenclature *(continued from previous page)*

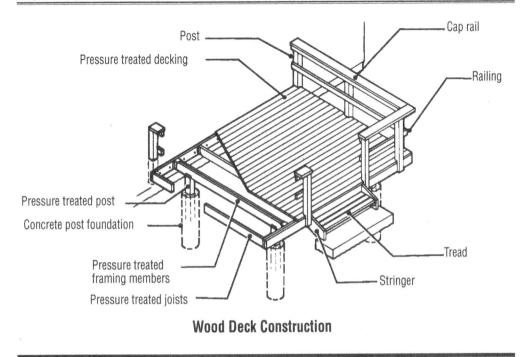

Wood Deck Construction

Figure 6-2   Lumber Industry Abbreviations

The following is a list of abbreviations used in connection with lumber in the construction industry. The abbreviations are commonly used in the form shown, but variations may occur.

AD	Air-dried
ADF	After deducting freight
ALS	American Lumber Standards
AV or AVG	Average
Bd	Board
Bd. ft.	Board foot or feet
Bdl	Bundle
Bev	Beveled
B/L	Bill of lading
BM	Board measure
Btr	Better
B&B or B& Btr	B and better
B&S	Beams and stringers
CB1S	Center bead one side
CB2S	Center bead two sides
CF	Cost and frieght
CG2E	Center groove and freight
CG2E	Center groove two edges
CIF	Cost, insurance, and freight
CIFE	Cost, insurance, freight, and exchange
Clg	Ceiling
Clr	Clear
CM	Center matched

Division 6

(continued on next page)

Figure 6-2 Lumber Industry Abbreviations *(continued from previous page)*

Com	Common
CS	Caulking seam
Csg	Casing
Cu. Ft.	Cubic foot or feet
CV1S	Center vee one side
CV2S	Center vee two sides
D&H	Dressed and headed
D&	Dressed and matched
DB. Clg	Double-headed ceiling (E&CB2S)
DB. Part	Double-beaded partition (E&CB2S)
DET	Double end trimmed
Dim	Dimension
Dkg	Decking
D/S or D/Sdg	Dropped siding
EB1S	Edge bead one side
EB2S	Edge bead two sides
E&CB1S	Edge and center bead one side
E&CB2S	Edge and center bead two sides
E&CV1S	Edge and center vee one side
E&CV2S	Edge and center vee two sides
EE	Eased edges
EG	Edge (vertical) grain
EM	End matched
EV1S	Edge vee one side
EV2S	Edge vee two sides
Fac	Factory
FAS	Free alongisde (named vessel)
FBM	Foot or feet board measure
FG	Flat (slash) grain
Fig	Flooring
FOB	Free on board (named point)
FOHC	Free of heart center or centers
FOK	Free of knots
Frt	Freight
Ft	Foot or feet
GM	Grade Marked
G/R or G/Rfg	Grooved roofing
HB	Hollow back
H&M	Hit-and-miss
Hrt	Heart
Hrt CC	Heart cubical content
Hrt FA	Heart facial area
Hrt G	Heart girth
IN	Inch or inches
J&P	Joists and planks
KD	Kiln-dried
LCLL	Less than carload
LFT or Lin. Ft.	Linear foot or feet
Lgr	Longer
Lgth	Length
Lin	Linear
Lng	Lining
M	Thousand
MBM	Thousand (feet) board measure

(continued on next page)

Figure 6-2 Lumber Industry Abbreviations *(continued from previous page)*

MC	Moisture content
Merch	Merchantable
Mldg	Moulding
No	Number
N1E	Nosed one edge
N2E	Nosed two edges
Og	Ogee
Ord	Order
Part	Paragraph
Part	Partition
Pat	Pattern
Pc	Piece
Pcs	Pieces
PE	Plain end
PO	Purchase order
P&T	Post and timbers
Reg	Regular
Res	Resawed or resawn
Rfg	Roofing
Rgh	Rough
R/L	Random lengths
R/W	Random widths
R/W&L	Random widths and lengths
Sdg	Siding
Sel	Select
S&E	Side and Edge (surfaced on)
SE Sdg	Square edge siding
SE & S	Square edge and sound
S/L or S/LAP	Shiplap
SL&C	Shipper's load and count
SM. or Std. M	Standard Matched
Spec	Specifications
Std	Standard
Stpg	Stepping
Str. or Struc	Structural
S1E	Surfaced one edge
S1S	Surfaced one side
S1S1E	Surfaced one side and one edge
S1S2E	Surfaced one side and two edges
S2E	Surfaced two edges
S2S	Surfaced two sides
S2S1E	Surfaced two sides and one edge
S2S&CM	Surfaced two sides and center matched
S2S&SM	Surfaced two sides and standard match
S4S	Surfaced four sides
S4S&CS	Surfaced four sides and caulking seam
T&G	Tongued and grooved
VG	Vertical grain
Wdr	Wider
Wt	Weight

(courtesy How to Estimate Building Losses and Construction Costs, *Paul I. Thomas, Prentice Hall)*

Division 6

Figure 6-3   Number of Board Feet in Various Sized Pieces of Lumber

This table shows how many board feet are in various pieces of lumber for the lengths indicated.

Size (in inches)	Length of Piece							
	8'	10'	12'	14'	16'	18'	20'	22'
2 x 3	4	5	6	7	8	9	10	11
2 x 4	5-1/3	6-2/3	8	9-1/3	10-2/3	12	13-1/3	14-2/3
2 x 6	8	10	12	14	16	18	20	22
2 x 8	10-2/3	13-1/3	16	18-2/3	21-1/3	24	26-2/3	29-1/3
2 x 12	16	20	24	28	32	36	40	44
3 x 4	8	10	12	14	16	18	20	22
3 x 6	12	15	18	21	24	27	30	33
3 x 8	16	20	24	28	32	36	40	44
3 x 10	20	25	30	35	40	45	50	55
3 x 12	24	30	36	42	48	54	60	66
4 x 4	10-2/3	13-1/3	16	18-2/3	21-1/3	24	26-2/3	29-1/3
4 x 6	16	20	24	28	32	36	40	44
4 x 8	21-1/3	26-2/3	32	37-1/3	42-2/3	48	53-1/3	58-2/3
6 x 6	24	30	36	42	48	54	60	66
6 x 8	32	40	48	56	64	72	80	88
6 x 10	40	50	60	70	80	90	100	110
8 x 8	42-2/3	53-1/3	64	74-2/3	85-1/2	96	106-2/3	117-1/3
8 x 10	53-1/3	66-2/3	80	93-1/3	106-2/3	120	133-1/3	146-2/3
8 x 12	64	80	96	112	128	144	160	176
10 x 10	66-2/3	83-1/3	100	116-2/3	133-1/3	150	166-2/3	183-1/3
10 x 12	80	100	120	140	160	180	200	220
12 x 14	112	140	168	196	224	252	280	308

Figure 6-4 Board Foot Multipliers

This table is used for computing the number of board feet in any length of dimension lumber. Find the size of lumber to be used, and read across for the multiplier. For example, to find out how many board feet of lumber are in an 8' length of 2 × 4 lumber, look for the multiplier for 2 × 4's (0.667), and multiply it by the length of lumber (8'). The result is 0.667 × 8, or 5.336 board feet.

Nominal Size (in inches)	Multiply length by	Nominal Size (in inches)	Multiply length by
2 x 2	0.333	4 x 4	1.333
2 x 3	0.500	4 x 6	2.000
2 x 4	0.667	4 x 8	2.667
2 x 6	1.000	4 x 10	3.333
2 x 8	1.333	4 x 12	4.000
2 x 10	1.667		
2 x 12	2.000	6 x 6	3.000
		6 x 8	4.000
3 x 3	.0750	6 x 10	5.000
3 x 4	1.000	6 x 12	6.000
3 x 6	1.500		
3 x 8	2.000	8 x 8	5.333
3 x 10	2.500	8 x 10	6.667
3 x 12	3.000	8 x 12	8.000

Figure 6-5 Factors for the Board Foot Measure of Floor or Ceiling Joists

This table allows you to compute the amount of lumber (in board feet) for the floor or ceiling joists of a known area, based on the size of lumber and the spacing. To use this chart, simply multiply the area of the room or building in question by the factor across from the appropriate board size and spacing.

Ceiling Joists			
Joist Size	Inches On Center	Board Feet per Square Foot of Ceiling Area	Nails Lbs. per MBM
2" x 4"	12"	.78	17
	16"	.59	19
	20"	.48	19
2" x 6"	12"	1.15	11
	16"	.88	13
	20"	.72	13
	24"	.63	13
2" x 8"	12"	1.53	9
	16"	1.17	9
	20"	.96	9
	24"	.84	9
2" x 10"	12"	1.94	7
	16"	1.47	7
	20"	1.21	7
	24"	1.04	7
3" x 8"	12"	2.32	6
	16"	1.76	6
	20"	1.44	6
	24"	1.25	6

Figure 6-6 Floor and Ceiling Joist Calculation Tables

This table provides the amount of lumber (in board feet) for floor or ceiling joists based on the size of lumber and the spacing.

Floor Area	Ceiling Joists														
	2" x 4" On Center			2" x 6" On Center			2" x 8" On Center			2" x 10" On Center			2" x 12" On Center		
	16"	20"	24"	16"	20"	24"	16"	20"	24"	16"	20"	24"	16"	20"	24"
200	110	88	73	170	136	114	220	176	147	270	216	180	330	264	220
300	165	132	110	255	204	170	330	264	220	405	324	270	495	396	330
400	220	176	147	340	272	227	440	352	293	540	432	360	660	528	440
500	275	220	183	425	340	283	550	440	367	695	540	450	825	660	550
600	330	264	220	510	408	340	660	528	440	810	648	540	990	792	660
700	385	308	257	595	476	397	770	616	514	945	756	630	1155	924	770
800	440	352	293	680	544	454	880	704	587	1080	864	720	1320	1056	880
900	495	396	330	765	612	510	990	792	660	1215	972	810	1485	1188	990
1000	550	440	367	850	680	567	1100	880	734	1350	1080	900	1650	1320	1100
1100	605	484	404	935	748	624	1210	968	807	1485	1188	990	1815	1415	1210
1200	660	528	440	1020	816	680	1320	1056	880	1620	1296	1080	1980	1584	1320
1300	715	572	477	1105	884	737	1430	1144	954	1755	1404	1170	2145	1716	1430
1400	770	616	514	1190	952	794	1540	1232	1027	1890	1512	1260	2310	1848	1540
1500	825	660	550	1275	1020	850	1650	1320	1100	2025	1620	1350	2475	1980	1650
1600	880	704	587	1360	1088	907	1760	1408	1177	2160	1728	1440	2640	2112	1760
1700	935	748	624	1445	1156	964	1870	1496	1247	2295	1836	1530	2805	2244	1870
1800	990	792	660	1530	1224	1021	1980	1584	1320	2430	1944	1620	2970	2376	1980
1900	1045	836	697	1615	1292	1077	2090	1672	1394	2565	2052	1710	3135	2508	2090
2000	1100	880	734	1700	1360	1134	2200	1760	1467	2700	2160	1800	3300	2640	2200
2100	1155	924	770	1785	1428	1190	2310	1848	1540	2835	2268	1890	3465	2772	2310
2200	1210	968	807	1870	1496	1247	2420	1936	1614	2970	2376	1980	3630	2904	2420
2300	1265	1012	844	1955	1564	1304	2530	2024	1688	3105	2484	2070	3795	3036	2530
2400	1320	1056	880	2040	1632	1360	2640	2112	1760	3240	2592	2160	3960	3168	2640
2500	1375	1100	917	2125	1700	1417	2750	2200	1834	3375	2700	2250	4125	3300	2750
2600	1430	1144	953	2210	1768	1474	2860	2288	1908	3510	2808	2340	4290	3432	2860
2700	1485	1188	990	2295	1836	1530	2970	2376	1980	3645	2916	2430	4455	3564	2970
2800	1540	1232	1027	2380	1904	1587	3080	2464	2054	3780	3024	2520	4620	3696	3080
2900	1595	1276	1064	2465	1972	1644	3190	2552	2128	3915	3132	2610	4785	3828	3190
3000	1650	1320	1100	2550	2040	1700	3300	2640	2200	4050	3240	2700	4950	3960	3300

(courtesy Estimating Tables for Home Building, *Paul I. Thomas, Craftsman Book Company.*)

Division 6

Figure 6-7   Pieces of Studding Required for Walls, Floors, or Ceilings

This table lists the number of pieces of studding or furring needed for framing walls, floors, ceilings, partitions, etc., based on the length of the item being framed and its required spacing. Note that no provisions have been made for the doubling of studs at corners, window openings, door openings, etc. These must be added as called for.

Length of Wall, Floor or Ceiling (in feet)	On Center Spacing			
	12"	16"	20"	24"
8	9	7	6	5
9	10	8	6	6
10	11	9	7	6
11	12	9	8	7
12	13	10	8	7
13	14	11	9	8
14	15	12	9	8
15	16	12	10	9
16	17	13	11	9
17	18	14	11	10
18	19	15	12	10
19	20	15	12	11
20	21	16	13	11
21	22	17	14	12
22	23	18	14	12
23	24	18	15	13
24	25	19	15	13
25	26	20	16	14
26	27	21	17	14
27	28	21	17	15
28	29	22	18	15
29	30	23	18	16
30	31	24	19	16
32	33	25	20	17
34	35	27	22	18
36	37	28	23	19

(courtesy How to Estimate Building Losses and Construction Costs, *Paul I. Thomas, Prentice Hall)*

Figure 6-8 Multipliers for Computing Required Studding or Furring

This table allows you to compute the number of pieces of studding or furring required for framing walls, floors, roofs, partitions, etc.

Spacing (in inches)	Factor by Which to Multiply the Width of Room, Length of Wall or Roof	
12	1.00	
16	.75	
18	.67	Add one for
20	.60	end in each
24	.50	case

Examples:	Answers:
A room 16' wide requires studs 16" O.C.	16' x .75 + 1 = 13 joists
A wall 30' long requires studs 18" O.C.	30' x .67 + 1 = 21 studs
A gable roof 40' long requires studs 24" O.C.	40' x .50 + 1 = 21 rafters

(courtesy How to Estimate Building Losses and Construction Costs, *Paul I. Thomas, Prentice Hall*)

Division 6

Figure 6-9 Partition Framing

This table can be used to compute the board feet of lumber required for each square foot of wall area to be framed, based on the lumber size and spacing design.

Stud Size	Inches on Center	Studs Including Sole and Cap Plates		Horizontal Bracing in All Partitions		Horizontal Bracing in Bearing Partitions Only	
		Board Feet per Square Foot of Partition Area	Lbs. Nails per MBM of Stud Framing	Board Feet per Square Foot of Partition Area	Lbs. Nails per MBM of Bracing	Board Feet per Square Foot of Partition Area	Lbs. Nails per MBM of Bracing
2" x 3"	12"	.91	25	.04	145	.01	145
	16"	.83	25	.04	111	.01	111
	20"	.78	25	.04	90	.01	90
	24"	.76	25	.04	79	.01	79
2" x 4"	12"	1.22	19	.05	108	.02	108
	16"	1.12	19	.05	87	.02	87
	20"	1.05	19	.05	72	.02	72
	24"	1.02	19	.05	64	.02	64
2" x 6"	16"	1.38	19			.04	59
	20"	1.29	16			.04	48
	24"	1.22	16			.04	43
2" x 4" Staggered	8"	1.69	22				
3" x 4"	16"	1.35	17				
2" x 4" 2" Way	16"	1.08	19				

Figure 6-10 Exterior Wall Stud Framing

This table allows you to compute the board feet of lumber required for each square foot of exterior wall to be framed based on the lumber size and spacing design.

Stud Size	Inches On Center	Studs Including Corner Bracing		Horizontal Bracing Midway Between Plates	
		Board Feet per Square Foot of Ext. Wall Area	Lbs. of Nails per MBM of Stud Framing	Board Feet per Square Foot of Ext. Wall Area	Lbs. of Nails per MBM of Bracing
2" x 3"	16"	.78	30	.03	117
	20"	.74	30	.03	97
	24"	.71	30	.03	85
2" x 4"	16"	1.05	22	.04	87
	20"	.98	22	.04	72
	24"	.94	22	.04	64
2" x 6"	16"	1.51	15	.06	59
	20"	1.44	15	.06	48
	24"	1.38	15	.06	43

Figure 6-11 Furring Quantities

This table provides multiplication factors for converting square feet of wall area requiring furring into board feet of lumber. These figures are based on lumber size and spacing.

Size	Board Feet per Square Feet of Wall Area				Lbs. Nails per MBM of Furring
	Spacing Center to Center				
	12"	16"	20"	24"	
1" x 2"	.18	.14	.11	.10	55
1" x 3"	.28	.21	.17	.14	37

Figure 6-12 Floor Framing

This table can be used to compute the board feet required for each square foot of the floor area, based on the size of the lumber to be used and spacing design.

		Floor Joists		Block Over Main Bearing	
Joist Size	Inches On Center	Board Feet per Square Foot of Floor Area	Nails Lbs. per MBM	Board Feet per Square Foot of Floor Area	Nails Lbs. per MBM Blocking
2" x 6"	12"	1.28	10	.16	133
	16"	1.02	10	.03	95
	20"	.88	10	.03	77
	24"	.78	10	.03	57
2" x 8"	12"	1.71	8	.04	100
	16"	1.36	8	.04	72
	20"	1.17	8	.04	57
	24"	1.03	8	.05	43
2" x 10"	12"	2.14	6	.05	79
	16"	1.71	6	.05	57
	20"	1.48	6	.06	46
	24"	1.30	6	.06	34
2" x 12"	12"	2.56	5	.06	66
	16"	2.05	5	.06	47
	20"	1.77	5	.07	39
	24"	1.56	5	.07	29
3" x 8"	12"	2.56	5	.04	39
	16"	2.05	5	.05	57
	20"	1.77	5	.06	45
	24"	1.56	5	.06	33
3" x 10"	12"	3.20	4	.05	72
	16"	2.56	4	.07	46
	20"	2.21	4	.07	36
	24"	1.95	4	.08	26

Figure 6-13 Lumber (in Board Feet) Required for Framing a
One-Story Structure

This table shows the total number of board feet required to frame a typical one-story structure
based on ground floor area, spacing, and building style. The figures include the shoe and plate.

| Ground Floor Area | 2" x 4" Exterior Wall Studs | | | | 2" x 4" Partition Studs | | Total FBM Wall & Partition Studs, 16" O.C. | |
| | Spaced 16" O.C. | | Spaced 24" O.C. | | Spacing On Center | | | |
	Gables Included	No Gables	Gables Included	No Gables	16"	24"	Gables Included	No Gables
200	419	402	313	300	132	88	551	534
300	521	496	389	370	198	132	719	694
400	603	570	450	425	264	176	867	834
500	678	637	506	475	330	220	1008	967
600	747	697	558	520	396	264	1143	1093
700	809	750	604	560	462	308	1271	1212
800	871	804	650	600	528	352	1400	1332
900	925	850	691	635	594	396	1520	1444
1000	981	898	732	670	660	440	1641	1558
1100	1037	945	774	705	726	484	1763	1671
1200	1085	985	810	735	792	528	1877	1777
1300	1134	1025	847	765	858	572	1992	1883
1400	1182	1065	883	795	924	616	2106	1990
1500	1225	1100	914	820	990	660	2215	2090
1600	1273	1140	950	850	1056	704	2329	2196
1700	1314	1173	981	875	1122	748	2436	2295
1800	1356	1206	1013	900	1188	792	2544	2394
1900	1400	1240	1044	925	1254	836	2654	2494
2000	1440	1273	1075	950	1320	880	2760	2593
2100	1475	1300	1100	970	1386	924	2861	2686
2200	1516	1333	1133	995	1452	968	2968	2785
2300	1550	1360	1158	1015	1518	1012	3068	2878
2400	1594	1394	1190	1040	1584	1056	3178	2978
2500	1628	1420	1216	1060	1650	1100	3278	3070
2600	1664	1447	1242	1080	1716	1144	3380	3163
2700	1700	1474	1269	1100	1782	1188	3482	3256
2800	1733	1500	1295	1120	1848	1232	3581	3348
2900	1768	1528	1320	1140	1914	1276	3682	3442
3000	1811	1561	1353	1165	1980	1320	3791	3541

(courtesy Estimating Tables for Home Building, *Paul I. Thomas, Craftsman Book Company)*

Division 6

Figure 6-14 Lumber (in Board Feet) Required for Framing a
One-and-a-Half Story Structure

This table shows the total number of board feet required to frame a typical one-and-a-half story structure based on ground floor area, spacing, and building style. The figures include the shoe and plate.

Ground Floor Area	2" x 4" Exterior Wall Studs				2" x 4" Partition Studs Spacing On Center		Total FBM Wall & Partition Studs, 16" O.C.	
	Spaced 16" O.C.		Spaced 24" O.C.					
	Gables Included	No Gables	Gables Included	No Gables	16"	24"	Gables Included	No Gables
200	713	696	541	528	234	154	947	930
300	883	858	670	651	351	231	1234	1209
400	1019	986	773	748	468	308	1487	1454
500	1143	1102	867	836	585	385	1728	1687
600	1256	1206	953	915	702	462	1958	1908
700	1360	1300	1053	986	819	539	2178	2119
800	1459	1392	1106	1056	936	616	2395	2328
900	1550	1475	1174	1118	1053	693	2603	2528
1000	1638	1555	1241	1179	1170	770	2808	2725
1100	1728	1636	1310	1241	1287	847	3015	2923
1200	1806	1706	1369	1294	1404	924	3210	3110
1300	1884	1775	1428	1346	1521	1000	3405	3296
1400	1961	1844	1488	1400	1638	1078	3600	3482
1500	2027	1902	1537	1443	1755	1155	3782	3657
1600	2105	1972	1596	1496	1872	1232	3977	3844
1700	2171	2030	1646	1540	1990	1310	4161	4020
1800	2238	2088	1697	1584	2106	1386	4344	4194
1900	2305	2146	1744	1628	2223	1463	4528	4369
2000	2371	2204	1797	1672	2340	1540	4711	4544
2100	2425	2250	1838	1707	2457	1617	4882	4707
2200	2492	2309	1890	1751	2574	1694	5066	4883
2300	2545	2355	1930	1786	2691	1771	5236	5046
2400	2613	2413	1980	1830	2808	1848	5421	5221
2500	2668	2460	2022	1866	2925	1925	5593	5385
2600	2723	2506	2062	1900	3042	2000	5765	5548
2700	2777	2552	2105	1936	3159	2080	5936	5711
2800	2831	2598	2146	1971	3276	2156	6107	5874
2900	2885	2645	2186	2006	3393	2233	6278	6038
3000	2953	2703	2238	2050	3510	2310	6463	6213

(courtesy Estimating Tables for Home Building, *Paul I. Thomas, Craftsman Book Company)*

Figure 6-15 Lumber (in Board Feet) Required for Framing a Two-Story Structure

This table shows the total number of board feet required to frame a typical two-story structure based on ground floor area, spacing, and building style. The figures include the shoe and plate.

| Ground Floor Area | 2" x 4" Exterior Wall Studs | | | | 2" x 4" Partition Studs | | Total FBM Wall & Partition Studs, 16" O.C. | |
| | Spaced 16" O.C. | | Spaced 24" O.C. | | Spacing On Center | | | |
	Gables Included	No Gables	Gables Included	No Gables	16"	24"	Gables Included	No Gables
200	821	804	613	600	264	176	1085	1068
300	1017	992	759	740	396	264	1413	1388
400	1172	1140	875	850	528	352	1700	1668
500	1315	1274	981	950	660	440	1975	1934
600	1444	1394	1078	1040	792	528	2236	2186
700	1559	1500	1164	1120	924	616	2483	2424
800	1675	1608	1250	1200	1056	704	2731	2664
900	1775	1700	1326	1270	1188	792	2963	2888
1000	1879	1796	1402	1340	1320	880	3200	3116
1100	1982	1890	1479	1410	1452	968	3434	3342
1200	2070	1970	1545	1470	1584	1056	3654	3554
1300	2159	2050	1612	1530	1716	1144	3875	3766
1400	2247	2130	1678	1590	1848	1232	4095	3980
1500	2325	2200	1734	1640	1980	1320	4305	4180
1600	2413	2280	1800	1700	2112	1408	4525	4392
1700	2487	2346	1856	1750	2244	1496	4731	4590
1800	2562	2412	1913	1800	2376	1584	4938	4788
1900	2639	2480	1969	1850	2508	1672	5147	4988
2000	2713	2546	2025	1900	2640	1760	5353	5186
2100	2775	2600	2071	1940	2772	1848	5547	5372
2200	2849	2666	2128	1990	2904	1936	5753	5570
2300	2910	2720	2173	2030	3036	2024	5946	5756
2400	2988	2788	2230	2080	3168	2112	6156	5956
2500	3048	2840	2276	2121	3300	2200	6348	6140
2600	3111	2894	2322	2160	3432	2288	6543	6326
2700	3173	2948	2369	2200	3564	2376	6737	6512
2800	3233	3000	2415	2240	3696	2464	6929	6696
2900	3296	3056	2460	2280	3828	2552	7124	6884
3000	3372	3122	2518	2330	3960	2640	7332	7082

(*courtesy* Estimating Tables for Home Building, *Paul I. Thomas, Craftsman Book Company*)

Division 6

Figure 6-16 Flat Roof Framing

This table can be used to compute the amount of lumber in board feet required to frame a flat roof, based on the size of lumber required.

Flat Roof Framing			
Joist Size	Inches On Center	Board Feet per Square Foot of Ceiling Area	Nails Lbs. per MBM
2" x 6"	12"	1.17	10
	16"	.91	10
	20"	.76	10
	24"	.65	10
2" x 8"	12"	1.56	8
	16"	1.21	8
	20"	1.01	8
	24"	.86	8
2" x 10"	12"	1.96	6
	16"	1.51	6
	20"	1.27	6
	24"	1.08	6
2" x 12"	12"	2.35	5
	16"	1.82	5
	20"	1.52	5
	24"	1.30	5
3" x 8"	12"	2.35	5
	16"	1.82	5
	20"	1.52	5
	24"	1.30	5
3" x 10"	12"	2.94	4
	16"	2.27	4
	20"	1.90	4
	24"	1.62	4

MBF, MFBM = Thousand feet board measure

Figure 6-17 Conversion Factors for Wood Joists and Rafters

This table is used to compute actual lengths or board feet of lumber for roofs of various inclines. To compute actual rafter quantities for inclined roofs, multiply the quantities figured for a flat roof by the factors as shown in the table. Please note that this table does not include quantities for cantilevered overhangs.

Roof Slope	Approximate Angle	Factor	Roof Slope	Approximate Angle	Factor
Flat	0°	1.000	12 in 12	45.0°	1.414
1 in 12	4.8°	1.003	13 in 12	47.3°	1.474
2 in 12	9.5°	1.014	14 in 12	49.4°	1.537
3 in 12	14.0°	1.031	15 in 12	51.3°	1.601
4 in 12	18.4°	1.054	16 in 12	53.1°	1.667
5 in 12	22.6°	1.083	17 in 12	54.8°	1.734
6 in 12	26.6°	1.118	18 in 12	56.3°	1.803
7 in 12	30.3°	1.158	19 in 12	57.7°	1.873
8 in 12	33.7°	1.202	20 in 12	59.0°	1.943
9 in 12	36.9°	1.250	21 in 12	60.3°	2.015
10 in 12	39.8°	1.302	22 in 12	61.4°	2.088
11 in 12	42.5°	1.357	23 in 12	62.4°	2.162

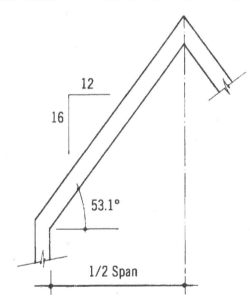

Inclined to Horizontal Sections

Figure 6-18   Allowance Factors for Roof Overhangs

Horizontal Span	Roof Overhang Measured Horizontally							
	0'-6"	1'-0"	1'-6"	2'-0"	2'-6"	3'-0"	3'-6"	4'-0"
6'	1.083	1.167	1.250	1.333	1.417	1.500	1.583	1.667
7'	1.071	1.143	1.214	1.286	1.357	1.429	1.500	1.571
8'	1.063	1.125	1.188	1.250	1.313	1.375	1.438	1.500
9'	1.056	1.111	1.167	1.222	1.278	1.333	1.389	1.444
10'	1.050	1.100	1.150	1.200	1.250	1.300	1.350	1.400
11'	1.045	1.091	1.136	1.182	1.227	1.273	1.318	1.364
12'	1.042	1.083	1.125	1.167	1.208	1.250	1.292	1.333
13'	1.038	1.077	1.115	1.154	1.192	1.231	1.269	1.308
14'	1.036	1.071	1.107	1.143	1.179	1.214	1.250	1.286
15'	1.033	1.067	1.100	1.133	1.167	1.200	1.233	1.267
16'	1.031	1.063	1.094	1.125	1.156	1.188	1.219	1.250
17'	1.029	1.059	1.088	1.118	1.147	1.176	1.206	1.235
18'	1.028	1.056	1.083	1.111	1.139	1.167	1.194	1.222
19'	1.026	1.053	1.079	1.105	1.132	1.158	1.184	1.211
20'	1.025	1.050	1.075	1.100	1.125	1.150	1.175	1.200
21'	1.024	1.048	1.071	1.095	1.119	1.143	1.167	1.190
22'	1.023	1.045	1.068	1.091	1.114	1.136	1.159	1.182
23'	1.022	1.043	1.065	1.087	1.109	1.130	1.152	1.174
24'	1.021	1.042	1.063	1.083	1.104	1.125	1.146	1.167
25'	1.020	1.040	1.060	1.080	1.100	1.120	1.140	1.160
26'	1.019	1.038	1.058	1.077	1.096	1.115	1.135	1.154
27'	1.019	1.037	1.056	1.074	1.093	1.111	1.130	1.148
28'	1.018	1.036	1.054	1.071	1.089	1.107	1.125	1.143
29'	1.017	1.034	1.052	1.069	1.086	1.103	1.121	1.138
30'	1.017	1.033	1.050	1.067	1.083	1.100	1.117	1.133
32'	1.016	1.031	1.047	1.063	1.078	1.094	1.109	1.125

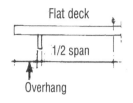

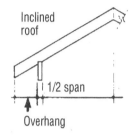

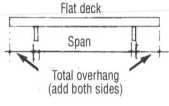

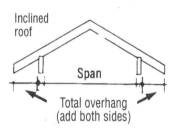

Overhang Sections

Figure 6-19 Roofing Factors When Rate of Rise is Known

No. Inches Rise per Foot of Run	*Pitch	Rafter Length in Inches per Foot of Run	To Obtain Rafter Length, Multiply Run (in feet) By:
3.0 3.5	1/8	12.37 ÷ 12 =	1.030 1.045
4.0 4.5	1/6	12.65	1.060 1.075
5.0 5.5	5/24	13.00	1.090 1.105
6.0 6.5	1/4	13.42	1.120 1.140
7.0 7.5	7/24	13.89	1.160 1.185
8.0 8.5	1/3	14.42	1.201 1.230
9.0 9.5	3/8	15.00	1.250 1.280
10.0 10.5	5/12	15.62	1.301 1.335
11.0 11.5	11/24	16.28	1.360 1.390
12.0	1/2	16.97	1.420

* Pitch is determined by dividing the inches rise-per-foot-of-run by 24. (Most roofs of dwellings are constructed with a pitch of 1/8, 1/6, and 1/4, 1/3, or 1/2.)

(courtesy How to Estimate Building Losses and Construction Costs, *Paul I. Thomas, Prentice Hall)*

Division 6

Figure 6-20 Pitched Roof Framing

This table is used to compute the board feet of lumber required per square foot of roof area, based on the required spacing and the lumber size to be used.

Rafters Including Collar Ties, Hip and Valley Rafters, Ridge Poles

Spacing Center to Center

Rafter Size	12" Board Feet per Square Foot of Roof Area	12" Nails Lbs. per MBM	16" Board Feet per Square Foot of Roof Area	16" Nails Lbs. per MBM	20" Board Feet per Square Foot of Roof Area	20" Nails Lbs. per MBM	24" Board Feet per Square Foot of Roof Area	24" Nails Lbs. per MBM
2" x 4"	.89	17	.71	17	.59	17	.53	17
2" x 6"	1.29	12	1.02	12	.85	12	.75	12
2" x 8"	1.71	9	1.34	9	1.12	9	.98	9
2" x 10"	2.12	7	1.66	7	1.38	7	1.21	7
2" x 12"	2.52	6	1.97	6	1.64	6	1.43	6
3" x 8"	2.52	6	1.97	6	1.64	6	1.43	6
3" x 10"	3.13	5	2.45	5	2.02	5	1.78	5

Figure 6-21   Board Feet Required for Roof Framing – Truss Construction

This table shows the number of board feet of lumber required for roof framing using "W" type truss construction. This table is based on a truss spacing of 24" on center, the roof pitch, and the area being covered. Braces and struts are included.

| Ground Floor Area | 2" x 4" Rafter and Chord | | | | | | 2" x 6" Rafter ... 2" x 4" Chord | | | | | |
| | Roof Pitch | | | | | | Roof Pitch | | | | | |
	1/8	1/6	5/24	1/4	1/3	1/2	1/8	1/6	5/24	1/4	1/3	1/2
200	210	212	214	218	224	240	246	248	252	256	264	288
300	315	318	321	327	336	360	369	372	378	384	396	432
400	420	424	428	436	448	480	492	496	504	512	528	576
500	525	530	535	545	560	600	615	620	630	640	660	720
600	630	636	642	654	672	720	738	744	756	768	792	864
700	735	742	749	763	784	840	861	868	882	896	924	1008
800	840	848	856	872	896	960	984	992	1008	1024	1056	1152
900	945	954	963	981	1008	1080	1107	1116	1134	1152	1188	1296
1000	1050	1060	1070	1090	1120	1200	1230	1240	1260	1280	1320	1440
1100	1155	1166	1177	1199	1232	1320	1353	1364	1386	1408	1452	1584
1200	1260	1272	1284	1308	1344	1440	1476	1488	1512	1536	1584	1728
1300	1365	1378	1391	1417	1456	1560	1599	1612	1638	1664	1716	1872
1400	1470	1484	1498	1526	1568	1680	1722	1736	1764	1792	1848	2016
1500	1575	1590	1605	1635	1680	1800	1845	1860	1890	1920	1980	2160
1600	1680	1696	1712	1744	1792	1920	1968	1984	2016	2048	2112	2304
1700	1785	1802	1819	1853	1904	2040	2091	2108	2142	2176	2244	2448
1800	1890	1908	1926	1962	2016	2160	2214	2232	2268	2304	2376	2592
1900	1995	2014	2033	2071	2128	2280	2337	2356	2394	2432	2508	2736
2000	2100	2120	2140	2180	2240	2400	2460	2480	2520	2560	2640	2880
2100	2205	2226	2247	2289	2352	2520	2583	2604	2646	2688	2772	3024
2200	2310	2332	2354	2398	2464	2640	2706	2728	2772	2816	2904	3168
2300	2415	2438	2461	2507	2576	2760	2829	2852	2898	2944	3036	3312
2400	2520	2544	2568	2616	2688	2880	2952	2976	3024	3072	3168	3456
2500	2625	2650	2675	2725	2800	3000	3075	3100	3150	3200	3300	3600
2600	2730	2756	2782	2834	2912	3120	3198	3224	3276	3328	3432	3744
2700	2835	2862	2889	2943	3024	3240	3321	3348	3402	3456	3564	3888
2800	2940	2968	2996	3052	3136	3360	3444	3472	3528	3584	3696	4032
2900	3045	3074	3103	3161	3248	3480	3567	3596	3654	3712	3828	4176
3000	3150	3180	3210	3270	3360	3600	3690	3720	3780	3840	3960	4320

(*courtesy* Estimating Tables for Home Building, *Paul I. Thomas, Craftsman Book Company*)

Division 6

Figure 6-22   Board Feet Required for Roof Framing – 2 x 4 Construction

This table shows the number of board feet of lumber required for roof framing using 2 x 4 rafters, and including the ridge pole. The table is based on the ground floor area being covered, spacing, and roof pitch.

Ground Floor Area	16" Center to Center						24" Center to Center					
	Roof Pitch						Roof Pitch					
	1/8	1/6	5/24	1/4	1/3	1/2	1/8	1/6	5/24	1/4	1/3	1/2
200	114	116	120	124	132	156	78	80	82	86	92	108
300	171	174	180	186	198	234	117	120	123	129	138	162
400	228	232	240	248	264	312	156	160	164	172	184	216
500	285	290	300	310	330	390	195	200	205	215	230	270
600	342	348	360	372	396	468	234	240	246	258	276	324
700	399	406	420	434	462	546	273	280	287	301	322	378
800	456	464	480	496	528	624	312	320	328	344	368	432
900	513	522	540	558	594	702	351	360	369	387	414	486
1000	570	580	600	620	660	780	390	400	410	430	460	540
1100	627	638	660	682	726	858	429	440	451	473	506	594
1200	684	696	720	744	792	936	468	480	492	516	552	648
1300	741	754	780	806	858	1014	507	520	533	559	598	702
1400	798	812	840	868	924	1092	546	560	574	602	644	756
1500	855	870	900	930	990	1170	585	600	615	645	690	810
1600	912	928	960	992	1056	1248	624	640	656	688	736	864
1700	969	986	1020	1054	1122	1326	663	680	697	731	782	918
1800	1026	1044	1080	1116	1188	1404	702	720	738	774	828	972
1900	1083	1102	1140	1178	1254	1482	741	760	779	817	874	1026
2000	1140	1160	1200	1240	1320	1560	780	800	820	860	920	1080
2100	1197	1218	1260	1302	1386	1638	819	840	861	903	966	1134
2200	1254	1276	1320	1363	1452	1716	858	880	902	946	1012	1188
2300	1311	1334	1380	1426	1518	1794	897	920	943	989	1058	1242
2400	1368	1392	1440	1488	1584	1872	936	960	984	1032	1104	1296
2500	1425	1450	1500	1550	1650	1950	975	1000	1025	1075	1150	1350
2600	1482	1508	1560	1612	1716	2028	1014	1040	1066	1118	1196	1404
2700	1539	1566	1620	1674	1782	2106	1053	1080	1107	1161	1242	1458
2800	1596	1624	1680	1736	1848	2184	1092	1120	1148	1204	1288	1512
2900	1653	1682	1740	1798	1914	2262	1131	1160	1189	1247	1334	1566
3000	1710	1740	1800	1860	1980	2340	1170	1200	1230	1290	1380	1620

(courtesy Estimating Tables for Home Building, *Paul I. Thomas, Craftsman Book Company*)

Figure 6-23   Board Feet Required for Roof Framing – 2 x 6 Construction

This table shows the number of board feet of lumber required for roof framing using 2 x 6 rafters, and including the ridge pole. The table is based on the ground floor area being covered, spacing, and roof pitch.

Ground Floor Area	16" Center to Center						24" Center to Center					
	Roof Pitch						Roof Pitch					
	1/8	1/6	5/24	1/4	1/3	1/2	1/8	1/6	5/24	1/4	1/3	1/2
200	164	170	174	180	192	228	114	116	120	124	132	156
300	246	255	261	270	288	342	171	174	180	186	198	234
400	328	340	348	360	384	456	228	232	240	248	264	312
500	410	425	435	450	480	570	285	290	300	310	330	390
600	492	510	522	540	576	684	342	348	360	372	396	468
700	574	595	609	630	672	798	399	406	420	434	462	546
800	656	680	696	720	768	912	456	464	480	496	528	624
900	739	765	783	810	864	1026	513	522	540	558	594	702
1000	820	850	870	900	960	1140	570	580	600	620	660	780
1100	902	935	957	990	1056	1254	627	638	660	682	726	858
1200	984	1020	1044	1080	1152	1368	684	696	720	744	792	936
1300	1066	1105	1131	1170	1248	1482	741	754	780	806	858	1014
1400	1148	1190	1218	1260	1344	1596	798	812	840	868	924	1092
1500	1230	1274	1305	1350	1440	1710	855	870	900	930	990	1170
1600	1312	1360	1392	1440	1536	1824	912	928	960	992	1056	1248
1700	1394	1445	1479	1530	1632	1938	969	986	1020	1054	1122	1326
1800	1476	1530	1566	1620	1728	2052	1026	1044	1080	1116	1188	1404
1900	1558	1615	1653	1710	1824	2166	1083	1102	1140	1178	1254	1482
2000	1640	1700	1740	1800	1920	2280	1140	1160	1200	1240	1320	1560
2100	1722	1785	1827	1890	2016	2394	1197	1218	1260	1302	1386	1638
2200	1804	1870	1914	1980	2112	2508	1254	1276	1320	1364	1452	1716
2300	1886	1955	2001	2070	2208	2622	1311	1334	1380	1426	1518	1794
2400	1968	2040	2088	2160	2304	2736	1368	1392	1440	1488	1584	1872
2500	2050	2125	2175	2250	2400	2850	1425	1450	1500	1550	1650	1950
2600	2132	2210	2262	2340	2496	2964	1482	1508	1560	1612	1716	2028
2700	2214	2295	2349	2430	2592	3078	1539	1566	1620	1674	1782	2106
2800	2296	2380	2436	2520	2688	3192	1596	1624	1680	1736	1848	2184
2900	2378	2465	2523	2610	2784	3306	1653	1682	1740	1798	1914	2262
3000	2460	2550	2610	2700	2880	3420	1710	1740	1800	1860	1980	2340

(courtesy Estimating Tables for Home Building, *Paul I. Thomas, Craftsman Book Company*)

Division 6

Figure 6-24 Material Required for On-the-Job Cut Bridging – Linear Feet

Based on the size of the lumber used for joists, the total amount of lumber (in board feet) can be obtained for the various spacing of the joists using this table.

Size Joist in Inches	Spacing in Inches	Linear Feet per Set (2)	Linear Feet per Foot-of-Row
2 x 6	16	2.57	1.92
2 x 8	16	2.70	2.02
2 x 10	16	2.87	2.15
2 x 12	16	3.06	2.30
2 x 8	20	3.31	2.00
2 x 10	20	3.45	2.07
2 x 12	20	3.61	2.17
2 x 8	24	3.94	2.00
2 x 10	24	4.05	1.97
2 x 12	24	4.19	2.10

Note: Add to the total lineal feet developed from the table at least 10% cutting waste.

Example: A room 20 feet wide has two rows of bridging. The 2" x 12" joists are 16" on center.

$$2 \times 20 \text{ L.F.} \times 2.30 \text{ ft. per foot-of-row} = 92.0 \text{ L.F.}$$
$$\text{Add 10\% waste} \quad 9.2$$
$$\text{Total L.F.} = 101.2$$
$$\text{Round out to 101} \quad \text{L.F.}$$

(Per set method would be, 30 sets x 3.06 = 91.8 to which must be added 10% waste.)

(*courtesy* How to Estimate Building Losses and Construction Costs, *Paul I. Thomas, Prentice Hall*)

Figure 6-25 Board Feet Required for On-the-Job Cut Bridging

Based on the size of the lumber used for joists, the total lengths of lumber for various sized bridging can be obtained from this chart.

Cross Bridging—Board Feet per Square Foot of Floors, Ceiling or Flat Roof Area Nails — Pounds Per MBM of Bridging							
		1" x 3"		1" x 4"		2" x 3"	
Joist Size	Spacing	B.F.	Nails	B.F.	Nails	B.F.	Nails
2" x 8"	12"	.04	147	.05	112	.08	77
	16"	.04	120	.05	91	.08	61
	20"	.04	102	.05	77	.08	52
	24"	.04	83	.05	63	.08	42
2" x 10"	12"	.04	136	.05	103	.08	71
	16"	.04	114	.05	87	.08	58
	20"	.04	98	.05	74	.08	50
	24"	.04	80	.05	61	.08	41
2" x 12"	12"	.04	127	.05	96	.08	67
	16"	.04	108	.05	82	.08	55
	20"	.04	94	.05	71	.08	48
	24"	.04	78	.05	59	.08	39
3" x 8"	12"	.04	160	.05	122	.08	84
	16"	.04	127	.05	96	.08	66
	20"	.04	107	.05	81	.08	54
	24"	.04	86	.05	65	.08	44
3" x 10"	12"	.04	146	.05	111	.08	77
	16"	.04	120	.05	91	.08	62
	20"	.04	102	.05	78	.08	52
	24"	.04	83	.05	63	.08	42

Division 6

Figure 6-26 Board Feet of Sheathing and Subflooring

This table is used to compute the board feet of sheathing or subflooring required per square foot of roof, ceiling, or flooring.

Type	Size	Board Feet per Square Foot of Area	Diagonal			
			Lbs. Nails per MBM Lumber			
			Joist, Stud or Rafter Spacing			
			12"	16"	20"	24"
Surface 4 Sides (S4S)	1" x 4"	1.22	58	46	39	32
	1" x 6"	1.18	39	31	25	21
	1" x 8"	1.18	30	23	19	16
	1" x 10"	1.17	35	27	23	19
Tongue and Groove (T&G)	1" x 4"	1.36	65	51	43	36
	1" x 6"	1.26	42	33	27	23
	1" x 8"	1.22	31	24	20	17
	1" x 10"	1.20	36	28	24	19
Shiplap	1" x 4"	1.41	67	53	45	37
	1" x 6"	1.29	43	33	28	23
	1" x 8"	1.24	31	24	20	17
	1" x 10"	1.21	36	28	24	19

Figure 6-27 Roof Decking for All Buildings with a Roof Pitch of 1/8

Ground Floor Area	Roof Surface Area	Shiplap 1" By		Tongue & Groove Boards 1" By			Square Edge Boards 1" By		Plywood Insulation & Particleboard
		6" & 8"	10"	4"	6" & 8"	10"	4"	6" & 8"	
S.F.	S.F.	B.F.	B.F.	B.F.	B.F.	B.F.	B.F.	B.F.	S.F.
200	206	252	244	274	252	244	246	236	226
300	309	378	366	411	378	366	369	354	339
400	412	504	488	548	504	488	492	472	452
500	515	630	610	685	630	610	615	590	565
600	618	756	632	822	756	732	738	708	678
700	721	882	854	959	882	854	861	826	791
800	824	1008	976	1096	1008	976	984	944	904
900	927	1134	1098	1233	1134	1098	1107	1062	1017
1000	1030	1260	1220	1370	1260	1220	1230	1180	1130
1100	1133	1386	1342	1507	1386	1342	1353	1298	1243
1200	1236	1512	1464	1644	1512	1464	1476	1416	1356
1300	1339	1638	1586	1781	1638	1586	1600	1534	1469
1400	1442	1764	1708	1918	1764	1708	1722	1652	1582
1500	1545	1890	1830	2055	1890	1830	1845	1770	1695
1600	1648	2016	1952	2192	2016	1952	1968	1888	1808
1700	1751	2142	2074	2329	2142	2074	2091	2006	1921
1800	1854	2268	2196	2466	2268	2196	2214	2124	2034
1900	1957	2394	2318	2603	2394	2318	2337	2242	2147
2000	2060	2520	2440	2740	2520	2440	2460	2360	2260
2100	2163	2646	2562	2877	2646	2562	2583	2478	2373
2200	2266	2772	2684	3014	2772	2684	2706	2596	2486
2300	2369	2898	2806	3151	2898	2806	2829	2714	2600
2400	2472	3024	2928	3288	3024	2928	2953	2832	2712
2500	2575	3150	3050	3425	3150	3050	3075	2950	2825
2600	2678	3276	3172	3562	3276	3172	3198	3068	2938
2700	2781	3402	3294	3699	3402	3294	3321	3186	3051
2800	2884	3528	3416	3836	3528	3416	3440	3304	3164
2900	2987	3654	3538	3973	3654	3538	3567	3422	3277
3000	3090	3780	3660	4110	3780	3660	3690	3540	3390

All figures include milling and cutting waste.

(*courtesy* Estimating Tables for Home Building, *Paul I. Thomas, Craftsman Book Company*)

Division 6

Figure 6-28 Roof Decking for All Buildings with a Roof Pitch of 1/6

Ground Floor Area	Roof Surface Area	Shiplap 1" By		Tongue & Groove Boards 1" By			Square Edge Boards 1" By		Plywood Insulation & Particleboard
		6" & 8"	10"	4"	6" & 8"	10"	4"	6" & 8"	
S.F.	S.F.	B.F.	B.F.	B.F.	B.F.	B.F.	B.F.	B.F.	S.F.
200	212	258	250	282	258	250	252	244	234
300	318	387	375	423	387	375	378	366	351
400	424	516	500	564	516	500	504	488	468
500	530	645	625	705	645	625	630	610	585
600	636	774	750	846	774	750	756	732	702
700	742	900	875	987	900	875	882	854	819
800	848	1030	1000	1128	1030	1000	1008	976	936
900	954	1160	1125	1269	1160	1125	1134	1098	1053
1000	1060	1290	1250	1410	1290	1250	1260	1220	1170
1100	1166	1420	1375	1551	1420	1375	1386	1342	1287
1200	1272	1551	1500	1692	1550	1500	1512	1464	1404
1300	1378	1680	1625	1833	1680	1625	1638	1586	1521
1400	1484	1810	1750	1974	1810	1750	1764	1708	1638
1500	1590	1935	1875	2115	1935	1875	1890	1830	1755
1600	1696	2064	2000	2256	2064	2000	2016	1952	1872
1700	1802	2195	2125	2397	2195	2125	2142	2074	1989
1800	1908	2322	2250	2538	2322	2250	2268	2196	2106
1900	2014	2450	2375	2679	2450	2375	2394	2318	2223
2000	2120	2580	2500	2820	2580	2500	2520	2440	2340
2100	2226	2710	2625	2961	2710	2625	2646	2562	2457
2200	2332	2840	2750	3102	2840	2750	2772	2684	2574
2300	2438	2970	2875	3243	2970	2875	2898	2806	2691
2400	2544	3100	3000	3384	3100	3000	3024	2928	2808
2500	2650	3225	3125	3525	3225	3125	3150	3050	2925
2600	2756	3354	3250	3666	3354	3250	3276	3172	3042
2700	2862	3483	3375	3807	3483	3375	3402	3294	3159
2800	2968	3610	3500	3948	3610	3500	3528	3416	3276
2900	3074	3740	3625	4089	3740	3625	3654	3528	3393
3000	3180	3870	3750	4230	3870	3750	3780	3660	3510

All figures include milling and cutting waste.

(*courtesy* Estimating Tables for Home Building, *Paul I. Thomas, Craftsman Book Company*)

Figure 6-29 Roof Decking for All Buildings with a Roof Pitch of 5/24

Ground Floor Area	Roof Surface Area	Shiplap 1" By		Tongue & Groove Boards 1" By			Square Edge Boards 1" By		Plywood Insulation & Particleboard
		6" & 8"	10"	4"	6" & 8"	10"	4"	6" & 8"	
S.F.	S.F.	B.F.	B.F.	B.F.	B.F.	B.F.	B.F.	B.F.	S.F.
200	218	266	258	290	266	258	260	250	240
300	327	400	387	435	400	387	390	375	360
400	426	532	516	580	532	516	520	500	480
500	546	665	645	725	665	645	650	625	600
600	654	800	774	870	800	774	780	750	720
700	763	930	900	1015	930	900	910	875	840
800	872	1064	1030	1160	1064	1030	1040	1000	960
900	981	1200	1160	1305	1200	1160	1170	1125	1080
1000	1090	1330	1290	1450	1330	1290	1300	1250	1200
1100	1199	1463	1420	1595	1463	1420	1430	1375	1320
1200	1308	1600	1550	1740	1600	1550	1560	1500	1440
1300	1417	1730	1680	1885	1730	1680	1690	1625	1560
1400	1526	1860	1810	2030	1860	1810	1820	1750	1680
1500	1635	2000	1935	2175	2000	1935	1950	1875	1800
1600	1744	2130	2064	2320	2130	2064	2080	2000	1920
1700	1853	2260	2195	2465	2260	2195	2210	2125	2040
1800	1962	2400	2322	2610	2400	2322	2340	2250	2160
1900	2071	2530	2450	2755	2530	2450	2470	2375	2280
2000	2180	2660	2580	2900	2660	2580	2600	2500	2400
2100	2289	2800	2710	3045	2800	2710	2730	2625	2520
2200	2398	2930	2840	3190	2930	2840	2860	2750	2640
2300	2507	3060	2970	3335	3060	2970	2990	2875	2760
2400	2616	3200	3100	3480	3200	3100	3120	3000	2880
2500	2725	3325	3225	3625	3325	3225	3250	3125	3000
2600	2834	3460	3354	3770	3460	3354	3380	3250	3120
2700	2943	3590	3483	3915	3590	3483	3510	3375	3240
2800	3052	3724	3610	4060	3724	3610	3640	3500	3360
2900	3161	3860	3740	4205	3860	3740	3770	3625	3480
3000	3270	3990	3870	4350	3990	3870	3900	3750	3600

All figures include milling and cutting waste.

(courtesy Estimating Tables for Home Building, *Paul I. Thomas, Craftsman Book Company*)

Division 6

Figure 6-30   Roof Decking for All Buildings with a Roof Pitch of 1/4

Ground Floor Area	Roof Surface Area	Shiplap 1" By		Tongue & Groove Boards 1" By			Square Edge Boards 1" By		Plywood Insulation & Particleboard
		6" & 8"	10"	4"	6" & 8"	10"	4"	6" & 8"	
S.F.	S.F.	B.F.	B.F.	B.F.	B.F.	B.F.	B.F.	B.F.	S.F.
200	224	274	264	298	274	264	266	258	246
300	336	411	396	447	411	396	400	387	369
400	448	548	528	600	548	528	532	516	492
500	560	685	660	745	685	660	665	645	615
600	672	822	792	894	822	792	800	774	738
700	784	959	924	1043	959	924	930	900	861
800	896	1096	1056	1192	1096	1054	1064	1030	984
900	1008	1233	1188	1340	1233	1188	1200	1160	1107
1000	1120	1370	1320	1490	1370	1320	1330	1290	1230
1100	1232	1507	1452	1639	1507	1452	1463	1420	1353
1200	1344	1644	1584	1788	1644	1584	1600	1550	1476
1300	1456	1781	1716	1937	1781	1716	1730	1680	1600
1400	1568	1918	1848	2086	1918	1848	1860	1810	1722
1500	1680	2055	1980	2235	2055	1980	2000	1935	1845
1600	1792	2192	2112	2384	2192	2112	2130	2064	1968
1700	1904	2329	2244	2533	2329	2244	2260	2195	2091
1800	2016	2466	2376	2682	2466	2376	2400	2322	2214
1900	2128	2603	2508	2830	2603	2508	2530	2450	2337
2000	2240	2740	2640	2980	2740	2640	2660	2580	2460
2100	2352	2877	2772	3130	2877	2772	2800	2710	2583
2200	2464	3014	2904	3278	3014	2904	2930	2840	2706
2300	2576	3151	3036	3427	3151	3036	3060	2970	2829
2400	2688	3288	3168	3576	3288	3168	3200	3100	2952
2500	2800	3425	3300	3725	3425	3300	3325	3225	3075
2600	2912	3562	3432	3874	3562	3432	3460	3354	3198
2700	3024	3700	3564	4023	3700	3564	3590	3483	3321
2800	3136	3836	3696	4172	3836	3696	3724	3610	3440
2900	3248	3973	3828	4321	3973	3828	3860	3740	3567
3000	3360	4110	3960	4470	4110	3960	3990	3870	3690

All figures include milling and cutting waste.

(courtesy Estimating Tables for Home Building, *Paul I. Thomas, Craftsman Book Company*)

Figure 6-31 Roof Decking for All Buildings with a Roof Pitch of 1/3

Ground Floor Area	Roof Surface Area	Shiplap 1" By		Tongue & Groove Boards 1" By			Square Edge Boards 1" By		Plywood Insulation & Particleboard
		6" & 8"	10"	4"	6" & 8"	10"	4"	6" & 8"	
S.F.	S.F.	B.F.	B.F.	B.F.	B.F.	B.F.	B.F.	B.F.	S.F.
200	240	292	284	320	292	284	286	276	264
300	360	438	426	480	438	426	429	414	396
400	480	584	569	640	584	568	572	552	528
500	600	730	710	800	730	710	715	690	660
600	720	876	852	960	876	852	858	828	792
700	840	1022	994	1120	1022	994	1000	966	924
800	960	1170	1136	1280	1170	1136	1144	1104	1056
900	1080	1314	1278	1440	1314	1278	1287	1242	1188
1000	1200	1460	1420	1600	1460	1420	1430	1380	1320
1100	1320	1610	1562	1760	1610	1562	1573	1518	1452
1200	1440	1752	1704	1920	1752	1704	1716	1656	1584
1300	1560	1900	1846	2080	1900	1846	1859	1794	1716
1400	1680	2044	1990	2240	2044	1990	2000	1932	1848
1500	1800	2190	2130	2400	2190	2130	2145	2070	1980
1600	1920	2336	2272	2560	2336	2272	2288	2208	2112
1700	2040	2480	2414	2720	2480	2414	2430	2346	2244
1800	2160	2630	2556	2880	2630	2556	2574	2484	2376
1900	2280	2774	2700	3040	2774	2700	2717	2622	2508
2000	2400	2920	2840	3200	2920	2840	2860	2760	2640
2100	2520	3066	2982	3360	3066	2982	3000	2898	2772
2200	2740	3212	3124	3520	3212	3124	3146	3036	2904
2300	2760	3358	3266	3680	3358	3266	3290	3174	3036
2400	2880	3504	3410	3840	3504	3410	3430	3312	3168
2500	3000	3650	3550	4000	3650	3550	3575	3450	3300
2600	3120	3800	3692	4160	3800	3692	3720	3588	3432
2700	3240	3942	3834	4320	3942	3834	3860	3726	3564
2800	3360	4090	3976	4480	4090	3976	4000	3864	3696
2900	3480	4234	4118	4640	4234	4118	4150	4000	3828
3000	3600	4380	4260	4800	4380	4260	4290	4140	3960

All figures include milling and cutting waste.

(courtesy Estimating Tables for Home Building, *Paul I. Thomas, Craftsman Book Company)*

Division 6

Figure 6-32   Roof Decking for All Buildings with a Roof Pitch of 1/2

Ground Floor Area	Roof Surface Area	Shiplap 1" By		Tongue & Groove Boards 1" By			Square Edge Boards 1" By		Plywood Insulation & Particleboard
		6" & 8"	10"	4"	6" & 8"	10"	4"	6" & 8"	
S.F.	S.F.	B.F.	B.F.	B.F.	B.F.	B.F.	B.F.	B.F.	S.F.
200	284	346	336	378	346	336	338	326	312
300	426	520	504	567	520	504	507	489	468
400	568	690	672	756	690	672	676	652	624
500	710	865	840	945	865	840	845	815	780
600	852	1040	1008	1134	1040	1008	1014	978	936
700	994	1210	1176	1323	1210	1176	1183	1141	1092
800	1136	1385	1344	1512	1385	1344	1352	1304	1248
900	1278	1560	1512	1700	1560	1512	1521	1467	1404
1000	1420	1730	1680	1890	1730	1680	1690	1630	1560
1100	1562	1900	1848	2080	1900	1848	1859	1793	1716
1200	1704	2076	2016	2268	2076	2016	2028	1956	1872
1300	1846	2250	2184	2457	2250	2184	2200	2119	2028
1400	1990	2422	2352	2646	2422	2352	2370	2282	2184
1500	2130	2595	2520	2835	2595	2520	2535	2445	2340
1600	2272	2770	2688	3024	2770	2688	2705	2608	2500
1700	2414	2940	2856	3213	2940	2856	2875	2771	2652
1800	2556	3115	3024	3400	3115	3024	3040	2934	2810
1900	2700	3290	3192	3590	3290	3192	3210	3097	2964
2000	2840	3460	3360	3780	3460	3360	3380	3260	3120
2100	2982	3633	3528	3970	3633	3528	3550	3423	3276
2200	3124	3810	3696	4158	3810	3696	3720	3586	3432
2300	3266	3980	3864	4350	3980	3864	3890	3749	3590
2400	3410	4150	4032	4536	4150	4032	4056	3912	3744
2500	3550	4325	4200	4725	4325	4200	4225	4075	3900
2600	3692	4500	4368	4914	4500	4368	4395	4238	4056
2700	3834	4670	4536	5100	4670	4536	4565	4400	4212
2800	3976	4845	4704	5292	4845	4704	4730	4564	4368
2900	4118	5020	4872	5480	5020	4872	4900	4727	4524
3000	4260	5190	5040	5670	5190	5040	5070	4890	4680

All figures include milling and cutting waste.

(courtesy Estimating Tables for Home Building, *Paul I. Thomas, Craftsman Book Company*)

Figure 6-33 Wood Siding Factors

This table shows the factor by which the area to be covered is multiplied to determine the exact amount of surface material needed.

Item	Nominal Size	Width Overall	Face	Area Factor
Shiplap	1" x 6"	5-1/2"	5-1/8"	1.17
	1 x 8	7-1/4	6-7/8	1.16
	1 x 10	9-1/4	8-7/8	1.13
	1 x 12	11-1/4	10-7/8	1.10
Tongue and Groove	1 x 4	3-3/8	3-1/8	1.28
	1 x 6	5-3/8	5-1/8	1.17
	1 x 8	7-1/8	6-7/8	1.16
	1 x 10	9-1/8	8-7/8	1.13
	1 x 12	11-1/8	10-7/8	1.10
S4S	1 x 4	3-1/2	3-1/2	1.14
	1 x 6	5-1/2	5-1/2	1.09
	1 x 8	7-1/4	7-1/4	1.10
	1 x 10	9-1/4	9-1/4	1.08
	1 x 12	11-1/4	11-1/4	1.07
Solid Paneling	1 x 6	5-7/16	5-7/16	1.19
	1 x 8	7-1/8	6-3/4	1.19
	1 x 10	9-1/8	8-3/4	1.14
	1 x 12	11-1/8	10-3/4	1.12
Bevel Siding*	1 x 4	3-1/2	3-1/2	1.60
	1 x 6	5-1/2	5-1/2	1.33
	1 x 8	7-1/4	7-1/4	1.28
	1 x 10	9-1/4	9-1/4	1.21
	1 x 12	11-1/4	11-1/4	1.17

Note: This area factor is strictly so-called milling waste. The cutting and fitting waste must be added.

* 1" lap

(courtesy Western Wood Products Association)

Figure 6-34 Milling and Cutting Waste Factors for Wood Siding

This table lists the milling waste, and the amount of nails required for various types of wood siding. Amounts are per 1,000 FBM of wood siding.

Type of Siding	Nominal Size in Inches	Lap in Inches 1" Lap	Pounds Nails per 1,000 FBM	Percentage of Waste
Bevel Siding	1 x 4	1	25-6d common	63
	1 x 6	1	25-6d common	35
	1 x 8	1-1/4	20-8d common	35
	1 x 10	1-1/2	20-8d common	30
Rustic and Drop Siding	1 x 4	Matched	40-8d common	33
	1 x 6	Matched	30-8d common	25
	1 x 8	Matched	25-8d common	20
Vertical Siding	1 x 6	Matched	25-8d finish	20
	1 x 8	Matched	20-8d finish	18
	1 x 10	Matched	20-8d finish	15
Batten Siding*	1 x 8	Rough	25	5
	1 x 10	Rough	20	5
	1 x 12	Rough	20	5
	1 x 8	Dressed	25	13
	1 x 10	Dressed	20	11
	1 x 12	Dressed	20	10
Plywood Siding	1/4	Sheets		5-10
	3/8	Sheets	15 per MSF	5-10
	5/8	Sheets		5-10

* For 1" x 10" boards, allow 1,334 linear feet 1" x 2" joint strips for each 1,000 FBM of batten siding. Add 12 pounds 8d common nails.

(courtesy How to Estimate Building Losses and Construction Costs, *Paul I. Thomas, Prentice Hall*)

Figure 6-35 Rough Hardware Allowances

Average Material Cost Allowances for Rough Hardware as a Percentage of Carpentry Material Costs	
Minimum	0.5%
Maximum	1.5%

Figure 6-36 Installation Time in Labor-Hours for Joists and Decking

Description	Labor-Hours Pneumatic Nailed	Labor-Hours Manual Nailed	Unit
Joist Framing			
2" x 6"	.011	.013	L.F.
2" x 8"	.013	.015	L.F.
2" x 10"	.015	.018	L.F.
2" x 12"	.016	.018	L.F.
2" x 14"	.018	.021	L.F.
3" x 8"	—	.017	L.F.
3" x 12"	—	.027	L.F.
4" x 8"	—	.026	L.F.
4" x 12"	—	.036	L.F.
Bridging Wood or Steel, Joists			
16" On Center	.047	.062	Pr.
24" On Center	—	.057	Pr.
Sub Floor Plywood CDX			
1/2" Thick	.009	.011	S.F.
5/8" Thick	.010	.012	S.F.
3/4" Thick	.010	.013	S.F.
Boards Diagonal			
1" x 8"	—	.019	S.F.
1" x 10"	—	.018	S.F.
Wood Fiber T & G			
2' x 8' Planks			
1" Thick	—	.016	S.F.
1-3/8" Thick	—	.018	S.F.

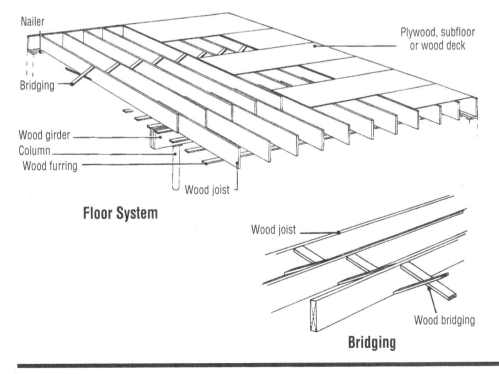

Nailer

Plywood, subfloor or wood deck

Bridging

Wood girder
Column
Wood furring

Wood joist

Floor System

Wood joist

Wood bridging

Bridging

Division 6

Figure 6-37 Installation Time in Labor-Hours for Columns, Beams, Girders, and Decking

Description	Labor-Hours	Unit
Columns		
6" x 6"	.074	L.F.
8" x 8"	.122	L.F.
10" x 10"	.178	L.F.
12" x 12"	.240	L.F.
Beams and Girders		
6" x 10"	.073	L.F.
8" x 16"	.142	L.F.
12" x 12"	.240	L.F.
10" x 16"	.213	L.F.
Wood Deck		
3" Nominal	.050	S.F.
4" Nominal	.064	S.F.
Laminated Wood Deck		
3" Nominal	.038	S.F.
4" Nominal	.049	S.F.

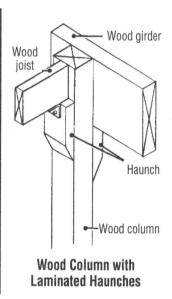

**Wood Column with
Laminated Haunches**

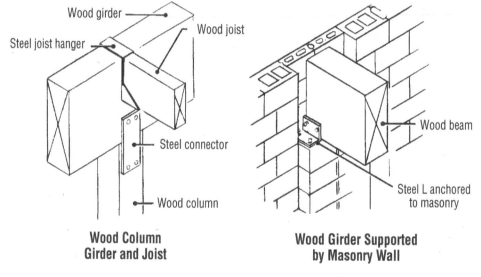

**Wood Column
Girder and Joist**

**Wood Girder Supported
by Masonry Wall**

(continued on next page)

Figure 6-37   Installation Time in Labor-Hours for Columns, Beams, Girders, and Decking *(continued from previous page)*

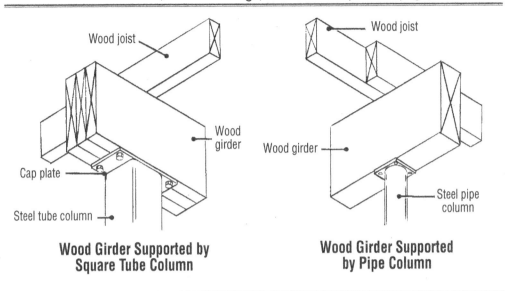

Wood Girder Supported by Square Tube Column

Wood Girder Supported by Pipe Column

Division 6

Figure 6-38 Installation Time in Labor-Hours for Wall Framing

Description	Labor-Hours	Unit
Wall Framing		
Headers over openings, 2 x 6	44.444	M.B.F.
2 x 6 pneumatic nailed	37.209	
2 x 8	35.556	
2 x 8 pneumatic nailed	29.630	
2 x 10	30.189	
2 x 10 pneumatic nailed	23.881	
2 x 12	26.667	
2 x 12 pneumatic nailed	22.222	
4 x 12	21.053	
4 x 12 pneumatic nailed	17.391	
6 x 12	19.048	
6 x 12 pneumatic nailed	15.873	
Plates, untreated, 2 x 3	37.209	
2 x 3 pneumatic nailed	30.769	
2 x 4	30.189	
2 x 4 pneumatic nailed	23.881	
2 x 6	21.333	
2 x 6 pneumatic nailed	17.778	
Studs, 8' high wall, 2 x 3	26.667	
2 x 3 pneumatic nailed	22.222	
2 x 4	17.391	
2 x 4 pneumatic nailed	14.493	
2 x 6	16.000	
2 x 6 pneumatic nailed	13.333	
3 x 4	20.000	
3 x 4 pneumatic nailed	16.667	

Figure 6-39 Installation Time in Labor-Hours for Structural
Insulated Panels (SIPs)

Description	Labor-Hours	Unit
Structural insulated panels, 7/16" OSB both faces, EPS insul. 3-5/8" T	.019	SF
5-5/8" thick	.023	
7-3/8" thick	.028	
9-3/8" thick	.036	
7/16" OSB one face, EPS insul, 3-5/8" thick	.018	
5-5/8" thick	.022	
7-3/8" thick	.026	
9-3/8" thick	.033	
7-16" OSB – 1/2" GWB faces, EPS insul, 3-5/8" T	.019	
5-5/8" thick	.023	
7-3/8" thick	.028	
9-3/8" thick	.036	
7/16" OSB – 1/2" MRGWB faces, EPS insul, 3-5/8" T	.019	
5-5/8" thick	.023	
7-3/8" thick	.028	
9-3/8" thick	.036	
Structural insulated panels, 7/16" OSB both sides, straw core	.017	

Division 6

Figure 6-40 Installation Time in Labor-Hours for Prefabricated
Wood Roof Trusses

Description	Labor-Hours	Unit
Wood Roof Trusses		
4/12 Pitch		
20' Span	.645	Ea.
28' Span	.755	Ea.
36' Span	.870	Ea.
8/12 Pitch		
20' Span	.702	Ea.
28' Span	.816	Ea.
36' Span	.976	Ea.
Fink or King Post 30' to 60' Span, 2' O.C.	.013	S.F. Floor
Plywood Roof Sheating		
1/2" Thick	.011	S.F.
5/8" Thick	.012	S.F.
3/4" Thick	.013	S.F.
Stressed Skin, Plywood Roof Panels 4' x 8'		
4-1/4" Deep	.019	S.F. Roof
6-1/2" Deep	.023	S.F. Roof
Wood Roof Decks		
3" Thick Nominal	.050	S.F.
4" Thick Nominal	.064	S.F.
Laminated Wood Roof Deck		
3" Thick Nominal	.038	S.F.
4" Thick Nominal	.049	S.F.

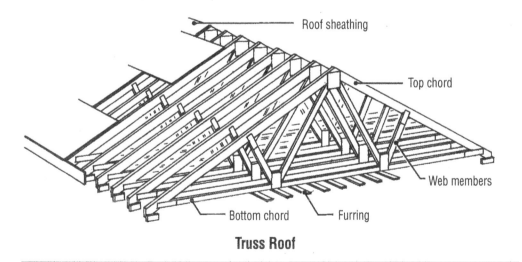

Truss Roof

Figure 6-41 Installation Time in Labor-Hours for Rafter Systems

Description	Labor-Hours	Unit
Rafters, Common, to 4 in 12 Pitch, 2" x 6"	.016	L.F.
2" x 8"	.017	L.F.
2" x 10"	.025	L.F.
2" x 12"	.028	L.F.
On Steep Roofs, 2" x 6"	.020	L.F.
2" x 8"	.021	L.F.
2" x 10"	.032	L.F.
2" x 12"	.035	L.F.
On Dormers or Complex Roofs, 2" x 6"	.027	L.F.
2" x 8"	.030	L.F.
2" x 10"	.038	L.F.
2" x 12"	.041	L.F.
Hip and Valley, to 4 in 12 Pitch, 2" x 6"	.021	L.F.
2" x 8"	.022	L.F.
2" x 10"	.028	L.F.
2" x 12"	.030	L.F.
Hip and Valley, on Steep Roofs, 2" x 6"	.027	L.F.
2" x 8"	.029	L.F.
2" x 10"	.036	L.F.
2" x 12"	.039	L.F.
Hip and Valley, on Dormers/Complex Roofs		
2" x 6"	.031	L.F.
2" x 8"	.034	L.F.
2" x 10"	.042	L.F.
2" x 12"	.045	L.F.
Hip and Valley Jacks to 4 in 12 Pitch		
2" x 6"	.027	L.F.
2" x 8"	.033	L.F.
2" x 10"	.036	L.F.
2" x 12"	.043	L.F.
Hip and Valley Jacks on Steep Roofs,		
2" x 6"	.034	L.F.
2" x 8"	.042	L.F.
2" x 10"	.046	L.F.
2" x 12"	.054	L.F.
Hip and Valley Jacks		
on Dormers/Complex Roofs, 2" x 6"	.039	L.F.
2" x 8"	.048	L.F.
2" x 10"	.052	L.F.
2" x 12"	.063	L.F.
Collar Ties, 1" x 4"	.020	L.F.
Ridge Board, 1" x 6"	.027	L.F.
1" x 8"	.029	L.F.
1" x 10"	.032	L.F.
2" x 6"	.032	L.F.
2" x 8"	.036	L.F.
2" x 10"	.040	L.F.
Sub-Fascia, 2" x 8"	.071	L.F.
2" x 10"	.089	L.F.

(continued on next page)

Division 6

Figure 6-41 Installation Time in Labor-Hours for Rafter Systems
(continued from previous page)

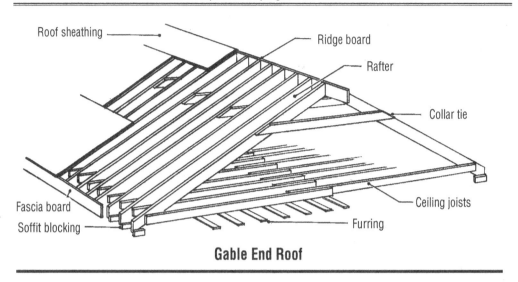

Gable End Roof

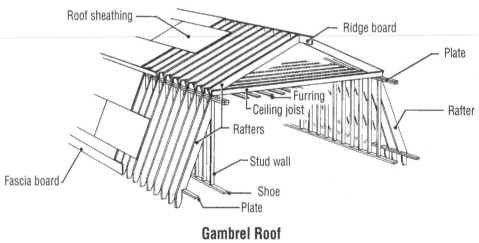

Gambrel Roof

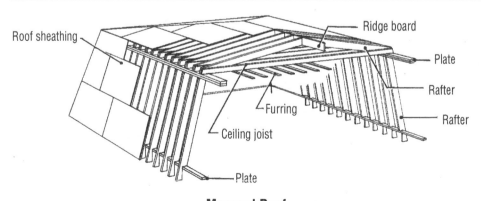

Mansard Roof

(continued on next page)

Figure 6-41   Installation Time in Labor-Hours for Rafter Systems
(continued from previous page)

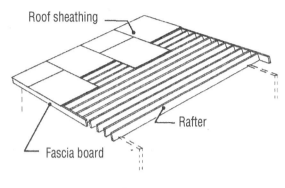

Roof sheathing

Rafter

Fascia board

Shed Roof

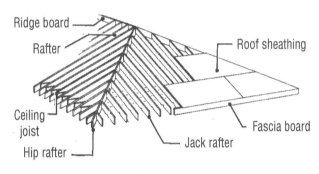

Ridge board

Rafter

Roof sheathing

Ceiling joist

Fascia board

Hip rafter

Jack rafter

Hip Roof

Division 6

Figure 6-42   Installation Time in Labor-Hours for Wood Siding

Description	Labor-Hours	Unit
Siding		
Clapboard		
Cedar Beveled		
1/2" x 6"	.032	S.F.
1/2" x 8"	.029	S.F.
3/4" x 10"	.027	S.F.
Redwood, Beveled		
1/2" x 4"	.040	S.F.
1/2" x 6"	.036	S.F.
1/2" x 8"	.032	S.F.
3/4" x 10"	.027	S.F.
Board		
Redwood		
Tongue and Groove		
1" x 4"	.053	S.F.
1" x 8"	.043	S.F.
Channel		
1" x 10"	.028	S.F.
Cedar		
Butted		
1" x 4"	.033	S.F.
Channel		
1" x 8"	.032	S.F.
Board and Batten		
1" x 12"	.031	S.F.
Pine		
Butted		
1" x 8"	.029	S.F.
Sheets		
Hardboard, Lapped 7/16" x 8"	.025	S.F.
Plywood		
MDO		
3/8" Thick	.021	S.F.
1/2" Thick	.023	S.F.
3/4" Thick	.025	S.F.
Texture 1-11		
5/8" Thick	.024	S.F.
Sheathing		
Plywood on Walls		
3/8" Thick	.013	S.F.
1/2" Thick	.014	S.F.
5/8" Thick	.015	S.F.
3/4" Thick	.016	S.F.
Wood Fiber	.013	S.F.
Trim Exterior, Up to 1" x 6"	.040	L.F.
Fascia		
1" x 6"	.032	L.F.
1" x 8"	.036	L.F.

(continued on next page)

Figure 6-42 Installation Time in Labor-Hours for Wood Siding
(continued from previous page)

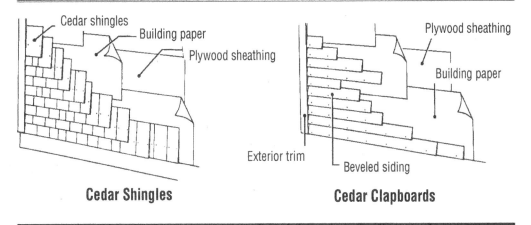

Cedar Shingles   **Cedar Clapboards**

Figure 6-43 Installation Time in Labor-Hours for Millwork

Description	Labor-Hours	Unit
Beams, Decorative 4" x 8"	.100	L.F.
Moldings, Base, 9/16" x 3-1/2"	.033	L.F.
Casing, Apron, 5/8" x 3-1/2"	.036	L.F.
Band, 11/16" x 1-3/4"	.032	L.F.
Casings, 11/16" x 3-1/2"	.037	L.F.
Ceilings, Bed, 9/16" x 2"	.033	L.F.
Cornice, 9/16" x 2-1/4"	.027	L.F.
Cove Scotia, 11/16" x 2-3/4"	.031	L.F.
Crown, 11/16" x 4-5/8"	.036	L.F.
Exterior Cornice Boards, 1" x 6"	.032	L.F.
Corner Board, 1" x 6"	.040	L.F.
Fascia, 1" x 6"	.032	L.F.
Back Band Trim	.032	L.F.
Verge Board, 1" x 6"	.040	L.F.
Trim, Astragal, 1-5/16" x 2-3/16"	.033	L.F.
Chair Rail, 5/8" x 3-1/2"	.033	L.F.
Handrail, 1-1/2" x 1-3/4"	.100	L.F.
Door Moldings	.471	Set
Door Trim Including Headers, Stops and Casing 4-1/2" Wide	1.509	Opng.
Stool Caps, 1-1/16" x 3-1/4"	.053	L.F.
Threshold, 3' Long, 5/8" x 3-5/8"	.250	Ea.
Window Trim Sets Including Casings, Header, Stops, Stool and Apron	.800	Opng.
Paneling, 1/4" Thick	.040	S.F.
3/4" Thick, Stock	.050	S.F.
Architectural Grade	.071	S.F.

Division 6

(continued on next page)

Figure 6-43 Installation Time in Labor-Hours for Millwork
(continued from previous page)

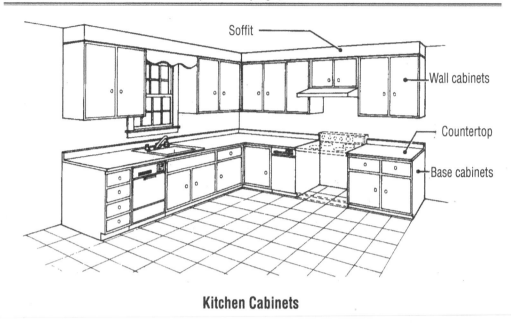

Kitchen Cabinets

Division Seven

Thermal & Moisture Protection

Introduction

This division covers all types of materials used to seal the outside of the building against moisture, thermal, and air penetration, plus the associated insulation and accessories. (Note, however, that doors and windows, including skylights, are covered in Division 8 – Openings.)

Dampproofing usually consists of one or two layers of coatings applied to a foundation wall up to the finished grade elevation. Dampproofing is used to inhibit the migration of moisture from the outside of the structure to the inside. Waterproofing usually consists of membrane sheeting used to prevent or stop the flow of water. Moisture protection measures improve the efficiency of thermal insulation while maintaining the proper humidity levels and preventing the growth of mold.

Insulation is primarily used to reduce the unwanted effects of heat transfer through the exterior enclosure of a building. The type, form, and material used for insulation depend on the location within the structure and the size of the space available.

green ☼ Asbestos was used extensively in the building industry for thermal insulation and fire protection before it was found to be carcinogenic and banned in the 1970s. Fortunately, there are replacements for asbestos that are not only safe, but actually environmentally friendly. In fact, Division 7 items are a big part of building green because of the multiple benefits of insulation. Proper thermal insulation is usually the most cost efficient way of reducing energy use. On top of these green building attributes, fiberglass, mineral wool, cellulose, and cotton insulation can be easily produced from recycled materials.

Roofing can be subdivided into various types depending on its application. Shingles are used mostly on residences. Commercial and/or industrial roofs can be constructed from single ply membranes, metal sheets, corrugated metal panels, copper sheets, copper panels, or corrugated fiberglass panels. When a roof consists of more than a single layer of asphalt felt and/or tar, it is called a built-up roof. Also included in Division 7 is sheet metal work associated with roofing: flashings, trim, gutters and downspouts, gravel stops, and the like.

A living (green) roof offers multiple benefits. These roofs are designed with plants, usually sedum, growing in soil over a waterproof membrane. In addition to providing a very high insulation value (soil is a good insulator), green roofs reduce the load on water treatment plants by cutting the amount of storm water runoff. They require little manufactured material, and the plants actually increase oxygen levels in the atmosphere. Reflective roofs are also green choices with much to offer. Simply by using a reflective coating, usually white, solar energy is reflected off of the roof reducing air conditioning loads and heat island effects, improving water quality, and prolonging the life of the roof.

The items addressed in this chapter include:

- Insulation and dampproofing
- Siding
- Roofing
- Flashings
- Roof specialties
- Fireproofing
- Joint sealants and caulking

Estimating Data

The following tables present estimating guidelines for items found in Division 7 – Thermal and Moisture Protection. Please note that these guidelines are intended as indicators of what may be expected, but that each project must be evaluated individually.

Table of Contents

Estimating Tips

Roof Walkways

When estimating the installation of a membrane roof, be sure to include the cost for walkway pavers to create a path to roof-mounted mechanical equipment. You will more than likely be required to install these in order to obtain a warranty.

Single Ply Roofing Sources

In many locations, manufacturers make their materials available through one distributor/contractor franchise. If only one type of roof is specified, you may not have much choice as far as acquiring costs or quotes.

Estimating the Specified System

Even though it may be less expensive to use a "sustainable substitute," the system that is specified is the one that should be estimated. If you include as part of your estimate a non-specified system and get the job, you might be required to furnish the originally specified, more expensive roof system.

Sheet Metal Work

Do not assume that all sheet metal items are "off-the-shelf" items. Gutters and downspouts as well as termite shields, gravel stops, expansion joints, and reglets may need to be shop-fabricated. Checking informational catalogs could save quite a tidy sum in extras.

Checklist ✓

For an estimate to be reliable, all items must be accounted for. A complete estimate can also limit contingencies. The following checklist can be used to help ensure that all items are included.

☐ **Waterproofing/ Dampproofing**
- ☐ Bentonite
- ☐ Bituminous coating
- ☐ Building paper
- ☐ Cementitious
- ☐ Elastomeric
- ☐ Liquid
- ☐ Membrane
- ☐ Preformed vapor barrier
- ☐ Sealants

☐ **Building Insulation**
- ☐ Rigid
  - ☐ Fiberglass
  - ☐ Foam glass
  - ☐ Polyisocyanurate
  - ☐ Polystyrene
- ☐ Non-rigid
  - ☐ Fiberglass
  - ☐ Mineral fiber
  - ☐ Vermiculite
  - ☐ Perlite
- ☐ Form board
  - ☐ Acoustical
  - ☐ Fiberglass
  - ☐ Gypsum
  - ☐ Mineral fiber
  - ☐ Wood fiber

☐ **Masonry Insulation**
- ☐ Poured
- ☐ Rigid
- ☐ Core
- ☐ Foam

☐ **Roof Deck Insulation**
- ☐ Fiberboard
- ☐ Fiberglass
- ☐ Foam glass
- ☐ Polyisocyanurate
- ☐ Polystyrene

☐ **Spray-on Insulation**

☐ **Special Insulation**

☐ **Vapor Retarders**

☐ **Exterior Insulation Finish System**

☐ **Shingles for Siding or Roof**
- ☐ Aluminum
- ☐ Asphalt
- ☐ Clay tile
- ☐ Concrete
- ☐ Slate
- ☐ Wood

☐ **Siding**

☐ **Fiberglass Panels**
- ☐ Roofing
- ☐ Siding

☐ **Metal Panels**

☐ **Steel Panels**
- ☐ Roofing
- ☐ Siding
- ☐ Insulated sandwich

Division 7

☐ *Vinyl Siding*

☐ *Membrane Roofing*

☐ *Built-Up Roofing*

☐ *Roll Roofing*

☐ *Metal Roofing*
- ☐ Lead
- ☐ Copper
- ☐ Stainless steel
- ☐ Zinc

☐ *Fluid Applied Roofing*

☐ *Roof Accessories*
- ☐ Downspouts
- ☐ Gutters
- ☐ Expansion joints
- ☐ Fascia
- ☐ Flashing

☐ Gravel stop
☐ Louvers
☐ Reglet
☐ Roof drains
☐ Soffit
☐ Hatches
- ☐ Roof
- ☐ Smoke vents
☐ Skyroof
☐ Ventilators
☐ Walkways

☐ *Fireproofing*
- ☐ Beams
- ☐ Columns
- ☐ Decking

☐ *Firestopping*

☐ *Joint Sealants and Caulking*

Figure 7-1 Common Insulating Materials Used in Construction

green | This table lists the common types of materials used in construction, their forms, and their major uses.

Physical Form	Type of Material	Major Uses
Powders	Diatomaceous earth; sawdust; silica aerogel	Filler
Loose fibrous materials	Cork granules; glass wool; mineral wool; shredded bark; vermiculite; perlite; mica	Flat areas such as air spaces above ceilings, adjacent to roof, to reduce conduction and convection
Batt or blanket insulation[a]	Glass wool; mineral wool; wood fibers, etc., enclosed by paper, cloth, wire mesh, or aluminum	Air spaces, particularly in vertical walls and flat surfaces, to reduce conduction and convection
Board; sheet, and integrant insulation[a]	Cork; fiber; paper pulp; cellular glass; glass fiber	Sheathing for walls, to increase strength as well as reduce heat loss; rigid insulation on roofs; perimeter insulation along edges of slab floors
Reflective insulation	Aluminum foil, often combined in layers with one or more adjoining air spaces or combined with sheets of paper	Used principally for all types of refrigerated or controlled environmental spaces
Special types of block and brick insulation and refractories	Insulation block made of cork, expanded glass, 85% magnesia, or vermiculite; insulating refractory block or brick made of diatomaceous earth or kaolin (clay); heavy refractories made of fire clay, magnesite, or silica	Special controlled temperature and high-temperature insulation problems; for example, refrigerator rooms, pipes, ducts, boilers, fire chambers of boilers, and chimneys
Foam-type insulation	Rigid boards; 2-component on-job application with special applicators; polystyrene and polyurethane	Interior-applied insulation for walls above or below grade; roof insulation; air space and perimeter insulation

[a] These are available with vapor barrier as part of insulation.

(courtesy Construction Materials, Caleb Hornbostel, John Wiley & Sons)

Division 7

Figure 7-2a Resistance Values of Construction Materials

This table lists the R-values for commonly used materials.

Material	R Value
Silver	0.00034
Copper	0.00039
Steel	0.0031
Granite	0.051
Slate shingles	0.096
Concrete per 1" (25.4 mm)	0.08
4" (101.6 mm) concrete block	0.71
8" (203.2 mm) concrete block	1.11
12" (304.8 mm) concrete block	1.28
Common brick per 1" (25.4 mm)	0.20
Face brick per 1" (25.4 mm)	0.11
Plate glass	0.18
Window glass	0.89
1/2" (12.7 mm) plasterboard	0.45
3/8" (11.03 mm) plywood	0.59
3/8" (11.03 mm) insulation board	2.06
1" (25.4 mm) polyurethane foam	6.25
2" (50.8 mm) polyurethane foam	14.29
Insulating glass, one air space	1.61
Insulating glass, two air spaces	2.13
Air space 3-1/2" (88.9 mm)	0.91
Air space 3/4" (19.05 mm)	0.92
Fiberglass, paper-faced	11.0
Built-up roofing, 4-ply slag	0.33
Asphalt strip shingles	0.44
White pine V-joint T & G	0.94
2" (50.8 mm) wood decking	2.03
3" (76.2 mm) wood decking	3.28
1" (25.4 mm) fiberboard	2.78
2" (50.8 mm) fiberboard	5.26
3" (76.2 mm) fiberboard	8.33
8" (203.2 mm) lapped beveled wood siding	0.81
10" (254.0 mm) lapped beveled wood siding	1.05
1" (25.4 mm) fiberglass perimeter insulation	4.30
Wood shingles	0.87

The resistance factor (R) is the reciprocal of conductivity (K), conductance (C), or overall heat transfer coefficient values (U):

$$R = \frac{1}{K} \text{ or } \frac{1}{C} \text{ or } \frac{1}{U}$$

(courtesy Construction Materials, *Caleb Hornbostel, John Wiley & Sons*)

(continued on next page)

Figure 7-2b Resistance Values of Construction Materials
(continued from previous page)

Type of Insulation	R-Value per Inch
Loose-fill:	
Cellulose	3.1–3.7
Fiberglass	2.2–4.2
Rock wool	2.2–2.9
Batts:	
Fiberglass	2.9–3.8
Cotton	3.0–3.7
Sprayed insulation:	
Polyurethane foam	5.6–6.2
Icynene foam	3.6–4.3
Wet-sprayed cellulose	2.9–3.4
Spray-in fiberglass	3.7–3.8
Foam board:	
Expanded polystyrene	3.9–4.2
Extruded polystyrene	5.0
Polyisocyanurate	5.6–7.0
Polyurethane	5.6–7.0
Phenolic (closed cell)	8.2
Phenolic (open cell)	4.4

(courtesy DOE Insulation Fact Sheet, www.eere.energy.gov/buildings)

Division 7

Figure 7-3 Characteristics of Roof Insulation Materials

This table lists the various types of roofing insulation (across the top) and the desirable characteristics (down the left side), and compares how each type is rated for those characteristics.

Characteristics	Type of Insulating Board								
	Polyiso-cyanurate Foam	Polyure-thane Foam	Extruded Poly-styrene	Molded Poly-styrene	Cellular Glass	Mineral Fiber	Phenolic Foam	Wood Fiber	Glass Fiber
Impact resistant	G	G	G	F	G	E	G	E	F
Moisture resistant	E	G	E	G	E	G	E		
Fire resistant	E				E	E	E		E
Compatible with bitumens	E	G	F	F	E	E	E	E	G
Durable	E	E	E	E	E	G	F	E	E
Stable "k" value			E	E	E	E		E	E
Dimensionally stable	E	E	E	E	E	E	E	E	G
High thermal resistance	E	E	E	G	F	F	E	F	G
Available tapered slabs	Y	Y	Y	Y	Y	Y	Y	Y	Y
"R" value per in. thickness	7.20	6.25	4.76	3.85–4.35	2.86	2.78	8.30	1.75–2.00	4.00
Thicknesses available	1"–3"	1"–4"	1"–3½"	½"–24"	1½"–4"	¾"–3"	1"–4"	1"–3"	¾"–2½"
Density (lb./ft.³)	2.0	1.5	1.8–3.5	1.0–2.0	8.5	16–17	1.5	22–27	49
Remarks	Prone to "thermal drift"	Prone to "thermal drift" Should be overlaid with a thin layer of wood fiber, glass fiber or perlite board, with staggered joints.	Somewhat sensitive to hot bitumen & adhesive vapors Should be overlaid with a thin layer of wood fiber, glass fiber or perlite board, with staggered joints.	Somewhat sensitive to hot bitumen & adhesive vapors Should be overlaid with a thin layer of wood fiber, glass fiber or perlite board, with staggered joints.			Prone to "thermal drift" Relatively new & untested.	Expands with moisture — holds moisture.	Prone to damage from moisture infiltration.

E = Excellent　　　G = Good　　　F = Fair

Figure 7-4 Factors for Converting Horizontal Roof Dimensions to Inclined

This table is useful when a pitched roof is called for and only the pitch and the horizontal dimension of the roof are known. Multiply the horizontal dimension by the conversion factor shown for the appropriate roof pitch.

Roof Slope	Approximate Angle	Factor
Flat	0	1.000
1 in 12	4.8	1.003
2 in 12	9.5	1.014
3 in 12	14.0	1.031
4 in 12	18.4	1.054
5 in 12	22.6	1.083
6 in 12	26.6	1.118
7 in 12	30.3	1.158
8 in 12	33.7	1.202
9 in 12	36.9	1.250
10 in 12	39.8	1.302
11 in 12	42.5	1.357
12 in 12	45.0	1.414
13 in 12	47.3	1.474
14 in 12	49.4	1.537
15 in 12	51.3	1.601
16 in 12	53.1	1.667
17 in 12	54.8	1.734
18 in 12	56.3	1.803
19 in 12	57.7	1.873
20 in 12	59.0	1.943
21 in 12	60.3	2.015
22 in 12	61.4	2.088
23 in 12	62.4	2.162

Figure 7-5 Roofing Quantities

The following tables show the quantities of roofing materials required for various pitched roofs for the building sizes described.

			Roofing for a Roof Pitch of 1-1/2:12 (Figures include cutting and fitting waste.)							
Ground Floor Area	Roof Surface Area	Asphalt Strip Shingles	Wood Shingles 18" Long		Wood Shingles 24" Long		Fiberglass Shingles	Clay Roofing Tile	Number of Rolls of Saturated Felt Roofing Paper	
			4" Expo.	5-1/2" Expo.	6" Expo.	7-1/2" Expo.				
S.F.	S.F.	S.F.	S.F.	S.F.	S.F.	S.F.	S.F.	S.F.	15 Lb.	30 Lb.
200	206	226	278	226	284	226	226	216	.57	1.14
300	309	339	417	339	426	339	339	324	.85	1.70
400	412	452	556	452	568	452	452	432	1.13	2.27
500	515	565	695	565	710	565	565	540	1.42	2.84
600	618	678	834	678	852	678	678	648	1 70	3.40
700	721	791	973	791	994	791	791	756	1.98	3.97
800	824	904	1112	904	1136	904	904	864	2.27	4.54
900	927	1010	1251	1010	1278	1010	1010	972	2.55	5.10
1000	1030	1130	1390	1130	1420	1130	1130	1080	2.83	5.66
1100	1133	1243	1529	1243	1562	1243	1243	1188	3.11	6.22
1200	1236	1356	1668	1356	1704	1356	1356	1296	3.40	6.80
1300	1339	1469	1810	1469	1846	1469	1469	1404	3.68	7.36
1400	1442	1582	1946	1582	1990	1582	1582	1512	3.96	7.92
1500	1545	1695	2085	1695	2130	1695	1695	1620	4.25	8.50
1600	1648	1808	2224	1808	2272	1808	1808	1728	4.53	9.06
1700	1751	1921	2363	1921	2414	1921	1921	1836	4.81	9.62
1800	1854	2034	2500	2034	2556	2034	2034	1944	5.09	10.18
1900	1957	2147	2640	2147	2700	2147	2147	2052	5.38	10.76
2000	2060	2260	2780	2260	2840	2260	2260	2160	5.66	11.32
2100	2163	2378	2920	2378	2982	2378	2378	2268	5.94	11.88
2200	2266	2486	3058	2486	3124	2486	2486	2376	6.23	12.46
2300	2369	2600	3200	2600	3266	2600	2600	2484	6.51	13.02
2400	2472	2712	3336	2712	3410	2712	2712	2592	6.80	13.58
2500	2575	2825	3475	2825	3550	2825	2825	2700	7.08	14.16
2600	2678	2938	3615	2938	3692	2938	2938	2808	7.36	14.72
2700	2781	3051	3753	3051	3834	3051	3051	2916	7.64	15.28
2800	2884	3164	3892	3164	3976	3164	3164	3024	7.92	15.84
2900	2987	3277	4030	3277	4118	3277	3277	3132	8.21	16.42
3000	3090	3390	4170	3390	4260	3390	3390	3240	8.49	16.98

(continued on next page)

Figure 7-5 Roofing Quantities *(continued from previous page)*

Ground Floor Area	Roof Surface Area	Asphalt Strip Shingles	Wood Shingles 18" Long 4" Expo.	Wood Shingles 18" Long 5-1/2" Expo.	Wood Shingles 24" Long 6" Expo.	Wood Shingles 24" Long 7-1/2" Expo.	Fiberglass Shingles	Clay Roofing Tile	Number of Rolls of Saturated Felt Roofing Paper 15 Lb.	Number of Rolls of Saturated Felt Roofing Paper 30 Lb.
S.F.	S.F.	S.F.	S.F.	S.F.	S.F.	S.F.	S.F.	S.F.	15 Lb.	30 Lb.
200	212	234	286	234	292	234	234	224	.59	1.17
300	318	351	429	351	438	351	351	336	.88	1 76
400	424	468	472	468	584	468	468	448	1 17	2.34
500	530	585	715	585	730	585	585	560	1.46	2.93
600	636	702	858	702	876	702	702	672	1 76	3.51
700	742	819	1000	819	1022	819	819	784	2.05	4.10
800	848	936	1144	936	1170	936	936	896	2.34	4.68
900	954	1053	1287	1050	1314	1050	1050	1008	2.63	5.27
1000	1060	1170	1430	1170	1460	1170	1170	1120	2.93	5.85
1100	1166	1287	1573	1287	1610	1287	1287	1232	3.22	6.44
1200	1272	1404	1716	1404	1752	1404	1404	1344	3.51	7.02
1300	1378	1521	1859	1521	1900	1521	1521	1456	3.80	7.61
1400	1484	1638	2000	1638	2044	1638	1638	1568	4 10	8.19
1500	1590	1755	2145	1755	2190	1755	1755	1680	4.39	8.78
1600	1696	1872	2288	1872	2336	1872	1872	1792	4.68	9.36
1700	1802	1989	2430	1989	2480	1989	1989	1904	4.97	9.95
1800	1908	2106	2574	2106	2630	2106	2106	2016	5.27	10.53
1900	2014	2223	2717	2223	2774	2223	2223	2128	5.56	11.12
2000	2120	2340	2860	2340	2920	2340	2340	2240	5.85	11 70
2100	2226	2457	3000	2457	3066	2457	2457	2352	6.14	12.29
2200	2332	2574	3146	2574	3212	2574	2574	2464	6.44	12.87
2300	2438	2691	3290	2691	3358	2691	2691	2576	6.73	13.46
2400	2544	2808	3430	2808	3504	2808	2808	2688	7.02	14.04
2500	2650	2925	3575	2925	3650	2925	2925	2800	7.31	14.63
2600	2756	3042	3720	3042	3800	3042	3042	2912	7.61	15.21
2700	2862	3159	3860	3159	3942	3159	3159	3024	7.90	15.80
2800	2968	3276	4000	3276	4090	3276	3276	3136	8.19	16.38
2900	3074	3393	4150	3393	4234	3393	3393	3248	8.48	16.96
3000	3180	3510	4290	3510	4380	3510	3510	3360	8.78	17.55

(continued on next page)

Division 7

Figure 7-5 Roofing Quantities *(continued from previous page)*

Ground Floor Area	Roof Surface Area	Asphalt Strip Shingles	Wood Shingles 18" Long		Wood Shingles 24" Long		Fiberglass Shingles	Clay Roofing Tile	Number of Rolls of Saturated Felt Roofing Paper	
			4" Expo.	5-1/2" Expo.	6" Expo.	7-1/2" Expo.			15 Lb.	30 Lb.
S.F.	S.F.	S.F.	S.F.	S.F.	S.F.	S.F.	S.F.	S.F.		
200	218	240	294	240	300	240	240	228	.60	1.20
300	327	360	441	360	450	360	360	342	.90	1.80
400	426	480	588	480	600	480	480	456	1.20	2.40
500	545	600	735	600	750	600	600	570	1.50	3.00
600	654	720	882	720	900	720	720	684	1.80	3.60
700	763	840	1029	840	1050	840	840	798	2.10	4.20
800	872	960	1176	960	1200	960	960	912	2.40	4.80
900	981	1080	1323	1080	1350	1080	1080	1026	2.70	5.40
1000	1090	1200	1470	1200	1500	1200	1200	1140	3.00	6.00
1100	1199	1320	1617	1320	1650	1320	1320	1254	3.30	6.60
1200	1308	1440	1764	1440	1800	1440	1440	1368	3.60	7.20
1300	1417	1560	1911	1560	1950	1560	1560	1482	3.90	7.80
1400	1526	1680	2058	1680	2100	1680	1680	1596	4.20	8.40
1500	1635	1800	2205	1800	2250	1800	1800	1710	4.50	9.00
1600	1744	1920	2352	1920	2400	1920	1920	1824	4.80	9.60
1700	1853	2040	2500	2040	2550	2040	2040	1938	5.10	10.20
1800	1962	2160	2646	2160	2700	2160	2160	2052	5.40	10.80
1900	2071	2280	2793	2280	2850	2280	2280	2166	5.70	11.40
2000	2180	2400	2940	2400	3000	2400	2400	2280	6.00	12.00
2100	2289	2520	3087	2520	3150	2520	2520	2394	6.30	12.60
2200	2398	2640	3234	2640	3300	2640	2640	2508	6.60	13.20
2300	2507	2760	3381	2760	3450	2760	2760	2622	6.90	13.80
2400	2616	2880	3528	2880	3600	2880	2880	2736	7.20	14.40
2500	2725	3000	3675	3000	3750	3000	3000	2850	7.50	15.00
2600	2834	3120	3822	3120	3900	3120	3120	2964	7.80	15.60
2700	2943	3240	3969	3240	4050	3240	3240	3078	8.10	16.20
2800	3052	3360	4116	3360	4200	3360	3360	3192	8.40	16.80
2900	3161	3480	4263	3480	4350	3480	3480	3306	8.70	17.40
3000	3270	3600	4410	3600	4500	3600	3600	3420	9.00	18.00

Table title: **Roofing for a Roof Pitch of 2-1/2:12 (Figures include cutting and fitting waste.)**

(continued on next page)

Figure 7-5   Roofing Quantities *(continued from previous page)*

Roofing for a Roof Pitch of 3:12 (Figures include cutting and fitting waste.)										
Ground Floor Area	Roof Surface Area	Asphalt Strip Shingles	Wood Shingles 18" Long		Wood Shingles 24" Long		Fiberglass Shingles	Clay Roofing Tile	Number of Rolls of Saturated Felt Roofing Paper	
			4" Expo.	5-1/2" Expo.	6" Expo.	7-1/2" Expo.				
S.F.	S.F.	S.F.	S.F.	S.F.	S.F.	S.F.	S.F.	S.F.	15 Lb.	30 Lb.
200	224	246	300	246	310	246	246	236	.62	1.24
300	336	369	450	369	465	369	369	354	.93	1.86
400	448	492	600	492	620	492	492	472	1.24	2.48
500	560	615	750	615	775	615	615	590	1.55	3.10
600	672	738	900	738	930	738	738	708	1.86	3.72
700	784	861	1050	861	1085	861	861	826	2.17	4.34
800	896	984	1200	984	1240	984	984	944	2.48	4.96
900	1008	1107	1350	1107	1395	1107	1107	1062	2.79	5.58
1000	1120	1230	1500	1230	1550	1230	1230	1180	3.10	6.20
1100	1232	1353	1650	1353	1705	1353	1353	1298	3.41	6.82
1200	1344	1476	1800	1476	1860	1476	1476	1416	3.72	7.44
1300	1456	1600	1950	1600	2015	1600	1600	1534	4.03	8.06
1400	1568	1722	2100	1722	2170	1722	1722	1652	4.34	8.68
1500	1680	1845	2250	1845	2325	1845	1845	1770	4.65	9.30
1600	1793	1968	2400	1968	2480	1968	1968	1888	4.96	9.92
1700	1904	2091	2550	2091	2635	2091	2091	2006	5.27	10.54
1800	2016	2214	2700	2214	2790	2214	2214	2124	5.58	11.16
1900	2128	2337	2850	2337	2945	2337	2337	2242	5.89	11.78
2000	2240	2460	3000	2460	3100	2460	2460	2360	6.20	12.40
2100	2352	2583	3150	2583	3255	2583	2583	2478	6.51	13.02
2200	2464	2706	3300	2706	3410	2706	2706	2596	6.82	13.64
2300	2576	2829	3450	2829	3565	2829	2829	2714	7.13	14.26
2400	2688	2952	3600	2952	3720	2952	2952	2832	7.44	14.88
2500	2800	3075	3750	3075	3875	3075	3075	2950	7.75	15.50
2600	2912	3198	3900	3198	4030	3198	3198	3068	8.06	16.12
2700	3024	3321	4050	3321	4185	3321	3321	3186	8.37	16.74
2800	3136	3440	4200	3440	4340	3440	3440	3304	8.68	17.36
2900	3248	3567	4350	3567	4495	3567	3567	3422	9.00	18.00
3000	3360	3690	4500	3690	4650	3690	3690	3540	9.30	18.60

(continued on next page)

Division 7

Figure 7-5　Roofing Quantities *(continued from previous page)*

Ground Floor Area	Roof Surface Area	Asphalt Strip Shingles	Wood Shingles 18" Long		Wood Shingles 24" Long		Fiberglass Shingles	Clay Roofing Tile	Number of Rolls of Saturated Felt Roofing Paper	
			4" Expo.	5-1/2" Expo.	6" Expo.	7-1/2" Expo.			15 Lb.	30 Lb.
S.F.	S.F.	S.F.	S.F.	S.F.	S.F.	S.F.	S.F.	S.F.		
200	240	264	324	264	332	264	264	252	.66	1.32
300	360	396	486	396	498	396	396	378	.99	1.98
400	480	528	648	528	664	528	528	504	1.32	2.64
500	600	660	810	660	830	660	660	630	1.65	3.30
600	720	792	972	792	1000	792	792	756	1.98	3.96
700	840	924	1134	924	1162	924	924	882	2.31	4.62
800	960	1056	1296	1056	1328	1056	1056	1008	2.64	5.28
900	1080	1188	1458	1188	1494	1188	1188	1134	2.97	5.94
1000	1200	1320	1620	1320	1660	1320	1320	1260	3.30	6.60
1100	1320	1452	1782	1452	1826	1452	1452	1386	3.63	7.26
1200	1440	1584	1944	1584	1992	1584	1584	1512	3.96	7.92
1300	1560	1716	2106	1716	2158	1716	1716	1638	4.29	8.58
1400	1680	1848	2268	1848	2324	1848	1848	1764	4.62	9.24
1500	1800	1980	2430	1980	2490	1980	1980	1890	4.95	9.90
1600	1920	2112	2592	2112	2656	2112	2112	2016	5.28	10.56
1700	2040	2244	2754	2244	2822	2244	2244	2142	5.61	11.22
1800	2160	2376	2916	2376	2990	2376	2376	2268	5.94	11.88
1900	2280	2508	3078	2508	3154	2508	2508	2394	6.27	12.54
2000	2400	2640	3240	2640	3320	2640	2640	2520	6.60	13.20
2100	2520	2772	3400	2772	3486	2772	2772	2646	6.93	13.86
2200	2640	2904	3564	2904	3652	2904	2904	2772	7.26	14.52
2300	2760	3036	3726	3036	3820	3036	3036	2898	7.59	15.18
2400	2880	3168	3888	3168	3984	3168	3168	3024	7.92	15.80
2500	3000	3300	4050	3300	4150	3300	3300	3150	8.25	16.50
2600	3120	3432	4212	3432	4316	3432	3432	3276	8.58	17.16
2700	3240	3564	4374	3564	4482	3564	3564	3402	8.91	17.82
2800	3360	3696	4536	3696	4648	3696	3696	3528	9.24	18.48
2900	3480	3828	4700	3828	4814	3828	3828	3654	9.57	19.14
3000	3600	3960	4860	3960	4980	3960	3960	3780	9.90	19.80

Roofing for a Roof Pitch of 4:12 (Figures include cutting and fitting waste.)

(continued on next page)

Figure 7-5 Roofing Quantities *(continued from previous page)*

Ground Floor Area	Roof Surface Area	Asphalt Strip Shingles	Wood Shingles 18" Long		Wood Shingles 24" Long		Fiberglass Shingles	Clay Roofing Tile	Number of Rolls of Saturated Felt Roofing Paper	
			4" Expo.	5-1/2" Expo.	6" Expo.	7-1/2" Expo.			15 Lb.	30 Lb.
S.F.	S.F.	S.F.	S.F.	S.F.	S.F.	S.F.	S.F.	S.F.		
200	284	312	384	312	392	312	312	298	78	1.55
300	426	468	576	468	588	468	468	447	1.16	2.32
400	568	624	768	624	784	624	624	600	1.55	3.10
500	710	780	960	780	980	780	780	745	1.94	3.88
600	852	936	1152	936	1176	936	936	894	2.33	4.65
700	994	1092	1344	1092	1372	1092	1092	1043	2.71	5.43
800	1136	1248	1536	1248	1568	1248	1248	1192	3.10	6.20
900	1278	1404	1728	1404	1764	1404	1404	1340	3.49	6.98
1000	1420	1560	1920	1560	1960	1560	1560	1490	3.88	7.75
1100	1562	1716	2112	1716	2156	1716	1716	1639	4.26	8.53
1200	1704	1872	2304	1872	2352	1872	1872	1788	4.65	9.30
1300	1846	2028	2496	2028	2548	2028	2028	1937	5.04	10.08
1400	1990	2184	2688	2184	2744	2184	2184	2086	5.43	10.85
1500	2130	2340	2880	2340	2940	2340	2340	2235	5.81	11.63
1600	2272	2500	3072	2500	3136	2500	2500	2384	6.20	12.40
1700	2414	2652	3264	2652	3332	2652	2652	2533	6.59	13.18
1800	2556	2810	3456	2810	3528	2810	2810	2682	6.98	13.95
1900	2700	2964	3648	2964	3724	2964	2964	2830	7.36	14.93
2000	2840	3120	3840	3120	3920	3120	3120	2980	7.75	15.50
2100	2982	3276	4032	3276	4116	3276	3276	3130	8.14	16.28
2200	3124	3432	4224	3432	4312	3432	3432	3278	8.53	17.05
2300	3266	3590	4416	3590	4508	3590	3590	3427	8.91	17.83
2400	3410	3744	4608	3744	4704	3744	3744	3576	9.30	18.60
2500	3550	3900	4800	3900	4900	3900	3900	3725	9.69	19.38
2600	3692	4056	4992	4056	5096	4056	4056	3874	10.08	20.15
2700	3834	4212	5184	4212	5292	4212	4212	4023	10.46	20.93
2800	3976	4368	5376	4638	5488	4368	4368	4172	10.85	21.70
2900	4118	4524	5568	4524	5684	4524	4524	4321	11.24	22.48
3000	4260	4680	5760	4680	5880	4680	4680	4470	11.63	23.25

Roofing for a Roof Pitch of 6:12 (Figures include cutting and fitting waste.)

(courtesy Estimating Tables for Home Building, *Paul I. Thomas, Craftsman Book Company)*

Division 7

Figure 7-6 Waste Values for Roof Styles

This table lists the approximate values for waste that should be added to the net area for materials of various roof types.

Roof Shape	Plain	Cut-up
Gable	10	15
Hip	15	20
Gambrel	10	20
Gothic	10	15
Mansard (sides)	10	15
Porches	10	—

Figure 7-7 Roofing Materials and Nomenclature

These graphics illustrate various types of roofing systems.

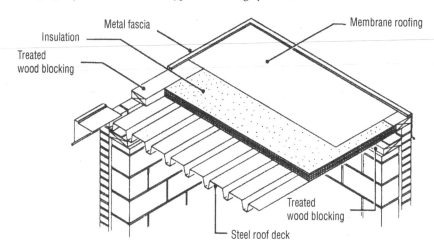

Single Ply Roofing

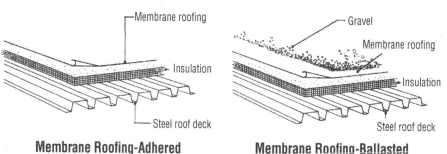

Membrane Roofing-Adhered   **Membrane Roofing-Ballasted**

(continued on next page)

Figure 7-7 Roofing Materials and Nomenclature
(continued from previous page)

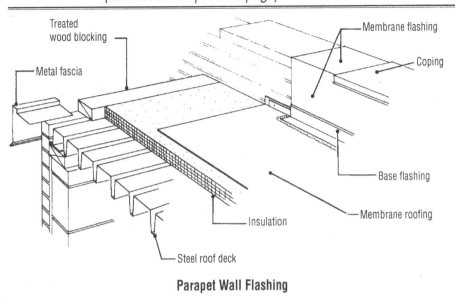

Parapet Wall Flashing

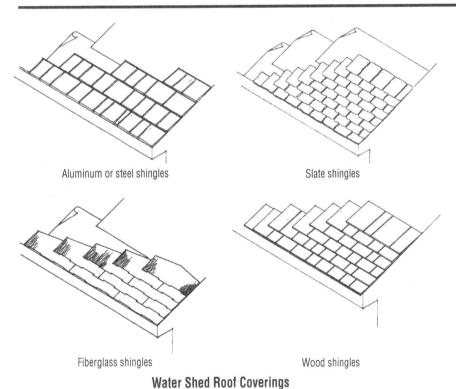

Water Shed Roof Coverings

(continued on next page)

Division 7

Figure 7-7 Roofing Materials and Nomenclature
(continued from previous page)

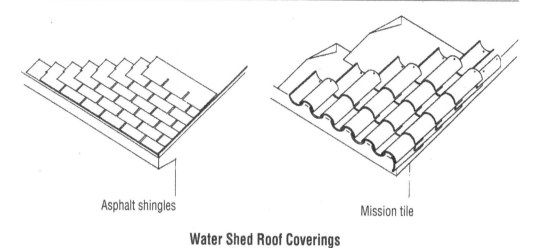

Asphalt shingles

Mission tile

Water Shed Roof Coverings

Figure 7-8 Commercial Roofing Systems

This table lists and compares the more common types of commercial roofing systems. It is not intended to be a comprehensive listing of available systems, but an overview of the most common types.

ROOF SYSTEM TYPE	Low initial cost	Many re-entrant corners & penetrations	High traffic roofs or areas	Resistant to oils, solvents	Mid-rise or high-rise use (over 5 stories)	Ease of maintenance	Compatible with light steel decks	Long record of serviceability	REMARKS
Built-up roof, Asphalt	3	5	5	3	1	5	1	5	Traditional materials
Built-up roof, Coal Tar Pitch	2	4	5	3	1	5	1	5	Traditional materials
Modified Bitumen (APP modifier)	2	5	4	4	3	4	2	3	Torch-applied materials
Modified Bitumen (SBS modifier)	2	5	4	4	3	5	2	3	Mop-applied materials
Single-ply, ballasted*	4	2	2	2	1	2	1	3	Simple & inexpensive
Single-ply, adhered*	2	4	2	2	4	3	4	3	High growth in this segment
Single-ply, mechanically attached*	3	3	2	2	4	3	4	3	Rapid development in this segment
Non-vulcanized elastomers, mechanically attached	2	2	2	4	4	3	4	3	Relatively small market segment
Preformed Metal	2	2	2	4	2	2	3	5	Durable materials
Spray-applied Polyurethane	2	5	1	2	3	1	2	1	Requires training and careful quality control

* Characteristics listed are for EPDM single ply sheet; other elastomers have similar characteristics.

A rating of:

     1 = poor; not recommended
     2 = fair
     3 = average
     4 = better than average
     5 = excellent; recommended

Division 7

Figure 7-9   Single Ply Roofing Guide

This table lists the common generic classification of single ply roofing materials, compatible substrates, methods of attachment to the roof deck, and methods for sealing the material.

	Generic Materials (Classification)	Compatible Substrates						Attachment Method				Sealing Method				
		Slip-sheet required	Concrete	Existing asphalt membrane	Insulation board	Plywood	Spray urethane foam	Adhesive	Fully adhered	Loosely laid/ballast	Partially-adhered	Adhesive	Hot air gun	Self sealing	Solvent	Torch heating
Thermo Setting	EPDM (Ethylene propylene diene monomer)	X	X	X	X	X	X	X	X	X	X	X		X	X	
	Neoprene (Synthetic rubber)	X	X		X	X		X	X	X		X				
	PIB (Polyisobutylene)	X	X	X	X	X	X	X		X		X	X		X	
Thermo Plastic	CSPE (Chlorosulfenated polyethylene)	X	X		X	X	X	X	X	X	X	X	X			
	CPE (Chlorinated polyethylene)	X	X		X	X			X	X	X	X				
	PVC (Polyvinyl chloride)	X	X		X	X	X			X	X	X			X	
Composites	Glass reinforced EPDM/neoprene	X	X		X	X	X			X		X				
	Modified bitumen/polyester	X		X	X	X			X			X	X			X
	Modified bitumen/polyethylene & aluminum	X	X		X	X		X	X			X				X
	Modified bitumen/polyethylene sheet	X	X		X	X				X			X			X
	Modified CPE				X	X			X			X				
	Non woven glass reinforced PVC							X	X	X		X				
	Nylon reinforced PVC		X		X	X				X				X	X	
	Nylon reinforced/butylorneoprene	X							X					X	X	
	Polyester reinforced CPE	X	X	X	X	X	X			X	X	X	X			
	Polyester reinforced PVC	X	X		X	X	X			X	X	X	X		X	
	Rubber asphalt/plastic sheet	X	X	X	X	X			X					X		

Single Ply Roofing Membrane Installation Guide

Figure 7-10 Roofing Materials Quantities – Asphalt Rolls and Sheets

This table shows some of the characteristics of asphalt roofing materials.

Product	Approximate Shipping Weight		Squares per Package	Length	Width	Selvage	Ex-posure	ASTM Fire & Wind Ratings
	per Roll	per Square						
Mineral Surface Roll	75# to 90#	75# to 90#	1	36' to 38'	36"	0" to 4"	32" to 34"	C
Mineral Surface Roll Double Coverage	55# to 70#	110# to 140#	1/2	36'	36"	19"	17"	C
Smooth Surface Roll	50# to 86#	40# to 65#	1 to 2	36' to 72'	36"	N/A	34"	None
Saturated Felt Underlayment (non-perforated)	35# to 60#	11# to 30#	2 to 4	72' to 144'	36"	N/A	17" to 34"	*

*May be a component in a complete fire-rated system. Check with the manufacturer.

(courtesy Asphalt Roofing Manufacturers Association)

Figure 7-11 Roofing Material Quantities – Built-Up Systems

This table shows typical quantities required to apply 100 SF (1 square) of built-up asphalt roofing.

Number of Plies	Rolls of 15 Lb. Saturated Felt (432 S.F.)	Pounds of Asphalt	Pounds of Slag	Pounds of Gravel
3	3/4	150	300	400
4	1	180	300	400
5	1-1/4	230	300	400
3 (1 dry ply)	3/4	120	300	400
4 (2 dry plies)	1	120	300	400
5 (2 dry plies)	1-1/4	150	300	400

For smooth-top finish without slag or gravel, deduct 20 lbs. of asphalt per 100 SF of roofing. When 30 lb. saturated felt is used as a first ply, add 1/4 of a roll of 15-lb. felt.

(courtesy How to Estimate Building Losses and Construction Costs, *Paul I. Thomas, Prentice-Hall*)

Division 7

Figure 7-12 Roofing Material Quantities – Shingles

This table describes various types of roofing shingles and some of their characteristics and expected coverage.

Product	Configuration	Approximate Shipping Weight per Square	Shingles per Square	Bundles per Square	Width	Length	Exposure	ASTM* Fire & Wind Ratings
Self-Sealing Random-Tab Strip Shingle  Multi-Thickness	Various Edge, Surface Texture & Application Treatments	240# to 360#	64 to 90	3, 4 or 5	11-1/2" to 14"	36" to 40"	4" to 6"	A or C - Many Wind Resistant
Self-Sealing Random-Tab Strip Shingle 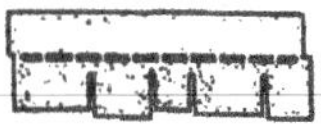 Single Thickness	Various Edge, Surface Texture & Application Treatments	240# to 300#	65 to 80	3 or 4	12" to 13-1/4"	36" to 40"	4" to 5-5/8"	A or C - Many Wind Resistant
Self-Sealing Square-Tab Strip Shingle  Three-Tab	3 Tab or 4 Tab	200# to 300#	65 to 80	3 or 4	12" to 13-1/4"	36" to 40"	5" to 5-5/8"	A or C - All Wind Resistant
Self-Sealing Square-Tab Strip Shingle 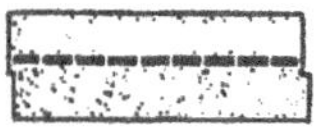 No Cut-Out	Various Edge and Surface Texture Treatments	200# to 300#	65 to 81	3 or 4	12" to 13-1/4"	36" to 40"	5" to 5 5/8"	A or C - All Wind Resistant
Individual Interlocking Shingle  Basic Design	Several Design Variations	180# to 250#	72 to 120	3 or 4	18" to 22-1/4"	20" to 22-1/2"	—	A or C - Many Wind Resistant

Other types [are] available from some manufacturers in certain areas of the country.

Consult your regional Asphalt Roofing Manufacturers Association.

(courtesy Asphalt Roofing Manufacturers Association)

Figure 7-13 Roofing Material Quantities – Red Cedar Shingles

This table shows the expected coverage and materials required for the installation of various sizes of red cedar shingles.

Covering Capacities and Approximate Nail Requirements of Certigrade Red Cedar Shingles

Shingle Exposure in Inches	No. 1 Grade Sixteen Inch Shingles — Nail Size: 3d, 1-1/4" Long				No. 1 Grade Eighteen Inch Shingles — Nail Size: 3d, 1-1/4" Long				No. 1 Grade Twenty-Four Inch Shingles — Nail Size: 4d, 1-1/2" Long			
	Four-Bundle Square		One Bundle		Four Bundle Square		One Bundle		Four-Bundle Square		One Bundle	
	Coverage in S.F.	Pounds Nails	Coverage in S.F.	Pounds Nails	Coverage in S.F.	Pounds Nails	Coverage in S.F.	Pounds Nails	Coverage in S.F.	Pounds Nails	Coverage in S.F.	Pounds Nails
3-1/2	70	2-7/8	17-1/2	3/4								
4	80	2-1/2	20	5/8	72-1/2	2-1/2	18	5/8				
4-1/2	90	2-1/4	22-1/2	5/8	81-1/2	2-1/4	20	5/8				
5	100*	2	25	1/2	90-1/2	2	22-1/2	1/2				
5-1/2	110	1-3/4	27-1/2	1/2	100*	1-3/4	25	1/2				
6	120	1-2/3	30	3/8	109	1-2/3	27	3/8	80	2-1/3	20	5/8
6-1/2	130	1-1/2	32-1/2	3/8	118	1-1/2	29-1/2	3/8	86-1/2	2-1/8	21-1/2	1/2
7	140	1-2/5	35	1/3	127	1-2/5	31-1/2	1/3	93	2	23	1/2
7-1/2	†150	1-1/3	37-1/2	1/3	136	1-1/3	34	1/3	100*	1-7/8	25	1/2
8	160		40		145-1/2	1-1/4	36	1/3	106-1/2	1-3/4	26-1/2	1/2
8-1/2	170		42-1/2		†154-1/2	1-1/4	38-1/2	1/4	113	1-2/3	28	1/2
9	180		45		163-1/2		40-1/2		120	1-1/2	30	3/8
9-1/2	190		47-1/2		172-1/2		43		126-1/2	1-1/2	31-1/2	3/8
10	200		50		181-1/2		45		133	1-1/2	33	3/8
10-1/2	210		52-1/2		191		47-1/2		140	1-1/3	35	1/3
11	220		55		200		50		146-1/2	1-1/4	36-1/2	1/3
11-1/2	230		57-1/2		209		52		†153	1-1/4	38	1/3
12	‡240		60		218		54-1/2		160		40	
12-1/2					227		56-1/2		166-1/2		41-1/2	
13					236		59		173		43	
13-1/2					245-1/2		61		180		45	
14					‡254-1/2		63-1/2		186-1/2		46-1/2	
14-1/2									193		48	
15									200		50	
15-1/2									206-1/2		51-1/2	
16									‡213		53	

* Maximum exposure recommended for roofs

† Maximum exposure recommended for single-coursing on side walls

‡ Maximum exposure recommended for double-coursing on side walls. Figures in italics are inserted for convenience in estimating quantity of shingles needed for wide exposures in double-coursing, with butt-nailing. In double-coursing, with any exposure chosen, the figures indicate the amount of shingles for the outer courses. Order an equivalent number of shingles for concealed courses. Approximately 1-1/2 lbs., 5d small-headed nails required per square (100 SF wall area) to apply the outer course of 16-inch shingles at 12-inch weather exposure. Plus 1/2 lb. 3d nails for under course shingles. Figure slightly fewer nails for 18-inch shingles at 14-inch exposure.

(courtesy American Institute of Steel Construction, Inc.) Excerpted from: Certigrade Handbook of Red Cedar Shingles, *published by Red Cedar Shingle Bureau (Red Cedar Shingle and Handsplit Shake Bureau), Seattle, Washington.*

Division 7

Figure 7-14   Roofing Material Quantities – Standard 3/16"-Thick Slate

This table shows the materials needed for installing various sizes of 3/16" thick roof slates.

Size of Slate (In.)	Slates per Square	Exposure with 3" Lap	Nails per Square Lbs.	Nails per Square Ozs.	Size of Slate (In.)	Slates per Square	Exposure with 3" Lap	Nails per Square Lbs.	Nails per Square Ozs.
26 x 14	89	11-1/2"	1	0	16 x 14	160	6-1/2"	1	13
					16 x 12	184	6-1/2"	2	2
24 x 16	86	10-1/2"	1	0	16 x 11	201	6-1/2"	2	5
24 x 14	98	10-1/2"	1	2	16 x 10	222	6-1/2"	2	8
24 x 13	106	10-1/2"	1	3	16 x 9	246	6-1/2"	2	13
24 x 12	114	10-1/2"	1	5	16 x 8	277	6-1/2"	3	2
24 x 11	125	10-1/2"	1	7					
					14 x 12	218	5-1/2"	2	8
22 x 14	108	9-1/2"	1	4	14 x 11	238	5-1/2"	2	11
22 x 13	117	9-1/2"	1	5	14 x 10	261	5-1/2"	3	3
22 x 12	126	9-1/2"	1	7	14 x 9	291	5-1/2"	3	5
22 x 11	138	9-1/2"	1	9	14 x 8	327	5-1/2"	3	12
22 x 10	152	9-1/2"	1	12	14 x 7	374	5-1/2"	4	4
20 x 14	121	8-1/2"	1	6	12 x 10	320	4-1/2"	3	10
20 x 13	132	8-1/2"	1	8	12 x 9	355	4-1/2"	4	1
20 x 12	141	8-1/2"	1	10	12 x 8	400	4-1/2"	4	9
20 x 11	154	8-1/2"	1	12	12 x 7	457	4-1/2"	5	3
20 x 10	170	8-1/2"	1	15	12 x 6	533	4-1/2"	6	1
20 x 9	189	8-1/2"	2	3					
					11 x 8	450	4"	5	2
18 x 14	137	7-1/2"	1	9	11 x 7	515	4"	5	14
18 x 13	148	7-1/2"	1	11					
18 x 12	160	7-1/2"	1	13	10 x 8	515	3-1/2"	5	14
18 x 11	175	7-1/2"	2	0	10 x 7	588	3-1/2"	7	4
18 x 10	192	7-1/2"	2	3	10 x 6	686	3-1/2"	7	13
18 x 9	213	7-1/2"	2	7					

Figure 7-15 Wood Siding Quantities

This table allows you to compute the amount of wood siding (in board feet required per square foot of wall area to receive siding) based on the style and dimensions of the siding.

Type of Siding	Size	Exposure	Board Feet per S.F. of Wall Area	Lbs. Nails per MBM of Siding		
				Stud Spacing		
				16"	20"	24"
Plain Bevel Siding	1/2" x 4"	2-1/2"	1.60	17	13	11
		2-3/4"	1.45			
	1/2" x 6"	4-1/2"	1.33	11	9	7
		4-3/4"	1.26			
		5"	1.20			
	1/2" x 8"	6-1/2"	1.23	8	7	6
		7"	1.14			
Plain Bevel Bungalow Siding	5/8" x 8"	6-1/2"	1.23	14	11	9
		7"	1.14			
	5/8" x 10"	8-1/2"	1.18	16	13	11
		9"	1.11			
	3/4" x 8"	6-1/2"	1.23	14	11	9
		7"	1.14			
	3/4" x 10"	8-1/2"	1.18	16	13	11
		9"	1.11			
	3/4" x 12"	10-1/2"	1.14	14	11	9
		11"	1.09			
Drop or Rustic Siding	3/4" x 4"	3-1/4"	1.23	27	22	18
	3/4" x 6"	5-1/6"	1.19	18	15	12
		5-3/16"	1.17			

Figure 7-16 Estimating Guide for Caulking

Linear Feet per Full Gallon (231 cu. in.)		Width of Joint						
		1/4"	3/8"	1/2"	5/8"	3/4"	7/8"	1"
Depth of Joint	1/4"	308	205	154	123	102	88	—
	3/8"	205	136	102	82	68	58	—
	1/2"	154	102	77	61	51	44	38
	3/4"	102	68	51	43	34	30	26

Example: One full gallon is sufficient material to fill a joint 1/2" wide, 3/8" deep, and 102' long.

Cartridges: When figuring feet per cartridge for a particular joint size, divide linear feet shown above by 12.

Division 7

Figure 7-17   Installation Time in Labor-Hours for Roof Deck Insulation

Description	Labor-Hours	Unit
Roof Deck Insulation		
Fiberboard to 2" Thick	.010	S.F.
Fiberglass	.008	S.F.
Foamglass		
1-1/2" and 2" Thick	.010	S.F.
3" and 4" Thick	.011	S.F.
Tapered	.013	B.F.
Perlite to 1-1/2" Thick	.010	S.F.
2" Thick	.011	S.F.
Polyisocyanurate		
3/4" Thick	.005	S.F.
1" and 1-1/2" Thick	.006	S.F.
2" Thick	.007	S.F.
2-1/2" to 3-1/2" Thick	.008	S.F.
Tapered	.006	B.F.
Polystyrene, Expanded		
1" Thick	.005	S.F.
2" and 3" Thick	.006	S.F.
4" and 6" Thick	.007	S.F.
Tapered	.005	B.F.
Polystyrene, Extruded		
1" Thick	.005	S.F.
2" Thick	.006	S.F.
3" and 4" Thick	.008	S.F.
Tapered	.005	B.F.
Sprayed Polystyrene or Urethane		
1" Thick	.031	S.F.
2" Thick	.051	S.F.
Lightweight Cellular Fill		
Portland Cement and Foaming Agent	1.120	C.Y.
Vermiculite or Perlite	1.120	C.Y.
Ready Mix		
2" Thick	.006	S.F.
3" Thick	.007	S.F.

(continued on next page)

Figure 7-17 Installation Time in Labor-Hours for Roof Deck Insulation
(continued from previous page)

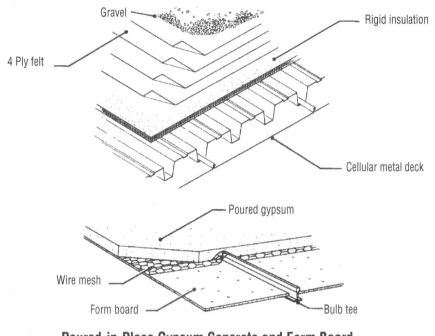

Gravel

Rigid insulation

4 Ply felt

Cellular metal deck

Poured gypsum

Wire mesh

Form board

Bulb tee

Poured-in-Place Gypsum Concrete and Form Board

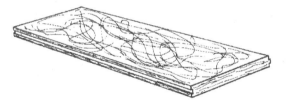

Tongue and Groove Fiberboard Insulation

Figure 7-18   Installation Time in Labor-Hours for Built-Up Roofing

Description	Labor-Hours	Unit
Built-up Roofing		
Asphalt Flood Coat with Gravel or Slag		
Fiberglass Base Sheet		
3 Plies Felt Mopped	2.545	Sq.
On Nailable Decks	2.667	Sq.
4 Plies Felt Mopped	2.800	Sq.
On Nailable Decks	2.947	Sq.
Coated Glass Fiber Base Sheet		
3 Plies Felt Mopped	2.800	Sq.
On Nailable Decks	2.947	Sq.
Organic Base Sheet and 3 Plies Felt	2.545	Sq.
On Nailable Decks	2.667	Sq.
Coal Tar Pitch with Gravel or Slag		
Coated Glass Fiber Base Sheet and		
2 Plies Glass Fiber Felt	2.947	Sq.
On Nailable Decks	3.111	Sq.
Asphalt Mineral Surface Roll Roofing	2.074	Sq.
Walkway		
Aslphalt Impregnated	.020	S.F.
Patio Blocks 2" Thick	.070	S.F.
Expansion Joints Covers	.048	L.F.

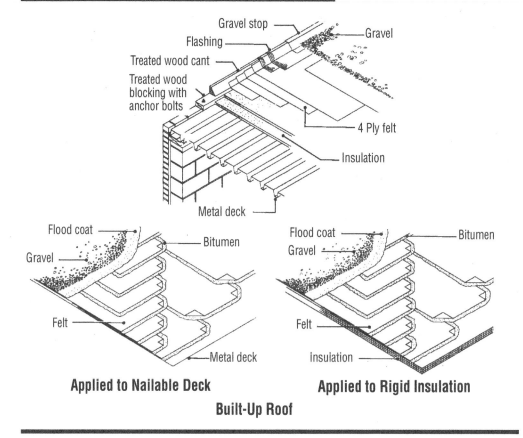

Applied to Nailable Deck   **Applied to Rigid Insulation**

Built-Up Roof

Figure 7-19 Installation Time in Labor-Hours for Single Ply Roofing

Description	Labor-Hours	Unit
Single Ply Membrane, General		
Loose-Laid and Ballasted	.784	Sq.
Mechanically Fastened	1.143	Sq.
Fully Adhered, All Types	1.538	Sq.
Modified Bitumen, Cap Sheet		
Fully Adhered Torch Welding	.019	S.F.
Asphalt Mopped	.028	S.F.
Cured Neoprene for Flashing	.028	S.F.

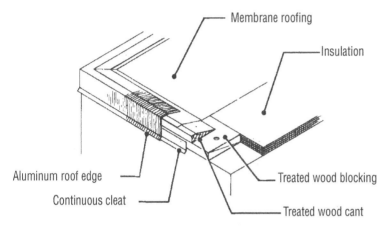

Roof Edge, Single Ply Roofing

Figure 7-20   Installation Time in Labor-Hours for Shingle and Tile Roofing

Description	Labor-Hours	Unit
Shingles, Manual Nailed		
Aluminium	1.600	Sq.
Ridge Cap or Valley, Manual Nailed	.047	L.F.
Fiber Cement		
500 lb. per Sq.	3.636	Sq.
Starters	2.667	C.L.F.
Hip and Ridge	8.000	C.L.F.
Asphalt Standard Strip		
Class A 210 to 235 lb. per Sq.	1.455	Sq.
Class C 235 to 240 lb. per Sq.	1.600	Sq.
Standard Laminated		
Class A 240 to 260 lb. per Sq.	1.778	Sq.
Class C 260 to 300 lb. per Sq.	2.000	Sq.
Premium Laminated		
Class A 260 to 300 lb. per Sq.	2.286	Sq.
Class C 300 to 385 lb. per Sq.	2.667	Sq.
Hip and Ridge Shingles	.024	L.F.
Slate Including Felt Underlay	4.571	Sq.
Steel	3.636	Sq.
Wood		
5" Exposure, 16" long	3.200	Sq.
5-1/2" Exposure, 18" long	2.090	Sq.
Panelized 8" Strips 7" Exposure	2.667	Sq.
Ridge	.023	L.F.
Shingles, Pneumatic Nailed		
Asphalt Standard Strip		
Class A 210 to 235 lb. per Sq.	1.143	Sq.
Class C 235 to 240 lb. per Sq.	1.280	Sq.
Standard Laminated		
Class A 240 to 260 lb. per Sq.	1.422	Sq.
Class C 260 to 300 lb. per Sq.	1.600	Sq.
Premium Laminated		
Class A 260 to 300 lb. per Sq.	1.831	Sq.
Class C 300 to 385 lb. per Sq.	2.133	Sq.
Hip and Ridge Shingles	.019	L.F.
Wood		
5" Exposure	2.462	Sq.
5-1/2" Exposure	2.241	Sq.
Panelized 8' Strips 7" Exposure	2.000	Sq.
Tiles		
Aluminum		
Mission	3.200	Sq.
Spanish	2.667	Sq.
Clay,		
Americana, 158 Pc/Sq.	4.848	Sq.
Spanish, 171 Pc/Sq.	4.444	Sq.
Mission, 192 Pc/ Sq.	6.957	Sq.
French, 133 Pc/Sq.	5.926	Sq.
Norman, 317 Pc/Sq.	8.000	Sq.
Williamsburg, 158 Pc/Sq.	5.926	Sq.
Concrete	5.926	Sq.

Figure 7-21 Installation Time in Labor-Hours for Gutters and Downspouts

Description	Labor-Hours	Unit
Gutters		
Aluminum	.067	L.F.
Copper Stock Units		
4" Wide	.067	L.F.
6" Wide	.070	L.F.
Steel Galvanized or Stainless	.067	L.F.
Vinyl	.073	L.F.
Wood	.080	L.F.
Downspouts		
Aluminum		
2" x 3"	.042	L.F.
3" Diameter	.042	L.F.
4" Diameter	.057	L.F.
Copper		
2" or 3" Diameter	.042	L.F.
4" Diameter	.055	L.F.
5" Diameter	.062	L.F.
2" x 3"	.042	L.F.
3" x 4"	.055	L.F.
Steel Galvanized		
2" or 3" Diameter	.042	L.F.
4" Diameter	.055	L.F.
5" Diameter	.062	L.F.
6" Diameter	.076	L.F.
2" x 3"	.042	L.F.
3" x 4"	.055	L.F.
Epoxy Painted		
2" x 3"	.042	L.F.
3" x 4"	.055	L.F.
Steel Pipe, Black, Extra Heavy		
4" Diameter	.400	L.F.
6" Diameter	.444	L.F.
Stainless Steel		
2" x 3" or 3" Diameter	.042	L.F.
3" x 4" or 4" Diameter	.055	L.F.
4" x 5" or 5" Diameter	.059	L.F.
Vinyl		
2" x 3"	.038	L.F.
2-1/2" Diameter	.036	L.F.

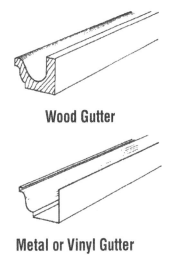

Wood Gutter

Metal or Vinyl Gutter

(continued on next page)

Figure 7-21   Installation Time in Labor-Hours for Gutters and
Downspouts *(continued from previous page)*

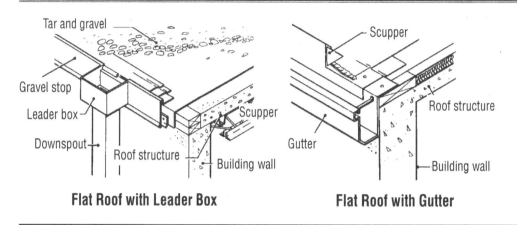

Flat Roof with Leader Box   **Flat Roof with Gutter**

Figure 7-22   Installation Time in Labor-Hours for Metal Siding

Description	Labor-Hours	Unit
Aluminum Siding		
On Steel Frame	.041	S.F.
On Wood Frame	.030	S.F.
Closure Strips and Flashing	.040	L.F.
Steel Siding, on Steel Frame	.040	S.F.
Metal Siding Panels, Insulated with Liner		
Factory Assembled	.084	S.F.
Metal Fascia, No Furring or Framing Incl.		
Long Panels	.055	S.F.
Short Panels	.070	S.F.
Mansard Roofing		
With Battens, Custom		
Straight Surfaces	.070	S.F.
Curved Surfaces	.107	S.F.
Framing, to 5' High	.070	S.F.
Soffits, Metal Panel	.064	S.F.
Furring		
Metal, 3/4" Channel		
16" on Center	.030	S.F.
24" on Center	.023	S.F.
Wood Strips		
On Masonry	.016	L.F.
On Concrete	.031	L.F.
Channel Framing, Overhead 24" on Center	.019	S.F.
Sheathing		
Drywall, 1/2" Thick	.015	S.F.
Plywood, 1/2" Thick	.014	S.f.
Channel Gifts, Steel		
Less than 8"	.072	Lb.
Greater than 8"	.048	Lb.
Slotted Channel Framing System		
Minimum	.007	Lb.
Maximum	.010	Lb.

(continued on next page)

Division 7

Figure 7-22   Installation Time in Labor-Hours for Metal Siding
(continued from previous page)

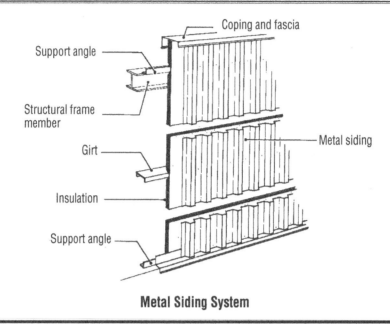

Metal Siding System

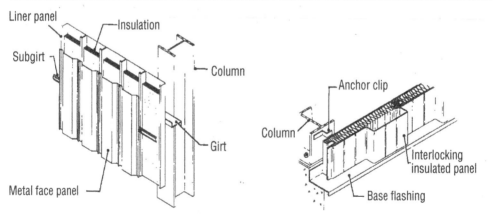

Field Assembled Insulated Metal Wall   **Factory Assembled Insulated Metal Wall**

Figure 7-23 Installation Time in Labor-Hours for Fireproofing
Structural Steel

Description	Labor-Hours	Unit
Fireproofing - 10" Column Encasements		
Perlite Plaster	.273	V.L.F.
1" Perlite on 3/8" Gypsum Lath	.345	V.L.F.
Sprayed Fiber	.131	V.L.F.
Concrete 1-1/2" Thick	.716	V.L.F.
Gypsum Board 1/2" Fire Resistant,		
1 Layer	.364	V.L.F.
2 Layer	.428	V.L.F.
3 Layer	.530	V.L.F.
Fireproofing - 16" x 7" Beam Encasements		
Perlite Plaster on Metal Lath	.453	L.F.
Gypsum Plaster on Metal Lath	.408	L.F.
Sprayed Fiber	.079	L.F.
Concrete 1-1/2" Thick	.554	L.F.
Gypsum Board 5/8" Fire Resistant	.488	L.F.

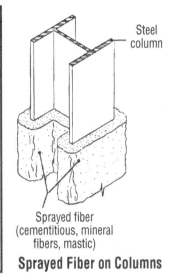

Steel column

Sprayed fiber
(cementitious, mineral
fibers, mastic)

Sprayed Fiber on Columns

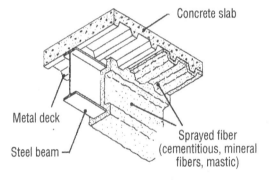

Concrete slab

Metal deck

Steel beam

Sprayed fiber
(cementitious, mineral
fibers, mastic)

Sprayed Fiber on Beams and Girders

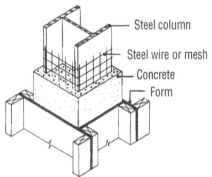

Steel column

Steel wire or mesh

Concrete

Form

Concrete Encasement on Columns

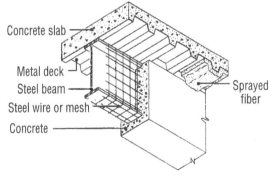

Concrete slab

Metal deck

Steel beam

Steel wire or mesh

Concrete

Sprayed fiber

Concrete Encasement on Beams and Girders

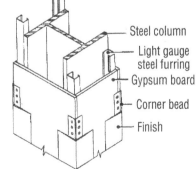

Steel column

Light gauge steel furring

Gypsum board

Corner bead

Finish

Gypsum Board on Columns

Division 7

(continued on next page)

Figure 7-23   Installation Time in Labor-Hours for Fireproofing
Structural Steel *(continued from previous page)*

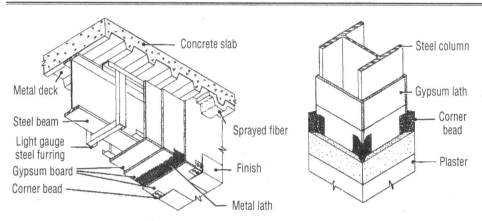

Gypsum Board on Beams and Girders   Plaster on Gypsum Lath — Columns

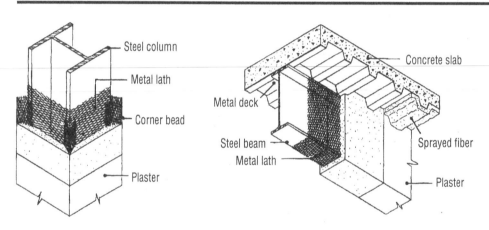

Plaster on Metal Lath — Columns   Plaster on Metal Lath — Beams and Girders

Figure 7-24 Installation Time in Labor-Hours for Exterior Insulation and Finish Systems

Description	Labor-Hours	Unit
Field applied, 1" EPS insulation	.136	S.F.
With 1/2" cement board sheathing	.182	
2" EPS insulation	.136	
With 1/2" cement board sheathing	.182	
3" EPS insulation	.136	
With 1/2" cement board sheathing	.182	
4" EPS insulation	.136	
With 1/2" cement board sheathing	.182	
Premium finish, add	.032	
Heavy duty reinforcement add	0.44	
2.5#/S.Y. metal lath substrate add	.107	S.Y.
3.4#/S.Y. metal lath substrate add	.107	"
Color Texture change	.032	S.F.
With substrate leveling base coat	.015	
With substrate sealing base coat	.007	

Figure 7-25 Installation Time in Labor-Hours for Spray Polyurethane Roof Systems

Description	Labor-Hours	Unit
Closed-cell, spray polyurethane foam, 2 pounds per cubic foot density		
1" thick	.004	S.F.
2" thick	.008	
3" thick	.012	
3-1/2" thick	.014	
4" thick	.016	
5" thick	.020	
5-1/2" thick	.022	
6" thick	.024	
Sprayed polyurethane foam roofing (SPF)		
Primer for metal substrate (when required)	.008	S.F.
Primer for non-metal substrate (when required)	.008	"
Closed cell spray, polyurethane foam, 3 lbs per		
CF density, 1", R6.7	.002	S.F.
2", R13.4	.002	
3", R20.1	.002	
Spray-on silicone coating	.010	

Division 7

Figure 7-26 Installation Time in Labor-Hours for Green Roof Systems

Description	Labor-Hours	Unit
Vegetated Roofing		
Soil mixture for green roof 30% sand, 55% gravel, 15% soil		
Hoist and spread soil mixture 4 inch depth, up to five stories tall roof	.014	S.F
6" depth	.021	
8" depth	.028	
10" depth	.035	
12" depth	.042	↓
Mobilization 55 ton crane to site	2.222	Ea.
Hoisting cost to five stories per day (Avg. 28 picks per day)	56.000	Day
Mobilization or demobilization, 100 ton crane to site driver & escort	6.400	Ea.
Hoisting cost six to ten stories per day (Avg. 21 picks per day)	56.000	Day
Hoist and spread soil mixture 4" depth, six to ten stories tall roof	.014	S.F.
6" depth	.021	
8" depth	.028	
10" depth	.035	
12" depth	.042	↓
Green roof edging treated lumber 4" x 4", no hoisting included	.040	L.F.
4" x 6"	.040	
4" x 8"	.044	
4" x 6" double stacked	.053	
Green roof edging redwood lumber 4" x 4", no hoisting included	.040	
4" x 6"	.040	
4" x 8"	.044	
4" x 6" double stacked	.053	↓
Planting sedum, light soil, potted, 2-1/4" diameter, two per S.F.	.019	S.F.
One per S.F.	.010	
Planting sedum mat per S.F. including shipping (4,000 S.F. minimum)	.008	
Installation sedum mat system (no soil required) per S.F.		
(4,000 S.F. minimum)	.008	↓

Division Eight
Openings

Introduction

Doors, windows, glass, and glazing can represent a considerable portion of a total construction contract. The more "upscale" a building or project, the more emphasis will be placed on the visible effects of the building, including doors, windows, and glass treatments. There is a wide range of door and window materials and hardware. Increasing the quality of these components can add a considerable amount to an estimate. The energy rating or material sustainability may also be a factor.

green | Division 8 works hand in hand with Division 7, because openings such as doors and windows are usually present in insulated walls and roofs. When doors and windows are closed, the walls should have the same level of thermal and moisture protection as if there were no openings at all. In the past, single pane glass for windows and solid wood for doors were about the only options, and neither were particularly well-sealed in their frames. With the greater focus on energy conservation and the green building movement, increasing insulating values while improving functionality has become a priority. More options for practical and affordable door and window frame sealing are available. Double pane glass, with argon filling the space between the panes, and coatings that limit specific spectrums of the sun's rays are commonly used. Skylights have come a long way, too. Reflectors are used to increase the light entering the skylight, and some can even be automated to track the sun. Devices called light tubes work by reflecting light through a flexible pipe, permitting an exterior opening to be a good distance, both vertically and horizontally, from where the light is desired.

Most architectural plans or specifications include door, window, and hardware schedules that indicate types, sizes, materials, and options. This takes most of the guesswork out of the estimating process. If these schedules are not available at the time the estimate is prepared, it's important to document the assumptions made for each of these components in the bid.

To assist in estimating this division, tables and charts cover the following areas:

- Metal doors/frames
- Wood/plastic doors
- Special doors
- Entrances/storefronts
- Windows – metal
- Windows – wood/plastic
- Hardware/operators/seals
- Glass/glazing
- Curtain walls

Estimating Data

The following tables present estimating guidelines for items found in Division 8 – Openings. Please note that these guidelines are intended as indicators of what may generally be expected, and that each project must be evaluated individually.

Table of Contents

Division 8

Estimating Tips

Interior Door Ratings

For walls to be considered fire-rated, any doors in those walls must be fire-rated. Most plans do not spell out which interior partitions are to be considered fire-rated. In commercial applications, a rule of thumb is that all partitions that have drywall (or masonry) from the floor to the above structure, and few, if any, penetrations should be considered fire-rated. The glass in a rated door should be fire-rated.

Door Schedule

If the drawings do not include a door schedule, it may be worth the estimator's time to develop one, especially if the project is large or complicated. The schedule should indicate the opening number, door type, size, material, glass or louver requirements, and remarks. The quantity take-off process can be expedited by making a copy of the schedule and noting the quantity and hardware requirements next to each door type.

The door schedule should include a frame schedule listing the frame material; type; and jamb, head, and sill details. Fire-rating, ballistic, and pressure requirements should be noted.

Hardware requirements should be listed on the door schedule as well. Keep in mind the fact that the hardware can, in some instances, be more costly than the door itself. The hardware schedule may be in the specification.

Handicap Access

While the drawings may not show it, local codes may require special hardware and opening systems to allow a structure to be accessible to the handicapped. Contact the local authorities for related codes and requirements.

Special Doors

Special attention should be given to any oversized or unusual type of door. Such items should not be priced on a prorated basis, as they are generally special ordered. The cost of special doors can skyrocket, especially if they involve exotic woods, special finishes, or special attention (which usually means higher labor costs). Note that special doors may require a considerable amount of lead time for ordering and shipping and may also require large equipment.

Window Schedule

As with all doors to be included in the project, all windows should be listed on the drawings in a window schedule. If none is included, it may be well worth the time to create one. The schedule should contain the opening number, window type, window size, glass type, frame material and details, and required accessories and hardware. In addition, energy ratings and window films should be reviewed. As with the door schedule, making a copy of the window schedule can expedite the quantity takeoff.

Building Hardware

As a rule of thumb, building hardware for an average quality building can be expected to run in the neighborhood of 2% of the entire building cost. However, security requirements and wiring of the hardware may drive up the cost.

Checklist ✓

For an estimate to be reliable, all items must be accounted for. A complete estimate can also limit contingencies. The following checklist can be used to help ensure that all items are included.

□ *Doors*
- □ Accordion
- □ Acoustical
- □ Aluminum
- □ Bi-parting
- □ Blast
- □ Cold storage
- □ Fire
  - □ Rated
  - □ Rolling industrial
- □ Folding
- □ Garage
- □ Glass
  - □ Sliding
  - □ Swing
- □ Grilles
  - □ Rolling
- □ Hollow metal
- □ Overhead
- □ Revolving
- □ Rolling
- □ Service
- □ Sliding
- □ Tin-clad
- □ Vault
- □ Wood
- □ Handicapped
- □ Security
- □ Ballistic
- □ Pressure

□ *Door Frames for above*

□ *Entrances*

□ *Storefronts*

□ *Windows*
- □ Casement
- □ Projected
- □ Single-hung
- □ Double-hung
- □ Picture
- □ Roof
- □ Screens
- □ Security
- □ Sliding
- □ Hurricane
- □ Ballistic

□ *Window Wall Systems*
- □ Curtain wall systems
- □ Blast protection

□ *Hardware*

□ *Hardware Allowance*
- □ Handicapped
- □ Security

□ *Automatic Openers*

□ *Glass*
- □ Acoustical
- □ Faceted
- □ Insulated
- □ Laminated
- □ Obscure
- □ Plexiglass
- □ Reflective
- □ Sheet
- □ Spandrel
- □ Stained

- ☐ Window
- ☐ Wire
- ☐ Fire Glass
- ☐ Ballistic
- ☐ Glazing
- ☐ Allowance for broken glass replacement
- ☐ Allowance for temporary glass
- ☐ Window films
- ☐ Energy rating

Figure 8-1 Standard Door Nomenclature

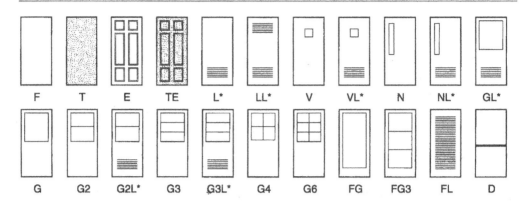

Consult SDI Technical Data Series SDI 106 for further information on standard design nomenclature for Standard Steel Doors.

Nomenclature Letter Symbols

F – Flush
T – Textured
E – Embossed
TE – Textured and Embossed
L* – Louvered (top or bottom)

LL* – Louvered (top and bottom)
V – Vision Lite
VL* – Vision Lite and Louvered
N – Narrow Lite
NL* – Narrow Lite and Louvered

GL* – Half Glass and Louvered
G – Half Glass (options G2, G3, G4 and G6)
FG – Full Glass (option FG3)
FL – Full Louver
D – Dutch Door

* Louvered door designs are further specified as inserted louver (I), pierced (P), or air condition grille (A). When ordering, specify design, louver size, and/or free area requirements.

(courtesy Steel Door Institute)

Figure 8-2   Door Density Guide

This table can be used to estimate the quantity of doors needed when no interior plan is available, such as in a conceptual estimating situation. To use this table, first pick out from the left-most column the type of structure to be estimated. Moving across that major row, from the second column, "Stories," pick out the number of stories in the building. Read across that line to the "SF Door" column. These figures are typical ratios of square feet of floor area to each door.

Example: To estimate the number of interior doors for a one-story office building of 7,000 square feet, find "Offices" in the left column; "200–500 SF/Door" is listed. If the building is average (no high density of small offices; no high density of large conference/meeting rooms), one would choose a density figure in the middle, say 350 SF. To find the total number of doors to be expected in the office, divide 7,000 SF by 350 SF = 20 doors.

Building Type	Stories	S.F./ Door	Building Type	Stories	S.F./ Door
Apartments	1	90	Nursing Home	1	80
	2	80		2–4	80
	3	90	Offices	1	200–500
	5	90		2	200–500
	6–15	80		3–5	200–500
Bakery	1	500		6–10	200–500
	2	500		11–20	200–500
Bank	1	200	Parking Ramp (Open)	2–8	600
	2–4	150	Parking Garage	2–8	600
Bottling Plant	1	500	Pre-Engineered Store	1	600
Bowling Alley	1	500	Office	1	150
			Shop	1	150
Bus Terminal	1	150	Radio & TV Broadcasting	1	250
Cannery	1	1000	& TV Transmitter	1	400
Car Wash	1	180	Self Service Restaurant	1	150
Dairy Plant	1	300	Cafe & Drive-In Restaurant	1	180
			Restaurant with seating	1	250
Department Store	1	600	Supper Club	1	250
	2–5	600	Bar or Lounge	1	240
Dormitory	2	90	Retail Store or Shop	1	600
	3–5	90	Service Station Masonry	1	150
	6–15	90	Metal panel	1	150
Funeral Home	1	150	Frame	1	150
	2	140	Shopping Center (strip)	1	300
Garage Sales & Service	1	300	(group)	1	400
Hotel	3–8	90		2	400
	9–15	90	Small Food Store	1	300
Laundromat	1	250	Store/Apt. above Masonry	2	100
Medical Clinic	1	60	Frame	2	100
	2–4	60	Frame	3	100
Motel	1	70	Supermarkets	1	400
	2–3	70	Truck Terminal	1	0
Movie Theater 200–600 seats	1	180	Warehouse	1	0
601–1400 seats		200			
1401–2200 seats		250			

Figure 8-3 Standard Steel Door Usage Guide

Building Types	Standard Steel Door Grades			Door Thickness		Door Design Nomenclature					General Remarks
	Grade I Standard Duty	Grade II Heavy Duty (1-3/4" only)	Grade III Extra Heavy Duty (1-3/4" only)	1-3/4"	1-3/4" or 1-3/8"	Flush (F)	Half Glass (G)	Vision Lite (V)	Full Glass (FG)	Narrow Lite (N)	
Apartment											
Main Entrance			•	•			•		•	•	Two way vision
Unit Entrance	•	•		•	•	•					Check for fire door rqmts.
Bedroom	•				•	•					
Bathroom	•				•	•					
Closet	•				•	•					
Stairwell		•	•	•				•			Fire door – two way vision
Dormitory											
Main Entrance			•	•			•		•	•	Two way vision
Unit Entrance	•	•		•		•					Check for fire door rqmts.
Bedroom	•				•	•					
Bathroom	•				•	•					
Closet	•				•	•					
Stairwell		•	•	•				•			Fire door – two way vision
Hotel – Motel											
Unit Entrance	•	•		•		•					Check for fire door rqmts.
Bathroom	•				•	•					
Closet	•				•	•					
Stairwell		•	•	•				•			Fire door – two way vision
Storage & Utility	•	•		•		•					Optional type selection

(courtesy Steel Door Institute)

(continued on next page)

Figure 8-3 Standard Steel Door Usage Guide
(continued from previous page)

Building Types	Standard Steel Door Grades			Door Thickness		Door Design Nomenclature					General Remarks
	Grade I Standard Duty	Grade II Heavy Duty (1-3/4" only)	Grade III Extra Heavy Duty (1-3/4" only)	1-3/4"	1-3/4" or 1-3/8"	Flush (F)	Half Glass (G)	Vision Lite (V)	Full Glass (FG)	Narrow Lite (N)	
Hospital—Nursing Home											
Main Entrance			•	•			•		•	•	Two way vision
Patient Room		•		•		•					
Stairwell		•	•	•				•			Fire door – two way vision
Opertg. & Exam.		•	•	•		•					Optional type selection
Bathroom	•				•	•					
Closet	•				•	•					
Recreation		•		•		•		•			Optional design selection
Kitchen		•	•	•				•			Two way vision
Industrial											
Entrance & Exit			•	•			•		•	•	Optional design selection
Office	•	•		•		•	•				Optional design selection
Production		•		•			•				
Toilet		•	•	•		•					Design option – louver door
Tool			•	•		•					Design option – Dutch door
Trucking		•		•			•				
Monorail		•		•		•	•				Optional design selection

(continued on next page)

Division 8

Figure 8-3 Standard Steel Door Usage Guide
(continued from previous page)

Building Types	Standard Steel Door Grades			Door Thickness		Door Design Nomenclature					General Remarks
	Grade I Standard Duty	Grade II Heavy Duty (1-3/4" only)	Grade III Extra Heavy Duty (1-3/4" only)	1-3/4"	1-3/4" or 1-3/8"	Flush (F)	Half Glass (G)	Vision Lite (V)	Full Glass (FG)	Narrow Lite (N)	
Office											
Entrance			•	•			•		•	•	Two way vision
Individual Office	•			•	•	•	•				Optional thickness & design
Closet	•				•	•					
Toilet		•	•	•		•					Design option-louver door
Stairwell		•	•	•				•			Fire door – two way vision
Equipment		•	•	•		•					Optional type selection
Boiler		•	•	•		•					Fire Door – flush design
School											
Entrance & Exit			•	•			•		•	•	Two way vision
Classroom		•		•			•			•	Two way vision
Toilet		•	•	•		•					Design option – louver door
Gymnasium		•	•	•		•		•			Optional design
Cafeteria		•	•	•			•				Two way vision
Stairwell		•	•	•				•			Fire door – two way vision
Closet	•				•	•					

Figure 8-4 Standard Steel Door Grades and Models

Building Types	Standard Steel Door Levels				Door Design Nomenclature					
	Level I Standard Duty 1 3/8"	Level II Heavy Duty 1 3/4" only	Level III Extra Heavy Duty 1 3/4" only	Level IV Maximum Duty 1 3/4" only	F	G	V	FG	N	L
Apartment										
Main Entrance		•	•		•	•		•	•	
Unit Entrance	•	•	•		•					
Bedroom	•				•					
Bathroom	•				•					•
Closet	•				•					•
Stairwell		•	•				•		•	
Dormitory										
Main Entrance		•	•	•		•		•	•	
Unit Entrance	•	•			•					
Bedroom	•	•			•					
Bathroom	•	•			•					•
Closet	•	•	✳		•					•
Stairwell		•	•				•		•	
Hotel - Motel										
Unit Entrance	•	•			•					
Bathroom	•				•					
Closet	•				•					•
Stairwell		•	•				•		•	
Storage & Utility		•	•		•					•
Hospital - Nursing Home										
Main Entrance		•				•		•	•	
Patient Room		•			•					
Stairwell		•	•				•		•	
Operating & Exam.		•	•		•					
Bathroom		•	•		•					•
Closet	•	•			•					•
Recreation		•			•		•			
Kitchen		•	•				•			
Industrial										
Entrance & Exit		•	•			•		•	•	
Office	•	•			•	•				
Production		•				•				
Toilet		•	•		•					•
Tool		•	•		•					
Trucking		•	•			•				
Monorail		•	•		•	•				
Office										
Entrance			•			•		•	•	
Individual Office	•				•	•				
Closet	•				•					•
Toilet		•	•		•					
Stairwell		•	•				•		•	
Equipment		•	•		•					
Boiler		•	•		•					•
School										
Entrance & Exit			•	•		•		•	•	
Classroom		•				•			•	
Toilet		•	•		•					•
Gymnasium			•	•	•	•	•			
Cafeteria		•	•			•				
Stairwell		•	•				•		•	
Closet	•	•	•	•	•					•

Note: Compliance with the local and national codes is the responsibility of the specifier.

(courtesy Steel Door Institute)

Division 8

Figure 8-5 Fire Door Classifications

This table lists the various fire door ratings by label, time, and temperature rating, plus allowable glass area for a fire-rated door.

Level		Model	Full Flush or Seamless			Construction
			Msg No.	IP in	SI mm	
I	Standard Duty	1	20	0.032	0.8	Full Flush
		2				Seamless
II	Heavy Duty	1	18	0.042	1	Full Flush
		2				Seamless
III	Extra Heavy Duty	1	16	0.053	1.3	Full Flush
		2				Seamless
		3				Stile & Rail
IV	Maximum Duty	1	14	0.067	1.6	Full Flush
		2				Seamless

(courtesy Steel Door Institute)

Figure 8-6 Hinge Requirements

All closer-equipped doors should have ball-bearing hinges. Lead-lined or extremely heavy doors require special strength hinges. Usually 1-1/2 pair of hinges are used per door for openings up to 7'-6" high. The table below shows typical hinge requirements.

Use Frequency	Type Hinge Required	Type of Opening	Type of Structure
High	Heavy Weight Ball Bearing	Entrances Toilet Rooms	Banks, Office Buildings, Schools, Stores and Theaters Office Buildings and Schools
Average	Standard Weight Ball Bearing	Entrances Corridors Toilet Rooms	Dwellings Office Buildings and Schools Stores
Low	Plan Bearing	Interior	Dwellings

Figure 8-7   Hardware Nomenclature

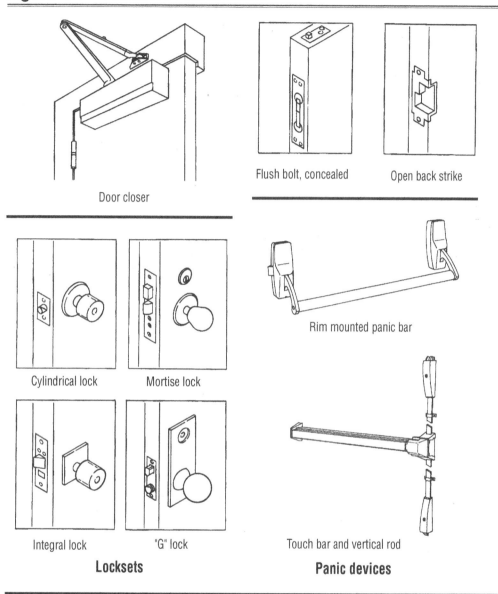

Door closer

Flush bolt, concealed

Open back strike

Cylindrical lock

Mortise lock

Rim mounted panic bar

Integral lock

"G" lock

Touch bar and vertical rod

Locksets

Panic devices

Figure 8-8   Finish Hardware Allowances

Average Allowances for Finish Hardware as a Percentage of Total Job Costs		
	Minimum	0.75%
	Maximum	3.50%

Average distribution of total finish hardware costs for a typical building is 85% material, 15% labor.

Figure 8-9 Properties of Different Types of Window Glazing

	Visible Transmittance	Solar Heat Gain Coefficient	U-Value (Heat Loss)
Double-Pane Clear	0.82	0.78	0.46
Low-E	0.78	0.58	0.25
Selective	0.72	0.37	0.27

Figure 8-10 Glazing Labor

Glass sizes are estimated by the "united inch" (height + width). The table below shows the number of lights glazed in an eight-hour period by the crew size indicated, for glass up to 1/4" thick. Square or nearly square lights are more economical on a SF basis. Long slender lights will have a high SF installation cost. For insulated glass reduce production by 33%. For 1/2" plate glass reduce production by 50%. Production time for glazing with two glazers per day averages: 1/4" plate glass 120 SF; 1/2" plate glass 55 SF; 1/2" insulated glass 95 SF; insulated glass 75 SF.

Glazing Method	United Inches per Light							
	40"	60"	80"	100"	135"	165"	200"	240"
Number of Men in Crew	1	1	1	1	2	3	3	4
Industrial sash, putty	60	45	24	15	18	—	—	—
With sops, putty bed	50	36	21	12	16	8	4	3
Wood stops, rubber	40	27	15	9	11	6	3	2
Metal stops, rubber	30	24	14	9	9	6	3	2
Structural glass	10	7	4	3	—	—	—	—
Corrugated glass	12	9	7	4	4	4	3	—
Storefronts	16	15	13	11	7	6	4	4
Skylights, putty glass	60	36	21	12	16	—	—	—
Thiokol set	15	15	11	9	9	6	3	2
Vinyl set, snap on	18	18	13	12	12	7	5	4
Maximum area per light	2.8 S.F.	6.3 S.F.	11.1 S.F.	17.4 S.F.	31.6 S.F.	47 S.F.	69 S.F.	100 S.F.

Figure 8-11 Installation Time in Labor-Hours for Exterior Doors and Entry Systems

Description	Labor-Hours	Unit
Glass Entrance Door, Including Frame and Hardware		
Balanced, Including Glass		
3' x 7'		
Economy	17.778	Ea.
Premium	22.857	Ea.
Hinged, Aluminum		
3' x 7'	8.000	Ea.
6' x 7', 3' Transom	8.889	Ea.
6' x 7'	12.308	Ea.
6' x 10', 3' Transom	14.545	Ea.
Stainless Steel, Including Glass		
3' x 7'		
Minimum	10.000	Ea.
Average	11.429	Ea.
Maximum	13.333	Ea.
Tempered Glass		
3' x 7'	8.00	Ea.
6' x 7'	11.429	Ea
Hinged, Automatic, Aluminum		
6' x 7'	22.860	Ea.
Revolving, Aluminum, 7'-0" Diameter x 7' High		
Minimum	42.667	Ea.
Average	53.333	Ea.
Maximum	71.111	Ea.
Stainless Steel, 7'-0" Diameter x 7' High	105.000	Ea.
Bronze, 7'-0" Diameter x 7' High	213.000	Ea.
Glass Storefront System, Including Frame and Hardware		
Hinged, Aluminum, Including Glass, 400 S.F.		
w/3' x 7' Door		
Commercial Grade	.107	S.F.
Institutional Grade	.123	S.F.
Monumental Grade	.139	S.F.
w/6' x 7' Door		
Commercial Grade	.119	S.F.
Institutional Grade	.130	S.F.
Monumental Grade	.160	S.F.
Sliding, Automatic, 12' x 7'-6" w/5' x 7' door	22.857	Ea.
Mall Front, Manual, Aluminum		
15' x 9'	12.308	Ea.
24' x 9'	22.857	Ea.
48' x 9' w/Fixed Panels	17.778	Ea.

(continued on next page)

Figure 8-11 Installation Time in Labor-Hours for Exterior Doors
and Entry Systems *(continued from previous page)*

Description	Labor-Hours	Unit
Tempered All-Glass w/Glass Mullions,		
up to 10' High	.185	S.F.
Up to 20' High, Minimum	.218	S.F.
Average	.240	S.F.
Maximum	.300	S.F.
Entrance Frames, Aluminum, 3' x 7'	2.286	Ea.
3' x 7', 3' Transom	2.462	Ea.
6' x 7'	2.667	Ea.
6' x 7', 3' Transom	2.909	Ea.
Glass, Tempered, 1/4" Thick	.133	S.F.
1/2" Thick	.291	S.F.
3/4" Thick	.457	S.F.
Insulating, 1" Thick	.213	S.F.
Overhead Commercial Doors		
Frames Not Included		
Stock Sectional Heavy Duty Wood		
1-3/4" Thick		
8' x 8'	8.000	Ea.
10' x 10'	8.889	Ea.
12' x 12'	10.667	Ea.
Fiberglass and Aluminum Heavy Duty Sectional		
12' x 12'	10.667	Ea.
20' x 20', Chain Hoist	32.000	Ea.
Steel 24' Gauge Sectional Manual		
8' x 8' High	8.000	Ea.
10' x 10' High	8.889	Ea.
12' x 12' High	10.667	Ea.
20' x 14' High, Chain Hoist	22.857	Ea.
For Electric Trolley Operator to 14' x 14'	4.000	Ea.
Over 14' x 14'	8.000	Ea.

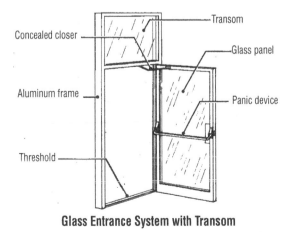

Glass Entrance System with Transom

Figure 8-12 Installation Time in Labor-Hours for Interior Doors
(incl. hinges, but not frames)

Description	Labor-Hours	Unit
Architectural, Flush, Interior, Hollow Core, Veneer Face		
Up to 3'-0" x 7'-0"	.941	Ea.
4'-0" x 7'-0"	1.000	Ea.
Plastic Laminate Face, Hollow Core		
Up to 2'-6" x 6'-8"	1.000	Ea.
3'-0" x 7'-0"	1.067	Ea.
4'-0" x 7'-0"	1.143	Ea.
Particle Core, Veneer Face		
2'-6" x 6'-9"	1.067	Ea.
3'-0" x 6'-8"	1.143	Ea.
3'-0" x 7'-0"	1.231	Ea.
4'-0" x 7'-0"	1.333	Ea.
M.D.O. on Hardboard Face		
3'-0" x 7'-0"	1.333	Ea.
4'-0" x 7'-0"	1.600	Ea.
Plastic Laminate Face, Solid Core		
3'-0" x 7'-0"	1.455	Ea.
4'-0" x 7'-0"	2.000	Ea.
Flush, Exterior, Solid Core, Veneer Face		
2'-6" x 7'-0"	1.067	Ea.
3'-0" x 7'-0"	1.143	Ea.
Decorator, Hand Carved Solid Wood		
Up to 3'-0" x 7'-0"	1.143	Ea.
3'-6" x 8'-0"	1.600	Ea.
Fire Door, Flush, Mineral Core		
B Label, 1 Hour, Veneer Face		
2'-6" x 6'-8"	1.143	Ea.
3'-0" x 7'-0"	1.333	Ea.
4'-0" x 7'-0"	1.333	Ea.
Plastic Laminate Face		
3'-0" x 7'-0"	1.455	Ea.
4'-0" x 7'-0"	1.600	Ea.
Residential, Interior		
Hollow Core or Panel		
Up to 2'-8" x 6'-8"	.889	Ea.
3'-0" x 6'-8"	.941	Ea.
Bi-folding Closet		
3'-0" x 6'-8"	1.231	Ea.
5'-0" x 6'-8"	1.455	Ea.
Interior Prehung, Hollow Core or Panel		
Up to 2'-8" x 6'-8"	.941	Ea.
3'-0" x 6'-8"	1.000	Ea.
Exterior, Entrance, Solid Core or Panel		
Up to 2'-8" x 6'-8"	1.000	Ea.
3'-0" x 6'-8"	1.067	Ea.
Exterior Prehung, Entrance		
Up to 3'-0" x 7'-0"	1.000	Ea.

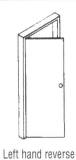

Left hand reverse

Right hand reverse

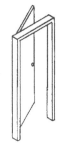

Left hand

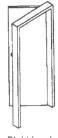

Right hand

Hand Designations

(continued on next page)

Figure 8-12   Installation Time in Labor-Hours for Interior Doors
(continued from previous page)

Description	Labor-Hours	Unit
Hollow Metal Doors Flush		
Full Panel, Commercial		
20 Gauge		
2'-0" x 6'-8"	.800	Ea.
2'-6" x 6'-8"	.889	Ea.
3'-0" x 6'-8" or 3'-0" x 7'-0"	.941	Ea.
4'-0" x 7'-0"	1.067	Ea.
18 Gauge		
2'-6" x 6'-8" or 2'-6" x 7'-0"	.941	Ea.
3'-0" x 6'-8" or 3'-0" x 7'-0"	1.000	Ea.
4'-0" x 7'-0"	1.067	Ea.
Residential		
24 Gauge, Prehung		
2'-8" x 6'-8"	1.000	Ea.
3'-0" x 7'-0"	1.067	Ea.
Bifolding		
3'-0" x 6'-9"	1.000	Ea.
5'-0" x 6'-8"	1.143	Ea.
Steel Frames		
16 Gauge		
3'-0" x 6'-9"	1.000	Ea.
5'-0" x 6'-8"	1.143	Ea.
14 Gauge		
4'-0" Wide	1.067	Ea.
8'-0" Wide	1.333	Ea.
Transom Lite Frames		
Fixed Add	.103	Ea.
Movable	.123	Ea.

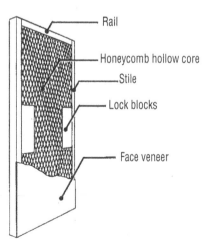

Hollow Core Door

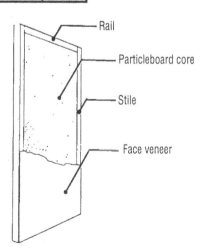

Solid Core Door

Figure 8-13 Installation Time in Labor-Hours for Special Doors

Description	Labor-Hours	Unit
Access Panels & Doors, Metal		
12" x 12"	.800	Ea.
18" x 18"	.889	Ea.
24" x 24"	.889	Ea.
36" x 36"	1.000	Ea.
Cold Storage Doors		
Single Galvanized Steel Horizontal Sliding		
5' x 7'		
Manual Operation	8.000	Ea.
Power Operation	8.421	Ea.
9' x 10'		
Manual Operation	9.412	Ea.
Power Operation	10.000	Ea.
Hinged Lightweight 3' x 6'-6"		
2" Thick	8.000	Ea.
4" Thick	8.421	Ea.
Bi-parting Electric Operated 6' x 8'	20.000	Opng.
For Door Buck Framing & Door Protection Add	6.400	Opng.
Galvanized Batten Door		
4' x 7'	8.000	Opng.
6' x 8'	8.889	Opng.
Fire Door 6' x 8'		
Single Slide	20.000	Opng.
Double Bi-parting	22.857	Opng.
Darkroom Doors Revolving		
2 Way 36" Diameter	5.161	Opng.
3 Way 51" Diameter	11.429	Opng.
4 Way 49" Diameter	11.429	Opng.
Hinged Safety		
2 Way 41" Diameter	6.957	Opng.
3 Way 51" Diameter	11.429	Opng.
Pop Out Safety		
2 Way 41" Diameter	5.161	Opng.
3 Way 51" Diameter	11.429	Opng.
Vault Front Door and Frame		
32" x 78", 1 Hour, 750 lbs.	10.667	Opng.
40" x 78", 2 Hours, 1130 lbs.	16.000	Opng.
Day Gate	10.667	Opng.
32" Wide	10.667	Opng.
40" Wide	11.429	Opng.

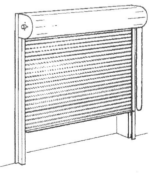

Roll Up Shutter

(continued on next page)

Figure 8-13   Installation Time in Labor-Hours for Special Doors
(continued from previous page)

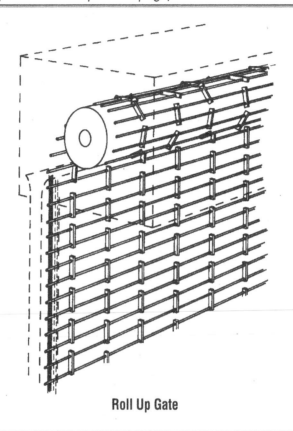

Roll Up Gate

Figure 8-14 Installation Time in Labor-Hours for Finish Hardware

Description	Labor-Hours	Unit
Astragals, 1/8" x 3"	.089	L.F.
Spring Hinged Security Seal with Cam	.107	L.F.
Two Piece Overlapping	.133	L.F.
Automatic Openers, Swing Doors, Single	20.000	Ea.
Single Operating, Pair	32.000	Pair
Sliding Doors, 3" Wide Including Track and Hanger, Single	26.667	Opng.
Bi-parting	40.000	Opng.
Handicap Opener Button, Operating	5.333	Pair
Bolts, Flush, Standard, Concealed	1.143	Ea.
Bumper Plates, 1-1/2" x 3/4" U Channel	.200	L.F.
Door Closer, Adjustable Backcheck, 3 Way Mount	1.333	Ea.
Doorstops	.250	Ea.
Kickplate	.533	Ea.
Lockset, Non-Keyed	.667	Ea.
Keyed	.800	Ea.
Dead Locks, Heavy Duty	.889	Ea.
Entrance Locks, Deadbolt	1.000	Ea.
Mortise Lockset, Non Keyed	.889	Ea.
Keyed	1,000	Ea.
Panic Device for Rim Locks		
Single Door, Exit Only	1.333	Ea.
Outside Key and Pull	1.600	Ea.
Bar and Vertical Rod, Exit Only	1.600	Ea.
Outside Key and Pull	2.000	Ea.
Push Pull Plate	.667	Ea.
Weatherstripping, Window, Double Hung	1.111	Opng.
Doors, Wood Frame, 3' x 7'	1.053	Opng.
6' x 7'	1.143	Opng.
Metal Frame, 3' x 7'	2.667	Opng.
6' x 7'	3.200	Opng.

Spring hinged security seal

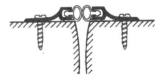

Two piece overlapping

Interlocking with bulb insert

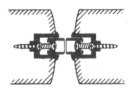

Two piece adjustable

Aluminum extrusion with sponge insert

Spring loaded locking bolt

Spring bronze strip

Astragals

Figure 8-15   Installation Time in Labor-Hours for Exterior Windows/Sash

Description	Labor-Hours	Unit
Custom Aluminum Sash, Glazing Not Included	.080	S.F.
Stock Aluminum Windows, Frame and		
Glazing Casement, 3'-1" x 3'-2" Opening	1.600	Ea
Combination Storm and Screen		
2'-0" x 3'-5" Opening	.533	Ea.
4'-0" x 6'-9" Opening	.640	Ea.
Projected, with Screen,		
3'-1" x 3'-2" Opening	1.600	Ea.
4'-5" x 5'-3" Opening	2.000	Ea.
Single Hung, 2' x 3' Opening	1.600	Ea.
3'-4" x 5'-0" Opening	1.778	Ea.
Sliding, 3' x 2' Opening	1.600	Ea.
5' x 3' Opening	1.778	Ea.
8' x 4' Opening	2.667	Ea.
Custom Steel Sash Units, Glazing and		
Trim Not Included	.080	S.F.
Stock Steel Windows, Frame, Trim and		
Glass Included		
Double Hung, 2'-8" x 4'-6" Opening	1.333	Ea.
Commercial Projected, 3'-9" x 5'-5" Opening	1.600	Ea.
6'-9" x 4'-1" Opening	2.286	Ea.
Intermediate Projected, 2'-9" x 4'-1" Opening	1.333	Ea.
Custom Wood Sash, Including Double/Triple		
Glazing but Not Trim		
5' x 4" Opening	3.200	Ea.
7' x 4'-6" Opening	3.721	Ea.
8'-6" x 5' Opening	4.571	Ea.
Stock Wood Windows, Frame, Trim and		
Glass Included		
Awning Type, 2'-10" x 1-10" Opening	800	Ea.
5'-0" x 3'-0" Opening	1.000	Ea.

(continued on next page)

Figure 8-15 Installation Time in Labor-Hours for Exterior Windows/Sash
(continued from previous page)

Description	Labor-Hours	Unit
Bow Bay, 8' x 5' Opening	1.600	Ea.
12'-0" x 6'-0" Opening	2.667	Ea.
Casement 2'-0" x 3'-0" Opening	.800	Ea.
4'-0" x 4'-0" Opening 2 Leaf	.920	Ea.
6'-0" x 4'-0" Opening 3 Leaf	1.040	Ea.
10'-0" x 4'-0" Opening 5 Leaf	1.280	Ea.
Double Hung 2-0" x 3'-0" Opening	.800	Ea.
3'-0" x 5'-0" Opening	1.000	Ea.
Picture Window 5'-0" x 4'-0" Opening	1.445	Ea.
Roof Window, Complete		
2'-9" x 4'-0" Opening	3.566	Ea.
Sliding 3'-0" x 3'-0" Opening	.800	Ea.
6'-0" x 5'-0" Opening	1.000	Ea.

Combination Storm and Screen Window

Single-Hung Window - Aluminum

Projected Window - Aluminum

(continued on next page)

Division 8

Figure 8-15   Installation Time in Labor-Hours for Exterior Windows/Sash
(continued from previous page)

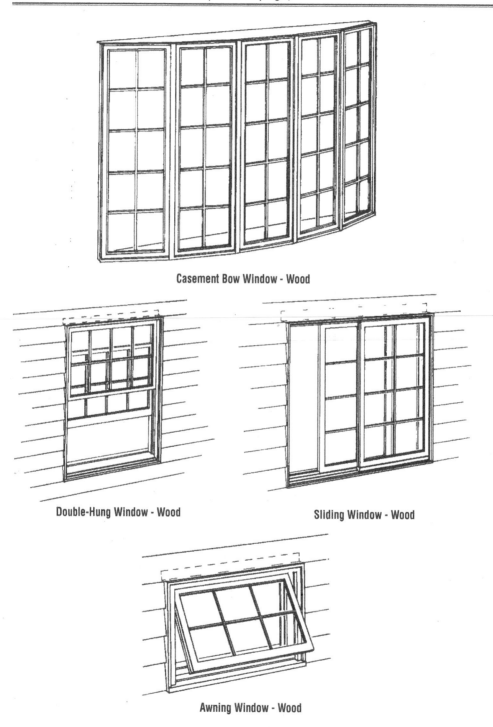

Casement Bow Window - Wood

Double-Hung Window - Wood

Sliding Window - Wood

Awning Window - Wood

(continued on next page)

Figure 8-15 Installation Time in Labor-Hours for Exterior Windows/Sash
(continued from previous page)

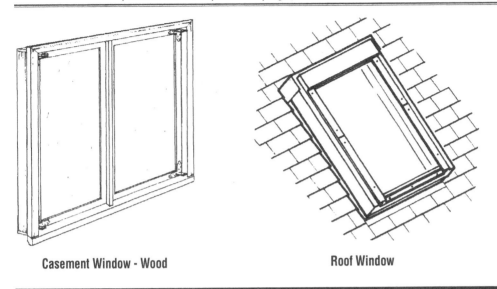

Casement Window - Wood   **Roof Window**

Figure 8-16 Demolition Time in Labor-Hours for Exterior
Windows/Glazing

Description	Labor-Hours	Unit
Aluminum Windows, to 12 S.F.	.500	Ea.
To 25 S.F.	.727	Ea.
To 50 S.F.	1.600	Ea.
Aluminum Storm Windows, to 12 S.F.	.296	Ea.
To 25 S.F.	.381	Ea.
To 50 S.F.	.500	Ea.
Steel Windows, to 12 S.F.	.615	Ea.
To 25 S.F.	.889	Ea.
To 50 S.F.	2.999	Ea.
Wood Windows, to 12 S.F.	.364	Ea.
To 25 S.F.	.444	Ea.
To 50 S.F.	.615	Ea.
Glazing, minimum	.040	S.F.
Maximum	.053	S.F.

Figure 8-17 Installation Time in Labor-Hours for Skylights

Description	Labor-Hours	Unit
Domed Unit Skylights		
Plastic domes, flush or curb mounted, ten or more units		
Nominal size under 10 S.F., double	.246	S.F.
Single	.200	
10 S.F. to 20 S.F., double	.102	
Single	.081	
20 S.F. to 30 S.F., double	.081	
Single	.069	
30 S.F. to 65 S.F., double	.069	
Single	.052	↓
Ventilating insulated plexiglass dome with curb mounting, 36" x 36"	2.667	Ea.
52" x 52"	2.667	
28" x 52"	3.200	
36" x 52"	3.200	↓
Field fabricated, factory type, aluminum and wire glass	.267	S.F.
Insulated safety glass with aluminum frame	.200	
Sandwich panels, fiberglass, for walls, 1-9/16" thick, to 250 S.F.	.160	
250 S.F. and up	.121	
As above, but for roofs, 2-3/4" thick, to 250 S.F.	.108	
250 S.F. and up	.097	↓
Metal-Framed Skylights		
Prefabricated glass block with metal frame		
Minimum	.121	S.F.
Maximum	.200	↓

Figure 8-18 Installation Time in Labor-Hours for Metal Wall Louvers

Metal Wall Louvers	Crew Makeup	Daily Output	Labor-Hours	Unit
LOUVERS Aluminum with screen, residential, 8" x 8"	1 Carpenter	38	.211	Ea.
12" x 12"		38	.211	Ea.
12" x 18"		35	.229	Ea.
14" x 24"		30	.267	Ea.
18" x 24"		27	.296	Ea.
30" x 24"		24	.333	Ea.
Triangle, adjustable, small		20	.400	Ea.
Large		15	.533	Ea.
Midget, aluminum, 3/4" deep, 1" diameter		85	.094	Ea.
3" diameter		60	.133	Ea.
4" diameter		50	.160	Ea.
6" diameter		30	.267	Ea.
Louvers, ridge vent strip, mill finish	1 Sheet Metal Worker	155	.052	L.F.
Soffit vent, continuous, 3" wide, aluminum, mill finish	1 Carpenter	200	.040	L.F.
Baked enamel finish		200	.040	L.F.
Under eaves vent, aluminum, mill finish, 16" x 4"		48	.167	Ea.
16" x 8"		48	.167	Ea.
Vinyl wall louvers, 1-1/2" deep, 8" x 8"		38	.211	Ea.
12" x 12"		38	.211	Ea.
12" x 18"		35	.229	Ea.
14" x 24"		30	.267	Ea.

Figure 8-19 Installation Time in Labor-Hours for Metal Casework

Metal Wall Louvers	Crew Makeup	Daily Output	Labor-Hours	Unit
KEY CABINETS Wall mounted, 60 key capacity	1 Carpenter	20	.400	Ea.
Drawer type, 600 key capacity	1 Building Laborer	15	.533	Ea.
2400 key capacity		20	.400	Ea.
Tray type, 20 key capacity		50	.160	Ea.
50 key capacity		40	.200	Ea.

Division 8

Notes

Division Nine
Finishes

Introduction

From an estimating standpoint, Division 9 – Finishes does not represent a heavy portion of the overall project cost. However, from the construction standpoint, it is one of the most important areas. Upon taking possession of the building, the owner cannot see the piles you had to drive in winter conditions while working on unstable ground, or the good quality of work that went into the structure of the building. The most visible components are those in the finish work: the paint job in the main office area, the carpeting in the executive area, the vinyl flooring in the production area. Therefore, when estimating the finishes for any structure, this is one category where corners should not be cut in an attempt to get the job or get back on budget.

Many buildings are built on a "spec" basis—that is, built before they have tenants committed, on pure speculation that someone will lease or purchase the structure, or at least part of it. In this case, the party financing the project may take one of two paths while completing the building. They might seek to attract a client by finishing the building in a handsome, attractive but generic manner, or they may leave the building unfinished, perhaps finishing only the lobby or common areas and leave tennant areas unfinished until leased. Either way, finishes are an important part of the overall project, even if they represent a relatively small portion of the whole picture.

If LEED® certification is a goal and the building is left unfinished, LEED CS (core and shell) is the appropriate version of the rating system for the builder to use.

This chapter contains many charts and tables for use in estimating Division 9 – Finishes. Included here are appropriate tables to estimate materials for finishing walls, ceilings, floors, doors, windows, and trim work, as well as installation time information. The areas that are covered include:

- Drywall and plaster construction
- Tile work
- Terrazzo
- Acoustical treatment
- Flooring
- Painting and wall coverings

Estimating Data

The following tables present estimating guidelines for items found in Division 9 – Finishes. Please note that these guidelines are intended as indicators of what may generally be expected, but that each project must be evaluated individually.

Table of Contents

Estimating Tips

General Note on Finishes

Interior finish creates the identity of the building. It makes a statement about the nature of the business within—whether utilitarian, adventurous, or staid and dignified. Therefore, selection of materials should be made with an eye toward purpose and function of occupancy.

The type of materials used will have an impact on LEED certification. Current trends encourage use of natural materials, particularly those sourced locally. Materials with high recycled content are also desirable. The selection of interior finishes will also have an impact on energy conservation, light reflectance, and thermal retention.

Room Finish Schedule

A complete set of plans should contain a room finish schedule. If one is not available, it would be well worth the time and effort to put one together. A room finish schedule should contain the room number, room name (for clarity), floor materials, base materials, wainscot materials, wainscot height, wall materials (for each wall), ceiling materials, and special instructions. It is handy and easier to work out a room finish schedule on a grid system.

Studs for Lathing and Plaster

If not specified, the studs for plaster partitions are usually spaced at no greater than 16" on center.

Lathing Waste Factor

Allow a 5% waste factor when figuring the amount of lathing.

Plaster Allowances

Deductions for openings in plaster partitions vary by the preference and experience of the estimator, from 0% for openings of less than 2 SF to 50% for openings greater than 2 feet in width.

For curved walls with a radius less than 6' allow twice as much plaster.

Pricing a plaster and lath job depends on the quality of work desired. A first class project will have variations in the wall surface of less than 1/16" in 10 feet. An ordinary plaster job (or commercial grade) may have variations of 1/8" to 3/16" in 10 feet. Overall, labor for first class work is approximately 20% higher than ordinary or commercial grade work.

Drywall Deductions

A rule of thumb when estimating materials for drywall is to not deduct for door or window openings of less than 32 SF.

Wall Coverings

A single roll of wallpaper will cover approximately 36 SF. Allow approximately 6 SF per roll for waste. This means take the total square foot area to be covered and divide by 30 to determine the number of rolls needed.

For vinyls and grass cloth coverings with no patterns to match, allow approximately 10% (3 SF per roll) for waste. For patterns requiring matching, allow up to 25% to 30%.

Waste can run as high as 50% to 60% for coverings with large, bold, or intricate patterns.

Always specify wall coverings from the same batch with identical batch numbers. This will help to ensure that the colors and patterns will match.

Flooring

Waste allowances for flooring depend on the material used, the room dimensions, and the pattern or texture of the chosen material. For tiles or strip flooring allow for 5% waste. For carpet, the amount of waste will range from 5% to 35% depending on the roll width, room dimensions, and pattern.

Access flooring, originally used exclusively for computer room applications, is now being used in other areas of the building as well. Raised access flooring, with air supply openings on the floor, take advantage of natural convection to circulate air. This approach can reduce HVAC energy costs by 20%.

Ceilings

Allow 5% minimum waste for acoustic ceiling grid and tile. The percent waste will increase for small or irregularly shaped areas. Be sure to include the cost of scaffolding where applicable.

Surplus Finishes

Review the specifications to determine if there are any requirements to provide certain amounts of extra floor tile, ceiling pads, paint, wall coverings, etc., for the owner's maintenance department. In some cases, the owner may require a substantial amount of materials, especially if it is a special order or long lead time item.

Checklist ✓

For an estimate to be reliable, all items must be accounted for. A complete estimate can also limit contingencies. The following checklist can be used to help ensure that all items are included.

☐ *Access Flooring*

☐ *Bases*
  ☐ Cove
  ☐ Sanitary

☐ *Ceilings*
  ☐ Acoustical
  ☐ Dropped
  ☐ Drywall
  ☐ Insulation
    ☐ Thermal
    ☐ Acoustic
  ☐ Plaster
  ☐ Suspension system

☐ *Drywall*

☐ *Flooring*
  ☐ Brick
  ☐ Carpet
  ☐ Carpet tile
  ☐ Ceramic tile
  ☐ Composition
  ☐ Concrete topping
  ☐ Metal tile
  ☐ Paint
  ☐ Epoxy
  ☐ Urethane
  ☐ Mosaic tile
  ☐ Plastic tile

☐ Quarry tile
☐ Raised access
☐ Resilient
  ☐ Asphalt
  ☐ Conductive
  ☐ Cork
  ☐ Linoleum
  ☐ Polyurethane
  ☐ Rubber tile
  ☐ Sheet vinyl
  ☐ Vinyl tile
☐ Stone
☐ Terrazzo
☐ Wood

☐ *Wall Finishes*
  ☐ Ceramic tile
  ☐ Planking
  ☐ Plastic tile
  ☐ Mosaic tile
  ☐ Quarry tile
  ☐ Metal tile
  ☐ Paint
  ☐ Paneling
  ☐ Plaster
  ☐ Stucco
  ☐ Wainscoting
  ☐ Wall coverings
    ☐ Cloth
    ☐ Paper
    ☐ Vinyl

Figure 9-1 Partition Density Guide

This table allows you to estimate the average quantities of partitions found in various types of buildings when no interior plans are available, such as in a conceptual estimating situation. To use this chart, pick out the type of structure to be estimated from the left-hand column. Moving across, choose from the second column the number of stories in the building. Read across to the "Partition Density" column. This figure represents the number of square feet of floor area for every linear foot of partition.

Example: The project to be estimated is a three-story office building of 10,000 square feet per floor (total of 30,000 square feet). To estimate the approximate quantity of interior partitions in the entire building, first look in column 1 for "Office" (on the next page). Reading across to the "Stories" column, find the "3–5 Story" line. The "Partition Density" column for that line indicates 20 SF/LF. To find the number of partitions, divide the building area by the density factor: 30,000 SF/20 = 1,500 LF of partition. Note that the right-most column gives a breakdown of average partition types. For our office building, the average mix is 30% concrete block (generally found at stairwells and service area partitions) and 70% drywall. Our drywall total would then be 1,050 LF. To determine the total square footage of partitions, multiply this linear footage by the specified (or assumed) partition height. Note that the remainder of the partitions (450 LF) must be added to the masonry portion of the estimate, or at least accounted for in the estimate.

Building Type	Stories	Partition Density	Description of Partition
Apartments	1 story	9 S.F./L.F.	Plaster, wood doors & trim
	2 story	8 S.F./L.F.	Drywall, wood studs, wood doors & trim
	3 story	9 S.F./L.F.	Plaster, wood studs, wood doors & trim
	5 story	9 S.F./L.F.	Plaster, wood studs, wood doors & trim
	6–15 story	8 S.F./L.F.	Drywall, wood studs, wood doors & trim
Bakery	1 story	50 S.F./L.F.	Conc. block, paint, door & drywall, wood studs
	2 story	50 S.F./L.F.	Conc. block, paint, door & drywall, wood studs
Bank	1 story	20 S.F./L.F.	Plaster, wood studs, wood doors & trim
	2-4 story	15 S.F./L.F.	Plaster, wood studs, wood doors & trim
Bottling Plant	1 story	50 S.F./L.F.	Conc. block, drywall, wood studs, wood trim
Bowling Alley	1 story	50 S.F./L.F.	Conc. block, wood & metal doors, wood trim
Bus Terminal	1 story	15 S.F./L.F.	Conc. block, ceramic tile, wood trim
Cannery	1 story	100 S.F./L.F.	Drywall on metal studs
Car Wash	1 story	18 S.F./L.F.	Concrete block, painted & hollow metal door
Dairy Plant	1 story	30 S.F./L.F.	Concrete block, glazed tile, insulated cooler doors
Department Store	1 story	60 S.F./L.F.	Drywall, wood studs, wood doors & trim
	2-5 story	60 S.F./L.F.	30% concrete block, 70% drywall, wood studs
Dormitory	2 story	9 S.F./L.F.	Plaster, concrete block, wood doors & trim
	3–5 story	9 S.F./L.F.	Plaster, concrete block, wood doors & trim
	6–15 story	9 S.F./L.F.	Plaster, concrete block, wood doors & trim
Funeral Home	1 story	15 S.F./L.F.	Plaster on concrete block & wood studs, paneling
	2 story	14 S.F./L.F.	Plaster, wood studs, paneling & wood doors
Garage Sales & Service	1 story	30 S.F./L.F.	50% conc. block, 50% drywall, wood studs
Hotel	3–8 story	9 S.F./L.F.	Plaster, conc. block, wood doors & trim
	9–15 story	9 S.F./L.F.	Plaster, conc. block, wood doors & trim
Laundromat	1 story	25 S.F./L.F.	Drywall, wood studs, wood doors & trim
Medical Clinic	1 story	6 S.F./L.F.	Drywall, wood studs, wood doors & trim
	2–4 story	6 S.F./L.F.	Drywall, wood studs, wood doors & trim
Motel	1 story	7 S.F./L.F.	Drywall, wood studs, wood doors & trim
	2–3 story	7 S.F./L.F.	Concrete block, drywall on wood studs, wood paneling

(continued on next page)

Figure 9-1 Partition Density Guide *(continued from previous page)*

Building Type	Stories	Partition Density	Description of Partition
Movie 200–600 seats	1 story	18 S.F./L.F.	Concrete block, wood, metal, vinyl trim
Theater 601–1400 seats		20 S.F./L.F.	Concrete block, wood, metal, vinyl trim
1401–2200 seats		25 S.F./L.F.	Concrete block, wood, metal, vinyl trim
Nursing Home	1 story	8 S.F./L.F.	Drywall, wood studs, wood doors & trim
	2–4 story	8 S.F./L.F.	Drywall, wood studs, wood doors & trim
Office	1 story	20 S.F./L.F.	30% concrete block, 70% drywall on wood studs
	2 story	20 S.F./L.F.	30% concrete block, 70% drywall on wood studs
	3–5 story	20 S.F./L.F.	30% concrete block, 70% movable partitions
	6–10 story	20 S.F./L.F.	30% concrete block, 70% movable partitions
	11–20 story	20 S.F./L.F.	30% concrete block, 70% movable partitions
Parking Ramp (Open)	2–8 story	60 S.F./L.F.	Stair and elevator enclosures only
Parking Garage	2–8 story	60 S.F./L.F.	Stair and elevator enclosures only
Pre-Engineered Store	1 story	60 S.F./L.F.	Drywall on wood studs, wood doors & trim
Office	1 story	15 S.F./L.F.	Concrete block, movable wood partitions
Shop	1 story	15 S.F./L.F.	Movable wood partitions
Radio & TV Broadcasting	1 story	25 S.F./L.F.	Concrete block, metal and wood doors
& TV Transmitter	1 story	40 S.F./L.F.	Concrete block, metal and wood doors
Self Service Restaurant	1 story	15 S.F./L.F.	Concrete block, wood and aluminum trim
Cafe & Drive-in Restaurant	1 story	18 S.F./L.F.	Drywall, wood studs, ceramic & plastic trim
Restaurant with seating	1 story	25 S.F./L.F.	Concrete block, paneling, wood studs & trim
Supper Club	1 story	25 S.F./L.F.	Concrete block, paneling, wood studs & trim
Bar or Lounge	1 story	24 S.F./L.F.	Plaster or gypsum lath, wood studs
Retail Store or Shop	1 story	60 S.F./L.F.	Drywall, wood studs, wood doors & trim
Service Station Masonry	1 story	15 S.F./L.F.	Concrete block, paint, door & drywall, wood studs
Metal panel	1 story	15 S.F./L.F.	Concrete block, paint door & drywall, wood studs
Frame	1 story	15 S.F./L.F.	Drywall, wood studs, wood doors & trim
Shopping Center (strip)	1 story	30 S.F./L.F.	Drywall, wood studs, wood doors & trim
(group)	1 story	40 S.F./L.F.	50% concrete block, 50% drywall, wood studs
	2 story	40 S.F./L.F.	50% concrete block, 50% drywall, wood studs
Small Food Store	1 story	30 S.F./L.F.	Concrete block, drywall, wood studs, wood trim
Store/Apt. Masonry	2 story	10 S.F./L.F.	Plaster, wood studs, wood doors & trim
above Frame	2 story	10 S.F./L.F.	Plaster, wood studs, wood doors & trim
Frame	3 story	10 S.F./L.F.	Plaster, wood studs, wood doors & trim
Supermarkets	1 story	40 S.F./L.F.	Concrete block, paint, drywall & porcelain panel
Truck Terminal	1 story	0	
Warehouse	1 story	0	

Figure 9-2 Surface Area of Rooms

This table provides the total surface area in square feet of room walls and ceilings. Select the room dimensions from the top row and left column. The square at which the appropriate column and row intersect is the total square footage for that room. This chart is especially helpful in determining quantities for painting or drywalling. Note that no deductions have been made for openings (windows or doors).

The last two sections of this table show square footages for single surface areas, such as a single wall or a floor area.

Rooms with 7' Ceilings—Total Area – 4 Walls and Ceiling (in square feet)

	3'	4'	5'	6'	7'	8'	9'	10'	11'	12'	13'	14'	15'	16'	17'	18'	19'	20'	21'	22'	23'	24'
22'	416	452	488	524	560	596	632	668	704	740	776	812	848	884	920	956	992	1028	1064	1100	1136	1172
21'	399	434	469	504	539	574	609	644	679	714	749	784	819	854	889	924	959	994	1029	1064	1099	1134
20'	382	416	450	484	518	552	586	620	654	688	722	756	790	824	858	892	926	960	994	1028	1062	1096
19'	365	398	431	464	497	530	563	596	629	662	695	728	761	794	827	860	893	926	959	992	1025	1058
18'	348	380	412	444	476	508	540	572	604	636	668	700	732	764	796	828	860	892	924	956	988	1020
17'	331	362	393	424	455	486	517	548	579	610	641	672	703	734	765	796	827	858	889	920	951	982
16'	314	344	374	404	434	464	494	524	554	584	614	644	674	704	734	764	794	824	854	884	914	944
15'	297	326	355	384	413	442	471	500	529	558	587	616	645	674	703	732	761	790	819	848	877	906
14'	280	308	336	364	392	420	448	476	504	532	560	588	616	644	672	700	728	756	784	812	840	868
13'	263	290	317	344	371	398	425	452	479	506	533	560	587	614	641	668	695	722	749	776	803	830
12'	246	272	298	324	350	376	402	428	454	480	506	532	558	584	610	636	662	688	714	740	766	792
11'	229	254	279	304	329	354	379	404	429	454	479	504	529	554	579	604	629	654	679	704	729	754
10'	212	236	260	284	308	332	356	380	404	428	452	476	500	524	548	572	596	620	644	668	692	716
9'	195	218	241	264	287	310	333	356	379	402	425	448	471	494	517	540	563	586	609	632	655	678
8'	178	200	222	244	266	288	310	332	354	376	398	420	442	464	486	508	530	552	574	596	618	640
7'	161	182	203	224	245	266	287	308	329	350	371	392	413	434	455	476	497	518	539	560	581	602
6'	144	164	184	204	224	244	264	284	304	324	344	364	384	404	424	444	464	484	504	524	544	564
5'	127	146	165	184	203	222	241	260	279	298	317	336	355	374	393	412	431	450	469	488	507	526
4'	110	128	146	164	182	200	218	236	254	272	290	308	326	344	362	380	398	416	434	452	470	488
3'	93	110	127	144	161	178	195	212	229	246	263	280	297	314	331	348	365	382	399	416	433	450
	3'	4'	5'	6'	7'	8'	9'	10'	11'	12'	13'	14'	15'	16'	17'	18'	19'	20'	21'	22'	23'	24'

(continued on next page)

Figure 9-2 Surface Area of Rooms *(continued from previous page)*

Division 9

Rooms with 7'6" Ceilings—Total Area – 4 Walls and Ceiling (in square feet)

	3'	4'	5'	6'	7'	8'	9'	10'	11'	12'	13'	14'	15'	16'	17'	18'	19'	20'	21'	22'
3'	99	117	135	153	171	189	207	225	243	261	279	297	315	333	351	369	387	405	423	441
4'	117	136	155	174	193	212	231	250	269	288	307	326	345	364	383	402	421	440	459	478
5'	135	155	175	195	215	235	255	275	295	315	335	355	375	395	415	435	455	475	495	515
6'	153	174	195	216	237	258	279	300	321	342	363	384	405	426	447	468	489	510	531	552
7'	171	193	215	237	259	281	303	325	347	369	391	413	435	457	479	501	523	545	567	589
8'	189	212	235	258	281	304	327	350	373	396	419	442	465	488	511	534	557	580	603	626
9'	207	231	255	279	303	327	351	375	399	423	447	471	495	519	543	567	591	615	639	663
10'	225	250	275	300	325	350	375	400	425	450	475	500	525	550	575	600	625	650	675	700
11'	243	269	295	321	347	373	399	425	451	477	503	529	555	581	607	633	659	685	711	737
12'	261	288	315	342	369	396	423	450	477	504	531	558	585	612	639	666	693	720	747	774
13'	279	307	335	363	391	419	447	475	503	531	559	587	615	643	671	699	727	755	783	811
14'	297	326	355	384	413	442	471	500	529	558	587	616	645	674	703	732	761	790	819	848
15'	315	345	375	405	435	465	495	525	555	585	615	645	675	705	735	765	795	825	855	885
16'	333	364	395	426	457	488	519	550	581	612	643	674	705	736	767	798	829	860	891	922
17'	351	383	415	447	479	511	543	575	607	639	671	703	735	767	799	831	863	895	927	959
18'	369	402	435	468	501	534	567	600	633	666	699	732	765	798	831	864	897	930	963	996
19'	387	421	455	489	523	557	591	625	659	693	727	761	795	829	863	897	931	965	999	1033
20'	405	440	475	510	545	580	615	650	685	720	755	790	825	860	895	930	965	1000	1035	1070
21'	423	459	495	531	567	603	639	675	711	747	783	819	855	891	927	963	999	1035	1071	1107
22'	441	478	515	552	589	626	663	700	737	774	811	848	885	922	959	996	1033	1070	1107	1144
23'	459	497	535	573	611	649	687	725	763	801	839	877	915	953	991	1029	1067	1105	1143	1181
24'	477	516	555	594	633	672	711	750	789	828	867	906	945	984	1023	1062	1101	1140	1179	1218

(continued on next page)

Figure 9-2 Surface Area of Rooms *(continued from previous page)*

Rooms with 8' Ceilings—Total Area – 4 Walls and Ceiling (in square feet)

	3'	4'	5'	6'	7'	8'	9'	10'	11'	12'	13'	14'	15'	16'	17'	18'	19'	20'	21'	22'
3'	105	124	143	162	181	200	219	238	257	276	295	314	333	352	371	390	409	428	447	466
4'	124	144	164	184	204	224	244	264	284	304	324	344	364	384	404	424	444	464	484	504
5'	143	164	185	206	227	248	269	290	311	332	353	374	395	416	437	458	479	500	521	542
6'	162	184	206	228	250	272	294	316	338	360	382	404	426	448	470	492	514	536	558	580
7'	181	204	227	250	273	296	319	342	365	388	411	434	457	480	503	526	549	572	595	618
8'	200	224	248	272	296	320	344	368	392	416	440	464	488	512	536	560	584	608	632	656
9'	219	244	269	294	319	344	369	394	419	444	469	494	519	544	569	594	619	644	669	694
10'	238	264	290	316	342	368	394	420	446	472	498	524	550	576	602	628	654	680	706	732
11'	257	284	311	338	365	392	419	446	473	500	527	554	581	608	635	662	689	716	743	770
12'	276	304	332	360	388	416	444	472	500	528	556	584	612	640	668	696	724	752	780	808
13'	295	324	353	382	411	440	469	498	527	556	585	614	643	672	701	730	759	788	817	846
14'	314	344	374	404	434	464	494	524	554	584	614	644	674	704	734	764	794	824	854	884
15'	333	364	395	426	457	488	519	550	581	612	643	674	705	736	767	798	829	860	891	922
16'	352	384	416	448	480	512	544	576	608	640	672	704	736	768	800	832	864	896	928	960
17'	371	404	437	470	503	536	569	602	635	668	701	734	767	800	833	866	899	932	965	998
18'	390	424	458	492	526	560	594	628	662	696	730	764	798	832	866	900	934	968	1002	1036
19'	409	444	479	514	549	584	619	654	689	724	759	794	829	864	899	934	969	1004	1039	1074
20'	428	464	500	536	572	608	644	680	716	752	788	824	860	896	932	968	1004	1040	1076	1112
21'	447	484	521	558	595	632	669	706	743	780	817	854	891	928	965	1002	1039	1076	1113	1150
22'	466	504	542	580	618	656	694	732	770	808	846	884	922	960	998	1036	1074	1112	1150	1188
23'	485	524	563	602	641	680	719	758	797	836	875	914	953	992	1031	1070	1109	1148	1187	1226
24'	504	544	584	624	664	704	744	784	824	864	904	944	984	1024	1064	1104	1144	1184	1224	1264

(continued on next page)

Figure 9-2 Surface Area of Rooms *(continued from previous page)*

Rooms with 8'6" Ceilings—Total Area – 4 Walls and Ceiling (in square feet)

	3'	4'	5'	6'	7'	8'	9'	10'	11'	12'	13'	14'	15'	16'	17'	18'	19'	20'	21'	22'
3'	111	131	151	171	191	211	231	251	271	291	311	331	351	371	391	411	431	451	471	491
4'	131	152	173	194	215	236	257	278	299	320	341	362	383	404	425	446	467	488	509	530
5'	151	173	195	217	239	261	283	305	327	349	371	393	415	437	459	481	503	525	547	569
6'	171	194	217	240	263	286	309	332	355	378	401	424	447	470	493	516	539	562	585	608
7'	191	215	239	263	287	311	335	359	383	407	431	455	479	503	527	551	575	599	623	647
8'	211	236	261	286	311	336	361	386	411	436	461	486	511	536	561	586	611	636	661	686
9'	231	257	283	309	335	361	387	413	439	465	491	517	543	569	595	621	647	673	699	725
10'	251	278	305	332	359	386	413	440	467	494	521	548	575	602	629	656	683	710	737	764
11'	271	299	327	355	383	411	439	467	495	523	551	579	607	635	663	691	719	747	775	803
12'	291	320	349	378	407	436	465	494	523	552	581	610	639	668	697	726	755	784	813	842
13'	311	341	371	401	431	461	491	521	551	581	611	641	671	701	731	761	791	821	851	881
14'	331	362	393	424	455	486	517	548	579	610	641	672	703	734	765	796	827	858	889	920
15'	351	383	415	447	479	511	543	575	607	639	671	703	735	767	799	831	863	895	927	959
16'	371	404	437	470	503	536	569	602	635	668	701	734	767	800	833	866	899	932	965	998
17'	391	425	459	493	527	561	595	629	663	697	731	765	799	833	867	901	935	969	1003	1037
18'	411	446	481	516	551	586	621	656	691	726	761	796	831	866	901	936	971	1006	1041	1076
19'	431	467	503	539	575	611	647	683	719	755	791	827	863	899	935	971	1007	1043	1079	1115
20'	451	488	525	562	599	636	673	710	747	784	821	858	895	932	969	1006	1043	1080	1117	1154
21'	471	509	547	585	623	661	699	737	775	813	851	889	927	965	1003	1041	1079	1117	1155	1193
22'	491	530	569	608	647	686	725	764	803	842	881	920	959	998	1037	1076	1115	1154	1193	1232
23'	511	551	591	631	671	711	751	791	831	871	911	951	991	1031	1071	1111	1151	1191	1231	1271
24'	531	572	613	654	695	736	777	818	859	900	941	982	1023	1064	1105	1146	1187	1228	1269	1310

(continued on next page)

Figure 9-2 Surface Area of Rooms *(continued from previous page)*

Rooms with 9' Ceilings—Total Area – 4 Walls and Ceiling (in square feet)

	3'	4'	5'	6'	7'	8'	9'	10'	11'	12'	13'	14'	15'	16'	17'	18'	19'	20'	21'	22'
3'	117	138	159	180	201	222	243	264	285	306	327	348	369	390	411	432	453	474	495	516
4'	138	160	182	204	226	248	270	292	314	336	358	380	402	424	446	468	490	512	534	556
5'	159	182	205	228	251	274	297	320	343	366	389	412	435	458	481	504	527	550	573	596
6'	180	204	228	252	276	300	324	348	372	396	420	444	468	492	516	540	564	588	612	636
7'	201	226	251	276	301	326	351	376	401	426	451	476	501	526	551	576	601	626	651	676
8'	222	248	274	300	326	352	378	404	430	456	482	508	534	560	586	612	638	664	690	716
9'	243	270	297	324	351	378	405	432	459	486	513	540	567	594	621	648	675	702	729	756
10'	264	292	320	348	376	404	432	460	488	516	544	572	600	628	656	684	712	740	768	796
11'	285	314	343	372	401	430	459	488	517	546	575	604	633	662	691	720	749	778	807	836
12'	306	336	366	396	426	456	486	516	546	576	606	636	666	696	726	756	786	816	846	876
13'	327	358	389	420	451	482	513	544	575	606	637	668	699	730	761	792	823	854	885	916
14'	348	380	412	444	476	508	540	572	604	636	668	700	732	764	796	828	860	892	924	956
15'	369	402	435	468	501	534	567	600	633	666	699	732	765	798	831	864	897	930	963	996
16'	390	424	458	492	526	560	594	628	662	696	730	764	798	832	866	900	934	968	1002	1036
17'	411	446	481	516	551	586	621	656	691	726	761	796	831	866	901	936	971	1006	1041	1076
18'	432	468	504	540	576	612	648	684	720	756	792	828	864	900	936	972	1008	1044	1080	1116
19'	453	490	527	564	601	638	675	712	749	786	823	860	897	934	971	1008	1045	1082	1119	1156
20'	474	512	550	588	626	664	702	740	778	816	854	892	930	968	1006	1044	1082	1120	1158	1196
21'	495	534	573	612	651	690	729	768	807	846	885	924	963	1002	1041	1080	1119	1158	1197	1236
22'	516	556	596	636	676	716	756	796	836	876	916	956	996	1036	1076	1116	1156	1196	1236	1276
23'	537	578	619	660	701	742	783	824	865	906	947	988	1029	1070	1111	1152	1193	1234	1275	1316
24'	558	600	642	684	726	768	810	852	894	936	978	1020	1062	1104	1146	1188	1230	1272	1314	1356

(continued on next page)

Division 9

Figure 9-2 Surface Area of Rooms (continued from previous page)

Rooms with 9'6" Ceilings — Total Area – 4 Walls and Ceiling (in square feet)

	3'	4'	5'	6'	7'	8'	9'	10'	11'	12'	13'	14'	15'	16'	17'	18'	19'	20'	21'	22'
3'	123	145	167	189	211	233	255	277	299	321	343	365	387	409	431	453	475	497	519	541
4'	145	168	191	214	237	260	283	306	329	352	375	398	421	444	467	490	513	536	559	582
5'	167	191	215	239	263	287	311	335	359	383	407	431	455	479	503	527	551	575	599	623
6'	189	214	239	264	289	314	339	364	389	414	439	464	489	514	539	564	589	614	639	664
7'	211	237	263	289	315	341	367	393	419	445	471	497	523	549	575	601	627	653	679	705
8'	233	260	287	314	341	368	395	422	449	476	503	530	557	584	611	638	665	692	719	746
9'	255	283	311	339	367	395	423	451	479	507	535	563	591	619	647	675	703	731	759	787
10'	277	306	335	364	393	422	451	480	509	538	567	596	625	654	683	712	741	770	799	828
11'	299	329	359	389	419	449	479	509	539	569	599	629	659	689	719	749	779	809	839	869
12'	321	352	383	414	445	476	507	538	569	600	631	662	693	724	755	786	817	848	879	910
13'	343	375	407	439	471	503	535	567	599	631	663	695	727	759	791	823	855	887	919	951
14'	365	398	431	464	497	530	563	596	629	662	695	728	761	794	827	860	893	926	959	992
15'	387	421	455	489	523	557	591	625	659	693	727	761	795	829	863	897	931	965	999	1033
16'	409	444	479	514	549	584	619	654	689	724	759	794	829	864	899	934	969	1004	1039	1074
17'	431	467	503	539	575	611	647	683	719	755	791	827	863	899	935	971	1007	1043	1079	1115
18'	453	490	527	564	601	638	675	712	749	786	823	860	897	934	971	1008	1045	1082	1119	1156
19'	475	513	551	589	627	665	703	741	779	817	855	893	931	969	1007	1045	1083	1121	1159	1197
20'	497	536	575	614	653	692	731	770	809	848	887	926	965	1004	1043	1082	1121	1160	1199	1238
21'	519	559	599	639	679	719	759	799	839	879	919	959	999	1039	1079	1119	1159	1199	1239	1279
22'	541	582	623	664	705	746	787	828	869	910	951	992	1033	1074	1115	1156	1197	1238	1279	1320
23'	563	605	647	689	731	773	815	857	899	941	983	1025	1067	1109	1151	1193	1235	1277	1319	1361
24'	585	628	671	714	757	800	843	886	929	972	1015	1058	1101	1144	1187	1230	1273	1316	1359	1402

(continued on next page)

Figure 9-2　Surface Area of Rooms *(continued from previous page)*

Rooms with 10' Ceilings — Total Area – 4 Walls and Ceiling (in square feet)

	3'	4'	5'	6'	7'	8'	9'	10'	11'	12'	13'	14'	15'	16'	17'	18'	19'	20'	21'	22'
3'	129	152	175	198	221	244	267	290	313	336	359	382	405	428	451	474	497	520	543	566
4'	152	176	200	224	248	272	296	320	344	368	392	416	440	464	488	512	536	560	584	608
5'	175	200	225	250	275	300	325	350	375	400	425	450	475	500	525	550	575	600	625	650
6'	198	224	250	276	302	328	354	380	406	432	458	484	510	536	562	588	614	640	666	692
7'	221	248	275	302	329	356	383	410	437	464	491	518	545	572	599	626	653	680	707	734
8'	244	272	300	328	356	384	412	440	468	496	524	552	580	608	636	664	692	720	748	776
9'	267	296	325	354	383	412	441	470	499	528	557	586	615	644	673	702	731	760	789	818
10'	290	320	350	380	410	440	470	500	530	560	590	620	650	680	710	740	770	800	830	860
11'	313	344	375	406	437	468	499	530	561	592	623	654	685	716	747	778	809	840	871	902
12'	336	368	400	432	464	496	528	560	592	624	656	688	720	752	784	816	848	880	912	944
13'	359	392	425	458	491	524	557	590	623	656	689	722	755	788	821	854	887	920	953	986
14'	382	416	450	484	518	552	586	620	654	688	722	756	790	824	858	892	926	960	994	1028
15'	405	440	475	510	545	580	615	650	685	720	755	790	825	860	895	930	965	1000	1035	1070
16'	428	464	500	536	572	608	644	680	716	752	788	824	860	896	932	968	1004	1040	1076	1112
17'	451	488	525	562	599	636	673	710	747	784	821	858	895	932	969	1006	1043	1080	1117	1154
18'	474	512	550	588	626	664	702	740	778	816	854	892	930	968	1006	1044	1082	1120	1158	1196
19'	497	536	575	614	653	692	731	770	809	848	887	926	965	1004	1043	1082	1121	1160	1199	1238
20'	520	560	600	640	680	720	760	800	840	880	920	960	1000	1040	1080	1120	1160	1200	1240	1280
21'	543	584	625	666	707	748	789	830	871	912	953	994	1035	1076	1117	1158	1199	1240	1281	1322
22'	566	608	650	692	734	776	818	860	902	944	986	1028	1070	1112	1154	1196	1238	1280	1322	1364
23'	589	632	675	718	761	804	847	890	933	976	1019	1062	1105	1148	1191	1234	1277	1320	1363	1406
24'	612	656	700	744	788	832	876	920	964	1008	1052	1096	1140	1184	1228	1272	1316	1360	1404	1448

(continued on next page)

Figure 9-2 Surface Area of Rooms *(continued from previous page)*

Rooms with 10'6" Ceilings—Total Area – 4 Walls and Ceiling (in square feet)

	3'	4'	5'	6'	7'	8'	9'	10'	11'	12'	13'	14'	15'	16'	17'	18'	19'	20'	21'	22'
3'	135	159	183	207	231	255	279	303	327	351	375	399	423	447	471	495	519	543	567	591
4'	159	184	209	234	259	284	309	334	359	384	409	434	459	484	509	534	559	584	609	634
5'	183	209	235	261	287	313	339	365	391	417	443	469	495	521	547	573	599	625	651	677
6'	207	234	261	288	315	342	369	396	423	450	477	504	531	558	585	612	639	666	693	720
7'	231	259	287	315	343	371	399	427	455	483	511	539	567	595	623	651	679	707	735	763
8'	255	284	313	342	371	400	429	458	487	516	545	574	603	632	661	690	719	748	777	806
9'	279	309	339	369	399	429	459	489	519	549	579	609	639	669	699	729	759	789	819	849
10'	303	334	365	396	427	458	489	520	551	582	613	644	675	706	737	768	799	830	861	892
11'	327	359	391	423	455	487	519	551	583	615	647	679	711	743	775	807	839	871	903	935
12'	351	384	417	450	483	516	549	582	615	648	681	714	747	780	813	846	879	912	945	978
13'	375	409	443	477	511	545	579	613	647	681	715	749	783	817	851	885	919	953	987	1021
14'	399	434	469	504	539	574	609	644	679	714	749	784	819	854	889	924	959	994	1029	1064
15'	423	459	495	531	567	603	639	675	711	747	783	819	855	891	927	963	999	1035	1071	1107
16'	447	484	521	558	595	632	669	706	743	780	817	854	891	928	965	1002	1039	1076	1113	1150
17'	471	509	547	585	623	661	699	737	775	813	851	889	927	965	1003	1041	1079	1117	1155	1193
18'	495	534	573	612	651	690	729	768	807	846	885	924	963	1002	1041	1080	1119	1158	1197	1236
19'	519	559	599	639	679	719	759	799	839	879	919	959	999	1039	1079	1119	1159	1199	1239	1279
20'	543	584	625	666	707	748	789	830	871	912	953	994	1035	1076	1117	1158	1199	1240	1281	1322
21'	567	609	651	693	735	777	819	861	903	945	987	1029	1071	1113	1155	1197	1239	1281	1323	1365
22'	591	634	677	720	763	806	849	892	935	978	1021	1064	1107	1150	1193	1236	1279	1322	1365	1408
23'	615	659	703	747	791	835	879	923	967	1011	1055	1099	1143	1187	1231	1275	1319	1363	1407	1451
24'	639	684	729	774	819	864	909	954	999	1044	1089	1134	1179	1224	1269	1314	1359	1404	1449	1494

Division 9

(continued on next page)

Figure 9-2 Surface Area of Rooms *(continued from previous page)*

Rooms with 11' Ceilings—Total Area – 4 Walls and Ceiling (in square feet)

	3'	4'	5'	6'	7'	8'	9'	10'	11'	12'	13'	14'	15'	16'	17'	18'	19'	20'	21'	22'
3'	141	166	191	216	241	266	291	316	341	366	391	416	441	466	491	516	541	566	591	616
4'	166	192	218	244	270	296	322	348	374	400	426	452	478	504	530	556	582	608	634	660
5'	191	218	245	272	299	326	353	380	407	434	461	488	515	542	569	596	623	650	677	704
6'	216	244	272	300	328	356	384	412	440	468	496	524	552	580	608	636	664	692	720	748
7'	241	270	299	328	357	386	415	444	473	502	531	560	589	618	647	676	705	734	763	792
8'	266	296	326	356	386	416	446	476	506	536	566	596	626	656	686	716	746	776	806	836
9'	291	322	353	384	415	446	477	508	539	570	601	632	663	694	725	756	787	818	849	880
10'	316	348	380	412	444	476	508	540	572	604	636	668	700	732	764	796	828	860	892	924
11'	341	374	407	440	473	506	539	572	605	638	671	704	737	770	803	836	869	902	935	968
12'	366	400	434	468	502	536	570	604	638	672	706	740	774	808	842	876	910	944	978	1012
13'	391	426	461	496	531	566	601	636	671	706	741	776	811	846	881	916	951	986	1021	1056
14'	416	452	488	524	560	596	632	668	704	740	776	812	848	884	920	956	992	1028	1064	1100
15'	441	478	515	552	589	626	663	700	737	774	811	848	885	922	959	996	1033	1070	1107	1144
16'	466	504	542	580	618	656	694	732	770	808	846	884	922	960	998	1036	1074	1112	1150	1188
17'	491	530	569	608	647	686	725	764	803	842	881	920	959	998	1037	1076	1115	1154	1193	1232
18'	516	556	596	636	676	716	756	796	836	876	916	956	996	1036	1076	1116	1156	1196	1236	1276
19'	541	582	623	664	705	746	787	828	869	910	951	992	1033	1074	1115	1156	1197	1238	1279	1320
20'	566	608	650	692	734	776	818	860	902	944	986	1028	1070	1112	1154	1196	1238	1280	1322	1364
21'	591	634	677	720	763	806	849	892	935	978	1021	1064	1107	1150	1193	1236	1279	1322	1365	1408
22'	616	660	704	748	792	836	880	924	968	1012	1056	1100	1144	1188	1232	1276	1320	1364	1408	1452
23'	641	686	731	776	821	866	911	956	1001	1046	1091	1136	1181	1226	1271	1316	1361	1406	1451	1496
24'	666	712	758	804	850	896	942	988	1034	1080	1126	1172	1218	1264	1310	1356	1402	1448	1494	1540

(continued on next page)

Figure 9-2 Surface Area of Rooms *(continued from previous page)*

Rooms with 12' Ceilings—Total Area – 4 Walls and Ceiling (in square feet)

	3'	4'	5'	6'	7'	8'	9'	10'	11'	12'	13'	14'	15'	16'	17'	18'	19'	20'	21'	22'
3'	153	180	207	234	261	288	315	342	369	396	423	450	477	504	531	558	585	612	639	666
4'	180	208	236	264	292	320	348	376	404	432	460	488	516	544	572	600	628	656	684	712
5'	207	236	265	294	323	352	381	410	439	468	497	526	555	584	613	642	671	700	729	758
6'	234	264	294	324	354	384	414	444	474	504	534	564	594	624	654	684	714	744	774	804
7'	261	292	323	354	385	416	447	478	509	540	571	602	633	664	695	726	757	788	819	850
8'	288	320	352	384	416	448	480	512	544	576	608	640	672	704	736	768	800	832	864	896
9'	315	348	381	414	447	480	513	546	579	612	645	678	711	744	777	810	843	876	909	942
10'	342	376	410	444	478	512	546	580	614	648	682	716	750	784	818	852	886	920	954	988
11'	369	404	439	474	509	544	579	614	649	684	719	754	789	824	859	894	929	964	999	1034
12'	396	432	468	504	540	576	612	648	684	720	756	792	828	864	900	936	972	1008	1044	1080
13'	423	460	497	534	571	608	645	682	719	756	793	830	867	904	941	978	1015	1052	1089	1126
14'	450	488	526	564	602	640	678	716	754	792	830	868	906	944	982	1020	1058	1096	1134	1172
15'	477	516	555	594	633	672	711	750	789	828	867	906	945	984	1023	1062	1101	1140	1179	1218
16'	504	544	584	624	664	704	744	784	824	864	904	944	984	1024	1064	1104	1144	1184	1224	1264
17'	531	572	613	654	695	736	777	818	859	900	941	982	1023	1064	1105	1146	1187	1228	1269	1310
18'	558	600	642	684	726	768	810	852	894	936	978	1020	1062	1104	1146	1188	1230	1272	1314	1356
19'	585	628	671	714	757	800	843	886	929	972	1015	1058	1101	1144	1187	1230	1273	1316	1359	1402
20'	612	656	700	744	788	832	876	920	964	1008	1052	1096	1140	1184	1228	1272	1316	1360	1404	1448
21'	639	684	729	774	819	864	909	954	999	1044	1089	1134	1179	1224	1269	1314	1359	1404	1449	1494
22'	666	712	758	804	850	896	942	988	1034	1080	1126	1172	1218	1264	1310	1356	1402	1448	1494	1540
23'	693	740	787	834	881	928	975	1022	1069	1116	1163	1210	1257	1304	1351	1398	1445	1492	1539	1586
24'	720	768	816	864	912	960	1008	1056	1104	1152	1200	1248	1296	1344	1392	1440	1488	1536	1584	1632

(continued on next page)

Figure 9-2 Surface Area of Rooms *(continued from previous page)*

Square Footage for Single Floor, Ceiling or Wall Area

	3'	4'	5'	6'	7'	8'	9'	10'	11'	12'	13'	14'	15'	16'	17'	18'	19'	20'	21'	22'
3'	9	12	15	18	21	24	27	30	33	36	39	42	45	48	51	54	57	60	63	66
4'	12	16	20	24	28	32	36	40	44	48	52	56	60	64	68	72	76	80	84	88
5'	15	20	25	30	35	40	45	50	55	60	65	70	75	80	85	90	95	100	105	110
6'	18	24	30	36	42	48	54	60	66	72	78	84	90	96	102	108	114	120	126	132
7'	21	28	35	42	49	56	63	70	77	84	91	98	105	112	119	126	133	140	147	154
8'	24	32	40	48	56	64	72	80	88	96	104	112	120	128	136	144	152	160	168	176
9'	27	36	45	54	63	72	81	90	99	108	117	126	135	144	153	162	171	180	189	198
10'	30	40	50	60	70	80	90	100	110	120	130	140	150	160	170	180	190	200	210	220
11'	33	44	55	66	77	88	99	110	121	132	143	154	165	176	187	198	209	220	231	242
12'	36	48	60	72	84	96	108	120	132	144	156	168	180	192	204	216	228	240	252	264
13'	39	52	65	78	91	104	117	130	143	156	169	182	195	208	221	234	247	260	273	286
14'	42	56	70	84	98	112	126	140	154	168	182	196	210	224	238	252	266	280	294	308
15'	45	60	75	90	105	120	135	150	165	180	195	210	225	240	255	270	285	300	315	330
16'	48	64	80	96	112	128	144	160	176	192	208	224	240	256	272	288	304	320	336	352
17'	51	68	85	102	119	136	153	170	187	204	221	238	255	272	289	306	323	340	357	374
18'	54	72	90	108	126	144	162	180	198	216	234	252	270	288	306	324	342	360	378	396
19'	57	76	95	114	133	152	171	190	209	228	247	266	285	304	323	342	361	380	399	418
20'	60	80	100	120	140	160	180	200	220	240	260	280	300	320	340	360	380	400	420	440
21'	63	84	105	126	147	168	189	210	231	252	273	294	315	336	357	378	399	420	441	462
22'	66	88	110	132	154	176	198	220	242	264	286	308	330	352	374	396	418	440	462	484
23'	69	92	115	138	161	184	207	230	253	276	299	322	345	368	391	414	437	460	483	506
24'	72	96	120	144	168	192	216	240	264	288	312	336	360	384	408	432	456	480	504	528

(continued on next page)

Figure 9-2 Surface Area of Rooms *(continued from previous page)*

Square Footage for Single Floor, Ceiling or Wall Area

	23'	24'	25'	26'	27'	28'	29'	30'	31'	32'	33'	34'	35'	36'	37'	38'	39'	40'	41'	42'
25'	575	600	625	650	675	700	725	750	775	800	825	850	875	900	925	950	975	1000	1025	1050
26'	598	624	650	676	702	728	754	780	806	832	858	884	910	936	962	988	1014	1040	1066	1092
27'	621	648	675	702	729	756	783	810	837	864	891	918	945	972	999	1026	1053	1080	1107	1134
28'	644	672	700	728	756	784	812	840	868	896	924	952	980	1008	1036	1064	1092	1120	1148	1176
29'	667	696	725	754	783	812	841	870	899	928	957	986	1015	1044	1073	1102	1131	1160	1189	1218
30'	690	720	750	780	810	840	870	900	930	960	990	1020	1050	1080	1110	1140	1170	1200	1230	1260
31'	713	744	775	806	837	868	899	930	961	992	1023	1054	1085	1116	1147	1178	1209	1240	1271	1302
32'	736	768	800	832	864	896	928	960	992	1024	1056	1088	1120	1152	1184	1216	1248	1280	1312	1344
33'	759	792	825	858	891	924	957	990	1023	1056	1089	1122	1155	1188	1221	1254	1287	1320	1353	1386
34'	782	816	850	884	918	952	986	1020	1054	1088	1122	1156	1190	1224	1258	1292	1326	1360	1394	1428
35'	805	840	875	910	945	980	1015	1050	1085	1120	1155	1190	1225	1260	1295	1330	1365	1400	1435	1470
36'	828	864	900	936	972	1008	1044	1080	1116	1152	1188	1224	1260	1296	1332	1368	1404	1440	1476	1512
37'	851	888	925	962	999	1036	1073	1110	1147	1184	1221	1258	1295	1332	1369	1406	1443	1480	1517	1554
38'	874	912	950	988	1026	1064	1102	1140	1178	1216	1254	1292	1330	1368	1406	1444	1482	1520	1558	1596
39'	897	936	975	1014	1053	1092	1131	1170	1209	1248	1287	1326	1365	1404	1443	1482	1521	1560	1599	1638
40'	920	960	1000	1040	1080	1120	1160	1200	1240	1280	1320	1360	1400	1440	1480	1520	1560	1600	1640	1680
41'	941	984	1025	1066	1107	1148	1189	1230	1271	1312	1353	1394	1435	1476	1517	1558	1599	1640	1681	1722
42'	966	1008	1050	1092	1134	1176	1218	1260	1302	1344	1386	1428	1470	1512	1554	1596	1638	1680	1722	1764
43'	989	1032	1075	1118	1161	1204	1247	1290	1333	1376	1419	1462	1505	1548	1591	1634	1677	1720	1763	1806
44'	1012	1056	1100	1144	1188	1232	1276	1320	1364	1408	1452	1496	1540	1584	1628	1672	1716	1760	1804	1848
45'	1035	1080	1125	1170	1215	1260	1305	1350	1395	1440	1485	1530	1575	1620	1665	1710	1755	1800	1845	1890
46'	1058	1104	1150	1196	1242	1288	1334	1380	1426	1472	1518	1564	1610	1656	1702	1748	1794	1840	1886	1932

Figure 9-3　Stud, Drywall, and Plaster Wall Systems

These figures show the components of common studding, drywall, and plaster partition systems.

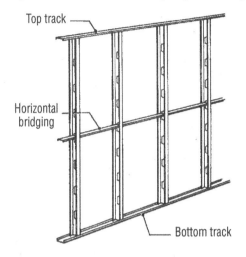

Load Bearing Steel Studs

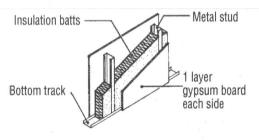

Fiberglass Batt Insulation

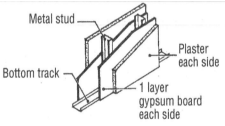

Plaster on Gypsum Lath

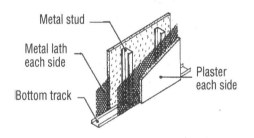

Plaster on Metal Lath

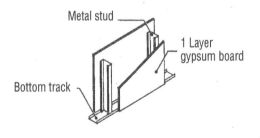

Gypsum Plasterboard

(continued on next page)

Figure 9-3 Stud, Drywall, and Plaster Wall Systems
(continued from previous page)

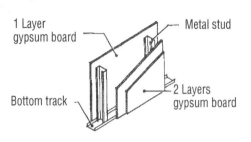

Gypsum Plasterboard, 2 Layers

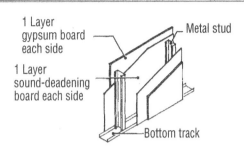

Sound-Deadening Board

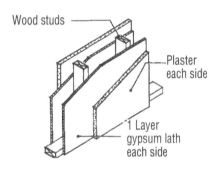

Plaster on Gypsum Lath

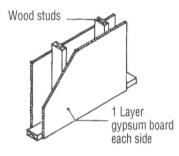

Gypsum Plasterboard

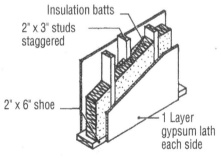

Staggered Stud Wall

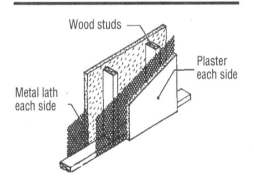

Plaster on Metal Lath

(continued on next page)

Figure 9-3 Stud, Drywall, and Plaster Wall Systems
(continued from previous page)

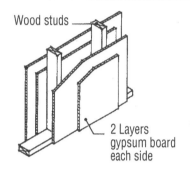

Gypsum Plasterboard, 2 Layers

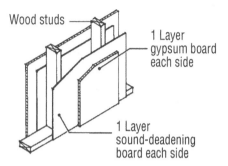

Sound-Deadening Board

Figure 9-4 Furring Quantities

This table provides multiplication factors for converting square feet of wall area requiring furring into board feet of lumber. These figures are based on lumber and size spacing.

Size	Board Feet per Square Feet of Wall Area				Lbs. Nails per MBM of Furring
	Spacing Center to Center				
	12"	16"	20"	24"	
1" x 2"	.18	.14	.11	.10	55
1" x 3"	.28	.21	.17	.14	37

Figure 9-5 Quantities for Perlite Gypsum Plaster

This table shows the approximate number of sacks of prepared perlite gypsum plaster required to cover 100 square yards of surface area. The coverage depends on the type of lathing or base used and the required thickness of plaster.

Type of Base	Total Plaster Thickness (inches)	Number of Sacks
Wood lath	5/8	22 (80 lbs. per sack)
Metal lath	5/8	35 (80 lbs. per sack)
Gypsum lath	1/2	20 (80 lbs. per sack)
Masonry walls	5/8	24 (67 lbs. per sack)

(courtesy How to Estimate Building Losses and Construction Costs, Paul I. Thomas, Prentice-Hall, Inc.)

Figure 9-6 Quantities for Job-Mixed Plaster

This table presents approximate quantities required to prepare enough job-mixed sanded or perlite plaster to cover 100 square yards of surface area. The coverage depends on the type of base materials used and the required thickness of plaster.

Type of Base	Total Plaster Thickness Inches	100 Lb. Sacks of Neat Gypsum	C.Y. Sand 1:2-1/2 Mix	Aggregate Perlite		
				C.F.	4 C.F. Sacks	Mix
Wood lath	5/8	11	1.1	27.5	12.5	1:2-1/2
Metal lath	5/8	20	2.0	50.0	12.5	1:2-1/2
Gypsum lath	1/2	10	1.0	25	6.5	1:2-1/2
Masonry walls	5/8	12	1.2	30	7.5	1:3

ASA permits 250 lbs. damp loose sand or 2-1/2 CF of vermiculite or perlite, provided this proportioning is used for both scratch and brown coats on three-coat work.

The number of 4 CF sacks of perlite are shown to the nearest 1/2 sack.

(*courtesy* How to Estimate Building Losses and Construction Costs, *Paul I. Thomas, Prentice-Hall, Inc.*)

Figure 9-7 Quantities for White Coat Finish Plaster

This table shows the approximate quantities of materials required to cover 100 square yards of surface area with 1/16" coat of white finish.

Type of Finish	100 Lb. Sacks Neat Gypsum	100 Lb. Sacks Keen's Cement	50 Lb. Sacks Hydrated Lime
Lime putty	2	—	8
Keene's cement	—	4	4

Based on ASA mix of 1 (100-lb.) sack neat gypsum gauging plaster to 4 (50-lb.) sacks hydrated lime and 1 (100-lb.) sack of Keene's cement to 1 (50-lb.) sack hydrated lime for medium hard finish.

(*courtesy* How to Estimate Building Losses and Construction Costs, *Paul I. Thomas, Prentice-Hall, Inc.*)

Figure 9-8 Quantities of Portland Cement Plaster

This table shows the approximate quantities of materials required to cover 100 square yards of surface area with Portland cement plaster. The coverage depends on the base material used and the required thickness of plaster.

Plaster Thickness (inches)	Cubic Yards on Masonry Base	Cubic Yards on Wire Lath Base
1/4	0.75	0.90
1/2	1.50	1.80
5/8	1.37	1.64
3/4	2.25	2.70
1	3.00	3.60

(courtesy How to Estimate Building Losses and Construction Costs, *Paul I. Thomas, Prentice-Hall, Inc.)*

Figure 9-9 Wallboard Quantities

This table lists factors that can be applied to predetermined area figures in order to quantify actual material requirements. Included are waste factors for the intended use of the materials. The table also lists the amount of screws required for the wallboard, based on stud spacing.

Includes Fiberboard, Gypsum Board, and Plywood, Used as Underflooring, Sheathing, Plaster Base, or as Drywall Finish					
Factors		Screws			
Used for Underflooring, Sheathing, and Plaster Base	Used for Exposed Drywall Finish	Pounds of Screws per 1000 Sq. Ft. of Wallboard			
		Joist, Stud or Rafter Spacing			
		12"	16"	20"	24"
1.05	1.10	7.4	5.8	5	4.1

Figure 9-10 Gypsum Panel Coverage

This table shows the expected coverage from the size panels indicated across the top of the table. Inversely, if the area to be covered with gypsum board panels is known (in square feet), you can quickly determine the required number of sheets of drywall. Note that no provision has been made for openings.

No. of Panels	4' x 7'	4' x 8'	4' x 9'	4' x 10'	4' x 11'	4' x 12'	4' x 13'	4' x 14'
10	280	320	360	400	440	480	520	560
11	308	352	396	440	484	528	572	616
12	336	384	432	480	528	576	624	672
13	364	416	468	520	572	624	676	728
14	392	448	504	560	616	672	728	784
15	420	480	540	600	660	720	780	840
16	448	512	576	640	704	768	832	896
17	476	544	612	680	748	816	884	952
18	504	576	648	720	792	864	936	1008
19	532	608	684	760	836	912	988	1064
20	560	640	720	800	880	960	1040	1120
21	588	672	756	840	924	1008	1092	1176
22	616	704	792	880	968	1056	1144	1232
23	644	736	828	920	1012	1104	1196	1288
24	672	768	864	960	1056	1152	1248	1344
25	700	800	900	1000	1100	1200	1300	1400
26	728	832	936	1040	1144	1248	1352	1456
27	756	864	972	1080	1188	1296	1404	1512
28	784	896	1008	1120	1232	1344	1456	1568
29	812	928	1044	1160	1276	1392	1508	1624
30	840	960	1080	1200	1320	1440	1560	1680
31	868	992	1116	1240	1364	1488	1612	1736

The header "Sizes and Numbers of Panels and S.F. of Area Covered" spans across the top of the table.

Figure 9-11   Drywall Accessories

This table provides the approximate quantities of screws, joint compound, and tape needed for the indicated amounts of drywall.

With this Amount of Sheetrock Gypsum Panel	Type GWB-54 Screws Required* USE	this Amount of Powder-Type Compound OR	this Amount of USG Ready-To-Use Compound-All Purpose AND	this Amount of Perf-A-Tape Reinforcing Tape
100 S.F.	.6 lbs.	6 lbs.	1 Gal.	37 Ft.
200 S.F.	1.2 lbs.	12 lbs.	2 Gals.	74 Ft.
300 S.F.	1.7 lbs.	18 lbs.	2 Gals.	111 Ft.
400 S.F.	2.3 lbs.	24 lbs.	3 Gals.	148 Ft.
500 S.F.	2.9 lbs.	30 lbs.	3 Gals.	185 Ft.
600 S.F.	3.5 lbs.	36 lbs.	4 Gals.	222 Ft.
700 S.F.	4.0 lbs.	42 lbs.	5 Gals.	259 Ft.
800 S.F.	4.6 lbs.	48 lbs.	5 Gals.	296 Ft.
900 S.F.	5.2 lbs.	54 lbs.	6 Gals.	333 Ft.
1000 S.F.	5.7 lbs.	60 lbs.	6 Gals.	370 Ft.

* Spaced 7" on ceiling; 8" on wall.

1-1/4" screw ≈ 266/lb.

Figure 9-12 Levels of Gypsum Drywall Finish

The finished appearance of drywall largely depends on the quality of framing and care exercised in applying the drywall panels. The better the framing and cladding, the easier it is to achieve a near-perfect wall. The last remaining challenge is to finish the joints to meet visual appearance expectations. This chapter helps you determine the level of finish quality needed and shows you how to obtain it.

Contract documents traditionally use nonspecific terms such as "industry standards" or "workmanlike finish" to describe how finished gypsum board walls and ceilings should look. This practice has often led to confusion about the degree or level of finishing required for a particular job.

Five leading industry trade associations—the Association of the Wall and Ceiling Industries International (AWCI), Ceilings and Interior Systems Construction Association (CISCA), Gypsum Association (GA), Drywall Finishing Council (DWFC) and Painting and Decorating Contractors of America (PDCA)—have combined efforts to collectively adopt a set of industry-wide recommended specifications for levels of gypsum board finish. This specification identifies five levels of finishing, enabling architects to more closely identify the quality of finish required and allowing for better competitive bidding among contractors. American Society for Testing & Materials (ASTM) recognizes this specification by including the levels of gypsum board finishing in ASTM C840.

Key factors in determining the quality level required include the following:

1. The location of the work to be done
2. The type and angle of surface illumination (both natural and artificial lighting)
3. The orientation of the panels during installation
4. The type of paint or wall covering to be used
5. The method of application

Critical lighting conditions, gloss paints, and thin wall coverings require a high level of finish, while heavily textured surfaces or those that will be decorated with heavy-gauge wall coverings require less attention to final surface quality.

Definitions of the five finishing levels are provided below, together with a matrix that details how each level of finishing should be achieved using SHEETROCK brand joint treatment and finishing products. The matrix also describes the appearance of the finished wall that may be anticipated for each level.

Finishing-Level Definitions

The following finishing-level definitions are based on GA-214-07, "Recommended Levels of Gypsum Board Finish," and are intended to provide an industry standard for drywall finishing.

Level 0 Used in temporary construction or wherever the final decoration has not been determined. Unfinished. No taping, finishing or corner beads are required. Also could be used where non-predecorated panels will be used in demountable-type partitions that are to be painted as a final finish.

Level 1 Frequently used in plenum areas above ceilings, in attics, in areas where the assembly would generally be concealed, or in building service corridors and other areas not normally open to public view. Some degree of sound and smoke control is provided; in some geographic areas,

(continued on next page)

Figure 9-12 Levels of Gypsum Drywall Finish
(continued from previous page)

this level is referred to as "fire-taping," although this level of finish does not typically meet fire-resistant assembly requirements. Where a fire resistance rating is required for the gypsum board assembly, details of construction should be in accordance with reports of fire tests of assemblies that have met the requirements of the fire rating imposed.

All joints and interior angles shall have tape embedded in joint compound. Accessories are optional at specifier discretion in corridors and other areas with pedestrian traffic. Tape and fastener heads need not be covered with joint compound. Surface shall be free of excess joint compound. Tool marks and ridges are acceptable.

Level 2 It may be specified for standard gypsum board surfaces in garages, warehouse storage, or other similar areas where surface appearance is not of primary importance.

All joints and interior angles shall have tape embedded in joint compound and shall be immediately wiped with a joint knife or trowel, leaving a thin coating of joint compound over all joints and interior angles. Fastener heads and accessories shall be covered with a coat of joint compound. Surface shall be free of excess joint compound. Tool marks and ridges are acceptable.

Level 3 Typically used in areas that are to receive heavy texture (spray or hand applied) finishes before final painting, or where commercial-grade (heavy duty) wall coverings are to be applied as the final decoration. This level of finish should not be used where smooth painted surfaces or where lighter weight wall coverings are specified. The prepared surface shall be coated with a drywall primer prior to the application of final finishes.

All joints and interior angles shall have tape embedded in joint compound and shall be immediately wiped with a joint knife or trowel, leaving a thin coating of joint compound over all joints and interior angles. One additional coat of joint compound shall be applied over all joints and interior angles. Fastener heads and accessories shall be covered with two separate coats of joint compound. All joint compounds shall be smooth and free of tool marks and ridges. The prepared surface shall be covered with a drywall primer prior to the application of the final decoration.

Level 4 This level should be used where residential grade (light duty) wall coverings, flat paints, or light textures are to be applied. The prepared surface shall be coated with a drywall primer prior to the application of final finishes. Release agents for wall coverings are specifically formulated to minimize damage if coverings are subsequently removed.

The weight, texture, and sheen level of the wall covering material selected should be taken into consideration when specifying wall coverings over this level of drywall treatment. Joints and fasteners must be sufficiently concealed if the wall covering material is lightweight, contains limited pattern, has a glossy finish, or has any combination of these features. In critical lighting areas, flat paints applied over light textures tend to reduce joint photographing. Gloss, semi-gloss, and enamel paints are not recommended over this level of finish.

All joints and interior angles shall have tape embedded in joint compound and shall be immediately wiped with a joint knife or trowel, leaving a thin coating of joint compound over all joints and interior angles. In addition, two separate coats of joint compound shall be applied over all flat joints and one separate coat of joint compound applied over interior angles. Fastener heads and accessories shall be covered with three separate coats of joint compound. All joint compounds

(continued on next page)

Figure 9-12 Levels of Gypsum Drywall Finish
(continued from previous page)

shall be smooth and free of tool marks and ridges. The prepared surface shall be covered with a drywall primer like SHEETROCK first coat prior to the application of the final decoration.

Level 5 The highest quality finish is the most effective method to provide a uniform surface and minimize the possibility of joint photographing and of fasteners showing through the final decoration. This level of finish is required where gloss, semi-gloss, or enamel is specified; when flat joints are specified over an untextured surface; or where critical lighting conditions occur. The prepared surface shall be coated with a drywall primer prior to the application of final decoration.

All joints and interior angles shall have tape embedded in joint compound and be immediately wiped with a joint knife or trowel, leaving a thin coating of joint compound over all joints and interior angles. Two separate coats of joint compound shall be applied over all flat joints and one separate coat of joint compound applied over interior angles. Fastener heads and accessories shall be covered with three separate coats of joint compound.

A thin skim coat of joint compound shall be trowel applied to the entire surface. Excess compound is immediately troweled off, leaving a film or skim coating of compound completely covering the paper. As an alternative to a skim coat, a material manufactured especially for this purpose may be applied such as SHEETROCK TUFF-HIDE primer surfacer. The surface must be smooth and free of tool marks and ridges. The prepared surface shall be covered with a drywall primer prior to the application of the final decoration.

(courtesy The Gypsum Construction Handbook, 6ᵗʰ Edition, by USG)

Figure 9-13 Wood Flooring Quantities

This table lists the amounts to be added to the measured floor areas for the materials required for installing strip wood flooring. Milling and cutting waste values are included.

Nominal Size in Inches	Actual Size in Inches	Milling + 5% Cutting Waste	Multiply Area by	Floor Needed per 1,000 FBM
1 x 2	25/32 x 1-1/2	55%	1.55	1,500
1 x 2-1/2	25/32 x 2	42	1.42	1,420
1 x 3	25/32 x 2-1/4	38	1.38	1,380
1 x 4	25/32 x 3-1/4	29	1.29	1,290
1 x 2	3/8 x 1-1/2	38	1.38	1,380
1 x 2-1/2	3/8 x 2	30	1.30	1,300
	1/2 x 1-1/2	38	1.38	1,380
	1/2 x 2	30	1.30	1,300

Figure 9-14 Approximate Waste Percentages to be Added for Resilient Tile

This table lists approximate waste figures to be added when estimating resilient tiling. These figures are based on expected averages for the size of the areas to receive the tile.

Area	Waste
Up to 75 S.F.	10–12%
75–150 S.F.	7–10%
150–300 S.F.	6–7%
300–1,000 S.F.	4–6%
1,000–5,000 S.F.	3–4%
5,000 S.F. and Up	2–3%

Figure 9-15   Painting Quantity Factors

The *Painting and Decorating Contractors of America's Estimating Guide* recommends that the following factors be used when estimating painting. For each type of area, the actual measured flat surface area of the item should be increased by the factors indicated. These factors are for adjusting painting rates and outputs of flat wall painting to account for the extra amount of detail and associated extra time needed for trim work.

Area		Factor
Balustrades:		1 side x 4
Blinds:	Plain	Actual area x 2
	Slotted	Actual area x 4
Cabinets:		Front area x 5
Clapboards and Drop Siding:		Actual area x 1.1
Cornices:	1 Story	Actual area x 2
	2 Story	Actual area x 3
	1 Story Ornamental	Actual area x 4
	2 Story Ornamental	Actual area x 6
Doors:	Flush	150% per side
	Two Panel	175% per side
	Four Panel	200% per side
	Six Panel	225% per side
Door Trim:		L.F. + 50% per side
Fences:	Chain Link	1 side x 3 for both sides
	Picket	1 side x 4 for both sides
Gratings:		1 side x 4/6
Grilles:	Plain	1 side x 200%
	Lattice	Area x 2 per side
	Moldings under 12" Wide	1 S.F./L.F.
Open Trusses:		Length x Depth x 2.5
Pipes:	Up to 4"	1 S.F. per L.F.
	4" to 8"	2 S.F. per L.F.
	8" to 12"	3 S.F. per L.F.
	12" to 16"	4 S.F. per L.F.
	Hangers Extra	
Radiator:		Face area x 7
Shingle Siding:		Area x 1.5
Stairs:		Number of risers x 8 widths
Tie Rods:		2 S.F. per L.F.
Wainscoting, Paneled:		Actual area x 2
Walls and Ceilings:		Length x Width, no deducts for less than 100 S.F.
Sanding and Puttying:	Quality Work	Actual area x 2
	Average Work	Actual area x 50%
	Industrial	Actual area x 25%
Downspouts and Gutters:		Actual area x 2
Window Sash:		1 L.F. of part = 1 S.F.

Figure 9-16 Paint Coverage

This table lists the average expected coverage for various painting tasks using different methods of application.

Item	Coat	One Gallon Covers			In 8 Hours a Laborer Covers			Labor-Hours per 100 S.F.		
		Brush	Roller	Spray	Brush	Roller	Spray	Brush	Roller	Spray
Paint wood siding	prime	250 S.F.	225 S.F.	290 S.F.	1150 S.F.	1300 S.F.	2275 S.F.	.695	.615	.351
	others	270	250	290	1300	1625	2600	.615	.492	.307
Paint exterior trim	prime	400	—	—	650	—	—	1.230	—	—
	1st	475	—	—	800	—	—	1.000	—	—
	2nd	520	—	—	975	—	—	.820	—	—
Paint shingle siding	prime	270	255	300	650	975	1950	1.230	.820	.410
	others	360	340	380	800	1150	2275	1.000	.695	.351
Stain shingle siding	1st	180	170	200	750	1125	2250	1.068	.711	.355
	2nd	270	250	290	900	1325	2600	.888	.603	.307
Paint brick masonry	prime	180	135	160	750	800	1800	1.066	1.000	.444
	1st	270	225	290	815	975	2275	.981	.820	.351
	2nd	340	305	360	815	1150	2925	.981	.695	.273
Paint interior plaster drywall	prime	400	380	495	1150	2000	3250	.695	.400	.246
	others	450	425	495	1300	2300	4000	.615	.347	.200
Paint interior doors and windows	prime	400	—	—	650	—	—	1.230	—	—
	1st	425	—	—	800	—	—	1.000	—	—
	2nd	450	—	—	975	—	—	.820	—	—

Figure 9-17 Painting Estimating Techniques

Proper estimating methodology is needed to obtain an accurate painting estimate. There is no known reliable shortcut or square foot method. These steps should be followed:

- List all surfaces to be painted, with an accurate quantity (area) of each. Items having similar surface condition, finish, application method, and accessibility may be grouped together.

- List all the tasks required for each surface to be painted, including surface preparation, masking, and protection of adjacent surfaces. Surface preparation may include minor repairs, washing, sanding, and puttying.

- Select the proper RSMeans cost data line for each task. Review and consider all adjustments to labor and materials for type of paint and location of work. Apply the height adjustment carefully. For instance, when applying the adjustment for work over 8' high to a wall that is 12' high, apply the adjustment only to the area between 8' and 12' high, and not to the entire wall.

- When applying more than one percent (1%) adjustment, apply each to the base cost of the data, rather than applying one percentage adjustment on top of the other.

- When estimating the cost of painting walls and ceilings, remember to add the brushwork for all cut-ins at inside corners and around windows and doors as a LF measure. One linear foot of cut-in with brush equals one square foot of painting.

- All items for spray painting include the labor for roll-back.

- Deduct for openings greater than 100 SF, or openings that extend from floor to ceiling and are greater than 5' wide. Do not deduct small openings.

- The cost of brushes, rollers, ladders, and spray equipment are considered to be part of a painting contractor's overhead, and should not be added to the estimate. The cost of rented equipment such as scaffolding and swing staging should be added to the estimate.

Figure 9-18 Wall Covering Quantities: Common Measuring Charts
for American Single Rolls

Wall and Border Estimating Chart			
Distance Of Wall Width in Feet	American Single Rolls for Wall Area —Height of Ceiling—		Number of Yards for Border
	8 feet	9 feet	10 feet	Border
4	2	2	2	2
5	2	2	2	2
6	2	2	2	3
7	2 to 3	3	3	3
8	2 to 3	3	3	3
9	3	3	3 to 4	4
10	3 to 4	4	4	4
11	3 to 4	4	4	4
12	4	4	4	5
13	4	4	5	5
14	4	5	5	5
16	4 to 5	5	6	6
18	5	6	6	7
20	6	6	7	7
22	6 to 7	7	8	8
24	7	7	8	9
26	7 to 8	8	9	9

Note: These charts should be used with caution because design repeat of pattern and width/length of pattern may increase your rollage. The first two charts allow for no door or window openings.

To determine number of Euro single rolls needed, you can take the above number of American rolls and multiply the rolls by 1.25. For example, if the number of American single rolls is 15, you multiply 15 x 1.25, and this equals 18.75 or 19 Euro single rolls.

(courtesy Painting and Decorating Contractors of America)

(continued on next page)

Figure 9-18 Wall Covering Quantities: Common Measuring Charts for American Single Rolls *(continued from previous page)*

Room and Border Estimating Chart					
Distance Around Room in Feet	American Single Rolls for Wall Area —Height of Ceiling—			Number of Yards for Border	American Single Rolls for Ceilings
	8 feet	9 feet	10 feet		
28	8	9	10	11	2
30	8	9	10	11	2
32	9	10	11	12	2
34	10	11	12	13	4
36	10	11	12	13	4
38	11	12	13	14	4
40	11	12	14	15	4
42	12	13	14	15	4
44	12	14	15	16	5
46	13	14	16	17	5
48	13	15	16	17	6
50	14	15	17	18	6
52	14	16	18	19	6
54	15	17	18	19	6
56	15	17	19	20	7
58	16	18	20	21	8
60	16	18	20	21	8

(continued on next page)

Figure 9-18 Wall Covering Quantities: Common Measuring Charts for American Single Rolls *(continued from previous page)*

Ceiling and Border Estimating Chart			
Room Size Length x Width In Feet	Square Footage	Number of American Single Rolls	Number of Yards for Border
4 x 4	16	1	6
5 x 5	25	1	7
6 x 6	36	2	9
7 x 7	49	2	11
8 x 8	64	2 to 3	12
9 x 9	81	4	13
10 x 10	100	4	15
11 x 11	121	5	16
12 x 12	144	6	17
13 x 13	169	6	19
14 x 14	196	7	20
15 x 15	225	8	21
16 x 16	256	9	23
17 x 17	289	10	24
18 x 18	324	12	25
19 x 19	361	12	27
20 x 20	400	14	28

Figure 9-19 Fire System Design Considerations

A large number of systems have been designed and tested for fire resistance. The systems vary greatly in both design and performance. Nevertheless, certain basic system designs are commonly used. As a frame of reference, several typical designs and their accompanying fire ratings are shown below for both wood- and steel-framed assemblies.

Below are a series of notes that apply to many of the fire tests:

1. Two recent tests, UL Design U419 for non-load-bearing partitions and UL Design U423 for load-bearing partitions, permit SHEETROCK brand gypsum panel products and IMPERIAL brand gypsum base products to be applied horizontally or vertically in partitions without compromising the fire rating. When either of these tests is listed with a USG system, the system can then be built with the panels oriented in either direction.

2. The two fire tests indicated above also demonstrated that when FIRECODE or FIRECODE C Core products are used, the horizontal joints on opposite sides of the studs need not be staggered (as was previously required).

3. In partition systems indicating the use of 5/8" SHEETROCK brand gypsum panels, FIRECODE Core, or 1/2" SHEETROCK brand gypsum panels, FIRECODE C Core, it is permissible to substitute 5/8" FIBEROCK brand panels without compromising the fire rating.

4. Where insulation is shown in assembly drawings, a specific product may be required to achieve the stated fire rating. Glass fiber insulation cannot be substituted in all cases for mineral wool insulation.

5. In fire-rated non-load-bearing partitions, steel studs should not be attached to floor and ceiling runners.

Wood Frame Partitions	**1-hr. Rating** **UL Design U305** **Drywall System**	
	Studs:	Wood 2 x 4
	Stud spacing:	16" o.c.
	Gypsum panel:	5/8" SHEETROCK brand FIRECODE Core gypsum panel, or 5/8" SHEETROCK brand FIRECODE Core MOLD TOUGH gypsum panel, each side.
	Panel orientation:	Vertical or horizontal.
	Attachment:	1-7/8" cement-coated nails spaced 7" o.c.
	Joints:	Exposed or taped and treated according to edge configuration.
	Insulation:	Optional.
	Perimeter:	Should be caulked With SHEETROCK brand acoustical sealant.

Veneer Plaster System

Studs:	Wood 2 x 4
Stud spacing:	16" o.c.
Gypsum panel:	5/8" IMPERIAL brand gypsum base FIRECODE Core, each side.
Panel orientation:	Vertical or horizontal.
Attachment:	1-7/8" cement-coated nails spaced 7" o.c.
Joints:	Taped.

(continued on next page)

Figure 9-19 Fire System Design Considerations
(continued from previous page)

Finish:	3/32" Diamond or Imperial veneer finish both sides.
Insulation:	Optional.
Perimeter:	Should be caulked with Sheetrock brand acoustical sealant.

**2-hr. Rating
UL Design U301
Drywall System**

6"

Studs:	Wood 2 x 4
Stud spacing:	16" o.c.
Gypsum panel:	Two layers of 5/8" Sheetrock brand Firecode Core gypsum panel, or 5/8" Sheetrock brand Firecode Core Mold Tough gypsum panel each side.
Panel orientation:	Horizontal or vertical—joints of face layer staggered over joints of base layer.
Attachment:	Base layer—1-7/8" cement-coated nails spaced 6" o.c. Face layer—2-3/8" nails 8" o.c.
Joints:	Exposed or taped and treated.
Perimeter:	Should be caulked with Sheetrock brand acoustical sealant.

Veneer Plaster System

Studs:	Wood 2 x 4
Stud spacing:	16" o.c.
Gypsum panel:	Two layers of 5/8" Imperial brand gypsum base Firecode Core.
Panel orientation:	Horizontal or vertical—joints of face layer staggered over joints of base layer.
Attachment:	Base layer—1-7/8" cement-coated nails spaced 6" o.c. Face layer—2-3/8" nails 8" o.c.
Joints:	Taped.
Finish:	3/32" Diamond or Imperial veneer finish both sides.
Perimeter:	Should be caulked with Sheetrock brand acoustical sealant.

**Steel Frame
Partitions**

**1-hr. Rating
UL Design U419
Drywall System**

4⅛"

(continued on next page)

Figure 9-19 Fire System Design Considerations
(continued from previous page)

Studs:	Steel 362S125-18 (minimum).
Stud spacing:	24" o.c.
Gypsum panel:	5/8" SHEETROCK brand FIRECODE Core gypsum panel or 5/8" SHEETROCK brand Mold-Tough FIRECODE Core gypsum panel each side.
Panel orientation:	Vertical or horizontal.
Attachment:	Type S screws 8" o.c.
Joints:	Taped and treated.
Insulation:	Optional.
Perimeter:	Should be caulked with SHEETROCK brand acoustical sealant.

Veneer Plaster System

Studs:	Steel 362S125-18 (minimum).
Stud spacing:	24" o.c.
Gypsum panel:	5/8" IMPERIAL brand gypsum Base FIRECODE Core, each side.
Panel orientation:	Vertical or horizontal.
Attachment:	Type S screws 8" o.c.
Joints:	Taped (paper) and treated.
Finish:	3/32" DIAMOND or IMPERIAL veneer finish both sides.
Insulation:	Optional.
Perimeter:	Should be caulked with SHEETROCK brand acoustical sealant.

2-hr. Rating
UL Design U411, U412 or U419
Drywall System

wall thickness varies with actual design

Studs:	Steel 250S125-18.
Stud spacing:	24" o.c.
Gypsum panel:	Two layers of 5/8" SHEETROCK brand FIRECODE Core gypsum panel, or 1/2" SHEETROCK brand FIRECODE C gypsum panel, Core, each side.
Panel orientation:	Vertical or horizontal—joints of face layer staggered over joints of base layer.
Attachment:	Base layer—1" Type S screws 8" o.c. Face layer—laminated with joint compound or attached with 1-5/8" Type S screws 12" o.c.

(continued on next page)

Figure 9-19 Fire System Design Considerations
(continued from previous page)

Joints:	U411, exposed or taped and treated; U412, outer layer taped and treated.
Perimeter:	Should be caulked with SHEETROCK brand acoustical sealant.

Veneer Plaster System

Studs:	Steel 250S125-18.
Stud spacing:	24" o.c.
Gypsum panel:	Two layers of 5/8" IMPERIAL brand gypsum base, FIRECODE Core, or 1/2" IMPERIAL gypsum base, FIRECODE C Core.
Panel orientation:	Vertical or horizontal—joints of face layer staggered over joints of base layer.
Attachment:	Base layer—1" Type S screws 8" o.c. Face layer—laminated with joint compound or attached with 1-5/8" Type S screws 12" o.c. Face layer—2-3/8" nails 8" o.c.
Joints:	Taped (paper) and treated.
Finish:	3/32" DIAMOND or IMPERIAL veneer finish both sides.
Perimeter:	Should be caulked with SHEETROCK brand acoustical sealant.

Wood Floor/ Ceilings

**1-hr. Rating
UL Design L501 or L512
Drywall System**

wall thickness varies with actual design

Floor:	1" nom. wood sub and finished floor.
Joists:	Wood 2 x 10 cross bridged with 1 x 3 lumber.
Joist spacing:	16" o.c.
Gypsum panel:	5/8" SHEETROCK brand gypsum panel, FIRECODE Core (L501), or 1/2" SHEETROCK brand gypsum panel, FIRECODE C Core (L512).
Panel orientation:	Perpendicular to joists.
Attachment:	1-7/8" cement-coated nails spaced 6" o.c.
Joints:	Taped and treated.

Veneer Plaster System

Floor:	1" nom. wood sub and finished floor.

(continued on next page)

Figure 9-19 Fire System Design Considerations
(continued from previous page)

	Joists:	Wood 2 x 10 cross bridged with 1 x 3 lumber.
	Joist spacing:	16" o.c.
	Gypsum panel:	5/8" IMPERIAL brand gypsum base, FIRECODE Core (L501), or 1/2" IMPERIAL brand gypsum base, FIRE CODE C Core (L512).
	Panel orientation:	Perpendicular to joists.
	Attachment:	1-7/8" cement-coated nails spaced 6" o.c.
	Joints:	Taped.
	Finish:	3/32" DIAMOND or IMPERIAL veneer finish both sides.

Steel Floor/ Ceilings	**3-hr. Rating UL Design G512 Drywall System**	
	Floor:	2-1/2" concrete on corrugated steel deck or riblath over bar joist—includes 3-hr. unrestrained beam.
	Joists:	Type 12J2 min. size, spaced 24" o.c. (riblath); Type 16J2 min. size, spaced 24" o.c. (corrugated steel deck).
	Furring channel:	25-ga. spaced 24" o.c. perpendicular to joists; 3" on each side of wallboard end joints—double-strand saddle tied.
	Gypsum panel:	5/8" SHEETROCK Brand Gypsum Panel, FIRECODE C Core.
	Panel orientation:	Perpendicular to furring.
	Attachment:	1" Type S screws 12 " o.c.
	Joints:	End joints backed with wallboard strips and attached to double channels.

Veneer Plaster System

Floor:	2-1/2" concrete on corrugated steel deck or riblath over bar joist—includes 3-hr. unrestrained beam.
Joists:	Type 12J2 min. size, spaced 24" o.c. (riblath); Type 16J2 min. size, spaced 24" o.c. (corrugated steel deck).
Furring channel:	25-ga. spaced 24" o.c. perpendicular to joists; 3" on each side of wallboard end joints—double-strand saddle tied.
Gypsum panel:	5/8" IMPERIAL brand gypsum base, FIRECODE C Core.
Panel orientation:	Perpendicular to furring.
Attachment:	1" Type S screws 12 " o.c.
Joints:	End joints backed with wallboard strips and attached to double channels.
Finish:	3/32" DIAMOND or IMPERIAL veneer finish both sides.

(courtesy of The Gypsum Construction Handbook, *6th Edition, by USG)*

Figure 9-20 Application Time in Labor-Hours for Painting

Description	Labor-Hours	Unit
Cabinets and casework		
Primer coat, oil base, brushwork	.012	SF
Paint, oil base, brushwork	.012	SF
1 coat	.012	SF
Stain, brushwork, wipe-off	.012	SF
Shellac, 1 coat, brushwork	.012	SF
Varnish, 3 coats, brushwork	.025	SF
Doors and windows, interior latex		
Doors, flush, both sides, incl. frame & trim		
Roll and brush primer	.800	Ea.
Finish coat, latex	.800	Ea.
Primer & 1 coat latex	1.143	Ea.
Primer & 2 coats latex	1.600	Ea.
Spray, both sides, primer	.400	Ea.
Finish coat, latex	.400	Ea.
Primer & 1 coat latex	.727	Ea.
Primer & 2 coats latex	1	Ea.
Doors, French, both sides, 10–15 lite, incl. frame & trim		
Roll & brush primer	1.333	Ea.
Finish coat, latex	1.333	Ea.
Primer & 1 coat latex	2.667	Ea.
Primer & 2 coats latex	4	Ea.
Doors, louvered, both sides, incl. frame & trim		
Roll & brush primer	1.143	Ea.
Finish coat, latex	1.143	Ea.
Primer & 1 coat latex	2	Ea.
Primer & 2 coats latex	2.667	Ea.
Spray, both sides, primer	.400	Ea.
Finish coat, latex	.400	Ea.
Primer & 1 coat latex	.727	Ea.
Primer & 2 coats latex	1	Ea.
Doors, panel, both sides, incl. frame & trim		
Roll & brush primer	1.333	Ea.
Finish coat, latex	1.333	Ea.
Primer & 1 coat latex	2.667	Ea.
Primer & 2 coats latex	3.200	Ea.
Spray, both sides, primer	.800	Ea.
Finish coat, latex	.800	Ea.
Primer & 1 coat latex	1.600	Ea.

(continued on next page)

Figure 9-20 Application Time in Labor-Hours for Painting
(continued from previous page)

Description	Labor-Hours	Unit
Primer & 2 coats latex	2	Ea.
Windows, per interior side, based on 15 SF		
1 to 6 lite		
Brushwork, primer	.615	Ea.
Finish coat, enamel	.615	Ea.
Primer & 1 coat enamel	1	Ea.
Primer & 2 coats enamel	1.333	Ea.
7 to 10 lite		
Brushwork, primer	.727	Ea.
Finish coat, enamel	.727	Ea.
Primer & 1 coat enamel	1.143	Ea.
Primer & 2 coats enamel	1.600	Ea.
12 lite		
Brushwork, primer	.800	Ea.
Finish coat, enamel	.800	Ea.
Primer & 1 coat enamel	1.333	Ea.
Primer & 2 coats enamel	1.600	Ea.
Fences		
Chain link or wire metal, one side, water base		
Roll & brush, first coat	.008	SF
Second coat	.006	SF
Spray, first coat	.004	SF
Second coat	.003	SF
Picket, water base		
Roll & brush, first coat	.009	SF
Second coat	.008	SF
Spray, first coat	.004	SF
Second coat	.003	SF
Stockade, water base		
Roll & brush, first coat	.008	SF
Second coat	.007	SF
Spray, first coat	.004	SF
Second coat	.003	SF
Floors		
Concrete paint, latex		
Brushwork		
first coat	.008	SF
second coat	.007	SF
third coat	.006	SF
Roll		
first coat	.003	SF
second coat	.002	SF
third coat	.002	SF

(continued on next page)

Figure 9-20 Application Time in Labor-Hours for Painting
(continued on previous page)

Description	Labor-Hours	Unit
Spray		
first coat	.003	SF
second coat	.002	SF
third coat	.002	SF
Acid stain and sealer		
Stain, 1 coat	.012	SF
2 coats	.014	SF
Acrylic sealer, 1 coat	.003	SF
2 coats	.006	SF
Floors, conc./wood, oil base, primer/sealer coat, brushwork		
Stain, wood floor, brushwork, 1 coat	.004	SF
Roller	.003	SF
Spray	.003	SF
Varnish, wood floor, brushwork	.004	SF
Roller	.003	SF
Spray	.003	SF
Grilles, per side, oil base, primer coat, brushwork		
Spray	.007	SF
Paint 2 coats, brushwork	.025	SF
Spray	.012	SF
Gutters and downspouts, wood		
Downspouts, 4", primer	.013	LF
Finish coat, exterior latex	.013	LF
Primer & 1 coat exterior latex	.020	LF
Primer & 2 coats exterior latex	.025	LF
Pipe, 1"–4" diameter, primer or sealer coat, oil base, brushwork	.013	LF
Spray	.007	LF
Paint 2 coats, brushwork	.021	LF
Spray	.012	LF
13"–16" diameter, primer or sealer coat, brushwork	.052	LF
Spray	.030	LF
Paint 2 coats, brushwork	.082	LF
Spray	.052	LF
Trim, wood, including puttying under 6" wide		
Primer coat, oil base, brushwork	.012	LF
Paint, 1 coat, brushwork	.012	LF
3 coats	.025	LF

(continued on next page)

Figure 9-20 Application Time in Labor-Hours for Painting
(continued from previous page)

Description	Labor-Hours	Unit
Over 6" wide, primer coat, brushwork	.012	LF
Paint, 1 coat, brushwork	.012	LF
3 coats	.025	LF
Cornice, simple design, primer coat, oil base, Brushwork	.012	SF
Paint, 1 coat	.012	SF
Ornate design, primer coat	.023	SF
Paint, 1 coat	.023	SF
Balustrades, primer coat, oil base, brushwork	.015	SF
Paint, 1 coat	.015	SF
Trusses and wood frames, primer boat, oil base, brushwork	.010	SF
Spray	.013	SF
Paint, 2 coats	.016	SF
Spray	.013	SF
Stain, brushwork, wipe off	.013	SF
Varnish, 3 coats, brushwork	.029	SF
Siding, exterior		
Steel siding, oil base, paint 1 coat, brushwork	.008	SF
Spray	.004	SF
Paint 2 coats, brushwork	.012	SF
Spray	.006	SF
Stucco, rough, oil base, paint 2 coats, brushwork	.012	SF
Roller	.010	SF
Spray	.005	SF
Texture 1–11 or clapboard, oil base, primer Coat, brushwork	.012	SF
Spray	.004	SF
Paint 2 coats, brushwork	.020	SF
Spray	.006	SF
Stain 2 coats, brushwork	.017	SF
Spray	.006	SF
Wood shingles, oil base primer coat, brushwork	.012	SF
Spray	.004	SF
Paint 2 coats, brushwork	.020	SF
Spray	.007	SF
Stain 2 coats, brushwork	.017	SF
Spray	.006	SF

(continued on next page)

Figure 9-20 Application Time in Labor-Hours for Painting
(continued from previous page)

Description	Labor-Hours	Unit
Wall coatings		
High build epoxy, 50 mil, minimum	.021	SF
Maximum	.084	SF
Laminated epoxy with fiberglass, min.	.027	SF
Maximum	.055	SF
Sprayed perlite or vermiculite, 1/16" thick,		
Minimum	.003	SF
Maximum	.013	SF
Vinyl plastic wall coating, minimum	.011	SF
Maximum	.033	SF
Urethane on smooth surface, 2 coats, min.	.007	SF
Maximum	.012	SF
Ceramic-like glazed coating, cementitious,		
Minimum	.018	SF
Maximum	.023	SF
Resin base, minimum	.013	SF
Maximum	.024	SF

Figure 9-21 Installation Time in Labor-Hours for Wood Stud Partition Systems

Description	Labor-Hours	Unit
Wood Partitions Studs with Shingle Bottom Plate and Double Top Plate		
2 x 3 or 2 x 4 Studs		
12" On Center	.020	S.F.
16" On Center	.016	S.F.
24" On Center	.013	S.F.
2 x 6 Studs		
12" On Center	.023	S.F.
16" On Center	.018	S.F.
24" On Center	.014	S.F.
Plates		
2 x 3	.019	L.F.
2 x 4	.020	L.F.
2 x 6	.021	L.F.
Studs		
2 x 3	.013	L.F.
2 x 4	.012	L.F.
2 x 6	.032	L.F.
Blocking	.032	L.F.
Grounds 1 x 2		
For Casework	.024	L.F.
For Plaster	.018	L.F.
Insulation Fiberglass Batts	.005	S.F.
Meth Lath Diamond Expanded		
2.5 lb. per S.Y.	.094	S.Y.
3.4 lb. per S.Y.	.100	S.Y.
Gypsum Lath		
3/8" Thick	.094	S.Y.
1/2" Thick	.100	S.Y.
Gypsum Plaster		
2 Coats	.381	S.Y.
3 Coats	.460	S.Y.
Perlite or Vermiculite Plaster		
2 Coats	.435	S.Y.
3 Coats	.541	S.Y.

(continued on next page)

Figure 9-21   Installation Time in Labor-Hours for Wood Stud
Partition Systems *(continued from previous page)*

Description	Labor-Hours	Unit
Drywall Gypsum Plasterboard Including Taping		
3/8" Thick	.015	S.F.
1/2" or 5/8" Thick	.017	S.F.
For Thin Coat Plaster Instead of Taping Add	.011	S.F.
Prefinished Vinyl Faced Drywall	.018	S.F.
Sound-deadening Board	.009	S.F.
Walls in Place		
2" x 4" Studs with 5/8"		
Gypsum Drywall Both Sides Taped	.053	S.F.
2" x 4" Studs with 2 Layers Gypsum Drywall		
Both Sides Taped	.078	S.F.

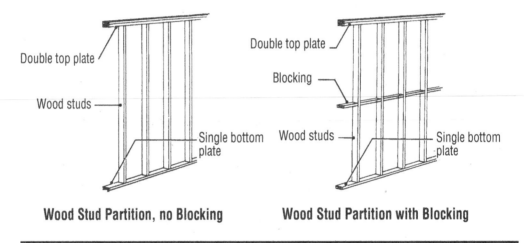

Wood Stud Partition, no Blocking   **Wood Stud Partition with Blocking**

Figure 9-22 Installation Time in Labor-Hours for Steel Stud
Partition Systems

Description	Labor-Hours	Unit
Non-Loadbearing Stud, Galv., 25 Gauge, 1-5/8" Wide,		
16" On Center	.019	S.F.
24" On Center	.016	S.F.
2-1/2" Wide,		
16" On Center	.020	S.F.
24" On Center	.016	S.F.
20 Gauge, 1-5/8" Wide,		
16" On Center	.018	S.F.
24" On Center	.016	S.F.
2-1/2" Wide,		
16" On Center	.019	S.F.
24" On Center	.016	S.F.
25 Gauge or 20 Gauge, 3-5/8" or 4" Wide		
16" On Center	.020	S.F.
24" On Center	.017	S.F.
6" Wide,		
16" On Center	.022	S.F.
24" On Center	.018	S.F.

(continued on next page)

Figure 9-22   Installation Time in Labor-Hours for Steel Stud
Partition Systems *(continued from the previous page)*

Description	Labor-Hours	Unit
Metal Lath Diamond Expanded		
2.5 lb. Screwed to Studs	.100	S.Y.
3.4 lb. Screwed to Studs or Wired to Framing	.107	S.Y.
Rib Lath Wired to Steel		
2.50 lb.	.107	S.Y.
3.40 lb.	.114	S.Y.
4.00 lb.	.123	S.Y.
Gypsum Lath		
3/8" Thick	.094	S.Y.
1/2" Thick	.100	S.Y.
Gypsum Plaster		
2 Coats	.381	S.Y.
3 Coats	.460	S.Y.
Perlite or Vermiculite Plaster		
2 Coats	.435	S.Y.
3 Coats	.541	S.Y.
Wood Fiber Plaster		
2 Coats	.556	S.Y.
3 Coats	.702	S.Y.
Drywall Gypsum Plasterboard Including Taping		
3/8" Thick	.015	S.F.
1/2" or 5/8" Thick	.017	S.F.
For Thincoat Plaster Instead of Taping, Add	.013	S.F.
Prefinished Vinyl Faced Drywall	.015	S.F.
Sound-deadening Board	.009	S.F.
Walls in Place		
3-5/8" Studs, NLB, 25 Gauge,		
16" On Center with 5/8" Gypsum Drywall		
Both Sides Taped	.047	S.F.
24" On Center	.044	S.F.
3-5/8" Studs, NLB, 25 Gauge,		
16" On Center with 2 Layers Gypsum Drywall		
Both Sides Taped	.065	S.F.
24" On Center	.060	S.F.

Figure 9-23 Installation Time in Labor-Hours for Ceiling Systems

Description	Labor-Hours	Unit
Ceiling Tile Stapled, Cemented or Installed on Suspension System, 12" x 12" or 12" x 24", Not Including Furring		
Mineral Fiber, Plastic Coated	.027	S.F.
Fire Rated, 3/4" Thick, Plain Faced	.027	S.F.
Plastic Coated Face	.027	S.F.
Aluminum Faced, 5/8" Thick, Plain	.027	S.F.
Metal Pan Units, 24 ga. Steel, Not Incl. Pads, Painted, 12" x 12"	.023	S.F.
12" x 36" or 12" x 24", 7% Open Area	.024	S.F.
Aluminum, 12" x 12"	.023	S.F.
12" x 24"	.022	S.F.
Stainless Steel, 12" x 24", 26 ga., Solid	.023	S.F.
5.2% Open Area	.024	S.F.
Suspended Acoustic Ceiling Boards Not Including Suspension System		
Fiberglass Boards, Film Faced, 2' x 2' or 2' x 4', 5/8" Thick	.012	S.F.
3/4" Thick	.016	S.F.
3" Thick, Thermal, R11	.018	S.F.
Glass Cloth Faced Fiberglass, 3/4" Thick	.016	S.F.
1" Thick	.016	S.F.
1-1/2" Thick, Nubby Face	.017	S.F.
Mineral Fiber Boards, 5/8" Thick, Aluminum Faced, 24" x 24"	.013	S.F.
24" x 48"	.012	S.F.
Plastic Coated Face	.020	S.F.
Mineral Fiber, 2 Hour Rating, 5/8" Thick	.012	S.F.
Mirror Faced Panels, 15/16" Thick	.016	S.F.
Air Distributing Ceilings, 5/8" Thick, F.R.D. Water Felted Board	.020	S.F.
Eggcrate, Acrylic, 1/2" x 1/2" x 1/2" Cubes	.016	S.F.
Polystyrene Eggcrate	.016	S.F.
Luminous Panels, Prismatic	.020	S.F.
Perforated Aluminum Sheets, .024" Thick, Corrugated, Painted	.016	S.F.
Mineral Fiber, 24" x 24" or 48", reveal edge, Painted, 5/8" Thick	.013	S.F.
3/4" Thick	.014	S.F.
Wood Fiber in Cementitious Binder, 2' x 2' or 4', Painted, 1" Thick	.013	S.F.
2" Thick	.015	S.F.
2-1/2" Thick	.016	S.F.
3" Thick	.018	S.F.
Access Panels, Metal, 12" x 12"	.400	Ea.
24" x 24"	.800	Ea.

(continued on next page)

Figure 9-23 Installation Time in Labor-Hours for Ceiling Systems
(continued from previous page)

Description	Labor-Hours	Unit
Suspension Systems		
Class A Suspension System, T Bar, 2' x 4' Grid	.010	S.F.
2' x 2' Grid	.012	S.F.
Concealed Z Bar Suspension System, 12"		
Module	.015	S.F.
1-1/2" Carrier Channels, 4' O.C., Add	.017	S.F.
Carrier Channels for Ceiling With Recessed		
Lighting Fixtures, Add	.017	S.F.
Hanging Wire, 12 Ga.	.002	S.F.
Suspended Ceilings, Complete Including		
Standard Suspension System but Not Including		
1-1/2" Carrier Channels		
Air Distributing Ceilings, Including Barriers,		
2' x 2' Board	.027	S.F.
12" x 12" Tile	.033	S.F.
Ceiling Board System, 2' x 4', Plain Faced,		
Supermarkets	.016	S.F.
Offices	.021	S.F.
Luminous Panels, Flat or Ribbed	.031	S.F.
Metal Pan with Acoustic Pad	.039	S.F.
Tile, Z Bar Suspension, 5/8" Mineral Fiber Tile	.034	S.F.
3/4" Mineral Fiber Tile	.035	S.F.
Reveal Tile With Drop, 2' x 2' Grid with Colored		
Suspension System	.023	S.F.
Gypsum Plaster Ceilings 2 Coats, No Lath		
Included	.435	S.Y.
2 Coats on and Incl. 3/8" Gypsum Lath on Steel	.578	S.Y.
3 Coats, No Lath Included	.513	S.Y.
3 Coats on and Including Painted Metal Lath	.627	S.Y.
Metal Lath 2.5 lb. Diamond Painted, on Wood		
Framing	.107	S.Y.
On Ceilings, 3.4 lb. Diamond Painted, on		
Wood Framing	.114	S.Y.
3.4 lb. Diamond Painted, Wired to Steel Framing	.133	S.Y.
Suspended Ceiling System, Incl. 3.4 lb.		
Diamond Lath	.533	S.Y.
Drywall Ceilings Gypsum Drywall, Fire Rated		
Finished		
Screwed to Grid, Channel of Joists	.021	S.F.
Over 8' High, 1/2" Thick	.026	S.F.
5/8" Thick	.026	S.F.
Grid Suspension System, Direct Hung		
1-1/2" C.R.C., With 7/8" Hi Hat Furring		
Channel, 16" O.C.	.016	S.F.
24" O.C.	.012	S.F.
3-5/8" Channel, 25 ga., with Track,		
16" O.C.	.016	S.F.
24" O.C.	.012	S.F.

(continued on next page)

Figure 9-23 Installation Time in Labor-Hours for Ceiling Systems
(continued from previous page)

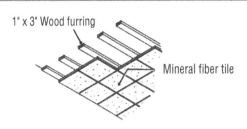

**Acoustical Mineral Fiber Tile
on 1" x 3" Wood Furring**

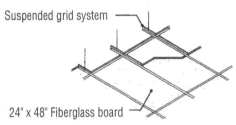

**Fiberglass Board
on Suspended Grid System**

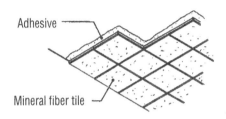

Mineral Fiber Tile Applied with Adhesive

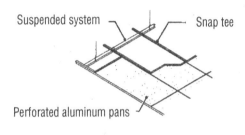

Acoustical Perforated Metal Pans

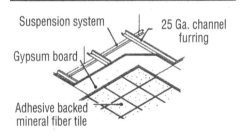

**Acoustical Mineral Fiber Tile
on Gypsum Board**

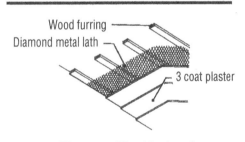

**Plaster on Metal Lath and
Wood Furring**

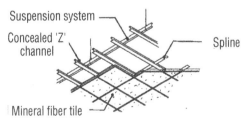

**Mineral Fiber Tile
on Concealed 'Z' Channel**

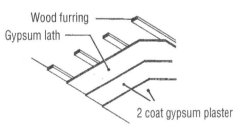

Plaster on Wood Furring

(continued on next page)

Figure 9-23 Installation Time in Labor-Hours for Ceiling Systems
(continued from previous page)

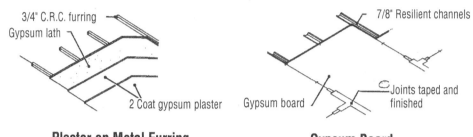

Plaster on Metal Furring

**Gypsum Board
on 7/8" Resilient Channel Furring**

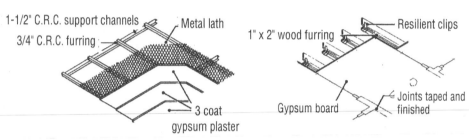

Plaster on Metal Lath

**Gypsum Board on 1" x 2",
Suspended, with Resilient Clips**

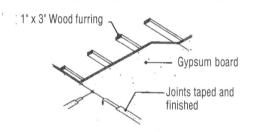

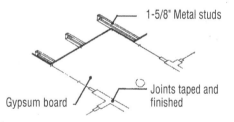

Gypsum Board on 1" x 3" Wood Furring

**Gypsum Board
on 1-5/8" Metal Stud Furring**

Division 9

Figure 9-24 Installation Time in Labor-Hours for Floor and Wall Tile Systems

Description	Labor-Hours	Unit
Flooring Cast Ceramic 4" x 8" x 3/4" Pressed	.160	S.F.
Hand Molded	.168	S.F.
8" x 3/4" Hexagonal	.188	S.F.
Heavy Duty Industrial Cement Mortar Bed	.200	S.F.
Ceramic Pavers 8" x 4"	.168	S.F.
Ceramic Tile Base, Using 1" x 1" Tiles, 4" High,		
Mud Set	.195	L.F.
Thin Set	.125	L.F.
Cove Base, 4-1/4" x 4-1/4" High, Mud Set	.176	L.F.
Thin Set	.125	L.F.
6" x 4-1/4" High, Mud Set	.160	L.F.
Thin Set	.117	L.F.
Sanitary Cove Base, 6" x 4-1/4" High, Mud Set	.172	L.F.
Thin Set	.129	L.F.
6" x 6" High, Mud Set	.190	L.F.
Thin Set	.137	L.F.
Bullnose Trim, 4-1/4" x 4-1/4", Mud Set	.195	L.F.
Thin Set	.125	L.F.
6" x 4-1/4" Bullnose Trim, Mud Set	.190	L.F.
Thin Set	.129	L.F.
Floors, Natural Clay, Random or Uniform,		
Thin Set, Color Group 1	.087	S.F.
Color Group 2	.087	S.F.
Porcelain Type, 1 Color, Color Group 2,		
1" x 1"	.087	S.F.
2" x 2" or 2" x 1", Thin Set	.084	S.F.
Conductive Tile, 1" Squares, Black	.147	S.F.
4" x 8" or 4" x 4", 3/8" Thick	.133	S.F.
Trim, Bullnose, Etc.	.080	L.F.
Specialty Tile, 3" x 6" x 1/2", Decorator Finish	.087	S.F.
Add For Epoxy Grout, 1/16" Joint,		
1" x 1" Tile	.020	S.F.
2" x 2" Tile	.020	S.F.
Pregrouted Sheets, Walls, 4-1/4", 6" x 4-1/4",		
and 8-1/2" x 4-1/4", S.F. Sheets,		
Silicone Grout	.067	S.F.
Floors, Unglazed, 2 S.F. Sheets		
Urethane Adhesive	.089	S.F.
Walls, Interior, Thin Set, 4-1/4" x 4-1/4" Tile	.084	S.F.
6" x 4-1/4" Tile	.084	S.F.
8-1/2" x 4-1/4" Tile	.084	S.F.
6" x 6" Tile	.080	S.F.

(continued on next page)

Figure 9-24 Installation Time in Labor-Hours for Floor and
Wall Tile Systems *(continued from previous page)*

Description	Labor-Hours	Unit
Decorated Wall Tile, 4-1/4"x 4-1/4"		
Minimum	.059	Ea.
Maximum	.089	Ea.
Exterior Walls, Frostproof, Mud Set,		
4-1/4" x 4-1/4"	.157	S.F.
1-3/8" x 1-3/8"	.172	S.F.
Crystalline Glazed, 4-1/4" x 4-1/4", Mud		
Set, Plain	.160	S.F.
4-1/4" x 4-1/4", Scored Tile	.160	S.F.
1-3/8" Squares	.172	S.F.
For Epoxy Grout, 1/16" Joints, 4-1/4" Tile,		
Add	.020	S.F.
For Tile Set in Dry Mortar, Add	.009	S.F.
For Tile Set in Portland Cement Mortar, Add	.055	S.F.
Regrout Tile 4-1/2" x 4-1/2", or Larger, Wall	.080	S.F.
Floor	.064	S.F.
Ceramic Tile Panels Insulated, Over 1000		
Square Feet,		
1-1/2" Thick	.073	S.F.
2-1/2" Thick	.073	S.F.
Glass Mosaics 3/4" Tile on 12" Sheets,		
Color Group 1 and 2 Minimum	.195	S.F.
Maximum (Latex Set)	.219	S.F.
Color Group 3	.219	S.F.
Color Group 4	.219	S.F.
Color Group 5	.219	S.F.
Color Group 6	.219	S.F.
Color Group 7	.219	S.F.
Color Group 8, Gold, Silvers and Specialties	.250	S.F.
Marble Thin Gauge Tile, 12" x 6", 9/32", White		
Carara	.250	S.F.
Filled Travertine	.250	S.F.
Synthetic Tiles, 12" x 12" x 5/8",		
Thin Set, Floors	.250	S.F.
On Walls	.291	S.F.
Metal Tile Cove Base, Standard Colors,		
4-1/4" Square	.053	L.F.
4-1/8" x 8-1/2"	.040	L.F.
Walls, Aluminum, 4-1/4" Square, Thin		
Set, Plain	.100	S.F.
Epoxy Enameled	.107	S.F.
Leather on Aluminum, Colors	.123	S.F.
Stainless Steel	.107	S.F.
Suede on Aluminum	.123	S.F.
Plastic Tile Walls, 4-1/4" x 4-1/4",		
.050" Thick	.064	S.F.
.110" Thick	.067	S.F.

(continued on next page)

Figure 9-24 Installation Time in Labor-Hours for Floor and
Wall Tile Systems *(continued from previous page)*

Description	Labor-Hours	Unit
Quarry Tile Base, Cove or Sanitary, 2" or 5" High, Mud Set		
1/2" Thick	.145	L.F.
Bullnose Trim, Red, Mud Set, 6" x 6" x 1/2"		
Thick	.133	L.F.
4" x 4" x 1/2" Thick	.145	L.F.
4" x 8" x 1/2" Thick, Using 8" as Edge	.123	L.F.
Floors, Mud Set, 1000 S.F. Lots, Red,		
4" x 4" x 1/2" Thick	.133	S.F.
6" x 6" x 1/2" Thick	.114	S.F.
4" x 8" x 1/2" Thick	.123	S.F.
Brown Tile, Imported, 6" x 6" x 7/8"	.133	S.F.
9" x 9" x 1-1/4"	.145	S.F.
For Thin Set Mortar Application, Deduct	.023	S.F.
Stair Tread and Riser, 6" x 6" x 3/4", Plain	.320	S.F.
Abrasive	.340	S.F.
Wainscot, 6" x 6" x 1/2", Thin Set, Red	.152	S.F.
Colors Other Than Green	.152	S.F.
Window Sill, 6" Wide, 3/4" Thick	.178	L.F.
Corners	.200	Ea.
Terra Cotta Tile on Walls, Dry Set, 1/2" Thick		
Square, Hexagonal or Lattice Shapes, Unglazed	.059	S.F.
Glazed, Plain Colors	.062	S.F.
Intense Colors	.064	S.F.

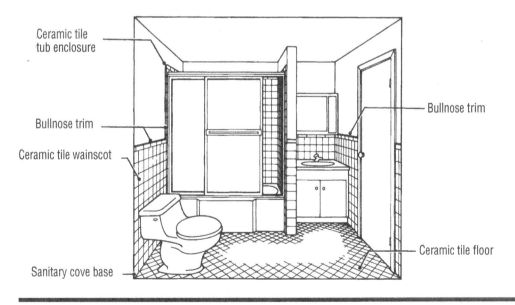

Ceramic tile tub enclosure

Bullnose trim

Bullnose trim

Ceramic tile wainscot

Sanitary cove base

Ceramic tile floor

Figure 9-25   Installation Time in Labor-Hours for Terrazzo

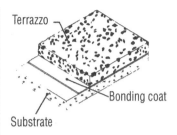

Description	Labor-Hours	Unit
Terrazzo Cast in Place Cove Base, 6" High	.400	L.F.
Curb, 6" High and 6" Wide	1.333	L.F.
Floor, Bonded to Concrete, 1-3/4" Thick,		
Gray Cement	.213	S.F.
White Cement	.213	S.F.
Not Bonded, 3" Total Thickness,		
Gray Cement	.229	S.F.
White Cement	.229	S.F.
Epoxy Terrazzo, 1/4" Thick, Minimum	.400	S.F.
Monolithic Terrazzo, 5/8" Thick, Including		
3-1/2" Base Slab, 10' Panels, Mesh and Felt	.128	S.F.
Stairs, Cast in Place, Pan Filled Treads	.533	L.F.
Treads and Risers	1.143	L.F.
Stair Landings, Add to Floor Prices	.258	S.F.
Stair Stringers and Fascia	.533	S.F.
For Abrasive Metal Nosings on Stairs, Add	.107	L.F.
For Abrasive Surface Finish, Add	.027	S.F.
For Flush Abrasive Strips, Add	.026	L.F.
Wainscot, Bonded, 1-1/2" Thick	.533	S.F.
Epoxy Terrazzo, 1/4" Thick	.400	S.F.
Terrazzo Precast Base, 6" High, Straight	.114	L.F.
Cove	.133	L.F.
8" High Base, Straight	.133	L.F.
Cove	.160	L.F.
Curbs, 4" x 4" High	.200	L.F.
8" x 8" High	.267	L.F.
Floor Tiles, Non-slip, 1" Thick, 12" x 12"	.267	S.F.
1-1/4" Thick, 12" x 12"	.267	S.F.
16" x 16"	.320	S.F.
1-1/2" Thick, 16" x 16"	.356	S.F.
Stair Treads, 1-1/2" Thick, Non-slip, Three Line	.229	L.F.
Nosing and Two Line	.229	L.F.
2" Thick Treads, Straight	.267	L.F.
Curved	.320	L.F.
Stair Risers, 1" Thick, to 6" High,		
Straight Sections	.267	L.F.
Cove	.320	L.F.
Curved, 1" Thick, to 6" High, Vertical	.333	L.F.
Cove	.421	L.F.
Stair Tread and Riser, Single Piece, Straight,		
Minimum	.267	L.F.
Maximum	.400	L.F.
Curved Tread and Riser, Minimum	.400	L.F.
Maximum	.500	L.F.
Stair Stringers, Notched, 1" Thick	.640	L.F.
2" Thick	.727	L.F.
Stair Landings, Structural, Non-slip,		
1-1/2" Thick	.188	S.F.
3" Thick	.213	S.F.
Wainscot, 12" x 12" x 1" Tiles	.667	S.F.
16" x 16" x 1-1/2" Tiles	.100	S.F.

Terrazzo

Bonding coat

Substrate

Figure 9-26 Installation Time in Labor-Hours for Wood Flooring

Description	Labor-Hours	Unit
Wood Floors Fir, Vertical Grain, 1" x 4", Not Including Finish	.031	S.F.
Gym Floor, in Mastic, Over 2 Ply Felt, #2 and Better 25/32" Thick Maple, Including Finish	.080	S.F.
33/32" Thick Maple, Including Finish	.082	S.F.
For 1/2" Corkboard Underlayment, Add	.011	S.F.
Maple Flooring, Over Sleepers, #2 and Better Including Finish, 25/32" Thick	.094	S.F.
33/32" Thick	.096	S.F.
For 3/4" Subfloor, Add	.023	S.F.
With Two 1/2" Subfloors, 25/32" Thick	.116	S.F.
Maple, Including Finish, #2 and better, 25/32" Thick, on Rubber Sleepers, with Two 1/2" Subfloors	.105	S.F.
With Steel Spline, Double Connection to Channels	.110	S.F.
Portable Hardwood, Prefinished Panels	.096	S.F.
Insulated with Polystyrene, Add	.048	S.F.
Running Tracks, Sitka Spruce Surface	.129	S.F.
3/4" Plywood Surface	.080	S.F.
Maple, Strip, Not Including Finish	.047	S.F.
Oak Strip, White or Red, Not Including Finish	.047	S.F.
Parquetry, Standard, 5/16" Thick, Not Including Finish, Minimum	.050	S.F.
13/16" Thick, Select Grade, Minimum	.050	S.F.
Maximum	.080	S.F.
Custom Parquetry, Including Finish, Minimum	.080	S.F.
Maximum	.160	S.F.
Prefinished White Oak, Prime Grade, 2-1/4" Wide	.047	S.F.
3-1/4" Wide	.043	S.F.
Ranch Plank	.055	S.F.
Hardwood Blocks, 9" x 9", 25/32" Thick	.050	S.F.

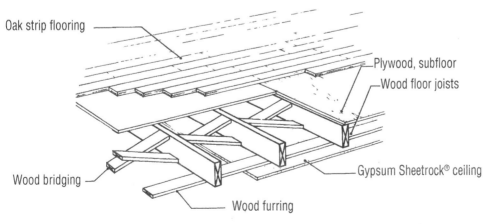

Oak strip flooring

Plywood, subfloor

Wood floor joists

Wood bridging

Gypsum Sheetrock® ceiling

Wood furring

(continued on next page)

Figure 9-26 Installation Time in Labor-Hours for Wood Flooring
(continued from previous page)

Description	Labor-Hours	Unit
Acrylic Wood Parquet Blocks, 12″ x 12″ x 5/16″, Irradiated, Set in Epoxy	.050	S.F.
Yellow Pine, 3/4″ x 3-1/8″, T & G, C and Better, Not Including Finish	.040	S.F.
Refinish Old Floors, Minimum	.020	S.F.
Maximum	.062	S.F.
Sanding and Finishing, Fill, Shellac, Wax	.027	S.F.
Wood Block Flooring End Grain Flooring, Natural Finish, 1″ Thick	.029	S.F.
1-1/2″ Thick	.031	S.F.
2″ Thick	.033	S.F.
Wood Composition Gym Floors 2-1/4″ x 6-7/8″ x 3/8″, on 2″ Grout Setting Bed	.107	S.F.
Thin Set, on Concrete	.064	S.F.
Sanding and Finishing, Add	.040	S.F.

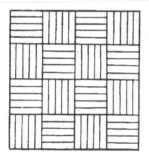
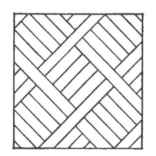
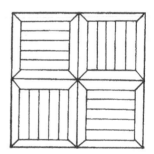

Random Patterns of Parquet Flooring

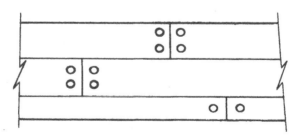

Strip Plank Flooring with Plugs

Figure 9-27 Installation Time in Labor-Hours for Carpeting

Description	Labor-Hours	Unit
Carpet Commercial Grades, Cemented		
Acrylic, 26 oz., Light to Medium Traffic	.216	S.F.
28 oz., Medium Traffic	.229	S.F.
35 oz., Medium to Heavy Traffic	.242	S.F.
Nylon, Non Anti-Static, 15 oz., Light Traffic	.229	S.F.
Nylon, With Anti-Static, 17 oz., Light to		
Medium Traffic	.211	S.F.
20 oz., Medium Traffic	.216	S.F.
22 oz., Medium Traffic	.216	S.F.
24 oz., Medium to Heavy Traffic	.229	S.F.
26 oz., Medium to Heavy Traffic	.229	S.F.
28 oz., Heavy Traffic	.229	S.F.
32 oz., Heavy Traffic	.242	S.F.
42 oz., Heavy Traffic	.258	S.F.
Needle Bonded, 20 oz., No Padding	.143	S.F.
Polypropylene, 15 oz., Light Traffic	.143	S.F.
22 oz., Medium Traffic	.182	S.F.
24 oz., Medium to Heavy Traffic	.182	S.F.
26 oz., Medium to Heavy Traffic	.182	S.F.
28 oz., Heavy Traffic	.205	S.F.
32 oz., Heavy Traffic	.205	S.F.
42 oz., Heavy Traffic	.216	S.F.
Scrim Installed, Nylon Sponge Back Carpet		
20 oz.	.242	S.Y.
60 oz.	.267	S.Y.
Tile, Foam-Backed, Needle Punch	.014	S.F.
Tufted Loop or Shag	.014	S.Y.
Wool, 30 oz., Medium Traffic	.229	S.Y.
Wool, 36 oz., Medium to Heavy Traffic	.242	S.Y.
Sponge Back, Wool, 36 oz., Medium to		
Heavy Traffic	.143	S.Y.
42 oz., Heavy Traffic	.229	S.Y.
Padding, Sponge Rubber Cushion, Minimum	.108	S.Y.
Maximum	.123	S.Y.
Felt, 32 oz. to 56 oz., Minimum	.108	S.Y.
Maximum	.123	S.Y.
Bonded Urethane, 3/8" Thick, Minimum	.094	S.Y.
Maximum	.107	S.Y.
Prime Urethane, 1/4" Thick, Minimum	.094	S.Y.
Maximum	.107	S.Y.

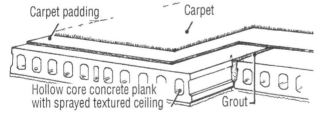

Carpet on Hollow Core Concrete Plank

Figure 9-28   Installation Time in Labor-Hours for Resilient Flooring

Description	Labor-Hours	Unit
Resilient Asphalt Tile, on Concrete, 1/8" Thick		
Color Group B	.015	S.F.
Color Group C and D	.015	S.F.
For Less Than 500 S.F., Add	.002	S.F.
For Over 5000 S.F., Deduct	.007	S.F.
Base, Cove, Rubber or Vinyl, .080" Thick		
Standard Colors, 2-1/2" High	.025	L.F.
4" High	.025	L.F.
6" High	.025	L.F.
1/8" Thick, Standard Colors,		
2-1/2" High	.025	L.F.
4" High	.025	L.F.
6" High	.025	L.F.
Corners, 2-1/2" High	.025	Ea.
4" High	.025	Ea.
6" High	.025	Ea.
Conductive Flooring, Rubber Tile,		
1/8" Thick	.025	S.F.
Homogeneous Vinyl Tile, 1/8" Thick	.025	S.F.
Cork Tile, Standard Finish, 1/8" Thick	.025	S.F.
3/16" Thick	.025	S.F.
5/16" Thick	.025	S.F.
1/2" Thick	.025	S.F.
Urethane Finish, 1/8" Thick	.025	S.F.
3/16" Thick	.025	S.F.
5/16" Thick	.025	S.F.
1/2" Thick	.025	S.F.
Polyethylene, in Rolls, No Base Incl.,		
Landscape Surfaces	.029	S.F.
Nylon Action Surface, 1/8" Thick	.029	S.F.
1/4" Thick	.029	S.F.
3/8" Thick	.029	S.F.
Golf Tee Surface with Foam Back	.033	S.F.
Practice Putting, Knitted Nylon Surface	.033	S.F.
Polyurethane, Thermoset, Prefabricated in Place, Indoor		
3/8" Thick for Basketball, Gyms, etc.	.080	S.F.

(continued on next page)

Figure 9-28 Installation Time in Labor-Hours for Resilient Flooring
(continued from previous page)

Description	Labor-Hours	Unit
Stair Treads and Risers		
Rubber, Molded Tread, 12" Wide, 5/16"		
Thick, Black	.070	L.F.
Colors	.070	L.F.
1/4" Thick, Black	.070	L.F.
Colors	.070	L.F.
Grit Strip Safety Tread, Colors 5/16" Thick	.070	L.F.
3/16" Thick	.067	L.F.
Landings, Smooth Sheet Rubber,		
1/8" Thick	.067	S.F.
3/16" Thick	.067	S.F.
Nosings, 1-1/2" Deep, 3" Wide, Residential	.057	L.F.
Commercial	.057	L.F.
Risers, 7" High, 1/8" Thick, Flat	.032	L.F.
Coved	.032	L.F.
Vinyl, Molded Tread, 12" Wide, Colors,		
1/8" Thick	.070	L.F.
1/4" Thick	.070	L.F.
Landing Material, 1/8" Thick	.040	S.F.
Riser, 7" High, 1/8" Thick, Coved	.046	L.F.
Threshold, 5-1/2" Wide	.080	L.F.
Tread and Riser Combined, 1/8" Thick	.100	L.F.

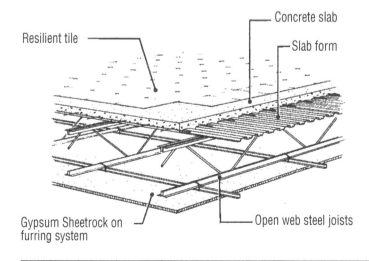

Concrete slab

Resilient tile

Slab form

Gypsum Sheetrock on
furring system

Open web steel joists

Figure 9-29   Installation Time in Labor-Hours for Composition Flooring

Description	Labor-Hours	Unit
Composition Flooring, Acrylic, 1/4" Thick	.092	S.F.
3/8" Thick	.107	S.F.
Cupric Oxychloride, on Bond Coat, Minimum	.100	S.F.
Maximum	.114	S.F.
Epoxy, with Colored Quartz Chips,		
Broadcast, Minimum	.071	S.F.
Maximum	.098	S.F.
Trowelled, Minimum	.086	S.F.
Maximum	.100	S.F.
Heavy Duty Epoxy Topping, 1/4" Thick		
500 to 1,000 S.F.	.114	S.F.
1,000 to 2,000 S.F.	.107	S.F.
Over 10,000 S.F.	.100	S.F.
Epoxy Terrazzo, 1/4" Thick, Chemical		
Resistant, Minimum	.080	S.F.
Maximum	.107	S.F.
Conductive, Minimum	.076	S.F.
Maximum	.100	S.F.
Mastic, Hot Laid, 2 Coat, 1-1/2" Thick,		
Standard, Minimum	.070	S.F.
Maximum	.092	S.F.
Acidproof, Minimum	.079	S.F.
Maximum	.137	S.F.
Neoprene, Trowelled on, 1/4" Thick,		
Minimum	.088	S.F.
Maximum	.112	S.F.
Polyacrylate Terrazzo, 1/4" Thick, Minimum	.065	S.F.
Maximum	.100	S.F.
3/8" Thick, Minimum	.077	S.F.
Maximum	.100	S.F.
Conductive Terrazzo, 1/4" Thick, Minimum	.107	S.F.
Maximum	.157	S.F.
3/8" Thick, Minimum	.132	S.F.
Maximum	.188	S.F.
Granite, Conductive, 1/4" Thick, Minimum	.069	S.F.
Maximum	.114	S.F.
3/8" Thick, Minimum	.069	S.F.
Maximum	.126	S.F.
Polyester, with Colored Quartz Chips,		
1/16" Thick, Minimum	.045	S.F.
Maximum	.086	S.F.
1/8" Thick, Minimum	.059	S.F.
Maximum	.071	S.F.
Polyester, Heavy Duty, Compared to		
Epoxy, Add	.019	S.F.
Polyurethane, with Suspended Vinyl Chips,		
Minimum	.045	S.F.
Maximum	.056	S.F.

Figure 9-30 Installation Time in Labor-Hours for Stone Flooring

Description	Labor-Hours	Unit
Marble Flooring Tiles, 3/8" Thick, Thinset	.267	S.F.
Mortar Bed	.369	S.F.
Marble Travertine; 1-1/4" Thick, Mortar Bed	.308	S.F.
Slate 1/2" Thick 6" x 6" x 1/2", Thinset	.267	S.F.
Mortar Bed	.320	S.F.
24" x 24" x 1/2" Mortar Bed	.267	S.F.

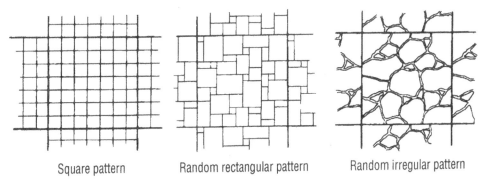

Square pattern   Random rectangular pattern   Random irregular pattern

Marble Flooring

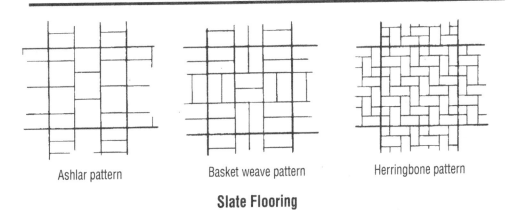

Ashlar pattern   Basket weave pattern   Herringbone pattern

Slate Flooring

Figure 9-31   Installation Time in Labor-Hours for Access Flooring

Description	Crew	Daily Output	Labor-Hours	Unit
Access floor package, including panel, pedestal, stringers & laminate cover				
Computer room greater than 6,000 SF	4 carp	750	.043	SF
Less than 6,000 SF	2 carp	375	.043	SF
Office greater than 6,000 SF	4 carp	1050	.030	SF
Panels, particleboard or steel, 1250# load, no covering, under 6,000 SF	2 carp	600	.027	SF
Over 6,000 SF	2 carp	640	.025	SF
Aluminum, 24" panels	2 carp	500	.032	SF
For snap-on stringer system, add	2 carp	1000	.016	SF
Office applications, steel or concrete panels, no covering, over 6,000 SF	2 carp	960	.017	SF
Machine cutouts after initial installation	1 carp	50	.160	Ea
Pedestals, 6" to 12"	2 carp	85	.188	Ea
Air conditioning grilles, 4" x 12"	1 carp	17	.471	Ea
4" x 18"	1 carp	14	.571	Ea
Approach ramps, minimum	2 carp	60	.267	SF
Maximum	2 carp	40	.400	SF
Handrail, 2 rail, aluminum	1 carp	15	.533	LF

Figure 9-32 Installation Time in Labor-Hours for Wall Coverings

Description	Labor-Hours	Unit
Wall Covering		
Aluminum Foil	.029	S.F.
Copper Sheets, .025" Thick		
Vinyl Backing	.033	S.F.
Phenolic Backing	.033	S.F.
Cork Tiles, Light or Dark, 12" x 12"		
3/16" Thick	.033	S.F.
5/16" Thick	.034	S.F.
1/4" Basketweave	.033	S.F.
1/2" Natural, Non-directional Pattern	.033	S.F.
Granular Surface, 12" x 36"		
1/2" Thick	.021	S.F.
1" Thick	.022	S.F.
Polyurethane Coated, 12" x 12"		
3/16" Thick	.033	S.F.
5/16" Thick	.034	S.F.
Cork Wallpaper, Paperbacked		
Natural	.017	S.F.
Colors	.017	S.F.
Flexible Wood Veneer, 1/32" Thick		
Plain Woods	.080	S.F.
Exotic Woods	.084	S.F.
Gypsum-based, Fabric-backed, Fire Resistant for Masonry Walls		
Minimum	.010	S.F.
Maximum	.013	S.F.
Acrylic, Modified, Semi-rigid PVC		
.028" Thick	.048	S.F.
.040" Thick	.050	S.F.
Vinyl Wall Covering, Fabric-backed		
Lightweight	.013	S.F.
Medium Weight	.017	S.F.
Heavy Weight	.018	S.F.
Grass Cloths with Lining Paper		
Minimum	.020	S.F.
Maximum	.023	S.F.

Notes

Division Ten

Specialties

Introduction

Specialties include prefinished, manufactured items that are usually installed at or near the end of a project—the finishing touches. Division 10 could be considered a catch-all type of division, containing miscellaneous items that do not quite fit into any other division. There is, of course, the potential for overlap; e.g., do toilet partitions belong in Division 8 – Openings, Division 9 – Finishes, or Division 10 – Specialties? (It has become accepted that they belong in Division 10, as they are a specialty item.)

The first step in estimating Division 10 is to check out the available specifications. Often, a list of items is provided, along with approved or accepted manufacturers of these items. The next step is to go through all drawings and to make a list of items to be priced.

Many of the items covered in Division 10 are priced on a delivered material only basis. It is important to keep this in mind, as the labor for the installation of these items will have to be figured separately.

Awnings and sun shades can be used as an inexpensive green building strategy to reduce cooling loads on a building. During the midday hours, the sun is overhead and its rays are intense. The position of awnings and shades above windows or openings blocks these intense rays, while allowing cooler indirect light to enter the building. During the early morning and late afternoon, when the sun's rays are less severe and daylight is more in demand, the position of the shades allows direct light into the building.

Portable and demountable partitions can also be part of a green building strategy if office layouts are subject to frequent changes. Manufacturing portable and demountable partitions is typically not an especially green process, but over the life cycle of the product, the embodied energy and pollution associated with manufacturing and waste is reduced because the partitions can be used over again many times. With fixed partitions, the new components need to be mined or harvested, processed, and shipped, and the old components must be demolished and disposed of.

Estimating Data

The following tables show the expected installation times for various specialty items. Each table lists the specialty item, the typical installation crew mix, the expected average daily output of that crew (the number of units the crew can be expected to install in one eight-hour day), and the labor-hours required for the installation of one unit.

Table of Contents

Estimating Tips

Item Takeoff

When taking off Division 10 items, list each item called for, along with the specified manufacturer. Often no substitutions are allowed for specified Division 10 items.

Labor Cost for Installation

When receiving bids for Division 10 items, make sure that the costs of installation are included. In many cases these items are sold per unit without installation. If installation is not included, inquire if delivery is included. Items sold by the piece without installation usually do not include any shipping costs.

Support Systems

Note that many Division 10 items require some type of support system not usually supplied with the item—such as support brackets, plates, or angles. These must be accounted for and may need to be added to the appropriate division. Phrases in the specifications that gloss over this subject and thus make it difficult to recover any added costs include, "The contractor shall install all products in accordance with the manufacturers' recommendations" and the like. Remember, it is much more costly to install a behind-the-wall support system after the wall is in place.

Preparation of Items

In some cases, Division 10 items may require some assembly before installation. This assembly time can often exceed the time required for installation.

Shop Smart

There can be a wide variance in costs for a specified item. Smart shopping for these items can help lower your overall bid.

Checklist ✓

For an estimate to be reliable, all items must be accounted for. A complete estimate can also limit contingencies. The following checklist can be used to help ensure that all items are included.

☐ *Canopies*

☐ *Coat Racks*

☐ *Control Boards*

☐ *Chair Rails*

☐ *Chutes*
   ☐ Laundry
   ☐ Trash
   ☐ Mail

☐ *Compartments and Cubicles*
   ☐ Shower and dressing compartments
   ☐ Storage and other cubicles
   ☐ Toilet compartments
      ☐ Handicapped

☐ *Corner Guards*

☐ *Directory Boards*

☐ *Disappearing Stairs*

☐ *Display Cases*

☐ *Fireplace Accessories*

☐ *Flagpoles*

☐ *Folding Grates*

☐ *Grilles and Screens — Decorative*

☐ *Hat Racks*

☐ *Hospital Equipment*
   ☐ Curtain rack
   ☐ Equipment mounting racks
   ☐ I.V. tracks

☐ *Identification Devices*
   ☐ Building sign
   ☐ Door signage
   ☐ Electronic signage
   ☐ Illuminated signs
   ☐ Plaques

☐ *Locker Room Equipment*
   ☐ Basket rack
   ☐ Benches
   ☐ Lockers

☐ *Louvers*
   ☐ Wall
   ☐ Soffit

☐ *Mail Items*
   ☐ Mail boxes
   ☐ Mail slots
   ☐ Mail chute system

☐ *Partitions*
   ☐ Demountable
   ☐ Folding
   ☐ Operable
   ☐ Portable
   ☐ Work station

☐ *Protection Screens*

☐ *Security Devices*
   ☐ Gates
   ☐ Miscellaneous

☐ *Shelving*
   ☐ Part bins
   ☐ Storage
   ☐ Pallet racks

☐ *Shower Doors*

☐ *Telephone Enclosures*

☐ *Toilet Accessories*
   ☐ Dispenser units
      ☐ Paper towel
      ☐ Napkin
      ☐ Soap
      ☐ Toilet paper
      ☐ Toilet seat covers
   ☐ Grab bars
   ☐ Hand dryers
   ☐ Handicapped bars
   ☐ Hat/coat rack
   ☐ Medicine chests

☐ Mirrors
☐ Mop holder
☐ Robe hook
☐ Shelving
☐ Towel bars
☐ Towel shelf
☐ Waste receptacles

☐ *Tub Enclosures*

☐ *Turnstiles*

☐ *Vending Machines*

☐ *Wall Guards*

☐ *Wardrobes*
   ☐ Coat rack
   ☐ Dormitory style
   ☐ Hat rack
   ☐ Hospital type

☐ *Waste Receptacles*

☐ *Visual Display Boards*
   ☐ Chalkboards
   ☐ Markerboards
   ☐ Tackboards

Figure 10-1   Installation Time for Toilet Compartments

Description	Crew	Daily Output	Labor-Hours	Unit
Metal Toilet Compartments				
Cubicles, ceiling hung, powder coated steel	2 Carpenters	4.00	4.000	Ea.
Stainless steel		4.00	4.000	
Floor and ceiling anchored, powder coated steel		5.00	3.200	
Stainless steel		5.00	3.200	
Floor anchored, powder coated steel		7.00	2.286	
Stainless steel		7.00	2.286	
Floor anchored, headrail braced, powder coated steel		6.00	2.667	
Stainless steel		4.60	3.478	
Wall-hung partitions, powder coated steel		7.00	2.286	
Stainless steel		7.00	2.286	
Screens, entrance, floor mounted, 58" high, 48" wide,				
powder coated steel		15.00	1.067	
Stainless steel		15.00	1.067	
Urinal screen, 18" wide, powder coated steel		6.15	2.602	
Stainless steel		6.15	2.602	
Floor mounted, head rail braced, powder coated steel		8.00	2.000	
Stainless steel		8.00	2.000	
Pilaster, flush, powder coated steel		10.00	1.600	
Stainless steel		10.00	1.600	
Post braced, powder coated steel		10.00	1.600	
Stainless steel		10.00	1.600	
Wall hung, bracket supported, powder coated steel		10.00	1.600	
Stainless steel		10.00	1.600	
Flange supported, powder coated steel		10.00	1.600	
Stainless steel		10.00	1.600	
Wedge type, powder coated steel		10.00	1.600	
Stainless steel		10.00	1.600	
Plastic-Laminate-Clad Toilet Compartments				
Cubicles, ceiling hung				
Plastic laminate on particleboard	2 Carpenters	4	4	Ea.
Floor and ceiling anchored				
Plastic laminate on particleboard	2 Carpenters	5	3.200	Ea.
Floor mounted				
Plastic laminate on particleboard	2 Carpenters	7	2.286	Ea.
Floor mounted, headrail braced				
Plastic laminate on particleboard	2 Carpenters	6	2.667	Ea.
Entrance screen, floor mtd. plastic laminate,				
58" high, 48" wide		15	1.067	
Urinal screen, 18" wide, ceiling braced, plastic				
laminate		8	2	
Floor mounted, headrail braced		8	2	

(continued on next page)

Division 10

Figure 10-1   Installation Time for Toilet Compartments
(continued from previous page)

Description	Crew	Daily Output	Labor-Hours	Unit
Pilaster, flush, plastic laminate		10	1.600	
Post braced, plastic laminate		10	1.600	
Wall hung, bracket supported				
Plastic laminate on particleboard	2 Carpenters	10	1.600	Ea.
Flange supported				
Plastic laminate on particleboard	2 Carpenters	10	1.600	Ea.
Plastic Toilet Compartments				
Cubicles, ceiling hung				
Phenolic	2 Carpenters	4	4	Ea.
Floor and ceiling anchored				
Phenolic	2 Carpenters	5	3.200	Ea.
Floor mounted				
Phenolic	2 Carpenters	7	2.286	Ea.
Floor mounted, headrail braced				
Phenolic	2 Carpenters	6	2.667	Ea.
Stone toilet comparments				
Cubicles, ceiling hung	2 Marblers	2	8	Ea.
Floor & ceiling anchored, marble		2.50	6.400	
Floor mounted, marble		3	5.333	
Floor mounted, headrail braced		3	5.333	
Entrance screen, floor mounted marble,				
58" high, 48" wide	↓	9	1.778	
Urinal screen, 18" wide, ceiling braced,				
marble	1 Bricklayer 1 Bricklayer Helper	6	2.667	Ea.
Floor mounted. Headrail braced,				
marble		6	2.667	
Pilaster, flush, marble		9	1.778	
Post braced, marble	↓	9	1.778	

Figure 10-2 Installation Time for Shower Compartments

Description	Crew Makeup	Daily Output	Labor-Hours	Unit
PARTITIONS, SHOWER Economy, painted steel, steel base, no door or plumbing included	2 Sheet Metal Workers	5	3.200	Ea.
Square, 32" x 32", stock, with receptor & door, fiberglass	↓	4.50	3.560	Ea.
Galvanized and painted steel		5	3.200	Ea.
Shower stall, double wall, incl. receptor but not including door or plumbing, enameled steel	2 Sheet Metal Workers	5	3.200	Ea.
Porcelain enamel steel		5	3.200	Ea.
Stainless steel		5	3.200	Ea.
Circular fiberglass, 36" diameter, no plumbing included		4	4.000	Ea.
One piece, 36" diameter, less door		4	4.000	Ea.
With door		3.50	4.570	Ea.
Curved shell shower, no door needed		3	5.330	Ea.
Glass stalls, with doors, no receptors, chrome on brass	↓	3	5.330	Ea.
Anodized aluminum		4	4.000	Ea.
Marble shower stall, stock design with shower door	2 Marble Setters	1.20	13.330	Ea.
With curtain		1.30	12.310	Ea.
Receptors, precast terrazzo, 32" x 32"		14	1.140	Ea.
48" x 34"		12	1.330	Ea.
Plastic, simulated terrazzo receptor, 32" x 32"		14	1.140	Ea.
32" x 48"		12	1.330	Ea.
Precast concrete, colors, 32" x 32"		14	1.140	Ea.
48" x 48"	↓	12	1.330	Ea.
Shower doors, economy plastic, 24" wide	1 Sheet Metal Worker	9	.889	Ea.
Tempered glass door, economy		8	1.000	Ea.
Folding, tempered glass, aluminum frame		6	1.330	Ea.
Sliding, tempered glass, 48" opening		6	1.330	Ea.
Deluxe, tempered glass, chrome on brass frame, min.		8	1.000	Ea.
Maximum		1	8.000	Ea.
On anodized aluminum frame, minimum		2	4.000	Ea.
Maximum		1	8.000	Ea.
Tub enclosure, plastic panels, economy, sliding panel		4	2.000	Ea.
Folding panel		4	2.000	Ea.
Deluxe, tempered glass, anodized alum. frame, min.		2	4.000	Ea.
Maximum		1.50	5.330	Ea.
On chrome-plated brass frame, minimum		2	4.000	Ea.
Maximum	↓	1.50	5.330	Ea.

Division 10

Figure 10-3 Installation Time for Wall and Corner Guards

Description	Crew Makeup	Daily Output	Labor-Hours	Unit
CORNER GUARDS Steel angle w/anchors, 1" x 1", 1.5#/L.F.	1 Struc. Steel Foreman 3 Struc. Steel Workers 1 Gas Welding Machine	320	.100	L.F.
2" x 2" angles, 3.5#/L.F.		300	.107	L.F.
3" x 3" angles, 6#/L.F.		275	.116	L.F.
4" x 4" angles, 8#/L.F.		240	.133	L.F.
Cast iron wheel guards, 3'-0" high		24	1.330	Ea.
5'-0" high		19	1.680	Ea.
Pipe bumper for truck doors, 8' long, 6" diameter		20	1.600	Ea.
8" diameter	↓	20	1.600	Ea.
CORNER PROTECTION				
Stainless steel, 16 ga., adhesive mount, 3-1/2" leg	1 Struc. Steel Worker	80	.100	L.F.
12 ga. Stainless, adhesive mount		80	.100	
For screw mount add 10%				
Vinyl acrylic, adhesive mount, 3" leg	1 Carpenter	128	0.63	
1-1/2" leg		160	.050	
Screw mounted, 3" leg		80	.100	
1-1/2" leg		100	.080	
Clear pastic, screw mounted, 2-1/2"		60	.133	
Vinyl cover, alum. Retainer, surface mount, 3" x 3"		48	.167	
2" x 2"		48	.167	
Flush mounted, 3" x 3"		32	.250	
2" x 2"	↓	32	.250	↓
WALLGUARD				
Neoprene with aluminum fastening strip, 1-1/2" x 2"	1 Carpenter	110	.073	L.F.
Trolley rail, PVC, clipped to wall, 5" high		185	.043	L.F.
8" high		180	.044	L.F.
Vinyl acrylic bed aligner and bumper, 37" long		10	.800	Ea.
43" long	↓	9	.889	Ea.

Figure 10-4 Installation Time for Prefabricated Fireplaces

Description	Crew Makeup	Daily Output	Labor-Hours	Unit
FIREPLACE, PREFABRICATED Free standing or wall hung with hood & screen, minimum	1 Carpenter	1.30	6.150	Ea.
Average		1	8.000	Ea.
Maximum		.90	8.890	Ea.
Double wall for chimney heights over 8'-6"				
7" diameter, add		33	.242	V.L.F.
10" diameter, add		32	.250	V.L.F.
12" diameter, add		31	.258	V.L.F.
14" diameter, add		30	.267	V.L.F.
Simulated brick chimney top, 4' high, 16" x 16"		10	.800	Ea.
24" x 24"		7	1.140	Ea.
Simulated logs, gas fired, 40,000 BTU, 2' long, min.		7	1.140	Set
Maximum		6	1.330	Set
Electric, 1500 BTU, 1'-6" long, minimum		7	1.140	Set
11,500 BTU, maximum		6	1.330	Set

Division 10

Figure 10-5 Installation Time for Fireplace Accessories

Description	Crew Makeup	Daily Output	Labor-Hours	Unit
FIREPLACE ACCESSORIES Chimney screens				
Galv., 13" x 13" flue	1 Bricklayer	8	1.000	Ea.
Galv., 24" x 24" flue		5	1.600	Ea.
Stainless steel, 13" x 13" flue		8	1.000	Ea.
20" x 20" flue		5	1.600	Ea.
Cleanout doors and frames, cast iron, 8" x 8"		12	.667	Ea.
12" x 12"		10	.800	Ea.
18" x 24"		8	1.000	Ea.
Cast iron frame, steel door, 24" x 30"		5	1.600	Ea.
Damper, rotary control, steel, 30" opening		6	1.330	Ea.
Cast iron, 30" opening		6	1.330	Ea.
36" opening		6	1.330	Ea.
48" opening		6	1.330	Ea.
60" opening		6	1.330	Ea.
72" opening		5	1.600	Ea.
84" opening		5	1.600	Ea.
96" opening		4	2.000	Ea.
Steel plate, poker control, 60" opening		8	1.000	Ea.
84" opening		5	1.600	Ea.
"Universal" type, chain operated, 32" x 20" opening		8	1.000	Ea.
48" x 24" opening		5	1.600	Ea.
Dutch Oven door and frame, cast iron, 12" x 15" opening		13	.615	Ea.
Copper plated, 12" x 15" opening		13	.615	Ea.
Fireplace forms with registers, 25" opening		3	2.670	Ea.
34" opening		2.50	3.200	Ea.
48" opening		2	4.000	Ea.
72" opening		1.50	5.330	Ea.
Squirrel and bird screens, galvanized, 8" x 8" flue		16	.500	Ea.
13" x 13" flue		12	.667	Ea.

Figure 10-6 Installation Time for Stoves

Description	Crew Makeup	Daily Output	Labor-Hours	Unit
WOODBURNING STOVES Cast iron, minimum	2 Carpenters Power Tools	1.30	12.310	Ea.
Average		1	16.000	Ea.
Maximum		.80	20.000	Ea.

Figure 10-7   Installation Time for Ground Set Flagpoles

Description	Crew Makeup	Daily Output	Labor-Hours	Unit
FLAGPOLE Not including base or foundation				
Aluminum, tapered, ground set 20' high	1 Carpenter 1 Truck Driver (light) 1 Truck w/Power Equip.	2	8.000	Ea.
25' high		1.70	9.410	Ea.
30' high		1.50	10.670	Ea.
35' high		1.40	11.430	Ea.
40' high		1.20	13.330	Ea.
50' high		1	16.000	Ea.
60' high		.90	17.780	Ea.
70' high		.80	20.000	Ea.
80' high		.70	22.860	Ea.
Counterbalanced, tapered, aluminum, 20' high		1.80	8.890	Ea.
30' high		1.50	10.670	Ea.
40' high		1.30	12.310	Ea.
50' high		1	16.000	Ea.
Aluminum, electronically operated, 30' high		1.40	11.430	Ea.
35' high		1.30	12.310	Ea.
40' high		1.10	14.550	Ea.
45' high		1	16.000	Ea.
50' high		.90	17.780	Ea.
Fiberglass, tapered, ground set, 25' high		2	8.000	Ea.
30' high		1.50	10.670	Ea.
35' high		1.40	11.430	Ea.
40' high		1.20	13.330	Ea.
50' high		1	16.000	Ea.
60' high		.90	17.780	Ea.
Yardarms and rigging for poles, 6' total length		1.90	8.420	Ea.
12' total length		1.80	8.890	Ea.
Steel, sectional, lightweight, ground set, 20' high		1.30	12.310	Ea.
25' high		1.20	13.330	Ea.
30' high		1.10	14.550	Ea.
35' high		1	16.000	Ea.
40' high		.90	17.780	Ea.
50' high		.80	20.000	Ea.
Tapered, heavyweight steel, ground set, 35' high		.80	20.000	Ea.
50' high		.70	22.860	Ea.
60' high		.70	22.860	Ea.
75' high		.60	26.670	Ea.
Bases, ornamental, minimum	1 Carpenter	6	1.330	Ea.
Average		4	2.000	Ea.
Maximum		2	4.000	Ea.
Wood poles, tapered, clear vertical grain fir with tilting base, not incl. foundation, 4" butt, 25' high	1 Carpenter 1 Truck Driver (light) 1 Truck w/Power Equip.	1.90	8.420	Ea.
6" butt, 30' high	" "	1.30	12.310	Ea.
Foundations for flagpoles, including excavation and concrete, to 35' high poles	3 Carpenters 1 Building Laborer Power Tools	10	3.200	Ea.
40' to 50' high		3.50	9.140	Ea.
Over 60' high		2	16.000	Ea.

Figure 10-8   Installation Time for Signs

Description	Crew Makeup	Daily Output	Labor-Hours	Unit
SIGNS Letters, individual, 2" high, cast aluminum	1 Carpenter	32	.250	Ea.
Cast bronze		32	.250	Ea.
4" high, 5/8" deep, cast aluminum		24	.333	Ea.
Cast bronze		24	.333	Ea.
6" high, 1" deep, cast aluminum		20	.400	Ea.
Cast bronze		20	.400	Ea.
12" high, 1-1/4" deep, cast aluminum		18	.444	Ea.
Cast bronze		18	.444	Ea.
18" high, 1-1/4" deep, cast aluminum		12	.667	Ea.
Cast bronze		12	.667	Ea.
Fabricated aluminum, 12" high, 3" deep		18	.444	Ea.
18" high, 3" deep		12	.667	Ea.
Fabricated stainless steel, 6" high, 3" deep		20	.400	Ea.
12" high, 3" deep		18	.444	Ea.
18" high, 3" deep		12	.667	Ea.
24" high, 4" deep		10	.800	Ea.
Painted sheet steel, 12" high, 2" deep		18	.444	Ea.
18" high, 3" deep		12	.667	Ea.
Plastic, 6" high, 1" deep		20	.400	Ea.
12" high, 2" deep		18	.444	Ea.
Plastic face, alum. frame, 20" high, 15" deep		5	1.600	Ea.
36" high, 24" deep		4	2.000	Ea.
Stainless steel frame, 12" high, 6" deep		5	1.600	Ea.
24" high, 8" deep		4	2.000	Ea.
Plaques, 20" x 30", for up to 450 letters, cast alum.	2 Carpenters	4	4.000	Ea.
Cast bronze		4	4.000	Ea.
30" x 40", up to 900 letters cast aluminum		3	5.330	Ea.
Plaques, 30" x 40", for up to 900 letters, cast bronze		3	5.330	Ea.
36" x 48", for up to 1300 letters, cast bronze		2	8.000	Ea.
Signs, cast aluminum steel signs, 2-way		30	.533	Ea.
4-way		30	.533	Ea.
Acrylic exit signs, 15" x 6", surface mounted				
Minimum		30	.533	Ea.
Maximum		20	.800	Ea.
Bracket mounted, double face, minimum		30	.533	Ea.
Maximum		20	.800	Ea.
Plexiglass, exterior, illuminated, single face		100	.160	S.F.
Double face		75	.213	S.F.
Interior, illuminated, single face		100	.160	S.F.
Double face		75	.213	S.F.
Painted plywood (MDO), over 4' x 8'		120	.133	S.F.
Under 4' x 8'		100	.160	S.F.

Figure 10-9 Installation Time for Metal Lockers

Description	Crew Makeup	Daily Output	Labor-Hours	Unit
LOCKERS Steel, baked enamel, 60" or 72", single tier				
Minimum	1 Sheet Metal Worker	14	.571	Opng.
Maximum		12	.667	Opng.
2 tier, 60" or 72" total height, minimum		26	.308	Opng.
Maximum		20	.400	Opng.
5 tier box lockers, minimum		30	.267	Opng.
Maximum		24	.333	Opng.
6 tier box lockers, minimum		36	.222	Opng.
Maximum		30	.267	Opng.
Basket rack with 32 baskets, 9" x 13" x 8" basket		50	.160	Basket
24 baskets, 12" x 13" x 8" basket		50	.160	Basket
Athletic, wire mesh, no lock, 18" x 18" x 72" basket		12	.667	Ea.
Overhead locker baskets on chains, 14" x 14" baskets	3 Sheet Metal Workers	96	.250	Basket
Overhead locker framing system, add		600	.040	Basket
Locking rail and bench units, add		120	.200	Basket
Locker bench, laminated maple, top only	1 Sheet Metal Worker	100	.080	L.F.
Pedestals, steel pipe	"	25	.320	Ea.
Teacher and pupil wardrobes, enameled				
22" x 15" x 61" high, minimum	1 Sheet Metal Worker	10	.800	Ea.
Average		9	.889	Ea.
Maximum		8	1.000	Ea.
Duplex lockers with 2 doors, 72" high, 15" x 15"		10	.800	Ea.
15" x 21"		10	.800	Ea.

Division 10

Figure 10-10   Installation Time for Fabric Awnings

Description	Crew Description	Daily Output	Labor-Hours	Unit
Awnings, Fabric				
Including acrylic canvas and frame, standard design				
Door and window, slope, 3' height, 4' wide	1 Carpenter	4.50	1.778	Ea.
6' wide		3.50	2.286	
8' wide		3.00	2.667	
Quarter round convex, 4' wide		3.00	2.667	
6' wide		2.25	3.556	
8' wide		1.80	4.444	
Dome, 4' wide		7.50	1.067	
6' wide		3.50	2.286	
8' wide		2.00	4.000	
Elongated dome, 4' wide		1.33	6.015	
6' wide		1.11	7.207	
8' wide	▼	1.00	8.000	
Entry or walkway, peak, 12' long, 4' wide	2 Carpenters	0.90	17.778	
6' wide		0.60	26.667	
8' wide		0.40	40.000	
Radius with dome end, 4' wide		1.10	14.545	
6' wide		0.70	22.857	
8' wide	▼	0.50	32.000	▼
Retractable lateral arm awning, manual				
To 12' wide, 8'-6" projection	2 Carpenters	1.70	9.412	Ea.
To 14' wide, 8'-6" projection		1.10	14.545	
To 19' wide, 8'-6" projection		0.85	18.824	
To 24' wide, 8'-6" projection	▼	0.67	23.881	▼
Patio/deck canopy with frame				
12' wide, 12' projection	2 Carpenters	2.00	8.00	Ea.
16' wide, 14' projection	"	1.20	13.333	"

Figure 10-11 Installation Time for Mail Boxes

Description	Crew Makeup	Daily Output	Labor-Hours	Unit
MAIL BOXES Horiz. key lock, 5"H x 6" W x 15"D, alum.				
Rear loading	1 Carpenter	34	.235	Ea.
Front loading		34	.235	Ea.
Double, 5"H x 12"W x 15"D, rear loading		26	.308	Ea.
Front loading		26	.308	Ea.
Quadruple, 10"H x 12"W x 15"D, rear loading		20	.400	Ea.
Front loading		20	.400	Ea.
Vertical, front loading, 15"H x 5" W x 6"D, aluminum		34	.235	Ea.
Bronze, duranodic finish		34	.235	Ea.
Steel, enameled		34	.235	Ea.
Alphabetical directories, 35 names		10	.800	Ea.
Letter collection box		6	1.330	Ea.
Letter slot, residential		20	.400	Ea.
Post office type		8	1.000	Ea.
Post office counter window, with grille		2	4.000	Ea.
Key keeper, single key, aluminum		26	.308	Ea.
Steel enameled	▼	26	.308	Ea.

Figure 10-12 Installation Time for Wire Mesh Partitions

Description	Crew Makeup	Daily Output	Labor-Hours	Unit
PARTITIONS, WOVEN WIRE For tool or stockroom enclosures				
Channel frame, 1-1/2" diamond mesh, 10 ga. wire painted				
Wall panels, 4' wide, 7' high	2 Carpenters	25	.640	Ea.
8' high		23	.696	Ea.
10' high		18	.889	Ea.
Ceiling panels, 10' long, 2' wide		25	.640	Ea.
4' wide		15	1.070	Ea.
Panel with service window & shelf, 5' long, 7' high		20	.800	Ea.
10' high		15	1.070	Ea.
Sliding doors, 3' wide, 7' full height		6	2.670	Ea.
10' full height		5	3.200	Ea.
6' wide, 7' full height		5	3.200	Ea.
10'full height		4	4.000	Ea.
Swinging doors, 3' wide, 7' high, no transom		6	2.670	Ea.
7' high, 3' transom	▼	5	3.200	Ea.

Division 10

Figure 10-13 Installation Time for Folding Gates

Description	Crew Makeup	Daily Output	Labor-Hours	Unit
SECURITY GATES				
Scissors type folding gate, painted steel				
Single, 6-1/2' high, 5-1/2' wide	2 Struc. Steel Workers	4.00	4.000	Opng.
6-1/2' wide		4.00	4.000	
7-1/2' wide		4.00	4.000	
Double gate, 8' high, 8' wide		2.50	6.400	
10' wide		2.50	6.400	
12' wide		2.00	8.000	
14' wide		2.00	8.000	

Figure 10-14 Installation Time for Demountable Partitions

Description	Crew Makeup	Daily Output	Labor-Hours	Unit
PARTITIONS, MOVABLE OFFICE Demountable, add for doors				
Do not deduct door openings from total L.F.				
Gypsum, laminated 2-1/4" thick, 9' high, unpainted	2 Carpenters	40	.400	L.F.
Painted		40	.400	L.F.
3" thick, acoustical, unpainted		40	.400	L.F.
Painted		40	.400	L.F.
Vinyl clad drywall on 2-1/2" metal studs, to 9' high		60	.267	L.F.
42" high, plus 10" glass		60	.267	L.F.
Vinyl clad gypsum with air space, 3" thick		60	.267	L.F.
Steel clad gypsum, as above		60	.267	L.F.
Hardboard, vinyl faced, 7' high, 1-9/16" thick		60	.267	L.F.
2-1/4" thick, painted		50	.320	L.F.
With 40" glass		40	.400	L.F.
10' high, vinyl faced, 1-9/16" thick		40	.400	L.F.
Enameled, 2-3/4" thick		30	.533	L.F.
Metal, to 9'-6" high, enameled steel, no glass		40	.400	L.F.
Steel frame, all glass		40	.400	L.F.
Vinyl covered, no glass		40	.400	L.F.
Steel frame with 52% glass		40	.400	L.F.
Free standing, 4'-6" high, steel with glass		100	.160	L.F.
Acoustical		100	.160	L.F.
Low rails, 3'-3" high, enameled steel		100	.160	L.F.
Vinyl covered		100	.160	L.F.
Plywood, prefin, 1-3/4" thick, rotary cut veneers				
Minimum		80	.200	L.F.
Maximum		80	.200	L.F.
Sliced veneers, book matched, minimum		80	.200	L.F.
Maximum		80	.200	L.F.
Sliced veneers, random matched, minimum		80	.200	L.F.
Maximum		80	.200	L.F.
Trackless wall, cork finish, semi-acoustic, 1-5/8" thick				
Minimum		325	.049	S.F.
Maximum		190	.084	S.F.
Acoustic, 2" thick, minimum		305	.052	S.F.
Maximum		225	.071	S.F.
For doors, not incl. hardware, hollow metal door, add		4.30	3.720	Ea.
Hardwood door, add		3.40	4.710	Ea.
Hardware for doors, not incl. closers, keyed		29	.552	Ea.
Non-keyed	▼	29	.552	Ea.

Figure 10-15 Installation Time for Portable Partitions

Description	Crew Makeup	Daily Output	Labor-Hours	Unit
PARTITIONS, HOSPITAL Curtain track, box channel				
Ceiling mounted	1 Carpenter	135	.059	L.F.
Suspended	"	100	.080	L.F.
Curtains, 8' to 9', nylon mesh tops, fire resistant				
Cotton, 2.85 lbs. per S.Y.	1 Carpenter	425	.019	L.F.
Anti-bacterial, thermoplastic		425	.019	L.F.
Fiberglass, 7 oz. per S.Y.	↓	425	.019	L.F.
I.V. track systems				
i.V. track, 4' x 7' oval	1 Carpenter	135	.059	L.F.
I.V. trolley		32	.250	Ea.
I.V. pendent, (tree, 5 hook)	↓	32	.250	Ea.
WALL SCREENS, divider panels, free standing, fiber core				
Fabric face straight				
3'-0" long, 4'-0" high	2 Carpenters	1.00	1.60	L.F.
5'-0" high		90	.178	L.F.
6'-0" high		75	.213	L.F.
5'-0" long, 4'-0" high		175	.091	L.F.
5'-0" high		150	.107	L.F.
6'-0" high		125	.128	L.F.
6'-0" long, 5'-0" high		162	.099	L.F.
Economical panels, fabric face, 5'-0" long, 5'-0" high		132	.121	L.F.
6'-0" high		112	.143	L.F.
5'-0" long, 5'-0" high		150	.107	L.F.
6'-0" high		125	.128	L.F.
Acoustical panels, 60 to 90 NRC, 3'-0" long, 5'-0" high		90	.178	L.F.
6'-0" high		75	.213	L.F.
5'-0" long, 5'-0" high		150	.107	L.F.
6'-0" high		125	.128	L.F.
5'-0" long, 5'-0" high		162	.099	L.f.
6'-0" high		138	.116	L.F.
Economy acoustical panels, 40 NRC, 4'-0" long, 5'-0" high		132	.121	L.F.
6'-0" high		112	.143	L.F.
5'-0" long, 6'-0" high		125	.128	L.F.
6'-0" long, 5'-0" high		162	.099	L.F.
Metal chalkboard, 6'-6" high, chalkboard, 1 side		125	.128	L.F.
Metal chalkboard, 2 sides		120	.133	L.F.
Tackboard, both sides	↓	123	.130	L.F.
PARTITIONS, WORK STATIONS Incl. top cabinet & desk top				
Multi-station units, 1 to 3 person seating capacity				
Minimum per person	1 Carpenter	6	1.330	Ea.
Average per person		5	1.600	Ea.
Maximum per person	↓	4	2.000	Ea.
4 to 6 person seating capacity				
Minimum per person	1 Carpenter	4	2.000	Ea.
Average per person		3	2.670	Ea.
Maximum per person	↓	2	4.000	Ea.

Figure 10-16 Installation Time for Panel Partitions

Description	Crew Makeup	Daily Output	Labor-Hours	Unit
PARTITIONS, FOLDING LEAF Acoustic, wood				
Vinyl faced, to 18' high, 6 psf, minimum	2 Carpenters	60	.267	S.F.
Average		45	.356	S.F.
Maximum		30	.533	S.F.
Formica or hardwood finish, minimum		60	.267	S.F.
Maximum		30	.533	S.F.
Wood, low acoustical type, 4.5 psf, to 14' high		50	.320	S.F.
Steel, acoustical, 9 to 12 lb. per S.F., vinyl faced				
Minimum		60	.267	S.F.
Maximum		30	.533	S.F.
Aluminum framed, acoustical, to 12' high, 5.5 psf				
Minimum		60	.267	S.F.
Maximum		30	.533	S.F.
6.5 lb. per S.F., minimum		60	.267	S.F.
Maximum		30	.533	S.F.
PARTITIONS, OPERABLE Acoustic air wall, 1-5/8" thick				
Minimum		375	.043	S.F.
Maximum		365	.044	S.F.
2-1/4" thick, minimum		360	.044	S.F.
Maximum		330	.048	S.F.
Overhead track type, acoustical, 3" thick, 11 psf				
Minimum		350	.046	S.F.
Maximum		300	.053	S.F.

Figure 10-17 Installation Time for Moveable Panel Systems

Description	Crew Makeup	Daily Output	Labor-Hours	Unit
PANELS AND DIVIDERS Free standing				
Fabric panel, class A fire rated				
Minimum NRC .95, STC 28, fabric edged				
40" high, 24" wide	2 Building Laborers	30	.533	Ea.
36" wide		28	.571	Ea.
48" wide		26	.615	Ea.
60" wide		25	.640	Ea.
50" high, 24" wide		30	.533	Ea.
36" wide		28	.571	Ea.
48" wide		26	.615	Ea.
60" wide		25	.640	Ea.
60" high, 24" wide		29	.552	Ea.
36" wide		27	.593	Ea.
48" wide		25	.640	Ea.
60" wide		24	.667	Ea.
72" high, 24" wide		29	.552	Ea.
36" wide		27	.593	Ea.
48" wide		25	.640	Ea.
60" wide		24	.667	Ea.
Hardwood edged, 40" high, 24" wide		30	.533	Ea.
36" wide		28	.571	Ea.
48" wide		26	.615	Ea.
60" wide		25	.640	Ea.
50" high, 24" wide		30	.533	Ea.
36" wide		28	.571	Ea.
48" wide		26	.615	Ea.
60" wide		25	.640	Ea.
60" high, 24" wide		29	.552	Ea.
36" wide		27	.593	Ea.
48" wide		25	.640	Ea.
60" wide		24	.667	Ea.
72" high, 24" wide		29	.552	Ea.
36" wide		27	.593	Ea.
48" wide		25	.640	Ea.
60" wide		24	.667	Ea.

Figure 10-18 Installation Time for Storage and Shelving

Description	Crew Makeup	Daily Output	Labor-Hours	Unit
PARTS BINS Metal, gray baked enamel finish				
6'-3" high, 3' wide				
12 bins, 18" wide x 12" high, 12" deep	2 Building Laborers	10	1.600	Ea.
24" deep		10	1.600	Ea.
72 bins, 6" wide x 6" high, 12" deep		8	2.000	Ea.
24" deep		8	2.000	Ea.
7'-3" high, 3' wide				
14 bins, 18" wide x 12" high, 12" deep	2 Building Laborers	10	1.600	Ea.
24" deep		10	1.600	Ea.
84 bins, 6" wide x 6" high, 12" deep		8	2.000	Ea.
24" deep		8	2.000	Ea.
SHELVING Metal, industrial, cross-braced, 3' wide				
12" deep	1 Struc. Steel Worker	175	.046	S.F. Shlf.
24" deep		330	.024	S.F. Shlf.
4' wide, 12" deep		185	.043	S.F. Shlf.
24" deep		380	.021	S.F. Shlf.
Enclosed sides, cross-braced back, 3' wide				
12" deep		175	.046	S.F. Shlf.
24" deep		290	.028	S.F. Shlf.
Fully enclosed, sides and back, 3' wide, 12" deep		150	.053	S.F. Shlf.
24" deep		255	.031	S.F. Shlf.
4' wide, 12" deep		150	.053	S.F. Shlf.
24" deep		290	.028	S.F. Shlf.
Wide span, 1600 lb. capacity per shelf, 7' wide				
24" deep		380	.021	S.F. Shlf.
36" deep		440	.018	S.F. Shlf.
8' wide, 24" deep		440	.018	S.F. Shlf.
36" deep		520	.015	S.F. Shlf.
Pallet racks, steel frame 2500 lb. capacity, 7' long				
30" deep.	2 Struc. Steel Workers	450	.036	S.F. Shlf.
36" deep		500	.032	S.F. Shlf.
42" deep		520	.031	S.F. Shlf.

Division 10

Figure 10-19   Installation Time for Bath Accessories

Description	Crew Makeup	Daily Output	Labor-Hours	Unit
COMMERCIAL TOILET ASSESORIES				
Curtain rod, stainless steel, 5' long, 1" diameter	1 Carpenter	13	.615	Ea.
1-1/4" diameter	"	13	.615	Ea.
Dispenser units, combined soap & towel dispensers,				
mirror and shelf, flush mounted	1 Carpenter	10	.800	Ea.
Towel dispenser and waste receptacle,				
flush mounted	1 Carpenter	10	.800	Ea.
Grab bar, straight, 1" diameter, stainless steel, 12" long		24	.333	Ea.
18" long		23	.348	Ea.
24" long		22	.364	Ea.
36" long		20	.400	Ea.
1-1/2" diameter, 18" long		23	.348	Ea.
36" long		20	.400	Ea.
Tub bar, 1" diameter, horizontal		14	.571	Ea.
Plus vertical arm		12	.667	Ea.
End tub bar, 1" diameter, 90° angle		12	.667	Ea.
Hand dryer, surface mounted, electric, 110 volt		4	2.000	Ea.
220 volt		4	2.000	Ea.
Hat and coat strip, stainless steel, 4 hook, 36" long		24	.333	Ea.
6 hook, 60" long		20	.400	Ea.
Mirror with stainless steel, 3/4" square frame				
18" x 24"		20	.400	Ea.
36" x 24"		15	.533	Ea.
48" x 24"		10	.800	Ea.
72" x 24"		6	1.330	Ea.
Mirror with 5" stainless steel shelf, 3/4" sq. frame				
18" x 24"		20	.400	Ea.
36" x 24"		15	.533	Ea.
48" x 24"		10	.800	Ea.
72" x 24"		6	1.330	Ea.
Mop holder strip, stainless steel, 6 holders, 60" long		20	.400	Ea.
Napkin/tampon dispenser, surface mounted		15	.533	Ea.
Robe hook, single, regular		36	.222	Ea.
Heavy duty, concealed mounting		36	.222	Ea.
Soap dispenser, chrome, surface mounted, liquid		20	.400	Ea.
Powder		20	.400	Ea.
Recessed stainless steel, liquid		10	.800	Ea.
Powder		10	.800	Ea.
Soap tank, stainless steel, 1 gallon		10	.800	Ea.
5 gallon		5	1.600	Ea.
Shelf, stainless steel, 5" wide, 18 ga., 24" long		24	.333	Ea.
72" long		16	.500	Ea.
8" wide shelf, 18 ga., 24" long		22	.364	Ea.
72" long		14	.571	Ea.
Toilet seat cover dispenser, stainless steel, recessed		20	.400	Ea.
Surface mounted		15	.533	Ea.
Toilet tissue dispenser, surface mounted, stainless steel				
Single roll		30	.267	Ea.
Double roll		24	.333	Ea.
Towel bar, stainless steel, 18" long		23	.348	Ea.
30" long		21	.381	Ea.

(continued on next page)

Figure 10-19 Installation Time for Bath Accessories
(continued from previous page)

Description	Crew Makeup	Daily Output	Labor-Hours	Unit
Towel dispenser, stainless steel, surface mounted	1 Carpenter	16	.500	Ea.
Flush mounted, recessed		10	.800	Ea.
Towel holder, hotel type, 2 guest size		20	.400	Ea.
Towel shelf, stainless steel, 24" long, 8" wide		20	.400	Ea.
Tumbler holder, tumbler only		30	.267	Ea.
Soap, tumbler & toothbrush		30	.267	Ea.
Wall urn ash receiver, recessed, 14" long		12	.667	Ea.
Surface, 8" long		18	.444	Ea.
Waste receptacles, stainless steel, with top, 13 gallon		10	.800	Ea.
36 Gallon		8	1.000	Ea.
MEDICINE CABINETS With mirror, stock, 16" x 22"				
Unlighted	1 Carpenter	14	.571	Ea.
Lighted		6	1.330	Ea.
Sliding mirror doors, 36" x 22", unlighted		7	1.140	Ea.
Lighted		5	1.600	Ea.
Center mirror, 2 end cabinets, unlighted, 48" long		7	1.140	Ea.
72" long		5	1.600	Ea.
For lighting, 48" long, add	1 Electrician	3.50	2.290	Ea.
72" long, add	"	3	2.670	Ea.
Hotel cabinets, stainless, with lower shelf, unlighted	1 Carpenter	10	.800	Ea.
Lighted	"	5	1.600	Ea.

Figure 10-20 Installation Time for Coat Racks/Wardrobes

Description	Crew Makeup	Daily Output	Labor-Hours	Unit
COAT RACKS & WARDROBES				
Utility hook strips 3/8" x 2-1/2" x 18", 6 hooks	1 Carpenter	48	.167	Ea.
34" long, 12 hooks	"	48	.167	"
Wall mounted racks, 16 gauge steel frame, 22 gauge steel shelves				
12" x 15" x 26", 6 hangers	1 Carpenter	32	.250	Ea.
12" x 15" x 50", 12 hangers	"	32	.250	"

Figure 10-21 Installation Time for Residential Casework

Description	Crew Makeup	Daily Output	Labor-Hours	Unit
IRONING CENTER Including cabinet, board & light				
Minimum	1 Carpenter	2	4.000	Ea.
Maximum	"	1.50	5.330	Ea.

Division Eleven
Equipment

Introduction

Division 11 covers the total package of equipment that is required for a particular project's type of construction. For example, equipment for a bank would include teller windows, package transfer units, automatic banking equipment, vault doors, safe deposit boxes, etc.

Energy Star, a voluntary labeling program jointly administered by the U.S. Environmental Protection Agency and the U.S. Department of Energy, is designed to identify and promote energy efficient products. Energy Star labels a multitude of products—many of which fall under Division 11—including products relating to residential appliances (11 31 00), food storage equipment (11 41 00), food cooking equipment (11 44 00), ice machines (11 47 00), and cleaning and disposal equipment (11 48 00).

Division 11 items are usually counted individually, as most are sold that way. To estimate equipment, begin with a review of the specifications. Items to be included may be listed singly or grouped. Approved manufacturers may also be listed. Note that some items may not be listed in the specifications but may appear on the drawings only. Review each drawing carefully for items that are included or hinted at (e.g., on the concrete drawings, an item such as "concrete base for book depository" may be included). Such items should be investigated to determine who is responsible for supplying them. The drawings may also designate items that are supplied "by others" or not in the contractor's scope of work.

Quite often, the equipment listed in Division 11 is purchased directly by the facility owner. In some cases, the contractor installs the equipment, while in others the owner arranges for the installation. When it is determined that the installation of the owner-supplied equipment is the contractor's responsibility, the labor should be figured separately and on a unit basis.

Estimating Data

The following tables show the expected installation time for various specialty items. Each table lists the specialty item, the typical installation crew mix, the expected average daily output of that crew (the number of units the crew can be expected to install in one eight-hour day), and the labor-hours required for the installation of one unit.

Table of Contents

Division 11

Estimating Tips

Faulty Assumptions

Do not assume that items covered in Division 11 will be purchased and installed by others outside of your contract. This can be a costly error. Check all drawings for these items and for terms such as "NIC" (Not in Contract) or "By Others." If these or similar terms are not in evidence, then it is safe to conclude that these are included in the scope of work.

Installation of Items

In many cases, Division 11 items are purchased by others, but their installation is the contractor's responsibility. Check all drawings and specifications carefully for these items.

Handling Charges

In cases where Division 11 items are purchased by others but installed by the contractor, contractors often add a handling charge (10% of the estimated material costs is common practice). This charge covers the receiving, handling, storage, protection, and final delivery of these items.

Support Systems

Note that Division 11 items may require some type of support system not usually supplied with the item—such as support brackets, plates, or angles. These need to be accounted for and may need to be added to the appropriate division. Phrases in the specifications that gloss over this subject and thus make it difficult to recover any added costs include, "The contractor shall install all products in accordance with the manufacturers' recommendations" and the like. Remember, it is much more costly to install a behind-the-wall support system after the wall is in place.

Preparation of Items

In some cases, Division 11 items may require assembly before installation. Be aware that the assembly time can often exceed the time required for installation.

Checklist ✓

For an estimate to be reliable, all items must accounted for. A complete estimate can also limit contingencies. The following checklist can be used to help ensure that all items are included.

☐ **Appliances**
- ☐ Cooktops
- ☐ Compactors
- ☐ Dehumidifiers – unit
- ☐ Dishwasher
- ☐ Dryer
- ☐ Exhaust hoods
- ☐ Garbage disposer
- ☐ Heaters – electric unit
- ☐ Humidifiers – unit
- ☐ Ice makers
- ☐ Ovens
- ☐ Refrigerators
- ☐ Sump pump – unit
- ☐ Washing machine
- ☐ Water heater – unit
- ☐ Water softener – unit

☐ **Automotive Equipment**
- ☐ Lifts
- ☐ Hoists
- ☐ Lube

☐ **Bank Equipment**
- ☐ Counters
- ☐ Safes
- ☐ Safe deposit boxes
- ☐ Vaults
- ☐ Teller stations/windows

☐ **Barber Shop Equipment**

☐ **Bowling Alley Equipment**

☐ **Check Room Equipment**

☐ **Church Equipment**
- ☐ Altar
- ☐ Baptistries
- ☐ Bells
- ☐ Carillons
- ☐ Confessionals
- ☐ Crosses
- ☐ Organ
- ☐ Pews
- ☐ Spires

☐ **Coin Stations**

☐ **Commercial Equipment**
- ☐ Cash register systems
- ☐ Check-out counters
- ☐ Display cases
- ☐ Refrigerated cases

☐ **Data Processing Equipment**

☐ **Dental Equipment**
- ☐ Chairs
- ☐ Drill/equipment stations
- ☐ Thruway equipment

☐ **Detention Equipment**

☐ **Dock Equipment**
- ☐ Bumpers
- ☐ Boards
- ☐ Door seals
- ☐ Levellers
- ☐ Lights
- ☐ Shelters

☐ *Food Service Equipment*
- ☐ Bar units
- ☐ Cooking equipment
- ☐ Dishwashing equipment
- ☐ Food preparation equipment
- ☐ Food serving equipment
- ☐ Refrigerated cases
- ☐ Tables

☐ *Gymnasium Equipment*

☐ *Industrial Equipment*

☐ *Laboratory Equipment*
- ☐ Casework
- ☐ Countertops
- ☐ Hoods
- ☐ Sinks
- ☐ Tables

☐ *Laundry Equipment*

☐ *Library Equipment*
- ☐ Bookshelves
- ☐ Book stacks
- ☐ Computer terminals
- ☐ Carrels
- ☐ Charging desks
- ☐ Racks

☐ *Medical Equipment*

☐ *Mortuary Equipment*

☐ *Musical Equipment*

☐ *Observatory Equipment*

☐ *Parking Equipment*
- ☐ Automatic gates
- ☐ Booths
- ☐ Card readers
- ☐ Control stations

- ☐ Ticket dispensers
- ☐ Traffic detectors

☐ *Playground Equipment*

☐ *Projection Equipment*

☐ *Prison Equipment*

☐ *Residential Equipment*
- ☐ Kitchen cabinets/appliances
- ☐ Lavatory cabinets
- ☐ Kitchen equipment
- ☐ Laundry equipment
- ☐ Unit kitchens
- ☐ Vacuum equipment

☐ *Safes*

☐ *Saunas*

☐ *School Equipment*
- ☐ Arts and crafts
- ☐ Audiovisual
- ☐ Language labs
- ☐ Vocational/shop equipment
- ☐ Wall benches
- ☐ Wall tables

☐ *Service Station Equipment*

☐ *Shop Equipment*

☐ *Shooting Ranges*

☐ *Stage Equipment*

☐ *Steam Baths*

☐ *Swimming Pool Equipment*
- ☐ Diving board
- ☐ Diving platform

- ☐ Filtration systems
- ☐ Heaters
- ☐ Ladders
- ☐ Lifeguard chair
- ☐ Lights
- ☐ Pool covers
- ☐ Slides

☐ **Unit Kitchens**

☐ **Vacuum Systems**

☐ **Waste Disposal**
- ☐ Compactors
- ☐ Incinerators

☐ **Special Equipment**

Figure 11-1   Installation Time for Maintenance Equipment

Description	Crew Makeup	Daily Output	Labor-Hours	Unit
VACUUM CLEANING Central, 3 inlet, residential	1 Skilled Worker	.90	8.890	Total
Commercial		.70	11.430	Total
5 inlet system, residential		.50	16.000	Total
7 inlet system		.40	20.000	Total
9 inlet system	↓	.30	26.670	Total

Figure 11-2   Installation Time for Teller and Bank Service Equipment

Description	Crew Makeup	Daily Output	Labor-Hours	Unit
BANK EQUIPMENT Alarm system, police	2 Electricians	1.60	10.000	Ea.
With vault alarm	"	.40	40.000	Ea.
Bullet-resistant teller window, 44" x 60"	1 Glazier	.60	13.330	Ea.
48" x 60"	"	.60	13.330	Ea.
Counters for banks, frontal only	2 Carpenters	1	16.000	Station
Complete with steel undercounter	"	.50	32.000	Station
Door and frame, bullet-resistant, with vision panel				
Minimum	2 Struc. Steel Workers	1.10	14.550	Ea.
Maximum		1.10	14.550	Ea.
Drive-up window, drawer & mike, not incl. glass				
Minimum		1	16.000	Ea.
Maximum		.50	32.000	Ea.
Night depository, with chest, minimum		1	16.000	Ea.
Maximum		.50	32.000	Ea.
Package receiver, painted		3.20	5.000	Ea.
Stainless steel	↓	3.20	5.000	Ea.
Partitions, bullet-resistant, 1-3/16" glass, 8' high	2 Carpenters	10	1.600	L.F.
Acrylic	"	10	1.600	L.F.
Pneumatic tube systems, 2 lane drive-up, complete	1 Carpenter .5 Electrician .5 Sheet Metal Worker	.25	64.000	Total
With T.V. viewer	"	.20	80.000	Total
Safety deposit boxes, minimum	1 Struc. Steel Worker	44	.182	Opng.
Maximum, 10" x 15" opening		19	.421	Opng.
Teller locker, average	↓	15	.533	Opng.
Pass thru, bullet-resistant window, painted steel				
24" x 36"	2 Struc. Steel Workers	1.60	10.000	Ea.
48" x 48"		1.20	13.330	Ea.
72" x 40"	↓	.80	20.000	Ea.
Surveillance system, 16mm film camera, complete	2 Electricians	1	16.000	Ea.
Surveillance system, video camera, complete	"	1	16.000	Ea.
Twenty-four hour teller, single unit,				
automated deposit, cash and memo	1 Carpenter .5 Electrician .5 Sheet Metal Worker	.25	64.000	Ea.

Figure 11-3 Installation Time for Religious Equipment

Description	Crew Makeup	Daily Output	Labor-Hours	Unit
CHURCH EQUIPMENT Altar, wood, custom design, plain	1 Carpenter	1.40	5.710	Ea.
Deluxe	"	.20	40.000	Ea.
Granite or marble, average	2 Marble Setters	.50	32.000	Ea.
Deluxe	"	.20	80.000	Ea.
Arks, prefabricated, plain	2 Carpenters	.80	20.000	Ea.
Deluxe, maximum	"	.20	80.000	Ea.
Baptistry, fiberglass, 3'-6" deep, x 13'-7" long,				
steps at both ends, incl. plumbing, minimum	2 Carpenters .5 Plumber	1	20.000	Ea.
Maximum	"	.70	28.570	Ea.
Confessional, wood, prefabricated, single, plain	1 Carpenter	.60	13.330	Ea.
Deluxe		.40	20.000	Ea.
Double, plain		.40	20.000	Ea.
Deluxe		.20	40.000	Ea.
Lecterns, wood, plain		5	1.600	Ea.
Deluxe		2	4.000	Ea.
Pews, bench type, hardwood, minimum	↓	20	.400	L.F.
Church equipment, pews, bench type, hardwood				
Maximum	1 Carpenter	15	.533	L.F.
Pulpits, hardwood, prefabricated, plain		2	4.000	Ea.
Deluxe		1.60	5.000	Ea.
Railing, hardwood, average		25	.320	L.F.
Seating, individual, oak, contour, laminated		21	.381	Person
Cushion seat		21	.381	Person
Fully upholstered		21	.381	Person
Combination, self-rising	↓	21	.381	Person
Steeples, translucent fiberglass, 30" square				
15' high	4 Carpenters 1 Equip. Oper. (crane) 1 Hyd. Crane, 12 Ton Power Tools	2	20.000	Ea.
25' high		1.80	22.220	Ea.
Painted fiberglass, 24" square, 14' high		2	20.000	Ea.
28' high		1.80	22.220	Ea.
Porcelain enamel steeples, custom, 40' high		.50	80.000	Ea.
60' high	↓	.30	133.000	Ea.
Wall cross, aluminum, extruded, 2" x 2" section	1 Carpenter	34	.235	L.F.
4" x 4" section		29	.276	L.F.
Bronze, extruded, 1" x 2" section		31	.258	L.F.
2-1/2" x 2-1/2" section		34	.235	L.F.
Solid bar stock, 1/2" x 3" section		29	.276	L.F.
Fiberglass, stock		34	.235	L.F.
Stainless steel, 4" deep, channel section		29	.276	L.F.
4" deep box section	↓	29	.276	L.F.

Division 11

Figure 11-4 Installation Time for Library Equipment

Description	Crew Makeup	Daily Output	Labor-Hours	Unit
LIBRARY EQUIPMENT Bookshelf, mtl., 90" high, 10" shelf				
Single face	1 Carpenter	12	.667	L.F.
Double face		12	.667	L.F.
Carrels, hardwood, 36" x 24", minimum		5	1.600	Ea.
Maximum		4	2.000	Ea.
Charging desk, built-in, with counter,				
plastic laminated top	↓	7	1.140	L.F.

Figure 11-5 Installation Time for Theater/Stage Equipment

Description	Crew Makeup	Daily Output	Labor-Hours	Unit
STAGE EQUIPMENT Control boards				
with dimmers & breakers, minimum	1 Electrician	1	8.000	Ea.
Average	↓	.50	16.000	Ea.
Maximum		.20	40.000	Ea.
Curtain track, straight, light duty	2 Carpenters	20	.800	L.F.
Heavy duty		18	.889	L.F.
Curved sections		12	1.330	L.F.
Curtains, velour, medium weight		600	.027	S.F.
Silica based yarn, fireproof	↓	50	.320	S.F.
Lights, border, quartz, reflector, vented,				
colored or white	1 Electrician	20	.400	L.F.
Spotlight, follow spot, with transformer, 2100 watt		4	2.000	Ea.
Stationary spot, fresnel quartz, 6" lens		4	2.000	Ea.
8" lens		4	2.000	Ea.
Ellipsoidal quartz, 1000W, 6" lens		4	2.000	Ea.
12" lens		4	2.000	Ea.
Strobe light, 1 to 15 flashes per second, quartz		3	2.670	Ea.
Color wheel, portable, five hole, motorized	↓	4	2.000	Ea.
Telescoping platforms, extruded alum., straight				
Minimum	4 Carpenters	157	.204	S.F. Stg.
Maximum		77	.416	S.F. Stg.
Pie-shaped, minimum		150	.213	S.F. Stg.
Maximum		70	.457	S.F. Stg.
Band risers, steel frame, plywood deck, minimum		275	.116	S.F. Stg.
Maximum	↓	138	.232	S.F. Stg.
Chairs for above, self-storing, minimum	2 Carpenters	43	.372	Ea.
Maximum	"	40	.400	Ea.
Rule of thumb: total stage equipment, minimum	4 Carpenters	100	.320	S.F. Stg.
Maximum	"	25	1.280	S.F. Stg.
MOVIE EQUIPMENT				
Lamphouses, incl. rectifiers, xenon, 1000 watt	1 Electrician	2	4.000	Ea.
1600 watt		2	4.000	Ea.
2000 watt		1.50	5.330	Ea.
4000 watt	↓	1.50	5.330	Ea.
Projection screens, rigid, in wall, acrylic, 1/4" thick	2 Glaziers	195	.082	S.F.
1/2" thick	"	130	.123	S.F.
Electric operated, heavy duty, 400 S.F.	2 Carpenters	1	16.000	Ea.
Sound systems, incl. amplifier, single system				
Minimum	1 Electrician	.90	8.890	Ea.
Dolby/Super Sound, maximum		.40	20.000	Ea.
Dual system, minimum		.70	11.430	Ea.
Dolby/Super Sound, maximum		.40	20.000	Ea.
Speakers, recessed behind screen, minimum	↓	2	4.000	Ea.
Maximum		1	8.000	Ea.
Seating, painted steel, upholstered, minimum	2 Carpenters	35	.457	Ea.
Maximum	"	28	.571	Ea.

Figure 11-6   Installation Time for Cash Register/Checkout

Description	Crew Makeup	Daily Output	Labor-Hours	Unit
CHECKOUT COUNTER Supermarket conveyor				
Single belt	2 Building Laborers	10	1.600	Ea.
Double belt		9	1.780	Ea.
Power take-away		7	2.290	Ea.
Warehouse or bulk type	↓	6	2.670	Ea.

Figure 11-7   Installation Time for Display Cases

Description	Crew Makeup	Daily Output	Labor-Hours	Unit
REFRIGERATED FOOD CASES Dairy, multi-deck				
12′ long	1 Steamfitter	3	5.330	Ea.
	1 Steamfitter Apprentice			
Delicatessen case, service deli, 12′ long				
Single deck		3.90	4.100	Ea.
Multi-deck, 18 S.F. shelf display		3	5.330	Ea.
Freezer, self-contained, chest-type, 30 C.F.		3.90	4.100	Ea.
Glass door, upright, 78 C.F.		3.30	4.850	Ea.
Frozen food, chest type, 12′ long		3.30	4.850	Ea.
Glass door, reach-in, 5 door		3	5.330	Ea.
Island case, 12′ long, single deck		3.30	4.850	Ea.
Multi-deck		3	5.330	Ea.
Meat case, 12′ long, single deck		3.30	4.850	Ea.
Multi-deck		3.10	5.160	Ea.
Produce, 12′ long, single deck		3.30	4.850	Ea.
Multi-deck	↓	3.10	5.160	Ea.

Figure 11-8 Installation Time for Turnstiles

Description	Crew Makeup	Daily Output	Labor-Hours	Unit
TURNSTILES One way, 4 arm, 46" diameter, economy				
Manual	2 Carpenters	5	3.200	Ea.
Electric		1.20	13.330	Ea.
High security, galv., 5'-5" diameter, 7' high, manual		1	16.000	Ea.
Electric		.60	26.670	Ea.
Three arm, 24" opening, light duty, manual		2	8.000	Ea.
Heavy duty		1.50	10.670	Ea.
Manual, with registering & controls, light duty		2	8.000	Ea.
Heavy duty		1.50	10.670	Ea.
Electric, heavy duty	↓	1.10	14.550	Ea.
One way gate with horizontal bars, 5'-5" diameter				
7" high, recreation or transit type	2 Carpenters	.80	20.000	Ea.

Figure 11-9 Installation Time for Projection Screens

Description	Crew Makeup	Daily Output	Labor-Hours	Unit
PROJECTION SCREENS Wall or ceiling hung,				
glass beaded, manually operated, economy	2 Carpenters	500	.032	S.F.
Intermediate		450	.036	S.F.
Deluxe		400	.040	S.F.
Electric operated, glass beaded, 25 S.F., economy		5	3.200	Ea.
Deluxe		4	4.000	Ea.
50 S.F., economy		3	5.330	Ea.
Deluxe		2	8.000	Ea.
Heavy duty, electric operated, 200 S.F.		1.50	10.670	Ea.
400 S.F.	↓	1	16.000	Ea.
Rigid acrylic in wall, for rear projection, 1/4" thick	2 Glaziers	30	.533	S.F.
1/2" thick (maximum size 10' x 20')	"	25	.640	S.F.

Division 11

Figure 11-10 Installation Time for Service Station Equipment

Description	Crew Makeup	Daily Output	Labor-Hours	Unit
AUTOMOTIVE Compressors, electric, 1-1/2 H.P.,				
Standard controls	2 Skilled Workers	1.50	16.000	Ea.
Dual controls	1 Helper	1.50	16.000	Ea.
5 H.P., 115/230 volt, standard controls		1	24.000	Ea.
Dual controls		1	24.000	Ea.
Gasoline pumps, conventional, lighted, single		2.50	9.600	Ea.
Double		2	12.000	Ea.
Hoists, single post, 8000# capacity, swivel arms		.40	60.000	Ea.
Two posts, adjustable frames, 11,000# capacity		.25	96.000	Ea.
24,000# capacity		.15	160.000	Ea.
7500# capacity, frame supports		.50	48.000	Ea.
Four post, roll on ramp		.50	48.000	Ea.
Lube equipment, 3 reel type, with pumps,				
not including piping		.50	48.000	Set
Spray painting booth, 26' long, complete		.40	60.000	Ea.

Figure 11-11 Installation Time for Parking Control Equipment

Description	Crew Makeup	Daily Output	Labor-Hours	Unit
PARKING EQUIPMENT Traffic, detectors, magnetic	2 Electricians	2.70	5.930	Ea.
Single treadle		2.40	6.670	Ea.
Automatic gates, 8' arm, one way		1.10	14.550	Ea.
Two way		1.10	14.550	Ea.
Fee indicator, 1" display		4.10	3.900	Ea.
Ticket printer and dispenser, standard		1.40	11.430	Ea.
Rate computing		1.40	11.430	Ea.
Card control station, single period		4.10	3.900	Ea.
4 period		4.10	3.900	Ea.
Key station on pedestal		4.10	3.900	Ea.
Coin station, multiple coins		4.10	3.900	Ea.

Figure 11-12 Installation Time for Loading Dock Equipment

Description	Crew Makeup	Daily Output	Labor-Hours	Unit
LOADING DOCK Bumpers, rubber blocks 4-1/2" thick				
10" high, 14" long	1 Carpenter	26	.308	Ea.
24" long		22	.364	Ea.
36" long		17	.471	Ea.
12" high, 14" long		25	.320	Ea.
24" long		20	.400	Ea.
36" long		15	.533	Ea.
Rubber blocks 6" thick, 10" high, 14" long		22	.364	Ea.
24" long		18	.444	Ea.
36" long		13	.615	Ea.
20" high, 11" long		13	.615	Ea.
Extruded rubber bumpers, T section				
22" x 22" x 3" thick		41	.195	Ea.
Molded rubber bumpers, 24" x 12" x 3" thick		20	.400	Ea.
Welded installation of above bumpers	1 Welder Foreman	8	1.000	Ea.
	1 Gas Welding Machine			
For drilled anchors, add per anchor	1 Carpenter	36	.222	Ea.
Door seal for door perimeter				
12" x 12", vinyl covered	"	26	.308	L.F.
Levellers, hinged for trucks, 10 ton capacity, 6' x 8'	1 Welder Foreman	1.08	14.815	Ea.
	1 Welder			
7' x 8'	1 Gas Welding Machine	1.08	14.815	Ea.
Hydraulic, 10 ton capacity, 6' x 8'		1.08	14.815	Ea.
7' x 8'		1.08	14.815	Ea.
Lights for loading docks, single arm, 24" long	1 Electrician	3.80	2.110	Ea.
Double arm, 60" long	"	3.80	2.110	Ea.
Shelters, fabric, for truck or train, scissor arms				
Minimum	1 Carpenter	1	8.000	Ea.
Maximum	"	.50	16.000	Ea.
DOCK BUMPERS Bolts not incl. 2" x 6" to 4" x 8"				
Average	1 Carpenter	.30	26.670	M.B.F.
	Power Tools			

Figure 11-13 Installation Time for Waste Handling Equipment

Description	Crew Makeup	Daily Output	Labor-Hours	Unit
WASTE HANDLING Compactors, 115 volt, 250#/hour, Chute fed	2 Skilled Workers 1 Helper	1	24.000	Ea.
Hand fed		2.40	10.000	Ea.
Multi-bag, hand or chute fed, 230 volt, 600#/hour		1	24.000	Ea.
Containerized, hand fed, 2 to 6 C.Y. containers				
250#/hour		1	24.000	Ea.
For chute fed, add per floor		1	24.000	Ea.
Heavy duty industrial compactor, 0.5 C.Y. capacity		1	24.000	Ea.
1.0 C.Y. capacity		1	24.000	Ea.
2.5 C.Y. capacity		.50	48.000	Ea.
5.0 C.Y. capacity		.50	48.000	Ea.
Combination shredder/compactor (5000#/hour)		.50	48.000	Ea.
Crematory, not including building, 1 place	1 Plumber Foreman (ins.) 2 Plumbers 1 Plumber Apprentice	.20	160.000	Ea.
2 place	"	.10	320.000	Ea.
Incinerator, electric, 100#/hour, minimum	1 Labor Foreman (inside) 2 Building Laborers 1 Struc. Steel Worker .5 Electrician	.75	48.000	Ea.
Maximum		.70	51.430	Ea.
400#/hour, minimum		.60	60.000	Ea.
Maximum		.50	72.000	Ea.
100#/hour, minimum		.25	114.000	Ea.
Maximum		.20	180.000	Ea.
Gas, not incl. chimney, electric or pipe				
50#/hour, minimum	1 Plumber Foreman (ins.) 2 Plumbers 1 Plumber Apprentice	.80	40.000	Ea.
Maximum		.70	45.710	Ea.
200#/hour, minimum (batch type)		.60	53.330	Ea.
Maximum (with feeder)		.50	64.000	Ea.
400#/hour, minimum (batch type)		.30	107.000	Ea.
Maximum (with feeder)		.25	128.000	Ea.
800#/hour, with feeder, minimum		.20	160.000	Ea.
Maximum		.17	188.000	Ea.
1200#/hour, with feeder, minimum		.15	213.000	Ea.
Maximum		.11	291.000	Ea.
2000#/hour, with feeder, minimum		.10	320.000	Ea.
Maximum		.05	640.000	Ea.
For heat recovery system, add, minimum		.25	128.000	Ea.
Add, maximum		.11	291.000	Ea.
For automatic ash conveyer, add		.50	64.000	Ea.
Large municipal incinerators, incl. stack, minimum		.25	128.000	Ton/Day
Maximum		.10	320.000	Ton/Day

Figure 11-14   Installation Time for Food Service Equipment

Description	Crew Makeup	Daily Output	Labor-Hours	Unit
APPLIANCES Cooking range, 30" free standing, 1 oven				
Minimum	2 Building Laborers	10	1.600	Ea.
Maximum		4	4.000	Ea.
2 oven, minimum		10	1.600	Ea.
Maximum		4	4.000	Ea.
Built-in, 30" wide, 1 oven, minimum	2 Carpenters	4	4.000	Ea.
Maximum		2	8.000	Ea.
2 oven, minimum		4	4.000	Ea.
Maximum		2	8.000	Ea.
Free-standing, 21" wide range, 1 oven, minimum	2 Building Laborers	10	1.600	Ea.
Maximum	"	4	4.000	Ea.
Countertop cooktops, 4 burner, standard				
Minimum	1 Electrician	6	1.330	Ea.
Maximum		3	2.670	Ea.
As above, but with grille and griddle attachment				
Minimum		6	1.330	Ea.
Maximum		3	2.670	Ea.
Induction cooktop, 30" wide		3	2.670	Ea.
Microwave oven, minimum		4	2.000	Ea.
Maximum		2	4.000	Ea.
Combination range, refrigerator and sink, 30" wide				
Minimum	1 Electrician 1 Plumber	2	8.000	Ea.
Maximum		1	16.000	Ea.
60" wide, average		1.40	11.430	Ea.
72" wide, average		1.20	13.330	Ea.
Office model, 48" wide		2	8.000	Ea.
Refrigerator and sink only		2.40	6.670	Ea.
Combination range, refrigerator, sink, microwave oven and ice maker	1 Electrician 1 Plumber	.80	20.000	Ea.
Compactor, residential size, 4 to 1 compaction				
Minimum	1 Carpenter	5	1.600	Ea.
Maximum	"	3	2.670	Ea.
Deep freeze, 15 to 23 C.F., minimum	2 Building Laborers	10	1.600	Ea.
Maximum		5	3.200	Ea.
30 C.F., minimum		8	2.000	Ea.
Maximum		3	5.330	Ea.
Dishwasher, built-in, 2 cycles, minimum	1 Electrician 1 Plumber	4	4.000	Ea.
Maximum		2	8.000	Ea.
4 or more cycles, minimum		4	4.000	Ea.
Maximum		2	8.000	Ea.
Dryer, automatic, minimum	1 Carpenter 1 Helper	3	5.330	Ea.
Maximum	"	2	8.000	Ea.
Garbage disposer, sink type, minimum	1 Electrician 1 Plumber	5	3.200	Ea.
Maximum	"	3	5.330	Ea.

(continued on next page)

Figure 11-14 Installation Time for Food Service Equipment
(continued from previous page)

Description	Crew Makeup	Daily Output	Labor-Hours	Unit
Appliances, Heater, electric, built-in, 1250 watt				
Ceiling type, minimum	1 Electrician	4	2.000	Ea.
Maximum		3	2.670	Ea.
Wall type, minimum		4	2.000	Ea.
Maximum		3	2.670	Ea.
1500 watt wall type, with blower		4	2.000	Ea.
3000 watt	↓	3	2.670	Ea.
Hood for range, 2 speed, vented, 30" wide				
Minimum	1 Carpenter .5 Electrician .5 Sheet Metal Worker	5	3.200	Ea.
Maximum		3	5.330	Ea.
42" wide, minimum		5	3.200	Ea.
Maximum	↓	3	5.330	Ea.
Icemaker, automatic, 13#/day	1 Plumber	7	1.140	Ea.
51#/day	"	2	4.000	Ea.
Refrigerator, no frost, 10 to 12 C.F., minimum	2 Building Laborers	10	1.600	Ea.
Maximum		6	2.670	Ea.
14 to 16 C.F., minimum		9	1.780	Ea.
Maximum		5	3.200	Ea.
18 to 20 C.F., minimum		8	2.000	Ea.
Maximum		4	4.000	Ea.
21 to 29 C.F., minimum	↓	7	2.290	Ea.
Maximum		3	5.330	Ea.
Sump pump cellar drainer, 1/3 H.P., minimum	1 Plumber	3	2.670	Ea.
Maximum		2	4.000	Ea.
Washing machine, automatic, minimum		3	2.670	Ea.
Maximum	↓	1	8.000	Ea.
Water heater, electric, glass lined, 30 gallon				
Minimum	1 Electrician 1 Plumber	5	3.200	Ea.
Maximum		3	5.330	Ea.
80 gallon	↓	2	8.000	Ea.
Maximum		1	16.999	Ea.
Water heater, gas, glass lined, 30 gallon, minimum	2 Plumbers	5	3.200	Ea.
Maximum		3	5.330	Ea.
50 gallon, minimum		2.50	6.400	Ea.
Maximum	↓	1.50	10.670	Ea.
Water softener, automatic, to 30 grains/gallon		5	3.200	Ea.
To 75 grains/gallon	↓	4	4.000	Ea.
Vent kits for dryers	1 Carpenter	10	.800	Ea.
KITCHEN EQUIPMENT Bake oven, single deck	1 Plumber 1 Plumber Apprentice	8	2.000	Ea.
Double deck		7	2.290	Ea.
Triple deck	↓	6	2.670	Ea.
Electric convection, 40" x 45" x 57"	2 Carpenters 1 Building Laborer .5 Electrician	4	7.000	Ea.

(continued on next page)

Figure 11-14 Installation Time for Food Service Equipment
(continued from previous page)

Description	Crew Makeup	Daily Output	Labor-Hours	Unit
KITCHEN EQUIPMENT Broiler, without oven				
Standard	1 Plumber	8	2.000	Ea.
	1 Plumber Apprentice			
Infra-red	2 Carpenters	4	7.000	Ea.
	1 Building Laborer			
	.5 Electrician			
Cooler, reach-in, beverage, 6' long	1 Plumber	6	2.670	Ea.
	1 Plumber Apprentice			
Dishwasher, commercial, rack type				
10 to 12 racks/hour	1 Plumber	3.20	5.000	Ea.
	1 Plumber Apprentice			
Semi-automatic 38 to 50 racks/hour	"	1.30	12.310	Ea.
Automatic 190 to 230 racks/hour	2 Plumbers	.70	34.290	Ea.
	1 Plumber Apprentice			
235 to 275 racks/hour	↓	.50	48.000	Ea.
8750 to 12,500 dishes/hour		.20	120.000	Ea.
Fast food equipment, total package, minimum	6 Skilled Workers	.08	600.000	Ea.
Maximum	"	.07	686.000	Ea.
Food mixers, 20 quarts	2 Carpenters	7	4.000	Ea.
	1 Building Laborer			
	.5 Electrician			
60 quarts	"	5	5.600	Ea.
Freezers, reach-in, 44 C.F.	1 Plumber	4	4.000	Ea.
	1 Plumber Apprentice			
68 C.F.		3	5.330	Ea.
Fryer, with submerger, single		7	2.290	Ea.
Double		5	3.200	Ea.
Griddle, 3' long		7	2.290	Ea.
4' long		6	2.670	Ea.
Ice cube maker, 50#/day		6	2.670	Ea.
500#/day	↓	4	4.000	Ea.
Kettles, steam-jacketed, 20 gallons	2 Carpenters	7	4.000	Ea.
	1 Building Laborer			
	.5 Electrician			
60 gallons	"	6	4.670	Ea.
Range, restaurant type, 6 burners and				
1 oven, 36"	1 Plumber	7	2.290	Ea.
	1 Plumber Apprentice			
2 ovens, 60"		6	2.670	Ea.
Heavy duty, single 34" oven, open top		5	3.200	Ea.
Fry top		6	2.670	Ea.
Hood fire protection system, minimum		3	5.330	Ea.
Maximum		1	16.000	Ea.
Refrigerators, reach-in type, 44 C.F.		5	3.200	Ea.
With glass doors, 68 C.F.	↓	4	4.000	Ea.

(continued on next page)

Figure 11-14 Installation Time for Food Service Equipment
(continued from previous page)

Description	Crew Makeup	Daily Output	Labor-Hours	Unit
KITCHEN EQUIPMENT Steamer, electric, 27 KW	2 Carpenters 1 Building Laborer .5 Electrician	7	4.000	Ea.
Electric, 10 KW or gas 100,000	"	5	5.600	Ea.
Rule of thumb: Equipment cost based on kitchen work area				
Office buildings, minimum	2 Carpenters 1 Building Laborer .5 Electrician	77	.364	S.F.
Maximum		58	.483	S.F.
Public eating facilities, minimum		77	.364	S.F.
Maximum		46	.609	S.F.
Hospitals, minimum		58	.483	S.F.
Maximum	↓	39	.718	S.F.
WINE VAULT Redwood, air conditioned, walk-in type 6'-8" high, incl. racks, 2' x 4' for 156 bottles	2 Carpenters	2	8.000	Ea.
4' x 6' for 614 bottles		1.50	10.670	Ea.
6' x 12' for 1940 bottles	↓	1	16.000	Ea.

Figure 11-15 Installation Time for Disappearing Stairs

Description	Crew Makeup	Daily Output	Labor-Hours	Unit
DISAPPEARING STAIRWAY No trim included				
Custom grade, pine, 8'-6" ceiling, minimum	1 Carpenter	4	2.000	Ea.
Average		3.50	2.290	Ea.
Maximum		3	2.670	Ea.
Heavy duty, pivoted, 8'-6" ceiling		3	2.670	Ea.
16' ceiling		2	4.000	Ea.
Economy folding, pine, 8'-6" ceiling		4	2.000	Ea.
9'-6" ceiling	↓	4	2.000	Ea.
Fire escape, galvanized steel, 8' to 10'-4" ceiling	2 Carpenters	1	16.000	Ea.
10'-6" to 13'-6" ceiling		1	16.000	Ea.
Automatic electric, aluminum, floor to floor height				
8' to 9'		1	16.000	Ea.
11' to 12'		.90	17.780	Ea.
14' to 15'	↓	.70	22.860	Ea.

Figure 11-16 Installation Time for Athletic/Recreational Equipment

Description	Crew Makeup	Daily Output	Labor-Hours	Unit
HEALTH CLUB EQUIPMENT				
Circuit training apparatus, 12 machines, minimum	2 Building Laborers	1.25	12.800	Set
Average		1	16.000	Set
Maximum		.75	21.330	Set
Squat racks	↓	5	3.200	Ea.
SCHOOL EQUIPMENT				
Basketball backstops, wall mounted, 6' extended, fixed				
Minimum	1 Carpenter 1 Helper	1	16.000	Ea.
Maximum		1	16.000	Ea.
Swing up, minimum		1	16.000	Ea.
Maximum		1	16.000	Ea.
Portable, manual, heavy duty, hydraulic		1.90	8.420	Ea.
Ceiling suspended, stationary, minimum		.78	20.510	Ea.
Fold up, with accessories, maximum	↓	1	16.000	Ea.
For electrically operated, add	1 Electrician	1	8.000	Ea.
Benches, folding, in wall, 14' table, 2 benches	2 Skilled Workers 1 Helper	2	12.000	Set
Bleachers, telescoping, manual to 15 tier, minimum	1 Carpenter Foreman 3 Carpenters Power Tools	65	.492	Seat
Maximum		60	.533	Seat
16 to 20 tier, minimum		60	.533	Seat
Maximum		55	.582	Seat
21 to 30 tier, minimum		50	.640	Seat
Maximum	↓	40	.800	Seat
For integral power operation, add, minimum	2 Electricians	300	.053	Seat
Maximum	"	250	.064	Seat
Exercise equipment				
Chinning bar, adjustable, wall mounted	1 Carpenter	5	1.600	Ea.
Exercise ladder, 16' x 1'-7", suspended	1 Carpenter 1 Helper	3	5.330	Ea.
High bar, floor plate attached	1 Carpenter	4	2.000	Ea.
Parallel bars, adjustable		4	2.000	Ea.
Uneven parallel bars, adjustable	↓	4	2.000	Ea.
Wall mounted, adjustable	1 Carpenter 1 Helper	1.50	10.670	Set
Rope, ceiling mounted, 18' long	1 Carpenter	10	.800	Ea.
Side horse, vaulting		5	1.600	Ea.
Treadmill, motorized, deluxe, training type	↓	5	1.600	Ea.
Weight lifting multi-station, minimum	2 Building Laborers	1	16.000	Ea.
Maximum	"	.50	32.000	Ea.
Scoreboards, baseball, minimum	1 Electrician Foreman 1 Electrician .5 Equip. Oper. (crane) .5 S.P. Crane, 5 Ton	2.40	8.330	Ea.
Maximum		.05	400.000	Ea.
Football, minimum		1.20	16.670	Ea.
Maximum		.20	100.000	Ea.
Basketball (one side), minimum		2.40	8.330	Ea.
Maximum		.30	66.670	Ea.
Hockey-basketball (four sides), minimum		.25	80.000	Ea.
Maximum	↓	.15	133.000	Ea.

Division 11

Figure 11-17 Installation Time for Industrial Equipment

Description	Crew Makeup	Daily Output	Labor-Hours	Unit
EQUIPMENT INSTALLATION Industrial equipment				
Minimum	1 Struc. Steel Foreman 4 Struc. Steel Workers 1 Equip. Oper. (crane) 1 Equp. Oper. Oiler 1 Crane 90 Ton	12	4.670	Ton
Maximum	"	2	28,000	Ton

Figure 11-18 Installation Time for Specialized Equipment

Description	Crew Makeup	Daily Output	Labor-Hours	Unit
VOCATIONAL SHOP EQUIPMENT Benches, work,				
Wood, average	2 Carpenters	5	3.200	Ea.
Metal, average		5	3.200	Ea.
Combination belt & disc sander, 6"		4	4.000	Ea.
Drill press, floor mounted, 12", 1/2 H.P.		4	4.000	Ea.
Dust collector, not incl. ductwork, 6' diameter	1 Sheet Metal Worker	1.10	7.270	Ea.
Grinders, double wheel, 1/2 H.P.	2 Carpenters	5	3.200	Ea.
Jointer, 4", 3/4 H.P.		4	4.000	Ea.
Kilns, 16 C.F., to 2000°		4	4.000	Ea.
Lathe, woodworking, 10", 1/2 H.P.		4	4.000	Ea.
Planer, 13" x 6"		4	4.000	Ea.
Potter's wheel, motorized		4	4.000	Ea.
Saws, band, 14", 3/4 H.P.		4	4.000	Ea.
Metal cutting band saw, 14"		4	4.000	Ea.
Radial arm saw, 10", 2 H.P.		4	4.000	Ea.
Scroll saw, 24"		4	4.000	Ea.
Table saw, 10", 3 H.P.		4	4.000	Ea.
Welder AC arc, 30 amp capacity		4	4.000	Ea.

Figure 11-19 Installation Time for Laboratory Equipment

Description	Crew Makeup	Daily Output	Labor-Hours	Unit
LABORATORY EQUIPMENT Cabinets, base, door units,				
metal	2 Carpenters	18	.889	L.F.
Drawer units		18	.889	L.F.
Tall storage cabinets, open, 7' high		20	.800	L.F.
With glazed doors		20	.800	L.F.
Wall cabinets, metal, 12-1/2" deep, open		20	.800	L.F.
With doors		20	.800	L.F.
Counter tops, not incl. base cabinets, acidproof,				
Minimum		82	.195	S.F.
Maximum		70	.229	S.F.
Stainless steel		82	.195	S.F.
Fume hood, with contertop & base, not including HVAC				
Simple, minimum	2 Carpenters	5.40	2.960	L.F.
Complex, including fixtures		2.40	6.670	L.F.
Special, maximum		1.70	9.410	L.F.
Ductwork, minimum	2 Sheet Metal Workers	1	16.000	Hood
Maximum	"	.50	32.000	Hood
For sink assembly with hot and cold water, add	1 Plumber	1.40	5.710	Ea.
Glassware washer, distilled water rinse, minimum	1 Electrician 1 Plumber	1.80	8.890	Ea.
Maximum	"	1	16.000	Ea.
Sink, one piece plastic, flask wash, hose, free standing	1 Plumber	1.60	5.000	Ea.
Epoxy resin sink, 25" x 16" x 10"	"	2	4.000	Ea.
Utility table, acid resistant top with drawers	2 Carpenters	30	.533	L.F.

Division 11

Figure 11-20 Installation Time for Medical Equipment

Description	Crew Makeup	Daily Output	Labor-Hours	Unit
MEDICAL EQUIPMENT Autopsy table, standard	1 Plumber	1	8.000	Ea.
Deluxe		.60	13.330	Ea.
Distiller, water, steam heated, 50 gal. capacity		1.40	5.710	Ea.
Automatic washer/sterilizer		2	4.000	Ea.
Steam generators, electric 10 KW to 180 KW				
Minimum	1 Electrician	3	2.670	Ea.
Maximum	"	.70	11.430	Ea.
Surgery table, minor minimum	1 Struc. Steel Worker	.70	11.430	Ea.
Maximum		.50	16.000	Ea.
Major surgery table, minimum		.50	16.000	Ea.
Maximum		.50	16.000	Ea.
Surgical lights, single arm	2 Electricians	.90	17.780	Ea.
Dual arm	"	.30	53.330	Ea.
Tables, physical therapy, walk off, electric	2 Carpenters	3	5.330	Ea.
Standard, vinyl top with base cabinets, minimum		3	5.330	Ea.
Maximum		2	8.000	Ea.
Utensil washer-sanitizer	1 Plumber	2	4.000	Ea.

Figure 11-21   Installation Time for Dental Equipment

Description	Crew Makeup	Daily Output	Labor-Hours	Unit
DENTAL EQUIPMENT Central suction system, minimum	1 Plumber	1.20	6.670	Ea.
Maximum	"	.90	8.890	Ea.
Air compressor, minimum	1 Skilled Worker	.80	10.000	Ea.
Maximum		.50	16.000	Ea.
Chair, electric or hydraulic, minimum		.50	16.000	Ea.
Maximum		.25	32.000	Ea.
Drill console with accessories, minimum		.50	16.000	Ea.
Maximum		.33	24.240	Ea.
Light, floor or ceiling mounted, minimum		3.60	2.220	Ea.
Maximum		1.20	6.670	Ea.
X-ray unit, wall		1.90	4.210	Ea.
Panoramic unit		.60	13.330	Ea.
Developers, X-ray, minimum		1	8.000	Ea.
Maximum		1	8.000	Ea.

Division Twelve
Furnishings

Introduction

Division 12 – Furnishings generally refers to the items brought into the building for the use and/or comfort of the occupants. Items in this category include furniture, artwork, window treatments (blinds, curtains, etc.), mats, rugs, plants, and the like. In the majority of cases, these items are supplied by the owner or are arranged for outside the scope of most construction-related contracts.

green: Since building occupants come into direct contact with furnishings, there is huge opportunity to make a green statement. Manufacturers understand this, and are now producing a broad range of environmentally friendly furnishings. Although not often thought of as green, ergonomic design aims to improve safety, comfort, performance, and aesthetics and therefore improves the indoor environment quality of a building for its occupants. Proper use of blinds and shades is a low-cost green strategy. By limiting excessive solar radiation into a building, air-conditioning loads can be reduced. For those that are involved in furnishing office, commercial, and residential buildings, LEED® CI (contract interiors) is the most appropriate version to use.

Items included within Division 12 are usually counted individually, as most items are bought/sold in this fashion. The procedures used to estimate this division are typically as follows:

- Review all contract documents to determine what is (and what is not) included in the scope of responsibility.
- Review the specifications (if applicable), all drawings, sketches, owner-supplied lists, and any other relevant documentation for the items required, specific models, and/or acceptable manufacturers.
- All items should be listed, accounted for, and have costs attached.
- Some items in this division require a support system that is not usually furnished with the item. Examples include blocking for the attachment of casework and heavy drapery rods. The required blocking must be added to the estimate in the appropriate division.

Many of the items can be delivered in place or have installation costs already included in the price of the item. Others, however, may need some attention or at least cleaning prior to installation. Costs must be estimated for these items.

Estimating Data

The following tables show the expected installation time for various specialty items. Each table lists the specialty item, the typical installation crew mix, the expected average daily output of that crew (the number of units the crew can be expected to install in one eight-hour day), and the labor-hours required for the installation of one unit.

Table of Contents

Division 12

Estimating Tips

Faulty Assumptions

Do not assume that items covered in Division 12 will be purchased and installed outside of your contract, especially if any furnishings or typical arrangements are shown on the drawings. This could be a costly assumption. Check all drawings and specifications for these items, and if any are found, check for the terms "NIC" (Not in Contract) or "By Others." If these or similar terms are not included, it is safe to conclude that these items are in the scope of work.

Installation of Items

In many cases, Division 12 items are purchased by others, but the contractor is responsible for installing them. Check all drawings and specifications for these items. Include receiving, storage, installation, and mechanical and electrical hookups as needed.

Handling Charges

In cases where Division 12 items are purchased by others but are to be installed by the contractor, many contractors add a handling charge. (Ten percent of the estimated material cost is common practice.) This charge covers the receiving, handling, storage, protection, and final delivery of these items.

Preparation of Items

In some cases, Division 12 items may require assembly before installation. Assembly time can often exceed the time required for installation.

Cleaning Time

When installing materials purchased by others, be sure to allow for cleaning time. Invariably, these items will need some cleaning, and this responsibility will fall on the installer.

Checklist ✓

For an estimate to be reliable, all items must be accounted for. A complete estimate can also limit contingencies. The following checklist can be used to help ensure that all items are included.

☐ *Artwork*

☐ *Interior Landscaping*
  ☐ Guarantee

☐ *Casework*

☐ *Furniture*
  ☐ Detention furniture
  ☐ Office furniture
  ☐ Hospital furniture
  ☐ Institutional furniture
  ☐ Residential furniture
  ☐ Retail furniture

☐ *Rugs & Mats*

☐ *Seating*
  ☐ Auditorium
  ☐ Booths and tables
  ☐ Classroom
  ☐ Stadium
  ☐ Theatre

☐ *Systems Furniture*

☐ *Window Treatments*
  ☐ Blinds
  ☐ Shades

Division 12

Figure 12-1   Installation Time for Artwork

Description	Crew Makeup	Daily Output	Labor-Hours	Unit
ARTWORK Framed				
Photography, minimum	1 Carpenter	36	.222	Ea.
Maximum		30	.267	Ea.
Posters, minimum		36	.222	Ea.
Maximum		30	.267	Ea.
Reproductions, minimum		36	.222	Ea.
Maximum	↓	30	.267	Ea.

Figure 12-2   Installation Time for Hospital Casework

Description	Crew Makeup	Daily Output	Labor-Hours	Unit
CABINETS				
Hospital, base cabinets, laminated plastic	2 Carpenters	10	1.600	L.F.
Enameled steel		10	1.600	L.F.
Stainless steel		10	1.600	L.F.
Cabinet base trim, 4" high, enameled steel		200	.080	L.F.
Stainless steel		200	.080	L.F.
Countertop, laminated plastic, no backsplash		40	.400	L.F.
With backsplash		40	.400	L.F.
For sink cutout, add		12.20	1.310	Ea.
Stainless steel counter top		40	.400	L.F.
Nurses station, door type, laminated plastic		10	1.600	L.F.
Enameled steel		10	1.600	L.F.
Stainless steel		10	1.600	L.F.
Wall cabinets, laminated plastic		15	1.070	L.F.
Enameled steel		15	1.070	L.F.
Stainless steel		15	1.070	L.F.
Kitchen, base cabinets, metal, minimum		30	.533	L.F.
Maximum		25	.640	L.F.
Wall cabinets, metal, minimum		30	.533	L.F.
Maximum		25	.640	L.F.
School, 24" deep		15	1.070	L.F.
Counter height units		20	.800	L.F.
Wood, custom fabricated, 32" high counter		20	.800	L.F.
Add for countertop		56	.286	L.F.
84" high wall units	↓	15	1.070	L.F.

Figure 12-3 Installation Time for Display Casework

Description	Crew Makeup	Daily Output	Labor-Hours	Unit
DISPLAY CASES Free standing, all glass				
Aluminum frame, 42" high x 36" x 12" deep	2 Carpenters	8	2.000	Ea.
70" high x 48" x 18" deep	"	6	2.670	Ea.
Wall mounted, glass front, aluminum frame				
Non-illuminated, one section 3' x 4' x 1'-4"	2 Carpenters	5	3.200	Ea.
5' x 4' x 1'-4"		5	3.200	Ea.
6' x 4' x 1'-4"		4	4.000	Ea.
Two sections, 8' x 4' x 1'-4"		2	8.000	Ea.
10' x 4' x 1'-4"		2	8.000	Ea.
Three sections, 16' x 4' x 1'-4"		1.50	10.670	Ea.
Table exhibit cases, 2' wide, 3' high, 4' long, flat top		5	3.200	Ea.
3' wide, 3' high, 4' long, sloping top		3	5.330	Ea.

Division 12

Figure 12-4 Installation Time for Window Treatments – Blinds

Description	Crew Makeup	Daily Output	Labor-Hours	Unit
BLINDS, INTERIOR Solid colors				
Horizontal, 1" aluminum slats, custom, minimum	1 Carpenter	590	.014	S.F.
Maximum		440	.018	S.F.
2" aluminum slats, custom, minimum		590	.014	S.F.
Maximum		440	.018	S.F.
Stock, minimum		590	.014	S.F.
Maximim		440	.018	S.F.
2" steel slats, stock, minimum		590	.014	S.F.
Maximum		440	.018	S.F.
Custom, minimum		590	.014	S.F.
Maximum		400	.020	S.F.
Alternate method of figuring:				
1" aluminum slats, 48" wide, 48" high	1 Carpenter	30	.267	Ea.
72" high		29	.276	Ea.
96" high		28	.286	Ea.
72" wide, 72" high		25	.320	Ea.
96" high		23	.348	Ea.
96" wide, 96" high		20	.400	Ea.
Vertical, 3" to 5" PVC or cloth strips, minimum		460	.017	S.F.
Maximum		400	.020	S.F.
4" aluminum slats, minimum		460	.017	S.F.
Maximum		400	.020	S.F.
Mylar mirror-finish strips, to 8" wide, minimum		460	.017	S.F.
Maximum		400	.020	S.F.
Alternate method of figuring:				
2" aluminum slats, 48" wide, 48" high	1 Carpenter	30	.267	Ea.
72" high		29	.276	Ea.
96" high		28	.286	Ea.
72" wide, 72" high		25	.320	Ea.
96" high		23	.348	Ea.
96" wide, 96" high		20	.400	Ea.
Mirror finish, 48" wide, 48" high		30	.267	Ea.
72" high		29	.276	Ea.
96" high		28	.286	Ea.
72" wide, 72" high		25	.320	Ea.
96" high		23	.348	Ea.
96" wide, 96" high		20	.400	Ea.
Decorative printed finish, 48" wide, 48" high		30	.267	Ea.
72" high		29	.276	Ea.
96" high		28	.286	Ea.
72" wide, 72" high		25	.320	Ea.
96" high		23	.348	Ea.
96" wide, 96" high		20	.400	Ea.

(continued on next page)

Figure 12-4 Installation Time for Window Treatments – Blinds
(continued from previous page)

Description	Crew Makeup	Daily Output	Labor-Hours	Unit
BLINDS, INTERIOR Wood folding panels with moveable louvers				
7" x 20" each	1 Carpenter	17	.471	Pr.
8" x 28" each		17	.471	Pr.
9" x 36" each		17	.471	Pr.
10" x 40" each		17	.471	Pr.
Fixed louver type, stock units, 8" x 20" each		17	.471	Pr.
10" x 28" each		17	.471	Pr.
12" x 36" each		17	.471	Pr.
18" x 40" each		17	.471	Pr.
Insert panel type, stock, 7" x 20" each		17	.471	Pr.
8" x 28" each		17	.471	Pr.
9" x 36" each		17	.471	Pr.
10" x 40" each		17	.471	Pr.
Raised panel type, stock, 10" x 24" each		17	.471	Pr.
12" x 26" each		17	.471	Pr.
14" x 30" each		17	.471	Pr.
16" x 36" each		17	.471	Pr.

Figure 12-5 Installation Time for Window Treatments – Shades

Description	Crew Makeup	Daily Output	Labor-Hours	Unit
SHADES Basswood, roll-up, stain finish, 3/8" slats	1 Carpenter	300	.027	S.F.
7/8" slats		300	.027	S.F.
Vertical side slide, stain finish, 3/8" slats		300	.027	S.F.
7/8" slats		300	.027	S.F.
Mylar, single layer, non-heat reflective		685	.012	S.F.
Double layered, heat reflective		685	.012	S.F.
Triple layered, heat reflective		685	.012	S.F.
Vinyl coated cotton, standard		685	.012	S.F.
Lightproof decorator shades		685	.012	S.F.
Vinyl, lightweight, 4 gauge		685	.012	S.F.
Heavyweight, 6 gauge		685	.012	S.F.
Vinyl laminated fiberglass, 6 gauge, translucent		685	.012	S.F.
Lightproof		685	.012	S.F.
Woven aluminum, 3/8" thick, lightproof and fireproof		350	.023	S.F.
Insulative shades		125	.064	S.F.
Solar screening, fiberglass		85	.094	S.F.
Interior insulative shutter, stock unit, 15" x 60"		17	.471	Pr.

Division 12

Figure 12-6 Installation Time for Window Treatments –
Draperies and Curtains

Description	Crew Makeup	Daily Output	Labor-Hours	Unit
DRAPERIES				
Drapery installation, hardware and drapes,				
Labor cost only, minimum	1 Building Laborer	75	.107	L.F.
Maximum	"	20	.400	L.F.

Figure 12-7 Installation Time for Window Treatments –
Drape/Curtain Hardware

Description	Crew Makeup	Daily Output	Labor-Hours	Unit
DRAPERY HARDWARE				
Standard traverse, per foot, minimum	1 Carpenter	59	.136	L.F.
Maximum		51	.157	L.F.
Decorative traverse, 28" to 48", minimum		22	.364	Ea.
Maximum		21	.381	Ea.
48" to 84", minimum		20	.400	Ea.
Maximum		19	.421	Ea.
66" to 120", minimum		18	.444	Ea.
Maximum		17	.471	Ea.
84" to 156", minimum		16	.500	Ea.
Maximum		15	.533	Ea.
130" to 240", minimum		14	.571	Ea.
Maximum		13	.615	Ea.
Ripplefold, snap-a-pleat system, 3' or less				
Minimum		15	.533	Ea.
Maximum		14	.571	Ea.
Traverse rods, adjustable, 28" to 48"		22	.364	Ea.
48" to 84"		20	.400	Ea.
66" to 120"		18	.444	Ea.
84" to 156"		16	.500	Ea.
120" to 220"		14	.571	Ea.
228" to 312"		13	.615	Ea.

Figure 12-8 Installation Time for Furniture

Description	Crew Makeup	Daily Output	Labor-Hours	Unit
DORMITORY				
Desk top, built-in, laminated plastic, 24" deep				
Minimum	2 Carpenters	50	.320	L.F.
Maximum		40	.400	L.F.
30" deep, minimum		50	.320	L.F.
Maximum		40	.400	L.F.
Dressing unit, built-in, minimum		12	1.330	L.F.
Maximum		8	2.000	L.F.
LIBRARY				
Attendant desk, 36" x 62" x 29" high	1 Carpenter	16	.500	Ea.
Book display, "A" frame display, both sides		16	.500	Ea.
Table with bulletin board		16	.500	Ea.
Book trucks, descending platform,				
Small, 14" x 30" x 35" high	1 Carpenter	16	.500	Ea.
Large, 14" x 40" x 42" high		16	.500	Ea.
Card catalogue, 30 tray unit		16	.500	Ea.
60 tray unit		16	.500	Ea.
72 tray unit	2 Carpenters	16	1.000	Ea.
120 tray unit	"	16	1.000	Ea.
Carrels, single face, initial unit	1 Carpenter	16	.500	Ea.
Additional unit	"	16	.500	Ea.
Double face, initial unit	2 Carpenters	16	1.000	Ea.
Additional unit		16	1.000	Ea.
Cloverleaf		11	1.450	Ea.
Chairs, sled base, arms, minimum	1 Carpenter	24	.333	Ea.
Maximum		16	.500	Ea.
No arms, minimum		24	.333	Ea.
Maximum		16	.500	Ea.
Standard leg base, arms, minimum		24	.333	Ea.
Maximum		16	.500	Ea.
No arms, minimum		24	.333	Ea.
Maximum		16	.500	Ea.
Charge desk, modular unit, 35" x 27" x 39" high				
Wood front and edges, plastic laminate tops				
Book return	1 Carpenter	16	.500	Ea.
Book truck port		16	.500	Ea.
Card file drawer, 5 drawers		16	.500	Ea.
10 drawers		16	.500	Ea.
15 drawers		16	.500	Ea.
Card and legal file		16	.500	Ea.
Charging machine		16	.500	Ea.
Corner		16	.500	Ea.
Cupboard		16	.500	Ea.
Detachable end panel		16	.500	Ea.
Gate		16	.500	Ea.
Knee space		16	.500	Ea.
Open storage		16	.500	Ea.
Station charge		16	.500	Ea.
Work station		16	.500	Ea.
Dictionary stand, stationary		16	.500	Ea.
Revolving		16	.500	Ea.

Division 12

(continued on next page)

Figure 12-8 Installation Time for Furniture *(continued from previous page)*

Description	Crew Makeup	Daily Output	Labor-Hours	Unit
LIBRARY (cont.)				
Exhibit case, table style, 60" x 28" x 36"	1 Carpenter	11	.727	Ea.
Globe stand		16	.500	Ea.
Magazine rack		16	.500	Ea.
Newspaper rack		16	.500	Ea.
Tables, card catalog reference, 24" x 60" x 42"		16	.500	Ea.
24" x 60" x 72"		16	.500	Ea.
Index, single tier, 48" x 72"		16	.500	Ea.
Double tier, 48" x 72"		16	.500	Ea.
Study, panel ends, plastic lam. surfaces 29" high				
36" x 60"	2 Carpenters	16	1.000	Ea.
36" x 72"		16	1.000	Ea.
36" x 90"		16	1.000	Ea.
48" x 72"		16	1.000	Ea.
Parsons, 29" high, plastic laminate top,				
wood legs and edges				
36" x 36"	2 Carpenters	16	1.000	Ea.
36" x 60"		16	1.000	Ea.
36" x 72"		16	1.000	Ea.
36" x 84"		16	1.000	Ea.
42" x 90"		16	1.000	Ea.
48" x 72"		16	1.000	Ea.
48" x 120"		16	1.000	Ea.
Round, leg or pedestal base, 36" diameter		16	1.000	Ea.
42" diameter		16	1.000	Ea.
48" diameter		16	1.000	Ea.
60" diameter		16	1.000	Ea.
RESTAURANT Bars, built-in, front bar	1 Carpenter	5	1.600	L.F.
Back bar	"	5	1.600	L.F.

Figure 12-9 Installation Time for Furniture Accessories

Description	Crew Makeup	Daily Output	Labor-Hours	Unit
ASH/TRASH RECEIVERS				
Ash urn, cylindrical metal				
8" diameter, 20" high	1 Building Laborer	60	.133	Ea.
8" diameter, 25" high		60	.133	Ea.
10" diameter, 26" high		60	.133	Ea.
12" diameter, 30" high	↓	60	.133	Ea.
Combination ash/trash urn, metal				
8" diameter, 20" high	1 Building Laborer	60	.133	Ea.
8" diameter, 25" high		60	.133	Ea.
10" diameter, 26" high		60	.133	Ea.
12" diameter, 30" high	↓	60	.133	Ea.
Trash receptacle, metal				
8" diameter, 15" high	1 Building Laborer	60	.133	Ea.
10" diameter, 18" high		60	.133	Ea.
16" diameter, 16" high		60	.133	Ea.
18" diameter, 32" high	↓	60	.133	Ea.
Trash receptacle, plastic, fire resistant				
Rectangular 11" x 8" x 12" high	1 Building Laborer	60	.133	Ea.
16" x 8" x 14" high	"	60	.133	Ea.
Trash receptacle, plastic, with lid				
35 gallon	1 Building Laborer	60	.133	Ea.
45 gallon	"	60	.133	Ea.

Figure 12-10 Installation Time for Miscellaneous Seating

Description	Crew Makeup	Daily Output	Labor-Hours	Unit
SEATING				
Lecture hall, pedestal type, minimum	2 Carpenters	35	.727	Ea.
Maximum		20	1.103	Ea.
Auditorium chair, all veneer construction		35	.727	Ea.
Veneer back, padded seat		35	.727	Ea.
Fully upholstered, spring seat	↓	35	.727	Ea.

Division 12

Figure 12-11 Installation Time for Booths and Tables

Description	Crew Makeup	Daily Output	Labor-Hours	Unit
BOOTHS				
Banquet, upholstered seat and back, custom				
Straight, minimum	2 Carpenters	40	.400	L.F.
Maximum		36	.444	L.F.
"L" or "U" shape, minimum		35	.457	L.F.
Maximum	↓	30	.533	L.F.
Upholstered outside finished backs for				
single booths and custom banquets				
Minimum	2 Carpenters	44	.364	L.F.
Maximum	"	40	.400	L.F.
Fixed seating, one piece plastic chair and				
plastic laminate table top				
Two seat, 24" x 24" table, minimum	2 Carpenters 2 Building Laborers Power Tools	30	1.070	Ea.
Maximum		26	1.230	Ea.
Four seat, 24" x 48" table, minimum		28	1.140	Ea.
Maximum		24	1.330	Ea.
Six seat, 24" x 76" table, minimum		26	1.230	Ea.
Maximum		22	1.450	Ea.
Eight seat, 24" x 102" table, minimum		20	1.600	Ea.
Maximum	↓	18	1.780	Ea.
Free standing, wood fiber core with				
plastic laminate face, single booth				
24" wide	2 Carpenters	38	.421	Ea.
48" wide		34	.471	Ea.
60" wide		30	.533	Ea.
Double booth, 24" wide		32	.500	Ea.
48" wide		28	.571	Ea.
60" wide	↓	26	.615	Ea.
Upholstered seat and back				
Foursome, single booth, minimum	2 Carpenters	38	.421	Ea.
Maximum		30	.533	Ea.
Double booth, minimum		32	.500	Ea.
Maximum	↓	26	.615	Ea.
Mount in floor, wood fiber core with				
plastic laminate face, single booth				
24" wide	2 Carpenters 2 Building Laborers Power Tools	30	1.070	Ea.
48" wide		28	1.140	Ea.
60" wide		26	1.230	Ea.
Double booth, 24" wide		26	1.230	Ea.
48" wide		24	1.330	Ea.
60" wide	↓	22	1.450	Ea.

Figure 12-12 Installation Time for Millwork

Description	Crew Makeup	Daily-Output	Labor-Hours	Unit
Cabinets, kitchen base, 24" x 24" x 35" high	2 Carpenters	22.30	.717	Ea.
Kitchen wall, 12" x 24" x 30" high		20.30	.788	"
Casework frames				
Base cabinets, 36" high, two bay, 36" wide	1 Carpenter	2.20	3.636	Ea.
Book cases, 7' high, two bay, 36" wide		1.60	5.000	
Coat racks, 7' high, two bay, 48" wide		2.75	2.909	
Wall mounted cabinets, 30" high, two bay, 36" wide		2.15	3.721	
Wardrobe, 7' high, 48" wide		1.70	4.706	
Cabinet doors				
Glass panel, hardware frame, 18" wide, 30" high	1 Carpenter	29.00	.276	Ea.
Hardwood, raised panel, 18" wide, 30" high		14.00	.571	"

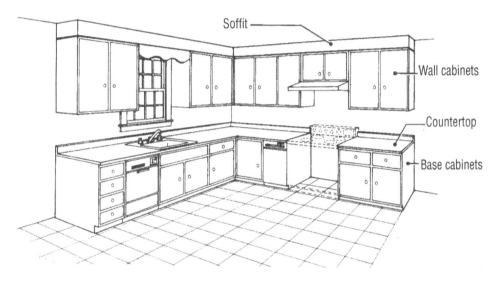

Kitchen Cabinets

Notes

Division Thirteen

Special Construction

Introduction

Division 13 – Special Construction deals with specialty subsystems. Items usually associated with this division can range from air-supported and pre-engineered structures to cold storage rooms.

The procedure for estimating this division starts with reading and reviewing the specifications. Usually if a special construction item is required, there will be some text about it in the specifications. The plans, as well as any other related construction document, should be thoroughly studied when special construction items are involved; they can contain important information.

Usually the items included in this division are purchased as a whole, complete, installed assembly, rather than installed in parts by one or more subcontractors. In many cases special construction systems are bought and installed by the owner. In some cases, the owner may purchase the unit and will have the contractor be responsible for the installation. In this case, the labor must be added to the overall estimate.

Greenhouse additions are used in passive solar applications. Greenhouse construction is ideal for letting the solar radiation into a building. Specially designed shades and blinds that cover the entire greenhouse area are used to control the amount of sun entering the structure.

Estimating Data

The following tables show the expected installation time for various specialty items. Each table lists the specialty item, the typical installation crew mix, the expected average daily output of that crew (the number of units the crew can be expected to install in one eight-hour day), and the labor-hours required for the installation of one unit.

Table of Contents

Division 13

Estimating Tips

Covering All Bases

If you take outside quotes as an aid to estimating Division 13, review the scope of work covered in those quotes. It may be that the outside agent has wrongfully assumed that you will provide traditionally supplied items such as excavation or an unloading crane. Check to make sure that it is covered somewhere in your estimate, and that it is not carried by both parties.

Interfaces

Review all documentation to ensure that all interfaces, such as electrical connections and control wiring, are accounted for. These items have a habit of falling through the cracks of an estimate.

Checklist ✓

For an estimate to be reliable, all items must be accounted for. A complete estimate can also limit contingencies. The following checklist can be used to help ensure that all items are included.

☐ *Access floors*

☐ *Acoustical chambers*
    ☐ Enclosures
    ☐ Panels

☐ *Air curtains*

☐ *Air-supported structures*

☐ *Bowling alleys*

☐ *Broadcast studio*

☐ *Chimneys*

☐ *Clean rooms*

☐ *Comfort stations*

☐ *Darkrooms*

☐ *Domes*

☐ *Garages*

☐ *Garden house, shed*

☐ *Grandstand*

☐ *Greenhouses*

☐ *Hangars*

☐ *Hyperbaric rooms*

☐ *Incinerators*

☐ *Insulated rooms*
    ☐ Coolers
    ☐ Freezers

☐ *Jacuzzis*

☐ *Music practice rooms*

☐ *Pre-engineered structures*

☐ *Pre-fabricated structures*

☐ *Radiation Protection*

Division 13

Figure 13-1 Installation Time for Air-Supported Structures

Description	Makeup	Daily Output	Labor-Hours	Unit
AIR-SUPPORTED STRUCTURES				
Site preparation, incl. anchor placement and utilities	1 Equipment Oper. (med.) 1 Building Laborer 1 Dozer, 200 H.P. 1 Air Powered Tamper 1 Air Compr. 365 C.F.M. 2-50' Air Hoses, 1-1/2" Dia.	1,000	.016	S.F. Flr.
Warehouse, polyester/vinyl fabric, 24 oz. over 10 yr. life, welded seams, tension cables, primary & auxiliary inflation system airlock, personnel doors and liner				
5000 S.F.	4 Building Laborers	5,000	.006	S.F. Flr.
12,000 S.F.	"	6,000	.005	S.F. Flr.
24,000 S.F.	8 Building Laborers	12,000	.005	S.F. Flr.
50,000 S.F.	"	12,500	.005	S.F. Flr.
12 oz. reinforced vinyl fabric, 5 yr. life, sewn seams, accordian door				
including liner, 3000 S.F.	4 Building Laborers	3,000	.011	S.F. Flr.
12,000 S.F.	"	6,000	.005	S.F. Flr.
24,000 S.F.	8 Building Laborers	12,000	.005	S.F. Flr.
Tedlar/vinyl fabric, 17 oz., with liner, over 10 yr. life, incl. overhead and				
personnel doors, 3000 S.F.	4 Building Laborers	3,000	.011	S.F. Flr.
12,000 S.F.	"	6,000	.005	S.F. Flr.
24,000 S.F.	8 Building Laborers	12,000	.005	S.F. Flr.
Greenhouse/shelter, woven polyethylene with liner, 2 yr. life, sewn seams, including doors				
3000 S.F.	4 Building Laborers	3,000	.011	S.F. Flr.
12,000 S.F.	"	6,000	.005	S.F. Flr.
24,000 S.F.	8 Building Laborers	12,000	.005	S.F. Flr.
Tennis/gymnasium, polyester/vinyl fabric 24 oz., over 10 yr. life, including thermal liner, heat and lights				
7200 S.F.	4 Building Laborers	6,000	.005	S.F. Flr.
13,000 S.F.	"	6,500	.005	S.F. Flr.
Over 24,000 S.F.	8 Building Laborers	12,000	.005	S.F. Flr.
Stadium/convention center, teflon coated fiberglass, heavy weight, over 20 yr. life incl. thermal liner and heating system				
Minimum	9 Building Laborers	26,000	.003	S.F. Flr.
Maximum	"	19,000	.004	S.F. Flr.
Doors, air lock, 15' long, 10' x 10'	2 Carpenters	.80	20.000	Ea.
15' x 15'		.50	32.000	Ea.
Revolving personnel door, 6' diameter, 6'- 6" high		.80	20.000	Ea.

(continued on next page)

Figure 13-1 Installation Time for Air-Supported Structures
(continued from previous page)

Description	Makeup	Daily Output	Labor-Hours	Unit

AIR-SUPPORTED STORAGE TANK COVERS
  Vinyl polyester scrim, double layer
    with hardware, blower, standby & controls
      Round, 75' diameter
	1 Labor Foreman (outside)
	4 Building Laborers
		4,500	.009	S.F.

100' diameter		5,000	.008	S.F.
150' diameter		5,000	.008	S.F.
Rectangular, 20' x 20'		4,500	.009	S.F.
30' x 40'		4,500	.009	S.F.
50' x 60'		4,500	.009	S.F.

Figure 13-2 Installation Time for Special Purpose Rooms

Description	Crew Makeup	Daily Output	Labor-Hours	Unit
DARKROOMS Shell, complete except for door, 64 S.F. 8' high	2 Carpenters	128	.125	S.F. Flr.
12' high		64	.250	S.F. Flr.
120 S.F., floor, 8' high		120	.133	S.F. Flr.
12' high		60	.267	S.F. Flr.
240 S.F. floor, 8' high		120	.133	S.F. Flr.
12' high		60	.267	S.F. Flr.
Mini-cylindrical, revolving, unlined, 4' diameter		3.50	4.570	Ea.
5'-6" diameter		2.50	6.400	Ea.

Division 13

Figure 13-3　Installation Time for Athletic Rooms

Description	Crew Makeup	Daily Output	Labor-Hours	Unit
SPORT COURT				
Rule of thumb for components:				
Walls	3 Carpenters	.15	160.000	Court
Floor	"	.25	96.000	Court
Lighting	2 Electricians	.60	26.670	Court
Handball, racquetball court in existing building				
Minimum	3 Carpenters	.20	160.000	Court
	1 Building Laborer			
	Power Tools			
Maximum	"	.10	320.000	Court
Rule of thumb for components: walls	3 Carpenters	.12	200.000	Court
Floor		.25	96.000	Court
Ceiling	↓	.33	72.730	Court
Lighting	2 Electricians	.60	26.670	Court

Figure 13-4　Installation Time for Audiometric Rooms

Description	Crew Makeup	Daily Output	Labor-Hours	Unit
AUDIOMETRIC ROOMS Under 500 S.F. surface area	4 Carpenters	98	.327	S.F. Surf.
Over 500 S.F. surface area	"	120	.267	S.F. Surf.

Figure 13-5 Installation Time for Cold Storage Rooms

Description	Crew Makeup	Daily Output	Labor-Hours	Unit
REFRIGERATORS Curbs, 12" high, 4" thick, concrete	2 Carpenters	58	.276	L.F.
Finishes, 2 coat Portland cement plaster, 1/2" thick	1 Plasterer	48	.167	S.F.
For galvanized reinforcing mesh, add	1 Lather	335	.024	S.F.
3/16" thick latex cement	1 Plasterer	88	.091	S.F.
For glass cloth reinforced ceilings, add	"	450	.018	S.F.
Fiberglass panels, 1/8" thick	1 Carpenter	149.45	.054	S.F.
Polystyrene, plastic finish ceiling, 1" thick		274	.029	S.F.
2" thick		274	.029	S.F.
4" thick		219	.037	S.F.
Floors, concrete, 4" thick	1 Cement Finisher	93	.086	S.F.
6" thick	"	85	.094	S.F.
Partitions, galv. sandwich panels, 4" thick, stock	2 Carpenters	219.20	.073	S.F.
Aluminum or fiberglass, 4" thick		219.20	.073	S.F.
Prefab walk-in, 7'-6" high, aluminum, incl. door & floors, not incl. partitions or refrigeration				
6' x 6' O.D. nominal		54.80	.292	S.F. Flr.
10' x 10' O.D. nominal		82.20	.195	S.F. Flr.
12' x 14' O.D. nominal		109.60	.146	S.F. Flr.
12' x 20' O.D. nominal		109.60	.146	S.F. Flr.
Rule of thumb for complete units, not incl. doors				
Cooler		146	.110	S.F. Flr.
Freezer		109.60	.146	S.F. Flr.
Shelving, plated or galvanized, steel wire type		360	.044	S.F. Hor.
Slat shelf type		375	.043	S.F. Hor.
Vapor barrier, on wood walls		1,644	.010	S.F.
On masonry walls		1,315	.012	S.F.

Figure 13-6 Installation Time for Saunas

Description	Crew Makeup	Daily Output	Labor-Hours	Unit
SAUNA Prefabricated, incl. heater & controls, 7' high				
6' x 4'	2 Carpenters 1 Building Laborer .5 Electrician	2.20	12.730	Ea.
6' x 5'		2	14.000	Ea.
6' x 6'		1.80	15.560	Ea.
6' x 9'		1.60	17.500	Ea.
8' x 12'		1.10	25.450	Ea.
8' x 8'		1.40	20.000	Ea.
8' x 10'		1.20	23.330	Ea.
10' x 12'		1	28.000	Ea.
Door only, with tempered insulated glass window	2 Carpenters	3.40	4.710	Ea.
Prehung, incl. jambs, pulls & hardware	"	12	1.330	Ea.

Figure 13-7 Installation Time for Steam Baths

Description	Crew Makeup	Daily Output	Labor-Hours	Unit
STEAM BATH Heater, timer & head, single, to 140 C.F.	1 Plumber	1.20	6.670	Ea.
To 300 C.F.		1.10	7.270	Ea.
Commercial size, to 800 C.F.		.90	8.890	Ea.
To 2500 C.F.		.80	10.000	Ea.
Multiple baths, motels, apartment, 2 baths	1 Plumber 1 Plumber Apprentice	1.30	12.310	Ea.
4 baths	"	.70	22.860	Ea.

Figure 13-8 Installation Time for Acoustical Enclosures

Description	Crew Makeup	Daily Output	Labor-Hours	Unit
ACOUSTICAL Enclosure, 4" thick wall and ceiling panels				
8# per S.F., up to 12' span	3 Carpenters	72	.333	S.F. Surf.
Better quality panels, 10.5# per S.F.		64	.375	S.F. Surf.
Reverb-chamber, 4" thick, parallel walls		60	.400	S.F. Surf.
Skewed walls, parallel roof, 4" thick panels		55	.436	S.F. Surf.
Skewed walls, skewed roof, 4" layers, 4" air space		48	.500	S.F. Surf.
Sound-absorbing panels, painted metal, 2'-6" x 8'				
Under 1000 S.F.		215	.112	S.F. Surf.
Over 2400 S.F.		240	.100	S.F. Surf.
Fabric faced		240	.100	S.F. Surf.
Flexible transparent curtain, clear	3 Sheet Metal Workers	215	.112	S.F. Surf.
50% foam		215	.112	S.F. Surf.
75% foam		215	.112	S.F. Surf.
100% foam		215	.112	S.F. Surf.
Audio masking system, including speakers				
amplification and signal generator				
Ceiling mounted, 5000 S.F.	2 Electricians	2,400	.007	S.F.
10,000 S.F.		2,800	.006	S.F.
Plenum mounted, 5000 S.F.		3,800	.004	S.F.
10,000 S.F.		4,400	.004	S.F.
MUSIC Practice room, modular, perforated steel				
Under 500 S.F.	2 Carpenters	70	.229	S.F. Surf.
Over 500 S.F.	"	80	.200	S.F. Surf.

Figure 13-9 Installation Time for Radiation Protection

Description	Crew Makeup	Daily Output	Labor-Hours	Unit
SHIELDING LEAD				
Lead lined door frame, not incl.				
hardware, 1/16" thick	1 Lather	2.40	3.330	Ea.
Lead sheets, 1/16" thick lead, prepped for hardware	2 Lathers	135	.119	S.F.
1/8" thick		120	.133	S.F.
Lead glass, 1/4" thick, 12" x 16", 2.0 mm LE		13	1.230	Ea.
24" x 36"		8	2.000	Ea.
36" x 60"		2	8.000	Ea.
Frame with 1/16" lead and voice passage				
36" x 60"		2	8.000	Ea.
24" x 36" frame		8	2.000	Ea.
Lead gypsum board, 5/8" thick with 1/16" lead		160	.100	S.F.
1/8" lead		140	.114	S.F.
1/32" lead		200	.080	S.F.
Butt joints in 1/8" lead or thicker, lead strip, add		240	.067	S.F.
X-ray protection, average radiography or				
fluoroscopy room, up to 300 S.F. floor, 1/16" lead				
Minimum	2 Lathers	.25	64.000	Total
Maximum, 7' walls	"	.15	107.000	Total
Deep therapy X-ray room, 250 KV capacity,				
up to 300 S.F. floor, 1/4" lead, minimum	2 Lathers	.08	200.000	Total
Maximum, 7' walls	"	.06	267.000	Total
SHIELDING, RADIO FREQUENCY				
Prefabricated or screen-type copper or steel				
Minimum	2 Carpenters	180	.089	S.F. Surf.
Average		155	.103	S.F. Surf.
Maximum		145	.110	S.F. Surf.

Division 13

Figure 13-10 Installation Time for Pre-Engineered Structures

Description	Crew Makeup	Daily Output	Labor-Hours	Unit
DOMES Revolving aluminum, electric drive, for astronomy observation, shell only,				
Stock units, 10' diameter, 800#, dome	2 Carpenters	.25	64.000	Ea.
Base		.67	23.880	Ea.
18' diameter, 2500#, dome		.17	94.120	Ea.
Base		.33	48.480	Ea.
24' diameter, 4500#, dome		.08	200.000	Ea.
Base		.25	64.000	Ea.
Bulk storage, shell only, dual radius hemispher. arch, steel framing, corrugated steel covering, 150' diameter	1 Struc. Steel Foreman 4 Struc. Steel Workers 1 Equip. Oper. (crane) 1 Equip. Oper. Oiler 1 Crane, 90 Ton	550	.102	S.F. Flr.
400' diameter	"	720	.078	S.F. Flr.
Wood framing, wood decking, to 400' diameter	4 Carpenters 1 Equip. Oper. (crane) 1 Equip. Oper. Oiler 1 Hyd. Crane, 55 Ton Power Tools	400	.120	S.F. Flr.
Radial framed wood (2" x 6"), 1/2" thick plywood, asphalt shingles 50' diameter	4 Carpenters 1 Equip. Oper. (crane) 1 Hyd. Crane, 12 Ton Power Tools	2,000	.020	S.F. Flr.
60' diameter		1,900	.021	S.F. Flr.
72' diameter		1,800	.022	S.F. Flr.
116' diameter		1,730	.023	S.F. Flr.
150' diameter		1,500	.027	S.F. Flr.

(continued on next page)

Figure 13-10 Installation Time for Pre-Engineered Structures
(continued from previous page)

Description	Crew Makeup	Daily Output	Labor-Hours	Unit
PRE-ENGINEERED STEEL BUILDINGS Clear span rigid frame, 26 ga. colored roofing and siding 20' wide, 10 ' eave height	1 Struc. Steel Foreman 4 Struc. Steel Workers 1 Equip. Oper. (crane) 1 Equip. Oiler 1 Crane, 90 Ton		.132	S.F. Flr.
14' eave height			.160	
16' eave height			.175	
20' eave height			.204	
24' eave height			.233	
30' to 40' wide, 10' eave height			.105	
14' eave height			.124	
16' eave height			.135	
20' eave height			.156	
24' eave height			.175	
50' to 100' wide, 10' eave height			.065	
14' eave height			.073	
16' eave height			.077	
20' eave height			.085	
24' eave height			.093	
Clear span tapered frame, 26 ga. colored roof & siding				
30' wide, 10' eave height			.105	
14' eave height			.124	
16' eave height			.135	
20' eave height			.156	
40' wide, 10' eave height			.093	
14' eave height			.110	
16' eave height			.118	
20' eave height			.135	
50' to 80' wide, 10' eave height			.073	
14' eave height			.083	
16' eave height			.088	
20' eave height	▼		.099	▼

(continued on next page)

Division 13

Figure 13-10 Installation Time for Pre-Engineered Structures
(continued from previous page)

Description	Crew Makeup	Daily Output	Labor-Hours	Unit
GARAGES Residential, prefab shell, stock, wood, single car				
Minimum	2 Carpenters Power Tools	1	16.000	Total
Maximum		.67	23.880	Total
Two car, minimum		.67	23.880	Total
Maximum	↓	.50	32.000	Total
SILOS Concrete stave industrial, not incl. foundations, conical or sloping bottoms				
12' diameter, 35' high	3 Bricklayers 2 Bricklayer Helpers	.11	363.000	Ea.
16' diameter, 45' high		.08	500.000	Ea.
25' diameter, 75' high	↓	.05	800.000	Ea.
Steel, factory fab., 30,000 gallon capacity, painted, minimum	1 Struc. Steel Foreman 5 Struc. Steel Workers 1 Equip. Oper. (crane) 1 Hyd. Crane, 25 Ton	1	56.000	Ea.
Maximum	↓	.50	112.000	Ea.
Epoxy lined, minimum		1	56.000	Ea.
Maximum	↓	.50	112.000	Ea.
TENSION STRUCTURES Rigid steel frame, vinyl coated polyester fabric shell, 72' clear span, not incl. foundations or floors				
4800 S.F.	1 Labor Foreman (outside) 4 Building Laborers .25 Equip. Oper. (crane) .25 Equip. Oper. Oiler .25 Crawler Crane, 40 Ton	1,000	.044	S.F. Flr.
12,000 S.F.		1,100	.040	S.F. Flr.
20,600 S.F.	↓	1,220	.036	S.F. Flr.
124' clear span, 11,000 S.F.	1 Struc. Steel Forman 5 Struc. Steel Workers 1 Equip. Oper. (crane) 1 Hyd. Crane, 25 Ton	2,175	.026	S.F. Flr.
25,750 S.F.		2,300	.024	S.F. Flr.
36,900 S.F.	↓	2,500	.022	S.F. Flr.
For roll-up door, 12' x 14' add	1 Carpenter 1 Helper	1	16.000	Ea.

Figure 13-11   Installation Time for Pre-Engineered Buildings

Description	Crew Makeup	Daily Output	Labor-Hours	Unit
HANGARS Prefabricated steel T hangars Galv. steel roof & walls, incl. electric bi-folding doors, 4 or more units, not including floors or foundations, minimum	1 Struc. Steel Foreman 4 Struc. Steel Workers 1 Equip. Oper. (crane) 1 Equip. Oper. Oiler 1 Crane, 90 Ton	1,275	.044	S.F. Flr.
Maximum		1,063	.053	S.F. Flr.
With bottom rolling doors, minimum		1,386	.040	S.F. Flr.
Maximum		966	.058	S.F. Flr.
Alternate pricing method: Galv. roof and walls, electric bi-folding doors, minimum	1 Struc. Steel Foreman 4 Struc. Steel Workers 1 Equip. Oper. (crane) 1 Equip. Oper. Oiler 1 Crane, 90 Ton	1.06	52.830	Plane
Maximum		.91	61.540	Plane
With bottom rolling doors, minimum		1.25	44.800	Plane
Maximum		.97	57.730	Plane
Circular type, prefab., steel frame, plastic skin, electric door, including foundations, 80′ diameter, for up to 5 light planes Minimum	1 Struc. Steel Foreman 4 Struc. Steel Workers 1 Equip. Oper. (crane) 1 Equip. Oper. Oiler 1 Crane, 90 Ton	.50	112.000	Total
Maximum	"	.25	224.000	Total

(continued on next page)

Division 13

Figure 13-11 Installation Time for Pre-Engineered Buildings
(continued from previous page)

Description	Labor-Hours	Unit
Pre-Engineered Building Shell		
Above Foundation Average	.116	S.F. floor
Eave Overhang 4' Wide with Soffit	.224	L.F.
End Wall Overhang 4' Wide with Soffit	.112	L.F.
Door 3' x 7'	3.200	opng
Door Framing Only 10' x 10'	2.667	opng
Sash		
3' x 3'	1.714	opng
4' x 3'	1.846	opng
6' x 4'	2.000	opng
Gutter	.050	L.F.
Skylight, Fiberglass Panel to 30 S.F.	2.400	Ea.
Roof Vents, Circular 20" Diameter	2.667	Ea.
Continuous 12" Wide 10' Long	4.000	Ea.

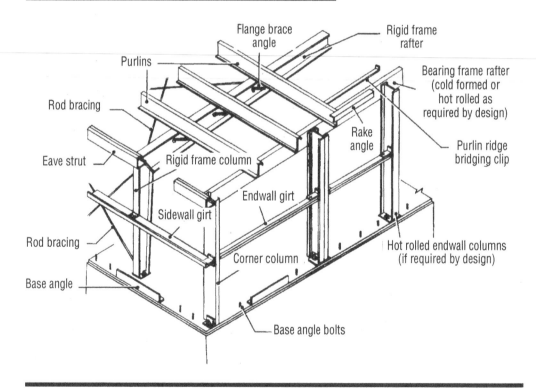

Figure 13-12 Installation Time for Metal Building Systems

Description	Crew Makeup	Daily Output	Labor-Hours	Unit
SHELTERS Aluminum frame, acrylic glazing				
3' x 9' x 8' high	2 Struc. Steel Workers	1.14	14.03	Ea.
9' x 12' x 8' high	"	.73	21.9	Ea.

Figure 13-13 Installation Time for Greenhouses

Description	Crew Makeup	Daily Output	Labor-Hours	Unit
GREENHOUSE Shell only, stock units				
not incl. 2' stud walls, foundation, floors				
heat or compartments				
Residential type, free standing, 8'-6" long				
7'-6" wide	2 Carpenters	59	.271	S.F. Flr.
10'-6" wide		85	.188	S.F. Flr.
13'-6" wide		108	.148	S.F. Flr.
17'-0" wide		160	.100	S.F. Flr.
Lean-to type, 3'-10" wide		34	.471	S.F. Flr.
6'-10" wide		58	.276	S.F. Flr.
Geodesic hemisphere, 1/8" plexiglass glazing				
8' diameter	2 Carpenters	2	8.000	Ea.
24' diameter		.35	45.710	Ea.
48' diameter		.20	80.000	Ea.

Figure 13-14 Installation Time for Portable Buildings

Description	Crew Makeup	Daily Output	Labor-Hours	Unit
COMFORT STATIONS Prefabricated, stock,				
with doors, windows & fixtures				
Not incl. interior finish or electrical				
Permanent, including concrete slab				
Minimum	1 Equip. Oper. (crane) 1 Equip. Oper. Oiler 1 Gradall, 3 Ton, .5 C.Y.	50	.320	S.F.
Maximum	"	43	.372	S.F.
GARDEN HOUSE Prefabricated wood, no floors or				
foundations, 48 to 200 S.F.				
Minimum	2 Carpenters	200	.080	S.F. Flr.
Maximum	"	48	.333	S.F. Flr.

Division 13

Figure 13-15   Installation Time for Swimming Pools

Description	Crew Makeup	Daily Output	Labor-Hours	Unit
SWIMMING POOL ENCLOSURE Translucent,				
free standing, not incl. foundations, heat or light				
Economy, minimum	2 Carpenters	200	.080	S.F. Hor.
Maximum		100	.160	S.F. Hor.
Deluxe, minimum		100	.160	S.F. Hor.
Maximum		70	.229	S.F. Hor.
SWIMMING POOL EQUIPMENT Diving stand,				
stainless steel, 3 meter		.40	40.000	Ea.
1 meter		2.70	5.930	Ea.
Diving boards, 16′ long, aluminum		2.70	5.930	Ea.
Fiberglass	↓	2.70	5.930	Ea.
Gutter system, stainless steel, with grating, stock,				
contains supply and drainage system	1 Welder Foreman	20	1.200	L.F.
	1 Welder			
	1 Equip. Oper. (light)			
	1 Gas Welding Mach.			
	″			
Integral gutter and 5′ high wall system,				
stainless steel		10	2.400	L.F.
Ladders, heavy duty, stainless steel, 2 tread	2 Carpenters	7	2.290	Ea.
4 tread		6	2.670	Ea.
Lifeguard chair, stainless steel, fixed	↓	2.70	5.930	Ea.
Lights, underwater, 12 volts, with transformer				
300 watt	1 Electrician	1	8.000	Ea.
110 volt, 500 watt, standard		1	8.000	Ea.
Low water cutoff type	↓	1	8.000	Ea.
Slides, fiberglass, aluminum handrails & ladder				
5′-0″, straight	2 Carpenters	1.60	10.000	Ea.
8′-0″, curved		3	5.330	Ea.
10′-0″, curved		1	16.000	Ea.
12′-0″, straight with platform	↓	1.20	13.330	Ea.
Hydraulic lift, movable pool bottom, single ram				
Under 1000 S.F. area	1 Labor Foreman (ins.)	.03	1200.000	Ea.
	2 Building Laborers			
	1 Struc. Steel Worker			
	.5 Electrician			
Four ram lift, over 1000 S.F.	1 Carpenter, 1 Helper	.02	1800.000	Ea.
Removable access ramp, stainless steel	2 Building Laborers	2	8.000	Ea.
Removable stairs, stainless steel, collapsible	″	2	8.000	Ea.

Figure 13-16 Installation Time for Ice Rinks

Description	Crew Makeup	Daily Output	Labor-Hours	Unit
ICE SKATING Dasher boards, polyethylene coated plywood 3' acrylic screen at sides, 5' acrylic ends, 85' x 200'	1 Carpenter Foreman 3 Carpenters Power Tools	.06	533.000	Ea.
Fiberglass & aluminum construction, same sides and ends	"	.06	533.000	Ea.
Subsoil heating system (recycled from compressor) 85' x 200'	1 Steamfitter Foreman (ins.) 2 Steamfitters 1 Steamfitter Apprentice	.27	118.000	Ea.
Subsoil insulation, 2 lb. polystyrene with vapor barrier 85' x 200'	2 Carpenters	.14	114.000	Ea.

Figure 13-17 Installation Time for Audio Masking

Description	Labor-Hours	Unit
Acoustical enclosure, 4" thick wall and ceiling		
8 lb. per S.F. up to 12" span	.333	S.F. Surf
Better quality panels, 10.5 lb. per S.F.	.375	S.F. Surf
Reverb-chamber, 4" thick panels	.400	S.F. Surf
Skewed wall, parallel roof, 4" thick panels	.436	S.F. Surf
Skewed wall, skewed roof, 4" layers, 4" air space	.500	S.F. Surf
Sound-absorbing panels, pntd mtl, 2'-6" x 8",		
under 1,000 S.F.	.112	S.F. Surf
Over 1,000 S.F.	.100	S.F. Surf
Fabric-faced	.100	S.F. Surf
Flexible transparent curtain, clear	.112	S.F. Surf
50% foam	.112	S.F. Surf
75% foam	.112	S.F. Surf
100% foam	.112	S.F. Surf
Audio masking system, incl speakers, amp, & signal generator		
Ceiling mounted, 5,000 S.F.	.007	S.F.
10,000 S.F.	.006	S.F.
Plenum mounted, 5,000 S.F.	.004	S.F.
10,000 S.F.	.004	S.F.

Division 13

Notes

Division Fourteen

Conveying Systems

Introduction

Division 14 – Conveying Systems is concerned with inter/intra-building transportation devices, such as elevators, handicap lifts, escalators, dumbwaiters, correspondence lifts, pneumatic tube systems, and moving walkways and ramps.

Passenger Elevators

Electric elevators are the most common, but hydraulic elevators can also be used for lifts up to 70' and where large capacities are required. Hydraulic speeds are limited to 200 FPM, but cars are self-leveling at the stops. On low rises, hydraulic installation runs about 15% less than standard electric elevators, but on higher rises this installation cost advantage is reduced. Maintenance of hydraulic elevators is about the same as for electric, but the underground portion is not included in the maintenance contract.

In standard electric elevators, there are two basic control systems: rheostatic systems for speeds up to 150 FPM and variable voltage systems for speeds over 150 FPM. The two types of drives are geared for low speeds and gearless for 450 FPM and over. As a rule of thumb, each added 100 FPM adds about 20% to the total cost.

Freight Elevators

Freight elevator capacities run from 1,500 lbs. to over 100,000 lbs., with 3,000- to 10,000-pound capacities the most common. Travel speeds are generally slower and control systems less intricate than on passenger elevators.

Escalators, Moving Stairs

Escalators are often used in buildings where 600 or more people will be traveling to the second floor or beyond on a daily basis. Freight cannot be carried on escalators; therefore, at least one elevator must be available for this function. The carrying capacity of an escalator is 5,000 to 8,000 people per hour. The power requirement is 2 to 3 kW per hour, and the required incline angle is 30°.

 To save electricity escalators can now be programmed to shut off during non-peak hours, and turn on rapidly when needed.

Estimating Data

The following tables and illustrations provide selection and installation data for conveying systems. Also included is a worksheet that can be used to help size and price elevators, and organize data for the estimate.

Table of Contents

Division 14

Estimating Tips

Dumbwaiters and Elevators
When developing an estimate for dumbwaiters and elevators, remember that some items needed by the installers may have to be included as part of the general contract. Examples are shaftways, rail support brackets, machine rooms, electrical supply and connections, sill angles, pits and ladders, and roof penthouses. Check the job specifications and drawings before pricing.

Elevator Doors
When preparing an estimate for elevators, check to make sure that doors have been included, not only for the elevator itself, but for each floor where the elevator stops.

Hydraulic Piston
When figuring the cost for a hydraulic elevator, make sure that the excavation for the elevator piston is included in the estimate. For each floor above grade that the elevator is to travel, there should be an equal length of piston below grade.

Fire Stops
When estimating any vertical conveying system, make sure all openings for doors, dumbwaiter access panels, etc., are fire-rated per the applicable code. In case of a fire emergency, their shaftways will otherwise act as a conduit for smoke and heat.

Support Systems
Many products in Division 14 will require some type of support or blocking for installation not included with the item itself. Examples are supports for tube systems, attachment points for lifts, and footings for hoists or cranes. Be sure to account for these supports.

Tie-in
Make sure that either the conveying system estimate, the electrical estimate, or mechanical estimate includes the tie-in to the rest of the electrical and control systems. This is another item that people often assume someone else has covered.

Historic Structures
Installation of elevators and handicapped lifts in historic structures can require significant additional costs. The associated structural requirements may involve cutting into and repairing finishes, moldings, flooring, etc.

Help
When in doubt about price, what is standard on elevators, material selection, availability, etc., do not hesitate to call on any of the elevator manufacturing companies directly. The competition for your business can be fierce, and they will try to assist you in any way they can.

Checklist ✓

For an estimate to be reliable, all items must be accounted for. A complete estimate can also limit contingencies. The following checklist can be used to help ensure that all items are included.

☐ **Conveyor Systems**
- ☐ Automatic

☐ **Manual Correspondence Lifts**

☐ **Dumbwaiters**
- ☐ Capacity
- ☐ Floors
- ☐ Size
- ☐ Speed
- ☐ Stops
- ☐ Finish

☐ **Elevators**
- ☐ Hydraulic
- ☐ Electric
- ☐ Capacity
- ☐ Floors
- ☐ Stops
- ☐ Finish
- ☐ Door type
- ☐ Geared
- ☐ Gearless
- ☐ Size
- ☐ Number required
- ☐ Speed
- ☐ Machinery location
- ☐ Signals
- ☐ Special requirements

☐ **Escalators**
- ☐ Capacity
- ☐ Floors
- ☐ Story height
- ☐ Finish
- ☐ Incline angle
- ☐ Size
- ☐ Number required
- ☐ Speed
- ☐ Machinery location
- ☐ Special requirements

☐ **Hoists and Cranes**
- ☐ Capacity
- ☐ Floors
- ☐ Story height
- ☐ Finish
- ☐ Incline angle
- ☐ Size
- ☐ Number required
- ☐ Speed
- ☐ Machinery location
- ☐ Special requirements

☐ **Lifts**

☐ **Wheelchair Lifts**

☐ **Platform Lifts**

Division 14

Figure 14-1   Elevator Nomenclature and Comparison of Elevator Types

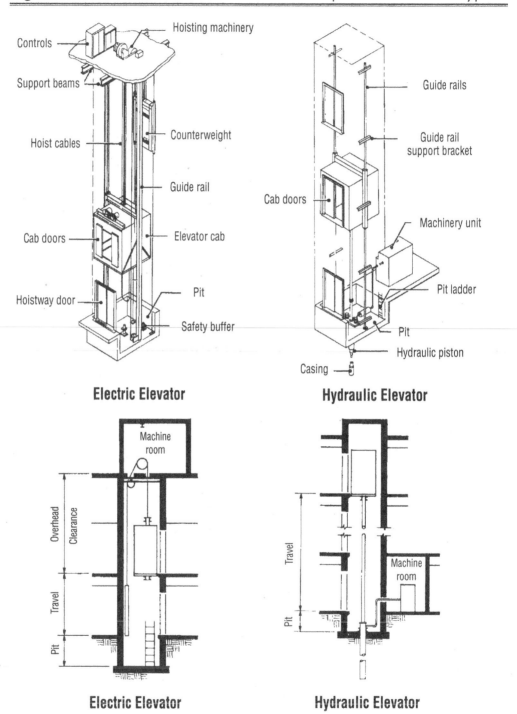

Electric Elevator   **Hydraulic Elevator**

Electric Elevator   **Hydraulic Elevator**

Figure 14-2 Cab Size Comparisons for Various Types of Elevators

This chart shows size comparisons for commonly encountered elevator cabs.

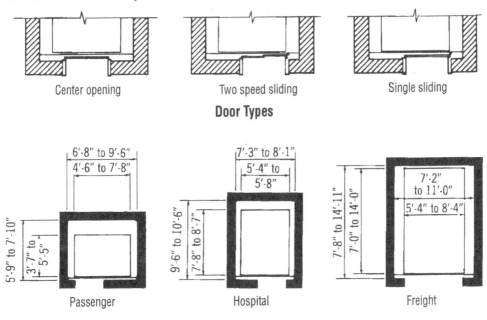

Door Types

Center opening   Two speed sliding   Single sliding

Passenger   Hospital   Freight

Figure 14-3 Elevator Speed versus Height Requirements

Building Type and Elevator Capacities	Travel Speeds in Feet per Minute								
	100 fpm	200 fpm	250 fpm	350 fpm	400 fpm	500 fpm	700 fpm	800 fpm	1000 fpm
Apartments 1200 lb. to 2500 lb.	to 70	to 100	to 125	to 150	to 250	to 350			
Department Stores 1200 lb. to 2500 lb.		100		125	175	250	350		
Hospitals 3500 lb. to 4000 lb.	to 70	100	125	150	175	250	350		
Office Buildings 2000 lb. to 4000 lb.		100	125	150		175	250	to 350	over 350

Note: Vertical transportation capacity may be determined by code occupancy requirements of the number of square feet per person divided into the total square feet of building type. If we are contemplating an office building, we find that the Occupancy Code Requirement is 100 S.F. per person. For a 20,000 S.F. building, we would have a legal capacity of two hundred people. Elevator handling capacity is subject to the five minute evacuation recommendation, but it may vary from 11% to 18%. Speed required is a function of the travel height, number of stops, and capacity of the elevator.

Division 14

Figure 14-4 Elevator Hoistway Size Requirements

This table lists and compares the common types of elevators, including their limitations, common usage, capacities, door/entry type, standard dimensions, and area per floor required.

Elevator Type	Floors	Building Type	Capacity Lbs.	Passengers	Entry*	Hoistway Width	Depth	S.F. Area per Floor
Hydraulic	5	Apt./Small Office	1500	10	S	6'-7"	4'-6"	29.6
			2000	13	S	7'-8"	4'-10"	37.4
	7	Av. Office/Hotel	2500	16	S	8'-4"	5'-5"	45.1
			3000	20	S	8'-4"	5'-11"	49.3
		Lg. Office/Store	3500	23	S	8'-4"	6'-11"	57.6
		Freight Light Duty	2500		D	7'-2"	7'-10"	56.1
		Heavy Duty	5000		D	10'-2"	10'-10"	110.1
			7500		D	10'-2"	12'-10"	131
			5000		D	10'-2"	10'-10"	110.1
			7500		D	10'-2"	12'-10"	131
			10,000		D	10'-4"	14'-10"	153.3
		Hospital	3500		D	6'-10"	9'-2"	62.7
			4000		D	7'-4"	9'-6"	69.6
Electric Traction, High Speed	High Rise	Apt./Small Office	2000	13	S	7'-8"	5'-10"	44.8
			2500	16	S	8'-4"	6'-5"	54.5
			3000	20	S	8'-4"	6'-11"	57.6
			3500	23	S	8'-4"	7'-7"	63.1
		Store	3500	23	S	9'-5"	6'-10"	64.4
		Large Office	4000	26	S	9'-4"	7'-6"	70
		Hospital	3500		D	7'-6"	9'-2"	69.4
			4000		D	7'-10"	9'-6"	58.8
Geared, Low Speed		Apartment	1200	8	S	6'-4"	5'-3"	33.2
			2000	13	S	7'-8"	5'-8"	43.5
			2500	16	S	8'-4"	6'-3"	52
		Office	3000	20	S	8'-4"	6'-9"	56
		Store	3500	23	S	9'-5"	6'-10"	64.4
						Add 4" width for multiple units		

* S = Single Door   * D = Double Door

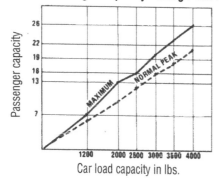

Elevator Passenger Capacity During Peak Periods

Passenger capacity

MAXIMUM NORMAL PEAK

Car load capacity in lbs.

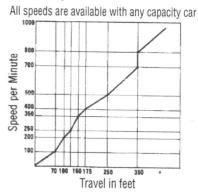

Speed & Travel Chart

All speeds are available with any capacity car

Speed per Minute

Travel in feet

Figure 14-5 Dumbwaiter Illustration and Nomenclature

This chart shows the workings of a common dumbwaiter and some of the terms used for its components.

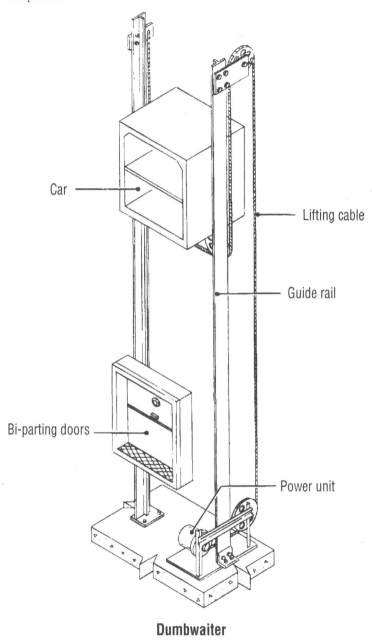

Dumbwaiter

Figure 14-6 Section Through an Escalator and Common Sizing Proportions

This illustration shows the workings of an escalator, common terms for its components, and some standard dimensional information.

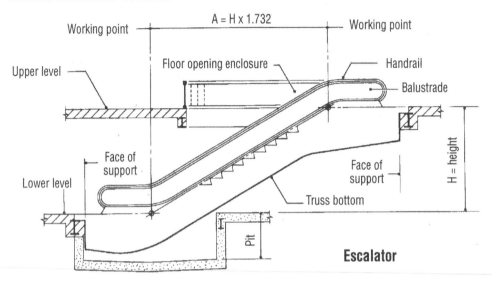

Figure 14-7 Installation Time for Elevators

Description	Crew Makeup	Daily Output	Labor-Hours	Units
ELECTRIC TRACTION FREIGHT ELEVATORS				
Electric freight, base unit, 4000 lb., 200 fpm, 4 stop, std. fin.	2 Elevator Constructors	.05	320	Ea.
For travel over 40 V.L.F, add		7.25	2.207	V.L.F.
For number of stops over 4, add		.27	59.259	Stop
HYDRAULIC FREIGHT ELEVATORS				
Hydraulic freight, base unit, 2000 lb, 50 fpm, 2 stop, std. fin.	2 Elevator Constructors	.10	160	Ea.
For travel over 40 V.L.F., add		7.25	2.207	V.L.F.
For number of stops over 2, add		.27	59.259	Stop
ELECTRIC TRACTION PASSENGER ELEVATORS				
Electric pass., base unit, 2000 lb, 200 fpm, 4 stop, std. fin	2 Elevator Constructors	.05	320	Ea.
For travel over 40 V.L.F., add		7.25	2.207	V.L.F.
For number of stops over 4, add		.27	59.259	Stop
Electric hospital, base unit, 4000 lb, 200 fpm, 4 stop, std. fin		.05	320	Ea.
For travel over 40 V.L.F., add		7.25	2.207	V.L.F.
For number of stops over 4, add		.27	59.259	Stop
HYDRAULIC PASSENGER ELEVATORS				
Hyd. Pass., base unit, 1500 lb, 100 fpm, 2 stop, std. fin.	2 Elevator Constructors	.10	160	Ea.
For travel over 12 V.L.F., add		7.25	2.207	V.L.F.
For number of stops over 2, add		.27	59.259	Stop
Hydraulic hospital, base unit, 4000 lb, 100 fpm, 2 stop, std. fin.		.10	160	Ea.
For travel over 12 V.L.F., add		7.25	2.207	V.L.F.
For number of stops over 2, add		.27	59.259	Stop

Figure 14-8 Installation Time for Electric Dumbwaiters

Description	Crew Makeup	Daily Output	Labor-Hours	Units
DUMBWAITERS 2 stop, electric, minimum	2 Elevator Constructors	.13	123.000	Ea.
Maximum		.11	145.000	Ea.
For each additional stop, add		.54	29.630	Stop

Figure 14-9 Installation Time for Manual Dumbwaiters

Description	Crew Makeup	Daily Output	Labor-Hours	Units
DUMBWAITERS 2 stop, hand, minimum	2 Elevator Constructors	.75	21.333	Ea.
Maximum		.50	32.000	Ea.
For each additional stop, add		.75	21.333	Stop

Division 14

Figure 14-10   Installation Time for Lifts

Description	Crew Makeup	Daily Output	Labor-Hours	Units
CORRESPONDENCE LIFT 1 floor, 2 stop,				
Electric, 25 lb. capacity	2 Elevator Constructors	.20	80.000	Ea.
Hand, 5 lb. capacity	"	.20	80.000	Ea.
PARCEL LIFT 20" x 20", 100 lb. capacity				
Electric, per floor	2 Millwrights	.25	64.000	Ea.

Figure 14-11   Installation Time for Escalators

Description	Crew Makeup	Daily Output	Labor-Hours	Units
ESCALATORS Per single unit, minimum	3 Elevator Constructors 1 Elevator Apprentice	.07	457.000	Ea.
Maximum	"	.04	800.000	Ea.

Figure 14-12   Installation Time for Moving Walkways

Description	Crew Makeup	Daily Output	Labor-Hours	Units
MOVING RAMPS AND WALKS Walk, 27" tread width				
Minimum	3 Elevator Constructors 1 Elevator Apprentice	6.50	4.920	L.F.
Maximum		4.43	7.220	L.F.
48" tread width walk, minimum		4.43	7.220	L.F.
Maximum		3.82	8.380	L.F.
Ramp, 12° incline, 36" tread width, minimum		5.27	6.070	L.F.
Maximum		3.82	8.380	L.F.
48" tread width, minimum		3.57	8.960	L.F.
Maximum		2.91	11.000	L.F.

Figure 14-13 Installation Time for Chutes

Description	Crew Makeup	Daily Output	Labor-Hours	Units
CHUTES Linen or refuse, incl. sprinklers				
Aluminized steel, 16 gauge, 18" diameter	2 Sheet Metal Workers	3.50	4.570	Floor
24" diameter		3.20	5.000	Floor
30" diameter		3	5.330	Floor
36" diameter		2.80	5.710	Floor
Galvanized steel, 16 gauge, 18" diameter		3.50	4.570	Floor
24" diameter		3.20	5.000	Floor
30" diameter		3	5.330	Floor
36" diameter		2.80	5.710	Floor
Stainless steel, 18" diameter		3.50	4.570	Floor
24" diameter		3.20	5.000	Floor
30" diameter		3	5.330	Floor
36" diameter		2.80	5.710	Floor
Linen bottom collector, aluminized steel		4	4.000	Ea.
Stainless steel		4	4.000	Ea.
Refuse bottom hopper, aluminized steel				
18" diameter		3	5.330	Ea.
24" diameter		3	5.330	Ea.
36" diameter		3	5.330	Ea.
Package chutes, spiral type, minimum		4.50	3.560	Floor
Maximum		1.50	10.670	Floor

Figure 14-14 Installation Time for Pneumatic Tube Systems

Description	Crew Makeup	Daily Output	Labor-Hours	Units
PNEUMATIC TUBE SYSTEM Single tube, 2 stations				
100' long, stock, economy				
3" diameter	2 Steamfitters	.12	133.000	Total
4" diameter	"	.09	177.000	Total
Twin tube, two stations or more,				
conventional system				
2-1/2" round	2 Steamfitters	62.50	.256	L.F.
3" round		46	.348	L.F.
4" round		49.60	.323	L.F.
4" x 7" oval		37.60	.426	L.F.
Add for blower		2	8.000	System
Plus for each round station, add		7.50	2.130	Ea.
Plus for each oval station, add		7.50	2.130	Ea.
Alternate pricing method: base cost, minimum		.75	21.330	Total
Maximum		.25	64.000	Total
Plus total system length, add, minimum		93.40	.171	L.F.
Maximum		37.60	.426	L.F.
Completely automatic system, 4" round,				
15 to 50 stations		.29	55.170	Station
51 to 144 stations		.32	50.000	Station
6" round or 4" x 7" oval, 15 to 50 stations		.24	66.670	Station
51 to 144 stations		.23	69.570	Station

Notes

Division Twenty-One
Fire Suppression

Introduction

Division 21, part of the Facility Services subgroup, addresses the instrumentation, control, operation, maintenance, and commissioning of fire protection systems. Formerly these items were part of Division 13 – Special Construction.

Normally this division is estimated by specialists, well-versed in this area, primarily because plans and specifications show only the general configuration and layout of these systems. Many of the methods and materials are standardized and understood by the installers and, therefore, not usually detailed in the contract documents.

Sprinkler systems: These systems vary, depending on the classification of the building occupancy, and level of hazard. The most common type is the wet-pipe system, which is filled with pressurized water from a municipal system. Quick response and low initial cost are the main reasons why wet-pipe systems are used. If the area will be subject to below-freezing temperatures, dry-pipe systems are recommended.

Standpipe systems: These are designed to bring water for fire control to remote areas of a building where laying a fire hose may be difficult. They consist of siamese connections, heavy pipe risers, valves, hoses, hose cabinets and racks, and alarms. Piping portions of these systems are taken off by the linear foot. All other components and accessories are taken off by units each.

Estimating Data

The following tables present guidelines for fire suppression systems.

Table of Contents

Division 21

Estimating Tips

Fire Pumps

These components need to be addressed within the larger fire suppression system to ensure that water can be moved to upper floors at sufficient pressure to supply sprinkler or standpipe systems. The unit for takeoff for these is "each" (Ea.).

Fire Suppression Water Storage

Be sure to address the relevant needed tanks for each project, including underground, ground, or elevated storage tanks that contain water for fire control.

Checklist ✓

The following tables present estimating guidelines for items found in Division 21 – Fire Suppression. Please note that these guidelines are intended as indicators of what may generally be expected, but that each project must be evaluated individually.

☐ *Fire Protection Systems*
- ☐ Supression systems
  - ☐ Hydrants
  - ☐ Fire department connections
- ☐ Sprinkler systems
  - ☐ Wet pipe
  - ☐ Dry pipe
  - ☐ Pre-action
  - ☐ Deluge
- ☐ Extinguishing systems
  - ☐ Carbon dioxide
  - ☐ Clean agent
- ☐ FM200
- ☐ Wet chemical
- ☐ Dry chemical
- ☐ Foam
- ☐ Firecycle
☐ Pumps
☐ Water storage tanks
☐ Standpipes
- ☐ Cabinets
- ☐ Hoses, reels, and racks
- ☐ Valves
- ☐ Wet
- ☐ Dry

Figure 21-1   Types of Automatic Sprinkler Systems

Sprinkler systems may be classified by type as follows:

1. **Wet Pipe System.** A system employing automatic sprinklers attached to a piping system containing water and connected to a water supply so that water discharges immediately from sprinklers opened by a fire.

2. **Dry Pipe System.** A system employing automatic sprinklers attached to a piping system containing air under pressure. When the pressure is released from the opening of sprinklers, the water pressure opens a valve known as a "dry pipe valve." The water then flows into the piping system and out the opened sprinklers.

3. **Pre-Action System.** A system employing automatic sprinklers attached to a piping system containing air that may or may not be under pressure. Pre-action systems have heat-activated devices that are more sensitive than the automatic sprinklers themselves, installed in the same areas as the sprinklers. Actuation of the heat responsive system, as from a fire, opens a valve which permits water to flow into the sprinkler piping system and to be discharged from any sprinklers which maybe open.

4. **Deluge System.** A dry system connected to a water supply through a valve which is opened by the operation of a heat responsive system (installed in the same areas as the sprinklers.) When this valve opens, water flows into the piping system and discharges from all attached sprinklers.

5. **Combined Dry Pipe and Pre-Action Sprinkler System.** A system employing automatic sprinklers attached to a piping system containing air under pressure. A supplemental heat responsive system of generally more sensitive characteristics than the automatic sprinklers themselves is installed in the same areas as the sprinklers. Operation of the heat responsive system, as from a fire, actuates tripping devices which open dry pipe valves simultaneously and without loss of air pressure in the system. Operation of the heat responsive system also opens approved air exhaust valves at the end of the feed main which facilitates the filling of the system with water (which usually precedes the opening of sprinklers). The heat responsive system also serves as an automatic fire alarm system.

6. **Limited Water Supply System.** A system employing automatic sprinklers and conforming to automatic sprinkler standards, but supplied by a pressure tank of limited capacity.

7. **Chemical Systems.** Systems using FM200, carbon dioxide, dry chemical or high expansion foam as selected for special requirements. The agent may extinguish flames by chemically inhibiting flame propagation, suffocate flames by excluding oxygen, interrupting chemical action of oxygen uniting with fuel or sealing and cooling the combusion center.

8. **Firecycle System.** Firecycle is a fixed fire protection sprinkler system utilizing water as its extinguishing agent. It is a time delayed, recycling, pre-action type which automatically shuts the water off when the heat is reduced below the detector operating temperature. Water is turned on again when that temperature is exceeded. The system senses a fire condition through a closed circuit electrical detector which controls water flow to the fire automatically. Batteries supply up to 90 hours emergency power supply for system operation. The piping system is dry (until water is required) and is monitored with pressurized air. Should any leak in the system piping occur, an alarm will sound, but water will not enter the system until heat is sensed by a Firecycle detector.

Area coverage sprinkler systems maybe laid out and fed from the supply in any one of several patterns as shown in the illustration. It is desirable, if possible, to utilize a central feed and achieve a shorter flow path from the riser to the farthest sprinkler. This permits use of the smallest sizes of pipe possible, with resulting savings.

(continued on next page)

Figure 21-1 Types of Automatic Sprinkler Systems
(continued from previous page)

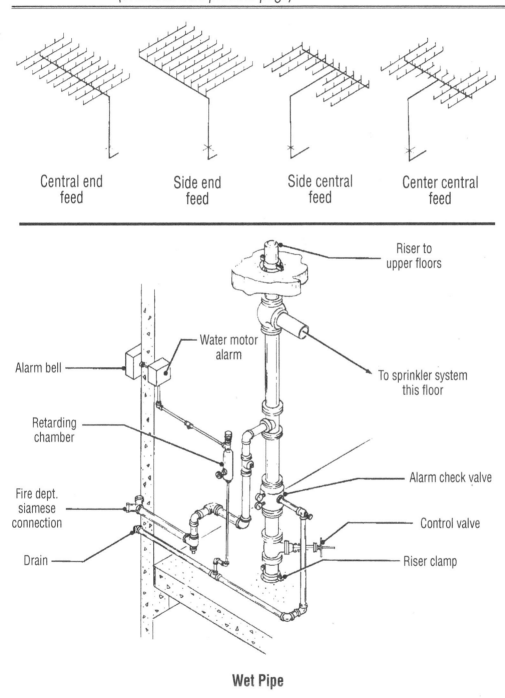

Central end feed

Side end feed

Side central feed

Center central feed

Riser to upper floors

Water motor alarm

Alarm bell

To sprinkler system this floor

Retarding chamber

Alarm check valve

Fire dept. siamese connection

Control valve

Drain

Riser clamp

Wet Pipe

Figure 21-2 Classification of Sprinkler Systems

LIGHT HAZARD OCCUPANCY
The protection area allotted per sprinkler should not exceed 200 S.F. with the maximum distance between lines and sprinklers on lines being 15'. The sprinklers do not need to be staggered. Branch lines should not exceed eight sprinklers on either side of a cross main. Each large area requring more than 100 sprinklers and without a sub-dividing partition should be supplied by feed mains or risers sized for ordinary hazard occupancy.

Included in this group are:

Auditoriums	Museums
Churches	Nursing homes
Clubs	Offices
Educational	Residential
Hospitals	Restaurants
Institutional	Schools
Libraries	Theaters
(except large stack rooms)	

ORINDARY HAZARD OCCUPANCY
The protection area allotted per sprinkler shall not exceed 130 S.F. of noncombustible ceiling and 120 S. F. of combustible ceiling. The maximum allowable distance between sprinkler lines and sprinklers on lines is 15'. Sprinklers shall be staggered if the distance between heads exceeds 12'. Branch lines should not exceed eight sprinklers on either side of a cross main.

Included in this group are:

Automative garages	Electric generating stations
Bakeries	Feed mills
Beverage manufacturing	Grain elevators
Bleacheries	Ice manufacturing
Boiler houses	Laundries
Canneries	Machine shops
Cement plants	Mercantiles
Clothing factories	Paper mills
Cold storage warehouses	Printing and publishing
Dairy products manufacturing	Shoe factories
	Warehouses
Distilleries	Wood product assembly
Dry cleaning	

EXTRA HAZARD OCCUPANCY
The protection area allotted per sprinkler shall not exceed 90 S.F. of of noncombustible ceiling and 80 S.F. of combustible ceiling. The maximum allowable distance between lines and between sprinklers on lines is 12'. Sprinklers on alternate lines shall be staggered if the distance between sprinklers on lines exceeds 8'. Branch lines should not exceed six sprinklers on either side of a cross main.

Included in this group are:

Aircraft hangars	Paint shops
Chemical works	Shade cloth manufacturing
Explosives manufacturing	Solvent extracting
Linoleum manufacturing	Varnish works
Linseed oil mills	Volatile, flammable liquid
Oil refineries	manufacturing and use

Figure 21-3 Sprinkler Head Quantities for Various Sizes
and Types of Pipe

Pipe Size Diameter	Light Hazard Occupancy		Ordinary Hazard Occupancy		Extra Hazard Occupancy	
	Steel Pipe Sprinklers	Copper Pipe Sprinklers	Steel Pipe Sprinklers	Copper Pipe Sprinklers	Steel Pipe Sprinklers	Copper Pipe Sprinklers
1"	2	2	2	2	1	1
1-1/4"	3	3	3	3	2	2
1-1/2"	5	5	5	5	5	5
2"	10	12	10	12	8	8
2-1/2"	30	40	20	25	15	20
3"	60	65	40	45	27	30
3-1/2"	100	115	65	75	40	45
4"			100	115	55	65
5"			160	180	90	100
6"			275	300	150	170

Dry Pipe Systems: A dry pipe system should be installed where a wet pipe system is impractical, as in buildings in areas of the country subject to subfreezing temperatures.

The use of an approved dry pipe system is more desirable than shutting off the water supply during cold weather.

Not more than 750 gallons of system capacity should be controlled by one dry pipe valve. Where two or more dry pipe valves are used, systems should preferably be divided horizontally.

While the above is a useful guide, sprinkler systems must be installed in accordance with the latest approved release of NFPA-13, the National Fire Protection Association code for "the installation of sprinkler systems."

Figure 21-4 Standpipe Systems

Class	Design-Use	Pipe Size Minimums	Water Supply Minimums
Class I	2-1/2" hose connection on each floor All areas within 30' of nozzle with 100' of hose Fire department trained personnel	Height to 100', 4" diam. Heights above 100', 6" diam. (275' max. except with pressure regulators 400' max.)	For each standpipe riser 500 GPM flow For common supply pipe allow 500 GPM for first standpipe plus 250 GPM for each additional standpipe (2500 GPM max. total), 30 min. duration, 65 PSI at 500 GPM
Class II	1-1/2" hose connection with hose on each floor All areas within 30' of nozzle with 100' of hose Occupant personnel	Height to 50', 2" diam. Height above 50', 2-1/2" diam.	For each standpipe riser 100 GPM flow For multiple riser common supply pipe 100 GPM 30 min. duration, 65 PSI at 100 GPM,
Class III	Both of above. Class I valved connections will meet Class III with addition of 2-1/2" by 1-1/2" adapter and 1-1/2" hose	Same as Class I	Same as Class I

Standpipe systems to be installed in accordance with the latest approved release of NFPA-14 "installation of standpipe and hose systems" published by the National Fire Protection Association.

Figure 21-5 Installation Time for Automatic Sprinkler Systems

Description	Labor-Hours	Unit
Valve, Gate, Iron Body, Flanged OS and Y, 125 lb. 4" Diameter	5.333	Ea.
Valve, Swing Check, w/Ball Drip, Flanged, 4" Diameter	5.333	Ea.
Valve, Swing Check, Bronze, Thread End, 2-1/2" Diameter	1.067	Ea.
Valve, Angle, Bronze, Thread End, 2" Diameter	.727	Ea.
Valve, Gate, Bronze, Thread End, 1" Diameter	.421	Ea.
Alarm Valve, 2-1/2" Diameter	5.333	Ea.
Alarm, Water Motor, with Gong	2.000	Ea.
Fire Alarm Horn, Electric	.308	Ea.
Pipe, Black Steel, Threaded, Schedule 40, 4" Diameter	.444	L.F.
2-1/2" Diameter	.320	L.F.
2" Diameter	.250	L.F.
1-1/4" Diameter	.180	L.F.
1" Diameter	.151	L.F.
Pipe Tee, 150 lb. Black Malleable 4" Diameter	4.000	Ea.
2-1/2" Diameter	1.778	Ea.
2" Diameter	1.455	Ea.
1-1/4" Diameter	1.143	Ea.
1" Diameter	1.000	Ea.
Pipe Elbow, 150 lb. Black Malleable 1" Diameter	.615	Ea.
Sprinkler Head, 135° to 286°, 1/2" Diameter	.500	Ea.
Sprinkler Head, Dry Pendent 1" Diameter	.571	Ea.
Dry Pipe Valve, w/Trim and Gauges, 4" Diameter	16.000	Ea.
Deluge Valve, w/Trim and Gauges, 4" Diameter	16.000	Ea.
Deluge System Monitoring Panel 120 Volt	.444	Ea.
Thermostatic Release	.400	Ea.
Heat Detector	.500	Ea.
Firecycle Controls w/Panel, Batteries, Valves and Switches	32.000	Ea.
Firecycle Package, Check and Flow Control Valves, Trim, 4" Diameter	16.000	Ea.
Air Compressor, Automatic, 200 gal. Sprinkler System 1/3 H.P.	6.154	Ea.

Division 21

(continued on next page)

Figure 21-5 Installation Time for Automatic Sprinkler Systems
(continued from previous page)

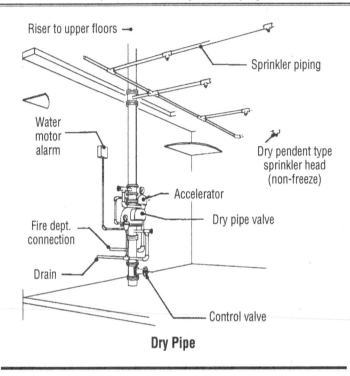

Dry Pipe

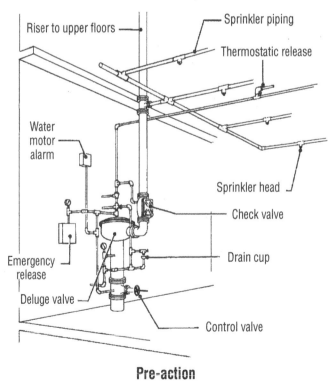

Pre-action

Figure 21-6   Installation Time for FM200 Fire Suppression Systems

Description	Labor-Hours	Unit
FM200 System, Filled, Including Mounting Bracket		
26 lb. Cylinder	2.000	Ea.
44 lb. Cylinder	2.286	Ea.
63 lb. Cylinder	2.667	Ea.
101 lb. Cylinder	3.200	Ea.
196 lb. Cylinder	4.000	Ea.
Electro/Mechanical Release	4.000	Ea.
Manual Pull Station	1.333	Ea.
Pneumatic Damper Release	1.000	Ea.
Discharge Nozzle	.570	Ea.
Control Panel Single Zone	8.000	Ea.
Control Panel Multi-zone (4)	16.000	Ea.
Battery Standby Power	2.810	Ea.
Heat Detector	1.000	Ea.
Smoke Detector	1.290	Ea.
Audio Alarm	1.194	Ea.

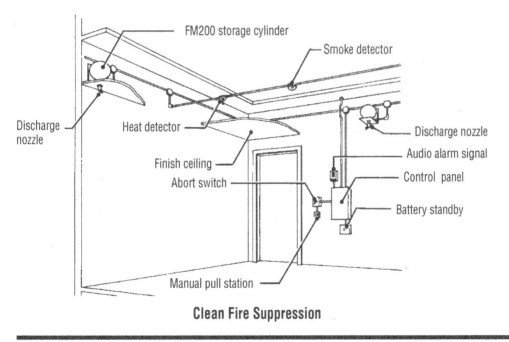

Clean Fire Suppression

Figure 21-7 Installation Time for Fire Standpipes

Description	Labor-Hours 4"	6"	8"	Unit
Black Steel Pipe	.444	.774	.889	L.F.
Pipe Tee	4.000	6.000	8.000	Ea.
Pipe Elbow	2.667	3.429	4.000	Ea.
Pipe Nipple, 2-1/2"	1.000	1.000	1.000	Ea.
Hose Valve, 2-1/2"	1.140	1.140	1.140	Ea.
Pressure Restricting Valve, 2-1/2"	1.140	1.140	1.140	Ea.
Check Valve with Ball Drip	5.333	8.000	10.667	Ea.
Siamese Inlet	3.200	3.478	3.478	Ea.
Roof Manifold with Valves	3.333	3.478	3.478	Ea.

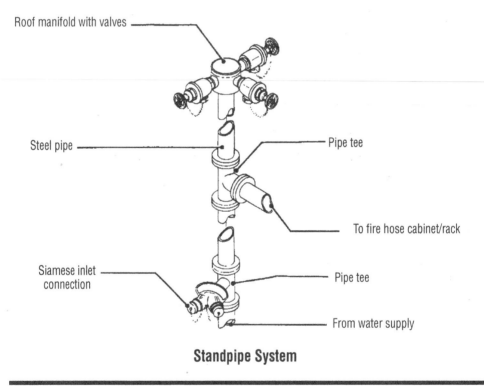

Roof manifold with valves

Steel pipe

Pipe tee

To fire hose cabinet/rack

Siamese inlet connection

Pipe tee

From water supply

Standpipe System

Figure 21-8 Installation Time in Labor-Hours for Pumps

Description	Labor-Hours	Unit
Fire Pumps with Controller and Fittings		
Electric		
4" Pipe Size 100 H.P. 500 GPM	51.619	Ea.
6" Pipe Size 250 H.P. 1000 GPM	88.889	Ea.
8" Pipe Size 300 H.P. 2000 GPM	114.286	Ea.
10" Pipe Size 450 H.P. 3500 GPM	133.333	Ea.
Diesel		
4" Pipe Size 111 H.P. 500 GPM	53.333	Ea.
6" Pipe Size 255 H.P. 1000 GPM	80.000	Ea.
8" Pipe Size 255 H.P. 2000 GPM	100.000	Ea.
10" Pipe Size 525 H.P. 3500 GPM	160.000	Ea.

Division 21

Notes

Division Twenty-Two
Plumbing

Introduction

Division 22 addresses piping and fixtures for potable water, domestic hot water, storm and sanitary waste, and interior gas piping systems. Plumbing work is shown on drawings indicated by fixtures on drawings, by riser diagram, and by schedules. Many details are not shown, however, and are left to the estimator to include on the estimate, based on personal knowledge and experience. In many cases, water control, pipe, and fittings may be estimated as percentages of the fixture cost. In addition to the work shown on the drawings, the plumbing contractor may be required to make connections for building equipment.

Fixture takeoff involves counting the various types, sizes, and styles. Be certain to include all costs, including trim, carrier, flush valves, and so forth if the "rough-in, supply, waste, and vent" line is not used. Include equipment costs such as pumps, water heaters, water softeners, and all items not counted.

Piping is taken off by the system. Pipe runs for any type of system consist of straight sections and fittings of various shapes, styles, and purposes. The estimator will start at one end of each system and measure and record the straight lengths of each size of pipe. Fittings are noted and recorded as they are encountered. Care should be taken to note all changes in size and material. This would only occur at a joint, so that it should become automatic habit to see that the material of piping going into a joint is the same as that leaving the joint. It is good practice to round off the totals of each size of pipe to the lengths normally available from the supplier.

green Division 22 is where many of the means, methods, and materials for achieving water efficiency can be found, including waterless urinals, ultra low flow toilets, and gray water recycling systems. Water efficiency is generally an important aspect of green building, and specifically LEED® certification. Reducing water use by 20% from the established baseline is a LEED NC prerequisite. Reductions of 30% can earn 2 points, 35% can earn 3 points, and 40% reduction can earn 4 points. Additional points can be earned by eliminating potable water use for landscaping and sewage conveyance, and treating wastewater on site.

Estimating Data

The following tables present estimating guidelines for items found in Division 22 – Plumbing. Please note that these guidelines are intended as indicators of what may generally be expected, and that each prospect must be evaluated individually.

Table of Contents

Estimating Tips

Plans and Specifications

Review all construction documentation before proceeding with the estimate
for Division 22. It is not unusual to find items on the plans, but not in the
specifications; or they may be in the specifications, but not on the drawings.
Also check any and all communications; the owner may have requested a change
or expressed a concern or a specific need, and not have had it included in your
drawings or specifications.

Plumbing Piping and Pumps

It is important to note that Division 22 addresses basic piping materials and
related materials. Pipe may be used by any of the mechanical disciplines, i.e.,
plumbing, fire protection, heating, and air conditioning.

Non-specialty estimators who are unfamiliar with the details of plumbing systems
may choose to take off fixtures only, and use general percentage markups for the
various elements of the plumbing system.

Most pipe is priced first as straight pipe with a joint (coupling, weld, etc.) every
10' and a hanger usually every 10'. There are exceptions with hanger spacing,
such as for cast iron pipe (5') and plastic pipe (3 per 10').

Fittings

When preparing an estimate, it may be necessary to approximate the fittings. They
usually run between 25% and 50% of the cost of the pipe. The lower percentage
is for simpler runs, and the higher for complex areas, such as mechanical rooms.

Historic Restoration Projects

For historic restoration projects, the systems must be as invisible as possible, and
pathways must be sought for pipes, conduit, and ductwork. While installation in
accessible spaces (such as basements and attics) is relatively straightforward to
estimate, labor costs may be more difficult to determine when delivery systems
must be concealed.

Checklist ✓

For an estimate to be reliable, all items must be accounted for. A complete estimate can also limit contingencies. The following checklist can be used to help ensure that all items are included.

☐ *Fixtures*
- ☐ Bathtubs
- ☐ Drinking fountains
- ☐ Hose bibs
- ☐ Lavatories
- ☐ Showers
- ☐ Sinks
  - ☐ Bar
  - ☐ Janitor
  - ☐ Kitchen
  - ☐ Laundry
  - ☐ Slop
- ☐ Toilets
- ☐ Urinals
- ☐ Wash centers
- ☐ Wash fountains
- ☐ Water closets
- ☐ Water coolers

☐ *Piping*
- ☐ Air chambers
- ☐ Concrete encasement
- ☐ Escutcheons
- ☐ Expansion joints
- ☐ Excavation for piping
- ☐ Fittings
- ☐ Hangers
- ☐ Insulation
- ☐ Materials

- ☐ Shock absorbers
- ☐ Specialty
  - ☐ Carbon dioxide
  - ☐ Compressed air
  - ☐ Industrial gases
  - ☐ Nitrous oxide
  - ☐ Oxygen
  - ☐ Process
  - ☐ Sanitary
  - ☐ Vacuum

☐ *Sanitary System*
- ☐ Sump pumps
- ☐ Bilge pumps
- ☐ Trash pumps
- ☐ Soils piping
- ☐ Stacks

☐ *Water Heaters*
- ☐ Electric
- ☐ Fuel-fired

☐ *Water Heat Exchangers*

☐ *Drinking Fountains and Water Coolers*

☐ *Swimming Pool Plumbing*

Figure 22-1 Capacity in Cubic Feet and U.S. Gallons of Pipes and Cylinders

Diameter in Inches	For 1 Foot Length		Diameter in Inches	For 1 Foot Length		Diameter in Inches	For 1 Foot Length	
	Cubic Feet	U.S. Gallons		Cubic Feet	U.S. Gallons		Cubic Feet	U.S. Gallons
1/4	0.0003	0.0025	6-3/4	0.2485	1.859	19	1.969	14.73
5/16	0.0005	0.0040	7	0.2673	1.999	19-1/2	2.074	15.51
3/8	0.0008	0.0057	7-1/4	0.2867	2.145	20	2.182	16.32
7/16	0.0010	0.0078	7-1/2	0.3068	2.295	20-1/2	2.292	17.15
1/2	0.0014	0.0102	7-3/4	0.3276	2.450	21	2.405	17.99
9/16	0.0017	0.0129	8	0.3491	2.611	21-1/2	2.521	18.86
5/8	0.0021	0.0159	8-1/4	0.3712	2.777	22	2.640	19.75
11/16	0.0026	0.0193	8-1/2	0.3941	2.948	22-1/2	2.761	20.66
3/4	0.0031	0.0230	8-3/4	0.4176	3.125	23	2.885	21.58
13/16	0.0036	0.0269	9	0.4418	3.305	23-1/2	3.012	22.53
7/8	0.0042	0.0312	9-1/4	0.4667	3.491	24	3.142	23.50
15/16	0.0048	0.0359	9-1/2	0.4922	3.682	25	3.409	25.50
1	0.0055	0.0408	9-3/4	0.5185	3.879	26	3.687	27.58
1-1/4	0.0085	0.0638	10	0.5454	4.080	27	3.976	29.74
1-1/2	0.0123	0.0918	10-1/4	0.5730	4.286	28	4.276	31.99
1-3/4	0.0167	0.1249	10-1/2	0.6013	4.498	29	4.587	34.31
2	0.0218	0.1632	10-3/4	0.6303	4.715	30	4.909	36.72
2-1/4	0.0276	0.2066	11	0.6600	4.937	31	5.241	39.21
2-1/2	0.0341	0.2550	11-1/4	0.6903	5.164	32	5.585	41.78
2-3/4	0.0412	0.3085	11-1/2	0.7213	5.396	33	5.940	44.43
3	0.0491	0.3672	11-3/4	0.7530	5.633	34	6.305	47.16
3-1/4	0.0576	0.4309	12	0.7854	5.875	35	6.681	49.98
3-1/2	0.0668	0.4998	12-1/2	0.8522	6.375	36	7.069	52.88
3-3/4	0.0767	0.5738	13	0.9218	6.895	37	7.467	55.86
4	0.0873	0.6528	13-1/2	0.9940	7.436	38	7.876	58.92
4-1/4	0.0985	0.7369	14	1.069	7.997	39	8.296	62.06
4-1/2	0.1104	0.8263	14-1/2	1.147	8.578	40	8.727	65.28
4-3/4	0.1231	0.9206	15	1.227	9.180	41	9.168	68.58
5	0.1364	1.020	15-1/2	1.310	9.801	42	9.621	71.97
5-1/4	0.1503	1.125	16	1.396	10.44	43	10.085	75.44
5-1/2	0.1650	1.234	16-1/2	1.485	11.11	44	10.559	78.99
5-3/4	0.1803	1.349	17	1.576	11.79	45	11.045	82.62
6	0.1963	1.469	17-1/2	1.670	12.49	46	11.541	86.33
6-1/4	0.2131	1.594	18	1.767	13.22	47	12.048	90.13
6-1/2	0.2304	1.724	18-1/2	1.867	13.96	48	12.566	94.00

Division 22

Figure 22-2 Gas Pipe Capacities

*(Gas-ft.^{3}/hr.) Maximum capacity of pipe in cubic feet of gas per hour for gas pressures of 0.5 PSIG or less and a pressure drop of 0.3" water column, based on a 0.60 specific gravity natural gas. If 1.5 L.P. gas is used, multiply capacity by 0.633.

Nominal Iron Pipe Size, Inches	Internal Diameter, Inches	Length of Pipe in Feet													
		10	20	30	40	50	60	70	80	90	100	125	150	175	200
1/4	.364	32	22	18	15	14	12	11	11	10	9	8	8	7	6
3/8	.493	72	49	40	34	30	27	25	23	22	21	18	17	15	14
1/2	.622	132	92	73	63	56	50	46	43	40	38	34	31	28	26
3/4	.824	278	190	152	130	115	105	96	90	84	79	72	64	59	55
1	1.049	520	350	285	245	215	195	180	170	160	150	130	120	110	100
1-1/4	1.380	1,050	730	590	500	440	400	370	350	320	305	275	250	225	210
1-1/2	1.610	1,600	1,100	890	760	670	610	560	530	490	460	410	380	350	320
2	2.067	3,050	2,100	1,650	1,450	1,270	1,150	1,050	990	930	870	780	710	650	610
2-1/2	2.469	4,800	3,300	2,700	2,300	2,000	1,850	1,700	1,600	1,500	1,400	1,250	1,130	1,050	980
3	3.068	8,500	5,900	4,700	4,100	3,600	3,250	3,000	2,800	2,600	2,500	2,200	2,000	1,850	1,700
4	4.026	17,500	12,000	9,700	8,300	7,400	6,800	6,200	5,800	5,400	5,100	4,500	4,100	3,800	3,500

*Per AGA and NFPA

Figure 22-3   Linear Expansion of Copper Tubing

Boiler Water Temp. °F	Total expansion in inches, based on initial tubing temperature of 70°F Note that amount of expansion is the same for all tubing sizes.									
	10 Ft.	20 Ft.	30 Ft.	40 Ft.	50 Ft.	60 Ft.	70 Ft	80 Ft.	90 Ft.	100 Ft.
80°F	.01	.02	.04	.05	.06	.07	.08	.09	.11	.12
100°F	.04	.07	.11	.14	.18	.22	.25	.28	.32	.35
120°F	.06	.12	.18	.24	.30	.35	.41	.47	.53	.59
140°F	.08	.16	.25	.33	.41	.50	.58	.66	.74	.82
160°F	.11	.21	.32	.42	.53	.64	.74	.85	.95	1.06
180°F	.13	.26	.39	.52	.65	.78	.91	1.04	1.17	1.29
200°F	.15	.31	.46	.61	.77	.92	1.07	1.22	1.38	1.53
220°F	.18	.35	.53	.71	.88	1.06	1.24	1.41	1.59	1.77
240°F	.20	.40	.60	.80	1.00	1.20	1.40	1.60	1.80	2.00
260°F	.22	.45	.67	.89	1.12	1.34	1.56	1.79	2.01	2.23
280°F	.25	.50	.74	.99	1.24	1.49	1.73	1.98	2.23	2.47

Note: calculations based on average expansion coefficient of 0.0000098 per °F for copper

Figure 22-4   Linear Expansion of Steel Pipe

Boiler Water Temp. °F	Total expansion in inches, based on initial tubing temperature of 70°F Note that amount of expansion is the same for all tubing sizes.									
	10 Ft.	20 Ft.	30 Ft.	40 Ft.	50 Ft.	60 Ft.	70 Ft.	80 Ft.	90 Ft.	100 Ft.
80°F	.01	.02	.03	.04	.04	.05	.06	.07	.07	.08
100°F	.03	.05	.07	.10	.12	.14	.17	.19	.22	.24
120°F	.04	.08	.12	.16	.20	.24	.28	.32	.36	.40
140°F	.06	.11	.17	.23	.28	.33	.39	.45	.50	.56
160°F	.07	.14	.22	.29	.36	.43	.50	.57	.65	.71
180°F	.09	.18	.27	.35	.44	.53	.61	.70	.79	.87
200°F	.10	.21	.31	.42	.52	.62	.72	.83	.93	1.03
220°F	.12	.24	.36	.48	.60	.72	.83	.95	1.07	1.19
240°F	.14	.27	.40	.57	.67	.81	.94	1.08	1.21	1.35
260°F	.15	.30	.45	.60	.75	.90	1.05	1.20	1.35	1.50
280°F	.17	.33	.50	.67	.83	1.00	1.16	1.33	1.50	1.67

Note: calculations based on average expansion coefficient of 0.0000066 per °F for steel

Division 22

Figure 22-5 General Pipe Material Considerations and
                 Background Data

1. Malleable fittings should be used for gas service.
2. Malleable fittings are used where there are stresses/strains due to expansion and vibration.
3. Cast fittings may be broken as an aid to disassembling of heating lines frozen by long use, temperature, and minerals.
4. Cast iron pipe is extensively used for underground and submerged service.
5. Type M (light wall) copper tubing is available in hard temper only and is used for non-pressure and less severe applications than K and L.
6. Type L (medium wall) copper tubing, available hard or soft for interior service.
7. Type K (heavy wall) copper tubing, available in hard or soft temper for use where conditions are severe. For underground and interior service.
8. Hard drawn tubing requires fewer hangers or supports but should not be bent. Silver brazed fittings are recommended, however, soft solder is usually used.

Figure 22-6 Estimated Quantity of Soft Solder Required to
                 Make 100 Joints

Size	Quantity in Pounds
3/8"	.5
1/2"	.75
3/4"	1.0
2"	1.5
1-1/4"	1.75
1-1/2"	2.0
2"	2.5
2-1/2"	3.4
3"	4.0
3-1/2"	4.8
4"	6.8
5"	8.0
6"	15.0
8"	32.0
10"	42.0

Notes:
1. The quantity of hard solder used is dependent on the skill of the operator, but for estimating purposes, 75% of the above figures may be used.
2. Two oz. of solder flux will be required for each pound of solder.
3. Includes an allowance for waste.
4. Drainage fittings consume 20% less.

Figure 22-7 Pipe Sizing for Heating

Heating Load, BTU/HR.	GPM Circulated (20°T.D.)	Recommended Connecting Tubing Size (Type M) for Various Heating Loads and Connecting Tubing Lengths. (Figures based on 10,000 BTU per GPM, or on temperature drop of 20° thru the circuit.)				
		Total Length Ft. Connecting Tubing				
		0–50	50–100	100–150	150–200	200–300
		Tubing, Nominal O.D., Type M				
5,000	0.5	3/8	3/8	1/2	1/2	1/2
10,000	1.0	3/8	3/8	1/2	1/2	1/2
15,000	1.5	1/2	1/2	1/2	1/2	3/4
20,000	2.0	1/2	1/2	1/2	3/4	3/4
30,000	3.0	1/2	1/2	3/4	3/4	3/4
40,000	4.0	1/2	3/4	3/4	3/4	3/4
50,000	5.0	3/4	3/4	3/4	1	1
60,000	6.0	3/4	3/4	1	1	1
75,000	7.5	3/4	1	1	1	1
100,000	10.0	1	1	1	1-1/4	1-1/4
125,000	12.5	1	1	1-1/4	1-1/4	1-1/4
150,000	15.0	1	1-1/4	1-1/4	1-1/4	1-1/2
200,000	20.0	1-1/4	1-1/4	1-1/2	1-1/2	1-1/2
250,000	25.0	1-1/4	1-1/2	1-1/2	2	2
300,000	30.0	1-1/2	1-1/2	2	2	2
400,000	40.0	2	2	2	2	2
500,000	50.0	2	2	2	2-1/2	2-1/2
600,000	60.0	2	2	2-1/2	2-1/2	2-1/2
800,000	80.0	2-1/2	2-1/2	2-1/2	2-1/2	3
1,000,000	100.0	2-1/2	2-1/2	3	3	3
1,250,000	125.0	2-1/2	3	3	3	3-1/2
1,500,000	150.0	3	3	3	3-1/2	3-1/2
2,000,000	200.0	3	3-1/2	3-1/2	4	4
2,500,000	250.0	3-1/2	3-1/2	4	4	5
3,000,000	300.0	3-1/2	4	4	4	5
4,000,000	400.0	4	4	5	5	5
5,000,000	500.0	5	5	5	6	6
6,000,000	600.0	5	6	6	6	8
8,000,000	800.0	8	6	8	8	8
10,000,000	1,000.0	8	8	8	8	10

Figure 22-8 Pipe Sizing for Cooling

Cooling Load, BTU/HR.	Cooling Load, Tons	GPM Circulated 3 GPM/Ton 8°Rise	Recommended Connecting Tubing Size (Type M) for Various Cooling Loads and Connecting Tubing Lengths. (Figures based on 3 GPM/Ton, or on temperature rise of 8° thru the circuit.)				
			Total Length Ft. Connecting Tubing				
			0–50	50–100	100–150	150–200	200–300
			Tubing, Nominal O.D., Type M				
6,000	0.5	1.5	1/2	1/2	1/2	1/2	3/4
9,000	0.75	2.25	1/2	1/2	1/2	3/4	3/4
12,000	1.0	3.0	1/2	1/2	3/4	3/4	3/4
18,000	1.5	4.5	3/4	3/4	3/4	1	1
24,000	2.0	6.0	3/4	3/4	1	1	1
30,000	2.5	7.5	3/4	1	1	1	1
36,000	3.0	9.0	1	1	1	1-1/4	1-1/4
48,000	4.0	12.0	1	1	1-1/4	1-1/4	1-1/4
60,000	5.0	15.0	1	1-1/4	1-1/4	1-1/4	1-1/2
72,000	6.0	18.0	1-1/4	1-1/4	1-1/2	1-1/2	1-1/2
96,000	8.0	24.0	1-1/4	1-1/2	1-1/2	2	2
120,000	10.0	30.0	1-1/2	1-1/2	2	2	2
144,000	12.0	36.0	1-1/2	2	2	2	2
180,000	15.0	45.0	2	2	2	2	2-1/2
240,000	20.0	60.0	2	2	2-1/2	2-1/2	2-1/2
300,000	25.0	75.0	2-1/2	2-1/2	2-1/2	2-1/2	3
360,000	30.0	90.0	2-1/2	2-1/2	2-1/2	3	3
480,000	40.0	120.0	2-1/2	3	3	3	3-1/2
600,000	50.0	150.0	3	3	3	3-1/2	3-1/2
720,000	60.0	180.0	3	3-1/2	3-1/2	4	4
900,000	75.0	225.0	3-1/2	3-1/2	4	4	5
1,200,000	100.0	300.0	3-1/2	4	4	4	5
1,500,000	125.0	375.0	4	4	5	5	5
1,800,000	150.0	450.0	4	5	5	5	6
2,400,000	200.0	600.0	5	5	6	6	6
3,000,000	250.0	750.0	5	6	6	6	8
3,600,000	300.0	900.0	6	6	8	8	8
4,800,000	400.0	1200.0	8	8	8	8	10
6,000,000	500.0	1500.0	8	8	10	10	10
7,200,000	600.0	1800.0	10	10	10	10	12

Figure 22-9 Labor Factors Due to Special Conditions

For installations higher than average of 15', labor costs may be increased by the following suggested percentages.

Ceiling Height	Labor Increase
Add to labor for elevated installation (Above floor level)	
10' to 14.5' high	10%
15' to 19.5'	20%
20' to 24.5'	25%
25' to 29.5'	35%
30' to 34.5'	40%
35' to 39.5'	50%
40' and higher	55%
Add to labor for crawl space	
3' high	40%
4' high	30%
Add to labor for multi-story building	
Add per floor floors 3 thru 19	2%
Add per floor for floors 20 and up	4%
Add to labor for working in existing occupied buildings	
Hospital	35%
Office building	25%
School	20%
Factory or warehouse	15%
Multi dwelling	15%
Add to labor, miscellaneous	
Cramped shaft	35%
Congested area	15%
Excessive heat or cold	30%
Labor factors, the above are reasonable suggestions, however each project should be evaluated for its own peculiarities.	
Other factors to be considered are:	
Movement of material and equipment through finished areas	
Equipment room	
Attic space	
No service road	
Poor unloading/storage area	
Congested site area/heavy traffic	

Figure 22-10   Steel Pipe/Size and Weight Data

Nom. Size	O.D.	Schedule 10		Schedule 20		Schedule 30		Standard (P.E.)	
		Wall	Lbs./Ft.	Wall	Lbs./Ft.	Wall	Lbs./Ft.	Wall	Lbs./Ft.
1/8"	.405							.068	.24
1/4"	.540							.088	.42
3/8"	.675							.091	.57
1/2"	.840							.109	.85
3/4"	1.050							.113	1.13
1"	1.315							.133	1.68
1-1/4"	1.660							.140	2.27
1-1/2"	1.900							.145	2.72
2"	2.375							.154	3.65
2-1/2"	2.875							.203	5.79
3"	3.500							.216	7.58
3-1/2"	4.000							.226	9.11
4"	4.500							.237	10.79
5"	5.563							.258	14.62
6"	6.625							.280	18.97
8"	8.625			.250	22.36	.277	24.70	.322	28.55
10"	10.750			.250	28.04	.307	34.24	.365	40.48
12"	12.750			.250	33.38	.330	43.77	.375	49.56
14"	14.000	.250	36.71	.312	45.61	.375	54.57	.375	54.57
16"	16.000	.250	42.05	.312	52.27	.375	62.58	.375	62.58
18"	18.000	.250	47.39	.312	58.94	.438	82.15	.375	70.59
20"	20.000	.250	52.73	.375	78.60	.500	104.13	.375	78.60
22"	22.000	.250	58.07	.375	86.61	.500	114.81	.375	86.61
24"	24.000	.250	63.41	.375	94.62	.562	140.68	.375	94.62
26"	26.000	.312	85.60	.500	136.17			.375	102.63
28"	28.000	.312	92.41	.500	146.85	.625	182.73	.375	110.64
30"	30.000	.312	98.93	.500	157.53	.625	196.08	.375	118.65
32"	32.000	.312	105.59	.500	168.21	.625	209.43	.375	126.66
34"	34.000	.312	112.25	.500	178.89	.625	222.78	.375	134.67
36"	36.000	.312	118.92	.500	189.57	.625	236.13	.375	142.68

(continued on next page)

Figure 22-10 Steel Pipe/Size and Weight Data
(continued from previous page)

Nom. Size	O.D.	Schedule 40		Schedule 60		Extra Strong		Schedule 80	
		Wall	Lbs./Ft.	Wall	Lbs./Ft.	Wall	Lbs./Ft.	Wall	Lbs./Ft.
1/8"	.405	.068	.24			.095	.31	.095	.31
1/4"	.540	.088	.42			.119	.54	.119	.54
3/8"	.675	.091	.57			.126	.74	.126	.74
1/2"	.840	.109	.85			.147	1.09	.147	1.09
3/4"	1.050	.113	1.13			.154	1.47	.154	1.47
1"	1.315	.133	1.68			.179	2.17	.179	2.17
1-1/4"	1.660	.140	2.27			.191	3.00	.191	3.00
1-1/2"	1.900	.145	2.72			.200	3.63	.200	3.63
2"	2.375	.154	3.65			.218	5.02	.218	5.02
2-1/2"	2.875	.203	5.79			.276	7.66	.276	7.66
3"	3.500	.216	7.58			.300	10.25	.300	10.25
3-1/2"	4.000	.226	9.11			.318	.12.50	.318	12.50
4"	4.500	.237	10.79			.337	14.98	.337	14.98
5"	5.563	.258	14.62			.375	20.78	.375	20.78
6"	6.625	.280	18.97			.432	28.57	.432	28.57
8"	8.625	.322	28.55	.406	35.64	.500	43.39	.500	43.39
10"	10.750	.365	40.48	.500	54.74	.500	54.74	.594	64.43
12"	12.750	.406	53.52	.562	73.15	.500	65.42	.688	88.63
14"	14.000	.438	63.44	.594	85.05	.500	72.09	.750	106.13
16"	16.000	.500	82.77	.656	107.50	.500	82.77	.844	136.61
18"	18.000	.562	104.67	.750	138.17	.500	93.45	.938	170.92
20"	20.000	.594	123.11	.812	166.40	.500	104.13	1.031	208.87
22"	22.000			.875	197.41	.500	114.81	1.125	250.81
24"	24.000	.688	171.29	.969	238.35	.500	125.49	1.219	296.53
26"	26.000					.500	136.17		
28"	28.000					.500	146.85		
30"	30.000					.500	157.53		
32"	32.000	.688	229.92			.500	168.21		
34"	34.000	.688	244.77			.500	178.89		
36"	36.000	.750	282.35			.500	189.57		

(continued on next page)

Figure 22-10 Steel Pipe/Size and Weight Data
(continued from previous page)

Nom. Size	O.D.	Schedule 100		Schedule 120		Schedule 140		Schedule 160		XX Strong	
		Wall	Lbs./Ft.	Wall	Lbs./Ft.	Wall	Lbs./Ft.	Wall	Lbs./Ft.	Wall	Lbs./Ft.
1/8"	.405										
1/4"	.540										
3/8"	.675										
1/2"	.840								.188	1.31	.294	1.71
3/4"	1.050								.219	1.94	.308	2.44
1"	1.315								.250	2.84	.358	3.66
1-1/4"	1.660								.250	3.76	.382	5.21
1-1/2	1.900								.281	4.86	.400	6.41
2"	2.375								.344	7.46	.436	9.03
2-1/2"	2.875								.375	10.01	.552	13.69
3"	3.500								.438	14.32	.600	18.58
3-1/2"	4.000										.636	22.85
4"	4.500			.438	19.00			.531	22.51	.674	27.54
5"	5.563			.500	27.04			.625	32.96	.750	38.55
6"	6.625			.562	36.39			.719	45.35	.864	53.16
8"	8.625	.594	50.95	.719	60.71	.812	67.76	.906	74.69	.875	72.42
10"	10.750	.719	77.03	.844	89.29	1.000	104.13	1.125	115.64	1.000	104.13
12"	12.750	.844	107.32	1.000	125.49	1.125	139.67	1.312	160.27	1.000	125.49
14"	14.000	.938	130.85	1.094	150.79	1.250	170.21	1.406	189.11		
16"	16.000	1.031	164.82	1.219	192.43	1.438	223.64	1.594	245.25		
18"	18.000	1.156	207.96	1.375	244.14	1.562	274.22	1.781	308.55		
20"	20.000	1.281	256.19	1.500	296.37	1.750	341.10	1.969	379.14		
22"	22.000	1.375	302.88	1.625	353.61	1.875	403.01	2.125	451.07		
24"	24.000	1.531	367.45	1.812	429.50	2.062	483.24	2.344	542.09		
26"	26.000										
28"	28.000										
30"	30.000										
32"	32.000										
34"	34.000										
36"	36.000										

Figure 22-11 Bolting Information for Standard Flanges

150 Lb. Steel Flanges				
Pipe Size	Diam. of Bolt Circle	Diam. of Bolts	No. of Bolts	Bolt Length
1/2	2-3/8	1/2	4	1-3/4
3/4	2-3/4	1/2	4	2
1	3-1/8	1/2	4	2
1-1/4	3-1/2	1/2	4	2-1/4
1-1/2	3-7/8	1/2	4	2-1/4
2	4-3/4	5/8	4	2-3/4
2-1/2	5-1/2	5/8	4	3
3	6	5/8	4	3
3-1/2	7	5/8	8	3
4	7-1/2	5/8	8	3
5	8-1/2	3/4	8	3-1/4
6	9-1/2	3/4	8	3-1/4
8	11-3/4	3/4	8	3-1/2
10	14-1/4	7/8	12	3-3/4
12	17	7/8	12	4
14	18-3/4	1	12	4-1/4
16	21-1/4	1	16	4-1/2
18	22-3/4	1-1/8	16	4-3/4
20	25	1-1/8	20	5-1/4
22	27-1/4	1-1/4	20	5-1/2
24	29-1/2	1-1/4	20	5-3/4
26	31-3/4	1-1/4	24	6
30	36	1-1/4	28	6-1/4
34	40-1/2	1-1/2	32	7
36	42-3/4	1-1/2	32	7
42	49-1/2	1-1/2	36	7-1/2

Steel, cast iron, bronze, stainless, etc. Flanges have identical bolting requirements per pipe size and pressure rating.

Figure 22-12 Support Spacing for Metal Pipe

Nominal Pipe Size Inches	Span Water Feet	Steam, Gas, Air Feet	Rod Size
1	7	9	
1-1/2	9	12	3/8"
2	10	13	
2-1/2	11	14	
3	12	15	1/2"
3-1/2	13	16	
4	14	17	5/8"
5	16	19	
6	17	21	3/4"
8	19	24	
10	20	26	7/8"
12	23	30	
14	25	32	
16	27	35	1"
18	28	37	
20	30	39	
24	32	42	1-1/4"
30	33	44	

Figure 22-13 Copper Tubing Dimensions and Weights

Nominal Size of Tube/In.	Actual Outside Diam. of Tube/In.	Type K Tube		Type L Tube		Type M Tube	
		Wall Thick-ness/In.	Weight Per Foot/Lbs.	Wall Thick-ness/In.	Weight Per Foot/Lbs.	Wall Thick-ness/In.	Weight Per Foot/Lbs.
1/4	3/8			.030	.126		
3/8	1/2	.049	.269	.035	.198		
1/2	5/8	.049	.344	.040	.285		
3/4	7/8	.065	.641	.045	.455		
1	1-1/8	.065	.839	.050	.655		
1-1/4	1-3/8	.065	1.04	.055	.884	.042	.68
1-1/2	1-5/8	.072	1.36	.060	1.14	.049	.94
2	2-1/8	.083	2.06	.070	1.75	.058	1.46
2-1/2	2-5/8	.095	2.93	.080	2.48	.065	2.03
3	3-1/8	.109	4.00	.090	3.33	.072	2.68
3-1/2	3-5/8	.120	5.12	.100	4.29	.083	3.58
4	4-1/8	.130	6.51	.110	5.38	.095	4.66
5	5-1/8	.160	9.67	.125	7.61	.109	6.66
6	6-1/8	.192	13.90	.140	10.20	.122	8.92
8	8-1/8	.271	25.90	.200	19.30	.170	16.50
10	10-1/8	.338	40.30	.250	30.10	.212	25.60
12	12-1/8	.405	57.80	.280	40.40	.254	36.70

(continued on next page)

Figure 22-13 Copper Tubing Dimensions and Weights
(continued from previous page)

Copper Pipe & Tubing

When measuring copper pipe, sweat fittings are measured by their inside diameter (ID) and compression fittings are measured by their outside diameter (OD). Hard temper comes in 20-foot straight lengths and soft temper comes in 20-foot straight lengths or 60-foot coils. Copper tubing is normally designed to conform with ASTM Designation B88. See the code for specific information on each type.

Use lead-free solid core solder (NOT ROSIN CORE) and a high quality flux when soldering sweat fittings.

Types of Copper Pipe

Type	Characteristics
DWV	DWV stands for "Drain, Waste, and Vent" and is recommended for above ground use only and no pressure applications. Sweat fittings only. Available only in hard type and in sizes from 1-1/4 inch to 6 inch.
K	A thick-walled, flexible copper tubing. Much thicker wall than Type L and M and is required for all underground installations. Typical uses include water services, plumbing, heating, steam, gas, oil, oxygen, and other applications where thick-walled tubing is required. Can be used with sweat, flared, and compression fittings. Available in hard and soft types.
L	Standard tubing used for interior, above ground plumbing. Uses include heating, air-conditioning, steam, gas and oil and for underground drainage lines. This is a flexible tubing but be very careful not to crimp the line when bending it. Special tools (inexpensive) are readily available to make bending much easier and safer. Although sweat, compression, and flare fittings are available, only sweat and flare fittings are legal for gas lines. Available in hard and soft types.
M	Typically used with interior heating and pressure line applications. Wall thickness is slightly less than types K and L. Normally used with sweat fittings. Available in hard and soft types.

(courtesy Pocket Ref, *3rd Edition, Thomas J. Glover, Sequoia Publishing, Inc.)*

Figure 22-14 Copper Tubing: Commercially Available Lengths

	Hard Drawn		Annealed	

Type K	Straight lengths up to:
	  8-inch diameter
	  10-inch diameter
	  12-inch diameter
		20 ft.
		18 ft.
		12 ft.
			Straight lengths up to:
			  8-inch diameter
			  10-inch diameter
			  12-inch diameter
			
			Coils up to:
			  1-inch diameter
			
			  1-1/4 and 1-1/2 inch diameter
			  2-inch diameter
				20 ft.
				18 ft.
				12 ft.
				
				60 ft.
				100 ft.
				60 ft.
				40 ft.
				45 ft.

Type L	Straight lengths up to:
	  10-inch diameter
	  12-inch diameter
		20 ft.
		18 ft.
			Straight Lengths up to:
			  10-inch diameter
			  12-inch diameter
			
			Coils up to:
			  1-inch diameter
			
			  1-1/4 and 1-1/2 inch diameter
			  2-inch diameter
				20 ft.
				18 ft.
				
				60 ft.
				100 ft.
				60 ft.
				40 ft.
				45 ft.

Type M	Straight lengths:
	  All diameters
		20 ft.	Straight lengths up to:
			  12-inch diameter
			
			Coils up to:
			  1-inch diameter
			
			  1-1/4 and 1-1/2-inch diameter
			  2-inch diameter
				20 ft.
				
				60 ft.
				100 ft.
				60 ft.
				40 ft.
				45 ft.

DWV	Straight lengths: All diameters	20 ft.	Not available	—
ACR	Straight lengths:	20 ft.	Coils:	50 ft.

Figure 22-15 Valve Selection Considerations

In any piping application, valve performance is critical. Valves should be selected to give the best performance at the lowest cost.

The following is a list of performance characteristics generally expected of valves:

1. Stopping flow or starting it
2. Throttling flow (modulation)
3. Changing the direction of flow
4. Checking backflow (permitting flow in only one direction)
5. Relieving or regulating pressure

In order to properly select the right valve, some facts must be determined, including the following:

1. What liquid or gas will flow through the valve?
2. Does the fluid contain suspended particles?
3. Does the fluid remain in liquid form at all times?
4. Which metals does the fluid corrode?
5. What are the pressure and temperature limits? (As temperature and pressure rise, so will the price of the valve.)
6. Is there constant line pressure?
7. Is the valve merely an on-off valve?
8. Will checking of backflow be required?
9. Will the valve operate frequently or infrequently?

Valves are classified by design type into such classifications as gate, globe, angle, check, ball, butterfly, and plug. They are also classified by end connection, stem, pressure restrictions, and material such as bronze, cast iron, etc. Each valve has a specific use. A quality valve used correctly will provide a lifetime of trouble-free service, but even a high quality valve installed in the wrong service may require frequent attention.

Figure 22-16 Valve Materials, Service Pressures, Definitions

Valve Materials

Bronze:

Bronze is one of the oldest materials used to make valves. It is most commonly used in hot and cold water systems and other noncorrosive services. It is often used as a seating surface in larger iron body valves to ensure tight closure.

Carbon Steel:

Carbon steel is a high strength material. Therefore, valves made from this metal are used in higher pressure services, such as steam lines up to 600 psi at 850°F. Many steel valves are available with butt-weld ends for economy and are generally used in high pressure steam service as well as other higher pressure noncorrosive services.

Forged Steel:

Valves from tough carbon steel are used in service up to 2000 psi and temperatures up to 1000°F in gate, globe, and check valves.

Iron:

Valves are normally used in medium to large pipe lines to control noncorrosive fluid and gases, where pressures do not exceed 250 psi at 450° or 500 psi cold water, oil, or gas.

Stainless Steel:

Developed steel alloys can be used in over 90% corrosive services.

Plastic PVC:

This is used in a great variety of valves, generally in high corrosive service with lower temperatures and pressures.

Valve Service Pressures

Pressure ratings on valves provide an indication of the safe operating pressure for a valve at some elevated temperature. This temperature is dependent upon the materials used and the fabrication of the valve. When specific data is not available, a good "rule-of-thumb" to follow is the temperature of saturated steam on the primary rating indicated on the valve body. Example: The valve has the number 150S printed on the side indicating 150 psi and hence, a maximum operating temperature of 367°F (temperature of saturated steam and 150 psi).

Valve Definitions

1. "WOG" – Water, oil, gas (cold working pressures).
2. "SWP" – Steam working pressure.
3. 100% area (full port) – Means the area through the valve is equal to or greater than the area of standard pipe.
4. "Standard Opening" – Means that the area through the valve is less than the area of standard pipe, and therefore these valves should be used only where restriction of flow is unimportant.
5. "Round Port" – Means the valve has a full round opening through the plug and body, of the same size and area as standard pipe.
6. "Rectangular Port" – Valves have rectangular shaped ports through the plug body. The area of the port is either equal to 100% of the area of standard pipe, or restricted (standard opening). In either case it is clearly marked.
7. "ANSI" – American National Standards Institute.

Figure 22-17a Stem Types

(O.S. & Y) - Rising Stem—Outside Screw and Yoke
Offers a visual indication of whether the valve is open or
closed. The stem threads are engaged by the yoke bushing so
the stem rises through the hand wheel as it is turned.

(R.S.) - Rising Stem—Inside Screw
Adequate clearance for operation must be provided because
both the hand wheel and the stem rise. The valve wedge
position is indicated by the position of the stem and hand
wheel.

(N.R.S.) - Non-Rising Stem—Inside Screw
A minimum clearance is required for operating this type of
valve. Excessive wear or damange to stem threads inside the
valve may be caused by heat, corrosion, and solids. Because
the hand wheel and stem do not rise, wedge position cannot be
visually determined.

Figure 22-17b Valve Types

Gate Valves
Provide full flow, minute pressure drop, minimum turbulence
and minimum fluid trapped in the line.
They are normally used where operation is infrequent.

Globe Valves
Globe valves are designed for throttling and/or frequent
operation with positive shut-off. Particular attention must be
paid to the several types of seating materials available to avoid
unnecessary wear. The seats must be compatible with the fluid
in service and may be composition or metal. The configuration
of the Globe valve opening causes turbulence which results
in increased resistance. Most bronze Globe valves are rising
stem-inside screw, but they are also available on O.S. & Y.

Angle Valves
The fundamental difference between the Angle valve and the
Globe valve is the fluid flow through the Angle valve. It makes
a 90° turn and offers less resistance to flow than the Globe
valve while replacing an elbow. An Angle valve thus reduces the
number of joints and installation time.

Check Valves
Check valves are designed to prevent backflow by automatically
seating when the direction of fluid is reversed.
Swing Check valves are generally installed with Gate valves, as
they provide comparable full flow.

Usually recommended for lines where flow velocities are
low and should not be used on lines with pulsating flow.
Recommended for horizontal installation, or in vertical lines
only where flow is upward.

(continued on next page)

Division 22

Figure 22-17b Valve Types *(continued from previous page)*

Lift Check Valves

These are commonly used with Globe and Angle valves since they have similar diaphragm seating arrangements and are recommended for preventing backflow of steam, air, gas, and water, and on vapor lines with high flow velocities. For horizontal lines, horizontal lift checks should be used and vertical lift checks for vertical lines.

Ball Valves

Ball valves are light and easily installed, yet because of modern elastomeric seats, provide light closure. Flow is controlled by rotating up to 90° a drilled ball which fits tightly against resilient seals. This ball seats with flow in either direction, and valve handle indicates the degree of opening. Recommended for frequent operation, readily adaptable to automation, ideal for installation where space is limited.

Butterfly Valves

Butterfly valves provide bubble-tight closure with excellent throttling characteristics. They can be used for full-open, closed, and for throttling applications.
The Butterfly valve consists of a disc within the valve body which is controlled by a shaft.
In its closed position, the valve disc seals against a resilient seat. The disc position throughout the full 90° rotation is visually indicated by the position of the operator.
A Butterfly valve is only a fraction of the weight of a Gate valve and requires no gaskets between flanges in most cases. Recommended for frequent operation and adaptable to automation where space is limited.
Wafer and Lug type bodies, when installed between two pipe flanges, can be easily removed from the line. The pressure of the bolted flanges holds the valve in place.
Locating lugs makes installation easier.

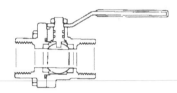

Plug Valves

Lubricated plug valves, because of the wide range of service to which they are adapted, may be classified as all purpose valves. They can be safely used at all pressures and vacuums, and at all temperatures up to the limits of available lubricants. They are the most satisfactory valves for the handling of gritty suspensions and many other destructive, erosive, corrosive and chemical solutions.

Figure 22-18 Roof Drain Sizing Chart

Pipe Diameter	Max. S.F. Roof Area*	Gal./Min.
2"	544	23
3"	1,610	67
4"	3,460	144
5"	6,280	261
6"	10,200	424
8"	22,000	913

*Design Assumptions: Vertical conductor size is based on a maximum rate of rainfall to 4" per hour. To convert roof area to other rates multiply "Max. S.F. Roof Area" shown by four and divide the result by desired local rate. The answer is the local roof area that may be handled by the indicated pipe diameter.

The above chart is meant to be used as a guide. Local building codes and requirements may vary from the values indicated and should be used when available.

Figure 22-19   Minimum Plumbing Fixture Requirements*

Type of Building or Occupancy (2)	Water Closets (14) (Fixtures per Person)		Urinals (6,10) (Fixtures per Person)		Lavatories (Fixtures per Person)		Bathtubs or Showers (Fixtures per Person)	Drinking Fountains (Fixtures per Person) (3, 13)
	Male	Female	Male	Female	Male	Female		
Assembly Places- Theatres, Auditoriums, Convention Halls, etc.-for permanent employee use	1: 1 - 15 2: 16 - 35 3: 36 - 55 Over 55, add 1 fixture for each additional 40 persons	1: 1 - 15 2: 16 - 35 3: 36 - 55	0: 1 - 9 1: 10 - 50 Add one fixture for each additional 50 males		1 per 40	1 per 40		
Assembly Places- Theatres, Auditoriums, Convention Halls, etc. - for public use	1: 1 - 100 2: 101 - 200 3: 201 - 400 Over 400, add 1 fixture for each additional 500 males and 1 for each additional 125 females	3: 1 - 50 4: 51 - 100 6: 101 - 200 11: 201 - 400	1: 1 - 100 2: 101 - 200 3: 201 - 400 4: 401 - 600 Over 600, add 1 fixture for each additional 300 males		1: 1 - 200 2: 201 - 400 3: 401 - 750 Over 750, add 1 fixture for each additional 500 persons	1: 1 - 200 2: 201 - 400 3: 401 - 750		1: 1 - 150 2: 151 - 400 3: 401 - 750 Over 750, add one fixture for each additional 500 persons
Dormitories (9) School or Labor	1 per 10 Add 1 fixture for each additional 25 males (over 10) and 1 for each additional 20 females (over 8)	1 per 8	1 per 25 Over 150, add 1 fixture for each additional 50 males		1 per 12 Over 12 add 1 fixture for each additional 20 males and 1 for each 15 additional females	1 per 12	1 per 8 For females add 1 bathtub per 30. Over 150, add 1 per 20	1 per 150 (12)
Dormitories- for Staff Use	1: 1 - 15 2: 16 - 35 3: 36 - 55 Over 55, add 1 fixture for each additional 40 persons	1: 1 - 15 3: 16 - 35 4: 36 - 55	1 per 50		1 per 40	1 per 40	1 per 8	
Dwellings: Single Dwelling Multiple Dwelling or Apartment House	1 per dwelling 1 per dwelling or apartment unit				1 per dwelling 1 per dwelling or apartment unit		1 per dwelling 1 per dwelling or apartment unit	
Hospital Waiting rooms	1 per room				1 per room			1 per 150 (12)
Hospitals- for employee use	1: 1 - 15 2: 16 - 35 3: 36 - 55 Over 55, add 1 fixture for each additional 40 persons	1: 1 - 15 3: 16 - 35 4: 36 - 55	0: 1 - 9 1: 10 - 50 Add 1 fixture for each additional 50 males		1 per 40	1 per 40		
Hospitals: Individual Room Ward Room	1 per room 1 per 8 patients				1 per room 1 per 10 patients		1 per room 1 per 20 patients	1 per 150 (12)
Industrial (6) Warehouses Workshops, Foundries and similar establishments- for employee use	1: 1 -10 2: 11 - 25 3: 26 - 50 4: 51 - 75 5: 76 - 100 Over 100, add 1 fixture for each additional 30 persons	1: 1 -10 2: 11 - 25 3: 26 - 50 4: 51 - 75 5: 76 - 100			Up to 100, per 10 persons Over 100, 1 per 15 persons (7, 8)		1 shower for each 15 persons exposed to excessive heat or to skin contamination with poisonous, infectious or irritating material	1 per 150 (12)
Institutional - Other than Hospitals or Penal Institutions (on each occupied floor)	1 per 25	1 per 20	0: 1 - 9 1: 10 - 50 Add 1 fixture for each additional 50 males		1 per 10	1 per 10	1 per 8	1 per 150 (12)
Institutional - Other than Hospitals or Penal Institutions (on each occupied floor)- for employee use	1: 1 - 15 2: 16 - 35 3: 36 - 55 Over 55, add 1 fixture for each additional 40 persons	1: 1 - 15 3: 16 - 35 4: 36 - 55	0: 1 - 9 1: 10 - 50 Add 1 fixture for each additional 50 males		1 per 40	1 per 40	1 per 8	1 per 150 (12)
Office or Public Buildings	1: 1 - 100 2: 101 - 200 3: 201 - 400 Over 400, add 1 fixture for each additional 500 males and 1 for each additional 150 females	3: 1 - 50 4: 51 - 100 8: 101 - 200 11: 201 - 400	1: 1 - 100 2: 101 - 200 3: 201 - 400 4: 401 - 600 Over 600, add 1 fixture for each additional 300 males		1: 1 - 200 2: 201 - 400 3: 401 - 750 Over 750, add 1 fixture for each additional 500 persons	1: 1 - 200 2: 201 - 400 3: 401 - 750		1 per 150 (12)
Office or Public Buildings - for employee use	1: 1 - 15 2: 16 - 35 3: 36 - 55 Over 55, add 1 fixture for each additional 40 persons	1: 1 - 15 3: 16 - 35 4: 36 - 55	0: 1 - 9 1: 10 - 50 Add 1 fixture for each additional 50 males		1 per 40	1 per 40		

*Note: Minimum plumbing fixture requirements vary in the several plumbing codes in use. The most recent plumbing code approved for the structure location should be reviewed prior to actual design.

(continued on next page)

Figure 22-19 Minimum Plumbing Fixture Requirements
(continued from previous page)

	Suggested Minimum Plumbing Fixture Requirements							
Type of Building or Occupancy	Water Closets (14) (Fixtures per Person)		Urinals (5, 10) (Fixtures per Person)		Lavatories (Fixtures per Person)		Bathtubs or Showers (Fixtures per Person)	Drinking Fountains (Fixtures per Person) (3, 13)
	Male	Female	Male	Female	Male	Female		
Penal Institutions - for employee use	1: 1 - 15 2: 18 - 35 3: 36 - 55 Over 55, add 1 fixture for each additional 40 persons	1: 1 - 15 3: 18 - 35 4: 36 - 55	0: 1 - 9 1: 10 - 50 Add 1 fixture for each additional 50 males		1 per 40	1 per 40		1 per 150 (12)
Penal Institutions - for prison use								
Cell	1 per cell				1 per cell			1 per cellblock floor
Exercise room	1 per exercise room		1 per exercise room		1 per exercise room			1 per exercise room
Restaurants, Pubs and Lounges (11)	1: 1 - 50 2: 51 - 150 3: 151 - 300 Over 300, add 1 fixture for each additional 200 persons	1: 1 - 50 2: 51 - 150 4: 151 - 300	1: 1 - 150 Over 150, add 1 fixture for each additional 150 males		1: 1 - 150 2: 151 - 200 3: 201 - 400 Over 400, add 1 fixture for each additional 400 persons	1: 1 - 150 2: 151 - 200 3: 201 - 400		
Schools - for staff use All Schools	1: 1 - 15 2: 18 - 35 3: 36 - 55 Over 55, add 1 fixture for each additional 40 persons	1: 1 - 15 3: 18 - 35 4: 36 - 55	1 per 50		1 per 40	1 per 40		
Schools - for student use: Nursery	1: 1 - 20 2: 21 - 50 Over 50, add 1 fixture for each additional 50 persons	1: 1 - 20 2: 21 - 50			1: 1 - 25 2: 26 - 50 Over 50, add 1 fixture for each additional 50 persons	1: 1 - 25 2: 26 - 50		1 per 150 (12)
Elementary	1 per 30	1 per 25	1 per 75		1 per 35	1 per 35		1 per 150 (12)
Secondary	1 per 40	1 per 30	1 per 35		1 per 40	1 per 40		1 per 150 (12)
Others (Colleges, Universities, Adult Centers, etc.	1 per 40	1 per 30	1 per 35		1 per 40	1 per 40		1 per 150 (12)
Worship Places Educational and Activities Unit	1 per 150	1 per 75	1 per 150		1 per 2 water closets			1 per 150 (12)
Worship Places Principal Assembly Place	1 per 150	1 per 75	1 per 150		1 per 2 water closets			1 per 150 (12)

Notes:

1. The figures shown are based upon one (1) fixture being the minimum required for the number of persons indicated or any fraction thereof.
2. Building categories not shown on this table shall be considered separately by the administrative authority.
3. Drinking fountains shall not be installed in toilet rooms.
4. Laundry trays. One (1) laundry tray or one (1) automatic washer standpipe for each dwelling unit or one (1) laundry tray or one (1) automatic washer standpipe, or combination thereof, for each twelve (12) apartments. Kitchen sinks, one (1) for each dwelling or apartment unit.
5. For each urinal added in excess of the minimum required, one water closet may be deducted. The number of water closets shall not be reduced to less than two-thirds (2/3) of the minimum requirement.
6. As required by ANSI Z4.1-1968. Sanitation in Places of Employment.
7. Where there is exposure to skin contamination with poisonous, infectious, or irritating materials, provide one (1) lavatory for each five (5) persons.
8. Twenty-four (24) linear inches of wash sink or eighteen (18) inches of a circular basin, when provided with water outlets for such space shall be considered equivalent to one (1) lavatory.
9. Laundry trays, one (1) for each fifty (50) persons. Service sinks, one (1) for each hundred (100) persons.
10. General. In applying this schedule of facilities, consideration shall be given to the accessibility of the fixtures. Conformity purely on a numerical basis may not result in an installation suited to the need of the individual establishment. For example, schools should be provided with toilet facilities on each floor having classrooms.
    a. Surrounding materials, wall and floor space to a point two (2) feet in front of urinal lip and four (4) feet above the floor, and at least two (2) feet to each side of the urinal shall be lined with non-absorbent materials.
    b. Trough urinals shall be prohibited.
11. A restaurant is defined as a business which sells food to be consumed on the premises.
    a. The number of occupants for a drive-in restaurant shall be considered as equal to the number of parking stalls.
    b. Employee toilet facilities shall not to be included in the above restaurant requirements. Hand washing facilities shall be available in the kitchen for employees.
12. Where food is consumed indoors, water stations may be substituted for drinking fountains. Offices, or public buildings for use by more than six (6) persons shall have one (1) drinking fountain for the first one hundred fifty (150) persons and one additional fountain for each three hundred (300) persons thereafter.
13. There shall be a minimum of one (1) drinking fountain per occupied floor in schools, theaters, auditoriums, dormitories, and offices of public buildings.
14. The total number of water closets for females shall be at least equal to the total number of water closets and urinals required for males.

Figure 22-20 Fixture Demands in Gallons per Hour per Fixture

The table below is based on 140°F final temperature except for dishwashers in public places where 180°F water is mandatory.

Fixture	Apt. House	Club	Gym	Hos-pital	Hotel	Indust. Plant	Office	Private Home	School
Bathtubs	20	20	30	20	20			20	
Dishwashers, automatic*	15	50-150		50-150	50-200	20-100		15	20-100
Kitchen sink	10	20		20	30	20	20	10	20
Laundry, stationary tubs	20	28		28	28			20	
Laundry, automatic wash	75	75		100	150			75	
Private lavatory	2	2	2	2	2	2	2	2	2
Public lavatory	4	6	8	6	8	12	6		15
Showers	30	150	225	75	75	225	30	30	225
Service sink	20	20		20	30	20	20	15	20
Demand factor	0.30	0.30	0.40	0.25	0.25	0.40	0.30	0.30	0.40
Storage capacity factor	1.25	0.90	1.00	0.60	0.80	1.00	2.00	0.70	1.00

*To obtain the probable maximum demand, multiply the total demands for the fixtures (gal./fixture/hour) by the demand factor. The heater should have a heating capacity in gallons per hour equal to this maximum. The storage tank should have a capacity in gallons equal to the probable maximum demand multiplied by the storage capacity factor.

Figure 22-21 Water Cooler Application/Capacities

Type of Service	Requirement
Office, School or Hospital	12 persons per gallon per hour
Office Lobby or Department Store	4 or 5 gallons per hour per fountain
Light manufacturing	7 persons per gallon per hour
Heavy manufacturing	5 persons per gallon per hour
Hot heavy manufacturing	4 persons per gallon per hour
Hotels	.08 gallons per hour per room
Theatre	1 gallon per hour per 100 seats

Figure 22-22 Drainage Requirements for Plumbing Fixtures

Drainage lines must be on a slope to maintain flow for proper operation. This slope should not be less than 1/4" per foot for 3" diameter or smaller pipe or not less than 1/8" per foot for 4" diameter or larger pipe. The capacity of building drainage systems is calculated in a basis of "drainage fixture units" (d.f.u.) as per the following chart.

Type of Fixture	d.f.u. Value	Type of Fixture	d.f.u. Value
Automatic clothes washer (2" standpipe)	3	Service sink (trap standard)	3
Bathroom group (water closet, lavatory and		Service sink (P trap)	2
bathtub or shower) tank type closet	6	Urinal, pedestal, syphon jet blowout	6
Bathtub (with or without overhead shower)	2	Urinal, wall hung	4
Clinic sink	6	Urinal, stall washout	4
Combination sink & tray with food disposal	4	Wash sink (circ. or mult.) per faucet set	2
Dental unit or cuspidor	1	Water closet, tank operated	4
Dental lavatory	1	Water closet, valve operated	6
Drinking fountain	1/2	Fixtures not listed above	
Dishwasher, domestic	2	Trap size 1-1/4" or smaller	1
Floor drains with 2" waste	3	Trap size 1-1/2"	2
Kitchen sink, domestic with one 1-1/2" trap	2	Trap size 2"	3
Kitchen sink, domestic with food disposal	2	Trap size 2-1/2"	4
Lavatory with 1-1/4" waste	1	Trap size 3"	5
Laundry tray (1 or 2 compartment)	2	Trap size 4"	6
Shower stall, domestic	2		

For continuous or nearly continuous flow into the system from a pump, air conditioning equipment, or other item, allow 2 fixture units for each gallon per minute of flow.

When the d.f.u. for each horizontal branch or vertical stack is computed from the table above, the appropriate pipe size for each branch or stack is determined from the table below.

Figure 22-23 Allowable Fixture Units (d.f.u.) for Branches and Stacks

Pipe Diam.	Horiz. Branch (not incl. drains)	Stack Size for 3 Stories or 3 Levels	Stack Size for Over 3 Levels	Maximum for 1 Story Building Stack
1-1/2"	3	4	8	2
2"	6	10	24	6
2-1/2"	12	20	42	9
3"	20*	48*	72*	20*
4"	160	240	500	90
5"	360	540	1100	200
6"	620	960	1900	350
8"	1400	2200	3600	600
10"	2500	3800	5600	1000
12"	3900	6000	8400	1500
15"	7000			

*Not more than two water closets or bathroom groups within each branch interval nor more than six water closets or bathroom groups on the stack. Stacks sized for the total may be reduced as load decreases at each story to a minimum diameter of ½ the maximum diameter.

Figure 22-24 Hot Water Consumption Rates for Commercial Applications

Type of Building	Size Factor	Maximum Hourly Demand	Average Day Demand
Apartment Dwellings	No. of apartments: Up to 20 21 to 50 51 to 75 76 to 100 101 to 200 201 up	 12.0 gal. per apt. 10.0 gal. per apt. 8.5 gal. per apt. 7.0 gal. per apt. 6.0 gal. per apt. 5.0 gal. per apt.	 42.0 gal. per apt. 40.0 gal. per apt. 38.0 gal. per apt. 37.0 gal. per apt. 36.0 gal. per apt. 35.0 gal. per apt.
Dormitories	Men Women	3.8 gal. per man 5.0 gal. per woman	13.1 gal. per man 12.3 gal. per woman
Hospitals	Per bed	23.0 gal. per Patient	90.0 gal. per Patient
Hotels	Single room with bath Double room with bath	17.0 gal. per unit 27.0 gal. per unit	50.0 gal. per unit 80.0 gal. per unit
Motels	No. of units: Up to 20 21 to 100 101 Up	 6.0 gal. per unit 5.0 gal. per unit 4.0 gal. per unit	 20.0 gal. per unit 14.0 gal. per unit 10.0 gal. per unit
Nursing Homes		4.5 gal. per bed	18.4 gal. per bed
Office Buildings		0.4 gal. per person	1.0 gal. per person
Restaurants	Full meal type Drive-in snack type	1.5 gal./max. meals/hr. 0.7 gal./max. meals/hr.	2.4 gal. per meal 0.7 gal. per meal
Schools	Elementary Secondary and high	0.6 gal. per student 1.0 gal. per student	0.6 gal. per student 1.8 gal. per student

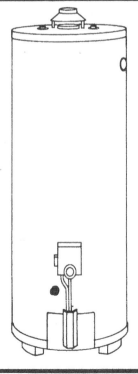

Residential Gas Fired Water Heater

Figure 22-25 Recommended Thickness of Insulation for Fiberglass and Rock Wool on Piping and Ductwork

Process Temperature (°F)											
Nominal Pipe Size (in inches)		150°	250°	350°	450°	550°	650°	750°	850°	950°	1050°
1/2	Thickness	1	1-1/2	2	2-1/2	3	3-1/2	4	4	4-1/2	5-1/2
	Heat Loss	8	16	24	33	43	54	66	84	100	114
	Surf. Temp.	72	75	76	78	79	81	82	86	87	87
1	Thickness	1	1-1/2	2	2-1/2	3-1/2	4	4	4-1/2	5	5-1/2
	Heat Loss	11	21	30	41	49	61	79	96	114	135
	Surf. Temp.	73	76	78	80	79	81	84	86	88	89
1-1/2	Thickness	1	2	2-1/2	3	4	4	4	5-1/2	5-1/2	6
	Heat Loss	14	22	33	45	54	73	94	103	128	152
	Surf. Temp.	73	74	77	79	79	82	86	84	88	90
2	Thickness	1-1/2	2	3	3-1/2	4	4	4	5-1/2	6	6
	Heat Loss	13	25	24	47	61	81	105	114	137	168
	Surf. Temp.	71	75	75	77	79	83	87	85	87	91
3	Thickness	1-1/2	2-1/2	3-1/2	4	4	4-1/2	4-1/2	6	6-1/2	7
	Heat Loss	16	28	39	54	75	94	122	133	154	184
	Surf. Temp.	72	74	75	77	81	83	87	86	87	90
4	Thickness	1-1/2	3	4	4	4	5	5-1/2	6	7	7-1/2
	Heat Loss	19	29	42	63	88	102	126	152	174	206
	Surf. Temp.	72	73	74	78	82	86	85	87	88	90
6	Thickness	2	3	4	4	4-1/2	5	5-1/2	6-1/2	7-1/2	8
	Heat Loss	21	38	54	81	104	130	159	181	208	246
	Surf. Temp.	71	74	75	79	82	84	87	88	89	91
8	Thickness	2	3-1/2	4	4	5	5	5-1/2	7	8	8-1/2
	Heat Loss	26	42	65	97	116	155	189	204	234	277
	Surf. Temp.	71	73	76	80	81	86	89	88	89	92
10	Thickness	2	3-1/2	4	4	5	5-1/2	5-1/2	7-1/2	8-1/2	9
	Heat Loss	32	50	77	115	136	170	220	226	259	307
	Surf. Temp.	72	74	77	81	82	85	90	87	89	91

Heat Loss = BTU/per ft./per hr.

Based on 65°F ambient temperature

Users are advised to consult manufacturer's literature for specific product temperature limitations.

(continued on next page)

Division 22

Figure 22-25 Recommended Thickness of Insulation for Fiberglass and Rock Wool on Piping and Ductwork

(continued from previous page)

Nominal Pipe Size (in inches)		Process Temperature (°F)									
		150°	250°	350°	450°	550°	650°	750°	850°	950°	1050°
12	Thickness	2	3-1/2	4	4	5	5-1/2	5-1/2	7-1/2	8-1/2	9-1/2
	Heat Loss	36	57	87	131	154	192	249	253	290	331
	Surf. Temp.	72	74	77	82	82	86	91	88	89	91
14	Thickness	2	3-1/2	4	4	5	5-1/2	6-1/2	7-1/2	9	9-1/2
	Heat Loss	40	61	94	141	165	206	236	271	297	352
	Surf. Temp.	72	74	77	82	83	86	87	89	89	91
16	Thickness	2-1/2	3-1/2	4	4	5-1/2	5-1/2	7	8	9	10
	Heat Loss	37	68	105	157	171	228	247	284	326	372
	Surf. Temp.	71	74	78	83	82	87	86	88	89	91
18	Thickness	2-1/2	3-1/2	4	4	5-1/2	5-1/2	7	8	9	10
	Heat Loss	41	75	115	173	187	250	270	310	354	404
	Surf. Temp.	71	74	78	83	83	87	87	88	90	91
20	Thickness	2-1/2	3-1/2	4	4	5-1/2	5-1/2	7	8	9	10
	Heat Loss	45	82	126	189	204	272	292	335	383	436
	Surf. Temp.	71	75	78	83	83	87	87	89	90	92
24	Thickness	2-1/2	4	4	4	5-1/2	6	7-1/2	8	9	10
	Heat Loss	53	86	147	221	237	295	320	386	439	498
	Surf. Temp.	71	74	78	83	83	86	86	89	91	93
30	Thickness	2-1/2	4	4	4	5-1/2	6-1/2	7-1/2	8-1/2	10	10
	Heat Loss	65	105	179	268	286	332	383	439	481	591
	Surf. Temp.	71	74	79	84	84	85	87	89	89	94
36	Thickness	2-1/2	4	4	4	5-1/2	7	8	9	10	10
	Heat Loss	77	123	211	316	335	364	422	486	556	683
	Surf. Temp.	71	74	79	84	84	84	86	88	90	94
Flat	Thickness	2	3-1/2	4	4-1/2	5-1/2	8-1/2	9-1/2	10	10	10
	Heat Loss	10	14	20	27	31	27	31	38	47	58
	Surf. Temp.	72	74	77	80	82	80	82	85	89	93

Heat Loss = BTU/per ft./per hr.

Based on 65°F ambient temperature

Users are advised to consult manufacturer's literature for specific product temperature limitations.

Figure 22-26 Standard Plumbing and Piping Symbols

Symbol	Description	Abbreviation
——— SD ———	Storm drain, rainwater drain	SD, ST
—— ——— SSD ———	Subsoil drain, footing drain	SSD
——— SS ———	Soil, waste, or sanitary sewer	S, W, SAN, SS
— — — — — — — —	Vent	V
——— AW ———	Acid waste	AW
— — — AV — — — —	Acid vent	AV
——— D ———	Indirect drain	D
——— PD ———	Pump discharge line	PD
——— — — ———	Cold water	CW
——— — — ———	Hot water supply (140°F)[a]	HW
——— — — — ———	Hot water recirculating (140°F)[a]	HWR
——— TW ———	Tempered hot water (temp. °F)[b]	TEMP. HW, TW
——— TWR ———	Tempered hot water recirculating (temp. °F)[b]	TEMP. HWR, TWR
——— DWS ———	(Chilled) drinking water supply	DWS
——— DWR ———	(Chilled) drinking water recirculating	DWR
——— SW ———	Soft water	SW
——— CD ———	Condensate drain	CD
——— DI ———	Distilled water	DI
——— DE ———	Deionized water	DE

(continued on next page)

Division 22

Figure 22-26 Standard Plumbing and Piping Symbols
(continued from previous page)

Symbol	Description	Abbreviation
——————— CWS —————	Chilled water supply	CWS
——————— CWR —————	Chilled water return	CWR
——————— LS —————	Lawn sprinkler supply	LS
——————— F —————	Fire protection water supply	F
——————— G —————	Gas–low-pressure	G
——————— MG —————	Gas–medium-pressure	MG
——————— HG —————	Gas–high-pressure	HG
— — — — GV — — — — —	Gas vent	GV
——————— FOS —————	Fuel oil supply	FOS
——————— FOR —————	Fuel oil return	FOR
— — — — FOV — — — — —	Fuel oil vent	FOV
——————— LO —————	Lubricating oil	LO
— — — — LOV — — — — —	Lubricating oil vent	LOV
——————— WO —————	Waste oil	WO
— — — — WOV — — — —	Waste oil vent	WOV
——————— OX —————	Oxygen	OX
——————— LOX —————	Liquid oxygen	LOX
——————— A —————	Compressed air[c]	A
——————— X#A —————	Compressed air–X#[c]	X#A
——————— MA —————	Medical compressed air	MA
——————— LA —————	Laboratory compressed air	LA
——————— HHWS —————	(Heating) hot water supply	HHWS
——————— HHWR —————	(Heating) hot water return	HHWR
——————— V —————	Vacuum	VAC

(continued on next page)

Figure 22-26 Standard Plumbing and Piping Symbols
(continued from previous page)

Symbol	Description	Abbreviation
——— MV ———	Medical vacuum	MV
——— SV ———	Surgical vacuum	SV
——— LV ———	Laboratory vacuum	LV
——— N ———	Nitrogen	N
——— N_2O ———	Nitrous oxide	N_2O
——— CO_2 ———	Carbon dioxide	CO_2
——— WVC ———	Wet vacuum cleaning	WVC
——— DVC ———	Dry vacuum cleaning	DVC
——— LPS ———	Low-pressure steam supply	LPS
— — — LPC — — —	Low-pressure condensate	LPC
——— MPS ———	Medium-pressure steam supply	MPS
— — — MPC — — —	Medium-pressure condensate	MPC
——— HPS ———	High-pressure steam supply	HPS
— — — HPC — — —	High-pressure condensate	HPC
— — — ATV — — —	Atmospheric vent (steam or hot vapor)	ATV
	Gate valve	GV
	Globe valve	GLV
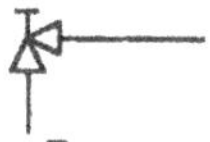	Angle valve	AV
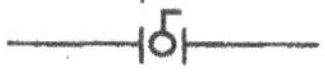	Ball valve	BV
	Butterfly valve	BFV
	Gas cock, gas stop	
	Balancing valve (specify type)	BLV
	Check valve	CV
	Plug valve	PV

(continued on next page)

Figure 22-26 Standard Plumbing and Piping Symbols
(continued from previous page)

Symbol	Description	Abbreviation
	Solenoid valve	
	Motor-operated valve (specify type)	
	Pressure-reducing valve	PRV
	Pressure-relief valve	RV
	Temperature-pressure-relief valve	TPV
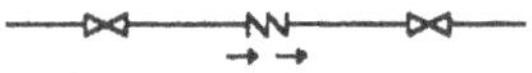	Reduced zone backflow preventer	RZBP
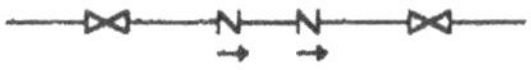	Double-check backflow preventer	DCBP
	Hose bibb	HB
	Recessed-box hose bibb or wall hydrant	WH
	Valve in yard box (valve type symbol as required for valve use)	YB
	Union (screwed)	
	Union (flanged)	
	Strainer (specify type)	
	Pipe anchor	PA
	Pipe guide	
	Expansion joint	EJ
	Flexible connector	FC
	Tee	
	Concentric reducer	
	Eccentric reducer	

(continued on next page)

Figure 22-26 Standard Plumbing and Piping Symbols
(continued from previous page)

Symbol	Description	Abbreviation
	Aquastat	
 FS	Flow switch	FS
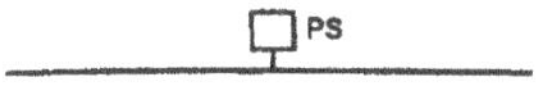 PS	Pressure switch	PS
	Water hammer arrester	WHA
	Pressure gauge with gauge cock	PG
	Thermometer (specify type)	
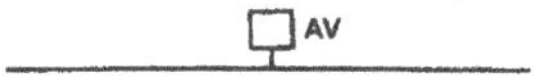 AV	Automatic air vent	AAV
	Valve in riser (type as specified or noted)	
	Riser down (elbow)	
	Riser up (elbow)	
	Air chamber	AC
	Rise or drop	
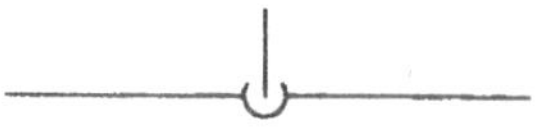	Branch–top connection	
	Branch–bottom connection	
	Branch–side connection	
	Cap on end of pipe	
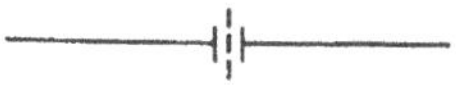	Flow indicator for stationary meter (orifice)	
	Flow indicator for portable meter (specify flow rate)	
	Upright fire sprinkler head	
	Fire hose rack	FHR

(continued on next page)

Division 22

Figure 22-26 Standard Plumbing and Piping Symbols
(continued from previous page)

Symbol	Description	Abbreviation
	Fire hose cabinet (surface-mounted)	FHC
	Fire hose cabinet (recessed)	FHC
	Cleanout plug	CO
	Floor cleanout	FCO
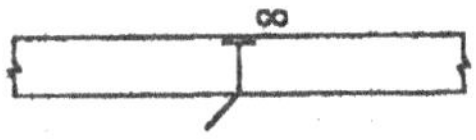	Wall cleanout	WCO
	Yard cleanout or cleanout to grade	CO
	Drain (all types) (specify)	D
	Pitch down or up–in direction of arrow	
	Flow–in direction of arrow	
	Point of connect	POC
	Outlet (specify type)	
	Steam trap (all types)	
	Floor drain with p-trap	FD

ªHot water (140°F) and hot water return (140°F). Use for normal hot water distribution system, usually but not necessarily (140°F). Change temperature designation if required.

ᵇHot water (temp. °F) and hot water return (temp. °F). Use for any domestic hot water system (e.g., tempered or sanitizing) required in addition to the normal system (see note "a" above). Insert system supply temperature where "temp." is indicated.

ᶜCompressed air and compressed air X#. Use pressure designations (X#) when compressed air is to be distributed at more than one pressure.

(continued on next page)

Figure 22-26 Standard Plumbing and Piping Symbols
(continued from previous page)

Referent (Synonym)	Symbol	Comments
Water supply and distribution symbols		
Mains, pipe		
Riser	⊗	
Hydrants		
Public hydrant, two hose outlets		Indicate size,[a] type of thread or connection.
Public hydrant, two hose outlets, and pumper connection		Indicate size,[a] type of thread or connection.
Wall hydrant, two hose outlets		Indicate size,[a] type of thread or connection.
Fire department connections		
Siamese fire department connection		Specify type, size, and angle.
Free-standing siamese fire department connection		Sidewalk or pit type, specify size.
Fire pumps		
Fire pump		Free-standing. Specify number and sizes of outlets.
Test header		Wall
Symbols for control panels		
Control panel	▭	Basic shape.
(a)	 FCP	Fire alarm control panel
Symbols for fire extinguishing systems		
Symbols for various types of extinguishing systems[b]		
Supplementary symbols		
Fully sprinklered space	(AS)	
Partially sprinklered space	(AS)	
Nonsprinklered space	 ⬨NS⬨	

(continued on next page)

Figure 22-26 Standard Plumbing and Piping Symbols
(continued from previous page)

Referent (Synonym)	Symbol	Comments

Symbols for fire sprinkler heads

Upright sprinkler[c]	$\int$---O---$\int$	
Pendent sprinkler[c, d]	$\int$---⊖---$\int$	
Upright sprinkler, nippled up[c]	$\int$---◎---$\int$	
Pendent sprinkler, on drop nipple[c, d]	$\int$---●---$\int$	
Sidewall sprinkler[c]	$\int$---▽---$\int$	

Symbols for piping, valves, control devices, and hangers.[e]

Pipe hanger	$\int$---\---$\int$	This symbol is a diagonal stroke imposed on the pipe that it supports.
Alarm check valve	$\int$---▷---$\int$	Specify size, direction of flow.
Dry pipe valve	$\int$---⋈---$\int$	Specify size.
Deluge valve	$\int$---◇---$\int$	Specify size and type.
Pre-action valve	$\int$---◇---$\int$	Specify size and type.

Symbols for portable fire extinguishers

| Portable fire extinguisher | △ | Basic shape. |

Symbols for firefighting equipment

| Hose station, dry standpipe | | |
| Hose station, changed standpipe | | |

Source: National Fire Protection Association (NFPA), Standard 170.

[a] Symbol element can be utilized in any combination to fit the type of hydrant.

[b] These symbols are intended for use in identifying the type of system installed to protect an area within a building.

[c] Temperature rating of sprinkler and other characteristics can be shown via legends where a limited number of an individual type of sprinkler is called for by the design.

[d] Can notate "DP" on drawing and/or in specifications where dry pendent sprinklers are employed.

[e] See also NFPA Standard 170, Section 5-4, for related symbols.

(courtesy ASPE)

Figure 22-27 Demolition (Selective vs. Removal for Replacement)

Demolition can be divided into two basic categories.

One type of demolition involves the removal of material with no concern for its replacement. The labor-hours to estimate this work are found under "Selective Demolition" in the Fire Protection, Plumbing and HVAC Divisions. It is selective in that individual items or all the material installed as a system or trade grouping such as plumbing or heating systems are removed. This may be accomplished by the easiest way possible, such as sawing, torch cutting, or sledge hammer as well as simple unbolting.

The second type of demolition is the removal of some item for repair or replacement. This removal may involve careful draining, opening of unions, disconnecting and tagging of electrical connections, capping of pipes/ducts to prevent entry of debris or leakage of the material contained as well as transport of the item away from its in-place location to a truck/dumpster. An approximation of the time required to accomplish this type of demolition is to use half of the time indicated as necessary to install a new unit. For example, installation of a new pump might be listed as requiring 6 labor-hours, so if we had to estimate the removal of the old pump we would allow an additional 3 hours for a total of 9 hours. That is, the complete replacement of a defective pump with a new pump would be estimated to take 9 labor-hours.

Figure 22-28   Installation Time in Labor-Hours for Cast Iron Soil Piping

Description	Labor-Hours	Unit
Cast Iron Soil Pipe Service Weight, Single Hub with Hangers Every Five Feet, Lead and Oakum Joints Every Ten Feet		
2" Pipe Size	.254	L.F.
3" Pipe Size	.267	L.F.
4" Pipe Size	.291	L.F.
5" Pipe Size	.316	L.F.
6" Pipe Size	.329	L.F.
8" Pipe Size	.542	L.F.
10" Pipe Size	.593	L.F.
12" Pipe Size	.667	L.F.
Push on Gasket Joints Every Ten Feet		
2" Pipe Size	.242	L.F.
3" Pipe Size	.254	L.F.
4" Pipe Size	.281	L.F.
5" Pipe Size	.304	L.F.
6" Pipe Size	.320	L.F.
8" Pipe Size	.516	L.F.
10" Pipe Size	.571	L.F.
12" Pipe Size	.653	L.F.
Cast Iron Soil Pipe Fittings Hub and Spigot Service Weight Bends or Elbows		
2" Pipe Size	1.000	Ea.
3" Pipe Size	1.140	Ea.
4" Pipe Size	1.230	Ea.
5" Pipe Size	1.330	Ea.
6" Pipe Size	1.410	Ea.
8" Pipe Size	2.910	Ea.
10" Pipe Size	3.200	Ea.
12" Pipe Size	3.560	Ea.
Tees or Wyes		
2" Pipe Size	1.600	Ea.
3" Pipe Size	1.780	Ea.
4" Pipe Size	2.000	Ea.
5" Pipe Size	2.000	Ea.
6" Pipe Size	2.180	Ea.
8" Pipe Size	4.570	Ea.
10" Pipe Size	4.870	Ea.
12" Pipe Size	5.330	Ea.

Eighth bend

Sanitary tee

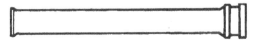

Single hub soil pipe

(continued on next page)

Figure 22-28   Installation Time in Labor-Hours for Cast Iron Soil Piping
(continued from previous page)

Description	Labor-Hours	Unit
Push on Gasket Joints		
Bends or Elbows		
2" Pipe Size	.800	Ea.
3" Pipe Size	.941	Ea.
4" Pipe Size	1.070	Ea.
5" Pipe Size	1.140	Ea.
6" Pipe Size	1.260	Ea.
8" Pipe Size	2.670	Ea.
10" Pipe Size	2.910	Ea.
12" Pipe Size	3.200	Ea.
Tees or Wyes		
2" Pipe Size	1.330	Ea.
3" Pipe Size	1.600	Ea.
4" Pipe Size	1.780	Ea.
5" Pipe Size	1.850	Ea.
6" Pipe Size	2.180	Ea.
8" Pipe Size	4.000	Ea.
10" Pipe Size	4.870	Ea.
12" Pipe Size	5.330	Ea.
Cleanouts		
Floor Type		
2" Pipe Size	.800	Ea.
3" Pipe Size	1.000	Ea.
4" Pipe Size	1.333	Ea.
5" Pipe Size	2.000	Ea.
6" Pipe Size	2.667	Ea.
8" Pipe Size	4.000	Ea.
Cleanout Tee		
2" Pipe Size	2.000	Ea.
3" Pipe Size	2.222	Ea.
4" Pipe Size	2.424	Ea.
5" Pipe Size	2.909	Ea.
6" Pipe Size	3.200	Ea.
8" Pipe Size	6.400	Ea.
Drains		
Heelproof Floor Drain		
2" to 4" Pipe Size	1.600	Ea.
5" and 6" Pipe Size	1.778	Ea.
8" Pipe Size	2.000	Ea.
Shower Drain		
1-1/2" to 3" Pipe Size	2.000	Ea.
4" Pipe Size	2.286	Ea.
Cast Iron Service Weight Traps		
Deep Seal		
2" Pipe Size	1.143	Ea.
3" Pipe Size	1.333	Ea.
4" Pipe Size	1.455	Ea.

Cleanout, floor type

Cleanout tee

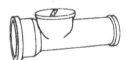

Heelproof floor drain

Shower drain

Deep seal trap

(continued on next page)

Division 22

Figure 22-28   Installation Time in Labor-Hours for Cast Iron Soil Piping
(continued from previous page)

Description	Labor-Hours	Unit
P Trap		
2" Pipe Size	1.000	Ea.
3" Pipe Size	1.143	Ea.
4" Pipe Size	1.231	Ea.
5" Pipe Size	1.333	Ea.
6" Pipe Size	1.412	Ea.
8" Pipe Size	2.909	Ea.
10" Pipe Size	3.200	Ea.
Running Trap with Vent		
3" Pipe Size	1.143	Ea.
4" Pipe Size	1.231	Ea.
5" Pipe Size	2.182	Ea.
6" Pipe Size	3.000	Ea.
8" Pipe Size	3.200	Ea.
S Trap		
2" Pipe Size	1.067	Ea.
3" Pipe Size	1.143	Ea.
4" Pipe Size	1.231	Ea.
No Hub with Couplings Every Ten Feet OC		
1-1/2" Pipe Size	.225	L.F.
2" Pipe Size	.239	L.F.
3" Pipe Size	.250	L.F.
4" Pipe Size	.276	L.F.
5" Pipe Size	.289	L.F.
6" Pipe Size	.304	L.F.
8" Pipe Size	.464	L.F.
10" Pipe Size	.525	L.F.
No Hub Couplings*		
1-1/2" Pipe Size	.333	Ea.
2" Pipe Size	.364	Ea.
3" Pipe Size	.421	Ea.
4" Pipe Size	.485	Ea.
5" Pipe Size	.545	Ea.
6" Pipe Size	.600	Ea.
8" Pipe Size	.970	Ea.
10" Pipe Size	1.230	Ea.

P trap

Running trap with vent

S trap

No hub coupling

*Note: In estimating labor for hub fittings, all the labor is included in the no hub couplings, one coupling per joint.

Figure 22-29   Installation Time in Labor-Hours for Steel Pipe

Description	Labor-Hours	Unit
Labor-hours to Install Black Schedule #40 Flanged with a Pair of Weld Neck Flanges and a Roll Type Hanger Every Ten Feet. The Pipe Hanger Is Oversized to Allow for Insulation.		
1-1/4" Pipe Size	.250	L.F.
1-1/2" Pipe Size	.276	L.F.
2" Pipe Size	.356	L.F.
2-1/2" Pipe Size	.444	L.F.
3" Pipe Size	.500	L.F.
3-1/2" Pipe Size	.552	L.F.
4" Pipe Size	.615	L.F.
5" Pipe Size	.762	L.F.
6" Pipe Size	.960	L.F.
8" Pipe Size	1.263	L.F.
10" Pipe Size	1.500	L.F.
12" Pipe Size	1.714	L.F.
Fittings for Use with Steel Pipe Flanged Elbows, Cast Iron 90° or 45°, 125 lb.		
1-1/2" Pipe Size	1.140	Ea.
2" Pipe Size	1.231	Ea.
2-1/2" Pipe Size	1.333	Ea.
3" Pipe Size	1.455	Ea.
3-1/2" Pipe Size	2.000	Ea.
4" Pipe Size	2.000	Ea.
5" Pipe Size	2.286	Ea.
6" Pipe Size	2.667	Ea.
8" Pipe Size	3.000	Ea.
10" Pipe Size	3.429	Ea.
12" Pipe Size	4.000	Ea.
Flanged Tees, Cast Iron 125 lb.		
1-1/2" Pipe Size	1.778	Ea.
2" Pipe Size	1.778	Ea.
2-1/2" Pipe Size	2.000	Ea.
3" Pipe Size	2.286	Ea.
3-1/2" Pipe Size	3.200	Ea.
4" Pipe Size	3.200	Ea.
5" Pipe Size	4.000	Ea.
6" Pipe Size	4.000	Ea.
8" Pipe Size	4.800	Ea.
10" Pipe Size	6.000	Ea.
12" Pipe Size	8.000	Ea.
Gasket and Bolt Sets Required at Each Flanged Joint		
1-1/2" Pipe Size	.267	Ea.
2" Pipe Size	.267	Ea.
2-1/2" Pipe Size	.267	Ea.
3" Pipe Size	.267	Ea.
3-1/2" Pipe Size	.286	Ea.
4" Pipe Size	.296	Ea.
5" Pipe Size	.308	Ea.
6" Pipe Size	.333	Ea.
8" Pipe Size	.400	Ea.
10" Pipe Size	.444	Ea.
12" Pipe Size	.500	Ea.

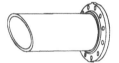

Flanged steel pipe

90° elbow — flanged

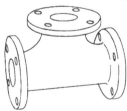

Tee — flanged

(continued on next page)

Division 22

Figure 22-29 Installation Time in Labor-Hours for Steel Pipe
(continued from previous page)

Description	Labor-Hours	Unit
Steel Pipe Labor-hours to Install Black, Schedule #10, Grooved Joint or Plain End with a Mechanical Joint Coupling and Pipe Hanger Every Ten Feet		
2" Pipe Size	.186	L.F.
2-1/2" Pipe Size	.262	L.F.
3" Pipe Size	.291	L.F.
3-1/2" Pipe Size	.302	L.F.
4" Pipe Size	.327	L.F.
5" Pipe Size	.400	L.F.
6" Pipe Size	.522	L.F.
8" Pipe Size	.585	L.F.
10" Pipe Size	.706	L.F.
12" Pipe Size	.800	L.F.
Black or Galvanized Schedule #40 Grooved Joint or Plain End with a Mechanical Joint Coupling and Pipe Hanger Every Ten Feet		
3/4" Pipe Size	.113	L.F.
1" Pipe Size	.127	L.F.
1-1/4" Pipe Size	.138	L.F.
1-1/2" Pipe Size	.157	L.F.
2" Pipe Size	.200	L.F.
2-1/2" Pipe Size	.281	L.F.
3" Pipe Size	.320	L.F.
3-1/2" Pipe Size	.340	L.F.
4" Pipe Size	.356	L.F.
5" Pipe Size	.432	L.F.
6" Pipe Size	.571	L.F.
8" Pipe Size	.649	L.F.
10" Pipe Size	.774	L.F.
12" Pipe Size	.889	L.F.
Fittings for Use with Grooved Joint or Plain End Steel Pipe Elbows 90° or 45°		
3/4" Pipe Size	.160	Ea.
1" Pipe Size	.160	Ea.
1-1/4" Pipe Size	.200	Ea.
1-1/2" Pipe Size	.242	Ea.
2" Pipe Size	.320	Ea.
2-1/2" Pipe Size	.400	Ea.
3" Pipe Size	.485	Ea.
4" Pipe Size	.640	Ea.
5" Pipe Size	.800	Ea.
6" Pipe Size	.960	Ea.
8" Pipe Size	1.143	Ea.
10" Pipe Size	1.333	Ea.
12" Pipe Size	1.600	Ea.

Flanged joint

Grooved joint steel pipe

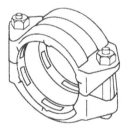

Grooved joint coupling

Mechanical joint elbow
90° — plain end pipe

(continued on next page)

Figure 22-29 Installation Time in Labor-Hours for Steel Pipe
(continued from previous page)

Description	Labor-Hours	Unit
Tees		
3/4" Pipe Size	.211	Ea.
1" Pipe Size	.242	Ea.
1-1/4" Pipe Size	.296	Ea.
1-1/2" Pipe Size	.364	Ea.
2" Pipe Size	.471	Ea.
2-1/2" Pipe Size	.593	Ea.
3" Pipe Size	.727	Ea.
4" Pipe Size	.941	Ea.
5" Pipe Size	1.231	Ea.
6" Pipe Size	1.412	Ea.
8" Pipe Size	1.714	Ea.
10" Pipe Size	2.000	Ea.
12" Pipe Size	2.400	Ea.
Labor-hours to Install Black or Galvanized Schedule #40 Threaded with a Coupling and Pipe Hanger Every Ten Feet. The Pipe Hanger is Oversized to Allow for Insulation.		
1/2" Pipe Size	.127	L.F.
3/4" Pipe Size	.131	L.F.
1" Pipe Size	.151	L.F.
1-1/4" Pipe Size	.180	L.F.
1-1/2" Pipe Size	.200	L.F.
2" Pipe Size	.250	L.F.
2-1/2" Pipe Size	.320	L.F.
3" Pipe Size	.372	L.F.
3-1/2" Pipe Size	.400	L.F.
4" Pipe Size	.444	L.F.
Fittings for Use with Steel Pipe. Threaded Fittings, Cast Iron, 125 lb. or Malleable Iron Rated at 150 lb. Elbows, 90° or 45°		
1/2" Pipe Size	.533	Ea.
3/4" Pipe Size	.571	Ea.
1" Pipe Size	.615	Ea.
1-1/4" Pipe Size	.727	Ea.
1-1/2" Pipe Size	.800	Ea.
2" Pipe Size	.889	Ea.
2-1/2" Pipe Size	1.143	Ea.
3" Pipe Size	1.600	Ea.
3-1/2" Pipe Size	2.000	Ea.
4" Pipe Size	2.667	Ea.

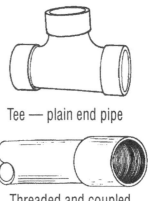

Tee — plain end pipe

Threaded and coupled steel pipe

45° elbow — malleable iron

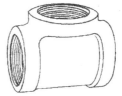

Tee — cast iron

(continued on next page)

Division 22

Figure 22-29   Installation Time in Labor-Hours for Steel Pipe
(continued from previous page)

Description	Labor-Hours	Unit
Tees		
1/2" Pipe Size	.889	Ea.
3/4" Pipe Size	.889	Ea.
1" Pipe Size	1.000	Ea.
1-1/4" Pipe Size	1.143	Ea.
1-1/2" Pipe Size	1.231	Ea.
2" Pipe Size	1.455	Ea.
2-1/2" Pipe Size	1.778	Ea.
3" Pipe Size	2.667	Ea.
3-1/2" Pipe Size	3.200	Ea.
4" Pipe Size	4.000	Ea.
Unions Malleable Iron 150 lb.*		
1/2" Pipe Size	.571	Ea.
3/4" Pipe Size	.615	Ea.
1" Pipe Size	.667	Ea.
1-1/4" Pipe Size	.762	Ea.
1-1/2" Pipe Size	.842	Ea.
2" Pipe Size	.941	Ea.
2-1/2" Pipe Size	1.231	Ea.
3" Pipe Size	1.778	Ea.
3-1/2" Pipe Size	3.200	Ea.
4" Pipe Size	3.200	Ea.
Steel Pipe Labor-hours to Install Black, Schedule #40 Welded with a Butt Joint and a Roll Type Hanger Every Ten Feet. The Pipe Hanger is Oversized to Allow for Insulation.		
1-1/4" Pipe Size	.190	L.F.
1-1/2" Pipe Size	.211	L.F.
2" Pipe Size	.262	L.F.
2-1/2" Pipe Size	.340	L.F.
3" Pipe Size	.372	L.F.
3-1/2" Pipe Size	.410	L.F.
4" Pipe Size	.432	L.F.
5" Pipe Size	.500	L.F.
6" Pipe Size	.667	L.F.
8" Pipe Size	.828	L.F.
10" Pipe Size	1.000	L.F.
12" Pipe Size	1.263	L.F.
Fittings for Use with Steel Pipe Butt Weld, Long Radius 90° Elbows 150 lb.		
1-1/4" Pipe Size	1.143	Ea.
1-1/2" Pipe Size	1.231	Ea.
2" Pipe Size	1.600	Ea.
2-1/2" Pipe Size	2.000	Ea.
3" Pipe Size	2.286	Ea.
3-1/2" Pipe Size	3.200	Ea.
4" Pipe Size	3.200	Ea.
5" Pipe Size	4.800	Ea.
6" Pipe Size	4.800	Ea.
8" Pipe Size	6.000	Ea.
10" Pipe Size	8.000	Ea.
12" Pipe Size	9.600	Ea.

*For unions larger than 4", flanges must be used.

Union — malleable iron

Bevel end steel pipe

90° elbow, butt weld, long radius — steel pipe

(continued on next page)

Figure 22-29 Installation Time in Labor-Hours for Steel Pipe
(continued from previous page)

Description	Labor-Hours	Unit
Forty-five Degree Elbows Require The Same Labor-hours for Installation as Ninety Degree Elbows. Since Forty-five Degree Ells are Often Double the Price of Ninety Degree Ells, a Cost-conscious Pipe Fitter Will Cut a Ninety in Half to Make Two Forty-fives.		
Butt Weld Tees 150 lb.		
1-1/4" Pipe Size	1.778	Ea.
1-1/2" Pipe Size	2.000	Ea.
2" Pipe Size	2.667	Ea.
2-1/2" Pipe Size	3.200	Ea.
3" Pipe Size	4.000	Ea.
3-1/2" Pipe Size	5.333	Ea.
4" Pipe Size	5.333	Ea.
5" Pipe Size	8.000	Ea.
6" Pipe Size	8.000	Ea.
8" Pipe Size	9.600	Ea.
10" Pipe Size	12.000	Ea.
12" Pipe Size	15.000	Ea.
Weld Neck Flanges 150 lb.		
1-1/4" Pipe Size	.552	Ea.
1-1/2" Pipe Size	.615	Ea.
2" Pipe Size	.800	Ea.
2-1/2" Pipe Size	1.000	Ea.
3" Pipe Size	1.143	Ea.
3-1/2" Pipe Size	1.333	Ea.
4" Pipe Size	1.600	Ea.
5" Pipe Size	2.000	Ea.
6" Pipe Size	2.400	Ea.
8" Pipe Size	3.429	Ea.
10" Pipe Size	4.000	Ea.
12" Pipe Size	4.800	Ea.
Slip-on Weld Flanges 150 lb.		
1-1/4" Pipe Size	1.000	Ea.
1-1/2" Pipe Size	1.067	Ea.
2" Pipe Size	1.333	Ea.
2-1/2" Pipe Size	1.600	Ea.
3" Pipe Size	1.778	Ea.
3-1/2" Pipe Size	2.286	Ea.
4" Pipe Size	2.667	Ea.
5" Pipe Size	3.200	Ea.
6" Pipe Size	4.000	Ea.
8" Pipe Size	4.800	Ea.
10" Pipe Size	6.000	Ea.
12" Pipe Size	8.000	Ea.

Tee, butt weld —
steel pipe

Weld neck flange —
steel pipe

Slip-on weld flange —
steel pipe

Figure 22-30 Installation Time in Labor-Hours for Copper Tubing

Description	Labor-Hours	Unit
Labor-hours to Install the Several Types of Copper Tubing Based on a Soft Soldered Coupling and a Clevis Hanger Every Ten Feet		
Type K Tubing		
1/2" Pipe Size	.103	L.F.
3/4" Pipe Size	.108	L.F.
1" Pipe Size	.121	L.F.
1-1/4" Pipe Size	.143	L.F.
1-1/2" Pipe Size	.160	L.F.
2" Pipe Size	.200	L.F.
2-1/2" Pipe Size	.267	L.F.
3" Pipe Size	.296	L.F.
3-1/2" Pipe Size	.381	L.F.
4" Pipe Size	.421	L.F.
5" Pipe Size	.500	L.F.
6" Pipe Size	.632	L.F.
8" Pipe Size	.706	L.F.
Type L Tubing		
1/2" Pipe Size	.099	L.F.
3/4" Pipe Size	.105	L.F.
1" Pipe Size	.118	L.F.
1-1/4" Pipe Size	.138	L.F.
1-1/2" Pipe Size	.154	L.F.
2" Pipe Size	.190	L.F.
2-1/2" Pipe Size	.258	L.F.
3" Pipe Size	.286	L.F.
3-1/2" Pipe Size	.372	L.F.
4" Pipe Size	.410	L.F.
5" Pipe Size	.471	L.F.
6" Pipe Size	.600	L.F.
8" Pipe Size	.667	L.F.
Type M Tubing		
1/2" Pipe Size	.095	L.F.
3/4" Pipe Size	.103	L.F.
1" Pipe Size	.114	L.F.
1-1/4" Pipe Size	.133	L.F.
1-1/2" Pipe Size	.148	L.F.
2" Pipe Size	.182	L.F.
2-1/2" Pipe Size	.250	L.F.
3" Pipe Size	.276	L.F.
3-1/2" Pipe Size	.356	L.F.
4" Pipe Size	.400	L.F.
5" Pipe Size	.444	L.F.
6" Pipe Size	.571	L.F.
8" Pipe Size	.632	L.F.
Type DWV Tubing		
1-1/4" Pipe Size	.133	L.F.
1-1/2" Pipe Size	.148	L.F.
2" Pipe Size	.182	L.F.
2-1/2" Pipe Size	.225	L.F.
3" Pipe Size	.276	L.F.
4" Pipe Size	.400	L.F.
5" Pipe Size	.444	L.F.
6" Pipe Size	.571	L.F.
8" Pipe Size	.632	L.F.

(continued on next page)

Figure 22-30 Installation Time in Labor-Hours for Copper Tubing
(continued from previous page)

Description	Labor-Hours	Unit
Fittings for Copper Tubing, Pressurized Systems, Solder Joint		
Elbows 90° or 45°		
1/2" Tubing Size	.400	Ea.
3/4" Tubing Size	.421	Ea.
1" Tubing Size	.500	Ea.
1-1/4" Tubing Size	.533	Ea.
1-1/2" Tubing Size	.615	Ea.
2" Tubing Size	.727	Ea.
2-1/2" Tubing Size	1.231	Ea.
3" Tubing Size	1.455	Ea.
3-1/2" Tubing Size	1.600	Ea.
4" Tubing Size	1.778	Ea.
5" Tubing Size	2.667	Ea.
6" Tubing Size	2.667	Ea.
8" Tubing Size	3.000	Ea.
Tees		
1/2" Tubing Size	.615	Ea.
3/4" Tubing Size	.667	Ea.
1" Tubing Size	.800	Ea.
1-1/4" Tubing Size	.889	Ea.
1-1/2" Tubing Size	1.000	Ea.
2" Tubing Size	1.143	Ea.
2-1/2" Tubing Size	2.000	Ea.
3" Tubing Size	2.286	Ea.
3-1/2" Tubing Size	2.667	Ea.
4" Tubing Size	3.200	Ea.
5" Tubing Size	4.000	Ea.
6" Tubing Size	4.000	Ea.
8" Tubing Size	4.800	Ea.
Drainage Waste and Vent Systems, Solder Joint		
Elbows 90° or 45°		
1-1/4" Tubing Size	.615	Ea.
1-1/2" Tubing Size	.667	Ea.
2" Tubing Size	.800	Ea.
3" Tubing Size	1.600	Ea.
4" Tubing Size	1.778	Ea.
5" Tubing Size	2.667	Ea.
6" Tubing Size	2.667	Ea.
8" Tubing Size	3.000	Ea.

90° elbow —
solder joint

45° elbow —
solder joint

Tee —
solder joint

Figure 22-31   Installation Time in Labor-Hours for Flexible Plastic Tubing

Description	Labor-Hours	Unit
Labor-hours to Install Flexible Plastic Piping Pipe Hangers and Couplings Are Not Included Since They Are Not Always Necessary Such as When This Piping is Laid in a Trench or Strung Through Joists or Clipped to Structural Members.		
1/2" Pipe Size	.020	L.F.
3/4" Pipe Size	.030	L.F.
1" Pipe Size	.040	L.F.
1-1/4" Pipe Size	.060	L.F.
1-1/2' Pipe Size	.080	L.F.
2" Pipe Size	.100	L.F.
3" Pipe Size	.130	L.F.
Fittings for Polybutylene or Polyethylene Pipe Acetal Insert Type with Crimp Rings (Hot and Cold Water)		
Elbows 90°		
1/2" Pipe Size	.350	Ea.
3/4" Pipe Size	.360	Ea.
Couplings		
1/2" Pipe Size	.350	Ea.
3/4" Pipe Size	.350	Ea.
Tees		
1/2" Pipe Size	.530	Ea.
3/4" Pipe Size	.570	Ea.
Nylon Insert Type with Stainless Steel Clamps (Cold Water Only)		
Elbows 90°		
3/4" Pipe Size	.360	Ea.
1" Pipe Size	.420	Ea.
1-1/4" Pipe Size	.440	Ea.
1-1/2" Pipe Size	.470	Ea.
2" Pipe Size	.500	Ea.
Couplings		
3/4" Pipe Size	.360	Ea.
1" Pipe Size	.420	Ea.
1-1/4" Pipe Size	.440	Ea.
1-1/2" Pipe Size	.470	Ea.
2" Pipe Size	.500	Ea.
Tees		
3/4" Pipe Size	.570	Ea.
1" Pipe Size	.620	Ea.
1-1/4" Pipe Size	.670	Ea.
1-1/2" Pipe Size	.730	Ea.
2" Pipe Size	.800	Ea.
Stainless Steel Clamps		
3/4" Pipe Size	.070	Ea.
1" Pipe Size	.070	Ea.
1-1/4" Pipe Size	.080	Ea.
1-1/2" Pipe Size	.080	Ea.
2" Pipe Size	.090	Ea.

90° elbow — acetal joint

Coupling — acetal insert

Tee — acetal insert

90° elbow — nylon insert

Coupling — nylon insert

Tee — nylon insert

Stainless steel clamp ring

(continued on next page)

Figure 22-31 Installation Time in Labor-Hours for Flexible Plastic Tubing
(continued from previous page)

Description	Labor-Hours	Unit
Acetal Flare Type (Hot and Cold Water)		
Elbows 90°		
1/2" Pipe Size	.360	Ea.
3/4" Pipe Size	.380	Ea.
1" Pipe Size	.440	Ea.
Couplings		
1/2" Pipe Size	.360	Ea.
3/4" Pipe Size	.380	Ea.
1" Pipe Size	.440	Ea.
Tees		
1/2" Pipe Size	.570	Ea.
3/4" Pipe Size	.620	Ea.
1" Pipe Size	.670	Ea.
Fusion Type (Hot and Cold Water)		
Elbows 90° or 45°		
1" Pipe Size	.420	Ea.
1-1/4" Pipe Size	.440	Ea.
1-1/2" Pipe Size	.470	Ea.
2" Pipe Size	.500	Ea.
3" Pipe Size	.800	Ea.
Couplings		
1" Pipe Size	.420	Ea.
1-1/4" Pipe Size	.440	Ea.
1-1/2" Pipe Size	.470	Ea.
2" Pipe Size	.500	Ea.
3" Pipe Size	.800	Ea.
Tees		
1" Pipe Size	.620	Ea.
1-1/4" Pipe Size	.670	Ea.
1-1/2" Pipe Size	.730	Ea.
2" Pipe Size	.800	Ea.
3" Pipe Size	1.140	Ea.
Brass Insert Type Fittings (Hot and Cold Water)		
Elbows 90°		
1/2" Pipe Size	.350	Ea.
3/4" Pipe Size	.360	Ea.
Couplings		
1/2" Pipe Size	.350	Ea.
3/4" Pipe Size	.360	Ea.
Tees		
1/2" Pipe Size	.530	Ea.
3/4" Pipe Size	.570	Ea.

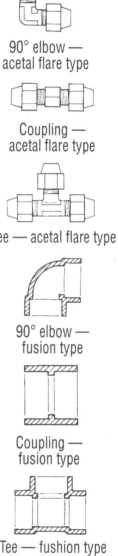

90° elbow —
acetal flare type

Coupling —
acetal flare type

Tee — acetal flare type

90° elbow —
fusion type

Coupling —
fusion type

Tee — fushion type

Tee —
brass insert type

Coupling —
brass insert type

90° elbow —
brass insert type

Figure 22-32 Installation Time in Labor-Hours for Plastic Piping

Description	Labor-Hours	Unit
Labor-hours to Install DMW Piping, ABS, or PVC with a Coupling and 3 Pipe Hangers Every Ten Feet		
1-1/4" Pipe Size	.190	L.F.
1-1/2" Pipe Size	.222	L.F.
2" Pipe Size	.271	L.F.
3" Pipe Size	.302	L.F.
4" Pipe Size	.333	L.F.
6" Pipe Size	.410	L.F.
Fittings for Use with Plastic Pipe DWV, ABS or PVC, with Socket Joints Bends - Quarter or Eighth		
1-1/4" Pipe Size	.471	Ea.
1-1/2" Pipe Size	.500	Ea.
2" Pipe Size	.571	Ea.
3" Pipe Size	.941	Ea.
4" Pipe Size	1.143	Ea.
6" Pipe Size	2.000	Ea.
Couplings		
1-1/4" Pipe Size	.471	Ea.
1-1/2" Pipe Size	.500	Ea.
2" Pipe Size	.571	Ea.
3" Pipe Size	.727	Ea.
4" Pipe Size	.941	Ea.
6" Pipe Size	1.333	Ea.
Tees and Wyes - Sanitary		
1-1/4" Pipe Size	.727	Ea.
1-1/2" Pipe Size	.800	Ea.
2" Pipe Size	.941	Ea.
3" Pipe Size	1.455	Ea.
4" Pipe Size	1.778	Ea.
6" Pipe Size	3.200	Ea.
Labor-hours to Install PVC Schedule 40 Piping with a Coupling and 3 Hangers Every Ten Feet. PVC Piping is Used for Water Service, Gas Service, Irrigation and Air Conditioning.		
1/2" Pipe Size	.148	L.F.
3/4" Pipe Size	.157	L.F.
1" Pipe Size	.174	L.F.
1-1/4" Pipe Size	.190	L.F.
1-1/2" Pipe Size	.222	L.F.
2" Pipe Size	.271	L.F.
2-1/2" Pipe Size	.286	L.F.
3" Pipe Size	.302	L.F.
4" Pipe Size	.333	L.F.
5" Pipe Size	.372	L.F.
6" Pipe Size	.410	L.F.
8" Pipe Size	.500	L.F.

Quarter bend —
socket joint

Eighth bend —
socket joint

(continued on next page)

Figure 22-32 Installation Time in Labor-Hours for Plastic Piping
(continued from previous page)

Description	Labor-Hours	Unit
Fittings for Use with PVC Schedule 40, with Socket Joints, Elbows, 90° or 45°		
1/2" Pipe Size	.364	Ea.
3/4" Pipe Size	.381	Ea.
1" Pipe Size	.444	Ea.
1-1/4" Pipe Size	.471	Ea.
1-1/2" Pipe Size	.500	Ea.
2" Pipe Size	.571	Ea.
2-1/2" Pipe Size	.727	Ea.
3" Pipe Size	.941	Ea.
4" Pipe Size	1.143	Ea.
5" Pipe Size	1.333	Ea.
6" Pipe Size	2.000	Ea.
8" Pipe Size	2.400	Ea.
Couplings		
1/2" Pipe Size	.364	Ea.
3/4" Pipe Size	.381	Ea.
1" Pipe Size	.444	Ea.
1-1/4" Pipe Size	.471	Ea.
1-1/2" Pipe Size	.500	Ea.
2" Pipe Size	.571	Ea.
2-1/2" Pipe Size	.700	Ea.
3" Pipe Size	.842	Ea.
4" Pipe Size	1.000	Ea.
5" Pipe Size	1.143	Ea.
6" Pipe Size	1.333	Ea.
8" Pipe Size	1.714	Ea.
Tees		
1/2" Pipe Size	.571	Ea.
3/4" Pipe Size	.615	Ea.
1" Pipe Size	.667	Ea.
1-1/4" Pipe Size	.727	Ea.
1-1/2" Pipe Size	.800	Ea.
2" Pipe Size	.941	Ea.
2-1/2" Pipe Size	1.143	Ea.
3" Pipe Size	1.455	Ea.
4" Pipe Size	1.778	Ea.
5" Pipe Size	2.000	Ea.
6" Pipe Size	3.200	Ea.
8" Pipe Size	4.000	Ea.
Labor-hours to Install PVC Schedule 80 Piping with a Coupling and 3 Hangers Every Ten Feet		
1/2" Pipe Size	.160	L.F.
3/4" Pipe Size	.170	L.F.
1" Pipe Size	.186	L.F.
1-1/4" Pipe Size	.205	L.F.
1-1/2" Pipe Size	.235	L.F.
2" Pipe Size	.291	L.F.

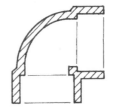

Vent tee — socket joint

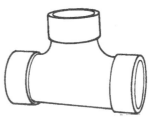

90° elbow — socket joint

Coupling — socket joint

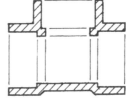

Tee — socket joint

(continued on next page)

Division 22

Figure 22-32 Installation Time in Labor-Hours for Plastic Piping
(continued from previous page)

Description	Labor-Hours	Unit
2-1/2" Pipe Size	.308	L.F.
3" Pipe Size	.320	L.F.
4" Pipe Size	.348	L.F.
5" Pipe Size	.381	L.F.
6" Pipe Size	.421	L.F.
8" Pipe Size	.511	L.F.
Fittings for Use with PVC Schedule 80, Socket or Threaded Joints, Elbows, 90° or 45°		
1/2" Pipe Size	.444	Ea.
3/4" Pipe Size	.471	Ea.
1" Pipe Size	.533	Ea.
1-1/4" Pipe Size	.571	Ea.
1-1/2" Pipe Size	.615	Ea.
2" Pipe Size	.727	Ea.
2-1/2" Pipe Size	.941	Ea.
3" Pipe Size	1.143	Ea.
4" Pipe Size	1.333	Ea.
5" Pipe Size	2.000	Ea.
6" Pipe Size	2.286	Ea.
8" Pipe Size	3.000	Ea.
Couplings		
1/2" Pipe Size	.444	Ea.
3/4" Pipe Size	.471	Ea.
1" Pipe Size	.533	Ea.
1-1/4" Pipe Size	.571	Ea.
1-1/2" Pipe Size	.615	Ea.
2" Pipe Size	.727	Ea.
2-1/2" Pipe Size	.800	Ea.
3" Pipe Size	.842	Ea.
4" Pipe Size	1.000	Ea.
5" Pipe Size	1.143	Ea.
6" Pipe Size	1.333	Ea.
8" Pipe Size	1.714	Ea.
Tees		
1/2" Pipe Size	.667	Ea.
3/4" Pipe Size	.727	Ea.
1" Pipe Size	.800	Ea.
1-1/4" Pipe Size	.889	Ea.
1-1/2" Pipe Size	1.000	Ea.
2" Pipe Size	1.143	Ea.
2-1/2" Pipe Size	1.455	Ea.
3" Pipe Size	1.778	Ea.
4" Pipe Size	2.000	Ea.
5" Pipe Size	3.200	Ea.
6" Pipe Size	3.200	Ea.
8" Pipe Size	4.000	Ea.

45° elbow — threaded joint

Coupling — threaded joint

Tee — threaded joint

Note: The labor-hours are the same to install CPVC schedule 80 piping (chlorinated polyvinyl chloride) with a coupling and 3 hangers every ten feet.

(continued on next page)

Figure 22-32   Installation Time in Labor-Hours for Plastic Piping
(continued from previous page)

Plastic Pipe

Although there are many plastic pipe types listed below, PVC and ABS are by far the most common types. It is imperative that the correct primers and solvents be used on each type of pipe or the joints will not seal properly, and the overall strength will be weakened.

Types of Plastic Pipe

Type	Characteristics
PVC	Polyvinyl Chloride, Type 1, Grade 1. This pipe is strong, rigid and resistant to a variety of acids and bases. Some solvents and chlorinated hydrocarbons may damage the pipe. PVC is very common, easy to work with and readily available at most hardware stores. Maximum useable temperature is 140°F (60°C) and pressure ratings start at a minimum of 125 to 200 psi (check for specific ratings on the pipe or ask the seller). PVC can be used with water, gas, and drainage systems, but NOT with hot water systems.
ABS	Acrylonitrile Butadiene Styrene, Type 1. This pipe is strong and rigid and resistant to a variety of acids and bases. Some solvents and chlorinated hydrocarbons may damage the pipe. ABS is very common, easy to work with and readily available at most hardware stores. Maximum useable temperature is 160°F (71°C) at low pressures. It is most common as a DWV pipe.
CPVC	Chlorinated polyvinyl chloride. Similar to PVC but designed specifically for piping water at up to 180°F (82°C) (can actually withstand 200°F for a limited time.) Pressure rating is 100 psi.
PE	Polyethylene. A flexible pipe for pressurized water systems such as sprinklers. Not for hot water.

(courtesy Pocket Ref, *3rd Edition, Thomas J. Glover, Sequoia Publishing, Inc.)*

Division 22

(continued on next page)

Figure 22-32 Installation Time in Labor-Hours for Plastic Piping
(continued from previous page)

PEX	Polyethylene cross-linked. A flexible pipe for pressurized water systems such as sprinklers.
PB	Polybutylene. A flexible pipe for pressurized water systems both hot and cold. ONLY compression and banded type joints can be used.
Polypropylene	Low pressure, lightweight material that is good up to 180°F (82°C). Highly resistant to acids, bases, and many solvents. Good for laboratory plumbing.
PVDF	Polyvinylidene fluoride. Strong, very tough, and resistant to abrasion, acids, bases, solvents, and much more. Good to 280°F (138°C). Good in lab.
FRP Epoxy	A thermosetting plastic over fiberglass. Very high strength and excellent chemical resistance. Good to 220°F (105°C). Excellent for labs.

Every product has certain unique terms and definitions associated with it. Listed below are some of the more common terms and definitions that relate to plastic pipe. A good understanding of these definitions is essential to understanding flexible pipe. ASTM D 883 and F 412 define additional terms relating to plastic pipe.

ACRYLONITRILE-BUTADIEN-STYRENE (ABS): Plastics containing polymers or blends of polymers, or both, in which the minimum butadiene content is 6%, the minimum acrylonitrile content is 15%, the minimum styrene or substituted content, or both, is 15%, and the maximum content of all other monomers is not more than 5%; plus lubricants, stabilizers, and colorants.

ADDITIVE: A substance added to another substance, usually to improve properties, such as plasticizers, initiators, light stabilizers, and flame retardants.

AGING: The effects on materials of exposure to an environment for an interval of time.

ASH CONTENT: Proportion of the solid residue remaining after a reinforcing substance has been incinerated (charred or intensely heated).

BURST STRENGTH: The internal pressure required to cause a pipe or fitting to fail.

CARBON BLACK: A black pigment produced by the incomplete burning of natural gas or oil. Because it possesses useful ultraviolet protective properties, it is also much used in molding compounds intended for outside weathering applications.

(continued on next page)

Figure 22-32 Installation Time in Labor-Hours for Plastic Piping
(continued from previous page)

CELL CLASSIFICATION: A rating of the primary properties of the resin material. To include density, melt index, flexural modulus, tensile strength at yield, environmental stress crack resistance, hydrostatic design basis, color, and UV stabilizer.

COEXTRUSION: A process whereby two or more heated or unheated plastic material streams forced through one or more shaping orifice(s) become one continuously formed piece.

CRACK: Any narrow opening or fissure in the surface.

CRATER: A small, shallow surface imperfection.

CRAZING: Fine cracks at or under the surface of a plastic.

CREEP: The time-dependent part of strain resulting from stress, that is, the dimensional change caused by the application of load over and above the elastic deformation and with respect to time.

DEFLECTION: Any change in the inside diameter of the pipe resulting from installation and imposed loads. Deflection may be either vertical or horizontal and is usually reported as a percentage of the base (undeflected) inside pipe diameter.

DEGRADATION: A deleterious change in the chemical structure, physical properties, or appearance of a plastic.

DETERIORATION: A permanent change in the physical properties of a plastic evidenced by impairment of these properties.

ENVIRONMENTAL STRESS CRACKING: The development of cracks in a material that is subjected to stress or strain in the presence of specific chemicals.

EXTRUSION: A process in which heated or unheated plastic is forced through a shaping orifice (a die) in one continuously formed shape, as in film sheet, rod, or tubing.

FILLER: A relatively inert material added to a plastic to modify its strength, permanence, working properties, or other qualities or to lower costs.

FLEXURAL STRENGTH: The stress that a specimen will withstand when subjected to a bending moment.

HOOP STRESS: The tensile stress acting on the pipe along the circumferential direction of the pipe wall when the pipe contains liquid or gas.

IMPACT STRENGTH: The characteristic that gives a material the ability to withstand shock loading. The work done in fracturing a test specimen in a specified manner under shock loading.

LONG-TERM HYDROSTATIC STRENGTH (LTHS): The hoop stress that when applied continuously will cause failure of the pipe at 100,000 hours (11.43 years).

MODULUS OF ELASTICITY: The ratio of stress to the strain produced in a material that is elastically deformed. Also called Young's modulus.

MONOMER: A low molecular weight substance consisting of molecules capable of reacting with like to unlike molecules to form a polymer.

(continued on next page)

Figure 22-32 Installation Time in Labor-Hours for Plastic Piping
(continued from previous page)

PERMANENT SET: The deformation remaining after a specimen has been stressed a prescribed amount in tension, compression, or shear for a specified time period and released for a specified time period.

For creep tests, the residual unrecoverable deformation after the load causing the creep has been removed for a substantial and specified period of time. Also, the increase in length, expressed as a percentage of the original length, by which an elastic material fails to return to its original length after being stressed for a standard period of time.

PIPE STIFFNESS: A measure of the inherent strength of the pipe as a function of applied load and resulting deformation (See ASTM Method D 2412).

PLASTIC: A material that contains as an essential ingredient one or more organic polymeric substances of large molecular weights, is solid in its finished state, and, at some stage in its manufacture or processing into finished articles, can be shaped by flow.

POLYETHYLENE: A plastic or resin prepared by the polymerization of ethylene as essentially the sole monomer.

POLYMER: A substance consisting of molecules characterized by the repetition of one or more types of monomeric units.

POLYMERIZATION: A chemical reaction in which the molecules of a monomer are linked together to form polymers.

POLYVINYLCHLORIDE: A polymer prepared by the polymerization of vinyl chloride with or without small amounts of other monomers.

RECYCLED PLASTIC: A plastic prepared from discarded articles that have been cleaned and reground.

REPROCESSED PLASTIC: A thermoplastic prepared from usually melt-processed scrap or reject parts by a plastics processor, or from non-standard or non-virgin material.

RESIN: Any polymer that is a basic material for plastics.

REWORKED PLASTIC: A plastic from a processor's own production that has been reground, pelletized, or solvated after having been previously processed by molding, extrusion, etc.

SET: Strain remaining after complete release of the force producing the deformation.

STABILIZERS: Chemicals used in plastics formulation to help maintain physical and chemical properties during processing and service life. A specific type of stabilizer, known as an ultraviolet stabilizer, is designed to absorb ultraviolet rays and prevent them from attacking the plastic.

STANDARD DIMENSION RATIO: The ratio of the outside pipe diameter to the wall thickness (Do/t).

STIFFNESS FACTOR: A physical property of plastic pipe that indicates the degree of flexibility of the pipe, when subjected to external loads. See ASTM D 2412.

(continued on next page)

Figure 22-32 Installation Time in Labor-Hours for Plastic Piping
(continued from previous page)

STRAIN: The change per unit of length in a linear dimension of a body, that accompanies a stress.

STRESS: When expressed with reference to pipe: the force per unit area in the wall of the pipe in the circumferential orientation due to internal hydrostatic pressure.

STRESS-CRACK: An external or internal crack in a plastic caused by tensile stresses less than its short-time mechanical strength.

STRESS-CRACKING FAILURE: The failure of a material by cracking or crazing some time after it has been placed under load. Time-to-failure may range from minutes to years. Causes include molded-in stresses, post-fabrication shrinkage or warpage, and hostile environment.

STRESS RELAXATION: The decrease of stress with respect to time in a piece of plastic that is subject to a constant strain.

TENSILE STRENGTH: The maximum pulling stress or force per unit cross-sectional area that the specimen can withstand before breaking.

THERMOPLASTIC: A plastic that repeatedly can be softened by heating and hardened by cooling through a temperature range characteristic of the plastic, and that in the softened state can be shaped by flow into articles by molding or extrusion.

THERMOSET: A plastic, after being reacted (i.e., cured) cannot be reversed to its original state.

VIRGIN MATERIAL: A plastic material in the form of pellets, granules, powder floc or liquid that has not been subjected to use or processing other than that required for its original manufacture.

(courtesy Rinker Materials)

Figure 22-33 Installation Time in Labor-Hours for Piping Supports

Description	Labor-Hours	Unit
Side Beam Brackets	.167	Ea.
Wall Brackets	.235	Ea.
C Clamps	.050	Ea.
I Beam Clamps		
2" Flange Size	.083	Ea.
3" Flange Size	.084	Ea.
4" Flange Size	.086	Ea.
5" Flange Size	.087	Ea.
6" Flange Size	.089	Ea.
7" Flange Size	.091	Ea.
8" Flange Size	.093	Ea.
Riser Clamps		
3/4" Pipe Size	.167	Ea.
1" Pipe Size	.170	Ea.
1-1/4" Pipe Size	.174	Ea.
1-1/2" Pipe Size	.178	Ea.
2" Pipe Size	.186	Ea.
2-1/2" Pipe Size	.195	Ea.
3" Pipe Size	.200	Ea.
3-1/2" Pipe Size	.205	Ea.
4" Pipe Size	.211	Ea.
5" Pipe Size	.216	Ea.
6" Pipe Size	.222	Ea.
8" Pipe Size	.235	Ea.
10" Pipe Size	.250	Ea.
12" Pipe Size	.286	Ea.
Split Ring Clamps		
1/2" Pipe Size	.117	Ea.
3/4" Pipe Size	.119	Ea.
1" Pipe Size	.121	Ea.
1-1/4" Pipe Size	.124	Ea.
1-1/2" Pipe Size	.127	Ea.
2" Pipe Size	.129	Ea.
2-1/2" Pipe Size	.133	Ea.
3" Pipe Size	.137	Ea.
3-1/2" Pipe Size	.140	Ea.
4" Pipe Size	.145	Ea.
5" Pipe Size	.151	Ea.
6" Pipe Size	.154	Ea.
8" Pipe Size	.160	Ea.
High Temperature Clamps		
4" Pipe Size	.151	Ea.
6" Pipe Size	.151	Ea.
8" Pipe Size	.165	Ea.
10" Pipe Size	.190	Ea.
12" Pipe Size	.222	Ea.

Side beam bracket

Medium welded steel bracket

Brackets

C-clamp

I-beam clamp

Extension pipe or riser clamp

Alloy steel pipe clamp

Medium pipe clamp

Split ring pipe clamp

Clamp Types

(continued on next page)

Figure 22-33 Installation Time in Labor-Hours for Piping Supports
(continued from previous page)

Description	Labor-Hours	Unit
Two Bolt Clamps		
1/2" Pipe Size	.117	Ea.
3/4" Pipe Size	.119	Ea.
1" Pipe Size	.121	Ea.
1-1/4" Pipe Size	.123	Ea.
1-1/2" Pipe Size	.127	Ea.
2" Pipe Size	.129	Ea.
2-1/2" Pipe Size	.133	Ea.
3" Pipe Size	.137	Ea.
3-1/2" Pipe Size	.140	Ea.
4" Pipe Size	.145	Ea.
5" Pipe Size	.151	Ea.
6" Pipe Size	.154	Ea.
8" Pipe Size	.160	Ea.
10" Pipe Size	.167	Ea.
12" Pipe Size	.180	Ea.
Pipe Alignment Guides		
1" Pipe Size	.308	Ea.
1-1/4" to 2" Pipe Size	.381	Ea.
2-1/2" to 3-1/2" Pipe Size	.444	Ea.
4" to 5" Pipe Size	.500	Ea.
6" Pipe Size	.762	Ea.
8" Pipe Size	1.000	Ea.
10" Pipe Size	1.333	Ea.
12" Pipe Size	1.412	Ea.
Band Hangers		
1/2" Pipe Size	.113	Ea.
3/4" Pipe Size	.114	Ea.
1" Pipe Size	.117	Ea.
1-1/4" Pipe Size	.119	Ea.
1-1/2" Pipe Size	.122	Ea.
2" Pipe Size	.124	Ea.
2-1/2" Pipe Size	.128	Ea.
3" Pipe Size	.131	Ea.
3-1/2" Pipe Size	.134	Ea.
4" Pipe Size	.140	Ea.
5" Pipe Size	.145	Ea.
6" Pipe Size	.148	Ea.
8" Pipe Size	.154	Ea.
Adjustable Swivel Rings		
1/2" Pipe Size	.117	Ea.
3/4" Pipe Size	.118	Ea.
1" Pipe Size	.121	Ea.
1-1/4" Pipe Size	.124	Ea.
1-1/2" Pipe Size	.127	Ea.
2" Pipe Size	.129	Ea.
2-1/2" Pipe Size	.133	Ea.
3" Pipe Size	.137	Ea.
3-1/2" Pipe Size	.140	Ea.
4" Pipe Size	.145	Ea.
5" Pipe Size	.151	Ea.
6" Pipe Size	.154	Ea.
8" Pipe Size	.160	Ea.

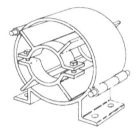

Pipe alignment
guide

Adjustable
band hanger

Adjustable pipe ring

(continued on next page)

Division 22

Figure 22-33 Installation Time in Labor-Hours for Piping Supports
(continued from previous page)

Description	Labor-Hours	Unit
Clevis		
1/2" Pipe Size	.117	Ea.
3/4" Pipe Size	.119	Ea.
1" Pipe Size	.121	Ea.
1-1/4" Pipe Size	.124	Ea.
1-1/2" Pipe Size	.127	Ea.
2" Pipe Size	.129	Ea.
2-1/2" Pipe Size	.133	Ea.
3" Pipe Size	.137	Ea.
3-1/2" Pipe Size	.140	Ea.
4" Pipe Size	.145	Ea.
5" Pipe Size	.151	Ea.
6" Pipe Size	.154	Ea.
8" Pipe Size	.160	Ea.
10" Pipe Size	.167	Ea.
12" Pipe Size	.180	Ea.
Insulation Protection Saddles		
3/4" Pipe Size	.118	Ea.
1" Pipe Size	.118	Ea.
1-1/4" Pipe Size	.118	Ea.
1-1/2" Pipe Size	.121	Ea.
2" Pipe Size	.121	Ea.
2-1/2" Pipe Size	.125	Ea.
3" Pipe Size	.125	Ea.
3-1/2" Pipe Size	.129	Ea.
4" Pipe Size	.129	Ea.
5" Pipe Size	.133	Ea.
6" Pipe Size	.133	Ea.
Adjustable Yoke with Roll		
2-1/2" Pipe Size	.117	Ea.
3" Pipe Size	.122	Ea.
3-1/2" Pipe Size	.129	Ea.
4" Pipe Size	.137	Ea.
5" Pipe Size	.145	Ea.
6" Pipe Size	.154	Ea.
8" Pipe Size	.167	Ea.
10" Pipe Size	.200	Ea.
12" Pipe Size	.235	Ea.
Two Rod Rolls		
2-1/2" Pipe Size	.136	Ea.
3" Pipe Size	.139	Ea.
3-1/2" Pipe Size	.142	Ea.
4" Pipe Size	.143	Ea.
5" Pipe Size	.145	Ea.
6" Pipe Size	.158	Ea.
8" Pipe Size	.178	Ea.
10" Pipe Size	.200	Ea.
12" Pipe Size	.235	Ea.

Adjustable clevis

Pipe covering
protection saddle

Adjustable steel yoke
pipe roll

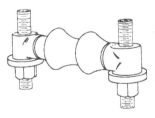

Adjustable two-rod
roller hanger

(continued on next page)

Figure 22-33 Installation Time in Labor-Hours for Piping Supports
(continued from previous page)

Description	Labor-Hours	Unit
Chair Roll		
2" Pipe Size	.118	Ea.
2-1/2" Pipe Size	.123	Ea.
3" Pipe Size	.129	Ea.
3-1/2" Pipe Size	.133	Ea.
4" Pipe Size	.138	Ea.
5" Pipe Size	.143	Ea.
6" Pipe Size	.151	Ea.
8" Pipe Size	.160	Ea.
10" Pipe Size	.167	Ea.
12" Pipe Size	.174	Ea.
Pipe Straps		
1/2" Pipe Size	.113	Ea.
3/4" Pipe Size	.114	Ea.
1" Pipe Size	.117	Ea.
1-1/4" Pipe Size	.119	Ea.
1-1/2" Pipe Size	.122	Ea.
2" Pipe Size	.124	Ea.
2-1/2" Pipe Size	.128	Ea.
3" Pipe Size	.131	Ea.
3-1/2" Pipe Size	.134	Ea.
4" Pipe Size	.140	Ea.
U-Bolts		
1/2" Pipe Size	.050	Ea.
3/4" Pipe Size	.051	Ea.
1" Pipe Size	.053	Ea.
1-1/4" Pipe Size	.054	Ea.
1-1/2" Pipe Size	.056	Ea.
2" Pipe Size	.058	Ea.
2-1/2" Pipe Size	.060	Ea.
3" Pipe Size	.063	Ea.
3-1/2" Pipe Size	.066	Ea.
4" Pipe Size	.068	Ea.
5" Pipe Size	.070	Ea.
6" Pipe Size	.072	Ea.
8" Pipe Size	.073	Ea.
10" Pipe Size	.075	Ea.
12" Pipe Size	.077	Ea.
U-Hooks		
3/4" to 2" Pipe Size	.083	Ea.

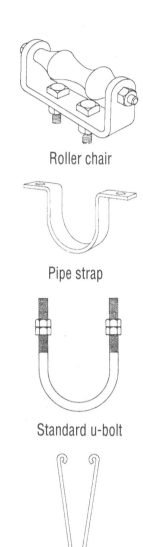

Roller chair

Pipe strap

Standard u-bolt

U-hook

Figure 22-34   Installation Time in Labor-Hours for Valves

Description	Labor-Hours	Unit
Valves		
Threaded or Solder End		
1/8" Diameter	.333	Ea.
1/4" Diameter	.333	Ea.
3/8" Diameter	.333	Ea.
1/2" Diameter	.333	Ea.
3/4" Diameter	.400	Ea.
1" Diameter	.421	Ea.
1-1/4" Diameter	.533	Ea.
1-1/2" Diameter	.615	Ea.
2" Diameter	.727	Ea.
2-1/2" Diameter	1.067	Ea.
3" Diameter	1.231	Ea.
Flanged End		
2" Diameter	1.600	Ea.
2-1/2" Diameter	3.200	Ea.
3" Diameter	3.556	Ea.
3-1/2" Diameter	5.333	Ea.
4" Diameter	5.333	Ea.
5" Diameter	7.059	Ea.
6" Diameter	8.000	Ea.
8" Diameter	9.600	Ea.
10" Diameter	10.909	Ea.
12" Diameter	14.118	Ea.
Plastic Threaded or Socket End		
1/4" Diameter	.308	Ea.
1/2" Diameter	.308	Ea.
3/4" Diameter	.320	Ea.
1" Diameter	.348	Ea.
1-1/4" Diameter	.381	Ea.
1-1/2" Diameter	.400	Ea.
2" Diameter	.471	Ea.
2-1/2" Diameter	.615	Ea.
3" Diameter	.667	Ea.
Backflow Preventers		
Threaded End		
3/4" Diameter	.500	Ea.
1" Diameter	.571	Ea.
1-1/2" Diameter	.800	Ea.
2" Diameter	1.143	Ea.
Flanged End		
2-1/2" Diameter	3.200	Ea.
3" Diameter	3.556	Ea.
4" Diameter	5.333	Ea.
6" Diameter	8.000	Ea.
8" Diameter	12.000	Ea.
10" Diameter	24.000	Ea.

Double check

Reduced pressure —
flanged

Reduced pressure —
threaded

Backflow Preventers

Figure 22-35   Installation Time in Labor-Hours for Roof Storm Drains

Description	Labor-Hours						Units
	2"	3"	4"	5"	6"	8"	
Roof Drain	1.143	1.231	1.333	1.600	2.000	2.290	Ea.
Cleanout Tee	2.000	2.222	2.424	2.909	3.200	6.400	Ea
C.I. Pipe (no hub)	.262	.276	.302	.324	.343	.533	L.F.
Elbow	1.230	1.334	1.454	1.714	2.000	3.368	Ea.
Tee	1.845	2.001	2.181	2.571	3.000	5.052	Ea.
Copper Tube	.348	.533	.696	.842	1.000	1.143	L.F.
Elbow	.727	1.455	1.778	2.667	2.667	3.000	Ea.
Tee	1.143	2.286	3.200	3.700	4.000	4.800	Ea.
Galvanized Steel Pipe	.250	.372	.444	.615	.774	.889	L.F.
Elbow	.889	1.600	2.667	3.200	3.429	4.000	Ea.
Tee	1.455	2.667	4.000	5.333	6.000	8.000	Ea.
DWV Pipe	.271	.302	.333	.370	.410	.500	L.F.
(ABS/PVC) Elbow	.571	.941	1.143	1.333	2.000	2.400	Ea.
Tee	.941	1.455	1.778	2.000	3.200	4.000	Ea.

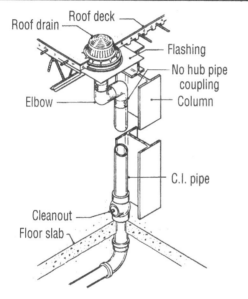

Riser Clamp

Figure 22-36   Installation Time in Labor-Hours for Rough-in of
Plumbing Fixtures

Description	Labor-Hours	Unit
For Roughing-In		
Bathtub	7.730	Ea.
Bidet	8.990	Ea.
Dental Fountain	6.900	Ea.
Drinking Fountain	4.370	Ea.
Lavatory		
Vanity Top	6.960	Ea.
Wall Hung	9.640	Ea.
Laundry Sinks	7.480	Ea.
Prison/Institution Fixtures		
Lavatory	10.670	Ea.
Service Sink	17.978	Ea.
Urinal	10.738	Ea.
Water Closet	13.445	Ea.
Combination Water Closet and Lavatory	16.000	Ea.
Shower Stall	7.800	Ea.
Sinks		
Corrosion Resistant	7.920	Ea.
Kitchen, Countertop	7.480	Ea.
Kitchen, Raised Deck	8.650	Ea.
Service, Floor	9.760	Ea.
Service, Wall	12.310	Ea.
Urinals		
Wall Hung	5.650	Ea.
Stall Type	8.040	Ea.
Wash Fountain, Group	8.790	Ea.
Water Closets		
Tank Type, Wall Hung	5.861	Ea.
Floor Mount, One Piece	5.246	Ea.
Bowl Only, Wall Hung	6.250	Ea.
Bowl Only, Floor Mount	5.634	Ea.
Gang, Side by Side, First	8.120	Ea.
Each Additional	7.480	Ea.
Gang, Back to Back, First Pair	9.090	Pair
Each Additional Pair	8.840	Pair
Water Conserving Type	8.250	Ea.
Water Cooler	3.620	Ea.

Figure 22-37 Installation Time in Labor-Hours for Plumbing Fixtures

Description	Labor-Hours	Unit
For Setting Fixture and Trim		
Bathtub	3.636	Ea.
Bidet	3.200	Ea.
Dental Fountain	2.000	Ea.
Drinking Fountain	2.500	Ea.
Lavatory		
Vanity Top	2.500	Ea.
Wall Hung	2.000	Ea.
Laundry Sinks	2.667	Ea.
Prison/Institution Fixtures		
Lavatory	2.000	Ea.
Service Sink	5.333	Ea.
Urinal	4.000	Ea.
Water Closet	2.759	Ea.
Combination Water Closet and Lavatory	3.200	Ea.
Shower Stall	2.909	Ea.
Sinks		
Corrosion Resistant	3.333	Ea.
Kitchen, Countertop	3.330	Ea.
Kitchen, Raised Deck	6.154	Ea.
Service, Floor	3.640	Ea.
Service, Wall	4.000	Ea.
Urinals		
Wall Hung	5.333	Ea.
Stall Type	6.400	Ea.
Wash Fountain, Group	9.600	Ea.
Water Closets		
Tank Type, Wall Hung	3.019	Ea.
Floor Mount, One Piece	3.019	Ea.
Bowl Only, Wall Hung	2.759	Ea.
Bowl Only, Floor Mount	2.759	Ea.
Gang, Side by Side, First	2.759	Ea.
Each Additional	2.759	Ea.
Gang, Back to Back, First Pair	5.520	Pair
Each Additional Pair	5.520	Pair
Water Conserving Type	2.963	Ea.
Water Cooler	4.000	Ea.

Figure 22-38   Installation Time in Labor-Hours for Commercial Gas Hot Water Heaters

Description	Labor-Hours	Unit
Gas fired, flush jacket, std. controls, vent not incl.		
75 MBH input, 73 GPH	5.714	Ea.
98 MBH input, 95 GPH	5.714	
120 MBH input, 110 GPH	6.667	
200 MBH input, 192 GPH	13.333	
250 MBH input, 245 GPH	16.000	

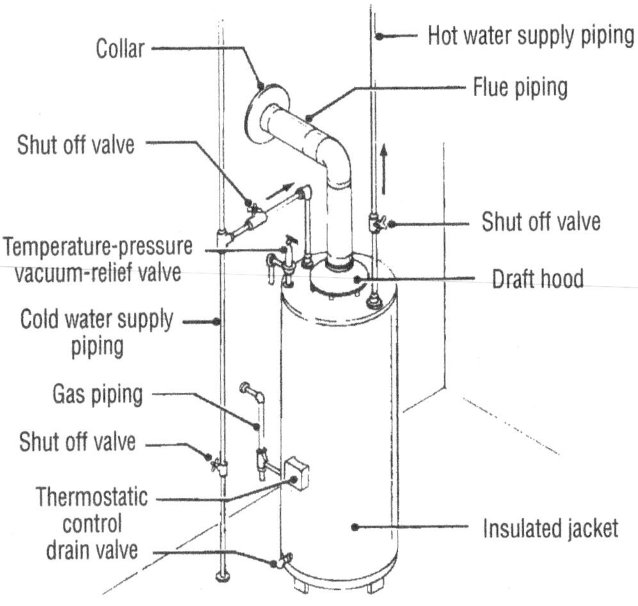

Collar

Hot water supply piping

Flue piping

Shut off valve

Temperature-pressure vacuum-relief valve

Shut off valve

Draft hood

Cold water supply piping

Gas piping

Shut off valve

Thermostatic control drain valve

Insulated jacket

Gas Fired Hot Water System

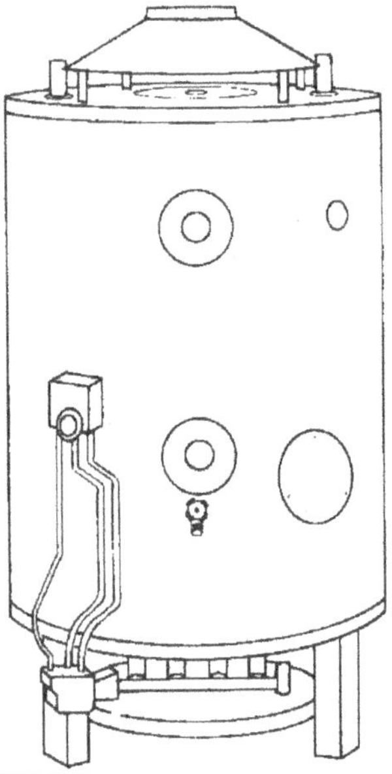

Gas Fired Water Heater — Commercial

Figure 22-39 Installation Time in Labor-Hours for Commercial Oil
Fired Hot Water Heaters

Description	Labor-Hours	Unit
Oil fired, glass lined, UL listed, std. controls, vent not incl.		
140 gal., 140 MBH input, 134 GPH	7.512	Ea.
140 gal., 199 MBH input, 191 GPH	8.000	Ea.
140 gal., 255 MBH input, 247 GPH	10.000	Ea.
140 gal., 540 MBH input, 519 GPH	16.667	Ea.
140 gal., 720 MBH input, 691 GPH	17.391	Ea.
201 gal., 1250 MBH input, 1200 GPH	19.672	Ea.
201 gal., 1500 MBH input, 1441 GPH	20.690	Ea.

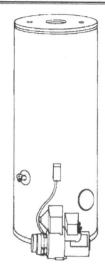

**Oil Fired Water Heater —
Commercial**

Figure 22-40 Installation Time in Labor-Hours for Commercial Electric
Hot Water Heaters

Description	Labor-Hours	Unit
Commercial, 100° rise		
Electric		
5 gal., 3 kW, 12 GPH, 208V	4.000	Ea.
10 gal., 6 kW, 25 GPH, 208V	4.000	Ea.
50 gal., 9 kW, 37 GPH, 208V	4.444	Ea.
50 gal., 36 kW, 148 GPH, 208V	4.444	Ea.
200 gal., 15 kW, 61 GPH, 480V	9.412	Ea.
200 gal., 120 kW, 490 GPH, 480V	9.412	Ea.
400 gal., 30 kW, 123 GPH, 480V	16.000	Ea.
Modulating step control, 2–5 steps	1.509	Ea.
6–10 steps	2.500	Ea.
11–15 steps	2.963	Ea.
16–20 steps	5.000	Ea.

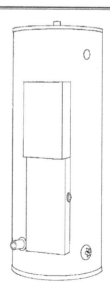

**Electric Water Heater —
Commercial**

Division 22

Figure 22-41   Installation Time in Labor-Hours for Commercial Hot
Water Heater Components

Description	Labor-Hours	Unit
Water Heater, Commercial, Gas, 75.5 MBH, 63 gal./hr	5.714	Ea.
Copper Tubing, Type L, Solder Joint, Hanger 10' On Center, 1" Diameter	.118	L.F.
Wrought Copper 90° Elbow for Solder Joints 1" Diameter	.500	Ea.
Wrought Copper Tee for Solder Joints 1" Diameter	.800	Ea.
Wrought Copper Union for Solder Joints 1" Diameter	.533	Ea.
Valve, Gate, Bronze, 125 lb., NRS, Soldered 1" Diameter	.421	Ea.
Relief Valve, Bronze, Press and Temp, Self Close, 3/4" IPS	.286	Ea.
Copper Tubing, Type L, Solder Joints, 3/4" Diameter	.105	L.F.
Wrought Copper 90° Elbow for Solder Joints 3/4" Diameter	.421	Ea.
Wrought Copper, Adapter, CTS to MPT, 3/4" IPS	.381	Ea.
Pipe Steel Black, Schedule 40, Threaded, 3/4" Diameter	.131	L.F.
Pipe, 90° Elbow, Malleable Iron Black, 150 lb. Threaded, 3/4" Diameter	.571	Ea.
Pipe, Union with Brass Seat, Malleable Iron Black, 3/4" Diameter	.615	Ea.
Valve, Gas Stop w/o Check, Brass, 3/4" IPS	.364	Ea.

Figure 22-42 Installation Time in Labor-Hours for Residential Hot
Water Heaters

Number of Bedrooms	H.W. Storage Capacity	Labor-Hours	Unit
1	20 gal.	3.810	Ea.
2	30 gal.	4.000	Ea.
3	40 gal.	4.211	Ea.
4	50 gal.	4.444	Ea.

For each additional bedroom go to the next larger size: 70, 85, and 100 gallon.

Figure 22-43 Installation Time in Labor-Hours for Pumps

Description	Labor-Hours	Unit
Pumps		
Sewage Ejector with Basin and Cover		
Single		
110 GPM 1/2 H.P.	6.400	Ea.
173 GPM 3/4 H.P.	8.000	Ea.
218 GPM 1 H.P.	10.000	Ea.
285 GPM 2 H.P.	12.308	Ea.
325 GPM 3 H.P.	17.143	Ea.
370 GPM 5 H.P.	24.000	Ea.
Duplex		
110 GPM 1/2 H.P.	8.000	Ea.
173 GPM 3/4 H.P.	10.000	Ea.
218 GPM 1 H.P.	20.000	Ea.
285 GPM 2 H.P.	24.000	Ea.
325 GPM 3 H.P.	30.000	Ea.
370 GPM 5 H.P.	48.000	Ea.
Sump Pumps, Submersible		
Pedestal Type Cellar Drainer with External Float		
42 GPM 1/3 H.P.	1.600	Ea.
Submersible with Built in Float		
22 GPM 1/4 H.P.	1.330	Ea.
68 GPM 1/2 H.P.	1.600	Ea.
94 GPM 1/2 H.P.	1.600	Ea.
105 GPM 1/2 H.P.	2.000	Ea.

**Submersible
Sump Pump**

Notes

Division Twenty-Three

Heating, Ventilating & Air Conditioning

Introduction

Setting HVAC units, running waste piping, and the many other functions of the mechanical contractor may involve the services of a multitude of other trades. For this reason, the full service mechanical contractor, much like the general contractor, may employ or subcontract for all of the individual mechanical trades, as well as temperature control specialists, balancing contractors, excavators, and even electricians. Although there are thousands of full service mechanical contractors, this number is far out-weighed by the number of individual plumbing shops, HVAC firms, sprinkler contractors, and sheet metal contractors that accomplish the bulk of mechanical work. This division is designed to be used and understood by not only the total mechanical contractor, but also by the individual trades.

The trade divisions included in this division are:

- Facility fuel systems, including piping and pumps
- HVAC piping and pumps
- HVAC air distribution
- HVAC air cleaning devices
- Central heating equipment
- Central cooling equipment
- Central HVAC equipment
- Decentralized HVAC equipment

The Energy Star program is a great source for identifying energy efficient HVAC units—both residential and commercial systems. For consumer HVAC units, the Energy Guide label compares each product with other comparable ones, and now also includes the Energy Star logo (if the unit qualifies).

For medium to large commercial installations it is important have the HVAC system commissioned to ensure that energy efficiency and occupant comfort is attained and maintained. The American Society of Heating, Refrigerating, and Air-Conditioning Engineers (ASHRAE) defines commissioning as the process of ensuring that systems are designed, installed, functionally tested, and capable of being operated and maintained to conform to the design intent. This is especially important in green buildings that may employ special high-tech systems, which the facility staff may be unfamiliar with and may require special training. Commissioning is so important that basic commissioning is a prerequisite for LEED® NC certification. More advanced commissioning will earn an additional credit.

Estimating Data

The following tables present estimating guidelines for items found in Division 23. Please note that these guidelines are intended as indicators of what may generally be expected, and that each project must be evaluated individually.

Table of Contents

Estimating Tips

Plans and Specifications

Review all construction documentation before proceeding with the estimate for Division 23. It is not unusual to find items on the plans, but not in the specifications; or they may be in the specifications but not on the drawings. Also check any and all communications; the owner may have requested a change or expressed a concern or a specific need, and not have had it included in your drawings or specifications.

Review All Drawing Divisions

HVAC systems are shown primarily in riser diagrams and schedules. It is important to review all of the drawings available to ensure that the systems called for will physically fit where they are supposed to. For example, sometimes the architect subcontracts out the HVAC design function as well as the structural design. If the drawings are not carefully coordinated, a conflict may result, such as inadequate space under deep trusses where main ductwork was supposed to be installed and a required finished ceiling height maintained.

Document All Inconsistencies

All inconsistencies with standard practice, conflicts, omissions, and other concerns should be addressed before the final estimate is resolved. Do not rely on memory, which can become quickly overloaded in a rush situation.

Temperature Control System

When estimating the cost of an HVAC system, check to see who is responsible for providing and installing the temperature control wiring system. In many cases, this item is overlooked, as it was assumed that it would be included in the electrical estimate. This can be a costly assumption.

Connecting to an Existing System

When tapping into an existing system for water, drainage, or waste, check with the local authorities (or building owner, in the case of interior work) to see if it is required that they or their designated contractor make the tap. Such surprises could lead to an unexpected expense or stoppage.

Non-Listed Items

It is important to include items that are not shown on the plans, but must be priced. These items include, but are not limited to, roof penetrations and pitchpots, dust protection, coring floors and walls, temporary water supply, testing and balancing HVAC systems, cleaning piping, purifying potable water systems, cleanup, and final adjustments.

Cutting and Patching

Not all projects will be done under ideal conditions, that is, wide open spaces with no interferences. Many times the mechanical contractor will have to run duct or piping through pre-existing partitions or partitions that have been recently installed. Allowances must be made for cutting into the partition and the ensuing patching. Otherwise, a statement must be included with the estimate that all cutting and patching will be the responsibility of others.

Electrical Connections

Clarify who is responsible for connecting the electrical wiring to mechanical items. It has been argued that the electrician is responsible only for bringing the wiring to the unit, while others maintain that it is the electrician's job to connect the wiring.

Checklist ✓

For an estimate to be reliable, all items must be accounted for. A complete estimate can also limit contingencies. The following checklist can be used to help ensure that all items are included.

☐ *Air Handling Units*
- ☐ Balance
- ☐ Cooling
- ☐ Heating
- ☐ Filtration
- ☐ Supply fans
- ☐ Supply system

☐ *Balance Air Systems*

☐ *Building Drainage*
- ☐ Area drains
- ☐ Balcony drains
- ☐ Floor drains
- ☐ Roof drains
- ☐ Shower drains
- ☐ Sump drains

☐ *Coils*
- ☐ Cooling
- ☐ Preheat
- ☐ Reheat

☐ *Cooling Plant*
- ☐ Chillers
- ☐ Compressors
- ☐ Condensers
- ☐ Cooling towers
- ☐ Pumps

☐ *Domestic Water – Hot*
- ☐ Boiler
- ☐ Conditioner
- ☐ Hose bibs
- ☐ Pumps
- ☐ Storage tanks

☐ *Ductwork System*
- ☐ Sheet metal
  - ☐ Double duct
  - ☐ High velocity
  - ☐ With coils
- ☐ Diffusers
- ☐ Grilles
- ☐ Hoods
- ☐ Registers
- ☐ Terminal boxes
  - ☐ Double duct
  - ☐ High velocity
  - ☐ With coils
- ☐ Volume dampers

☐ *Fans*
- ☐ Exhaust
- ☐ Return

☐ *Gas Supply System*
- ☐ Bottled gas
- ☐ Natural gas
- ☐ Laboratory gas systems
- ☐ Oxygen
- ☐ Industrial gases
- ☐ Control valves
- ☐ Fittings
- ☐ Meters
- ☐ Shutoffs

☐ *Heating*
- ☐ Boilers
- ☐ Forced air
- ☐ Fuel storage tanks
- ☐ Heat pumps
- ☐ Rooftop units
- ☐ Unit heaters

- ☐ Water storage tanks
- ☐ Systems
  - ☐ Multi-zone
  - ☐ Perimeter radiant
  - ☐ Radiant panels
  - ☐ Single zone
  - ☐ Terminal reheat
  - ☐ VAV box system

☐ *Insulation*

- ☐ Duct
- ☐ Piping
  - ☐ Cold water
  - ☐ Hot water
  - ☐ Process
  - ☐ Steam
  - ☐ Tanks
- ☐ Sound
- ☐ Water heaters

☐ *Oil Supply System*

- ☐ Filters
- ☐ Meters
- ☐ Pumps
- ☐ Safety valves
- ☐ Tanks

☐ *Piping*

- ☐ Air chambers
- ☐ Concrete encasement
- ☐ Escutcheons
- ☐ Expansion joints
- ☐ Excavation for piping
- ☐ Fittings
- ☐ Hangers
- ☐ Materials
- ☐ Shock absorbers
- ☐ Specialty
  - ☐ Carbon dioxide
  - ☐ Compressed air
  - ☐ Industrial gases
  - ☐ Nitrous oxide
  - ☐ Oxygen
  - ☐ Process
  - ☐ Sanitary
  - ☐ Vacuum

☐ *Temperature Control Systems*

- ☐ Devices
- ☐ Wiring

☐ *Special Systems*

Figure 23-1 Conversion Factors and Equivalents

Conversion Factors and Equivalents		
Multiply	**By**	**To Obtain**
Acres	43.560	Square Feet
B.T.U.s	0.2530	Kilogram-calories
B.T.U.s	778.2	Foot-pounds
B.T.U.s	.0002928	Kilowatt-hours
B.T.U.s per min.	12.97	Foot-pounds/sec.
B.T.U.s per min.	0.02356	Horsepower
B.T.U.s per min.	0.01757	Kilowatts
B.T.U.s per min.	17.57	Watts
Centimeters	0.3937	Inches
Cubic Feet	7.481	Gallons
Cubic feet/min.	0.1247	Gallons/sec.
Cubic feet/min.	62.43	Lbs. of water/min.
Cubic feet water	62.43	Lbs. of water
EDR	240	B.T.U.s
Ft. head	2.31	Lbs.
Ft. of Water	62.43	Lbs./Sq. Ft.
Ft. of Water	0.8826	Inches of mercury
Ft. of Water	0.4335	Lbs./Sq. In.
Ft./min.	0.01136	Miles/Hr.
Foot-pounds	.001285	B.T.U.s
Gallons	231	Cubic inches
Gallons/min.	.002228	Cu Ft. /sec.
Gallons water	8.345	Lbs. of water
Grains (troy)	0.06480	Grams
Grams	0.03527	Ounces
Horsepower	42.44	B.T.U.s/min.
Horsepower	33000	Ft. Lbs./min.
Horsepower	550	Ft. Lbs./sec.
Horsepower	745.7	Watts
Horsepower (Boiler)	33479	B.T.U.s/Hr.
Horsepower (Boiler)	9.804	Kilowatts
Horsepower-hours	2547	B.T.U.s
Inches	2.540	Centimeters
Inches of mercury	1.113	Ft. of water
Inches of mercury	0.4912	Lbs./Sq. In.
Inches of water	0.03613	Lbs./Sq. In.
Kilograms	2.2046	Pounds
Kilowatts	56.92	B.T.U.s/min.
Kilowatts	1.341	Horsepower
Kilowatt-hours	3415	B.T.U.s
Liters	0.2642	Gallons
Miles	5280	Feet
Ounces	437.5	Grains
Ounces	28.35	Grams
Pounds	7000	Grains
Pounds	453.6	Grams
Pounds of water	27.68	Cubic inches
Pounds of water	0.1198	Gallons
Pounds of water	7000	Grains
Pounds of water/min.	0.2669	Cu.Ft./sec.
Square feet	144	Square inches
Square inches	1.273×10^{-6}	Circular mils
Square miles	640	Acres
Temp. (degs. C) + 273	1	Abs. Temp. (degs. K)
Temp. (degs. F) + 17.8	1.8	Temp (degs. Fahr.)
Temp. (degs. F) -32	5/9	Temp. (degs. Cent.)
Ton (Refrigeration)	200	B.T.U.s/min.
Watts	.001341	Horsepower
Watt-hours	3.415	B.T.U.s

Figure 23-2a Air Requirements for Air Cooled Condenser

Refrigeration:
750 CFM per HP, 1000 CFM per ton
Air Conditioning:
1000 CFM per HP, 400 CFM per ton

Figure 23-2b Water Requirements for Condensing Units

City Water — Tons refrigeration x 1.5 = GPM
Cooling Tower — Tons refrigeration x 3.0 = GPM

Division 23

Figure 23-3 Units of Measure and Their Equivalents

Length
1 in. = 25.4 mm
1 mm = .03937 in.
1 ft. = 30.48 cm
1 meter = 3.28083 ft.
1 micron = .001 mm

Area
1 sq. in. = 6.4516 sq. cm
1 sq. ft. = 929.03 sq. cm.
1 sq. cm. = 0.155 sq. in.
1 sq. cm. = 0.0010764 sq. ft.

Volume
1 cu. in. = 16.387 cu. cm
1 cu. ft. = 1728 cu. in.
1 cu. ft. = 7.4805 U.S. gal.
1 cu. ft. = 6.229 British gal.
1 cu. ft. = 28.317 liters
1 U.S. gal. = 0.1337 cu. ft.
1 U.S. gal. = 231 cu. in.
1 U.S. gal. = 3.785 liters
1 British gal. = 1.20094 U.S. gal.
1 British gal. = 277.3 cu. in.
1 British gal. = 4.546 liters
1 liter = 61.023 cu. in.
1 liter = 0.03531 cu. ft.
1 liter = 0.2642 U.S. gal.

Weight
1 ounce av. = 28.35 g
1 lb. av. = 453.59 g
1 gram = 0.03527 oz. av.
1 kg. = 2.205 lb. av.
1 cu. ft. of water = 62.425 lb.
1 U.S. gal. of water = 8.33 lb.
1 cu. in. of water = 0.0361 lb.
1 British gal. of water = 10.04 lb.
1 cu. ft. of air at 32° F & 1 ATM = 0.080728 lb.

Velocity
1 ft. per sec. = 30.48 cm per sec.
1 cm. per sec. = .032808 ft. per sec.

Flow
1 cu. ft. per sec. = 448.83 gal. per min.
1 cu. ft. per sec. = 1699.3 liters per min.
1 U.S. gal. per min. = 0.002228 cu. ft. per sec.
1 U.S. gal. per min. = 0.06308 liters per sec.
1 cu. cm. per sec. = 0.0021186 cu. ft. per min.

Density
1 lb. per cu. ft. = 16.018 kg. per cu. meter
1 lb. per cu. ft. = .0005787 lb. per cu. in.
1 kg. per cu. meter = 0.06243 lb. per cu. ft.
1 g. per cu. cm. = 0.03613 lb. per cu. in.

Viscosity
1 Centipoise = .000672 lb. per ft. sec.
1 Centistoke = .000001076 sq. ft. per sec.

Pressure
1 in. of water = 0.03613 lb. per sq. in.
1 in. of water = 0.07355 in. of Hg.
1 ft. of water = 0.4335 lb. per sq. in.
1 ft. of water = 0.88265 in. of Hg.
1 in. of mercury = 0.49116 lb. per sq. in.
1 in. of mercury = 13.596 in. of water
1 in. of mercury = 1.13299 ft. of water
1 atmosphere = 14.696 lb. per sq. in.
1 atmosphere = 760 mm of Hg.
1 atmosphere = 29.921 in. of Hg.
1 atmosphere = 33.899 ft. of water
1 lb. per sq. in. = 27.70 in. of water
1 lb. per sq. in. = 2.036 in. of Hg.
1 lb. per sq. in. = .0703066 kg. per sq. cm
1 kg. per sq. cm. = 14.223 lb. per sq. in.
1 dyne per sq. cm. = .0000145 lb. per sq. in.
1 micron = .00001943 lb. per sq. in.

Energy
1 B.T.U.* = 777.97 ft. lbs.
1 erg = 9.4805×10^{-11} B.T.U.
1 erg = 7.3756×10^{-8} ft. lbs.
1 kilowatt hour = 2.655×10^6 ft. lbs.
1 kilowatt hour = 1.3410 h.p. hr.
1 kg. calorie = 3.968 B.T.U.

Power
1 horsepower = 33,000 ft. lb. per min.
1 horsepower = 550 ft. lb. per sec.
1 horsepower = 2,546.5 B.T.U. per hr.
1 horsepower = 745.7 watts
1 watt = 0.00134 horsepower
1 watt = 44.26 ft. lbs. per min.

Temperature
Temperature Fahrenheit (F) = 9/5 Centigrade + 32
  = 9/4 R +32
Temperature Centigrade (C) = 5/9 Fahrenheit (F) – 32
  = 5/4 R
Temperature Reaumur (°R) = 4/9 Fahrenheit (F) – 32 = 4/5 C
Absolute Temperature Centigrade or Kelvin (K)
  = Degrees C + 273.16
Absolute Temperature Fahrenheit or Rankine (R)
  = Degrees F + 459.69

Heat Transfer
1 B.T.U. per sq. ft. = .2712g cal. per sq. cm
1 g calorie per sq. cm. = 3.687 B.T.U. per sq. ft.
1 B.T.U. per hr. per sq. ft. per °F = 4.88 kg cal. per hr. per sq. m per °C.
1 Kg. cal. per hr. per sq. m. per °C = .205 B.T.U. per hr. per sq. ft. per °F
1 Boiler Horsepower = 33,479 B.T.U. per hr.

*BTU Formula (energy required to heat any substance): BTU/Hr. = Matl. Wt. (lbs.) x Temp. Rise (°F) x Spec. Heat.

Figure 23-4 Mechanical Equipment Service Life*

Service life is a time value established by ASHRAE that reflects the expected life of a specific component. Service life should not be confused with useful life or depreciation period used for income tax purposes. It is the life expectancy of system components. Equipment life is highly variable because of the diverse equipment applications, the preventive maintenance given, the environment, technical advancements of new equipment, and personal opinions. The values in this table are a median listing of replacement time of the components. Service life can be used to establish an amortization period; or, if an amortization period is given, service life can give insight into adjusting the maintenance or replacement costs of components.

Equipment Item	Median Years	Equipment Item	Median Years
Air conditioners		Coils	
Window unit	10	DX, water, or steam	20
Residential single or split package	15	Electric	15
Commercial through-the-wall	15	Heat exchangers	
Water-cooled package	15	Shell-and-tube	24
Computer room	15	Reciprocating compressors	20
Heat pumps		Package chillers	
Residential air-to-air	**	Reciprocating	20
Commercial air-to-air	15	Centrifugal	23
Commercial water-to-air	19	Absorption	23
Roof-top air conditioners		Cooling towers	
Single-zone	15	Galvanized metal	20
Multizone	15	Wood	20
Boilers, hot water (steam)		Ceramic	34
Steel water-tube	24 (30)	Air-cooled condensers	20
Steel fire-tube	25 (25)	Evaporative condensers	20
Cast iron	35 (30)	Insulation	
Electric	15	Molded	20
Burners	21	Blanket	24
Furnaces		Pumps	
Gas- or oil-fired	18	Base-mounted	20
Unit heaters		Pipe-mounted	10
Gas or electric	13	Sump and well	10
Hot water or steam	20	Condensate	15
Radiant heaters		Reciprocating engines	20
Electric	10	Steam turbines	30
Hot water or steam	25	Electric motors	18
Air terminals		Motor starters	17
Diffusers, grilles, and registers	27	Electric transformers	30
Induction and fan-coil units	20	Controls	
VAV and double-duct boxes	20	Pneumatic	20
Air washers	17	Electric	16
Duct work	30	Electronic	15
Dampers	20	Valve actuators	
Fans		Hydraulic	15
Centrifugal	25	Pneumatic	20
Axial	20	Self-contained	10
Propeller	15		
Ventilating roof-mounted	20		

*Obtained from a nation-wide survey by ASHRAE Technical Committee TC 1.8
**Data removed by TC 1.8 because of changing technology.

Figure 23-5 Storage Tank Capacities

To calculate the capacity (in gallons) of rectangular tanks, reduce all dimensions to inches, then multiply length by width by height and divide the resulting figure by 231.

Length in Feet	Diameter in Inches																
	12"	18"	24"	30"	36"	42"	48"	54"	60"	66"	72"	78"	84"	90"	96"	102"	108"
1	6	13	24	37	53	72	94	120	145	180	210	250	290	330	375	425	475
2	12	26	48	74	106	144	188	240	290	360	420	500	580	660	750	850	950
3	18	39	72	111	159	216	282	360	435	540	630	750	870	990	1125	1275	1425
4	24	52	96	148	212	288	376	480	580	720	840	1000	1160	1320	1500	1700	1900
5	30	65	120	185	265	360	470	600	725	900	1050	1250	1450	1650	1875	2125	2375
6	36	78	144	222	318	432	564	720	870	1080	1260	1500	1740	1980	2250	2550	2850
7	42	91	168	259	371	504	658	840	1015	1260	1470	1750	2030	2310	2625	2975	3325
8	48	104	192	296	424	576	752	960	1160	1440	1680	2000	2320	2640	3000	3400	3800
9	54	117	216	333	477	648	846	1080	1305	1620	1890	2250	2610	2970	3375	3825	4275
10	60	130	240	370	530	720	940	1200	1450	1800	2100	2500	2900	3300	3750	4250	4750
11	66	143	264	407	583	792	1034	1320	1595	1980	2310	2750	3190	3630	4125	4675	5225
12	72	156	288	444	636	864	1128	1440	1740	2160	2520	3000	3480	3960	4500	5100	5700
13	78	169	312	481	689	936	1222	1560	1885	2340	2730	3250	3770	4290	4875	5525	6175
14	84	182	336	518	742	1008	1316	1680	2030	2520	2940	3500	4060	4620	5250	5950	6650
15	90	195	360	555	795	1080	1410	1800	2175	2700	3150	3750	4350	4950	5625	6375	7125
16	96	208	384	592	848	1152	1504	1920	2320	2880	3360	4000	4640	5280	6000	6800	7600

Figure 23-6 Basics of a Heating System

The function of a heating system is to achieve and maintain a desired temperature in a room or building by replacing the amount of heat being dissipated. There are four kinds of heating systems: hot water, steam, warm air, and electric resistance. Each has certain essential and similar elements, with the exception of electric resistance heating.

The basic elements of a heating system are:

    A. A **combustion chamber** in which fuel is burned and heat transferred to a conveying medium

    B. The **"fluid"** used for conveying the heat (water, steam or air)

    C. **Conductors** or pipes for transporting the fluid to specific desired locations

    D. A means of disseminating the heat, sometimes called **terminal units**

A. The **combustion chamber** in a furnace heats air which is then distributed. This is called a *warm air system*.

The combustion chamber in the boiler heats water, which is either distributed as hot water or steam. This is termed a *hydronic system*.

The maximum allowable working pressures are limited by ASME "Code for Heating Boilers" to 15 PSI for steam and 160 PSI for hot water heating boilers, with a maximum temperature limitation of 250°F. Hot water boilers are generally rated for a working pressure of 30 PSI. High pressure boilers are governed by the ASME "Code for Power Boilers" which is used almost universally for boilers operating over 15 PSIG. High pressure boilers used for a combination of heating/process loads are usually designed for 150 PSIG.

Boiler ratings are usually indicated as either Gross or Net Output. The Gross Load is equal to the Net Load plus a piping and pickup allowance. When this allowance cannot be determined, divide the gross output rating by 1.25 for a value equal to or greater than the net heat loss requirement of the building.

B. Of the three **fluids** used, steam carries the greatest amount of heat per unit volume. This is because it gives up its latent heat of vaporization at a temperature considerably above room temperature. Another advantage is that the pressure to produce a positive circulation is readily available. Piping conducts the steam to terminal units and returns condensate to the boiler.

The **steam system** is well adapted to large buildings because of its positive circulation, its comparatively economical installation, and its ability to deliver large quantities of heat. Nearly all large office buildings, stores, hotels, and industrial buildings are so heated, in addition to many residences.

Hot water, when used as the heat carrying fluid, gives up a portion of its sensible heat and then returns to the boiler or heating apparatus for reheating. As the heat conveyed by each pound of water is about one-fiftieth of the heat conveyed by a pound of steam, it is necessary to circulate about fifty times as much water as steam by weight (although only one-thirtieth as much by volume). The hot water system is usually, although not necessarily, designed to operate at temperatures below that of the ordinary steam system and so the amount of heat transfer surface must be correspondingly greater. A temperature of 190°F to 200°F is normally the maximum. Circulation in small buildings may depend on the difference in density between hot water and the cool water returning to the boiler. Circulating pumps are normally used to maintain a desired rate of flow. Pumps permit a greater degree of flexibility and better control.

In **warm air** furnace systems, cool air is taken from one or more points in the building, passed over the combustion chamber and flue gas passages, and then distributed through a duct system. A disadvantage of this system is that the ducts take up much more building volume than steam or hot water pipes. Advantages of this system are the relative ease with which humidification can be accomplished by the evaporation of water as the air circulates through the heater, and the lack of need for expensive disseminating units as the warm air supply becomes part of the interior atmosphere of the building.

(continued on next page)

Figure 23-6 Basics of a Heating System *(continued from previous page)*

C. Conductors (pipes and ducts) have been lightly treated in the discussion of conveying fluids. For more detailed information such as sizing and distribution methods, the reader is referred to technical publications such as the American Society of Heating, Refrigerating, and Air Conditioning Engineers "Handbook of Fundamentals."

D. Terminal units come in an almost infinite variety of sizes and styles, but the basic principles of operation are very limited. As previously mentioned, warm air systems require only a simple register or diffuser to mix heated air with that present in the room. Special application items such as radiant coils and infrared heaters are available to meet particular conditions, but are not usually considered for general heating needs. Most heating is accomplished by having air flow over coils or pipes containing the heat transporting medium (steam, hot water, electricity). These units, while varied, may be separated into two general types: (1) radiator/convectors and (2) unit heaters.

Radiator/convectors may be cast, fin-tube or pipe assemblies. They may be direct, indirect, exposed, concealed or mounted within a cabinet enclosure, upright or baseboard style. These units are often collectively referred to as "radiators" or "radiation," although none gives off heat either entirely by radiation or by convection, but rather by a combination of both. The air flows over the units as a gravity "current." It is necessary to have one or more heat emitting units in each room. The most efficient placement is low along an outside wall or under a window to counteract the cold coming into the room and achieve an even distribution.

In contrast to radiator/convectors which operate most effectively against the walls of smaller rooms, unit **heaters** utilize a fan to move air over heating coils and are very effective in locations of relatively large volume. Unit heaters, while usually suspended overhead, may also be floor-mounted. They also may take in fresh outside air for ventilation. The heat distributed by unit heaters may be from a remote source and conveyed by a fluid, or it may be from the combustion of fuel in each individual heater. In the latter case, the only piping required would be for fuel. However, a vent for the products of combustion would be necessary. The following list gives many of the advantages of unit heaters for applications other than office or residential:

a. large capacity, so smaller number of units are required, **b.** piping system simplified, **c.** space saved, where they are located overhead out of the way, **d.** rapid heating directed where needed with effective wide distribution, **e.** difference between floor and ceiling temperature reduced, **f.** circulation of air obtained, and ventilation with introduction of fresh air possible, and **g.** heat output flexible and easily controlled.

Figure 23-7   Determining Heat Loss for Various Types of Buildings

General: While the most accurate estimates of heating requirements would naturally be based on detailed information about the building under consideration, it is possible to arrive at a reasonable approximation using the following procedure:

Calculate the cubic volume of the room or building.

Select the appropriate factor from the table below. Note that the factors apply only to inside temperatures listed in the first column and to 0°F outside temperature.

If the building has bad north and west exposures, multiply the heat loss factor by 1.1.

If the outside design temperature is other than 0°F, multiply the factor from the table below by the factor from Figure 23-8.

Multiply the cubic volume by the factor selected from the table below. This will give the estimated BTUH heat loss, which must be made up to maintain inside temperature.

Building Type	Conditions	Qualifications	Loss Factor*
Factories & Industrial Plants General Office Areas 70°F	One Story	Skylight in Roof	6.2
		No Skylight in Roof	5.7
	Multiple Story	Two Story	4.6
		Three Story	4.3
		Four Story	4.1
		Five Story	3.9
		Six Story	3.6
	All Walls Exposed	Flat Roof	6.9
		Heated Space Above	5.2
	One Long Warm Common Wall	Flat Roof	6.3
		Heated Space Above	4.7
	Warm Common Walls on Both Long Sides	Flat Roof	5.8
		Heated Space Above	4.1
Warehouses 60°F	All Walls Exposed	Skylights in Roof	5.5
		No Skylights in Roof	5.1
		Heated Space Above	4.0
	One Long Warm Common Wall	Skylight in Roof	5.0
		No Skylight in Roof	4.9
		Heated Space Above	3.4
	Warm Common Walls on Both Long Sides	Skylight in Roof	4.7
		No Skylight in Roof	4.4
		Heated Space Above	3.0

*Note: This table tends to be conservative, particularly for new buildings designed for minimum energy consumption.

Figure 23-8 Outside Design Temperature Correction Factor
(for Degrees Fahrenheit)

Outside Design Temperature	50	40	30	20	10	0	−10	−20	−30
Correction Factor	0.29	0.43	0.57	0.72	0.86	1.00	1.14	1.28	1.43

Figure 23-9 Basic Properties of Fuels

Fuel Type	Unit	Heating Value BTU/Unit	Overall System Efficiency	Remarks
Electricity	Kilowatt	3,412	100	
Steam	Pounds (at atmospheric pressure)	1,000	—	
Oil #2	Gallon	138,500	60–88	
Oil #4	Gallon	145,000	60–88	
Oil #6	Gallon	152,000	—	Preheat
Natural Gas	CCF (100 C.F.)*	103,000	65–92	
Propane	Gallon	95,500	65–90	
Coal	Pound	13,000	45–75	
Solar	S.F. of collector—varies with location and collector			

*Note: 1 therm = 1.013 CCF

Figure 23-10　Boiler Selection Chart

Several types of boilers are available to meet the hot water and heating needs of both residential and commercial buildings. Some different boiler types are shown in this table.

Most hot water boilers operate at less than 30 psig (low-pressure boilers). Low-pressure steam boilers operate at less than 15 psig. Above 30 psig, high-pressure boilers must conform to stricter requirements. Water is lost to a heating system through minor leaks and evaporation; therefore, a makeup water line to feed the boiler with fresh makeup water must be provided.

Boiler Type	Output Capacity Range – MBH Efficiency Range	Fuel Types	Uses
Cast Iron Sectional	80–14,500 80–92%	Oil, Gas, Coal, Wood/Fossil	Steam/Hot Water
Steel	1,200–18,000 80–92%	Oil, Gas, Coal, Wood/Fossil, Electric	Steam/Hot Water
Scotch Marine	3,400–24,000 80–92%	Oil, Gas	Steam/Hot Water
Pulse Condensing	40–150 90–95%	Gas	Hot Water
Residential/Wall Hung	15–60 90–95%	Gas, Electric	Hot Water

Efficiencies shown are averages and will vary with specific manufacturers. For existing equipment, efficiencies may be 60%–75%.

Figure 23-11 Electric Heaters

Description	Labor-Hours	Unit
Baseboard heaters		
2' long, 375 watt	1	Ea.
3' long, 500 watt	1	Ea.
4' long, 750 watt	1.190	Ea.
5' long, 935 watt	1.400	Ea.
6' long, 1125 watt	1.600	Ea.
7' long, 1310 watt	1.820	Ea.
8' long, 1500 watt	2	Ea.
9' long, 1680 watt	2.220	Ea.
10' long, 1875 watt	2.420	Ea.
Wall heaters with fan, 120 to 277 volt		
750 watt	1.140	Ea.
1000 watt	1.140	Ea.
1250 watt	1.330	Ea.
1500 watt	1.600	Ea.
2000 watt	1.600	Ea.
2500 watt	1.780	Ea.
3000 watt	2	Ea.
4000 watt	2.290	Ea.
Recessed heaters		
750 watt	1.330	Ea.
1000 watt	1.330	Ea.
1250 watt	1.600	Ea.
1500 watt	2	Ea.
2000 watt	2	Ea.
2500 watt	2.290	Ea.
3000 watt	2.670	Ea.
4000 watt	2.960	Ea.

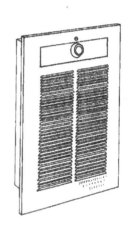

Rule of thumb: For a quick estimate of installation time for baseboard units, including controls, allow 1.820 labor-hours per kW. For duct heaters, including controls, allow 1.510 labor-hours per kW.

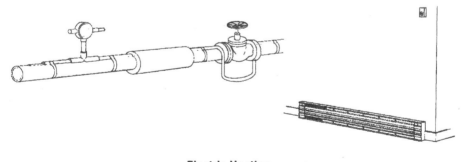

Electric Heating

Figure 23-12   Air Conditioning Basics

The purpose of air conditioning is to control the environment of a space so that comfort is provided for the occupants, and/or conditions are suitable for the processes or equipment contained therein. The several items which should be evaluated to define system objectives are:

- Temperature control
- Humidity control
- Cleanliness
- Odor, smoke, and fumes
- Ventilation

Efforts to control the above parameters must also include consideration of the degree or tolerance of variation, the noise level introduced, the velocity of air motion, and the energy requirements to accomplish the desired results.

The variation in temperature and humidity is a function of the sensor and the controller. The controller reacts to a signal from the sensor and produces the appropriate response in either the terminal unit, the conductor of the transporting medium (air, steam, chilled water, etc.) or the source (boiler, evaporating coils, etc.).

The noise level is a by-product of the energy supplied to moving components of the system. Those items which usually contribute the most noise are pumps, blowers, fans, compressors, and diffusers. The level of noise can be partially controlled through use of vibration pads, isolators, proper sizing, shields, baffles, and sound absorbing liners.

Some air motion is necessary to prevent stagnation and stratification. The maximum acceptable velocity varies with the degree of heating or cooling which is taking place. Most people feel air moving past them at velocities in excess of 25 FPM as an annoying draft. However, velocities up to 45 FPM may be acceptable in certain cases. Ventilation, expressed as air changes per hour and percentage of fresh air, is usually an item regulated by local codes.

Selection of the system to be used for a particular application is usually a trade-off. In some cases, the building size, style, or room available for mechanical use, limits the range of possibilities. Prime factors influencing the decision are first cost and total life (operating, maintenance, and replacement costs). The accuracy with which each parameter is determined will be an important measure of the reliability of the decision and subsequent satisfactory operation of the installed system.

Heat delivery may be desired from an air conditioning system. Heating capability usually is added as follows: A gas-fired burner or hot water/steam/electric coils may be added to the air handling unit directly and heat all air equally. For limited or localized heat requirements the water/steam/electric coils may be inserted into the duct branch supplying the cold areas. Gas-fired duct furnaces are also available. Note: when water or steam coils are used, the cost of the piping and boiler must also be added. For a rough estimate, use the cost per square foot of the appropriate sized hydronic system with unit heater. This will provide a cost of the boiler and piping, and the unit heaters of the system would equate to the approximate cost of the heating coils.

(continued on next page)

Figure 23-12 Air Conditioning Basics *(continued from previous page)*

Air Conditioning Requirements

BTUs per Hour per S.F. of Floor Area and S.F. per Ton of Air Conditioning

Type Building	BTU per S.F.	S.F. per Ton	Type Building	BTU per S.F.	S.F. per Ton	Type Building	BTU per S.F.	S.F. per Ton
Apartments, Individual			Dormitory, Rooms	40	300	Libraries	50	240
Corridors	26	450	Corridors	30	400	Low Rise Office, Exterior	38	320
	22	550	Dress Shops	43	280	Interior	33	360
Auditoriums & Theaters	40	300/18*	Drug Stores	80	150	Medical Centers	28	425
Banks	50	240	Factories	40	300	Motels	28	425
Barber Shops	48	250						
			High Rise Office —			Office (small suite)	43	280
Bars & Taverns	133	90	Exterior Rooms	46	263			
			Interior Rooms	37	325	P. O., Individual Office	42	285
Beauty Parlors	66	180	Hospitals, Core	43	280	Central Area	46	260
Bowling Alleys	68	175	Perimeter	46	260	Residences	20	600
Churches	36	330/20*	Hotel, Guest Rooms	44	275	Restaurants	60	200
Cocktail Lounges	68	175						
			Corridors	30	400	Schools & Colleges	46	260
Computer Rooms	141	85	Public Spaces	55	220	Shoe Stores	55	220
Dental Offices	52	230	Industrial Plants,			Shopping Centers,		
Department Stores,			Offices	38	320	Super Markets	34	350
Basement	34	350	General Offices	34	350	Retail Stores	48	250
Main Floor	40	300	Plant Areas	40	300	Specialty	60	200
Upper Floor	30	400						

* Persons per ton

12,000 BTU = 1 ton of air conditioning

(continued on next page)

Figure 23-12 Air Conditioning Basics *(continued from previous page)*

Type of Building	S.F. per Ton A/C	Ductwork # / SF of Building Area			
		Rooftop Unit Single Zone	Rooftop Unit Multizone	Self-contained Air- or Water-cooled	Split system Air- or Water-cooled
Apartments - Individual	0.0022	0.267	0.533	0.240	0.227
Corridors	0.0018	0.218	0.436	0.196	0.185
Auditoriums and Theaters	0.0033	0.400	0.800	0.360	0.340
Banks	0.0042	0.500	1.000	0.450	0.425
Barber Shops	0.0040	0.480	0.960	0.432	0.408
Bars and Taverns	0.0111	1.333	2.667	1.200	1.133
Beauty Parlors	0.0056	0.667	1.333	0.600	0.567
Bowling Alleys	0.0057	0.686	1.371	0.617	0.583
Churches	0.0030	0.364	0.727	0.327	0.309
Cocktail Lounges	0.0057	0.686	1.371	0.617	0.583
Computer Rooms	0.0118	1.412	2.824	1.271	1.200
Dental Offices	0.0043	0.522	1.043	0.470	0.443
Department Store, Basement	0.0029	0.343	0.686	0.309	0.291
Main Floor	0.0033	0.400	0.800	0.360	0.340
Upper Floor	0.0025	0.300	0.600	0.270	0.255
Dormitory Rooms	0.0033	0.400	0.800	0.360	0.340
Corridors	0.0025	0.300	0.600	0.270	0.255
Dress Shops	0.0036	0.429	0.857	0.386	0.364
Drug Stores	0.0067	0.800	1.600	0.720	0.680
Factories	0.0033	0.400	0.800	0.360	0.340
High Rise Office, Ext. Rooms	0.0038	0.456	0.913	0.411	0.388
Int. Rooms	0.0031	0.369	0.738	0.332	0.314
Hospitals, Core	0.0036	0.429	0.857	0.386	0.364
Perimeter	0.0038	0.462	0.923	0.415	0.392
Hotel, Guest Rooms	0.0036	0.436	0.873	0.393	0.371
Corridors	0.0025	0.300	0.600	0.270	0.255
Public Spaces	0.0045	0.545	1.091	0.491	0.464
Industrial Plants, Offices	0.0031	0.375	0.750	0.338	0.319
General Offices	0.0029	0.343	0.686	0.309	0.291
Plant Areas	0.0033	0.400	0.800	0.360	0.340
Libraries	0.0042	0.500	1.000	0.450	0.425
Low Rise Office, Exterior	0.0031	0.375	0.750	0.338	0.319
Interior	0.0028	0.333	0.667	0.300	0.283
Medical Centers	0.0024	0.282	0.565	0.254	0.240
Motels	0.0024	0.282	0.565	0.254	0.240
Office (small suite)	0.0036	0.429	0.857	0.386	0.364
Post Office, Individual Office	0.0035	0.421	0.842	0.379	0.358
Central Area	0.0038	0.462	0.923	0.415	0.392
Residences	0.0017	0.200	0.400	0.180	0.170
Restaurants	0.0050	0.600	1.200	0.540	0.510
Schools and Colleges	0.0038	0.462	0.923	0.415	0.392
Shoe Stores	0.0045	0.545	1.091	0.491	0.464
Shopping Centers, Supermarkets	0.0029	0.343	0.686	0.309	0.291
Retail Stores	0.0040	0.480	0.960	0.432	0.408
Specialty	0.0050	0.600	1.200	0.540	0.510

Note: In addition to the ductwork an allowance should be made for diffusers, registers, insulation and accessories such as turning vanes, volume dampers, fire dampers and access doors.

(continued on next page)

Figure 23-12 Air Conditioning Basics *(continued from previous page)*

Steel Sheet Gauges					
	Steel	Thickness Inches		Weight lbs/sq ft	
Gauge Number	Weight lbs per sq ft	US Standard Gauge	Manufac-turers Standard	Galvan-ized Sheet	Stain-less Steel
7/0	20.00	0.5000	...	...	...
6/0	18.75	0.4687	...	...	...
5/0	17.50	0.4375	...	...	...
4/0	16.25	0.4062	...	...	...
3/0	15.00	0.3750	...	...	...
2/0	13.75	0.3437	...	...	...
0	12.50	0.3125	...	...	...
1	11.25	0.2812	...	...	...
2	10.62	0.2656	...	...	...
3	10.00	0.2500	0.2391	...	...
4	9.37	0.2344	0.2242	...	...
5	8.75	0.2187	0.2092	...	...
6	8.12	0.2031	0.1943	...	...
7	7.50	0.1875	0.1793	...	...
8	6.87	0.1719	0.1644	...	...
9	6.25	0.1562	0.1495	...	...
10	5.62	0.1406	0.1345	5.7812	5.7937
11	5.00	0.1250	0.1196	5.1562	5.1500
12	4.37	0.1094	0.1046	4.5312	4.5063
13	3.75	0.0937	0.0897	3.9062	3.8625
14	3.12	0.0781	0.0747	3.2812	3.2187
15	2.81	0.0703	0.0673	2.9687	2.8968
16	2.50	0.0625	0.0598	2.6562	2.5750
17	2.25	0.0562	0.0538	2.4062	2.3175
18	2.00	0.0500	0.0478	2.1562	2.0600
19	1.75	0.0437	0.0418	1.9062	1.8025
20	1.50	0.0375	0.0359	1.6562	1.5450
21	1.37	0.0344	0.0329	1.5312	1.4160
22	1.25	0.0312	0.0299	1.4062	1.2875
23	1.12	0.0281	0.0269	1.2812	1.1587
24	1.00	0.0250	0.0239	1.1562	1.0300
25	0.875	0.0219	0.0209	1.0312	0.9013
26	0.750	0.0187	0.0179	0.9062	0.7725
27	0.687	0.0172	0.0164	0.8437	0.7081
28	0.625	0.0156	0.0149	0.7812	0.6438
29	0.562	0.0141	0.0135	0.7187	0.5794
30	0.500	0.0125	0.0120	0.6562	0.5150
31	0.437	0.0109	0.0105	...	...
32	0.406	0.0102	0.0097	...	...
33	0.375	0.0094	0.0090	...	...
34	0.344	0.0086	0.0082	...	...
35	0.312	0.0078	0.0075	...	...
36	0.281	0.0070	0.0067	...	...
37	0.266	0.0066	0.0064	...	...
38	0.250	0.0062	0.0060	...	...
39	0.234	0.0059	...	...	...
40	0.219	0.0055	...	...	...
41	0.211	0.0053	...	...	...
42	0.203	0.0051	...	...	...
43	0.195	0.0049	...	...	...
44	0.187	0.0047	...	...	...

(continued on next page)

Figure 23-12 Air Conditioning Basics *(continued from previous page)*

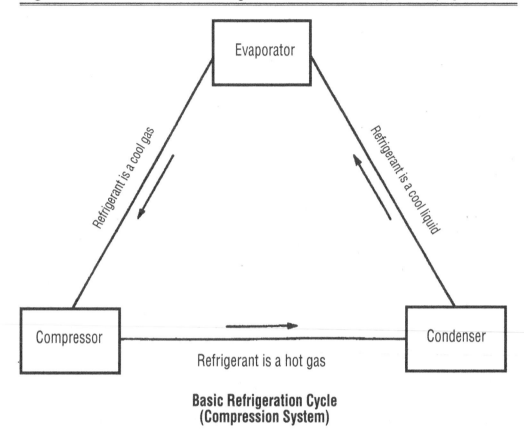

Basic Refrigeration Cycle
(Compression System)

All compression systems are composed of three basic components plus controls, regulators, etc.

1. When the compressor and condenser are combined, the assembly is called a *condensing unit.*
2. The condenser may be air cooled or water cooled. (Water cooled systems require a water source, usually a water tower or pond.)
3. Evaporators may also be called *air handlers*, *water chillers*, or *cooling coils.*

Figure 23-13 Recommended Ventilation Air Changes

Range of Time in Minutes per Change for Various Types of Facilities					
Assembly Halls	2–10	Dance Halls	2–10	Laundries	1–3
Auditoriums	2–10	Dining Rooms	3–10	Markets	2–10
Bakeries	2–3	Dry Cleaners	1–5	Offices	2–10
Banks	3–10	Factories	2–5	Pool Rooms	2–5
Bars	2–5	Garages	2–10	Recreation Rooms	2–10
Beauty Parlors	2–5	Generator Rooms	2–5	Sales Rooms	2–10
Boiler Rooms	1–5	Gymnasiums	2–10	Theaters	2–8
Bowling Alleys	2–10	Kitchens – Hospitals	2–5	Toilets	2–5
Churches	5–10	Kitchens – Restaurants	1–3	Transformer Rooms	1–5

CFM air required for changes = Volume of room in cubic feet/Minutes per change.

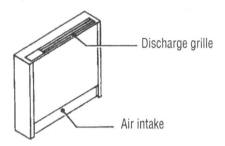

Discharge grille

Air intake

Room Size Fan Coil Unit

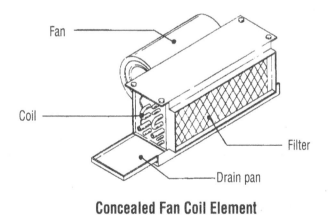

Fan

Coil

Filter

Drain pan

Concealed Fan Coil Element

Figure 23-14   Sheet Metal Calculator – Weight in Pounds per
Linear Foot

Example: If duct is 34" × 20" × 15' long, 34" is greater than 30' maximum for 24 ga. so 22 ga. must be used. 34" + 20' = 54". Going across from 54', find 13.5 lbs. per foot in 22 ga. column. 13.5 × 15' = 202.5 lbs. For SF of surface area 202.5/1.406 = 144 SF.

Note: Figures include an allowance for scrap.

Gauge	26	24	22	20	18	16	Gauge	26	24	22	20	18	16
Wt.–Lb./S.F.	.906	1.156	1.406	1.656	2.156	2.656	Wt.–Lb./S.F.	.906	1.156	1.406	1.656	2.156	2.656
SMACNA Max. Dimension Long Side		30"	54"	84"	85" Up		SMACNA Max. Dimension Long Side		30"	54"	84"	85" Up	
Sum–2 Sides							Sum–2 Sides						
2	.30	.40	.50	.60	.80	.90	42	7.0	9.0	10.5	12.2	16.0	18.9
3	.50	.65	.80	.90	1.1	1.4	43	7.2	9.2	10.8	12.5	16.3	19.4
4	.70	.85	1.0	1.2	1.5	1.8	44	7.3	9.5	11.0	12.8	16.7	19.8
5	.80	1.1	1.3	1.5	1.9	2.3	45	7.5	9.7	11.3	13.1	17.1	20.3
6	1.0	1.3	1.5	1.7	2.3	2.7	46	7.7	9.9	11.5	13.3	17.5	20.7
7	1.2	1.5	1.8	2.0	2.7	3.2	47	7.8	10.1	11.8	13.6	17.9	21.2
8	1.3	1.7	2.0	2.3	3.0	3.6	48	8.0	10.3	12.0	13.9	18.2	21.6
9	1.5	1.9	2.3	2.6	3.4	4.1	49	8.2	10.5	12.3	14.2	18.6	22.1
10	1.7	2.2	2.5	2.9	3.8	4.5	50	8.3	10.7	12.5	14.5	19.0	22.5
11	1.8	2.4	2.8	3.2	4.2	5.0	51	8.5	11.0	12.8	14.8	19.4	23.0
12	2.0	2.6	3.0	3.5	4.6	5.4	52	8.7	11.2	13.0	15.1	19.8	23.4
13	2.2	2.8	3.3	3.8	4.9	5.9	53	8.8	11.4	13.3	15.4	20.1	23.9
14	2.3	3.0	3.5	4.1	5.3	6.3	54	9.0	11.6	13.5	15.7	20.5	24.3
15	2.5	3.2	3.8	4.4	5.7	6.8	55	9.2	11.8	13.8	16.0	20.9	24.8
16	2.7	3.4	4.0	4.6	6.1	7.2	56	9.3	12.0	14.0	16.2	21.3	25.2
17	2.8	3.7	4.3	4.9	6.5	7.7	57	9.5	12.3	14.3	16.5	21.7	25.7
18	3.0	3.9	4.5	5.2	6.8	8.1	58	9.7	12.5	14.5	16.8	22.0	26.1
19	3.2	4.1	4.8	5.5	7.2	8.6	59	9.8	12.7	14.8	17.1	22.4	26.6
20	3.3	4.3	5.0	5.8	7.6	9.0	60	10.0	12.9	15.0	17.4	22.8	27.0
21	3.5	4.5	5.3	6.1	8.0	9.5	61	10.2	13.1	15.3	17.7	23.2	27.5
22	3.7	4.7	5.5	6.4	8.4	9.9	62	10.3	13.3	15.5	18.0	23.6	27.9
23	3.8	5.0	5.8	6.7	8.7	10.4	63	10.5	13.5	15.8	18.3	24.0	28.4
24	4.0	5.2	6.0	7.0	9.1	10.8	64	10.7	13.7	16.0	18.6	24.3	28.8
25	4.2	5.4	6.3	7.3	9.5	11.3	65	10.8	13.9	16.3	18.9	24.7	29.3
26	4.3	5.6	6.5	7.5	9.9	11.7	66	11.0	14.1	16.5	19.1	25.1	29.7
27	4.5	5.8	6.8	7.8	10.3	12.2	67	11.2	14.3	16.8	19.4	25.5	30.2
28	4.7	6.0	7.0	8.1	10.6	12.6	68	11.3	14.6	17.0	19.7	25.8	30.6
29	4.8	6.2	7.3	8.4	11.0	13.1	69	11.5	14.8	17.3	20.0	26.2	31.1
30	5.0	6.5	7.5	8.7	11.4	13.5	70	11.7	15.0	17.5	20.3	26.6	31.5
31	5.2	6.7	7.8	9.0	11.8	14.0	71	11.8	15.2	17.8	20.6	27.0	32.0
32	5.3	6.9	8.0	9.3	12.2	14.4	72	12.0	15.4	18.0	20.9	27.4	32.4
33	5.5	7.1	8.3	9.6	12.5	14.9	73	12.2	15.6	18.3	21.2	27.7	32.9
34	5.7	7.3	8.5	9.9	12.9	15.3	74	12.3	15.8	18.5	21.5	28.1	33.3
35	5.8	7.5	8.8	10.2	13.3	15.8	75	12.5	16.1	18.8	21.8	28.5	33.8
36	6.0	7.8	9.0	10.4	13.7	16.2	76	12.7	16.3	19.0	22.0	28.9	34.2
37	6.2	8.0	9.3	10.7	14.1	16.7	77	12.8	16.5	19.3	22.3	29.3	34.7
38	6.3	8.2	9.5	11.0	14.4	17.1	78	13.0	16.7	19.5	22.6	29.6	35.1
39	6.5	8.4	9.8	11.3	14.8	17.6	79	13.2	16.9	19.8	22.9	30.0	35.6
40	6.7	8.6	10.0	11.6	15.2	18.0	80	13.3	17.1	20.0	23.2	30.4	36.0
41	6.8	8.8	10.3	11.9	15.6	18.5	81	13.5	17.3	20.3	23.5	30.8	36.5

(continued on next page)

Figure 23-14 Sheet Metal Calculator — Weight in Pounds per Linear Foot *(continued from previous page)*

Gauge	26	24	22	20	18	16	Gauge	26	24	22	20	18	16
Wt.–Lb./S.F.	.906	1.156	1.406	1.656	2.156	2.656	Wt.–Lb./S.F.	.906	1.156	1.406	1.656	2.156	2.656
SMACNA Max. Dimension Long Side		30"	54"	84"	85" Up		SMACNA Max. Dimension Long Side		30"	54"	84"	85" Up	
Sum–2 Sides							Sum–2 Sides						
82	13.7	17.5	20.5	23.8	31.2	36.9	97	16.2	20.8	24.3	28.1	36.9	43.7
83	13.8	17.8	20.8	24.1	31.5	37.4	98	16.3	21.0	24.5	28.4	37.2	44.1
84	14.0	18.0	21.0	24.4	31.9	37.8	99	16.5	21.2	24.8	28.7	37.6	44.6
85	14.2	18.2	21.3	24.7	32.3	38.3	100	16.7	21.4	25.0	29.0	38.0	45.0
86	14.3	18.4	21.5	24.9	32.7	38.7	101	16.8	21.6	25.3	29.3	38.4	45.5
87	14.5	18.6	21.8	25.2	33.1	39.2	102	17.0	21.8	25.5	29.6	38.8	45.9
88	14.7	18.8	22.0	25.5	33.4	39.6	103	17.2	22.0	25.8	29.9	39.1	46.4
89	14.8	19.0	22.3	25.8	33.8	40.1	104	17.3	22.3	26.0	30.2	39.5	46.8
90	15.0	19.3	22.5	26.1	34.2	40.5	105	17.5	22.5	26.3	30.5	39.9	47.3
91	15.2	19.5	22.8	26.4	34.6	41.0	106	17.7	22.7	26.5	30.7	40.3	47.7
92	15.3	19.7	23.0	26.7	35.0	41.4	107	17.8	22.9	26.8	31.0	40.7	48.2
93	15.5	19.9	23.3	27.0	35.3	41.9	108	18.0	23.1	27.0	31.3	41.0	48.6
94	15.7	20.1	23.5	27.3	35.7	42.3	109	18.2	23.3	27.3	31.6	41.4	49.1
95	15.8	20.3	23.8	27.6	36.1	42.8	110	18.3	23.5	27.5	31.9	41.8	49.5
96	16.0	20.5	24.0	27.8	36.5	43.2							

Figure 23-15 Ductwork Packages (per Ton of Cooling)

System	Sheet Metal	Insulation	Diffusers	Return Register
Rooftop Unit Single Zone	120 lbs.	52 S.F.	1	1
Rooftop Unit Multizone	240 lbs.	104 S.F.	2	1
Self-contained Air or Water Cooled	108 lbs.	—	2	—
Split System Air Cooled	102 lbs.	—	2	—

Systems reflect most common usage

Note: See Figure 23-12 for SF/Ton for various building types.

Figure 23-16 Ductwork Fabrication and Installation Guidelines

Ductwork Installation Adds	Labor
Add to labor for elevated installation of fabricated ductwork	
10' to 15' high	6%
15' to 20' high	12%
20' to 25' high	15%
25' to 30' high	21%
30' to 35' high	24%
35' to 40' high	30%
Over 40' high	33%
Add to labor for elevated installation of prefabricated (purchased) ductwork	
10' to 15' high	10%
15' to 20' high	20%
20' to 25' high	25%
25' to 30' high	35%
30' to 35" high	40%
35' to 40' high	50%
Over 40' high	55%

The labor cost for sheet metal duct includes both the cost of fabrication and installation of the duct. The split is approximately 60% for fabrication, 40% for installation. It is for this reason that the percentage add for elevated installation is less than the percentage add for prefabricated duct.

Example: assume a piece of duct cost $100 installed (labor only).

$$\text{Sheet Metal Fabrication} = 40\% = \$40$$
$$\text{Installation} = 60\% = \$60$$

The add for elevated installation is:
$$\text{Based on total labor} = \$100 \times 6\% = \$6.00$$
(fabrication & installation)

Based on installation cost only
$$\text{(material purchased prefabricated)} = \$60 \times 10\% = \$6.00$$

The $6.00 markup (10' to 15' high) is the same.

Figure 23-17 Standard HVAC Symbols

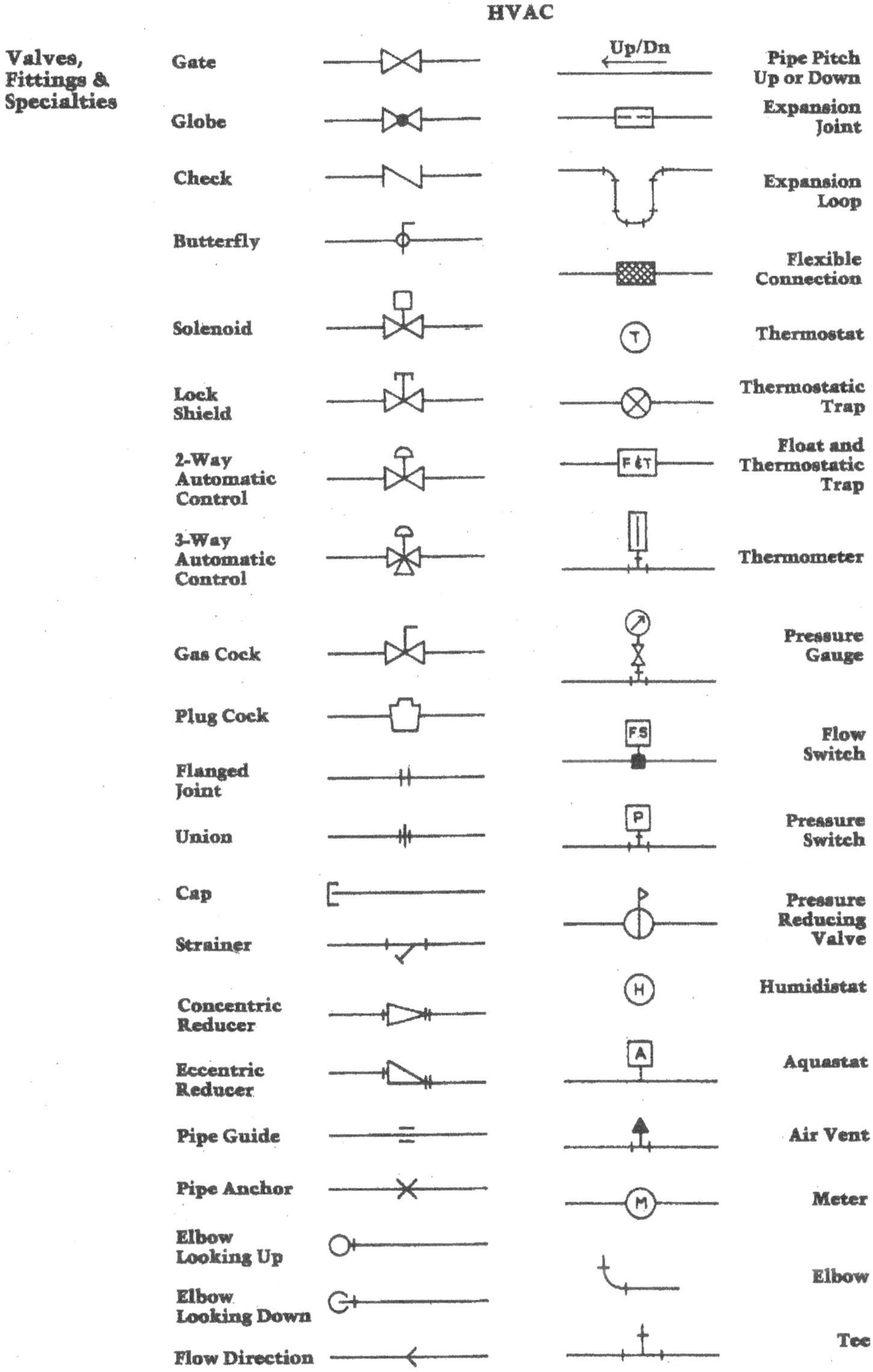

HVAC

Valves, Fittings & Specialties				
Gate	—▷◁—		←_Up/Dn_	Pipe Pitch Up or Down
Globe	—▷●◁—		—[ ]—	Expansion Joint
Check	—|\—		⊔	Expansion Loop
Butterfly	—φ—		▓▓▓	Flexible Connection
Solenoid	—▷◁—		(T)	Thermostat
Lock Shield	—▷◁—		—⊗—	Thermostatic Trap
2-Way Automatic Control	—▷◁—		—[F&T]—	Float and Thermostatic Trap
3-Way Automatic Control	—▷◁—		▯	Thermometer
Gas Cock	—▷◁—		⊘	Pressure Gauge
Plug Cock	—▢—		[FS]	Flow Switch
Flanged Joint	—||—		[P]	Pressure Switch
Union	—|||—		○	Pressure Reducing Valve
Cap	E—		(H)	Humidistat
Strainer	—↘—		[A]	Aquastat
Concentric Reducer	—▷||—		▲	Air Vent
Eccentric Reducer	—▷||—		(M)	Meter
Pipe Guide	—=—		└	Elbow
Pipe Anchor	—✕—			
Elbow Looking Up	O—		⊥	Tee
Elbow Looking Down	C—			
Flow Direction	—←—			

(continued on next page)

Figure 23-17   Standard HVAC Symbols *(continued from previous page)*

Plumbing		HVAC		HVAC (Cont.)
Floor Drain	▢	—— FOG ——	Fuel Oil Gauge Line	
Indirect Waste	—— W ——	—o— PD —o—	Pump Discharge	
Storm Drain	—— SD ——	— — — — — — —	Low Pressure Condensate Return	
Combination Waste & Vent	—— CWV ——	—— LPS ——	Low Pressure Steam	
Acid Waste	—— AW ——			
Acid Vent	— — AV — —	—— MPS ——	Medium Pressure Steam	
Cold Water	—— CW ——	—— HPS ——	High Pressure Steam	
Hot Water	—— HW ——			
Drinking Water Supply	—— DWS ——	—— BD ——	Boiler Blow-Down	
Drinking Water Return	—— DWR ——	—— F ——	Fire Protection Water Supply	Fire Protection
Gas-Low Pressure	—— G ——			
Gas-Medium Pressure	—— MG ——	—— WSP ——	Wet Standpipe	
Compressed Air	—— A ——	—— DSP ——	Dry Standpipe	
Vacuum	—— V ——	—— CSP ——	Combination Standpipe	
Vacuum Cleaning	—— VC ——	—— SP ——	Automatic Fire Sprinkler	
Oxygen	—— O ——	—o——o—	Upright Fire Sprinkler Heads	
Liquid Oxygen	—— LOX ——	—●——●—	Pendent Fire Sprinkler Heads	
Liquid Petroleum Gas	—— LPG ——			
Hot Water Heating Supply	—— HWS ——	⊲	Fire Hydrant	
Hot Water Heating Return	—— HWR ——	K	Wall Fire Dept. Connection	
Chilled Water Supply	—— CHWS ——	⊂<	Sidewalk Fire Dept. Connection	
Chilled Water Return	—— CHWR ——			
Drain Line	—— D ——	O—IIIIIIII FHR	Fire Hose Rack	
City Water	—— CW ——			
Fuel Oil Supply	—— FOS ——	FHC	Surface Mounted Fire Hose Cabinet	
Fuel Oil Return	—— FOR ——			
Fuel Oil Vent	—— FOV ——	FHC	Recessed Fire Hose Cabinet	

HVAC (left margin label)

(continued on next page)

Figure 23-17 Standard HVAC Symbols *(continued from previous page)*

HVAC Ductwork Symbols

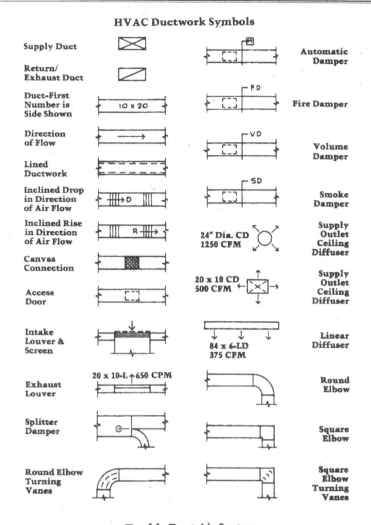

Supply Duct

Return/ Exhaust Duct

Duct-First Number is Side Shown — 10 x 20

Direction of Flow

Lined Ductwork

Inclined Drop in Direction of Air Flow — D

Inclined Rise in Direction of Air Flow — R

Canvas Connection

Access Door

Intake Louver & Screen

Exhaust Louver — 20 x 10-L ↑ 650 CPM

Splitter Damper

Round Elbow Turning Vanes

Automatic Damper

Fire Damper — FD

Volume Damper — VD

Smoke Damper — SD

24" Dia. CD 1250 CFM — Supply Outlet Ceiling Diffuser

20 x 10 CD 500 CFM — Supply Outlet Ceiling Diffuser

84 x 6-LD 375 CFM — Linear Diffuser

Round Elbow

Square Elbow

Square Elbow Turning Vanes

Double Duct Air System

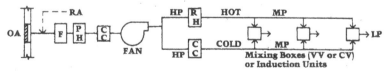

OA — F — PH — CC — FAN — HP R H — HOT — MP — LP
HP C C — COLD — MP
Mixing Boxes (VV or CV) or Induction Units

OA = Outside Air
RA = Return Air
F = Filter
PH = Preheat Coil

CC = Cooling Coil
RH = Reheat Coil
HP = High Pressure Duct
MP = Medium Pressure Duct

LP = Low Pressure Duct
VV = Variable Volume
CV = Constant Volume

Figure 23-18   Installation Time in Labor-Hours for Boilers

Description	Labor-Hours	Unit
Electric Fired, Steel, Output		
60 MBH	20.000	Ea.
500 MBH	36.293	Ea.
1,000 MBH	60.000	Ea.
2,000 MBH	94.118	Ea.
3,000 MBH	114.286	Ea.
7,000 MBH	117.778	Ea.
Gas Fired, Cast Iron, Output		
80 MBH	21.918	Ea.
500 MBH	53.333	Ea.
1,000 MBH	64.000	Ea.
2,000 MBH	99.250	Ea.
3,000 MBH	133.333	Ea.
7,000 MBH	320.000	Ea.
Oil Fired, Cast Iron, Output		
100 MBH	26.667	Ea.
500 MBH	64.000	Ea.
1,000 MBH	76.000	Ea.
2,000 MBH	114.286	Ea.
3,000 MBH	139.130	Ea.
7,000 MBH	360.000	Ea.
Scotch Marine Packaged Units, Gas or Oil Fired, Output		
1,300 MBH	80.000	Ea.
3,350 MBH	152.000	Ea.
4,200 MBH	170.000	Ea.
5,000 MBH	192.000	Ea.
6,700 MBH	223.000	Ea.
8,370 MBH	230.000	Ea.
10,000 MBH	250.000	Ea.
20,000 MBH	380.000	Ea.
23,400 MBH	484.000	Ea.

(continued on next page)

Figure 23-18 Installation Time in Labor-Hours for Boilers
(continued from previous page)

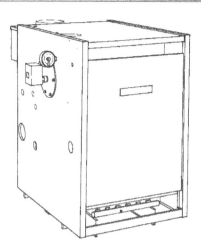

Packaged, Cast Iron Sectional, Gas Fired Boiler-Residential

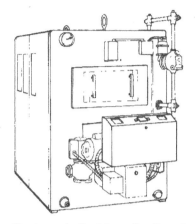

Packaged, Cast Iron Sectional, Gas/Oil Fired Boiler-Commercial

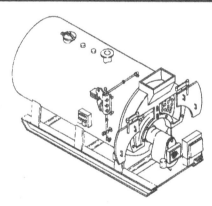

Packaged, Oil Fired, Modified Scotch Marine Boiler-Commercial

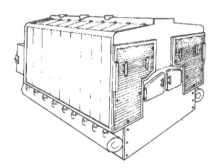

Cast Iron Sectional Boiler-Commercial

Figure 23-19 Installation Time in Labor-Hours for Forced Hot
                  Water Heating Systems

Description	Labor-Hours	Unit
Hot Water Heating System, Area to 2400 S.F.		
Boiler Package, Oil Fired, 225 MBH	17.143	Ea.
Oil Piping System	4.584	Ea.
Oil Tank, 550 Gallon, with Black Steel		
Fill Pipe	4.000	Ea.
Supply Piping, 3/4" Copper Tubing	.182	L.F.
Supply Fittings, Copper	.421	L.F.
Baseboard Radiation	.667	L.F.

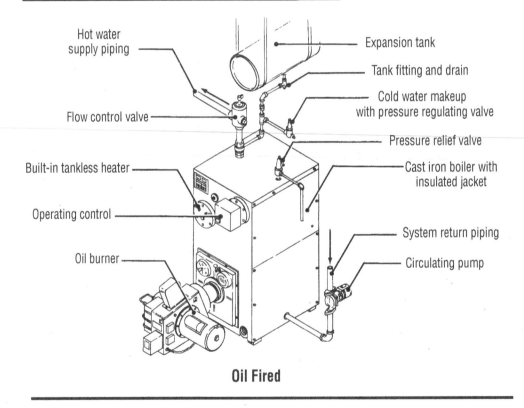

Oil Fired

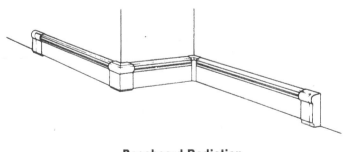

Baseboard Radiation

Figure 23-20   Installation Time in Labor-Hours for Forced Hot Air
Heating Systems

Description	Labor-Hours	Unit
Heating Only, Gas Fired Hot Air,		
One Zone, 1200 S.F. Building		
Furnace, Gas, Up Flow	4.710	Ea.
Intermittent Pilot	4.710	Ea.
Supply Duct, Rigid Fiberglass	.007	L.F.
Return Duct, Sheet Metal, Galvanized	.102	lb.
Lateral Ducts, 6" Flexible Fiberglass	.062	L.F.
Register, Elbows	.267	Ea.
Floor Registers, Enameled Steel	.250	Ea.
Floor Grille, Return Air	.364	Ea.
Thermostat	1.000	Ea.
Plenum	1.000	Ea.
Ductwork		
Fabricated Rectangular, Includes		
Fittings, Joists, Supports,		
Allowance for Flexible Connections,		
No Insulation.		
Aluminum, Alloy 3003-H14,		
Under 300 lbs.	.320	lb.
300 to 500 lbs.	.300	lb.
500 to 1000 lbs.	.253	lb.
1000 to 2000 lbs.	.200	lb.
2000 to 10,000 lbs.	.185	lb.
Over 10,000 lbs.	.166	lb.
Galvanized Steel, Under 400 lbs..		
400 to 1000 lbs.	.094	lb.
1000 to 2000 lbs.	.091	lb.
2000 to 5000 lbs.	.087	lb.
5000 to 10,000 lbs.	.084	lb.
Over 10,000 lbs.	.080	lb.

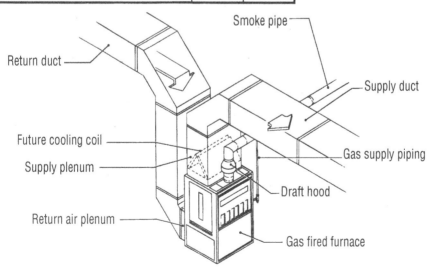

Gas Fired, Warm Air System

(continued on next page)

Figure 23-20 Installation Time in Labor-Hours for Forced Hot Air
Heating Systems *(continued from previous page)*

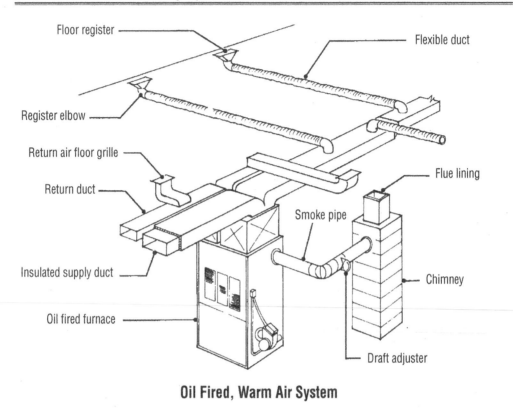

Oil Fired, Warm Air System

Figure 23-21 Installation Time in Labor-Hours for Air Handling Units

Description	Labor-Hours	Unit
Fan Coil Unit, Free Standing Finished Cabinet, 3 Row Cooling or Heating Coil, Filter		
200 CFM	2.000	Ea.
400 CFM	2.667	Ea.
600 CFM	2.909	Ea.
1000 CFM	3.200	Ea.
1200 CFM	4.000	Ea.
4000 CFM	8.571	Ea.
6000 CFM	16.000	Ea.
8000 CFM	30.000	Ea.
12,000 CFM	40.000	Ea.
Direct Expansion Cooling Coil, Filter		
2000 CFM	5.333	Ea.
3000 CFM	5.333	Ea.
4000 CFM	9.231	Ea.
8000 CFM	34.286	Ea.
12,000 CFM	40.000	Ea.
16,000 CFM	53.333	Ea.
20,000 CFM	63.158	Ea.
Central Station Unit, Factory Assembled, Modular, 4, 6, or 8 Row Coils, Filter and Mixing Box		
1500 CFM	13.333	Ea.
2200 CFM	14.545	Ea.
3800 CFM	20.000	Ea.
5400 CFM	30.000	Ea.
8000 CFM	40.000	Ea.
12,100 CFM	52.174	Ea.
18,400 CFM	72.727	Ea.
22,300 CFM	82.759	Ea.
33,700 CFM	126.316	Ea.
52,500 CFM	200.000	Ea.
63,000 CFM	246.154	Ea.

(continued on next page)

Figure 23-21 Installation Time in Labor-Hours for Air Handling Units
(continued from previous page)

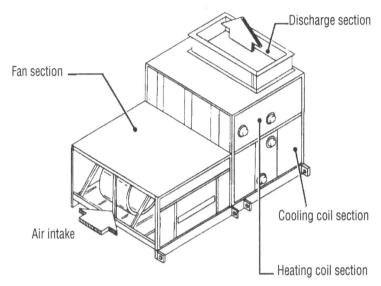

Discharge section

Fan section

Air intake

Cooling coil section

Heating coil section

Central Station Air Handling Unit

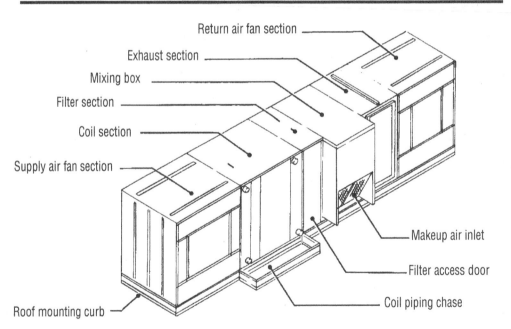

Return air fan section

Exhaust section

Mixing box

Filter section

Coil section

Supply air fan section

Makeup air inlet

Filter access door

Coil piping chase

Roof mounting curb

Central Station Air Handling Unit for Rooftop Location

Figure 23-22 Installation Time in Labor-Hours for Packaged Rooftop
Air Conditioner Units

Description	Labor-Hours	Unit
Single Zone, Electric Cool, Gas Heat		
2 Ton Cooling, 60 M BTU/Hr. Heating	17.204	Ea.
4 Ton Cooling, 95 M BTU/Hr. Heating	26.403	Ea.
5 Ton Cooling, 112 M BTU/Hr. Heating	28.571	Ea.
10 Ton Cooling, 200 M BTU/Hr. Heating	35.982	Ea.
15 Ton Cooling, 270 M BTU/Hr. Heating	42.032	Ea.
20 Ton Cooling, 360 M BTU/Hr. Heating	47.976	Ea.
30 Ton Cooling, 540 M BTU/Hr. Heating	68.376	Ea.
40 Ton Cooling, 675 M BTU/Hr. Heating	91.168	Ea.
Multi-Zone, Electric Cool, Gas Heat, Economizer		
15 Ton Cooling, 360 M BTU/Hr. Heating	52.545	Ea.
20 Ton Cooling, 360 M BTU/Hr. Heating	60.038	Ea.
25 Ton Cooling, 450 M BTU/Hr. Heating	71.910	Ea.
28 Ton Cooling, 450 M BTU/Hr. Heating	79.012	Ea.
30 Ton Cooling, 540 M BTU/Hr. Heating	85.562	Ea.
37 Ton Cooling, 540 M BTU/Hr. Heating	113.000	Ea.
70 Ton Cooling, 1,500 M BTU/Hr. Heating	198.000	Ea.
80 Ton Cooling, 1,500 M BTU/Hr. Heating	228.000	Ea.
90 Ton Cooling, 1,500 M BTU/Hr. Heating	256.000	Ea.
105 Ton Cooling, 1,500 M BTU/Hr. Heating	290.000	Ea.

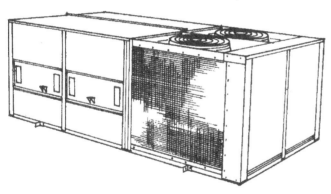

Packaged Rooftop Air Conditioner

Figure 23-23 Installation Time in Labor-Hours for Packaged
Water Chillers

Description	Labor-Hours	Unit
Packaged Water Chillers		
Centrifugal, Water Cooled, Hermetic		
400 Ton	283.000	
1,000 Ton	372.000	
1,300 Ton	410.000	
1,500 Ton	426.000	
Reciprocating, Air Cooled		
20 Ton	95.000	Ea.
40 Ton	108.000	
65 Ton	118.000	
100 Ton	129.000	
110 Ton	132.000	
125 Ton	135.000	
Water Cooled Multiple Compressor		
Semi Hermetic		
15 Ton	65.934	Ea.
20 Ton	78.049	
25 Ton	89.888	
30 Ton	101.000	
40 Ton	108.000	
50 Ton	113.000	
60 Ton	125.000	
80 Ton	151.000	
100 Ton	179.000	
120 Ton	196.000	
140 Ton	202.000	
Air Cooled for Remote Condenser		
40 Ton	144.000	Ea.
60 Ton	164.000	
80 Ton	183.000	
100 Ton	203.000	
120 Ton	223.000	
160 Ton	243.000	
Scroll Water Chillers		
Packaged w/integral air cooled condenser		
15 ton cooling	86.486	Ea.
20 ton cooling	94.118	
25 ton cooling	94.118	
30 ton cooling	101.000	
35 ton cooling	104.000	
40 ton cooling	108.000	
45 ton cooling	109.000	
50 ton cooling	113.000	
Scroll water cooled, single compressor, hermetic, tower not incl.		
2 ton cooling	28.070	Ea.
5 ton cooling	28.070	
6 ton cooling	38.005	
8 ton cooling	52.117	
10 ton cooling	67.039	
15 ton cooling	72.727	
20 ton cooling	83.990	
30 ton cooling	96.096	
Absorption		
Gas Fired, Air Cooled		
5 Ton	26.667	Ea.
10 Ton	40.000	
Steam or Hot Water, Water Cooled		
100 Ton	240.000	Ea.
600 Ton	336.000	
1,100 Ton	421.000	
1,500 Ton	463.000	

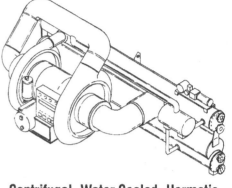

Centrifugal, Water Cooled, Hermetic

Figure 23-24 Installation Time in Labor-Hours for Cooling Towers

Description	Labor-Hours	Unit
Labor to Set in Place - Rigging not Included		
60 Ton Single Flow	8	Ea.
90 Ton Single Flow	16	Ea.
100 Ton Single Flow	16	Ea.
125 Ton Double Flow	24	Ea.
150 Ton Double Flow	30	Ea.
300 Ton Double Flow	40	Ea.
600 Ton Double Flow	80	Ea.
840 Ton Double Flow	136	Ea.
1000 Ton Double Flow	152	Ea.

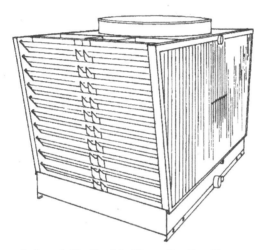

Induced Air, Double Flow, Cooling Tower

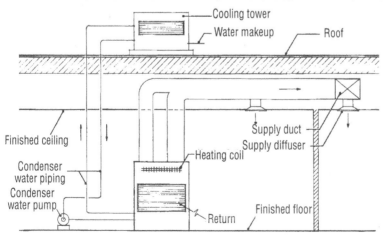

Cooling Tower and Condenser Water System

Figure 23-25 Installation Time in Labor-Hours for Computer Room Cooling Systems

Description	Labor-Hours	Unit
Computer Room Unit, Air Cooled, Includes Remote Condenser		
3 Ton	32.000	Ea.
5 Ton	35.556	Ea.
8 Ton	59.259	Ea.
10 Ton	64.000	Ea.
15 Ton	72.727	Ea.
20 Ton	82.759	Ea.
23 Ton	85.714	Ea.
Chilled Water, for Connection to Existing Chiller System		
5 Ton	21.622	Ea.
8 Ton	32.000	Ea.
10 Ton	32.653	Ea.
15 Ton	33.333	Ea.
20 Ton	34.783	Ea.
23 Ton	38.095	Ea.
Glycol System, Complete Except for Interconnecting Tubing		
3 Ton	40.000	Ea.
5 Ton	42.105	Ea.
8 Ton	69.565	Ea.
10 Ton	76.190	Ea.
15 Ton	92.308	Ea.
20 Ton	100.000	Ea.
23 Ton	109.091	Ea.
Water Cooled, Not Including Condenser Water Supply or Cooling Tower		
3 Ton	25.806	Ea.
5 Ton	29.630	Ea.
8 Ton	44.485	Ea.
15 Ton	59.259	Ea.
20 Ton	63.158	Ea.
23 Ton	70.588	Ea.

Note: The labor-hours do not include ductwork or piping.

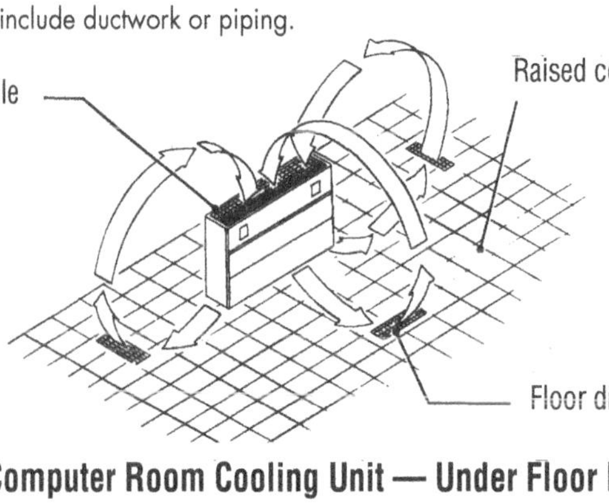

Cooling unit return air grille

Raised computer room floor

Floor distribution grilles

Self-Contained Computer Room Cooling Unit — Under Floor Distribution

Figure 23-26 Installation Time in Labor-Hours for Fans

Description	Labor-Hours	Unit
Fans		
Belt Drive, In-Line Centrifugal		
3800 CFM	5.882	Ea.
6400 CFM	7.143	Ea.
10,500 CFM	8.333	Ea.
15,600 CFM	12.500	Ea.
23,000 CFM	28.571	Ea.
28,000 CFM	50.000	Ea.
Direct Drive Ceiling Fan		
95 CFM	1.000	Ea.
210 CFM	1.053	Ea.
385 CFM	1.111	Ea.
885 CFM	1.250	Ea.
1650 CFM	1.538	Ea.
2960 CFM	1.818	Ea.
Direct Drive Paddle Blade Fan		
36", 4000 CFM	3.333	Ea.
52", 7000 CFM	5.000	Ea.
Direct Drive Roof Fan		
420 CFM	2.857	Ea.
675 CFM	3.333	Ea.
770 CFM	4.000	Ea.
1870 CFM	4.762	Ea.
2150 CFM	5.000	Ea.
Belt Drive Roof Fan		
1660 CFM	3.333	Ea.
2830 CFM	4.000	Ea.
4600 CFM	5.000	Ea.
8750 CFM	6.667	Ea.
12,500 CFM	10.000	Ea.
21,600 CFM	20.000	Ea.
Direct Drive Utility Set		
150 CFM	3.125	Ea.
485 CFM	3.448	Ea.
1950 CFM	4.167	Ea.
2410 CFM	4.545	Ea.
3328 CFM	6.667	Ea.
Belt Drive Utility Set		
800 CFM	3.333	Ea.
1300 CFM	4.000	Ea.
2000 CFM	4.348	Ea.
2900 CFM	4.762	Ea.
3600 CFM	5.000	Ea.
4800 CFM	5.714	Ea.
6700 CFM	6.667	Ea.
11,000 CFM	10.000	Ea.
13,000 CFM	12.500	Ea.
15,000 CFM	20.000	Ea.
17,000 CFM	25.000	Ea.
20,000 CFM	25.000	Ea.

(continued on next page)

Figure 23-26 Installation Time in Labor-Hours for Fans
(continued from previous page)

Description	Labor-Hours	Unit
Belt Drive Propeller Fan		
12", 1000 CFM	.571	Ea.
14", 1500 CFM	.667	Ea.
16", 2000 CFM	.889	Ea.
30", 4800 CFM	1.143	Ea.
36", 7000 CFM	1.333	Ea.
42", 10,000 CFM	1.600	Ea.
48", 16,000 CFM	2.000	Ea.
Belt Drive Airfoil Centrifugal		
12,420 CFM	7.273	Ea.
18,620 CFM	8.000	Ea.
27,580 CFM	8.889	Ea.
40,980 CFM	10.667	Ea.
60,920 CFM	16.000	Ea.
74,520 CFM	20.000	Ea.
90,160 CFM	22.857	Ea.
110,300 CFM	32.000	Ea.
134,960 CFM	40.000	Ea.

Figure 23-27   Installation Time in Labor-Hours for Roof Ventilators

Description	Labor-Hours	Unit
Rotary Syphons		
6" Diameter 185 CFM	1.000	Ea.
8" Diameter 215 CFM	1.143	Ea.
10" Diameter 260 CFM	1.333	Ea.
12" Diameter 310 CFM	1.600	Ea.
14" Diameter 500 CFM	1.600	Ea.
16" Diameter 635 CFM	1.778	Ea.
18" Diameter 835 CFM	1.778	Ea.
20" Diameter 1080 CFM	2.000	Ea.
24" Diameter 1530 CFM	2.000	Ea.
30" Diameter 2500 CFM	2.286	Ea.
36" Diameter 3800 CFM	2.667	Ea.
42" Diameter 4500 CFM	4.000	Ea.
Spinner Ventilators		
4" Diameter 180 CFM	.800	Ea.
5" Diameter 210 CFM	.889	Ea.
6" Diameter 250 CFM	1.000	Ea.
8" Diameter 360 CFM	1.143	Ea.
10" Diameter 540 CFM	1.333	Ea.
12" Diameter 770 CFM	1.600	Ea.
14" Diameter 830 CFM	1.600	Ea.
16" Diameter 1200 CFM	1.778	Ea.
18" Diameter 1700 CFM	1.778	Ea.
20" Diameter 2100 CFM	2.000	Ea.
24" Diameter 3100 CFM	2.000	Ea.
30" Diameter 4500 CFM	2.286	Ea.
36" Diameter 5500 CFM	2.667	Ea.
Stationary Gravity Syphons		
3" Diameter 40 CFM	.667	Ea.
4" Diameter 50 CFM	.800	Ea.
5" Diameter 58 CFM	.889	Ea.
6" Diameter 66 CFM	1.000	Ea.
7" Diameter 86 CFM	1.067	Ea.
8" Diameter 110 CFM	1.143	Ea.
10" Diameter 140 CFM	1.333	Ea.
12" Diameter 160 CFM	1.600	Ea.
14" Diameter 250 CFM	1.600	Ea.
16" Diameter 380 CFM	1.778	Ea.
18" Diameter 500 CFM	1.778	Ea.
20" Diameter 625 CFM	2.000	Ea.
24" Diameter 900 CFM	2.000	Ea.
30" Diameter 1375 CFM	2.286	Ea.
36" Diameter 2000 CFM	2.667	Ea.
42" Diameter 3000 CFM	4.000	Ea.

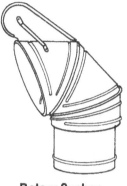

Rotary Syphon

Spinner Ventilator

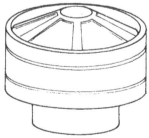

**Stationary
Gravity Syphon**

(continued on next page)

Figure 23-27 Installation Time in Labor-Hours for Roof Ventilators
(continued from previous page)

Description	Labor-Hours	Unit
Rotating Chimney Caps		
4" Diameter	.800	Ea.
5" Diameter	.889	Ea.
6" Diameter	1.000	Ea.
7" Diameter	1.067	Ea.
8" Diameter	1.143	Ea.
10" Diameter	1.333	Ea.
Relief Hoods, Intake/Exhaust		
500 CFM 12" x 16"	2.000	Ea.
750 CFM 12" x 20"	2.222	Ea.
1000 CFM 12" x 24"	2.424	Ea.
1500 CFM 12" x 36"	2.759	Ea.
3000 CFM 20" x 42"	4.000	Ea.
6000 CFM 20" x 84"	6.154	Ea.
8000 CFM 24" x 96"	6.957	Ea.
10,000 CFM 48" x 60"	8.889	Ea.
12,500 CFM 48" x 72"	10.000	Ea.
15,000 CFM 48" x 96"	12.308	Ea.
20,000 CFM 48" x 120"	13.333	Ea.
25,000 CFM 60" x 120"	17.778	Ea.
30,000 CFM 72" x 120"	22.857	Ea.
40,000 CFM 96" x 120"	26.667	Ea.
50,000 CFM 96" x 144"	32.000	Ea.

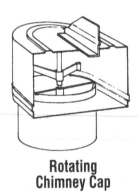

**Rotating
Chimney Cap**

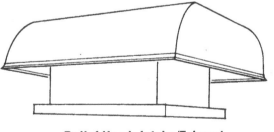

Relief Hood, Intake/Exhaust

Figure 23-28   Installation Time in Labor-Hours for Ductwork

Description	Labor-Hours	Unit
Ductwork		
Fabricated Rectangular, Includes Fittings, Joints, Supports		
Allowance for Flexible Connections, No Insulation		
Aluminum, Alloy 3003-H14, Under 300 lbs.	.320	lb.
300 to 500 lbs.	.300	lb.
500 to 1000 lbs.	.253	lb.
1000 to 2000 lbs.	.200	lb.
2000 to 10,000 lbs.	.185	lb.
Over 10,000 lbs.	.166	lb.
Galvanized Steel, Under 400 lbs.	.102	lb.
400 to 1000 lbs.	.094	lb.
1000 to 2000 lbs.	.091	lb.
2000 to 5000 lbs.	.087	lb.
5000 to 10,000 lbs.	.084	lb.
Over 10,000 lbs.	.080	lb.
Stainless Steel, Type 304, Under 400 lbs.	.145	lb.
400 to 1000 lbs.	.130	lb.
1000 to 2000 lbs.	.120	lb.
2000 to 10,000 lbs.	.107	lb.
Over 10,000 lbs.	.102	lb.
Flexible, Vinyl Coated Spring Steel or Aluminum, Pressure to 10" (WG) UL-181		
Non-Insulated, 3" Diameter	.040	L.F.
4" Diameter	.044	L.F.
5" Diameter	.050	L.F.
Ductwork, Flexible, Non-Insulated		
6" Diameter	.057	L.F.
7" Diameter	.067	L.F.
8" Diameter	.080	L.F.
9" Diameter	.089	L.F.
10" Diameter	.100	L.F.
12" Diameter	.133	L.F.
14" Diameter	.200	L.F.
16" Diameter	.267	L.F.
Insulated, 4" Diameter	.047	L.F.
5" Diameter	.053	L.F.
6" Diameter	.062	L.F.
7" Diameter	.073	L.F.
8" Diameter	.089	L.F.
9" Diameter	.100	L.F.
10" Diameter	.114	L.F.
12" Diameter	.160	L.F.
14" Diameter	.200	L.F.
16" Diameter	.267	L.F.
18" Diameter	.356	L.F.
20" Diameter	.369	L.F.

(continued on next page)

Figure 23-28 Installation Time in Labor-Hours for Ductwork

(continued from previous page)

Description	Labor-Hours	Unit
Fiberglass, Aluminized Jacket, 1-1/2" Blanket		
4" Diameter	.047	L.F.
5" Diameter	.053	L.F.
6" Diameter	.062	L.F.
7" Diameter	.073	L.F.
8" Diameter	.089	L.F.
9" Diameter	.100	L.F.
10" Diameter	.114	L.F.
12" Diameter	.160	L.F.
14" Diameter	.200	L.F.
16" Diameter	.267	L.F.
18" Diameter	.356	L.F.
Rigid Fiberglass, Round, .003" Foil Scrim Jacket		
4" Diameter	.052	L.F.
5" Diameter	.058	L.F.
6" Diameter	.067	L.F.
7" Diameter	.073	L.F.
8" Diameter	.089	L.F.
9" Diameter	.100	L.F.
10" Diameter	.114	L.F.
12" Diameter	.160	L.F.
14" Diameter	.200	L.F.
16" Diameter	.267	L.F.
18" Diameter	.356	L.F.
20" Diameter	.369	L.F.
22" Diameter	.400	L.F.
24" Diameter	.436	L.F.
26" Diameter	.480	L.F.
28" Diameter	.533	L.F.
30" Diameter	.600	L.F.
Rectangular, 1" Thick, Aluminum Faced, No Additional Insulation Required	.069	S.F. surf.

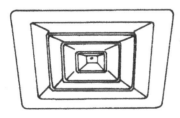

Supply Diffuser

(continued on next page)

Figure 23-28   Installation Time in Labor-Hours for Ductwork
(continued from previous page)

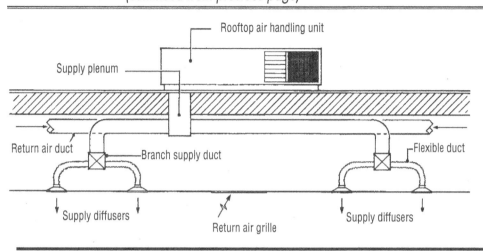

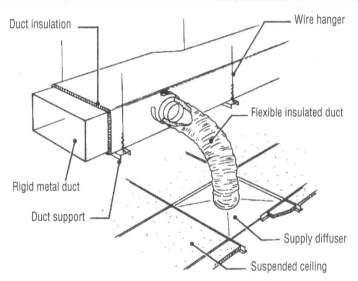

Ductwork System

Figure 23-29 Installation Time in Labor-Hours for Pumps

Description	Labor-Hours	Unit
Circulating Pumps, Heating, Cooling, Potable Water		
In Line Type		
3/4" Pipe Size 1/40 H.P.	1.000	Ea.
3/4" Through 1-1/2" Pipe Size 1/3 H.P.	2.667	Ea.
2" Pipe Size 1/6 H.P.	3.200	Ea.
2-1/2" Pipe Size 1/4 H.P.	3.200	Ea.
3" Pipe Size 1/3 H.P. through 1 H.P.	4.000	Ea.
Close Coupled Type		
1-1/2" Pipe Size 1-1/2 H.P. 50 GPM	5.333	Ea.
2" Pipe Size 3 H.P. 90 GPM	6.957	Ea.
2-1/2" Pipe Size 3 H.P. 150 GPM	8.000	Ea.
3" Pipe Size 5 H.P. 225 GPM	8.889	Ea.
4" Pipe Size 7-1/2 H.P. 350 GPM	10.000	Ea.
5" Pipe Size 15 H.P. 1000 GPM	14.118	Ea.
6" Pipe Size 25 H.P. 1550 GPM	16.000	Ea.
Base Mounted Type		
2-1/2" Pipe Size 3 H.P. 150 GPM	8.889	Ea.
3" Pipe Size 5 H.P. 225 GPM	10.000	Ea.
4" Pipe Size 7-1/2 H.P. 350 GPM	10.667	Ea.
5" Pipe Size 15 H.P. 1000 GPM	15.000	Ea.
6" Pipe Size 25 H.P. 1550 GPM	17.143	Ea.
8" Pipe Size 30 H.P. 1660 GPM	20.000	Ea.
10" Pipe Size 40 H.P. 1800 GPM	20.000	Ea.
12" Pipe Size 50 H.P. 2200 GPM	24.000	Ea.

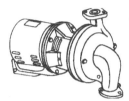

**In-Line
Centrifugal Pump**

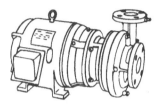

**Close Coupled,
Centrifugal Pump**

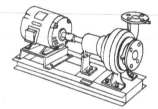

**Base Mounted
Centrifugal Pump**

Divisions 26, 27 & 28
Electrical, Communications & Security

Introduction

The electrical systems in building construction are now located in three different divisions of MasterFormat 2004 and are combined here in this chapter:

Division 26 – Electrical

Division 27 – Communications

Division 28 – Electronic Safety & Security

Like mechanical systems, electrical systems have three distinct portions: source, conductor/connection, and terminal unit. For example, take lighting. The source is a wall switch, the conductor/connector is wiring in conduit, and the terminal unit is the fixture. The first two components, the source and the conductor/connector, represent the rough-in phase of the project, and the terminal unit represents the finish portion of the electrical work.

Electrical systems share common components: conduit or raceways, conductors, devices, switches, panelboards, and fixtures. Raceways and conduits are channels that carry and protect conductors from the source to the terminal unit. Raceways consist of conduit, ducts, cable trays, and surface raceways.

A conductor is a wire or metal bar with a low level of resistance to the flow of electric current. Conductors are generally made of copper or aluminum, and are enclosed in an insulating jacket. Cables are conductors made up of heavy or multiple wires contained in a common jacket. Cable contained in a metallic sheathing (armored) is used where physical protection of the cable is necessary, but the use of conduit is not practical. Other types of conductors include high voltage shielded cable; flat wire or cable for undercarpet installations; low voltage wiring for telephones; and ethernet, coaxial, or optical fiber cable for data transmission.

Boxes are used in electrical wiring at each junction point or at each electrical device to provide easy and safe access for connections. Simply stated, devices control or conduct the flow of electricity without consuming any. Devices include wall switches, outlets, and light dimmers.

Switches are devices that are used to open, close, or change the condition of an electric circuit.

Boards serve as a means of grouping and mounting electric components, switches, safety devices, or controls. A load center or panelboard is a specialized type of board used principally for containing circuit breakers.

green Division 26 is an important section of the specifications for fulfilling the prerequisite level of energy efficiency requirements and additional credits in the LEED® Energy & Atmosphere category.

Electrical Sources

In addition to stand alone systems, photovoltaic collectors (CSI # 26-31-00) can be mounted on building roofs or integrated into the curtain wall system. Roof-mounted systems have the advantage of good to optimal positioning for capturing the sun's rays, but are limited to the size of the roof. Systems that lay flat on the roof are not as efficient as angled or tracking systems, but are less expensive. In addition, the lay-flat variety offers additional roof protection and insulation value. Curtain wall integral photovoltaic systems, which replace spandrel glass with photovoltaic, on the other hand, offer aesthetic and other advantages. However, the vertical position is not optimal for gathering the sun's rays. Wind energy equipment (CSI #26-32-23), like solar collectors, can be stand-alone systems or building-mounted. Because of noise, vibration, and structural issues, building-located systems tend to be small. Wind turbines benefit greatly from economies of scale, which is why large stand-alone turbines and wind farms are becoming increasingly common.

Terminal Units

Lighting fixtures are by far the most elemental portion of the electrical work. Lighting, in building construction, is still the largest single electrical cost center. LED and compact fluorescent lights (CFLs) are initially more expensive than incandescent lightbulbs, but save energy and money in the long run. Energy Star performance criteria ensure quality, safety, and durability. Energy Star rated exit lights (CSI #26-53-00) require only five (or fewer) watts to run. Full cut-off exterior luminaires (CSI #26-56-00) offer the multiple green advantages of directing light down where it is needed, saving energy, and also reducing the light escaping up to the night sky. These fixtures can earn credits in the LEED Sustainable Sites, Light Pollution Reduction category. There is little or no cost premium associated with this design.

The following topics are covered by the estimating charts in this section:

- Cable trays
- Conduits
- Conductors
- Boxes
- Wiring devices
- Starters and controls
- Boards
- Switchgear
- Transformers
- Lighting

Estimating Data

The following tables present effective estimating guidelines for items found in Divisions 26, 27, and 28 of MasterFormat 2004. Please note that these guidelines can be used as indicators of what may be expected, but that each project must be evaluated individually.

Table of Contents

Division 26

Estimating Tips

Conduit

Conduit should be taken off in three main categories: power distribution, branch power, and branch lighting. In this way, all of the conduit does not have to be taken off in one session. Instead, the estimator can concentrate on systems and components, thereby making it easier to ensure that all items have been accounted for.

Switchgear

When estimating costs for the installation of switchgear (especially large items), factors to review include access to the site, access and setting at the installation site, required connections, uncrating pads, anchors, leveling, final assembly of the components, and temporary protection from physical damage, including from exposure to the environment.

Pads, Supports, and Panel Backing

While supports and pads may be shown on drawings for the larger equipment, in many cases nothing is shown for smaller pieces, such as panelboards and area transformers. Whether a special floor-to-ceiling support system is required, or just a piece of plywood for the back of the panel, it must be included in the costs.

Installation Equipment

Do not overlook the costs for equipment used in the installation. If high lifts, scaffolding, or cherry pickers are available, the field will use them in lieu of the proposed ladders.

Material Weights

The estimator should take the weights of materials into consideration when completing a takeoff. Topics to consider include: How will the materials be supported? What methods of support are available? How high will the support structure have to reach? Will the final support structure be able to withstand the total burden? Is the support material included or separate from the fixture/equipment/material specified?

Non-Listed Items

It is important to include items that are not documented in the plans but must be priced. These items include, but are not limited to, testing, equipment hookups, motor controls, disconnect switches, special systems, dust protection, roof penetrations, pitch pots for the roof, coring concrete floors and walls, cleanup, and final adjustments.

Temporary Light and Power

Examine all contract documents to determine if providing temporary power and lighting is included in your scope of work. In many cases, temporary lighting and power requirements are not specified in contract documents. Either way, you will probably be required at some point to provide costs for these items. Depending on your bidding strategy, you may or may not want to list this item as an alternate to your estimate. However, if it is stated in the contract documents, then you will be fully responsible for providing this service.

Cutting and Patching

Not all projects will be performed under ideal conditions—that is, wide open spaces with no interferences. Many times the electrical contractor will have to run conduit through pre-existing partitions or partitions that were recently installed. Allowance must be made for this cutting into the partition and the ensuing patching. Otherwise, a statement must be included with the estimate that all cutting and patching will be the responsibility of others.

Mechanical Connections

Clarify who is responsible for connecting the electrical wiring to mechanical items. It has been argued that the electrician is responsible only for bringing the wiring to the unit, with others saying it is the electrician's job to connect the wiring.

Fixture and Device Counts

Performing a fixture takeoff is a good way to become familiar with a proposed project. Fixtures should be taken off room-by-room, using the fixture schedule (if provided), specifications, and the ceiling plan. While performing the count, it is a good idea to also take off the controlling devices and accessories (plaster ring, outlet boxes, cover plates, etc.). A spreadsheet works well for this purpose. When finished, this takeoff can be used for purchasing, accounting and billing, and cost control.

Special Systems

When estimating material costs for special systems, it is always prudent to obtain manufacturers' quotes for equipment prices. Also, some installations will require special accessories. Often, the sales engineers are a good source of information on these requirements.

Checklist ✓

For an estimate to be reliable, all items must be accounted for. A complete estimate can also limit contingencies. The following checklist can be used to help ensure that all items are included.

☐ **Building Service**
- ☐ Main switchboard
- ☐ Panels
- ☐ Distribution
- ☐ Lighting
- ☐ Power

☐ **Bus Duct Systems**

☐ **Cable**

☐ **Cable Trays**

☐ **Circuit Breakers**

☐ **Computer Power Supplies**

☐ **Conductors**

☐ **Conduit**

☐ **Controls**

☐ **Devices**
- ☐ Boxes for outlets, receptacles, light switches, dimmers
- ☐ Cover plates
- ☐ Dimmers
- ☐ Lighting control switches
- ☐ Outlets
- ☐ Receptacles

☐ **Disconnect Switches**

☐ **Emergency Systems**
- ☐ Area protection relay system
- ☐ Exit lighting
- ☐ Battery emergency lighting
- ☐ Generator
- ☐ Lighting
- ☐ Power
- ☐ Power connection to second service entrance
- ☐ Transfer switch

☐ **Equipment Connections**

☐ **Fixtures**
- ☐ Incandescent
- ☐ Fluorescent
- ☐ Metal halide
- ☐ High pressure sodium
- ☐ Area lighting
- ☐ Bollards
- ☐ Exit lights
- ☐ Floodlighting
- ☐ Fountain lighting
- ☐ High intensity discharge (HID)
- ☐ Site lighting
- ☐ Step lighting

☐ **Incoming Service**
- ☐ Connection to power company system
- ☐ Overhead
- ☐ Underground
- ☐ Primary
- ☐ Secondary
- ☐ Service feeds
- ☐ Main transformer

- ☐ Unit substation
  - ☐ Manholes
  - ☐ Feeder
  - ☐ Primary
  - ☐ Secondary
- ☐ Meters

☐ *Junction Boxes*

☐ *Motors*

☐ *Motor Control Centers*

☐ *Paging Systems*

☐ *Public Address Systems*

☐ *Pull Boxes*

☐ *Raceways*

☐ *Safety Devices*

☐ *Sound Systems*

☐ *Starters*

☐ *Substation*

☐ *Supports*

☐ *Switches*

☐ *Television Systems*
  - ☐ Antennae
  - ☐ Cables
  - ☐ Cameras
  - ☐ Satellite dishes

☐ *Transformers*

☐ *Underfloor Cabling Systems*

☐ *Undercarpet Cabling/ Wiring Systems*

☐ *Wiring*
  - ☐ Wire
  - ☐ Cable
    - ☐ Armored
    - ☐ Shielded
    - ☐ Insulated
    - ☐ Metallic
    - ☐ Nonmetallic
    - ☐ Coaxial
    - ☐ Ethernet
    - ☐ Optical fiber
  - ☐ Connections
  - ☐ Grounding

Figure 26-1 Electrical Symbols

This table contains an extensive list of legend symbols as an aid to the electrical estimator. These designations are commonly used and accepted on electrical plan drawings, schematic diagrams, wiring diagrams, and takeoff forms.

Frequently, drawings do not contain complete, detailed information for the estimator. The legend symbol must therefore be used to interpret sizes, capacities, and/or equipment requirements in order to identify and select unit prices for the estimate.

Division 26

Lighting Outlets

- Ceiling Surface Incandescent Fixture
- Wall Surface Incandescent Fixture
- Ceiling Recess Incandescent Fixture
- Wall Recess Incandescent Fixture
- Standard Designation for All Lighting Fixtures — A = Fixture Type, 3 = Circuit Number, b = Switch Control
- Ceiling Blanked Outlet
- Wall Blanked Outlet
- Ceiling Electrical Outlet
- Wall Electrical Outlet
- Ceiling Junction Box
- Wall Junction Box
- Ceiling Lamp Holder with Pull Switch
- Wall Lamp Holder with Pull Switch
- Ceiling Outlet Controlled by Low Voltage Switching When Relay Is Installed in Outlet Box
- Wall Outlet — Same as Above
- Outlet Box with Extension Ring
- Exit Sign with Arrow as Indicated
- Surface Fluorescent Fixture
- Pendant Fluorescent Fixture
- Recessed Fluorescent Fixture
- Wall Surface Fluorescent Fixture
- Channel Mounted Fluorescent Fixture
- Emergency Battery Pack Light Fixture
- Emergency Light Remote Head

- Surface or Pendant Continuous Row Fluorescent Fixtures
- Recessed Continuous Row Fluorescent Fixtures
- Incandescent Fixture on Emergency Circuit
- Fluorescent Fixture on Emergency Circuit

Receptacle Outlets

- Single Receptacle Outlet
- Duplex Receptacle Outlet
- Duplex Receptacle Outlet "X" Indicates Above Counter Max. Height = 42" or Above Counter
- Receptacle, Equipped with Ground Fault Circuit Interrupter
- Weatherproof Receptacle Outlet
- Triplex Receptacle Outlet
- Quadruplex Receptacle Outlet
- Duplex Receptacle Outlet — Split Wired
- Triplex Receptacle Outlet — Split Wired
- Single Special Purpose Receptacle Outlet
- Duplex Special Purpose Receptacle Outlet
- Range Outlet
- Special Purpose Connection — Dishwasher

(continued on next page)

Figure 26-1   Electrical Symbols (continued from previous page)

Receptacle Outlets (continued)

Explosion-Proof Receptacle Outlet
Max. Height = 36" to C_L

Multi-Outlet Assembly

Clock Hanger Receptacle

Fan Hanger Receptacle

Floor Single Receptacle Outlet

Floor Duplex Receptacle Outlet

Floor Special Purpose Outlet

Floor Telephone Outlet — Public

Floor Telephone Outlet — Private

Underfloor Duct and Junction Box for Triple, Double, or Single Duct System as Indicated by Number of Parallel Lines

Cellular Floor Header Duct

Switch Outlets

S   Single Pole Switch
Max. = 42" to C_L

S_2 Double Pole Switch

S_3 Three-Way Switch

S_4 Four-Way Switch

S_D Automatic Door Switch

S_K Key Operated Switch

S_P Switch & Pilot Lamp

S_{CB} Circuit Breaker

S_{WCB} Weatherproof Circuit Breaker

S_{MC} Momentary Contact Switch

S_{RC} Remote Control Switch (Receiver)

S_{WP} Weatherproof Switch

S_F Fused Switch

S_L Switch for Low Voltage Switching System

S_{LM} Master Switch for Low Voltage Switching System

S_T Time Switch

S_{TH} Thermal Rated Motor Switch

S_{DM} Incandescent Dimmer Switch

S_{FDM} Fluorescent Dimmer Switch

Switch & Single Receptacle

Switch & Double Receptacle

Special Outlet Circuits

Institutional, Commercial & Industrial System Outlets

Nurses Call System Devices — Any Type

Paging System Devices — Any Type

Fire Alarm System Devices — Any Type

Fire Alarm Manual Station — Max. Height = 48" to C_L

Fire Alarm Horn with Integral Warning Light

Fire Alarm Thermodetector, Fixed Temperature

Smoke Detector

Fire Alarm Thermodetector, Rate of Rise

(continued on next page)

Figure 26-1 Electrical Symbols *(continued from previous page)*

Institutional, Commercial & Industrial System Outlets (continued)

Fire Alarm Master Box — Max. Height per Fire Department

Magnetic Door Holder

Fire Alarm Annunciator

Staff Register System — Any Type

Electrical Clock System Devices — Any Type

Public Telephone System Devices

Private Telephone system Devices — Any Type

Watchman System Devices

Sound System, L = Speaker, V = Volume Control

Other Signal System Devices — CTV = Television Antenna, DP = Data Processing

Signal Central Station

Telephone Interconnection Box

Pneumatic/Electric Switch

Electric/Pneumatic Switch

Operating Room Grounding Plate

Patient Ground Point — 6 = Number of Jacks

Panelboards

Flush Mounted Panelboard & Cabinet

Surface Mounted Panelboard & Cabinet

Lighting Panel

Power Panel

Heating Panel

Controller (Starter)

Externally Operated Disconnect Switch

Busducts & Wireways

Trolley Duct

Busway (Service, Feeder, or Plug-In)

Cable Through Ladder or Channel

Wireway

Bus Duct Junction Box

Electrical Distribution or Lighting System, Aerial, Lightning Protection

Pole

Street Light & Bracket

Transformer

Primary Circuit

Secondary Circuit

Auxiliary System Circuits

Down Guy

Head Guy

Sidewalk Guy

Service Weather Head

Lightning Rod

Lightning Protection System Conductor

Residential Signaling System Outlets

Push Button

Buzzer

Bell

(continued on next page)

Figure 26-1   Electrical Symbols *(continued from previous page)*

Residential Signaling System Outlets (continued)

Bell and Buzzer Combination

Annunciator

Outside Telephone

Interconnecting Telephone

Telephone Switchboard

[BT]   Bell Ringing Transformer

[D]   Electric Door Opener

[M]   Maid's Signal Plug

[R]   Radio Antenna Outlet

[CH]   Chime

[TV]   Television Antenna Outlet

(T)   Thermostat

Underground Electrical Distribution or Lighting System

[M]   Manhole

[H]   Handhole

[TM]   Transformer — Manhole or Vault

[TP]   Transformer Pad

Underground Direct Burial Cable

Underground Duct Line

Street Light Standard Fed from Underground Circuit

Panel Circuits & Miscellaneous

Conduit Concealed in Floor or Walls

Wiring Exposed

Home Run to Panelboard — Number of Arrows Indicates Number of Circuits

Home Run to Panelboard — Two-Wire Circuit

Home Run to Panelboard — Number of Slashes Indicates Number of Wires (When more than two)

LS – L1,3,5   Home Run to Panelboard — 'LS' Indicates Panel Designation; LI, 3, 5, Indicates Circuit Breaker No.

Ground Connection

Grounded Wye

30 A   OR   30   Fuse, 30 A Type

225/125
3 P   Circuit Breaker, Molded Case, 3 Pole, 225 A Frame/125 A Trip

3 P
100/60 A   OR   3 P
100 A   60 A   Fused Disconnect Switch, 3 Pole, 100 A, 60 A Fuse

— C —   Clock Circuit, Conduit and Wire

— E —   Emergency Conduit and Wiring

— T —   Telephone Conduit and Wiring

Feeders

Conduit Turned Up

Conduit Turned Down

(G)   Generator

(M)   Motor

(5)   Motor — Numeral Indicates Horsepower

(I)   Instrument (Specify)

(continued on next page)

Figure 26-1 Electrical Symbols *(continued from previous page)*

Panel Circuits & Miscellaneous (continued)

$\boxed{\text{T}}$ Transformer

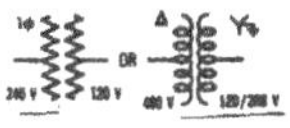

Transformer, 2 Winding, 1ϕ Indicates Single Phase; Δ-Y Indicates 3 Phase and Type of Connection; Numbers Indicate Voltages.

Remote Start-Stop Push Button Station

Remote Start-Stop Push Button Station w/Pilot Light

$\boxed{\text{HTR}}$ Electric Heater Wall Unit (Plan View)

Division 26

Figure 26-2 Electrical Formulas

The information in this section can be used as a rule or method for performing estimates, either before or after the preliminary design stage.

OHM'S LAW

Ohm's law is the mathematical relationship between voltage, current, and resistance in an electrical circuit. It is practically the basis of all electrical calculations. The term "electromotive force" is often used to designate pressure in volts. This formula can be expressed in various forms.

To find the current in amperes:

$$\text{Current} = \frac{\text{Voltage}}{\text{Resistance}} \quad \text{or} \quad \text{Amperes} = \frac{\text{Volts}}{\text{Ohms}} \quad \text{or} \quad I = \frac{E}{R}$$

The flow of current in amperes through any circuit is equal to the voltage or electromotive force divided by the resistance of that circuit.

To find the pressure or voltage:

$$\text{Voltage} = \text{Current} \times \text{Resistance} \quad \text{or} \quad \text{Volts} = \text{Amperes} \times \text{Ohms}$$
$$\text{or} \quad E = I \times R$$

The voltage required to force a current through a circuit is equal to the resistance of the circuit multiplied by the current.

To find the resistance:

$$\text{Resistance} = \frac{\text{Voltage}}{\text{Current}} \quad \text{or} \quad \text{Ohms} = \frac{\text{Volts}}{\text{Amperes}} \quad \text{or} \quad R = \frac{E}{I}$$

The resistance of a circuit is equal to the voltage divided by the current flowing through that circuit.

(continued on next page)

Figure 26-2 Electrical Formulas *(continued from previous page)*

POWER FORMULAS

One horsepower = 746 watts   One kilowatt = 1000 watts

The power factor of electric motors varies from 80% to 90% in the larger size motors.

SINGLE-PHASE ALTERNATING CURRENT CIRCUITS

Power in Watts = Volts x Amperes x Power Factor

To find current in amperes:

$$\text{Current} = \frac{\text{Watts}}{\text{Volts x Power Factor}} \quad \text{or}$$

$$\text{Amperes} = \frac{\text{Watts}}{\text{Volts x Power Factor}} \quad \text{or} \quad I = \frac{W}{E \times PF}$$

To find current of a motor, single phase:

$$\text{Current} = \frac{\text{Horsepower x 746}}{\text{Volts x Power Factor x Efficiency}} \quad \text{or}$$

$$I = \frac{HP \times 746}{E \times PF \times Eff.}$$

To find horsepower of a motor, single phase:

$$\text{Horsepower} = \frac{\text{Volts x Current x Power Factor x Efficiency}}{\text{746 Watts}} \quad \text{or}$$

$$HP = \frac{E \times I \times PF \times Eff.}{746}$$

To find power in watts of a motor, single phase:

Watts = Volts x Current x Power Factor x Efficiency   or

Watts = E x I x PF x Eff.

To find single phase kVA:

$$\text{1 Phase kVA} = \frac{\text{Volts x Amps}}{1000}$$

(continued on next page)

Figure 26-2   Electrical Formulas *(continued from previous page)*

POWER FORMULAS (continued)

THREE-PHASE ALTERNATING CURRENT CIRCUITS
Power in Watts = Volts x Amperes x Power Factor x 1.73

To find current in amperes in each wire:

$$\text{Current} = \frac{\text{Watts}}{\text{Voltage} \times \text{Power Factor} \times 1.73} \quad \text{or}$$

$$\text{Amperes} = \frac{\text{Watts}}{\text{Volts} \times \text{Power Factor} \times 1.73} \quad \text{or} \quad I = \frac{W}{E \times PF \times 1.73}$$

To find current of a motor, 3 phase:

$$\text{Current} = \frac{\text{Horsepower} \times 746}{\text{Volts} \times \text{Power Factor} \times \text{Efficiency} \times 1.73} \quad \text{or}$$

$$I = \frac{HP \times 746}{E \times PF \times Eff. \times 1.73}$$

To find horsepower of a motor, 3 phase:

$$\text{Horsepower} = \frac{\text{Volts} \times \text{Current} \times 1.73 \times \text{Power Factor}}{746 \text{ Watts}} \quad \text{or}$$

$$HP = \frac{E \times I \times 1.73 \times PF}{746}$$

To find power in watts of a motor, 3 phase:

Watts = Volts x Current x 1.73 x Power Factor x Efficiency   or

Watts = E x I x 1.73 x PF x Eff.

To find 3 phase kVA:

$$3 \text{ phase kVA} = \frac{\text{Volts} \times \text{Amps} \times 1.73}{1000} \quad \text{or}$$

$$kVA = \frac{V \times A \times 1.73}{1000}$$

Power Factor (PF) is the percentage ratio of the measured watts (effective power) to the volt-amperes (apparent watts).

$$\text{Power Factor} = \frac{\text{Watts}}{\text{Volts} \times \text{Amperes}} \times 100\%$$

Figure 26-3 Typical Commercial Service Entrance

This figure illustrates the basic components of a typical commercial service entrance.

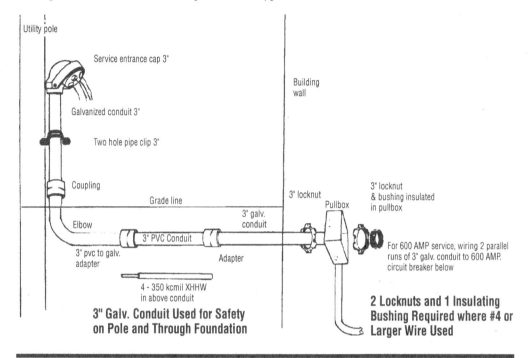

Utility pole

Service entrance cap 3"

Building wall

Galvanized conduit 3"

Two hole pipe clip 3"

Coupling

Grade line

3" locknut

3" locknut & bushing insulated in pullbox

Pullbox

Elbow

3" galv. conduit

3" PVC Conduit

3" pvc to galv. adapter

Adapter

For 600 AMP service, wiring 2 parallel runs of 3" galv. conduit to 600 AMP. circuit breaker below

4 - 350 kcmil XHHW in above conduit

3" Galv. Conduit Used for Safety on Pole and Through Foundation

2 Locknuts and 1 Insulating Bushing Required where #4 or Larger Wire Used

Figure 26-4 Typical Commercial Electrical System

This figure shows the basic lighting and power components used for the interior of a typical commercial project.

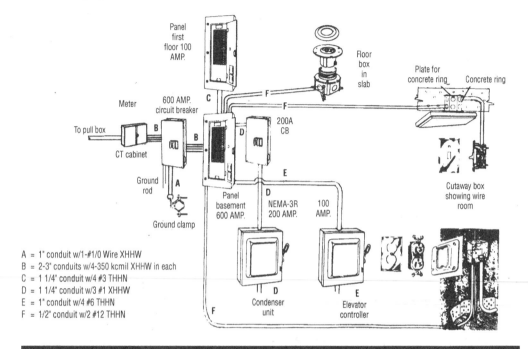

A = 1" conduit w/1-#1/0 Wire XHHW
B = 2-3" conduits w/4-350 kcmil XHHW in each
C = 1 1/4" conduit w/4 #3 THHN
D = 1 1/4" conduit w/3 #1 XHHW
E = 1" conduit w/4 #6 THHN
F = 1/2" conduit w/2 #12 THHN

Figure 26-5   kW Value/Cost Determination

General: Lighting and electric heating loads are expressed in watts and kilowatts.

Cost Determination: The proper ampere values can be obtained as follows:

1. Convert watts to kilowatts (watts ÷ 1000 = kilowatts).
2. Determine voltage rating of equipment.
3. Determine whether equipment is single phase or three phase.
4. Refer to Figure 26.6 to find ampere value from kW, ton and BTU/hr. values.
5. Determine type of wire insulation – TW, THW, THWN.
6. Determine if wire is copper or aluminum.
7. Refer to Figure 26.14 to obtain copper or aluminum wire size to accommodate the number and size of wires in each particular case.
8. Next refer to Figure 26.10 for the proper conduit size to accommodate the number and size of wires in each particular case.
9. Next refer to the per linear foot cost of the conduit.
10. Next refer to the per linear foot cost of the wire. Multiply cost of wire LF × number of wires in the circuits to obtain total wire cost per LF
11. Add values obtained in steps 9 and 10 for total cost per linear foot for conduit and wire × length of circuit = total cost.

Notes:

1. 1 Phase refers to single phase, 2 wire circuits.
2. 3 Phase refers to three phase, 3 wire circuits.
3. For circuits which operate continuously for 3 hours or more, multiply the ampere values by 1.25 for a given kW requirement.
4. For kW ratings not listed, add ampere values.
   For example: find the ampere value of 9 kW at 208 volt, single phase

$$\begin{array}{l} 4\ kW = 19.2A \\ \underline{5\ kW = 24.0A} \\ 9kW = 43.2A \end{array}$$

5. "Length of circuit" refers to the one-way distance of the run, not to the total sum of wire lengths.

Figure 26-6 Ampere Values as Determined by kW Requirements, BTU/Hr. or Ton, Voltage, and Phase Values

KW	Ton	BTU/Hr.	Ampere Values						
			120V	208V		240V		277V	480V
			1 Phase	1 Phase	3 Phase	1 Phase	3 Phase	1 Phase	3 Phase
0.5	.1422	1,707	4.2A	2.4A	1.4A	2.1A	1.2A	1.8A	0.6A
0.75	.2133	2,560	6.2	3.6	2.1	3.1	1.9	2.7	0.9
1.0	.2844	3,413	8.3	4.9	2.8	4.2	2.4	3.6	1.2
1.25	.3555	4,266	10.4	6.0	3.5	5.2	3.0	4.5	1.5
1.5	.4266	5,120	12.5	7.2	4.2	6.3	3.1	5.4	1.8
2.0	.5688	6,826	16.6	9.7	5.6	8.3	4.8	7.2	2.4
2.5	.7110	8,533	20.8	12.0	7.0	10.4	6.1	9.1	3.1
3.0	.8532	10,239	25.0	14.4	8.4	12.5	7.2	10.8	3.6
4.0	1.1376	13,652	33.4	19.2	11.1	16.7	9.6	14.4	4.8
5.0	1.4220	17,065	41.6	24.0	13.9	20.8	12.1	18.1	6.1
7.5	2.1331	25,598	62.4	36.0	20.8	31.2	18.8	27.0	9.0
10.0	2.8441	34,130	83.2	48.0	27.7	41.6	24.0	36.5	12.0
12.5	3.5552	42,663	104.2	60.1	35.0	52.1	30.0	45.1	15.0
15.0	4.2662	51,195	124.8	72.0	41.6	62.4	37.6	54.0	18.0
20.0	5.6883	68,260	166.4	96.0	55.4	83.2	48.0	73.0	24.0
25.0	7.1104	85,325	208.4	120.2	70.0	104.2	60.0	90.2	30.0
30.0	8.5325	102,390		144.0	83.2	124.8	75.2	108.0	36.0
35.0	9.9545	119,455		168.0	97.1	145.6	87.3	126.0	42.1
40.0	11.3766	136,520		192.0	110.8	166.4	96.0	146.0	48.0
45.0	12.7987	153,585			124.8	187.5	112.8	162.0	54.0
50.0	14.2208	170,650			140.0	208.4	120.0	180.4	60.0
60.0	17.0650	204,780			166.4		150.4	216.0	72.0
70.0	19.9091	238,910			194.2		174.6		84.2
80.0	22.7533	273,040			221.6		192.0		96.0
90.0	25.5975	307,170					225.6		108.0
100.0	28.4416	341,300							120.0

Figure 26-7 Electric Circuit Voltages

The following method provides the user with a simple nontechnical means of obtaining comparative costs of wiring circuits. The circuits considered serve the electrical loads of motors, electric heating, lighting and transformers, for example, those that require low voltage 60 hertz alternating current.

The method used here is suitable only for obtaining estimated costs. It is not intended to be used as a substitute for electrical engineering design applications.

Conduit and wire circuits can represent from twenty to thirty percent of the total building electrical cost. By following the described steps and using the tables, the user can translate the various types of electric circuits into estimated costs.

Wire Size: Wire size is a function of the electric load which is usually listed in one of the following units:
1. Amperes (A)
2. Watts (W)
3. Kilowatts (kW)
4. Volt amperes (kVA)
5. Kilovolt amperes (kVA)
6. Horsepower (HP)

The units of electric load must be converted to amperes in order to obtain the size of wire necessary to carry the load. To convert electric load units to amperes, one must have an understanding of the voltage classification of the power source and the voltage characteristics of the electrical equipment or load to be energized. The seven A.C. circuits commonly used are illustrated in the following figures showing the transformer load voltage and the point of use voltage at the point on the circuit where the load is connected. The difference between the source and point of use voltages is attributed to the circuit voltage drop and is considered to be approximately 4%.

Motor Voltages: Motor voltages are listed by their point of use voltage and not the power source voltage and not the point of wire voltage.

For example: 460 volts instead of 480 volts
200 volts instead of 208 volts
115 volts instead of 120 volts

Lighting and Heating Voltages: Lighting and heating equipment voltages are listed by the power source voltage and not the point of wire voltage.

For example: 480, 277, 120 volt lighting
480 volt heating or air conditioning unit
208 volt heating unit

Transformer Voltages: Transformer primary (input) and secondary (output) voltages are listed by the power source voltage.

For example: Single phase 10 kVA
Primary 240/480 volts
Secondary 120/240 volts

In this case, the primary voltage may be 240 volts with a 120 volts secondary or may be 480 volts with either a 120 V or a 240 V secondary.

For example: Three phase 10 kVA
Primary 480 volts
Secondary 208Y/120 volts

In this case, the transformer is suitable for connection to a circuit with a 3 phase 3 wire or 3 phase 4 wire circuit with a 480 voltage. This application will provide a secondary circuit of 3 phase 4 wire with 208 volts between the phase wires and 120 volts between any phase wire and the neutral (white) wire.

(continued on next page)

Figure 26-7 Electric Circuit Voltages *(continued from previous page)*

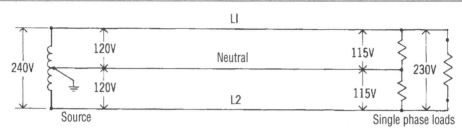

3 Wire, 1 Phase, 120/240 Volt System

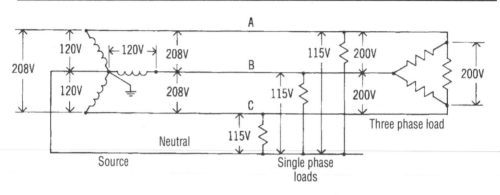

4 Wire, 3 Phase, 208Y/120 Volt System

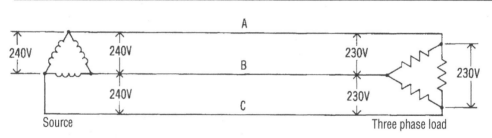

3 Wire, 3 Phase, 240 Volt System

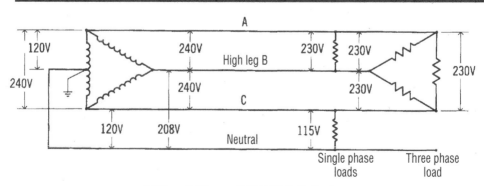

4 Wire, 3 Phase, 240/120 Volt System

(continued on next page)

Figure 26-7   Electric Circuit Voltages *(continued from previous page)*

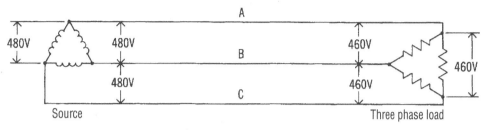

3 Wire, 3 Phase, 480 Volt System

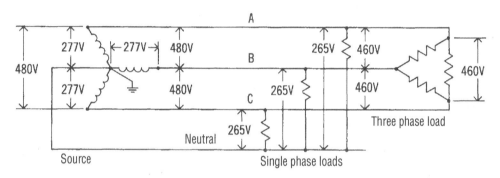

4 Wire, 3 Phase, 480Y/277 Volt System

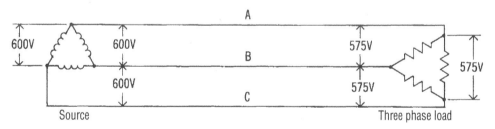

3 Wire, 3 Phase, 600 Volt System

Figure 26-8 Nominal Watts per Square Foot for Electrical Systems by Various Building Types

This table is used in the preliminary design stage to develop the total load of the building in watts, based on nominal watts per SF for electric systems for various building types.

The rule of thumb formula (see below) can be used to determinine the service current required, based on the available voltage.

Type Construction	1. Lighting	2. Devices	3. HVAC	4. Misc.	5. Elevator	Total Watts
Apartment, luxury high-rise	2	2.2	3	1		
Apartment, low-rise	2	2	3	1		
Auditorium	2.5	1	3.3	.8		
Bank, branch office	3	2.1	5.7	1.4		
Bank, main office	2.5	1.5	5.7	1.4		
Church	1.8	.8	3.3	.8		
College, science building	3	3	5.3	1.3		
College, library	2.5	.8	5.7	1.4		
College, physical education center	2	1	4.5	1.1		
Department store	2.5	.9	4	1		
Dormitory, college	1.5	1.2	4	1		
Drive-in donut shop	3	4	6.8	1.7		
Garage, commercial	.5	.5	0	.5		
Hospital, general	2	4.5	5	1.3		
Hospital, pediatric	3	3.8	5	1.3		
Hotel, airport	2	1	5	1.3		
Housing for the elderly	2	1.2	4	1		
Manufacturing, food processing	3	1	4.5	1.1		
Manufacturing, apparel	2	1	4.5	1.1		
Manufacturing, tools	4	1	4.5	1.1		
Medical clinic	2.5	1.5	3.2	1		
Nursing home	2	1.6	4	1		
Office building, hi-rise	3	2	4.7	1.2		
Office building, low-rise	3	2	4.3	1.2		
Radio-TV studio	3.8	2.2	7.6	1.9		
Restaurant	2.5	2	6.8	1.7		
Retail store	2.5	.9	5.5	1.4		
School, elementary	3	1.9	5.3	1.3		
School, junior high	3	1.5	5.3	1.3		
School, senior high	2.3	1.7	5.3	1.3		
Supermarket	3	1	4	1		
Telephone exchange	1	.6	4.5	1.1		
Theater	2.5	1	3.3	.8		
Town Hall	2	1.9	5.3	1.3		
U.S. Post Office	3	2	5	1.3		
Warehouse, grocery	1	.6	0	.5		

Rule of thumb: 1 kVA = 1 HP (single phase)

Three phase:

Watts = 1.73 x Volts x Current x Power Factor x Efficiency

$$\text{Horsepower} = \frac{\text{Volts x Current x Power Factor x Efficiency}}{746 \text{ Watts}}$$

Figure 26-9 Economy of Scale for Electrical Installations

For large installations, the economy of scale may have a definite impact on the electrical costs. If large quantities of a particular item are installed in the same general area, certain deductions can be made for labor. This table lists some suggested deductions for larger scale electrical installations.

Description	Quantity	Labor Deduction
Under floor ducts, bus ducts, conduit, cable systems:	150 to 250 L.F.	-10%
	251 to 350 L.F.	-15%
	351 to 500 L.F.	-20%
	Over 500 L.F.	-25%
Outlet boxes:	25 to 50 Ea.	-15%
	51 to 75 Ea.	-20%
	76 to 100 Ea.	-25%
	Over 100 Ea.	-30%
Wiring devices:	10 to 25 Ea.	-20%
	26 to 50 Ea.	-25%
	51 to 100 Ea.	-30%
	Over 100 Ea.	-35%
Lighting fixtures:	25 to 50 Ea.	-15%
	51 to 75 Ea.	-20%
	76 to 100 Ea.	-25%
	Over 100 Ea.	-30%

Figure 26-10 Maximum Number of Wires (Insulations Noted) for Various Conduit Sizes

The table below lists the maximum number of conductors for various conduit using THW, TW, or THWN insulations.

Copper Wire Size	1/2" TW	1/2" THW	1/2" THWN	3/4" TW	3/4" THW	3/4" THWN	1" TW	1" THW	1" THWN	1-1/4" TW	1-1/4" THW	1-1/4" THWN	1-1/2" TW	1-1/2" THW	1-1/2" THWN	2" TW	2" THW	2" THWN	2-1/2" TW	2-1/2" THW	2-1/2" THWN	3" TW	3" THW	3" THWN	3-1/2" TW	3-1/2" THW	3-1/2" THWN	4" TW	4" THW	4" THWN
#14	9	6	13	15	10	24	25	16	39	44	29	69	60	40	94	99	65	154	142	93		143			192					
#12	7	4	10	12	8	18	19	13	29	35	24	51	47	32	70	78	53	114	111	76	164	117			157					
#10	5	4	6	9	6	11	15	11	18	26	19	32	36	26	44	60	43	73	85	61	104	95		160	127			163		
#8	2	1	3	4	3	5	7	5	9	12	10	16	17	13	22	28	22	36	40	32	51	49		79	66		106	85		136
#6		1	1		2	4		4	6		7	11		10	15		16	26		23	37		36	57		48	76		62	98
#4		1	1		1	2		3	4		5	7		7	9		12	16		17	22		27	35		36	47		47	60
#3		1	1		1	1		2	3		4	6		6	8		10	13		15	19		23	29		31	39		40	51
#2		1	1		1	1		2	3		4	5		5	7		9	11		13	16		20	25		27	33		34	43
#1					1	1		1	1		3	3		4	5		6	8		9	12		14	18		19	25		25	32
1/0					1	1		1	1		2	3		3	4		5	7		8	10		12	15		16	21		21	27
2/0					1	1		1	1		1	2		3	3		5	6		7	8		10	13		14	17		18	22
3/0					1	1		1	1		1	1		2	3		4	5		6	7		9	11		12	14		15	18
4/0						1		1	1		1	1		1	2		3	4		5	6		7	9		10	12		13	15
250MCM								1	1		1	1		1	1		2	3		4	4		6	7		8	10		10	12
300								1	1		1	1		1	1		2	3		3	4		5	6		7	8		9	11
350									1		1	1		1	1		1	1		3	3		4	5		6	7		8	9
400								1	1		1	1					1	1		2	3		3	4		5	6		7	8
500								1	1		1	1					1	1		1	2		3	4		4	5		6	7
600									1			1					1	1		1	1		3	3		4	4		5	5
700												1					1	1		1	1		2	3		3	4		4	5
750												1					1	1		1	1		2	2		3	3		4	4

Figure 26-11 Conduit Weight Comparisons (lbs. per 100 ft.) Empty

Type	1/2"	3/4"	1"	1-1/4"	1-1/2"	2"	2-1/2"	3"	3-1/2"	4"	5"	6"
Rigid Aluminum	28	37	55	72	89	119	188	246	296	350	479	630
Rigid Steel	79	105	153	201	249	332	527	683	831	972	1314	1745
Intermediate Steel (IMC)	60	82	116	150	182	242	401	493	573	638		
Electrical Metallic Tubing (EMT)	29	45	65	96	111	141	215	260	365	390		
Polyvinyl Chloride Schedule 40	16	22	32	43	52	69	109	142	170	202	271	350
Polyvinyl Chloride Encased Burial						38		67	88	105	149	202
Fiber Duct Encased Burial						127		164	180	206	400	511
Fiber Duct Direct Burial						150		251	300	354		
Transite Encased Burial						160		240	290	330	450	550
Transite Direct Burial						220		310		400	540	640

Figure 26-12 Conduit Weight Comparisons (lbs. per 100 ft.) Filled*
with Maximum Conductors Allowed

Type	1/2"	3/4"	1"	1-1/4"	1-1/2"	2"	2-1/2"	3"	3-1/2"	4"	5"	6"
Rigid Galvanized Steel (RGS)	104	140	235	358	455	721	1022	1451	1749	2148	3083	4343
Intermediate Steel (IMC)	84	113	186	293	379	611	883	1263	1501	1830		
Electrical Metallic Tubing (EMT)	54	116	183	296	368	445	641	930	1215	1540		

*Conduit and heaviest conductor combination

Figure 26-13 Elevated Bus Ducts and Cable Tray Installation
Adjustments

For the installation of bus ducts and cable trays elevated more than 15' above the floor, the labor costs should be adjusted to allow for the added complexity. This table lists suggested adjustment factors.

Installation Height Above Floor	Labor Adjustments
15' to 20'	+ 10%
20' to 25'	+ 20%
25' to 30'	+ 25%
30' to 35'	+ 30%
35' to 40'	+ 35%
Over 40'	+ 40%

Division 26

Figure 26-14   Minimum Wire Sizes Allowed for Various Powered Circuits

The table below lists the minimum copper and aluminum wire size allowed for various types of insulation and design.

Minimum Wire Sizes									
	Copper		Aluminum			Copper		Aluminum	
Amperes	THW THWN or XHHW	THHN XHHW *	THW XHHW	THHN XHHW *	Amperes	THW THWN or XHHW	THHN XHHW *	THW XHHW	THHN XHHW *
15A	#14	#14	#12	#12	195	3/0	2/0	250kcmil	4/0
20	#12	#12	#10	#10	200	3/0	3/0	250kcmil	4/0
25	#10	#10	#10	#10	205	4/0	3/0	250kcmil	4/0
30	#10	#10	#8	#8	225	4/0	3/0	300kcmil	250kcmil
40	#8	#8	#8	#8	230	4/0	4/0	300kcmil	250kcmil
45	#8	#8	#6	#8	250	250kcmil	4/0	350kcmil	300kcmil
50	#8	#8	#6	#6	255	250kcmil	4/0	400kcmil	300kcmil
55	#6	#8	#4	#6	260	300kcmil	4/0	400kcmil	350kcmil
60	#6	#6	#4	#6	270	300kcmil	250kcmil	400kcmil	350kcmil
65	#6	#6	#4	#4	280	300kcmil	250kcmil	500kcmil	350kcmil
75	#4	#6	#3	#4	285	300kcmil	250kcmil	500kcmil	400kcmil
85	#4	#4	#2	#3	290	350kcmil	250kcmil	500kcmil	400kcmil
90	#3	#4	#2	#2	305	350kcmil	300kcmil	500kcmil	400kcmil
95	#3	#4	#1	#2	310	350kcmil	300kcmil	500kcmil	500kcmil
100	#3	#3	#1	#2	320	400kcmil	300kcmil	600kcmil	500kcmil
110	#2	#3	1/0	#1	335	400kcmil	350kcmil	600kcmil	500kcmil
115	#2	#2	1/0	#1	340	500kcmil	350kcmil	600kcmil	500kcmil
120	#1	#2	1/0	1/0	350	500kcmil	350kcmil	700kcmil	500kcmil
130	#1	#2	2/0	1/0	375	500kcmil	400kcmil	700kcmil	600kcmil
135	1/0	#1	2/0	1/0	380	500kcmil	400kcmil	750kcmil	600kcmil
150	1/0	#1	3/0	2/0	385	600kcmil	500kcmil	750kcmil	600kcmil
155	2/0	1/0	3/0	3/0	420	600kcmil	500kcmil		700kcmil
170	2/0	1/0	4/0	3/0	430		500kcmil		750kcmil
175	2/0	2/0	4/0	3/0	435		600kcmil		750kcmil
180	3/0	2/0	4/0	4/0	475		600kcmil		

* Dry locations only.

Notes:

1. Size #14 to 4/0 is in AWG units (American wire guage).
2. Size 250 to 750 is in kcmil units (thousand circular mils).
3. Use next higher ampere value if exact value is not listed in table.
4. For loads that operate continuously, increase ampere value by 25% to obtain proper wire size.
5. Refer to Figure 26-15 for the maximum circuit length for the various size wires.
6. This figure is for estimating purposes only, based on ambient temperature of 30° C (86°F); for ambient temperature other than 30° C, ampacity correction factors will be applied.

Figure 26-15 Maximum Circuit Length (Approximate) for Various Power Requirements

This figure assumes the use of THW, copper wire @ 75°C and a 4% voltage drop.

Maximum circuit length: This figure indicates the typical maximum installed length a circuit can have and still maintain an adequate voltage level at the point of use. The circuit length is similar to the conduit length.

If the circuit length for an ampere load and a copper wire size exceeds the length obtained from this table, use the next largest wire size to compensate for voltage drop.

Example: A 130 ampere load at 480 volts, 3 phase, 3 wire with No. 1 wire can be run a maximum length of 555 LF and provide satisfactory operation. If the same load is to be wired at the end of a 625 LF circuit, then a larger wire must be used.

Amperes	Wire Size	Maximum Circuit Length in Feet				
		2 Wire, 1 Phase		3 Wire, 3 Phase		
		120V	240V	240V	480V	600V
15	14*	50	105	120	240	300
	14	50	100	120	235	295
20	12*	60	125	145	290	360
	12	60	120	140	280	350
30	10*	65	130	155	305	380
	10	65	130	150	300	375
50	8	60	125	145	285	355
65	6	75	150	175	345	435
85	4	90	185	210	425	530
115	2	110	215	250	500	620
130	1	120	240	275	555	690
150	1/0	130	260	305	605	760
175	2/0	140	285	330	655	820
200	3/0	155	315	360	725	904
230	4/0	170	345	395	795	990
255	250	185	365	420	845	1055
285	300	195	395	455	910	1140
310	350	210	420	485	975	1220
380	500	245	490	565	1130	1415

*Solid Conductor.

Note: The circuit length is the one way distance between the origin and the load.

Figure 26-16   Electrical Demolition (Removal for Replacement)

A guide for electrical "removal for replacement" is to apply the rule of thumb: 1/3 of new installation time (typical range from 20% to 50%) for removal. Remember to use reasonable judgment when applying the suggested percentage factor.

For example: an electrical contractor has been asked to remove an existing fluorescent lighting fixture and replace it with a new fixture using energy-saver lamps and electronic ballast.

In order to fully understand the extent of the project, the contractor should visit the job site and estimate the time needed to perform the renovation work in accordance with applicable national, state, and local regulations and codes.

The contractor may need to add extra labor-hours to his estimate if he discovers conditions such as a contaminated asbestos ceiling or broken acoustical ceiling tile, both of which would need to be repaired, patched, and touched up with paint—tasks normally assigned to general contractors. Furthermore, the owner could request that the contractor either salvage the materials removed and turn them over to the owner or dispose of them in a reclamation station.

The normal estimate times are 0.5 labor-hour for removing a lighting fixture and 1.5 labor-hours for installing a new one. In this case, the contractor might need to revise the estimate time from 2 labor-hours up to a minimum of 4 labor-hours for replacing one fluorescent lighting fixture.

For removal of large concentrations of lighting fixtures in the same area, apply an "economy of scale" to reduce estimating labor-hours.

Figure 26-17 Hazardous Area Wiring

All hazardous locations require special material and wiring methods as determined by the classifications defined in the *National Electric Code*. The descriptions of hazardous locations listed below have been abbreviated. The *National Electric Code* should be consulted for complete descriptions.

Class 1 Locations	Class 2 Locations	Class 3 Locations
Locations in which flammable gases or vapors are or may be present in the air in quantities sufficient to produce explosive or ignitable mixtures.	Locations in which hazardous conditions exist because of the presence of combustible dust.	Locations in which the presence of easily flammable fibers or flyings are present in the air, but not in sufficient quantities to produce ignitable mixtures under normal conditions.
Division 1: This classification includes locations where a hazardous atmosphere is expected during normal operations; locations where a breakdown in operation of processing equipment results in release of hazardous vapors and the simultaneous failure of electrical equipment.	**Division 1:** This classification includes locations where combustible dust may be suspended in the air under normal conditions in sufficient quantities to produce explosive or ignitable mixtures; locations where a breakdown in operation of machinery or equipment might cause a hazardous condition to exist while creating a source of ignition with the simultaneous failure of electrical equipment; locations in which combustible dust of any electrically conductive nature may be present.	**Division 1:** This classification includes locations where easily ignitable fibers or materials producing combustible flyings are handled, manufactured, or used.
Division 2: This classification includes locations in which volatile flammable liquids or gases are handled, processed, or used, but will normally be confined to closed containers or systems from which they can escape only by accidental rupture or breakdown; locations where hazardous conditions will occur only under abnormal circumstances.	**Division 2:** This classification includes locations where air-suspended combustible dust is not at hazardous levels, but where an accumulation of dust may interfere with the safe dissipation of heat from electrical equipment, or may be ignited by arcs, sparks, or burning material located near electrical equipment.	**Division 2:** This classification includes locations where easily ignitable fibers are stored or handled.

Figure 26-18 Standard Electrical Enclosure Types

NEMA Enclosures

Electrical enclosures serve two basic purposes: they protect people from accidental contact with enclosed electrical devices and connections, and they protect the enclosed devices and connections from specified external conditions. The National Electrical Manufacturers Association (NEMA) has established the following standards. These brief summaries are not intended to be complete representations of NEMA listings, and consultation of NEMA literature is advised for detailed information.

The following definitions and descriptions pertain to NON-HAZARDOUS locations:

NEMA Type 1: General purpose enclosures intended for use indoors, primarily to prevent accidental contact of personnel with the enclosed equipment in areas that do not involve unusual conditions.

NEMA Type 2: Drip-proof indoor enclosures intended to protect the enclosed equipment against dripping noncorrosive liquids and falling dirt.

NEMA Type 3: Dustproof, rain-tight and sleet-resistant (ice-resistant) enclosures intended for use outdoors to protect the enclosed equipment against wind-blown dust, rain, sleet, and external ice formation.

NEMA Type 3R: Rainproof and sleet-resistant (ice-resistant) enclosures intended for use outdoors to protect the enclosed equipment against rain. These enclosures are constructed so that the accumulation and melting of sleet (ice) will not damage the enclosure and its internal mechanisms.

NEMA Type 3S: Enclosures intended for outdoor use to provide limited protection against wind-blown dust, rain, and sleet (ice) and to allow operation of external mechanisms when ice-laden.

NEMA Type 4: Watertight and dust-tight enclosures intended for use indoors and out to protect the enclosed equipment against splashing water, seepage of water, falling or hose-directed water, and severe external condensation.

NEMA Type 4X: Watertight, dust-tight, and corrosion-resistant indoor and outdoor enclosures featuring the same provisions as Type 4 enclosures, plus corrosion resistance.

NEMA Type 5: Indoor enclosures intended primarily to provide limited protection against dust and falling dirt.

NEMA Type 6: Enclosures intended for indoor and outdoor use, primarily to provide limited protection against the entry of water during occasional temporary submersion at a limited depth.

NEMA Type 6R: Enclosures intended for indoor and outdoor use, primarily to provide limited protection against the entry of water during prolonged submersion at a limited depth.

NEMA Type 11: Enclosures intended for indoor use, primarily to provide, by means of oil immersion, limited protection to enclosed equipment against the corrosive effects of liquids and gases.

NEMA Type 12: Dust-tight and drip-tight indoor enclosures intended for use indoors in industrial locations to protect the enclosed equipment against fibers, flyings, lint, dust, and dirt, as well as light splashing, seepage, dripping, and external condensation of non-corrosive liquids.

(continued on next page)

Figure 26-18 Standard Electrical Enclosure Types
(continued from previous page)

NEMA Type 13: Oil-tight and dust-tight indoor enclosures intended primarily to house pilot devices, such as limit switches, foot switches, push buttons, selector switches, and pilot lights, and to protect these devices against lint and dust, seepage, external condensation, and sprayed water, oil, and non-corrosive coolant.

The following definitions and descriptions pertain to HAZARDOUS, or CLASSIFIED locations:

NEMA Type 7: Enclosures intended to use in indoor locations classified as Class 1, Groups A, B, C, or D, as defined in the *National Electric Code*.

NEMA Type 9: Enclosures intended for use in indoor locations classified as Class 2, Groups E, F, or G, as defined in the *National Electric Code*.

Figure 26-19 Elevated Equipment Installation Factors

Installation Height Above Floor	Labor Adjustments
10' to 15'	+ 15%
15' to 25'	+ 30%
Over 25'	+ 35%

For the installation of electrical equipment elevated more than 10' above the floor, the labor costs should be adjusted to allow for the added complexity. This table lists suggested adjustments.

Figure 26-20 Maximum Horsepower for Starter Size by Voltage

Starter Size	Maximum HP (3Φ)			
	208V	240V	480V	600V
00	1-1/2	1-1/2	2	2
0	3	3	5	5
1	7-1/2	7-1/2	10	10
2	10	15	25	25
3	25	30	50	50
4	40	50	100	100
5		100	200	200
6		200	300	300
7		300	600	600
8		450	900	900
8L		700	1500	1500

Division 26

Figure 26-21 Wattages for Motors

The power factor of electric motors varies from 80% to 90% in larger size motors.

90% Power Factor & Efficiency @ 200 or 460V			
HP	Watts	HP	Watts
10	9024	30	25784
15	13537	40	33519
20	17404	50	41899
25	21916	60	49634

Figure 26-22 Horsepower Requirements for Elevators with Three-Phase Motors

Elevator Type	Maximum Travel Height in Ft.	Travel Speeds in FPM	Capacity of Cars in Lbs.								
			1200	1500	1800	2000	2500	3000	3500	4000	4500
Hydraulic	70	70	10	15	15	15	20	20	20	25	30
		85	15	15	15	20	20	25	25	30	30
		100	15	15	20	20	25	30	30	40	40
		110	20	20	20	20	25	30	40	40	50
		125	20	20	20	25	30	40	40	50	50
		150	25	25	25	30	40	50	50	50	60
		175	25	30	30	40	50	50	60		
		200	30	30	40	40	50	60	60		
Geared Traction	300	200				10	10	15	15		23
		350				15	15	23	23		35

Figure 26-23 Motor Control Center Standard Classifications

Motor control centers are available in two standard classes, with several types of units making up each class. These classifications and types are defined as follows:

Class 1

Type A: This type consists of a control unit with a circuit breaker or fusible disconnect wired to the line side of the starter only.

Type B: This type is the same as Type A, but the control circuit leads are wired to a fixed terminal block on the control unit.

Type C: This type is the same as Type B, but the leads are brought to the control unit terminal boards which are located at the top or bottom of the motor control.

Interwiring and interlocking do not exist between starters or cubicles in any type in this class of unit.

Class 2

Type B: This type is the same as Class 1, Type B, but it contains wiring between control units in the same or adjacent cubicles.

Type C: This type is the same as Class 2, Type B, but it provides interwiring from the master terminal boards at the top and bottom of the control center.

Figure 26-24 Transformer Weights (lbs.) by kVA

Oil Filled 3 Phase 5/15 KV to 480/277			
KVA	**Lbs.**	**KVA**	**Lbs.**
150	1800	1000	6200
300	2900	1500	8400
500	4700	2000	9700
750	5300	3000	15000
Dry 240/480 to 120/240 Volt			
1 Phase		3 Phase	
KVA	**Lbs.**	**KVA**	**Lbs.**
1	23	3	90
2	36	6	135
3	59	9	170
5	73	15	220
7.5	131	30	310
10	149	45	400
15	205	75	600
25	255	112.5	950
37.5	295	150	1140
50	340	225	1575
75	550	300	1870
100	670	500	2850
167	900	750	4300

Figure 26-25 Generator Weights (lbs.) by kW

3 Phase 4 Wire 277/480 Volt			
Gas		Diesel	
KW	Lbs.	KW	Lbs.
7.5	600	30	1800
10	630	50	2230
15	960	75	2250
30	1500	100	3840
65	2350	125	4030
85	2570	150	5500
115	4310	175	5650
170	6530	200	5930
		250	6320
		300	7840
		350	8220
		400	10750
		500	11900

Figure 26-26 Lighting: Calculating Footcandles and Watts per Square Foot

1. Initial footcandles = number of fixtures × lamps per fixture × lumens per lamp × coefficient of utilization ÷ SF.

2. Maintained footcandles = initial footcandles × maintenance factor.

3. Watts per SF = number of fixtures × lamps × (lamp watts + ballast watts) ÷ SF.

Example: To find footcandles and watts per SF for an office 20' × 20' with 11 fluorescent fixtures each having four 40 watt CW lamps.

Based on good reflectance and clean conditions:
Lumens per lamp = 40 watt cool white at 3150 lumens per lamp.
Coefficient of utilization = .42 (varies from .62 for light colored areas to .27 for dark).
Maintenance factor = .75 (varies from .80 for clean areas with good maintenance to .50 for poor).
Ballast loss = 9 watts per lamp.

1. Initial footcandles:

$$\frac{11 \times 4 \times 3150 \times .42}{400} = \frac{58212}{400} = 145 \text{ footcandles}$$

2. Maintained footcandles: 145 × .75 = 109 footcandles

3. Watts per SF

$$\frac{11 \times 4 \times (40+8)}{400} = \frac{2112}{400} = 5.3 \text{ watts per SF}$$

Figure 26-27 IESNA* Recommended Illumination Levels in Footcandles

Commercial Buildings			Industrial Buildings		
Type	Description	Foot-Candles	Type	Description	Foot-Candles
Banks	Lobby	50	Assembly Areas	Rough bench & machine work	50
	Customer Areas	70		Medium bench & machine work	100
	Teller Stations	150		Fine bench & machine work	500
	Accounting Areas	150	Inspection Areas	Ordinary	50
Offices	Routine Work	100		Difficult	100
	Accounting	150		Highly Difficult	200
	Drafting	200	Material Handling	Loading	20
	Corridors, Halls, Washrooms	30		Stock Picking	30
Schools	Reading or Writing	70		Packing, Wrapping	50
	Drafting, Labs, Shops	100	Stairways Washrooms	Service Areas	20
	Libraries	70		Service Areas	20
	Auditoriums, Assembly	15	Storage Areas	Inactive	5
	Auditoriums, Exhibition	30		Active, Rough, Bulky	10
Stores	Circulation Areas	30		Active, Medium	20
	Stock Rooms	30		Active, Fine	50
	Merchandise Areas, Service	100	Garages	Active Traffic Areas	20
	Self-Service Areas	200		Service & Repair	100

*IESNA: Illuminating Engineering Society of North America

Figure 26-28 Energy Efficiency Rating for Luminaires

The energy efficiency program for luminaires recommends the use of a metric called the Luminaire Efficacy Rating (LER). The LER value expresses the total lumens generated by a lamp to the watts consumed by the lamp.

Lamp Type	Lumens per Watt
Incandescent	17
Tungsten haologen	14–20
Fluorescent	50–104
Metal halide	64–96
High Pressure sodium	76–116

Figure 26-29   General Lighting Loads by Occupancies

This data is for estimating purposes only.

Type of Facility	Unit Load per S.F. (Watts)
Armories and Auditoriums	1
Banks	5
Barber Shops and Beauty Parlors	3
Churches	1
Clubs	2
Court Rooms	2
Dwelling Units (1)	3
Garages — Commercial (storage)	1/2
Hospitals	2
Hotels and Motels, including apartment houses without provisions for cooking by tenants (1)	2
Industrial Commercial (Loft) Buildings	2
Lodge Rooms	1-1/2
Office Buildings	5
Restaurants	2
Schools	3
Stores	3
Warehouses (storage)	1/4
(1) In any of the above occupancies except one-family dwellings and individual dwelling units of multi-family dwellings:	
Assembly Halls and Auditoriums	1
Halls, Corridors, Closets	1/2
Storage Spaces	1/4

Figure 26-30 Lighting Limit for Listed Occupancies

This data is for estimating purposes only.

Type of Use	Maximum Watts per S.F.
INTERIOR	
Category A: Classrooms, office areas, automotive mechanical areas, museums, conference rooms, drafting rooms, clerical areas, laboratories, merchandising areas, kitchens, examining rooms, book stacks, athletic facilities.	3.00
Category B: Auditoriums, waiting areas, spectator areas, restrooms, dining areas, transportation terminals, working corridors in prisons and hospitals, book storage areas, active inventory storage, hospital bedrooms, hotel and motel bedrooms, enclosed shopping mall concourse areas, stairways.	1.00
Category C: Corridors, lobbies, elevators, inactive storage areas.	0.50
Category D: Indoor parking.	0.25
EXTERIOR	
Category E: Building perimeter: wall-wash, facade, canopy.	5.00 (per L.F.)
Category F: Outdoor parking.	0.10

Figure 26-31 Approximate Watts per SF for Popular Fixture Types

This figure provides the wattage required to maintain a given number of footcandles for a known type of lighting fixture.

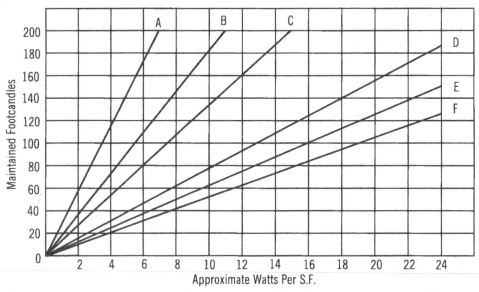

Due to the many variables involved, this information is recommended for preliminary estimating only.
  A. Fluorescent – industrial system
  B. Fluorescent – lens unit system
  C. Fluorescent – louvered unit system
  D. Incandescent – open reflector system
  E. Incandescent – lens unit system
  F. Incandescent – downlight system

Figure 26-32 Site Floodlighting Lamp Comparison (Includes Floodlight, Ballast, and Lamp for Pole Mounting)

This chart compares typical types of lamps used for floodlighting by required watts, intensity (measured in lumens, lumens per watt, and lumens at 40% life), and actual life of the lamp.

Type	Watts	Initial Lumens	Lumens per Watt	Lumens @ 40% Life	Life (Hours)
Incandescent	150	2880	19	85	750
	300	6360	21	84	750
	500	10,850	22	80	1000
	1000	23,740	24	80	1000
	1500	34,400	23	80	1000
Tungsten Halogen	500	10,950	22	97	2000
	1500	35,800	24	97	2000
Fluorescent Cool White	40	3150	79	88	20,000
	110	9200	84	87	12,000
	215	16,000	74	81	12,000
Metal Halide	175	14,000	80	77	7500
	400	34,000	85	75	15,000
	1000	100,000	100	83	10,000
	1500	155,000	103	92	1500
High Pressure Sodium	70	5800	83	90	20,000
	100	9500	95	90	20,000
	150	16,000	107	90	24,000
	400	50,000	125	90	24,000
	1000	140,000	140	90	24,000
Low Pressure Sodium	55	4600	131	98	18,000
	90	12,750	142	98	18,000
	180	33,000	183	98	18,000

Color: High Pressure Sodium – Slightly Yellow   Mercury Vapor – Green Blue
       Low Pressure Sodium – Yellow   Metal Halide – Blue-White   Note: Pole not included

Figure 26-33 Television Systems

Master TV antenna systems, part of Division 27, are used in schools, dormitories, and apartment buildings. Each system consists of the antenna, lightning arrester, splitters, and outlets. The signal is received by the antenna and increased by the amplifier. It then goes through the main cable to the splitter. Here, the signal is split among several branch cables.

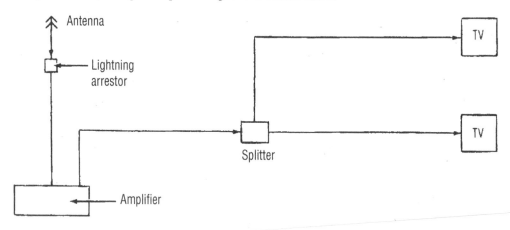

A closed-circuit TV system, part of Division 28, consists of a TV camera and monitor. These systems are used indoors and outdoors for security surveillance. Some applications require pan, tilt, and zoom (PTZ) mechanisms for remote control of the camera.

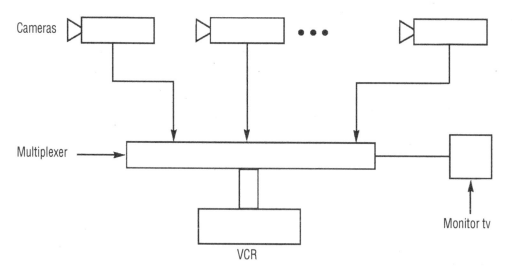

(continued on next page)

Figure 26-33 Television Systems *(continued from previous page)*

Television Equipment	Crew	Daily Output	Labor-Hours	Unit
T.V. Systems not including rough-in wires, cables & conduits				
Master TV antenna system				
VHF & UHF reception & distribution, 12 outlets	1 Elec.	6	1.333	Outlet
30 outlets		10	.800	
100 outlets		13	.615	
Amplifier		4	2	Ea.
Antenna		2	4	"
Closed circuit, surveillance, one station (camera & monitor)	2 Elec.	2.60	6.154	Total
For additional camera stations, add	1 Elec.	2.70	2.963	Ea.
Industrial quality, one station (camera & monitor)	2 Elec	2.60	6.154	Total
For additional camera stations, add	1 Elec.	2.70	2.963	Ea.
For low light, add		2.70	2.963	
For very low light, add		2.70	2.963	
For weatherproof camera station add		1.30	6.154	
For pan and tilt, add		1.30	6.154	
For zoom lens - remote control, add, minimum		2	4	
Maximum		2	4	
For automatic iris for low light, add		2	4	

Figure 26-34 Fire Alarm Systems

Fire alarm systems, part of Division 28, consist of control panels, annunciator panels, battery with rack and charger, various sensing devices (such as smoke and heat detectors), and alarm horn and light signals. Some fire alarm systems are very sophisticated and include speakers, telephone lines, door closer controls, and other components. Some are connected directly to the fire station. Requirements for the fire alarm systems are generally regulated by codes and by local authorities.

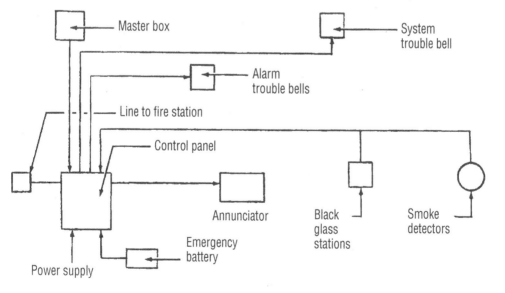

Fire Alarm

Figure 26-35 Burglar Alarm Systems

Burglar alarm systems, part of Division 28, consist of control panels, indicator panels, various types of alarm devices, and switches. The control panel is usually line powered with a battery backup supply. Some systems have a direct connection to the police or protection company, while others have auto-dial telephone capabilities. Most, however, simply have local control monitors and an annunciator. The sensing devices are various pressure switches, magnetic door switches, glass break sensors, infrared sensors, microwave detectors, and ultrasonic motion detectors. The alarms are sirens, horns, and/or flashing lights.

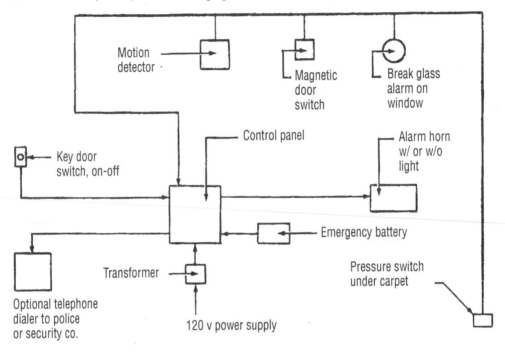

Burglar Alarm

Figure 26-36 Installation Time in Labor-Hours for Conduit

Conduit to 15' high, includes couplings, fittings, and support.

Description	Labor-Hours	Unit
Rigid Galvanized Steel 1/2" Diameter	.089	L.F.
1-1/2" Diameter	.145	L.F.
3" Diameter	.320	L.F.
6" Diameter	.800	L.F.
Aluminum 1/2" Diameter	.080	L.F.
1-1/2" Diameter	.123	L.F.
3" Diameter	.178	L.F.
6" Diameter	.400	L.F.
IMC 1/2" Diameter	.080	L.F.
1-1/2" Diameter	.133	L.F.
3" Diameter	.267	L.F.
4" Diameter	.320	L.F.
Plastic Coated Rigid Steel 1/2" Diameter	.100	L.F.
1-1/2" Diameter	.178	L.F.
3" Diameter	.364	L.F.
6" Diameter	.800	L.F.
EMT 1/2" Diameter	.047	L.F.
1-1/2" Diameter	.089	L.F.
3" Diameter	.160	L.F.
4" Diameter	.200	L.F.
PVC Nonmetallic 1/2" Diameter	.042	L.F.
1-1/2" Diameter	.080	L.F.
3" Diameter	.145	L.F.
6" Diameter	.267	L.F.

**Rigid Steel,
Plastic Coated Coupling**

PVC Conduit

PVC Elbow

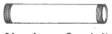

Aluminum Conduit

EMT Set Screw Connector

Aluminum Elbow

EMT Connector

Rigid Steel, Plastic Coated Conduit

EMT to Conduit Adapter

Rigid Steel, Plastic Coated Elbow

EMT to Greenfield Adapter

Division 26

Figure 26-37 Installation Time in Labor-Hours for Cable
Tray Systems

Description	Labor-Hours	Unit
Cable Tray		
Ladder Type 36" Wide	.267	L.F.
Elbows Vertical 36"	3.810	Ea.
Elbows Horizontal 36"	3.810	Ea.
Tee Vertical 36"	4.440	Ea.
Tee Horizontal 36"	5.330	Ea.
Drop-Out 36"	1.000	Ea.
Reducer 36" to 12"	2.290	Ea.
Wall Bracket 12"	.364	Ea.
Cover Straight 36"	.100	L.F.
Cover Elbow 36"	.320	Ea.

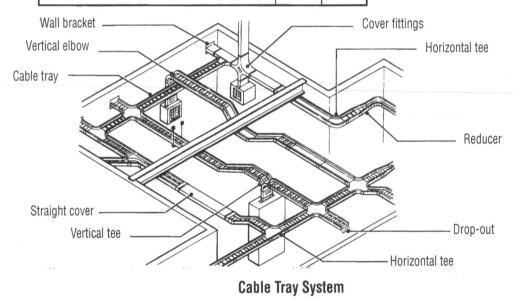

Cable Tray System

Figure 26-38 Installation Time in Labor-Hours for Underfloor
Raceway Systems

Description	Labor-Hours	Unit
Blank Duct	.100	L.F.
Insert Duct	.110	L.F.
Elbow Vertical	.800	Ea.
Elbow Horizontal	.300	Ea.
Panel Connector	.250	Ea.
Junction Box, Single Duct	2.000	Ea.
Double Duct	2.500	Ea.
Triple Duct	2.950	Ea.
Saddle Support, Single Duct	.290	Ea.
Double Duct	.500	Ea.
Triple Duct	.720	Ea.
Insert to Conduit Adapter	.250	Ea.
Outlet, Low Tension (Telephone and Signal)	1.000	Ea.
Outlet, High Tension (Power)	1.000	Ea.
Offset Duct Type	.300	Ea.

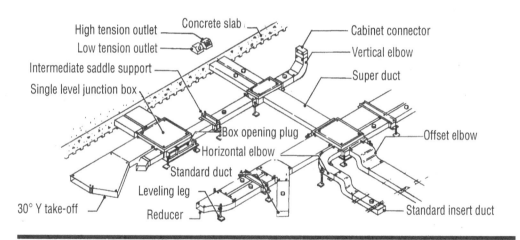

Figure 26-39 Installation Time in Labor-Hours for Cellular Concrete Floor Raceway Systems

Description	Labor-Hours	Unit
Underfloor Header Duct 3-1/8" Wide	.100	L.F.
7-1/4" Wide	.133	L.F.
Header Duct, Single Compartment 9" Wide	.400	L.F.
24" Wide	.727	L.F.
Double Compartment, 24" Wide	.800	L.F.
Triple Compartment, 36" Wide	1.330	L.F.
Location Market Plug	.250	Ea.

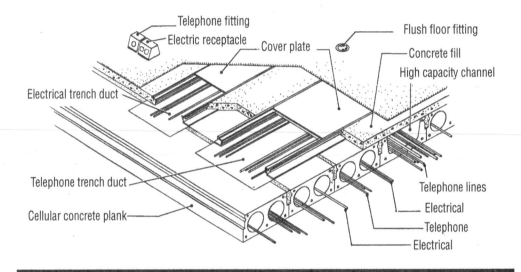

Figure 26-40 Installation Time in Labor-Hours for Electrical Conductors: Wire and Cable

Description	Labor-Hours	Unit
600V Copper #14 AWG	.610	CLF
#12 AWG	.720	CLF
#10 AWG	.800	CLF
#8 AWG	1.000	CLF
#6 AWG	1.230	CLF
#4 AWG	1.510	CLF
#3 AWG	1.600	CLF
#2 AWG	1.780	CLF
#1 AWG	2.000	CLF
#1/0	2.420	CLF
#2/0	2.760	CLF
#3/0	3.200	CLF
#4/0	3.640	CLF
250 kcmil	4.000	CLF
500 kcmil	5.000	CLF
1000 kcmil	9.000	CLF

(continued on next page)

Division 26

Figure 26-40   Installation Time in Labor-Hours for Electrical Conductors: Wire and Cable *(continued from previous page)*

Crimp, 1-hole lug   Terminal lug, solderless   Crimp, 2-way connector

Cable Terminations

Description	Labor-Hours	Unit
CABLE TERMINATIONS		
Wire connectors, screw type, #22 to #14	.031	Ea.
#18 to #12	.033	
#18 to #10	.033	
Crimp 1 hole lugs, copper or aluminum, 600 volt		
#14	.133	
#12	.160	
#10	.178	
#8	.222	
#6	.267	
#4	.296	
#2	.333	
#1	.400	
1/0	.457	
2/0	.533	
3/0	.667	
4/0	.727	
250 kcmil	.889	
300 kcmil	1.000	
350 kcmil	1.143	
400 kcmil	1.231	
500 kcmil	1.333	
600 kcmil	1.379	
700 kcmil	1.455	
750 kcmil	1.538	

PVC jacket connector   SER, insulated, aluminum

600 volt, armored   5 kv armored

Cable Connectors

Figure 26-41 Installation Time in Labor-Hours for Undercarpet
Power Systems

Description	Labor-Hours	Unit
Cable 3 Conductor #12 with Bottom Shield	.008	L.F.
Cable 5 Conductor #12 with Bottom Shield	.010	L.F.
Splice 3 Conductor with Insulating Patch	.334	Ea.
Splice 5 Conductor with Insulating Patch	.334	Ea.
Tap 3 Conductor with Insulating Patch	.367	Ea.
Tap 5 Conductor with Insulating Patch	.367	Ea.
Receptacle with Floor Box Pedestal Type	.500	Ea.
Receptacle Direct Connect	.320	Ea.
Top Shield	.005	L.F.
Transition Block	.104	Ea.
Transition Box, Flush Mount with Cover	.400	Ea.

Division 26

Figure 26-42 Installation Time in Labor-Hours for Undercarpet Telephone Systems

Description	Labor-Hours	Unit
Cable Assembly 25 Pair with Connectors 50'	.670	Ea.
3 Pair with Connectors 50'	.340	Ea.
4 Pair with Connectors 50'	.350	Ea.
Cable (Bulk) 3 Pair	.006	L.F.
4 Pair	.007	L.F.
Bottom Shield for 25 Pair Cable	.005	L.F.
3-4 Pair Cable	.005	L.F.
Top Shield for all Cable	.005	L.F.
Transition Box, Flush Mount	.330	Ea.
In Floor Service Box	2.000	Ea.
Floor Fitting with Duplex Jack and Cover	.380	Ea.
Floor Fitting Miniature with Duplex Jack	.150	Ea.
Floor Fitting with 25 Pair Kit	.380	Ea.
Floor Fitting Call Director Kit	.420	Ea.

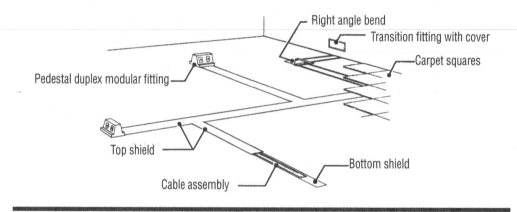

Figure 26-43 Installation Time in Labor-Hours for Undercarpet Data Systems

Description	Labor-Hours	Unit
Cable Assembly with Connectors 40'		
Single Lead	.360	Ea.
Dual Lead	.380	Ea.
Cable (Bulk) Single Lead	.010	L.F.
Dual Lead	.010	L.F.
Cable Notching 90 Degree	.080	Ea.
180 Degree	.130	Ea.
Connectors BNC Coax	.200	Ea.
Connectors TNC Coax	.200	Ea.
Transition Box, Flush Mount	.330	Ea.
In Floor Service Box	2.000	Ea.
Floor Fitting with Slotted Cover	.380	Ea.
With Blank Cover	.380	Ea.

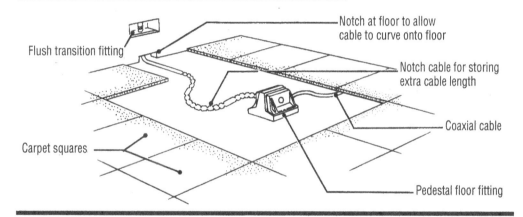

Flush transition fitting

Notch at floor to allow cable to curve onto floor

Notch cable for storing extra cable length

Coaxial cable

Carpet squares

Pedestal floor fitting

Figure 26-44   Installation Time in Labor-Hours for Unshielded
                   Twisted Cable

Description	Labor-Hours	Unit
High performance unshielded twisted pair (UTP)		
Category 3, #24, 2 pair solid, PVC jacket	.800	C.L.F.
4 pair solid	1.143	
25 pair solid	2.667	
2 pair solid, plenum	.800	
4 pair solid	1.143	
25 pair solid	2.667	
4 pair stranded, PVC jacket	1.143	
Category 5, #24, 4 pair solid, PVC jacket	1.143	
4 pair solid, plenum	1.143	
4 pair stranded, PVC jacket	1.143	
Category 5e, #24, 4 pair solid, PVC jacket	1.143	
4 pair solid, plenum	1.143	
4 pair stranded, PVC jacket	1.143	
Category 6, #24, 4 pair solid, PVC jacket	1.143	
4 pair solid, plenum	1.143	
4 pair stranded, PVC jacket	1.143	
Category 5, connector, UTP RJ-45	.100	Ea.
shielded RJ-45	.111	
Category 3, jack, UTP RJ-45	.111	
Category 5	.123	
Category 5e	.123	
Category 6	.123	
Category 5, jack, shielded RJ-45	.133	
Category 5e	.133	
Category 6	.133	

There are several categories used to describe high performance cable. The following information includes a description of categories CAT 3, 5, 5e, 6, and 7, and details classification of frequency and specific standards. The category standards have evolved under the sponsorship of organizations such as the Telecommunication Industry Association (TIA), the Electronic Industries Alliance (EIA), the American National Standards Institute (ANSI), the International Organization for Standardization (IOS), and the International Electrotechnical Commission (IEC), all of which have catered to the increasing complexities of modern network technology. For network cabling, users must comply with national or international standards. A breakdown of these categories is as follows:

Category 3: Designed to handle frequencies up to 16 MHz

Category 5: (TIA/EIA 568A) Designed to handle frequencies up to 100 MHz

Category 5e: Additional transmission performance to exceed Category 5

Category 6: Developed by TIA and other international groups to handle frequencies of 250 MHz

Category 7 (draft): Under development to handle a frequency range from 1 to 600 MHz

Figure 26-45 Installation Time in Labor-Hours for Hazardous Area Wiring

Description	Labor-Hours	Unit
Sealing Fitting 1/2" Diameter	.550	Ea.
2" Diameter	1.000	Ea.
3" Diameter	1.400	Ea.
4" Diameter	2.000	Ea.
Flexible Coupling 3/4" Diameter x 12" Long	.800	Ea.
2" Diameter x 12" Long	1.740	Ea.
3" Diameter x 12" Long	2.670	Ea.
4" Diameter x 12" Long	3.330	Ea.
Pulling Elbow 3/4" Diameter	1.000	Ea.
2" Diameter	2.000	Ea.
3" Diameter	2.670	Ea.
Conduit LB 3/4" Diameter	1.000	Ea.
T 3/4" Diameter	1.330	Ea.
Cast Box NEMA 7		
6" Long x 6" Wide x 6" Deep	4.000	Ea.
12" Long x 12" Wide x 6" Deep	8.000	Ea.
18" Long x 18" Wide x 8" Deep	16.000	Ea.

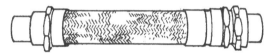

Explosion-proof Flexible Coupling

Explosion-proof Sealing Fitting

Explosion-proof Round Box with Cover, 3 Threaded Hubs

Explosion-proof NEMA 7, Surface Mounted, Pull Box

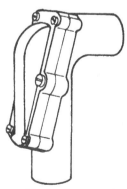

Explosion-proof Pulling Elbow

Figure 26-46 Installation Time in Labor-Hours for NEMA Enclosures

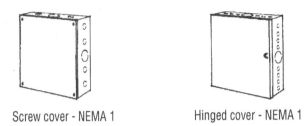

Screw cover - NEMA 1   Hinged cover - NEMA 1

Rainproof and weatherproof, screw cover - NEMA 3R

Sheet Metal Pull Boxes

Description	Labor-Hours	Unit
NEMA 1		
12" Long × 12" Wide × 4" Deep	1.330	Ea.
20" Long × 20" Wide × 8" Deep	2.500	Ea.
NEMA 3R		
12" Long × 12" Wide × 6" Deep	1.600	Ea.
24" Long × 24" Wide × 10" Deep	3.200	Ea.
NEMA 4		
12" Long × 12" Wide × 6" Deep	3.480	Ea.
24" Long × 24" Wide × 10" Deep	16.000	Ea.
NEMA 7		
12" Long × 12" Wide × 6" Deep	8.000	Ea.
24" Long × 18" Wide × 8" Deep	20.000	Ea.
NEMA 9		
12" Long × 12" Wide × 6" Deep	5.000	Ea.
24" Long × 24" Wide × 10" Deep	20.000	Ea.
NEMA 12		
12" Long × 14" Wide × 6" Deep	1.510	Ea.
24" Long × 30" Wide × 6" Deep	2.500	Ea.

Figure 26-47 Installation Time in Labor-Hours for Wiring Devices

Description	Labor-Hours	Unit
Receptacle 20A 250V	.290	Ea.
Receptacle 30A 250V	.530	Ea.
Receptacle 50A 250V	.720	Ea.
Receptacle 60A 250V	1.000	Ea.
Box, 4" Square	.400	Ea.
Box, Single Gang	.290	Ea.
Box, Cast Single Gang	.660	Ea.
Cover, Weatherproof	.120	Ea.
Cover, Raised Device	.150	Ea.
Cover, Brushed Brass	.100	Ea.

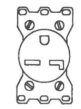

30 amp, 125 volt, NEMA 5

50 amp, 125 volt, NEMA 5

20 amp, 250 volt, NEMA 6

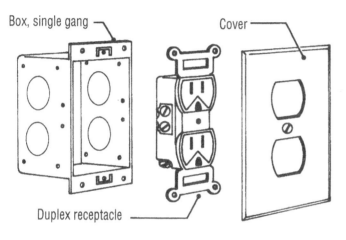

Box, single gang

Cover

Duplex receptacle

Receptacle, Including Box and Cover

Receptacles

Figure 26-48 Installation Time in Labor-Hours for Starters

Description	Labor-Hours	Unit
Starter 3-Pole 2 HP Size 00	2.290	Ea.
5 HP Size 0	3.480	Ea.
10 HP Size 1	5.000	Ea.
25 HP Size 2	7.270	Ea.
50 HP Size 3	8.890	Ea.
100 HP Size 4	13.330	Ea.
200 HP Size 5	17.780	Ea.
400 HP Size 6	20.000	Ea.
Control Station Stop/Start	1.000	Ea.
Stop/Start, Pilot Light	1.290	Ea.
Hand/Off/Automatic	1.290	Ea.
Stop/Start/Reverse	1.510	Ea.

Figure 26-49 Installation Time in Labor-Hours for Starters in
Hazardous Areas

Description	Labor-Hours	Unit
Circuit Breaker NEMA 7 600 Volts 3 Pole		
50 Amps	3.480	Ea.
150 Amps	8.000	Ea.
400 Amps	13.330	Ea.
Control Station Stop/Start	1.330	Ea.
Stop/Start Pilot Light	2.000	Ea.
Magnetic Starter FVNR 480 Volts 5 HP Size 0	5.000	Ea.
25 HP Size 2	8.890	Ea.
Combination 10 HP Size 1	8.000	Ea.
50 HP Size 3	20.000	Ea.
Panelboard 225 Amps M.L.O. 120/208 Volts		
24 Circuit	40.000	Ea.
Main Breaker	53.330	Ea.
Wall Switch, Single Pole 15 Amps	1.510	Ea.
Receptacle 15 Amps	1.510	Ea.

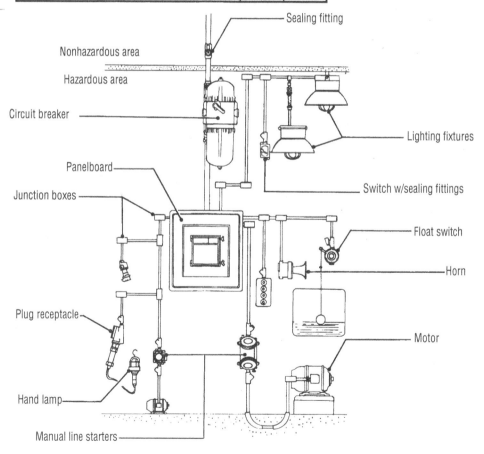

Starters for Class I, Division 2, Power and Lighting Installation

Division 26

Figure 26-50   Installation Time in Labor-Hours for Motor Control Systems

Description	Labor-Hours	Unit
Heavy Duty Fused Disconnect 30 Amps	2.500	Ea.
60 Amps	3.480	Ea.
100 Amps	4.210	Ea.
200 Amps	6.150	Ea.
600 Amps	13.330	Ea.
1200 Amps	20.000	Ea.
Starter 3-pole 2 HP Size 00	2.290	Ea.
5 HP Size 0	3.480	Ea.
10 HP Size 1	5.000	Ea.
25 HP Size 2	7.270	Ea.
50 HP Size 3	8.890	Ea.
100 HP Size 4	13.330	Ea.
200 HP Size 5	17.780	Ea.
400 HP Size 6	20.000	Ea.
Control Station Stop/Start	1.000	Ea.
Stop/Start, Pilot Light	1.290	Ea.
Hand/Off/Automatic	1.290	Ea.
Stop/Start/Reverse	1.510	Ea.

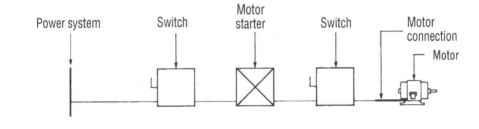

Figure 26-51 Installation Time in Labor-Hours for Motor
Control Centers

Description	Labor-Hours	Unit
Structures 300 Amps 72" High	10.000	Ea.
Structures 300 Amps 72" High		
Back to Back Type	13.300	Ea.
Starters Class 1 Type B Size 1	3.000	Ea.
Size 2	4.000	Ea.
Size 3	8.000	Ea.
Size 4	11.400	Ea.
Size 5	16.000	Ea.
Size 6	20.000	Ea.
Pilot Light Wiring in Starter	.500	Ea.
Push Button Wiring in Starter	.500	Ea.
Auxilliary Contacts in Starter	.500	Ea.

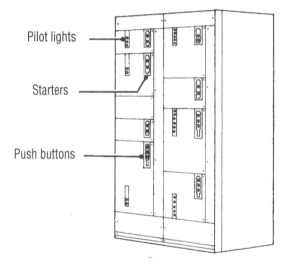

Pilot lights

Starters

Push buttons

Figure 26-52 Installation Time in Labor-Hours for Panelboards

Description	Labor-Hours	Unit
Panelboard 3-Wire 225 Amps Main Lugs		
38 Circuit	22.220	Ea.
4 Wire 225 Amps Main Lugs 42 Circuit	23.530	Ea.
3 Wire 400 Amps Main Circuit Breaker		
42 Circuit	32.000	Ea.
4 Wire 400 Amps Main Circuit Breaker		
42 Circuit	33.330	Ea.
3 Wire 100 Amps Main Lugs 20 Circuit	12.310	Ea.
4 Wire 100 Amps Main Circuit Breaker		
24 Circuit	17.020	Ea.

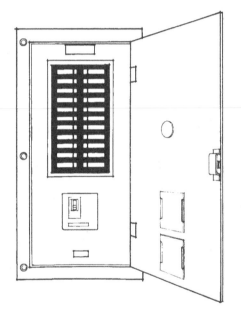

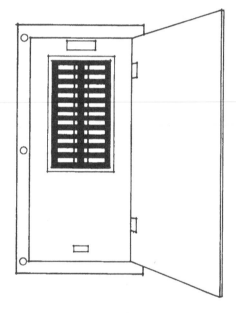

Main Circuit Breaker Panelboard   **Main Lugs Only Panelboard**

Figure 26-53   Installation Time in Labor-Hours for Safety Switches

Description	Labor-Hours	Unit
Safety Switch NEMA 1 600V 3P 200 Amps	6.150	Ea.
NEMA 3R	6.670	Ea.
NEMA 7	10.000	Ea.
NEMA 12	6.670	Ea.

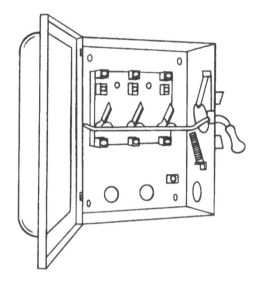

NEMA 1, non-fusible, 600 volt

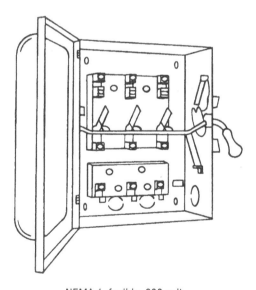

NEMA 1, fusible, 600 volt

Safety Switches

Figure 26-54 Installation Time in Labor-Hours for Metering
Switchboards

Description	Labor-Hours	Unit
Main Breaker Section 800 Amps	17.800	Ea.
1200 Amps	21.000	Ea.
1600 Amps	23.500	Ea.
Main Section 100 Amps 6 Meters	26.700	Ea.
8 Meters	30.800	Ea.
10 Meters	33.300	Ea.

Meter Socket

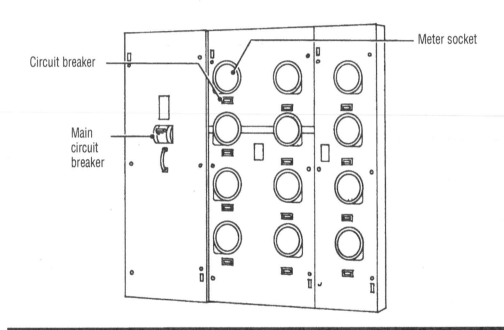

Figure 26-55   Installation Time in Labor-Hours for Multi-Section
Switchboards

Description	Labor-Hours	Unit
Main Switchboard Section 1200 Amps	18.000	Ea.
1600 Amps	19.000	Ea.
2000 Amps	20.000	Ea.
Main Ground Fault Protector 1200-2000 Amps	2.960	Ea.
Bus Way Connections 1200 Amps	6.150	Ea.
1600 Amps	6.670	Ea.
2000 Amps	8.000	Ea.
Auxilliary Pull Section	8.000	Ea.
Distribution Section 1200 Amps	22.220	Ea.
1600 Amps	24.240	Ea.
2000 Amps	25.810	Ea.
Breakers, 1 Pole 60 Amps	1.000	Ea.
2 Pole 60 Amps	1.150	Ea.
3 Pole 60 Amps	1.500	Ea.

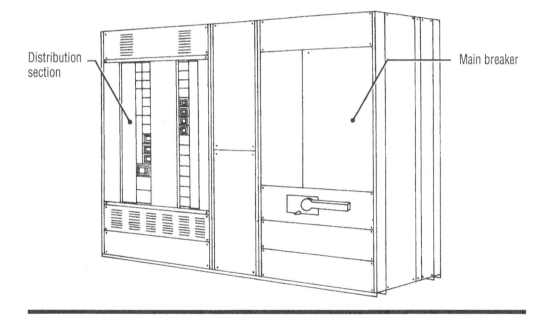

Distribution section

Main breaker

Figure 26-56   Installation Time in Labor-Hours for Metal-Clad
Switchgear

Description	Labor-Hours	Unit
Metal-Clad Structures 1200 Amps	17.500	Ea.
2000 Amps	19.500	Ea.
3000 Amps	24.000	Ea.
Breakers 1200 Amps	10.000	Ea.
2000 Amps	13.000	Ea.
3000 Amps	18.000	Ea.
Instrument Wiring	3.500	Ea.
Bus Bar Connections per Structure 1200 Amps	10.000	Ea.
2000 Amps	15.000	Ea.
3000 Amps	19.000	Ea.
Ground Bus Connection per Structure	6.000	Ea.

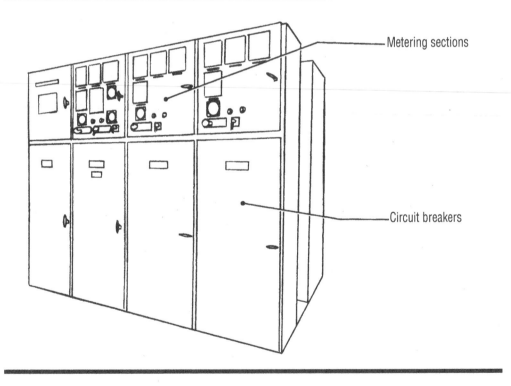

Metering sections

Circuit breakers

Figure 26-57 Installation Time in Labor-Hours for Low Voltage Metal Enclosed Switchgear

Description	Labor-Hours	Unit
Structures 1600 Amps	17.500	Ea.
4000 Amps	26.000	Ea.
Breakers Draw-Out 225 Amps	4.000	Ea.
600 Amps	8.000	Ea.
1600 Amps	11.000	Ea.
4000 Amps	23.000	Ea.
Bus Bar Connections 1600 Amps per Structure	12.000	Ea.
4000 Amps per Structure	21.000	Ea.

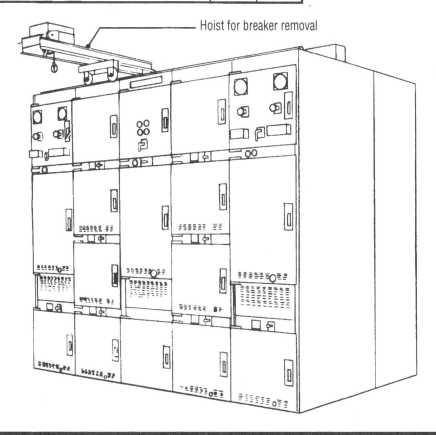

Hoist for breaker removal

Figure 26-58   Installation Time in Labor-Hours for Transformers

Description	Labor-Hours	Unit
Oil Filled 5 KV Primary 277/480 Volt Secondary		
3 Phase 150 KVA	30.770	Ea.
1000 KVA	76.920	Ea.
3750 KVA	125.000	Ea.
Liquid Filled 5 KV Primary 277/480 Volt		
Secondary 3 Phase 225 KVA	36.360	Ea.
1000 KVA	76.920	Ea.
2500 KVA	105.000	Ea.
Dry 480 Volt Primary 120/208 Volt Secondary		
3 Phase 15 KVA	14.550	Ea.
112 KVA	23.530	Ea.
500 KVA	44.440	Ea.

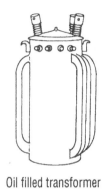

Oil filled transformer

Dry type transformer, 3 Phase

Dry type transformer, single phase

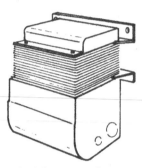

Buck-boost transformer

Transformer Types

Figure 26-59 Installation Time in Labor-Hours for Secondary
Unit Substations

Description	Labor-Hours	Unit
Load Interrupter Switch, 300 KVA and below	60.000	Ea.
400 KVA and above	63.000	Ea.
Transformer Section 112 KVA	37.000	Ea.
300 KVA	59.000	Ea.
500 KVA	67.000	Ea.
750 KVA	83.000	Ea.
Low Voltage Breakers		
2 Pole, 15 to 60 Amps, Type FA	1.430	Ea.
3 Pole, 15 to 60 Amps, Type FA	1.510	Ea.
2 Pole, 125 to 225 Amps, Type KA	2.350	Ea.
3 Pole, 125 to 225 Amps, Type KA	2.500	Ea.
2 Pole, 700 and 800 Amps, Type MA	5.330	Ea.
3 Pole, 700 and 800 Amps, Type MA	6.150	Ea.

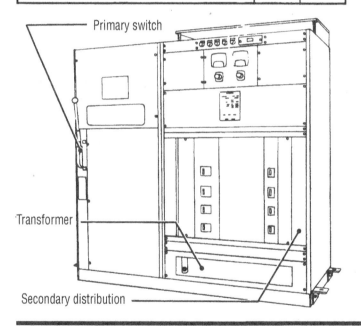

Figure 26-60 Installation Time in Labor-Hours for Emergency/Standby
Power Systems

Description	Labor-Hours	Unit
Battery Light Unit		
6 Volt Lead Battery and 2 Lights	2.000	Ea.
12 Volt Nickel Cadmium and 2 Lights	2.000	Ea.
Remote Mount Sealed Beam Light, 25W, 6 Volts	.300	Ea.
Self Contained Fluorescent Lamp Pack	.800	Ea.
Engine Generator 10kW Gas/Gasoline		
277/480 Volts Complete	34.000	System
175kW Complete	96.000	System
Engine Generator 500kW Diesel		
277/480 Volts Complete	133.000	System
1000kW Complete	180.000	System

Figure 26-61 Installation Time in Labor-Hours for Incandescent Lighting

Wire or cable termination of light fixtures are included in the installation of light fixtures. *Means Electrical Cost Data* provides a unique reference number with detailed labor task items.

Description	Hours	Unit
Ceiling, Recess Mounted Alzak Reflector		
150W	1.000	Ea.
300W	1.190	Ea.
Surface Mounted Metal Cylinder		
150W	.800	Ea.
300W	1.000	Ea.
Opal Glass Drum 10" 2-60W	1.000	Ea.
Pendant Mounted Globe 150W	1.000	Ea.
Vaportight 200W	1.290	Ea.
Chandelier 24" Diameter x 42" High		
6 Candle	1.330	Ea.
Track Light Spotlight 75W PAR Halogen	.500	Ea.
Wall Washer Quartz 250W	.500	Ea.
Exterior Wall Mounted Quartz 500W	1.510	Ea.
1500W	1.900	Ea.
Ceiling, Surface Mounted Vaportight		
100W	2.650	Ea.
150W	2.950	Ea.
175W	2.950	Ea.
250W	2.950	Ea.
400W	3.350	Ea.
1000W	4.450	Ea.

Track lighting spotlight

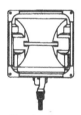

Exterior fixture,
wall mounted, quartz

Round ceiling fixture
with concentric louver

Round ceiling fixture
with reflector, no lens

Round ceiling fixture,
recessed, with Alzak reflector

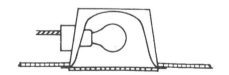

Square ceiling fixture, recessed, with glass lens, metal trim

Fixtures

Figure 26-62 Installation Time in Labor-Hours for Fluorescent
Lighting

Description	Labor-Hours	Unit
Troffer with Acrylic Lens 4-32W RS 2' x 4'	1.700	Ea.
2-32W URS 2' x 2'	1.400	Ea.
Surface Mounted Acrylic Wrap-around Lens		
4-40W RS 16" x 48"	1.500	Ea.
Industrial Pendant Mounted		
4' Long, 2-32W RS	1.400	Ea.
8' Long, 2-75W SL	1.820	Ea.
2-110W HO	2.000	Ea.
2-215W VHO	2.110	Ea.
Surface Mounted Strip, 4' Long, 1-40W RS	.940	Ea.
8' Long, 1-75W SL	1.190	Ea.

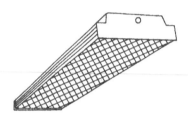

**Surface or Pendant Mounted Fixture with
Wrap Around Acrylic Lens, 4 Tube**

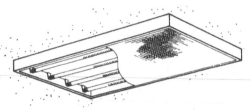

**Surface Mounted Fixture with
Acrylic Lens, 4 Tube**

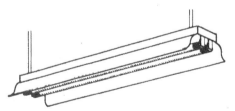

Pendant Mounted Industrial Fixture

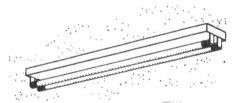

**Surface Mounted Strip Fixture,
2 Tube**

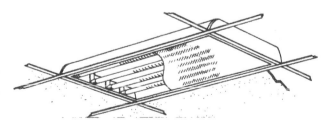

Troffer Mounted Fixture with Acrylic Lens, 4 Tube

Figure 26-63   Installation Time in Labor-Hours for High
Intensity Lighting

Description	Labor-Hours	Unit
Ceiling Recessed Mounted Prismatic Lens		
Integral Ballast 2' x 2' HID 150W	2.500	Ea.
250W	2.500	Ea.
400W	2.760	Ea.
Surface Mounted 250W	2.960	Ea.
400W	3.330	Ea.
High Bay Aluminum Reflector 400W	3.480	Ea.
1000W	4.000	Ea.

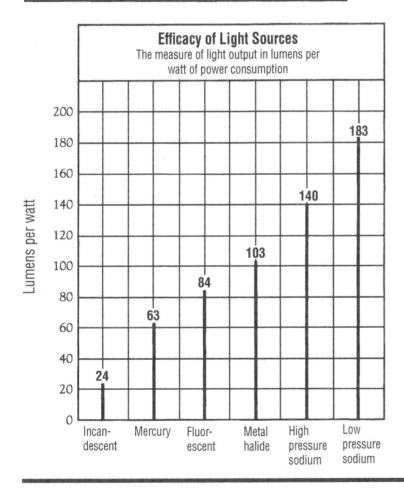

Figure 26-64   Installation Time in Labor-Hours for Hazardous Area Lighting

Description	Labor-Hours	Unit
Fixture, Pendant Mounted, Fluorescent 4' Long		
2-40W RS	3.480	Ea.
4-40W RS	4.710	Ea.
Incandescent 200W	2.290	Ea.
Ceiling Mounted, Incandescent 200W	2.000	Ea.
Ceiling, HID, Surface Mounted 100W	2.670	Ea.
150W	2.960	Ea.
250W	2.960	Ea.
400W	3.330	Ea.
Pendant Mounted 100W	2.960	Ea.
150W	3.330	Ea.
250W	3.330	Ea.
400W	3.810	Ea.

Explosion-proof Fluorescent Fixture, Pendant Mounted, 3 Tube

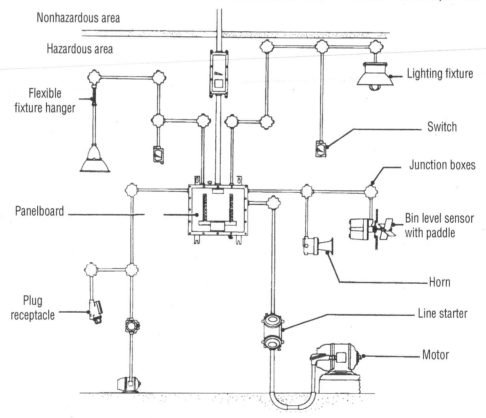

Class II Lighting Installation

Figure 26-65 Installation Time in Labor-Hours for Parking Area Lighting

Description	Labor-Hours	Unit
Luminaire HID 100 Watt	2.960	Ea.
150W	2.960	Ea.
175W	2.960	Ea.
250W	3.330	Ea.
400W	3.640	Ea.
1000W	4.000	Ea.
Bracket Arm 1 Arm	1.000	Ea.
2 Arm	1.000	Ea.
3 Arm	1.510	Ea.
4 Arm	1.510	Ea.
Aluminum Pole 20' High	6.900	Ea.
30' High	7.690	Ea.
40' High	10.00	Ea.
Steel Pole 20' High	7.690	Ea.
30' High	8.700	Ea.
40' High	11.77	Ea.
Fiberglass Pole 20' High	5.000	Ea.
30' High	5.560	Ea.
40' High	7.100	Ea.
Transformer Base	2.670	Ea.

Large Luminaire Light Fixture, 1000 Watt

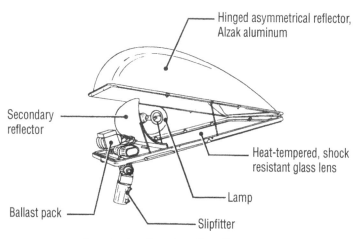

Hinged asymmetrical reflector, Alzak aluminum

Secondary reflector

Heat-tempered, shock resistant glass lens

Lamp

Ballast pack

Slipfitter

Luminaire Construction Features

Figure 26-66 Installation Time in Labor-Hours for Nurses' Call Systems

Nurse call systems are part of Division 27 – Communications.

Description	Labor-Hours	Unit
Call station, single bedside	1	Ea.
Double bedside	2	Ea.
Ceiling speaker station	1	Ea.
Emergency call station	1	Ea.
Pillow speaker	1	Ea.
Duty station	2	Ea.
Standard call button	1	Ea.
Lights, corridor, dome or zone indicator	1	Ea.
Master control station for 20 stations	24.620	Total

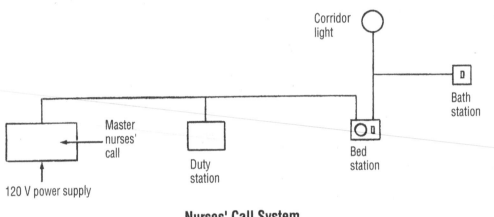

Nurses' Call System

Figure 26-67   Installation Time in Labor-Hours for Lightning Protection

Lightning protection for the rooftop of buildings is achieved by a series of lightning rods or air terminals joined together by either copper or aluminum cable. The cable size is determined by the height of the building. The lightning cable system is connected through a download to a ground rod that is a minimum of 2' below grade and 1-1/2' to 3' out from the foundation wall.

All equipment used in the fundamental grounding system within and on the structure should be UL approved and completely installed to proper electrical standards.

Description	Labor-Hours	Unit
Air Terminals, copper		
3/8" diameter x 10" (to 75' high)	1	Ea.
1/2" diameter x 12" (over 75' high)	1	Ea.
Aluminum, 1/2" diameter x 12" (to 75' high)	1	Ea.
5/8" diameter x 12" (over 75' high)	1	Ea.
Cable, copper		
220 lb. per thousand ft. (to 75' high)	.025	L.F.
375 lb. per thousand ft. (over 75' high)	.034	L.F.
Aluminum		
101 lb. per thousand ft. (to 75' high)	.028	L.F.
199 lb. per thousand ft. (over 75' high)	.033	L.F.
Arrestor, 175 volt AC to ground	1	Ea.
650 volt AC to ground	1.190	Ea.

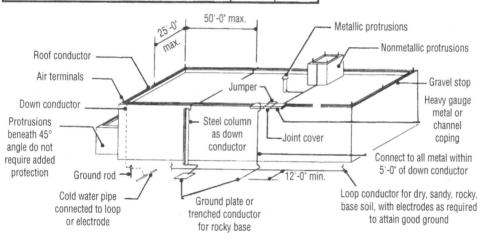

Figure 26-68 Installation Time in Labor-Hours for Clock Systems

Description	Labor-Hours	Unit
Clock Systems		
Time system components, master controller	24.240	Ea.
Program bell	1	Ea.
Combination clock & speaker	2.500	Ea.
Frequency generator	4	Ea.
Job time automatic stamp recorder, minimum	2	Ea.
Maximum	2	Ea.
Master time clock system, clocks & bells		
20 room	160	Ea.
50 room	400	Ea.
Time clock, 100 cards in & out, 1 color	2.500	Ea.
2 colors	2.500	Ea.
With 3 circuit program device, minimum	4	Ea.
Maximum	4	Ea.
Metal rack for 25 cards	1.140	Ea.
Watchman's tour station	1	Ea.
Annunciator with zone indication	8	Ea.

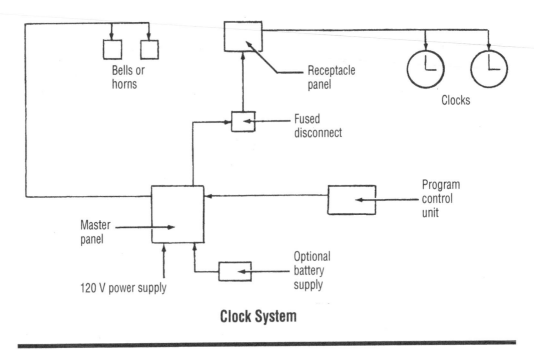

Clock System

Figure 26-69 Installation Time in Labor-Hours for BX Cable

Note: BX cable, technically known as armored cable (Type AC), has a flexible aluminum or steel sheath over the conductors.

Description	Labor-Hours	Unit
Armored cable		
600 bolt (BX), #14, 2 conductor, solid	3.333	CLF
3 conductor, solid	3.636	
4 conductor, solid	4	
#12, 2 conductor, solid	3.478	
3 conductor, solid	4	
4 conductor, solid	4.444	
#12, 19 conductor, stranded	7.273	
#10, 2 conductor, solid	4	
3 conductor, solid	5	
4 conductor, solid	5.714	
#8, 3 conductor, solid	6.154	
4 conductor, stranded	7.273	
#6, 2 conductor, stranded	6.154	▼

Figure 26-70 Installation Time in Labor-Hours for Romex (Non-Metallic Sheathed) Cable

Note: Romex is a brand name for a type of plastic insulated wire, sometimes called non-metallic sheathed.

Description	Labor-Hours	Unit
Non-metallic sheathed cable 600 volt		
Copper with ground wire (Romex)		
#14, 2 conductor	2.963	CLF
3 conductor	3.333	
4 conductor	3.636	
#12, 2 conductor	3.200	
3 conductor	3.636	
4 conductor	4	
#10, 2 conductor	3.636	
3 conductor	4.444	
4 conductor	5	
#8, 3 conductor	5.333	
4 conductor	5.714	
#6, 3 conductor	5.714	
#4, 3 conductor	6.667	
#2, 3 conductor	7.273	▼

Notes

Division Thirty-One
Earthwork

Introduction

Division 31 covers earthwork related items. Under the previous MasterFormat 95, these items were located in Division 2.

Division 31 contains several sections for earthwork, beginning with general site work materials. This is followed by clearing and grubbing, tree and shrub removal, and soil stripping and stockpiling. These are usually the first items of work that are accomplished.

Other major earthwork items include grading, excavation, dewatering, backfill, compaction, hauling, erosion control, soil stabilization, shoring, sheet piling, driven piles, drilled piles, and some earthwork specialty items.

Excavation and fill includes several subsections. The first, trench excavation and backfill, is specific to utilities and footings, and covers trench excavation by plow, trenching machine, and excavator/backhoes. The next category is minor structure excavation for small structures. Finally, there is excavation for larger buildings and large volume excavation projects.

In addition, there are some special categories including rock removal, which addresses drilling, blasting, and ripping. Also included are excavation methods by specific equipment including dozer, dragline, and scrapers. Backfilling has it own subsets including large backfill projects, structural backfill, combined loading and spreading, and utility bedding. Other support items within the earthwork division include dewatering, hauling, grading, and compaction.

The next sections are erosion and sedimentation control, soil treatment, and soil stabilization. Soil stabilization has several major subsections including soil mixing, geosynthetic stabilization, soil nailing, gabions, and riprap.

Lack of erosion and sediment control during and after construction can cause significant harm to the environment. Silt can clog up rivers and steams, disrupting aquatic wildlife. It can also harm local air quality by allowing exposed earth to generate airborne dust. Measures to control erosion and sediment include mulching, seeding, earth dikes, sediment traps, sediment basins hay bales, and silt fences. Erosion and sediment control plans are often required by federal and local regulations. They are also a prerequisite for obtaining LEED® NC (new construction) certification.

Shoring is another category worth noting. It includes timber shoring and sheet piling. Other categories of work are cofferdams, slurry walls, driven piles, and bored piles. The driven piles have several types including prestressed concrete, steel, wood, plastic, and micropiles. Some are composite piles that include plastic with steel reinforcement, and concrete filled steel piles.

There are some special categories in the heavy data sections of Division 31, including backfill and compaction for airport subgrades, tunnel excavation, tunnel grouting, and cast-in-place tunnel linings.

Division 31

Estimating Data

The following tables present estimating guidelines for items found in Division 31 – Earthwork. Please note that these guidelines are intended as indicators of what may generally be expected, but that each project must be evaluated individually.

Table of Contents

Division 31

Estimating Tips

Common Work Results

Estimating the actual cost of performing earthwork requires careful consideration of the variables involved. These include type of soil, whether water will be encountered, dewatering, whether banks need bracing, disposal of excavated earth, and length of haul to fill or spoil sites, etc. If the project has large quantities of cut or fill, consider raising or lowering the site to reduce costs, while paying close attention to the effect this has on site drainage and utilities.

It is very important to consider what time of year the project is scheduled for completion. Bad weather can cause large cost overruns from dewatering, site repair, and lost productivity.

Checklist ✓

For an estimate to be reliable, all items must be accounted for. A complete estimate can also limit contingencies. The following checklist can be used to help ensure that all items are included.

☐ **Excavation**
- ☐ Mass
- ☐ Trench

☐ **Borrow pit**
- ☐ Groundwater expected
  - ☐ None
  - ☐ Pump
  - ☐ Wellpoint system
  - ☐ Trench away from excavation
- ☐ Sheeting/shoring
- ☐ Disposal
  - ☐ Off site
  - ☐ On site
  - ☐ Use for fill

☐ **Backfill**
- ☐ Mass
- ☐ Trench
- ☐ Berms
- ☐ Topsoil
- ☐ Gravel
- ☐ Stone
- ☐ Source

- ☐ Borrow pit
- ☐ Commercial pit
- ☐ On site

☐ **Compaction**
- ☐ Hand
- ☐ Mechanical

☐ **Water Control**
- ☐ Dams
- ☐ Ditching
- ☐ Pumping
- ☐ Wellpoint system
- ☐ Sheet piling

☐ **Pestilence Control**

☐ **Piles/Piling**

☐ **Caissons**

☐ **Pressure-Injected Footings**

☐ **Special Footings**

Figure 31-1 Soil Bearing Capacity in Kips per SF

This table can be used to determine foundation footing size. Once the load on a footing has been determined, the footing size is determined by dividing the load by the allowable bearing capacity in this table.

Bearing Material	Typical Allowable Bearing Capacity
Hard sound rock	120 KSF
Medium hard rock	80
Hardpan overlaying rock	24
Compact gravel and boulder–gravel; very compact sandy gravel	20
Soft rock	16
Loose gravel; sandy gravel; compact sand; very compact sand–inorganic silt	12
Hard dry consolidated clay	10
Loose coarse to medium sand; medium compact fine sand	8
Compact sand–clay	6
Loose fine sand; medium compact sand–inorganic silts	4
Firm or stiff clay	3
Loose saturated sand–clay; medium soft clay	2

One kip = 1,000 lbs.

Figure 31-2 Weights and Characteristics of Materials

Approximate Material Characteristics*				
Material	**Loose (Lbs./C.Y.)**	**Bank (Lbs./C.Y.)**	**Swell (%)**	**Load Factor**
Clay, dry	2,100	2,650	26	0.79
Clay, wet	2,700	3,575	32	0.76
Clay and gravel, dry	2,400	2,800	17	0.85
Clay and gravel, wet	2,600	3,100	17	0.85
Earth, dry	2,215	2,850	29	0.78
Earth, moist	2,410	3,080	28	0.78
Earth, wet	2,750	3,380	23	0.81
Gravel, dry	2,780	3,140	13	0.88
Gravel, wet	3,090	3,620	17	0.85
Sand, dry	2,900	3,250	12	0.89
Sand, wet	3,200	3,600	13	0.89
Sand and gravel, dry	2,900	3,250	12	0.89
Sand and gravel, wet	3,400	3,750	10	0.91

*Exact values will vary with grain size, moisture content, compaction, etc. Test to determine exact values for specific soils.

Typical Soil Volume Conversion Factors				
Soil Type	**Initial Soil Condition**	**Bank**	**Converted to:**	
			Loose	**Compacted**
Clay	Bank	1.00	1.27	0.90
	Loose	0.79	1.00	0.71
	Compacted	1.11	1.41	1.00
Common earth	Bank	1.00	1.25	0.90
	Loose	0.80	1.00	0.72
	Compacted	1.11	1.39	1.00
Rock (blasted)	Bank	1.00	1.50	1.30
	Loose	0.67	1.00	0.87
	Compacted	0.77	1.15	1.00
Sand	Bank	1.00	1.12	0.95
	Loose	0.89	1.00	0.85
	Compacted	1.05	1.18	1.00

$$\text{Swell (\%)} = \left(\frac{\text{Wt/bank C.Y.}}{\text{Wt./loose C.Y.}} - 1 \right) \times 100$$

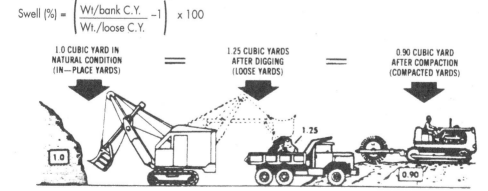

1.0 CUBIC YARD IN NATURAL CONDITION (IN—PLACE YARDS) == 1.25 CUBIC YARDS AFTER DIGGING (LOOSE YARDS) == 0.90 CUBIC YARD AFTER COMPACTION (COMPACTED YARDS)

Division 31

Figure 31-3 Material Weights

Material	Wt. in Bank per C.Y.	Percent of Swell	Swell Factor	Loose Wt. per C.Y.
Ashes, Hard Coal	700-1000 lbs.	8%	.93	650-930 lbs.
Ashes, Soft Coal with Clinkers	1000-1515 lbs.	8%	.93	930-1410 lbs.
Ashes, Soft Coal, Ordinary	1080-1215 lbs.	8%	.93	1000-1130 lbs.
Bauxite	2700-4325 lbs.	33%	.75	2020-3240 lbs.
Brick				2700 lbs.
Cement, Portland	94 lbs. per bag			
Cement, Portland	2970 lbs. (packed)	20%	.83	2450 lbs.
Coke, Lump, Loose				620-865 lbs.
Coke, Solvay, Egg, Chestnut or Pea				840 lbs.
Coke, Gas, Egg, Chestnut or Pea				785 lbs.
Coke, Gas Furnace				730 lbs.
Concrete	3240-4185 lbs.	40%	.72	2330-3000 lbs.
Concrete Mix, Wet				3500-3750 lbs.
Copper Ore	3800 lbs.	35%	.74	2800 lbs.
Gasoline, 56° Baumé	6.3 lbs. per gallon			
Granite	4500 lbs.	50 to 80%	.67 to .56	1520-3000 lbs.
Iron Ore, Hematite	6500-8700 lbs.		.45	3900 lbs.
Iron Ore, Limonite	6400 lbs.			
Iron Ore, Magnetite	8500 lbs			
Kaolin	2800 lbs.	30%	.77	2160 lbs.
Lead Ore, Galina	12,550 lbs.			
Lime				1400 lbs.
Limestone, Blasted	4200 lbs.	67 to 75%	.60 to .57	2400-2520 lbs.
Limestone, Loose, Crushed				2600-2700 lbs.
Limestone, Marble	4600 lbs.	67 to 75%	.60 to .57	2620-2760 lbs.
Mud, Dry (Close)	2160-2970 lbs.	20%	.83	1790-2460 lbs.
Mud, Wet (Moderately packed)	2970-3510 lbs.	20%	.83	2470-2910 lbs.
Oil, Crude	6.42 lbs. per gallon			
Phosphate Rock	5400 lbs.			
Sand, Dry	3250 lbs.	12%	.89	2900 lbs.
Sand, Wet	3600 lbs.	13%	.89	3200 lbs.
Sandstone	4140 lbs.	40 to 60%	.72 to .63	2610-2980 lbs.
Shale, Riprap	2800 lbs.	33%	.75	2100 lbs.
Slag, Sand	1670 lbs.	12%	.89	1485 lbs.
Slag, Solid	4320-4860 lbs.	33%	.75	2640-3240 lbs.
Slag, Crushed				1900 lbs.
Slag, Furnace, Granulated	1600 lbs.	12%	.89	1430 lbs.
Slate	4590-4860 lbs.	30%	.77	3530-3740 lbs.
Trap Rock	5075 lbs.	50%	.67	3400 lbs.

(continued on next page)

Figure 31-3 Material Weights *(continued from previous page)*

Wood and Lumber Weights per Cord		
Material	Lb./C.F.	Lb. per Cord
Beechwood		3250
Cedar, red	24.2	
Cedar, white	22.2	
Chestnut		2350
Douglas Fir	32.7	2350
Elm.		
Fir, commercial white	27	
Hemlock	28 - 29	2200
Hickory		4500
Maple		
Oak, red or white	47.3	
Pine: Norway or White	25	2000
Pine: Southern Yellow	37.3	
Poplar	29.4	2350
Redwood	26	
Walnut, black	38	

Figure 31-4 Quantities for Wellpoint Systems

Description for 200' System with 8" Header		Quantities
Equipment & Material	Wellpoints 25' long, 2" diameter @ 5' O.C.	40 Ea.
	Header pipe, 8" diameter	200 L.F.
	Discharge pipe, 8" diameter	100 L.F.
	8" Valves	3 Ea.
	Combination Jetting and Wellpoint pump (standby)	1 Ea.
	Wellpoint pump, 8" diameter	1 Ea.
	Transportation to and from site	1 day
	Fuel 30 days x 60 gal./day	1800 gal.
	Lubricants for 30 days x 16 lbs./day	480 Lbs.
	Sand for points	40 C.Y.
Labor	Technician to supervise installation	1 week
	Labor for installation and removal of system	300 labor-hours
	4 Operators straight time 40 hrs./wk. for 4.33 wks.	693 hrs.
	4 Operators overtime 2 hrs./wk. for 4.33 wks.	35 hrs.

Figure 31-5 Area Clearing Equipment Selection Table by Size of Area, Vegetation to be Cleared, and Method

This table suggests equipment requirements for light, intermediate, and heavy clearing. The productivity of the equipment will depend on the the density and type of growth.

Light Clearing, Vegetation up to 2 in. (5 cm) Diameter				
	Uprooting Vegetation	Cutting Vegetation At or Above Ground Level	Knocking the Vegetation to the Ground	Incorporation of Vegetation into the Soil
Small areas 10 acres (4.0 hectares)	Bulldozer blade, axes, grub hoes and mattocks	Axes, machetes, brush hooks, grub hoes and mattocks, wheel-mounted circular saws	Bulldozer blade	Moldboard plows, disc plows, disc harrows
Medium areas 100 acres (40 hectares)	Bulldozer blade,	Heavy-duty sickle mowers (up to 1-1/2" (3.7 cm) diameter), tractor-mounted circular saws; suspended rotary mowers	Bulldozer blade, rotary mowers, flail-type rotary cutters, rolling brush cutters	Moldboard plows, disc plows, disc harrows
Large areas 1,000 acres (400 hectares)	Bulldozer blade, root rake, grubber, root plow, anchor chain drawn between two crawler tractors, rails		Rolling brush cutter, flail-type cutter, anchor chain drawn between two crawler tractors, rails	Undercutter with disc, moldboard plows, disk plows, disk harrows

(continued on next page)

Figure 31-5 Area Clearing Equipment Selection Table by Size of Area, Vegetation to be Cleared, and Method

(continued from previous page)

Intermediate Clearing, Vegetation 2 to 8 in. (5 - 20 cm) Diameter				
	Uprooting Vegetation	Cutting Vegetation at or Above Ground Level	Leveling Vegetation	Tilling Vegetation into Soil
Small areas 10 acres (4.0 hectares)	Bulldozer blade	Axes, crosscut saws, power chain saws, wheel-mounted circular saws	Bulldozer blade	Heavy-duty disc plow; disc harrow
Medium areas 100 acres (40 hectares)	Bulldozer blade	Power chain saws, tractor-mounted circular saws	Bulldozer blade, rolling brush cutter (up to 5 in. (12 cm) diameter), rotary mower (up to 4 in. (10 cm) diameter)	Heavy-duty disc plow, disc harrow
Large areas 1000 acres (400 hectares)	Shearing blade, angling (tilted), bulldozer blade, rakes, anchor chain drawn between two crawler tractors, root plow	Shearing blade (angling or V-type)	Bulldozer blade, flail-type rotary cutter, anchor chain	Bulldozer blade with heavy-duty harrow

Heavy Clearing, Vegetation 8 in. (20 cm) Diameter or Larger			
	Uprooting Vegetation	Cutting Vegetation at or Above Ground Level	Knocking the Vegetation to the Ground
Small areas 10 acres (4.0 hectares)	Bulldozer blade	Axes, crosscut saws, power chain saws	Bulldozer blade
Medium areas 100 acres (40 hectares)	Shearing blade, angling (tilted) knockdown beam, rakes, tree stumper	Shearing blade (angling or V-type), tree shear (up to 26 in. (65 cm) softwood; 14 in. (35 cm) hardwood), shearing blade - power saw combination	Bulldozer blade
Large areas 1,000 acres (400 hectares)	Shearing blade, angling (tilted), knockdown beam, rakes, tree stumper, anchor chain with ball drawn between two crawler tractors	Shearing, blade (angling or V-type) shearing blade-power saw combination	Anchor chain with ball drawn between two crawler tractors

Figure 31-6 Excavating

The selection of equipment used for structural excavation and bulk excavation or for grading is determined by the following factors:

1. Quantity of material
2. Type of material
3. Depth or height of cut
4. Length of haul
5. Condition of haul road
6. Accessibility of site
7. Moisture content and dewatering requirements
8. Availability of excavating and hauling equipment

Some additional costs must be allowed for hand trimming the sides and bottom of concrete pours and other excavation below the general excavation.

When planning excavation and fill, the following should also be considered.

1. Swell factor
2. Compaction factor
3. Moisture content
4. Density requirements

A typical example for scheduling and estimating the cost of excavation of a 15' deep basement on a dry site when the material must be hauled off the site, is outlined below.

Assumptions:

1. Swell factor, 18%
2. No mobilization or demobilization
3. Allowance included for idle time and moving on job
4. No dewatering, sheeting, or bracing
5. No truck spotter or hand trimming

Number of B.C.Y. per truck

$$= 1.5 \text{ C.Y. bucket} \times 8 \text{ passes} = 12 \text{ loose C.Y.}$$

$$= 12 \times \frac{100}{118} = 10.2 \text{ B.C.Y. per truck}$$

Truck haul cycle:

Load truck 8 passes	= 4 minutes
Haul distance 1 mile	= 9 minutes
Dump time	= 2 minutes
Return 1 mile	= 7 minutes
Spot under machine	= 1 minute
	23 minute cycle

Fleet Haul Production per Day in B.C.Y.

4 trucks x $\dfrac{50 \text{ min. hr.}}{23 \text{ min. haul cycle}}$ x 8 hrs. x 10.2 B.C.Y. = 4 x 2.2 x 8 x 10.2 = 718 B.C.Y./day

Note: B.C.Y. = Bank Measure Cubic Yards

Figure 31-7 Excavation Equipment

The table below lists theoretical hourly production in CY/hr. bank measure for some typical excavation equipment. Figures assume 50-minute hours, 83% job efficiency, 100% operator efficiency, 90° swing, and properly sized hauling units, which must be modified for adverse digging and loading conditions. Actual production costs average about 50% of the theoretical values listing here.

Equipment	Soil Type	B.C.Y. Wt.	% Swell	1 C.Y.	1-1/2 C.Y.	2 C.Y.	2-1/2 C.Y.	3 C.Y.	3-1/2 C.Y.	4 C.Y.
Hydraulic Excavator "Backhoe" 15' deep cut	Moist loam, sandy clay	3400 lbs.	40%	85	125	175	220	275	330	380
	Sand and gravel	3100	18	80	120	160	205	260	310	365
	Common earth	2800	30	70	105	150	190	240	280	330
	Clay, hard, dense	3000	33	65	100	130	170	210	255	300
Power Shovel Optimum Cut (Ft.)	Moist loam, sandy clay	3400	40	170 (6.0)	245 (7.0)	295 (7.8)	335 (8.4)	385 (8.8)	435 (9.1)	475 (9.4)
	Sand and gravel	3100	18	165 (6.0)	225 (7.0)	275 (7.8)	325 (8.4)	375 (8.8)	420 (9.1)	460 (9.4)
	Common earth	2800	30	145 (7.8)	200 (9.2)	250 (10.2)	295 (11.2)	335 (12.1)	375 (13.0)	425 (13.8)
	Clay, hard, dense	3000	33	120 (9.0)	175 (10.7)	220 (12.2)	255 (13.3)	300 (14.2)	335 (15.1)	375 (16.0)
Drag line Optimum Cut (Ft.)	Moist loam, sandy clay	3400	40	130 (6.6)	180 (7.4)	220 (8.0)	250 (8.5)	290 (9.0)	325 (9.5)	385 (10.0)
	Sand and gravel	3100	18	130 (6.6)	175 (7.4)	210 (8.0)	245 (8.5)	280 (9.0)	315 (9.5)	375 (10.0)
	Common earth	2800	30	110 (8.0)	160 (9.0)	190 (9.9)	220 (10.5)	250 (11.0)	280 (11.5)	310 (12.0)
	Clay, hard, dense	3000	33	90 (9.3)	130 (10.7)	160 (11.8)	190 (12.3)	225 (12.8)	250 (13.3)	280 (12.0)

Equipment	Soil Type	B.C.Y. Wt.	% Swell	Wheel Loaders				Track Loaders		
				3 C.Y.	4 C.Y.	6 C.Y.	8 C.Y.	2-1/4 C.Y.	3 C.Y.	4 C.Y.
Loading Tractors	Moist loam, sandy clay	3400	40	260	340	510	690	135	180	250
	Sand and gravel	3100	18	245	320	480	650	130	170	235
	Common earth	2800	30	230	300	460	620	120	155	220
	Clay, hard, dense	3000	33	200	270	415	560	110	145	200
	Rock, well blasted	4000	50	180	245	380	520	100	130	180

Figure 31-8 Material Excavation

Volume of Excavated Material*	
Depth in Inches and Feet	Cubic Yards per Square Surface Foot
2"	.006
4"	.012
6"	.019
8"	.025
10"	.031
1'	.037
2'	.074
3'	.111
4'	.148
5'	.185
6'	.222
7'	.259
8'	.296
9'	.333
10'	.370

*No swellage factor applied

Example: Excavation required: 20' x 30' = 600 x .148 = 88.8 CY.

Figure 31-9 Hand Excavation

Task	C.Y. per Hour	Labor-Hours per C.Y.
Excavate sandy loam	1 – 2	0.5 – 1
Shovel loose earth into truck, dry	1/2 – 1	1.0 – 2.0
Shovel loose earth into truck, wet	1/4 – 1/2	2.0 – 4.0
Loosen soil with pick	1/4 – 1/2	2.0 – 4.0
Shovel from trench to 6'-0" deep, dry	1/2 – 1	1.0 – 2.0
Shovel from trench to 6'-0" deep, dry	1/4 – 1/2	2.0 – 4.0
Shovel from pits to 6'-0" deep, dry	1/2 – 1	1.0 – 2.0
Shovel from pits to 6'-0" deep, wet	1/4 – 1/2	2.0 – 4.0
Backfill, dry soil	1-1/2 – 2-1/2	0.4 – 0.7
Backfill, wet soil	3/4 – 1-1/2	0.7 – 1.3
Spread loose earth, dry	4 – 7	0.15 – 0.25
Spread loose earth, wet	2 – 3-1/2	0.3 – 0.5

Note: The lower values in the CY per hour column and the higher values in the Labor-Hours per CY column relate to heavy soils such as clay.

Figure 31-10 Compacting Backfill

Compaction of fill in embankments, around structures, in trenches, and under slabs is important to control settlement. Factors affecting compaction are:

1. Soil gradation
2. Moisture content
3. Equipment used
4. Depth of fill per lift
5. Density required

> Example: Compact granular fill around a building foundation using a 21" wide x 24" vibratory plate in 8" lifts. Operator moves at 50 FPM working a 50 minute hour to develop 95% Modified Proctor Density with 4 passes

Production Rate:

$$\frac{1.75' \text{ plate width x 50 FPM x 50 min./hr. x .67' lift}}{27 \text{ C.F. per C.Y.}} = 108.5 \text{ C.Y./hr.}$$

Production Rate for 4 Passes:

$$\frac{108.5 \text{ C.Y.}}{4 \text{ Passes}} = 27 \text{ C.Y./hr. x 8 hrs.} = 216 \text{ C.Y./day}$$

Figure 31-11 Trenching Machine Data

Type of Trenching Machine	Trench Depth (in Feet)	Trench Width (in Inches)	Digging Speed (Feet/Hour)
Wheel Type	2–4	16, 18, 20	150/600
		22, 24, 26	90/300
		28, 30	60/180
	4–6	16, 18, 20	40/120
		22, 24, 26	25/90
		28, 30	15/40
Ladder Type	4–6	16, 20, 24	100/300
		22, 26, 30	75/200
		28, 32, 36	40/125
	6–8	16, 20, 24	40/125
		22, 26, 30	30/60
		28, 32, 36	25/50
	8–12	18, 24, 30	30/75
		30, 33, 36	15/40
Chair Boom Type	2	4, 6	100/250
	3	4, 8, 12	50/200
	5	4, 8, 12	50/100
	6	6, 12, 18	30/90
	8	6, 12, 24	30/75

Figure 31-12 Trench Bottom Widths for Various Outside Diameters of Buried Pipe

Proper bedding of buried pipe is important. When figuring trench excavation and bedding material, use this table to estimate quantities. The side slopes will depend on type of soil and whether or not sheeting is used.

Outside Diameter in Inches	Trench Bottom Width in Feet
24	4.1
30	4.9
36	5.6
42	6.3
48	7.0
60	8.5
72	10.0
84	11.4

Figure 31-13 OSHA Soil Type Categories

OSHA Soil Type Categories

STABLE ROCK is natural solid mineral matter that can be excavated with vertical sides and remain intact while exposed. It is usually identified by a rock name such as granite or sandstone. Determining whether a deposit is of this type may be difficult unless it is known whether cracks exist and whether or not the cracks run into or away from the excavation.

TYPE A SOILS are cohesive soils with an unconfined compressive strength of 1.5 tons per square foot (tsf) (144 kPa) or greater. Examples of Type A cohesive soils are often: clay, silty clay, sandy clay, clay loam and, in some cases, silty clay loam and sandy clay loam. (No soil is Type A if it is fissured, is subject to vibration of any type, has previously been disturbed, is part of a sloped, layered system where the layers dip into the excavation on a slope of 4 horizontal to 1 vertical (4H:1V) or greater, or has seeping water.

TYPE B SOILS are cohesive soils with an unconfined compressive strength greater than 0.5 tsf (48 kPa) but less than 1.5 tsf (144 kPa). Examples of other Type B soils are: angular gravel; silt; silt loam; previously disturbed soils, unless otherwise classified as Type C; soils that meet the unconfined compressive strength or cementation requirements of Type A soils but are fissured or subject to vibration; dry unstable rock; and layered systems sloping into the trench at a slope less than 4H:1V (only if the material would be classified as a Type B soil).

TYPE C SOILS are cohesive soils with an unconfined compressive strength of 0.5 tsf (48 kPa) or less. Other Type C soils include granular soils such as gravel, sand and loamy sand, submerged soil, soil from which water is freely seeping, and submerged rock that is not stable. Also included in this classification is material in a sloped, layered system where the layers dip into the excavation or have a slope of four horizontal to one vertical (4H:1V) or greater.

LAYERED GEOLOGICAL STRATA. Where soils are configured in layers, i.e., where a layered geologic structure exists, the soil must be classified on the basis of the soil classification of the weakest soil layer. Each layer may be classified individually if a more stable layer lies below a less stable layer, i.e., where a Type C soil rests on top of stable rock.

(courtesy OSHA)

Figure 31-14 Shoring and Shielding Considerations

Shoring Types

Shoring is the provision of a support system for trench faces used to prevent movement of soil, underground utilities, roadways, and foundations. Shoring or shielding is used when the location or depth of the cut makes sloping back to the maximum allowable slope impractical. Shoring systems consist of posts, wales, struts, and sheeting. There are two basic types of shoring: timber, and aluminum hydraulic.

The trend today is toward the use of hydraulic shoring, a prefabricated strut and/or wale system manufactured of aluminum or steel. Hydraulic shoring provides a critical safety advantage over timber shoring, because workers do not have to enter the trench to install or remove hydraulic shoring. Other advantages of most hydraulic systems are that they:

- are light enough to be installed by one worker;
- are gauge-regulated to ensure even distribution of pressure along the trench line;
- can have their trench faces "preloaded" to use the soil's natural cohesion to prevent movement; and
- can be adapted easily to various trench depths and widths.

All shoring should be installed from the top down and removed from the bottom up. Hydraulic shoring should be checked at least once per shift for leaking hoses and/or cylinders, broken connections, cracked nipples, bent bases, and any other damaged or defective parts.

Shielding Types

Trench boxes are different from shoring because, instead of shoring up or otherwise supporting the trench face, they are intended primarily to protect workers from cave-ins and similar incidents. The excavated area between the outside of the trench box and the face of the trench should be as small as possible. The space between the trench boxes and the excavation side are backfilled to prevent lateral movement of the box. Shields may not be subjected to loads exceeding those which the system was designed to withstand.

Trench boxes are generally used in open areas, but they also may be used in combination with sloping and benching. The box should extend at least 18 in (0.45 m) above the surrounding area if there is sloping toward excavation. This can be accomplished by providing a benched area adjacent to the box.

Earth excavation to a depth of 2 ft (0.61 m) below the shield is permitted, but only if the shield is designed to resist the forces calculated for the full depth of the trench and there are no indications while the trench is open of possible loss of soil from behind or below the bottom of the support system. Conditions of this type require observation on the effects of bulging, heaving, and boiling as well as surcharging, vibration, adjacent structures, etc., on excavating below the bottom of a shield. Careful visual inspection of the conditions mentioned above is the primary and most prudent approach to hazard identification and control.

(courtesy OSHA)

Figure 31-15 Slope and Shield Configurations

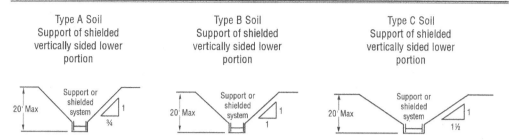

Maximum allowable slopes for excavations less than 20 feet (6.09 m) based on soil type and angle to the horizontal are as follows:

Allowable Slopes

Soil Type	Height/Depth Ratio	Slope Angle
Stable Rock	Vertical	90
Type A	3/4:1	53
Type B	1:1	45
Type C	1-1/2:1	34
Type A (Short Term)	1/2:1	63
(For a maximum excavation depth of 12 ft)		

(courtesy OSHA)

Figure 31-16 Data for Drilling and Blasting Rock

Hole Size (in Inches)	Hole Dimensions (in Feet)	Hole Area (in S.F.)	Rock Volume (per L.F. of Hole, in C.Y.)	Lbs. of Explosive Required (per L.F. of Hole)	Lbs. of Explosive per C.Y. of Rock % of Hole Filled		
					100%	75%	50%
1-1/2"	4 x 4	16	0.59	0.9	1.52	1.14	0.76
	5 x 5	25	0.93	0.9	0.97	0.73	0.48
	6 x 6	36	1.33	0.9	0.68	0.51	0.34
	7 x 7	49	1.81	0.9	0.50	0.38	0.25
2"	5 x 5	25	0.93	1.7	1.83	1.37	0.92
	6 x 6	36	1.33	1.7	1.28	0.96	0.64
	7 x 7	49	1.81	1.7	0.94	0.71	0.47
	8 x 8	64	2.37	1.7	0.72	0.54	0.36
3"	7 x 7	49	1.81	3.9	2.15	1.61	1.08
	8 x 8	64	2.37	3.9	1.65	1.24	0.83
	9 x 9	81	3.00	3.9	1.30	0.97	0.65
	10 x 10	100	3.70	3.9	1.05	0.79	0.53
	11 x 11	121	4.48	3.9	0.87	0.65	0.44
4"	8 x 8	64	2.37	7.5	3.16	2.37	1.58
	10 x 10	100	3.70	7.5	2.03	1.52	1.02
	12 x 12	144	5.30	7.5	1.42	1.06	0.71
	14 x 14	196	7.25	7.5	1.03	0.77	0.52
	16 x 16	256	9.50	7.5	0.79	0.59	0.40
5"	12 x 12	144	5.30	10.9	2.05	1.54	1.02
	14 x 14	196	7.25	10.9	1.50	1.13	0.75
	16 x 16	256	9.50	10.9	1.15	0.86	0.58
	18 x 18	324	12.00	10.9	0.91	0.68	0.46
	20 x 20	400	14.85	10.9	0.73	0.55	0.37
6"	12 x 12	144	5.30	15.6	2.94	2.20	1.47
	14 x 14	196	7.25	15.6	2.05	1.54	1.02
	16 x 16	256	9.50	15.6	1.64	1.23	0.82
	18 x 18	324	12.00	15.6	1.30	0.97	0.65
	20 x 20	400	14.85	15.6	1.05	0.89	0.53
	24 x 24	576	21.35	15.6	0.73	0.55	0.37
9"	20 x 20	400	14.85	35.0	2.36	1.77	1.18
	24 x 24	576	21.35	35.0	1.64	1.23	0.82
	28 x 28	784	29.00	35.0	1.21	0.91	0.61
	30 x 30	900	33.30	35.0	1.05	0.79	0.53
	32 x 32	1,024	37.90	35.0	0.92	0.69	0.46

Figure 31-17 Tunnel Excavation

Bored tunnel excavation is common in rock for diameters from 4 feet for sewer and utilities, to 60 feet for vehicles. Production varies from a few linear feet per day to over 200 linear feet per day. In the smaller diameters, the productivity is limited by the restricted area for mucking or the removal of excavated material.

Most of the tunnels in rock today are excavated by boring machines called moles. Preparation for starting the excavation or setting up the mole is very costly. Shafts must be excavated to the invert of the proposed tunnel and the mole must be lowered into the shaft. If excavating a portal tunnel that is starting at an open face, the cost is reduced considerably, both for mobilization and mucking.

In soft ground and mixed material, special bucket excavators and rotary excavators are used inside a shield. Tunnel liners must follow directly behind the shield to support the earth and prevent cave-ins.

Traditional muck haulage operations are performed by rail with locomotives and muck cars. Sometimes conveyors are more economical and require less ventilation of the tunnel.

Ventilation and air compression are other important cost factors to consider in tunnel excavation. Continuous ventilation ducts are sometimes fabricated at the tunnel site.

Tunnel linings are steel, cast-in-place reinforced concrete, shotcrete, or a combination of these. When required, contact grouting is performed by pumping grout between the lining and the excavation. Intermittent holes are drilled into the lining and separate costs are determined for drilling per hole, grout pump connecting per hole, and grout per cubic foot. Consolidation grouting and roof bolts may also be required where the excavation is unstable or faulting occurs.

Tunnel boring is usually done 24 hours per day. A typical crew for rock boring is:

Tunneling Crew based on three 8-hour shifts

1 Shifter
1 Walker
1 Machine operator for mole
1 Oiler
1 Mechanic
3 Locomotives with operators
5 Miners for rails, vent ducts, and roof bolts
1 Electrician
2 Pumps
2 Laborers for hoisting
1 Hoist operator for muck removal

Surface Crew based on normal 8-hour shift

2 Shop mechanics
1 Electrician
1 Shifter
2 Laborers
1 Operator with 18 ton cherry picker
1 Operator with front end loader

Figure 31-18 Piles

Piles are used to transmit foundation loads to strata of adequate bearing capacity and to eliminate settlement from the consolidation of overlying materials. This table lists nine principal pile categories of the three structural materials: wood, steel, and concrete. No exact criteria for the applicability of the various pile types can be given. The selection of types should be based on factors listed in the figures and on comparative costs.

Pile Type	Timber	Steel
Consider for length of	30–60 ft.	40–100 ft
Applicable material specifications	TS-2P3	TS-P67
Maximum stresses	Measured at most critical point, 1,200 psi for southern pine and Douglas fir. See U.S.D.A. Wood Handbook No. 72 for stress values of other species.	12,000 psi
Consider for design loads of	10–50 tons	40–120 tons
Disadvantages	Difficult to splice. Vulnerable to damage in hard driving. Vulnerable to decay unless treated, when piles are intermittently submerged.	Vulnerable to corrosion where exposed. BP section may be damaged or deflected by major obstructions.
Advantages	Comparatively low initial cost. Permanently submerged piles are resistant to decay. Easy to handle.	Easy to splice. High capacity. Small displacement. Able to penetrate through light obstructions.
Remarks	Best suited for friction pile in granular material.	Best suited for endbearing on rock. Reduce allowable capacity for corrosive locations.
Typical illustrations	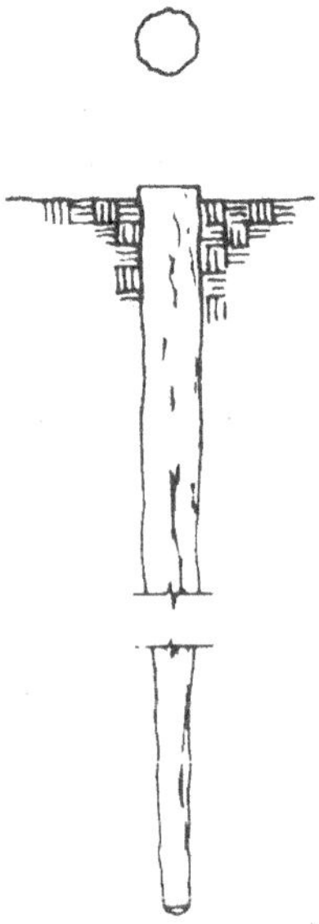	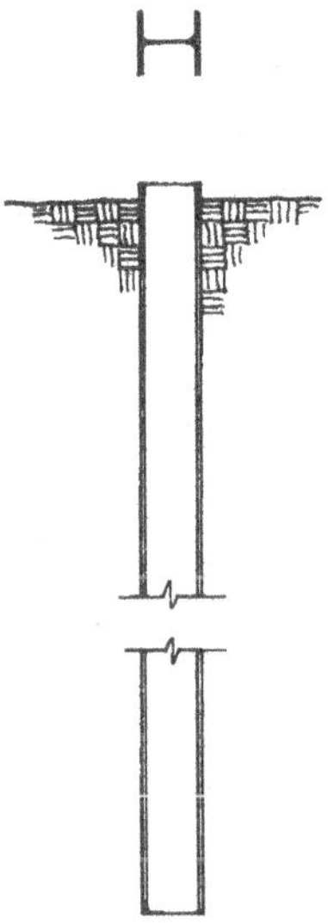

(continued on next page)

Figure 31-18 Piles *(continued from previous page)*

Pile Type	Precast Concrete (including prestressed)	Cast-in-Place Concrete (thin shell driven with mandrel)
Consider for length of	40–50 ft. for precast. 60–100 ft. for prestressed.	100 ft.
Applicable material specifications	TS-P57	ACI Code 318–for concrete
Maximum stresses	For precast—15% of 28-day strength of concrete, but no more than 700 psi. For prestressed—20% of 28-day strength of concrete, but no more than 1,000 psi in excess of prestress.	25% of 28-day strength of concrete with 1,000 psi maximum, measured at midpoint of length in bearing stratum.
Specifically designed for a wide range of loads		
Disadvantages	Unless prestressed, vulnerable to handling. High initial cost. Considerable displacement. Prestressed difficult to splice.	Difficult to splice after concreting. Redriving not recommended. Thin shell vulnerable during driving. Considerable displacement.
Advantages	High load capacities. Corrosion resistance can be attained. Hard driving possible.	Initial economy. Tapered sections provide higher bearing resistance in granular stratum.
Remarks	Cylinder piles in particular are suited for bending resistance.	Best suited for medium load friction piles in granular materials.
Typical illustrations		

(continued on next page)

Figure 31-18 Piles *(continued from previous page)*

Pile Type	Cast-in-place piles (shells driven without mandrel)	Pressure Injected Footings
Consider for length of	30–80 ft.	10–60 ft.
Applicable material specifications	ACI Code 318	TS-F16
Maximum stresses	25% of 28-day strength of concrete with maximum of 1,000 psi measured at midpoint of length in bearing stratum. 9,000 psi in shell.	25% of 28-day strength of concrete with a minimum of 1,000 psi. 9,000 psi for pipe shell if thickness greater than 1/8".
Consider for design loads of	50–70 tons	60–120 tons
Disadvantages	Hard to splice after concreting. Considerable displacement.	Base of footing cannot be made in clay. When clay layers must be penetrated to reach suitable material, special precautions are required for shafts if in groups.
Advantages	Can be redriven. Shell not easily damaged.	Provides means of placing high capacity footing on bearing stratum without necessity for excavation or dewatering. Required depths can be predicted accurately. High blow energy available for overcoming obstructions. Great uplift resistance if suitably reinforced.
Remarks	Best suited for friction piles of medium length.	Best suited for granular soils where bearing achieved through compaction around base. Minimum spacing 4'-6" on center. For further design requirements see your local building code.
Typical illustrations		

(continued on next page)

Figure 31-18 Piles *(continued from previous page)*

Pile Type	Concrete Filled Steel Pipe Piles	Composite Piles
Consider for length of	40–120 ft.	60–120 ft.
Applicable material specifications	ASTM A7—for Core ASTM A252—for Pipe ACI Code 318—for Concrete	ACI Code 318—for Concrete ASTM-36—for Structural Section. ASTM A252—for Steel Pipe TS-P2—for Timber.
Maximum stresses	9,000 psi for pipe shell. 25% of 28-day strength of concrete with a maximum of 1,000 psi. 12,000 psi on Steel Cores.	25% of 28-day strength of concrete with 1,000 psi maximum. 9,000 psi for structural and pipe sections. Same as timber piles for wood composite.
Consider for design loads of	80–120 tons without cores. 500–1,500 tons with cores.	30–80 tons
Disadvantages	High initial cost. Displacement for closed end pipe.	Difficult to attain good joint between two materials.
Advantages	Best control during installation. No displacement for open end installation. Open end pipe best against obstructions. High load capacities. Easy to splice.	Considerable length can be provided at comparatively low cost.
Remarks	Provides high bending resistance where unsupported length is loaded laterally.	The weakest of any material used shall govern allowable stresses and capacity.
Typical illustrations		

(continued on next page)

Division 31

Figure 31-18   Piles *(continued from previous page)*

Pile Type	Concrete Filled Steel Pipe Piles	General Notes
Consider for length of	30–60 ft.	1. Stresses given for steel piles are for noncorrosive locations. For corrosive locations, estimate possible reduction in steel cross section or provide protection from corrosion.
Applicable material specifications	TS-2P69	
Maximum stresses	25% of 28-day strength of concrete with a maximum of 1,000 psi.	
Consider for design loads of	35–70 tons	2. Lengths and loads indicated are for feasibility guidance only. They generally represent current practice.
Disadvantages	More than average dependence on quality workmanship. Not suitable through peat or similar highly compressible material.	3. Design load capacity should be determined by soil mechanics principles limiting stresses in piles and type and function of structure.
Advantages	Economy. Completely nondisplacement. No driving vibration to endanger adjacent structures. High skin friction. Good contact on rock for end bearing. Convenient for low-headroom underpinning work. Visual inspection of augered material. No splicing required.	
Remarks	Process patented	
Typical illustrations		

Figure 31-19 Caissons – General

General: Caissons, as covered in this section, are drilled cylindrical foundation shafts which function primarily as short column-like compression members. They transfer superstructure loads through inadequate soils to bedrock or hard stratum. They may be either reinforced or unreinforced and either straight or belled out at the bearing level.

Shaft diameters range in size from 20" to 84" with the most usual sizes beginning at 34". If inspection of bottom is required, the minimum diameter practical is 30". If handwork is required (in addition to mechanical belling, etc.) the minimum diameter is 32". The most frequently used shaft diameter is probably 36" with a 5' or 6' bell diameter. The maximum bell diameter practical is three times the shaft diameter.

Plain concrete is commonly used, poured directly against the excavated face of soil. Permanent casings add to cost and economically should be avoided. Wet or loose strata are undesirable. The associated installation sometimes involves a mudding operation with bentonite clay slurry to keep walls of excavation stable (costs not included here).

Reinforcement is sometimes used, especially for heavy loads. It is required if uplift, bending moment, or lateral loads exist. A small amount of reinforcement is desirable at the top portion of each caisson, even if the above conditions theoretically are not present. This will provide for construction eccentricities and other possibilities. Reinforcement, if present, should extend below the soft strata. Horizontal reinforcement is not required for belled bottoms.

There are three basic types of caisson bearing details:

1. Belled, which are generally recommended to provide reduced bearing pressure on soil. These are not for shallow depths or poor soils. Good soils for belling include most clays, hardpan, soft shale, and decomposed rock.

   Soils requiring handwork include hard shale, limestone, and sandstone.

   Soils not recommended include sand, gravel, silt, and igneous rock. Compact sand and gravel above water table may stand. Water in the bearing strata is undesirable.

2. Straight shafted, which have no bell but the entire length is enlarged to permit safe bearing pressures. They are most economical for light loads on high bearing capacity soil.

3. Socketed (or keyed), which are used for extremely heavy loads. They involve sinking the shaft into rock for combined friction and bearing support action. Reinforcement of shaft is usually necessary. Wide flange cores are frequently used here.

Advantages include:
- Shafts can pass through soils that piles cannot
- No soil heaving or displacement during installation
- No vibration during installation
- Less noise than pile driving
- Bearing strata can be visually inspected and tested

Uses include:
- Situations where unsuitable soil exists to moderate depth
- Tall structures
- Heavy structures
- Underpinning (extensive use)

(continued on next page)

Figure 31-19 Caissons – General *(continued from previous page)*

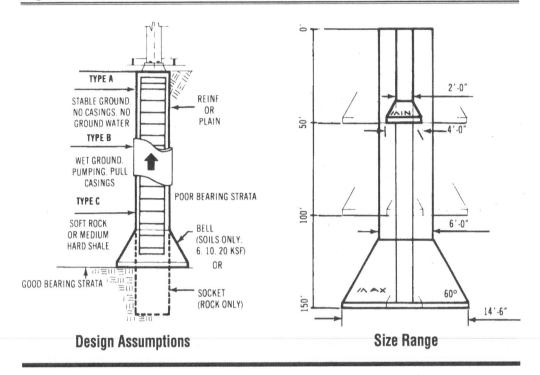

Design Assumptions   **Size Range**

Figure 31-20 Caissons – Types

The three principal types of caissons are:

1. Belled caissons, which, except for shallow depths and poor soil conditions, are generally recommended. They provide more bearing than shaft area. Because of its conical shape, no horizontal reinforcement of the bell is required.

2. Straight shaft caissons are used where relatively light loads are to be supported by a caisson that rests on high value bearing strata. While the shaft is larger in diameter than for belled types, this is more than offset by the saving in time and labor.

3. Keyed caissons are used when extremely heavy loads are to be carried. A keyed or socketed caisson transfers its load into rock by a combination of end-bearing and shear reinforcing of the shaft. The most economical shaft often consists of a steel casing, a steel wide flange core, and concrete. Allowable compressive stesses of 0.225 f'c for concrete, 16,000 psi for the wide flange core, and 9,000 psi for the steel casing are commonly used. The usual range of shaft diameter is from 18" to 84". The number of sizes specified for any one project should be limited due to the problems of casing and auger storage.

When handwork is to be performed, shaft diameters should not be less than 32". When inspection of borings is required, a minimum shaft diameter of 30" is recommended. Concrete caissons are intended to be poured against earth excavation so permanent forms that add to cost should not be used in the excavation if the excavation is clean and the earth sufficiently impervious to prevent excessive loss of concrete.

Soil Conditions for Belling		
Good	**Requires Handwork**	**Not Recommended**
Clay	Hard shale	Silt
Sandy clay	Limestone	Sand
Silty clay	Sandstone	Gravel
Clayey silt	Weathered mica	Igneous rock
Hard-Pan		
Soft shale		
Decomposed rock		

Division 31

Figure 31-21 Table of Shaft Diameters, Perimeters, Areas, and Volumes; and Shaft or Bell End-Bearing Areas

Shaft Diameter		Shaft Perimeter		Shaft or Bell End Area		Shaft Volume	
in.	cm.	ft.	m	ft.^{2}	m^2	yd^3/ft.	m^3/m
18	45.7	4.71	1.44	1.77	0.164	0.06	0.16
20	50.8	5.24	1.60	2.18	0.203	0.08	0.20
22	55.9	5.76	1.75	2.64	0.245	0.10	0.24
24	61.0	6.28	1.92	3.14	0.292	0.12	0.29
26	66.0	6.81	2.08	3.69	0.343	0.14	0.34
28	71.1	7.33	2.23	4.28	0.397	0.16	0.40
30	76.2	7.85	2.39	4.91	0.456	0.18	0.46
32	81.3	8.38	2.55	5.58	0.519	0.21	0.52
34	86.4	8.90	2.71	6.30	0.586	0.23	0.59
36	91.4	9.42	2.87	7.07	0.657	0.26	0.66
38	96.5	9.95	3.03	7.88	0.732	0.29	0.73
40	102	10.47	3.19	8.73	0.811	0.32	0.81
42	107	11.00	3.35	9.62	0.894	0.34	0.89
44	112	11.52	3.51	10.56	0.981	0.39	0.98
46	117	12.04	3.67	11.54	1.072	0.43	1.07
48	122	12.57	3.83	12.57	1.167	0.46	1.17
50	127	13.09	3.99	13.64	1.267	0.50	1.27
52	132	13.61	4.15	14.75	1.370	0.55	1.37
54	137	14.14	4.31	15.90	1.478	0.59	1.48
56	142	14.66	4.47	17.10	1.589	0.63	1.59
58	147	15.18	4.63	18.35	1.705	0.68	1.70
60	152	15.71	4.79	19.64	1.824	0.73	1.82
62	158	16.23	4.95	20.97	1.948	0.78	1.95
64	163	16.76	5.11	22.34	2.075	0.83	2.08
66	168	17.28	5.27	23.76	2.207	0.88	2.21
68	173	17.80	5.43	25.22	2.343	0.93	2.34
70	178	18.33	5.59	26.72	2.483	0.99	2.48
72	183	18.85	5.74	28.27	2.627	1.05	2.63
74	188	19.37	5.90	29.87	2.775	1.11	2.78
76	193	19.90	6.06	31.50	2.927	1.17	2.93
78	198	20.42	6.22	33.18	3.083	1.23	3.08
80	203	20.94	6.38	34.91	3.243	1.29	3.24
82	208	21.47	6.54	36.67	3.407	1.36	3.41
84	213	21.99	6.70	38.48	3.575	1.42	3.58

Figure 31-22 Bell or Underream Volumes – Dome Type

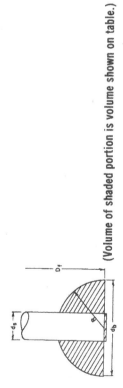

(Volume of shaded portion is volume shown on table.)

Bell Diameter	18" Diameter Shaft ft.³	18" Diameter Shaft m³	24" Diameter Shaft ft.³	24" Diameter Shaft m³	30" Diameter Shaft ft.³	30" Diameter Shaft m³	36" Diameter Shaft ft.³	36" Diameter Shaft m³	42" Diameter Shaft ft.³	42" Diameter Shaft m³	48" Diameter Shaft ft.³	48" Diameter Shaft m³
2'-0" 61 cm	2.5	0.07	4.0	0.11								
2'-6" 76 cm	4.0	0.11	6.0	0.17								
3'-0" 91 cm	6.5	0.18	9.0	0.25	6.0	0.17						
3'-6" 1.07 m	10.0	0.28			8.5	0.22	8.5	0.22				
4'-0" 1.22 m	15.0	0.42	14.0	0.40	12.5	0.35	11.5	0.33	13.0	0.37		
4'-6" 1.37 m			20.5	0.58	18.5	0.52	16.0	0.45	15.0	0.42	18.0	0.51
5'-0" 1.52 m			28.5	0.81	26.0	0.74	23.5	0.67	20.5	0.58	21.0	0.59
5'-6" 1.68 m					35.5	1.00	32.5	0.92	29.0	0.82	24.0	0.68
6'-0" 1.83 m					47.0	1.33	43.5	1.23	39.0	1.10	34.5	0.98
6'-6" 1.98 m					61.5	1.74	57.0	1.61	52.0	1.47	46.5	1.32
7'-0" 2.13 m					78.0	2.21	72.5	2.05	67.5	1.91	61.0	1.73
7'-6" 2.29 m					97.5	2.76	92.0	2.60	85.0	2.41	78.0	2.21
8'-0" 2.44 m							114.0	3.23	106.5	3.01	98.5	2.79
8'-6" 2.59 m							138.5	3.92	131.0	3.71	122.0	3.45
9'-0" 2.74 m							167.0	4.73	158.5	4.49	149.0	4.22
9'-6" 2.90 m									189.5	5.36	179.5	5.08
10'-0" 3.05 m									224.5	6.35	213.0	6.03
10'-6" 3.20 m									263.5	7.46	251.5	7.12
11'-0" 3.35 m											293.5	8.31
11'-6" 3.51 m											340.0	9.62
12'-0" 3.66 m											391.0	11.07

Division 31

Figure 31-23   Bell or Underream Volumes – 45° Type

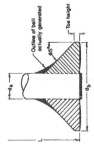

Bell Diameter	18" Shaft ft.³	18" m³	24" Shaft ft.³	24" m³	30" Shaft ft.³	30" m³	36" Shaft ft.³	36" m³	42" Shaft ft.³	42" m³	48" Shaft ft.³	48" m³	52" Shaft ft.³	52" m³	60" Shaft ft.³	60" m³
2'-0" 61 cm	0.9	0.02														
2'-6" 76 cm	2.3	0.06	1.1	0.03												
3'-0" 91 cm	4.4	0.13	2.9	0.08	1.3	0.04										
3'-6" 1.07 m	7.3	0.21	5.4	0.15	3.5	0.10	1.6	0.04								
4'-0" 1.22 m	11.1	0.31	8.9	0.25	6.5	0.18	4.1	0.11	1.8	0.05						
4'-6" 1.37 m	15.9	0.45	13.3	0.38	10.5	0.30	7.5	0.21	4.6	0.13	2.1	0.06				
5'-0" 1.52 m			18.8	0.53	15.5	0.44	12.0	0.34	8.5	0.24	5.2	0.15	2.3	0.07		
5'-6" 1.68 m			25.5	0.72	21.8	0.62	17.7	0.50	13.6	0.39	9.6	0.27	5.8	0.16	2.6	0.07
6'-0" 1.83 m			33.5	0.95	29.3	0.83	24.7	0.70	20.0	0.57	15.2	0.43	10.6	0.30	6.4	0.18
6'-6" 1.98 m					38.2	1.08	33.1	0.94	27.7	0.78	22.2	0.63	16.7	0.47	11.6	0.33
7'-0" 2.13 m					48.6	1.38	42.9	1.22	36.9	1.04	30.6	0.87	24.4	0.69	18.3	0.52
7'-6" 2.29 m					60.5	1.71	54.3	1.54	47.6	1.35	40.6	1.15	33.6	0.95	26.6	0.75
8'-0" 2.44 m							67.4	1.91	60.1	1.70	52.3	1.48	44.4	1.26	36.5	1.03
8'-6" 2.59 m							82.2	2.33	74.2	2.10	65.8	1.86	57.0	1.62	48.2	1.36
9'-0" 2.74 m							98.9	2.80	90.3	2.56	81.1	2.30	71.5	2.03	61.8	1.75
9'-6" 2.90 m									108.0	3.07	98.4	2.74	88.0	2.49	77.3	2.19
10'-0" 3.05 m									128.0	3.64	118.0	3.33	106.0	3.02	94.9	2.69
11'-0" 3.35 m											163.0	4.62	150.0	4.25	137.0	3.87
12'-0" 3.66 m											218.0	6.17	203.0	5.75	188.0	5.32
13'-0" 3.96 m													266.0	7.54	249.0	7.05
14'-0" 4.27 m															322.0	9.10
15'-0" 4.57 m															406.0	11.50

Figure 31-24 Bell or Underream Volumes – 30° Type

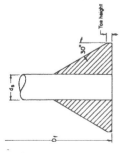

Bell Diameter	18" Diameter Shaft ft.³	18" Diameter Shaft m³	24" Diameter Shaft ft.³	24" Diameter Shaft m³	30" Diameter Shaft ft.³	30" Diameter Shaft m³	36" Diameter Shaft ft.³	36" Diameter Shaft m³	42" Diameter Shaft ft.³	42" Diameter Shaft m³	48" Diameter Shaft ft.³	48" Diameter Shaft m³
2'-0" 61 cm	1.0	0.03										
2'-6" 76 cm	2.8	0.08	1.2	0.04								
3'-0" 91 cm	5.7	0.16	3.6	0.10	1.5	0.04						
3'-6" 1.07 m	9.8	0.28	7.1	0.20	4.3	0.12	1.8	0.05				
4'-0" 1.22 m	15.3	0.43	12.0	0.33	8.4	0.24	5.0	0.14	2.1	0.06		
4'-6" 1.37 m	22.4	0.63	18.4	0.52	14.1	0.40	9.8	0.28	5.8	0.16	2.4	0.06
5'-0" 1.52 m			26.6	0.75	21.5	0.61	16.3	0.46	11.1	0.31	6.5	0.18
5'-6" 1.68 m			36.7	1.04	30.8	0.87	24.6	0.70	18.4	0.52	12.5	0.35
6'-0" 1.83 m			48.8	1.38	42.2	1.19	35.1	0.99	27.8	0.79	20.6	0.58
6'-6" 1.98 m					55.8	1.58	47.8	1.35	39.3	1.11	30.8	0.87
7'-0" 2.13 m					71.9	2.01	62.9	1.78	53.3	1.51	43.6	1.23
7'-6" 2.29 m					90.5	2.56	80.5	2.28	70.0	1.98	58.8	1.66
8'-0" 2.44 m							101.0	2.86	89.2	2.52	76.9	2.18
8'-6" 2.59 m							124.5	3.52	111.0	3.14	97.8	2.77
9'-0" 2.74 m							151.0	4.26	137.0	3.87	122.0	3.45
9'-6" 2.90 m									165.0	4.68	149.0	4.22
10'-0" 3.05 m									197.0	5.57	180.0	5.09
10'-6" 3.20 m									233.0	6.60	214.0	6.06
11'-0" 3.35 m											252.0	7.13
12'-0" 3.66 m											340.0	9.61

(continued on next page)

Figure 31-24 Bell or Underream Volumes – 30° Type
(continued from previous page)

	54" Diameter Shaft		60" Diameter Shaft		66" Diameter Shaft		72" Diameter Shaft		78" Diameter Shaft		84" Diameter Shaft	
5'-0" 1.52 m	2.7	0.08										
5'-6" 1.68 m	7.2	0.20	2.9	0.09								
6'-0" 1.83 m	13.8	0.39	8.0	0.23	3.2	0.09						
6'-6" 1.98 m	22.7	0.64	15.2	0.43	8.7	0.25	3.5	0.10				
7'-0" 2.13 m	34.0	0.96	24.8	0.70	16.6	0.47	9.4	0.27	3.8	0.11		
8'-0" 2.44 m	64.4	1.82	52.0	1.47	40.2	1.14	29.1	0.82	19.2	0.54	10.9	0.31
9'-0" 2.74 m	106.0	3.02	90.0	2.57	75.5	2.14	60.5	1.71	46.4	1.31	33.4	0.95
10'-0" 3.05 m	162.0	4.57	143.0	4.04	124.0	3.50	105.0	2.97	86.6	2.45	69.0	1.95
11'-0" 3.35 m	231.0	6.54	209.0	5.92	187.0	5.28	164.0	4.64	141.0	4.00	119.0	3.36
12'-0" 3.66 m	316.0	8.95	291.0	8.24	265.0	7.50	238.0	6.75	211.0	5.97	185.0	5.24

Figure 31-25 Pile Type, Classification, Use, and Support Determination

Pile Type (Material)	Pile Type (Form)	Classification* Displacement	Classification* Non-Displacement	Use	Support* End Brng.	Support* Friction	Comments
General	Preformed	√	√	Above ground. Through water.	√	√	Determines some pile types.
Timber General	Preformed	√	?	Structures with moderate load. Waterfront structures and protection. Trestles and bents. Temporary structures.	?	√	Relatively inexpensive per foot. Good impact absorption. May be jetted in pure sand. Easy to handle and cut off. Use only single length needed (extension is hard and expensive). Prone to hard driving damage. Not driven thru hard stratum or boulders. Limited in size and capacity. Scour and ice action injure pile. without concrete protection. Test piles determine length.
Treated				Above permanent water level. Moist soil (>20%).			Creosote pressure treated. Treat field cuts. 50-yr. life.
Untreated				Below permanent water level. Dry soil (<20%).			Lower material cost. 25-yr. life. Vulnerable to marine borers.
Concrete General				Building structures of moderate to heavy load. Bridge foundations.			Bracing is not easily attached.
C.I.P.	General						Permanent. Can be treated for sea water. Easy to alter lengths. Damage due to handling eliminated. Easily bonded into pile cap. Specialist contractor.

* √ = Most frequent application

? = Least frequent or questionable application

Division 31

(continued on next page)

Figure 31-25   Pile Type, Classification, Use, and Support Determination
(continued from previous page)

Pile Type		Classification*		Use	Support*		Comments
Material	Form	Displacement	Non-Displacement		End Brng.	Friction	
Concrete (cont.) / C.I.P. (cont.)	Uncased		√	Firm soils. Length, less than 25'	√	?	No storage space required. Can be made before excavation. Can eliminate vibration and noise. Soft crumbly soils cave in. Inspection difficult. Concrete completed can be damaged by subsequent driving. Bored or extracted casing used.
	Shell left in ground		Open end	Soft soils. All lengths.	√	?	Allows inspection before concreting. Easy to cut off or extend. Thin casings may be damaged by handling or soil pressure. Less economical as steel cost increases. Mandrel may be required.
		Closed end			?	√	Same as open end plus below. Increased lateral soil pressures. Thick casings carry part of load. Thin casings support concrete. Mandrel may be required.
Pre-cast (pre-formed)	General			Trestles and bents. Waterfront.			Reinforced for handling. Space required for casting and storage.
	Solid	√	?		√	√	Requires heavy equipment for handling and driving.
	Cylinder		Open end		√	√	Durable, can be treated for salt water. Difficult and expensive to cut off, extend or bond into pile cap.
		Plugged end	?		?		Convenient with concrete superstructure.

* √ = Most frequent application

? = Least frequent or questionable application

(continued on next page)

Figure 31-25 Pile Type, Classification, Use, and Support Determination
(continued from previous page)

| Pile Type | | Classification* | | Use | Support* | | Comments |
Material	Form	Displacement	Non-Displacement		End Brg.	Friction	
Steel — General				Large structures of heavy loads. Trestles and bents.			Easy to handle, cut off, extend or bond into pile cap. Available any length by size. Convenient to combine with steel superstructure. Can stand rough handling. Possible damage from corrosion. Requires marine environment protection. Relatively expensive except where bearing stratum can develop large pile capacity. General contractor installs.
HP	Preformed	Small		Driving through dense layers. When bottom is irregular.	√	?	Seldom used as friction pile. Add 1/16" to thickness for corrosion or .2% copper added. Reinforce bottom for driving into rock or hardpan for stability when muddy soil above. If rock is too hard for substantial penetration with little support above, don't use. Hard driving with no lateral support tends to bend pile. Boulders tend to force piles out of plumb. Do not use in cinder or ash fill. May need to batter for lateral resistance in water.
Pipe	General (preformed)			Good in battered applications.			Considerable structural strength. Flexurally strong. Driven with standard hammer rather than mandrel. Normally concrete filled.

* √ = Most frequent application

? = Least frequent or questionable application

Division 31

(continued on next page)

Figure 31-25 Pile Type, Classification, Use, and Support Determination
(continued from previous page)

Pile Type		Classification*		Use	Support*		Comments
Material	Form	Displacement	Non-Displacement		End Brng.	Friction	
Steel (cont.) — Pipe (cont.)	Concrete fill		Open end	Decrease disturbance to adj. structures. Avoid heaving when driving next piles. Driving thru obstructions.		?	Strength and rigidity. Sometimes classified as "composite."
		Closed end		No firm strata at reasonable depth. Desire to compact soil. Water bearing strata punctured or ended in.	?	√	Lower capacity than above if not driven to refusal. May be preferred without concrete fill, if friction support.

* √ = Most frequent application

? = Least frequent or questionable application

Figure 31-26 Wood Sheet Piling

Wood sheet piling may be used for depths to 20' where there is no groundwater. If moderate groundwater is encountered, tongue-and-groove sheeting will help to keep it out. When considerable groundwater is present, steel sheeting must be used.

For estimating purposes on trench excavation, sizes are as follows:

Sheeting is to be toed in at least 2' depending on the soil conditions. A five-person crew with an air compressor and sheeting driver can drive and brace 440 SF/day at 8' deep, 360 SF/day at 12' deep, and 320 SF/day at 16' deep. For normal soils, piling can be pulled in 1/3 the time to install. Pulling difficulty increases with time in the ground. Production can be increased by high pressure jetting. Assume 50% of lumber is salvaged, and include pulling costs. Some jurisdictions require an additional equipment operator.

Depth	Sheeting	Wales	Braces	B.F. per S.F.
To 8'	3 x 12's	6 x 8's, 2 Line	6 x 8's, @ 10'	4.0 @ 8'
8' x 12'	3 x 12's	10x 10's, 2 Line	10 x 10's, @ 9'	5.0 average
12' x 20'	3 x 12's	12x 12's, 3 Line	12 x 12's, @ 8'	7.0 average

Figure 31-27 Installation Time in Labor-Hours for Treated Wood Piles

Description	Labor-Hours	Unit
Wood Piles Treated		
12" Butts 8" Points up to 30' Long	.102	V.L.F.
Boot for Pile Tip	.300	Ea.
Point for Pile Tip	.450	Ea.
Mobilization for 10,000 L.F. Job	.019	V.L.F.

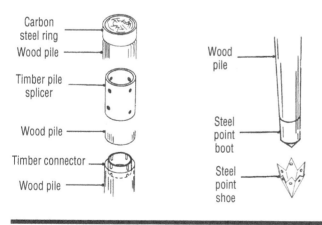

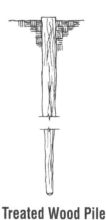

Treated Wood Pile

Figure 31-28   Installation Time in Labor-Hours for Step-Tapered Piles

Description	Labor-Hours	Unit
Cast-in-Place Step-Tapered Pile 12" Diameter	.107	V.L.F.
Mobilization		
Small Job	142.000	
Large Job	237.000	

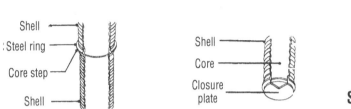

Step-Tapered Pile

Figure 31-29a   Installation Time in Labor-Hours for Steel Pipe Piles

Description	Labor-Hours	Unit
Pipe Piles 44 lb. per L.F.		
No Concrete	.135	V.L.F.
Concrete Filled	.154	V.L.F.
Splices	2.111	Ea.
Standard Points	1.975	Ea.
Heavy Duty Points	3.960	Ea.
Mobilization		
Small Job	142.000	
Large Job	237.000	

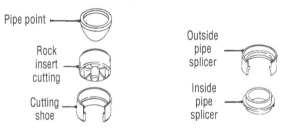

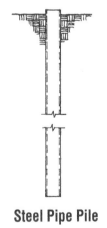

Steel Pipe Pile

Figure 31-29b Installation Time in Labor-Hours for Steel HP Piles

Description	Labor-Hours	Unit
H Sections 12″ x 12″, 53 lb. per L.F.	.108	V.L.F.
Splice or Standard Points	2.000	Ea.
Heavy Duty Points	2.286	Ea.
Mobilization		
Small Job	142.000	
Large Job	237.000	

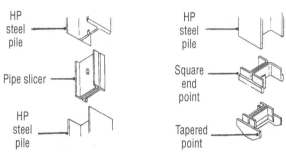

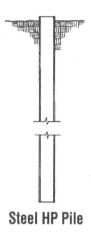

Steel HP Pile

Figure 31-30　Installation Time in Labor-Hours for Dewatering

Description	Labor-Hours	Unit
Excavate Drainage Trench, 2' Wide		
2' Deep	.178	C.Y.
3' Deep	.160	C.Y.
Sump Pits, by Hand		
Light Soil	1.130	C.Y.
Heavy Soil	2.290	C.Y.
Pumping 8 Hours, Diaphragm or Centrifugal Pump		
Attended 2 hours per day	3.000	Day
Attended 8 hours per day	12.000	Day
Pumping 24 Hours, Attended 24 Hours, 4 Men at		
6 Hour Shifts, 1 Week Minimum	25.140	Day
Relay Corrugated Metal Pipe, Including		
Excavation, 3' Deep		
12" Diameter	.209	L.F.
18" Diameter	.240	L.F.
Sump Hole Construction, Including Excavation,		
with 12" Gravel Collar		
Corrugated Pipe		
12" Diameter	.343	L.F.
18" Diameter	.480	L.F.
Wood Lining, Up to 4'x4'	.080	SFCA
Wellpoint System, Single Stage, Install and		
Remove, per Length of Header		
Minimum	.750	L.F.
Maximum	2.000	L.F.
Wells, 10' to 20' Deep with Steel Casing		
2' Diameter		
Minimum	.145	V.L.F.
Average	.245	V.L.F.
Maximum	.490	V.L.F.

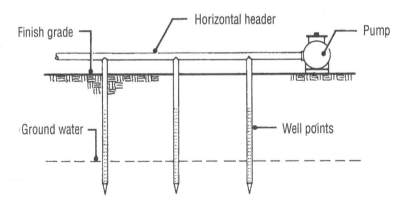

Figure 31-31 Installation Time in Labor-Hours for Caissons

Description	Labor-Hours	Unit
Caissons Open Style Machine Drilled to 50' Deep in Stable Ground, No Casings or Ground Water		
36" Diameter	.384	V.L.F.
8' Bell Diameter Add	20.000	Ea.
Open Style Machine Drilled to 50' Deep in Wet Ground Pulled Casing and Pumping		
36" Diameter	.933	V.L.F.
8' Bell Diameter Add	23.333	Ea.
Open Style Machine Drilled to 50' Deep in Soft Rock and Medium Hard Shale		
36" Diameter	5.867	V.L.F.
8' Bell Diameter Add	67.692	Ea.
Mobilization	65.000	
Bottom Inspection	6.667	Ea.

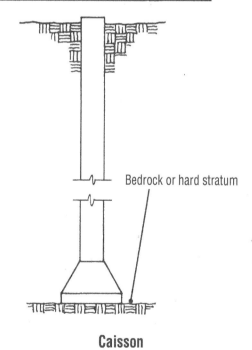

Bedrock or hard stratum

Caisson

Division 31

Figure 31-32 Installation Time in Labor-Hours for Pressure-Injected Footings

Description	Labor-Hours	Unit
Pressure Injected Footings 12" Diameter Shaft	.400	V.L.F.
Pressure Injected Footing 12" to 18"		
Diameter to 40'	.278	V.L.F.
Pile Cutoff Concrete Pile with Thin Shell	.211	Ea.
Mobilization		
Small Job	142.000	
Large Job	237.000	

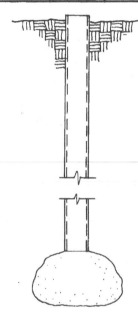

Pressure Injected Footing

Figure 31-33   Installation Time in Labor-Hours for Sheet Piling (Sheeting)

Description	Labor-Hours	Unit
Wood Sheeting Including Wales, Braces and Spacers		
8' Deep Excavation Pull and Salvage	.121	S.F.
Left in Place	.091	S.F.
12' Deep Pull and Salvage	.148	S.F.
Left in Place	.111	S.F.
16' Deep Pull and Salvage	.167	S.F.
Left in Place	.125	S.F.
20' Deep Pull and Salvage	.190	S.F.
Left in Place	.143	S.F.
Steel Sheet Piling		
15' Deep Excavation Pull and Salvage	.098	S.F.
Left in Place	.065	S.F.
20' Deep Pull and Salvage	.100	S.F.
Left in Place	.067	S.F.
25' Deep Pull and Salvage	.096	S.F.
Left in Place	.064	S.F.
Tieback, Based on Total Length		
Minimum	.553	L.F.
Maximum	1.250	L.F.

Normal

1-7/8"

Alternate

Steel Sheet Piling Interlocking Connections

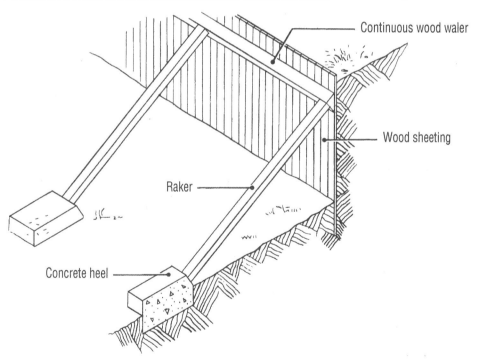

Wood Sheet Piling System

(continued on next page)

Division 31

Figure 31-33 Installation Time in Labor-Hours for Sheet Piling (Sheeting)
(continued from previous page)

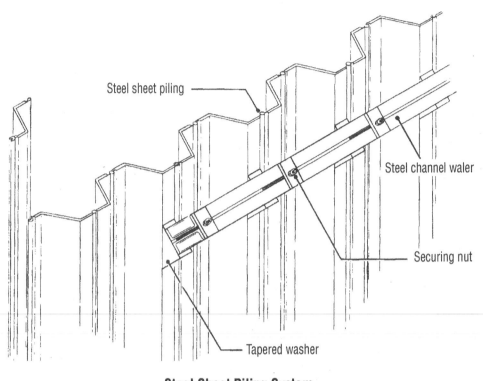

Steel Sheet Piling System

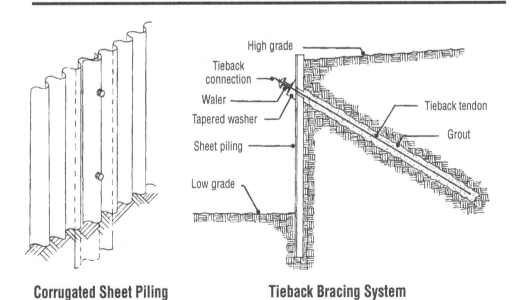

Corrugated Sheet Piling   **Tieback Bracing System**

Figure 31-34 Installation Time in Labor-Hours for Erosion Control

Description	Labor-Hours	Unit
Mulch		
Hand Spread		
Wood Chips, 2" Deep	.004	S.F.
Oat Straw, 1" Deep	.002	S.F.
Excelsior w/Netting	.001	S.F.
Polyethytlene Film	.001	S.F.
Shredded Bark, 3" Deep	.009	S.F.
Pea Stone	.643	C.Y.
Marble Chips	2.400	C.Y.
Polypropylene Fabric	.001	S.F.
Jute Mesh	.001	S.F.
Machine Spread		
Wood Chips, 2" Deep	1.970	MSF
Oat Straw, 1" Deep	.089	MSF
Shredded Bark, 3" Deep	2.960	MSF
Pea Stone	.047	C.Y.
Hydraulic Spraying		
Wood Cellulose	.200	MSF
Rip Rap		
Filter Stone, Machine Placed	.258	C.Y.
1/3 C.Y. Pieces, Crane Set, Grouted	.700	S.Y.
18" Thick, Crane Set, Not Grouted	1.060	S.Y.
Gabion Revetment Mats, Stone Filled,		
12" Deep	.366	S.Y.
Precast Interlocking Concrete Block Pavers	.078	S.F.

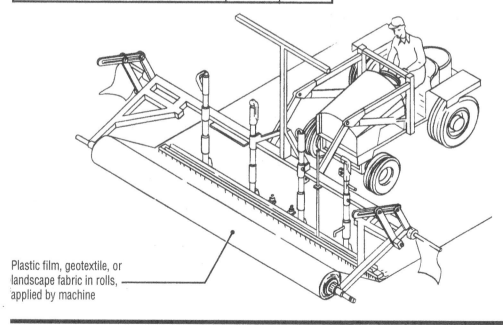

Plastic film, geotextile, or
landscape fabric in rolls,
applied by machine

Figure 31-35   Installation Time in Labor-Hours for Erosion Control:
Gabion Retaining Walls

Description	Labor-Hours	Unit
Gabion		
3 x 6 x 1	0.496	Ea.
3 x 6 x 2	4.308	
3 x 9 x 1	1.120	
3 x 9 x 3	9.333	
3 x 12 x 1	2.000	
3 x 12 x 3	18.667	↓
Drainage Stone, 3/4 Diameter	.092	C.Y.
Backfill Dozer	.010	
Compaction, Roller, 12 Lifts	.014	↓

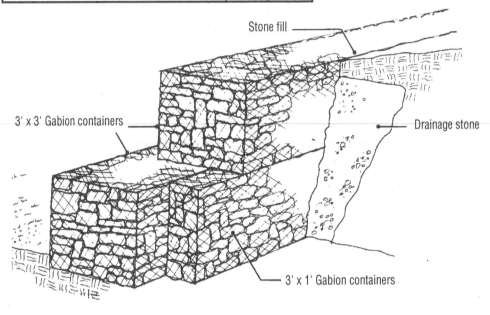

Division Thirty-Two
Exterior Improvements

Introduction

Division 32 covers exterior improvements, which were previously categorized in MasterFormat 95 Division 2. Addressed in Division 32 is site improvement maintenance data including pavement maintenance, snow removal, and landscape maintenance. Also included is data for sand seal, fog seal, slurry seal, and cold milling of asphalt pavement.

In addition to maintenance, Division 32 addresses sidewalks, driveways, and patios. Coverage includes base course gravel and crushed stone, bituminous-stabilized base courses, asphalt paving of roads, rigid concrete paving, rigid paving for airports, and rigid paving for canals. Unit pavers are also addressed, including precast concrete, brick, asphalt blocks, and stone, in addition to paving specialties such as cast-in-place, precast, stone, and bituminous concrete curbs. Other paving specialties include parking bumpers, pavement marking, and athletic surfacing.

Other site improvements include various types of fencing including wood, chain link, specialty metals, and plastic. The retaining wall section addresses cast-in-place concrete, segmental walls, metal cribs, gabion, and stone. Some specialty items include fabricated bridges and an extensive landscaping section.

The landscaping section of MasterFormat Division 32 includes data on mulching, planting, seeding, sodding, sprigging, and soil structure improvement. An extensive list of plants, trees, and ground covers are also included in the data. There are also support items including tree guying, edging, and transplanting of landscape items.

Reducing the amount of impervious surfaces is an important method of controlling storm water runoff. This can be accomplished by limiting the building and paved surfaces footprint, using pervious concrete for paved surfaces, or capturing the rainwater in cisterns for non-potable water needs. A successful strategy for reducing stormwater runoff can earn a Sustainable Sites point in LEED® NC (new construction).

Landscaping can be planned so that potable water use can be reduced or eliminated. Strategies include using captured stormwater runoff (as mentioned above), recycled (and treated) wastewater, efficient irrigation methods, and draught resistant plants. Reducing potable water for landscaping can earn a LEED NC Water Efficiency point, and eliminating potable water entirely for landscaping can earn an additional point.

Estimating Data

The following tables present estimating guidelines for items found in Division 32 – Exterior Improvements. Please note that these guidelines are intended as indicators of what may generally be expected, but that each prospect must be evaluated individually.

Table of Contents

Division 32

Estimating Tips

Paving

When estimating paving, keep in mind the overall project schedule. Although it is common to wait until the end of the project before paving, consider what time of the year the project is scheduled for completion. In colder climates, many concrete plants close for the season in late November and are thus very busy just before then. The supply or availability may be limited, and the prices may be higher.

Recycling of asphalt pavement is becoming very popular and is an alternative to removal and replacement of asphalt pavement. It can be a good value engineering proposal if removed pavement can be recycled either at the site or another site that is reasonably close to the project site.

Quick Quantity for Asphaltic Concrete Paving

A quick rule of thumb for asphaltic concrete is that for each inch of pavement thickness, one square yard is equivalent to 110 pounds.

Bases, Ballasts, and Paving

When estimating paving, keep in mind the project schedule. If an asphaltic paving project is in a colder climate and runs through to the spring, consider placing the base in the autumn and then topping it in the spring just prior to completion. This could save considerable costs in spring repair. Keep in mind that prices for asphalt and concrete are generally higher in the cold seasons.

Planting

The timing of planting and the guarantee specifications often determine the costs for establishing tree and shrub growth and a stand of grass or ground cover. Plan the work schedule to coincide with the local planting season. Maintenance and growth guarantees can add from 20% to 100% to the total landscaping cost. The cost to replace trees and shrubs can be as high as 5% of the total cost, depending on the planting zone, soil conditions, and the time of year.

Checklist ✓

For an estimate to be reliable, all items must be accounted for. A complete estimate can also limit contingencies. The following checklist can be used to help ensure that all items are included.

☐ **Flexible Paving Surface Treatment**
   ☐ Slurry seal
   ☐ Sand seal
   ☐ Fog seal

☐ **Driveways**

☐ **Sidewalks**

☐ **Steps**

☐ **Curbs**
   ☐ Paving
   ☐ Roads
   ☐ Walks
   ☐ Patios
   ☐ Other

☐ **Flexible Paving**
   ☐ Asphalt paving

☐ **Rigid Paving**

☐ **Unit Paving**
   ☐ Interlocking precast concrete units
   ☐ Precast concrete unit paving slabs
   ☐ Brick unit
   ☐ Asphalt unit
   ☐ Stone paving

☐ **Athletic and Recreational Surfacing**

☐ **Retaining Walls**

☐ **Parking Bumpers**

☐ **Pavement Markings**

☐ **Fences and Gates**

☐ **Fabricated Bridges**

☐ **Screening Devices**

☐ **Planting Irrigation**

☐ **Planting Preparation**
   ☐ Lawn
   ☐ Seeding
   ☐ Sodding
   ☐ Trees
   ☐ Bushes
   ☐ Shrubs
   ☐ Plantings
   ☐ Ground cover
   ☐ Mulch
   ☐ Edgings
   ☐ Maintenance agreements

☐ **Planting Accessories**
   ☐ Edging
   ☐ Tree guying

☐ **Transplanting**

☐ **Special Considerations**

Figure 32-1 Costs per Square Yard of Granular Base and Subbase Courses – per Ton

Where cost per ton is the pay unit and assumed density is 118.5 lb./CF (88.9 lb. SY-in):

Cost per Ton	Thickness in Inches									
	3	4	5	6	7	8	9	10	11	12
15	2.00	2.67	3.33	4.00	4.67	5.33	6.00	6.67	7.33	8.00
16	2.13	2.84	3.56	4.27	4.98	5.69	6.40	7.11	7.82	8.53
17	2.27	3.02	3.78	4.53	5.29	6.05	6.80	7.56	8.31	9.07
18	2.40	3.20	4.00	4.80	5.60	6.40	7.20	8.00	8.80	9.60
19	2.53	3.38	4.22	5.07	5.91	6.76	7.60	8.45	9.29	10.13
20	2.67	3.56	4.45	5.33	6.22	7.11	8.00	8.89	9.78	10.67
21	2.80	3.73	4.67	5.60	6.53	7.47	8.40	9.33	10.27	11.20
22	2.93	3.91	4.89	5.87	6.85	7.82	8.80	9.78	10.76	11.73
23	3.07	4.09	5.11	6.13	7.16	8.18	9.20	10.22	11.25	12.27
24	3.20	4.27	5.33	6.40	7.47	8.53	9.60	10.67	11.73	12.80
25	3.33	4.45	5.56	6.67	7.78	8.89	10.00	11.11	12.22	13.34
26	3.47	4.62	5.78	6.93	8.09	9.25	10.40	11.56	12.71	13.87
27	3.60	4.80	6.00	7.20	8.40	9.60	10.80	12.00	13.20	14.40
28	3.73	4.98	6.22	7.47	8.71	9.96	11.20	12.45	13.69	14.94
29	3.87	5.16	6.45	7.73	9.02	10.31	11.60	12.89	14.18	15.47
30	4.00	5.33	6.67	8.00	9.33	10.67	12.00	13.34	14.67	16.00
31	4.13	5.51	6.89	8.27	9.65	11.02	12.40	13.78	15.16	16.54
32	4.27	5.69	7.11	8.53	9.96	11.38	12.80	14.22	15.65	17.07
33	4.40	5.87	7.33	8.80	10.27	11.73	13.20	14.67	16.14	17.60
34	4.53	6.05	7.56	9.07	10.58	12.09	13.60	15.11	16.62	18.14
35	4.67	6.22	7.78	9.33	10.89	12.45	14.00	15.56	17.11	18.67

Example:
10,000 SY of base, 6" thick at $30 per ton
10,000 SY x $8.00/SY = $80,000

Division 32

Figure 32-2 Costs per Square Yard of Granular Base and Subbase Courses – per CY

Where Cost per cubic yard is the pay unit:

Cost per C.Y.	3	4	5	6	7	8	9	10	11	12
16.00	1.33	1.78	2.22	2.67	3.11	3.56	4.00	4.44	4.89	5.33
17.00	1.42	1.89	2.36	2.83	3.31	3.78	4.25	4.72	5.19	5.67
18.00	1.50	2.00	2.50	3.00	3.50	4.00	4.50	5.00	5.50	6.00
19.00	1.58	2.11	2.64	3.17	3.69	4.22	4.75	5.28	5.81	6.33
20.00	1.67	2.22	2.78	3.33	3.89	4.44	5.00	5.56	6.11	6.67
21.00	1.75	2.33	2.92	3.50	4.08	4.67	5.25	5.83	6.42	7.00
22.00	1.83	2.44	3.06	3.67	4.28	4.89	5.50	6.11	6.72	7.33
23.00	1.92	2.56	3.19	3.83	4.47	5.11	5.75	6.39	7.03	7.67
24.00	2.00	2.67	3.33	4.00	4.67	5.33	6.00	6.67	7.33	8.00
25.00	2.08	2.78	3.47	4.17	4.86	5.56	6.25	6.94	7.64	8.33
26.00	2.17	2.89	3.61	4.33	5.06	5.78	6.50	7.22	7.94	8.67
27.00	2.25	3.00	3.75	4.50	5.25	6.00	6.75	7.50	8.25	9.00
28.00	2.33	3.11	3.89	4.67	5.44	6.22	7.00	7.78	8.56	9.33
29.00	2.42	3.22	4.03	4.83	5.64	6.44	7.25	8.06	8.86	9.67
30.00	2.50	3.33	4.17	5.00	5.83	6.67	7.50	8.33	9.17	10.00
31.00	2.58	3.44	4.31	5.17	6.03	6.89	7.75	8.61	9.47	10.33
32.00	2.67	3.56	4.44	5.33	6.22	7.11	8.00	8.89	9.78	10.67
33.00	2.75	3.67	4.58	5.50	6.42	7.33	8.25	9.17	10.08	11.00
34.00	2.83	3.78	4.72	5.67	6.61	7.56	8.50	9.44	10.39	11.33
35.00	2.92	3.89	4.86	5.83	6.81	7.78	8.75	9.72	10.69	11.67
36.00	3.00	4.00	5.00	6.00	7.00	8.00	9.00	10.00	11.00	12.00
37.00	3.08	4.11	5.14	6.17	7.19	8.22	9.25	10.28	11.31	12.33
38.00	3.17	4.22	5.28	6.33	7.39	8.44	9.50	10.56	11.61	12.67
39.00	3.25	4.33	5.42	6.50	7.58	8.67	9.75	10.83	11.92	13.00
40.00	3.33	4.44	5.56	6.67	7.78	8.89	10.00	11.11	12.22	13.33

Thickness in Inches

Example:

10,000 SY of base, 6" thick at $24.00/CY

10,000 SY x $4.00/SY = $40,000

Figure 32-3 Costs per Square Yard of Prime Coats, Tack Coats – Gallons

Where cost per gallon is the pay unit:

Cost per Gal.	Application Rate, Gal. Per S.Y.					
	0.05	0.10	0.15	0.20	0.25	0.30
4.00	0.20	0.40	0.60	0.80	1.00	1.20
4.20	0.21	0.42	0.63	0.84	1.05	1.26
4.40	0.22	0.44	0.66	0.88	1.10	1.32
4.60	0.23	0.46	0.69	0.92	1.15	1.38
4.80	0.24	0.48	0.72	0.96	1.20	1.44
5.00	0.25	0.50	0.75	1.00	1.25	1.50
5.20	0.26	0.52	0.78	1.04	1.30	1.56
5.40	0.27	0.54	0.81	1.08	1.35	1.62
5.60	0.28	0.56	0.84	1.12	1.40	1.68
5.80	0.29	0.58	0.87	1.16	1.45	1.74
6.00	0.30	0.60	0.90	1.20	1.50	1.80
6.20	0.31	0.62	0.93	1.24	1.55	1.86
6.40	0.32	0.64	0.96	1.28	1.60	1.92
6.60	0.33	0.66	0.99	1.32	1.65	1.98
6.80	0.34	0.68	1.02	1.36	1.70	2.04
7.00	0.35	0.70	1.05	1.40	1.75	2.10
7.20	0.36	0.72	1.08	1.44	1.80	2.16
7.40	0.37	0.74	1.11	1.48	1.85	2.22
7.60	0.38	0.76	1.14	1.52	1.90	2.28
7.80	0.39	0.78	1.17	1.56	1.95	2.34
8.00	0.40	0.80	1.20	1.60	2.00	2.40
8.20	0.41	0.82	1.23	1.64	2.05	2.46
8.40	0.42	0.84	1.26	1.68	2.10	2.52
8.60	0.43	0.86	1.29	1.72	2.15	2.58
8.80	0.44	0.88	1.32	1.76	2.20	2.64
9.00	0.45	0.90	1.35	1.80	2.25	2.70
9.20	0.46	0.92	1.38	1.84	2.30	2.76
9.40	0.47	0.94	1.41	1.88	2.35	2.82
9.60	0.48	0.96	1.44	1.92	2.40	2.88
9.80	0.49	0.98	1.47	1.96	2.45	2.94

Division 32

Example:

10,000 SY of prime coat, 0.15 Gal. per SY at $5.00 per Gal.

10,000 SY x $0.75/SY = $7,500

Figure 32-4 Costs per Square Yard of Prime Coats, Tack Coats – Ton

Where cost per ton is the pay unit (assume 1 ton = 241 gal.):

Cost per Ton	Application Rate, Gal. Per S.Y.					
	0.05	0.10	0.15	0.20	0.25	0.30
1,000.00	0.21	0.41	0.62	0.83	1.04	1.24
1,020.00	0.21	0.42	0.63	0.85	1.06	1.27
1,040.00	0.22	0.43	0.65	0.86	1.08	1.29
1,060.00	0.22	0.44	0.66	0.88	1.10	1.32
1,080.00	0.22	0.45	0.67	0.90	1.12	1.34
1,200.00	0.25	0.50	0.75	1.00	1.24	1.49
1,220.00	0.25	0.51	0.76	1.01	1.27	1.52
1,240.00	0.26	0.51	0.77	1.03	1.29	1.54
1,260.00	0.26	0.52	0.78	1.05	1.31	1.57
1,280.00	0.27	0.53	0.80	1.06	1.33	1.59
1,300.00	0.27	0.54	0.81	1.08	1.35	1.62
1,320.00	0.27	0.55	0.82	1.10	1.37	1.64
1,340.00	0.28	0.56	0.83	1.11	1.39	1.67
1,360.00	0.28	0.56	0.85	1.13	1.41	1.69
1,380.00	0.29	0.57	0.86	1.15	1.43	1.72
1,400.00	0.29	0.58	0.87	1.16	1.45	1.74
1,420.00	0.29	0.59	0.88	1.18	1.47	1.77
1,440.00	0.30	0.60	0.90	1.20	1.49	1.79
1,460.00	0.30	0.61	0.91	1.21	1.51	1.82
1,480.00	0.31	0.61	0.92	1.23	1.54	1.84
1,500.00	0.31	0.62	0.93	1.24	1.56	1.87
1,520.00	0.32	0.63	0.95	1.26	1.58	1.89
1,540.00	0.32	0.64	0.96	1.28	1.60	1.92
1,560.00	0.32	0.65	0.97	1.29	1.62	1.94
1,580.00	0.33	0.66	0.98	1.31	1.64	1.97

Example:

10,000 SY of prime coat, 0.15 gal. per SY at $1200 per ton

10,000 SY x $0.75/SY = $7500

Figure 32-5 Costs per Square Yard of Asphalt Concrete Pavement Courses

Where cost per ton is the pay unit and assumed density is 145 lbs per ft.[3]:

Cost per Ton	Thickness of Pavement Course in Inches												
	.5	1	2	3	4	5	6	7	8	9	10	11	12
50	1.36	2.72	5.44	8.16	10.88	13.59	16.31	19.03	21.75	24.47	27.19	29.91	32.63
55	1.50	2.99	5.98	8.97	11.96	14.95	17.94	20.93	23.93	26.92	29.91	32.90	35.89
60	1.63	3.26	6.53	9.79	13.05	16.31	19.58	22.84	26.10	29.36	32.63	35.89	39.15
62	1.69	3.37	6.74	10.11	13.49	16.86	20.23	23.60	26.97	30.34	33.71	37.08	40.46
64	1.74	3.48	6.96	10.44	13.92	17.40	20.88	24.36	27.84	31.32	34.80	38.28	41.76
66	1.79	3.59	7.18	10.77	14.36	17.94	21.53	25.12	28.71	32.30	35.89	39.48	43.07
68	1.85	3.70	7.40	11.09	14.79	18.49	22.19	25.88	29.58	33.28	36.98	40.67	44.37
70	1.90	3.81	7.61	11.42	15.23	19.03	22.84	26.64	30.45	34.26	38.06	41.87	45.68
72	1.96	3.92	7.83	11.75	15.66	19.58	23.49	27.41	31.32	35.24	39.15	43.07	46.98
74	2.01	4.02	8.05	12.07	16.10	20.12	24.14	28.17	32.19	36.21	40.24	44.26	48.29
76	2.07	4.13	8.27	12.40	16.53	20.66	24.80	28.93	33.06	37.19	41.33	45.46	49.59
78	2.12	4.24	8.48	12.72	16.97	21.21	25.45	29.69	33.93	38.17	42.41	46.65	50.90
80	2.18	4.35	8.70	13.05	17.40	21.75	26.10	30.45	34.80	39.15	43.50	47.85	52.20
82	2.23	4.46	8.92	13.38	17.84	22.29	26.75	31.21	35.67	40.13	44.59	49.05	53.51
84	2.28	4.57	9.14	13.70	18.27	22.84	27.41	31.97	36.54	41.11	45.68	50.24	54.81
86	2.34	4.68	9.35	14.03	18.71	23.38	28.06	32.73	37.41	42.09	46.76	51.44	56.12
87	2.37	4.73	9.46	14.19	18.92	23.65	28.38	33.11	37.85	42.58	47.31	52.04	56.77
88	2.39	4.79	9.57	14.36	19.14	23.93	28.71	33.50	38.28	43.07	47.85	52.64	57.42
89	2.42	4.84	9.68	14.52	19.36	24.20	29.04	33.88	38.72	43.55	48.39	53.23	58.07
90	2.45	4.89	9.79	14.68	19.58	24.47	29.36	34.26	39.15	44.04	48.94	53.83	58.73

Example:

10,000 SY of pavement, 4" thick at $70 per ton

10,000 SY x $15.23/SY = $152,300

Division 32

Figure 32-6 Costs per Square Yard of Cement Treated Bases

Where cost per cubic yard is the pay unit:

Cost per C.Y.	Thickness in Inches							
	3	4	5	6	7	8	9	10
14.00	1.17	1.56	1.94	2.33	2.72	3.11	3.50	3.89
15.00	1.25	1.67	2.08	2.50	2.92	3.33	3.75	4.17
16.00	1.33	1.78	2.22	2.67	3.11	3.56	4.00	4.44
17.00	1.42	1.89	2.36	2.83	3.31	3.78	4.25	4.72
18.00	1.50	2.00	2.50	3.00	3.50	4.00	4.50	5.00
19.00	1.58	2.11	2.64	3.17	3.69	4.22	4.75	5.28
20.00	1.67	2.22	2.78	3.33	3.89	4.44	5.00	5.56
25.00	2.08	2.78	3.47	4.17	4.86	5.56	6.25	6.94
30.00	2.50	3.33	4.17	5.00	5.83	6.67	7.50	8.33
35.00	2.92	3.89	4.86	5.83	6.81	7.78	8.75	9.72
40.00	3.33	4.44	5.56	6.67	7.78	8.89	10.00	11.11
45.00	3.75	5.00	6.25	7.50	8.75	10.00	11.25	12.50
50.00	4.17	5.56	6.94	8.33	9.72	11.11	12.50	13.89
55.00	4.58	6.11	7.64	9.17	10.69	12.22	13.75	15.28
60.00	5.00	6.67	8.33	10.00	11.67	13.33	15.00	16.67
65.00	5.42	7.22	9.03	10.83	12.64	14.44	16.25	18.06
70.00	5.83	7.78	9.72	11.67	13.61	15.56	17.50	19.44
75.00	6.25	8.33	10.42	12.50	14.58	16.67	18.75	20.83

Example:

10,000 SY of base, 8" thick at $30.00 per CY

10,000 SY x $6.67/SY = $66,700

Figure 32-7 Costs per Square Yard of Portland Cement Concrete Pavement Courses

Where cost per cubic yard is the pay unit:

Cost per C.Y.	Pavement Thickness in Inches								
	4	5	6	7	8	9	10	11	12
65.00	7.22	9.03	10.83	12.64	14.44	16.25	18.06	19.86	21.67
70.00	7.78	9.72	11.67	13.61	15.56	17.50	19.44	21.39	23.33
75.00	8.33	10.42	12.50	14.58	16.67	18.75	20.83	22.92	25.00
80.00	8.89	11.11	13.33	15.56	17.78	20.00	22.22	24.44	26.67
85.00	9.44	11.81	14.17	16.53	18.89	21.25	23.61	25.97	28.33
90.00	10.00	12.50	15.00	17.50	20.00	22.50	25.00	27.50	30.00
95.00	10.56	13.19	15.83	18.47	21.11	23.75	26.39	29.03	31.67
100.00	11.11	13.89	16.67	19.44	22.22	25.00	27.78	30.56	33.33
105.00	11.67	14.58	17.50	20.42	23.33	26.25	29.17	32.08	35.00
110.00	12.22	15.28	18.33	21.39	24.44	27.50	30.56	33.61	36.67
115.00	12.78	15.97	19.17	22.36	25.56	28.75	31.94	35.14	38.33
120.00	13.33	16.67	20.00	23.33	26.67	30.00	33.33	36.67	40.00
125.00	13.89	17.36	20.83	24.31	27.78	31.25	34.72	38.19	41.67
130.00	14.44	18.06	21.67	25.28	28.89	32.50	36.11	39.72	43.33
135.00	15.00	18.75	22.50	26.25	30.00	33.75	37.50	41.25	45.00
140.00	15.56	19.44	23.33	27.22	31.11	35.00	38.89	42.78	46.67

Example:
10,000 SY of pavement 8" thick at $100 per CY
10,000 SY x $22.22 SY = $222,200

Division 32

Figure 32-8 Area of Road Surface for Various Road Widths

Area in Square Yards

Road Width	Per Linear Foot	Per 100 Feet	Per Mile
6 Ft.	0.67	66.67	3,520
7	0.78	77.78	4,107
8	0.89	88.89	4,693
9	1.00	100.00	5,280
10	1.11	111.11	5,867
11	1.22	122.22	6,453
12	1.33	133.33	7,040
13	1.44	144.44	7,627
14	1.56	155.56	8,213
15	1.67	166.67	8,800
16	1.78	177.78	9,387
17	1.89	188.89	9,973
18	2.00	200.00	10,560
20	2.22	222.22	11,733
22	2.44	244.44	12,907
24	2.67	266.67	14,080
25	2.78	277.78	14,667
26	2.89	288.89	15,253
28	3.11	311.11	16,427
30	3.33	333.33	17,600
32	3.56	355.56	18,773
34	3.78	377.78	19,947
36	4.00	400.00	21,120
38	4.22	422.22	22,293
40	4.44	444.44	23,467
50	5.56	555.56	29,333
60	6.67	666.67	35,200
70	7.78	777.78	41,067
75	8.33	833.33	44,000
80	8.89	888.89	46,933

(continued on next page)

Figure 32-8 Area of Road Surface for Various Road Widths
(continued from previous page)

Area in Square Meters

Road Width	Per Linear Meter	Per 50 m	Per Kilometer
2 m	2 m²	100 m²	2,000 m²
2.5	2.5	125	2,500
3	3	150	3,000
3.5	3.5	175	3,500
4	4	200	4,000
4.5	4.5	225	4,500
5	5	250	5,000
5.5	5.5	275	5,500
6	6	300	6,000
6.5	6.5	325	6,500
7	7	350	7,000
7.5	7.5	375	7,500
8	8	400	8,000
8.5	8.5	425	8,500
9	9	450	9,000
9.5	9.5	475	9,500
10	10	500	10,000
10.5	10.5	525	10,500
11	11	550	11,000
11.5	11.5	575	11,500
12	12	600	12,000
15	15	750	15,000
20	20	1,000	20,000
25	25	1,250	25,000

Division 32

Figure 32-9 Cubic Yards of Material Required for Various Widths and Depths per 100 Linear Feet and per Mile

	Width in Feet	\multicolumn{12}{c}{Depth in Inches}

Width in Feet	1	2	3	4	5	6	7	8	9	10	11	12
Per 100 Linear Feet												
1	0.31	0.62	0.93	1.23	1.54	1.85	2.16	2.47	2.78	3.09	3.40	3.70
2	0.62	1.23	1.85	2.47	3.09	3.70	4.32	4.94	5.56	6.17	6.79	7.41
3	0.93	1.85	2.78	3.70	4.63	5.56	6.48	7.41	8.33	9.26	10.20	11.10
4	1.23	2.47	3.70	4.94	6.17	7.41	8.64	9.88	11.10	12.30	13.60	14.80
5	1.54	3.09	4.63	6.17	7.72	9.26	10.80	12.30	13.90	15.40	17.00	18.50
6	1.85	3.70	5.56	7.41	9.26	11.10	13.00	14.80	16.70	18.50	20.40	22.20
7	2.16	4.32	6.48	8.64	10.80	13.00	15.10	17.30	19.40	21.60	23.80	25.90
8	2.47	4.94	7.41	9.88	12.30	14.80	17.30	19.80	22.20	24.70	27.20	29.60
9	2.78	5.56	8.33	11.10	13.90	16.70	19.40	22.20	25.00	27.80	30.60	33.30
10	3.09	6.17	9.26	12.30	15.40	18.50	21.60	24.70	27.80	30.90	34.00	37.00
20	6.17	12.30	18.50	24.70	30.90	37.00	43.20	49.40	55.60	61.70	67.90	74.10
30	9.26	18.50	27.80	37.00	46.30	55.60	64.80	74.10	83.30	92.60	102.00	111.00
40	12.30	24.70	37.00	49.40	61.70	74.10	86.40	98.80	111.00	123.00	136.00	148.00
50	15.40	30.90	46.30	61.70	77.20	92.60	108.00	123.00	139.00	154.00	170.00	185.00
60	18.50	37.00	55.60	74.10	92.60	111.00	130.00	148.00	167.00	185.00	204.00	222.00
70	21.60	43.20	64.80	86.40	108.00	130.00	151.00	173.00	194.00	216.00	238.00	259.00
80	24.70	49.40	74.10	98.80	123.00	148.00	173.00	198.00	222.00	247.00	272.00	296.00
90	27.80	55.60	83.30	111.00	139.00	167.00	194.00	222.00	250.00	278.00	306.00	333.00
100	30.90	61.70	92.60	123.00	154.00	185.00	216.00	247.00	278.00	309.00	340.00	370.00
Per Mile												
1	16.30	32.60	48.90	65.20	81.50	97.80	114.00	130.00	147.00	163.00	179.00	196.00
2	32.60	65.20	97.80	130.00	163.00	196.00	228.00	261.00	293.00	326.00	359.00	391.00
3	48.90	97.80	147.00	196.00	244.00	293.00	342.00	391.00	440.00	489.00	538.00	587.00
4	65.20	130.00	196.00	261.00	326.00	391.00	456.00	521.00	587.00	652.00	717.00	782.00
5	81.50	163.00	244.00	326.00	407.00	489.00	570.00	652.00	733.00	815.00	896.00	978.00
6	97.80	196.00	293.00	391.00	489.00	587.00	684.00	782.00	880.00	978.00	1,076.00	1,173.00

(continued on next page)

Figure 32-9 Cubic Yards of Material Required for Various Widths and Depths per 100 Linear Feet and per Mile

(continued from previous page)

Width in Feet	Depth in Inches											
	1	2	3	4	5	6	7	8	9	10	11	12
7	114.00	228.00	342.00	456.00	570.00	684.00	799.00	913.00	1,027.00	1,141.00	1,255.00	1,369.00
8	130.00	261.00	391.00	521.00	652.00	782.00	913.00	1,043.00	1,173.00	1,304.00	1,434.00	1,564.00
9	147.00	293.00	440.00	587.00	733.00	880.00	1,027.00	1,173.00	1,320.00	1,467.00	1,613.00	1,760.00
10	163.00	326.00	489.00	652.00	815.00	978.00	1,141.00	1,304.00	1,467.00	1,630.00	1,793.00	1,956.00
20	326.00	652.00	978.00	1,304.00	1,630.00	1,956.00	2,281.00	2,607.00	2,933.00	3,259.00	3,585.00	3,911.00
30	489.00	978.00	1,467.00	1,956.00	2,444.00	2,933.00	3,422.00	3,911.00	4,440.00	4,889.00	5,378.00	5,867.00
40	652.00	1,304.00	1,956.00	2,607.00	3,259.00	3,911.00	4,563.00	5,215.00	5,867.00	6,519.00	7,170.00	7,822.00
50	815.00	1,630.00	2,444.00	3,259.00	4,074.00	4,889.00	5,704.00	6,519.00	7,333.00	8,148.00	8,963.00	9,778.00
60	978.00	1,956.00	2,933.00	3,911.00	4,889.00	5,867.00	6,844.00	7,822.00	8,800.00	9,778.00	10,756.00	11,733.00
70	1,141.00	2,281.00	3,422.00	4,563.00	5,704.00	6,844.00	7,985.00	9,126.00	10,267.00	11,407.00	12,548.00	13,689.00
80	1,304.00	2,607.00	3,911.00	5,215.00	6,519.00	7,822.00	9,126.00	10,430.00	11,733.00	13,037.00	14,341.00	15,644.00
90	1,467.00	2,933.00	4,400.00	5,867.00	7,333.00	8,800.00	10,267.00	11,733.00	13,200.00	14,667.00	16,133.00	17,600.00
100	1,630.00	3,259.00	4,889.00	6,519.00	8,148.00	9,778.00	11,407.00	13,037.00	14,667.00	16,296.00	17,926.00	19,556.00

Per Mile

Where: q = Quantity of material, cubic yards
D = Depth, inches
W = Width, feet
L = Length

Formulas used for calculations:

$$100\ \text{L.F.:}\ q = \frac{D}{36}\ \frac{W}{3}\ \frac{100}{3} = 0.3086\ DW$$

$$\text{Mile:}\ q = \frac{D}{35}\ \frac{W}{3}\ \frac{5{,}280}{3} = 16.2963\ DW$$

Division 32

Figure 32-10 Excavation and Fill Volume Chart

This table is a handy quick reference for estimators when calculating base course gravel or stone for walks, driveways, or parking lots.

Compacted cubic yards of run-of-bank gravel = square yards × depth in inches × .0333*

Square Yards	2"	3"	4"	5"	6"	8"	10"
1	.066	.099	.132	.165	.198	.264	.330
2	.132	.198	.264	.330	.396	.528	.660
3	.198	.297	.396	.495	.594	.792	.990
4	.264	.396	.528	.660	.792	1.056	1.320
5	.330	.495	.660	.825	.990	1.320	1.650
6	.396	.594	.792	.990	1.188	1.584	1.980
7	.412	.693	.924	1.155	1.386	1.848	2.310
8	.528	.792	1.056	1.320	1.584	2.112	2.640
9	.594	.891	1.188	1.485	1.782	2.376	2.970
10	.660	.990	1.320	1.650	1.980	2.640	3.300
20	1.32	1.98	2.64	3.30	3.96	5.28	6.60
30	1.98	2.97	3.96	4.95	5.94	7.92	9.90
40	2.64	3.96	5.28	6.60	7.92	10.56	13.20
50	3.30	4.95	6.60	8.25	9.90	13.20	16.50
60	3.96	5.94	7.92	9.90	11.88	15.84	19.80
70	4.62	6.93	9.24	11.55	13.86	18.48	23.10
80	5.28	7.92	10.56	13.20	15.84	21.12	26.40
90	5.94	8.91	11.88	14.85	17.82	23.76	29.70
100	6.60	9.90	13.20	16.50	19.80	26.40	33.00
200	13.20	19.80	26.40	33.00	39.60	52.80	66.00
300	19.80	29.70	39.60	49.50	59.40	79.20	99.00
400	26.40	39.60	52.80	66.00	79.20	105.60	132.00
500	33.00	49.50	66.00	82.50	99.00	132.00	165.00
600	39.60	59.40	79.20	99.00	118.00	158.40	198.00
700	46.20	69.30	92.40	115.50	138.60	184.80	231.00
800	52.80	79.20	105.60	132.00	158.40	211.20	264.00
900	59.40	89.10	118.80	148.50	178.20	237.60	297.00
1000	66.00	99.00	132.00	165.00	198.00	264.00	330.00
2000	132.00	198.00	264.00	330.00	396.00	528.00	660.00
3000	198.00	297.00	396.00	495.00	594.00	792.00	990.00
4000	264.00	396.00	528.00	660.00	792.00	1056.00	1320.00
5000	330.00	495.00	660.00	825.00	990.00	1320.00	1650.00
6000	396.00	594.00	792.00	990.00	1180.00	1584.00	1980.00
7000	462.00	693.00	924.00	1155.00	1386.00	1848.00	2310.00
8000	528.00	792.00	1056.00	1320.00	1584.00	2112.00	2640.00
9000	594.00	891.00	1188.00	1485.00	1782.00	2376.00	2970.00
10,000	660.00	990.00	1320.00	1650.00	1980.00	2640.00	3300.00

* Includes 20% compaction factor

Example: 546 SY x 6 inches deep: 500 SY   99.000 CY
                                    40 SY   7.920 CY
                                     6 SY   1.188 CY

For crushed stone, add 10% to figures in table.   108.108, say 108 CY

For loose yards of gravel (truck measure) add 30% to figures in table.

(courtesy of Peckham Industries)

Figure 32-11   Options to be Added for Roadway in Design

Description	Quantity*
2" Bituminous concrete mix for sidewalk, incl. fine grade & roll	1.1 S.Y.
5" x 16" Granite vertical curbing	2.0 L.F.
9" Loam for tree lawn	.21 C.Y.
1" Sod, in the East	.7 S.Y.
Paint 4" centerline strip	1.0 L.F.
Lighting 30' aluminum pole, 140' on center, 400 watt mercury vapor	.007 Ea.

*Per L.F. of roadway

Figure 32-12   Typical Parking Lot Plan (50 Cars)

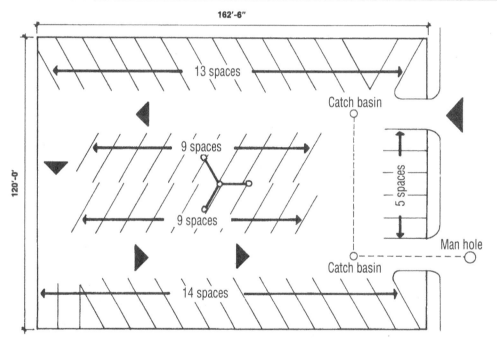

Preliminary Design Data: The space required for parking and maneuvering is between 300 and 400 SF per car, depending on engineering layout and design.

Ninety degree (90°) parking, with a central driveway and two rows of parked cars, will provide the best economy.

Diagonal parking is easier than 90° for the driver and reduces the necessary driveway width, but requires more total space.

Division 32

Figure 32-13 Parking Lot Layout Data Based on 9' x 19'
Parking Stall Size

Φ	P	A	C	W	N	G	D	L	P'	W'
Angle of Stall	Parking Depth	Aisle Width	Curb Length	Width Overall	Net Car Area	Gross Car Area	Distance Last Car	Lost Area	Parking Depth	Width Overall
90°	19'	24'	9	62'	171 S.F.	171 S.F.	9'	0	19'	62'
60°	21'	18'	10.4'	60'	171 S.F.	217 S.F.	7.8'	205 S.F.	18.8'	55.5'
45°	19.8'	13'	12.8'	52.7'	171 S.F.	252 S.F.	6.4'	286 S.F.	16.6'	46.2'

Note: Square foot per car areas do not include the area of the travel lane.

90° Stall Angle: The main reason for use of this stall angle is to achieve the highest car capacity. This may be sound reasoning for employee lots with all day parking, but in most (in and out) lots, there will be difficulty entering the stalls and no traffic lane direction. This may outweigh the advantage of high capacity.

60° Stall Angle: This layout is used most often due to the ease of entering and backing out. Also, the traffic aisle may be smaller.

45° Stall Angle: Requires a small change of direction from the traffic aisle to the stall, so the aisle may be reduced in width.

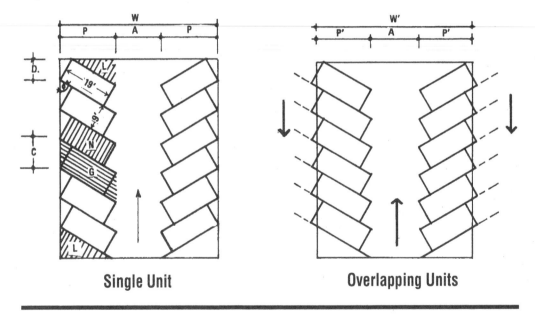

Single Unit   **Overlapping Units**

Figure 32-14 Options to be Added for Parking Lot Appurtenances

Description	Unit
8" x 6" Asphalt berm curbing	L.F.
5" x 16" Vertical granite curbing	L.F.
4-1/2" x 12" Sloped granite edging	L.F.
6" x 18" Precast concrete curb	L.F.
24" Wide 6" high concrete curb and gutter	L.F.
4'-0" Wide bituminous concrete walk, 2" thick, 6" gravel base	S.Y.
4'-0" Wide concrete walk 4" thick, 6" gravel base	S.Y.
2-Coat sealcoating, petroleum resistant, under 1000 S.Y.	S.Y.
1/4 to 3/8 C.Y. rip rap slope protection, 18" thick, not grouted	S.Y.
35"-0' High aluminum pole with (3) - 1,000 watt mercury vapor roadway type fixtures	Ea.
6'0" Long concrete car bumpers, 6" x 10"	Ea.
6'-0" High chain link fence, 1-5/8" top rail, 2" line posts, 10'-0" O.C., 6 gal.	L.F.
3'-0" High corrugated steel guardrail post, 6'-3" O.C.	L.F.
18" Deep perimeter planter bed with shrubs @ 3" O.C.	S.F.
6" Topsoil, including fine grading and seeding	S.Y.

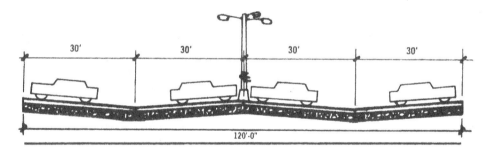

Division 32

Figure 32-15 Costs per Square Yard for Steel – 1-Inch Pavement Thickness

This table reflects continuously reinforced concrete pavement courses where cost per pound is the pay unit (based on cross-sectional area of steel expressed as a percentage of cross-sectional area of pavement course).

Cost per Pound	For 1-Inch Pavement Thickness			
	Cross-Sectional Area of Steel			
	0.5%	0.6%	0.7%	0.8%
1.00	1.84	2.21	2.57	2.94
1.05	1.93	2.32	2.70	3.09
1.10	2.02	2.43	2.83	3.23
1.15	2.11	2.54	2.96	3.38
1.20	2.21	2.65	3.09	3.53
1.25	2.30	2.76	3.22	3.68
1.30	2.39	2.87	3.34	3.82
1.35	2.48	2.98	3.47	3.97
1.40	2.57	3.09	3.60	4.12
1.45	2.66	3.20	3.73	4.26
1.50	2.76	3.31	3.86	4.41
1.55	2.85	3.42	3.99	4.56
1.60	2.94	3.53	4.12	4.70
1.65	3.03	3.64	4.25	4.85
1.70	3.12	3.75	4.37	5.00
1.75	3.22	3.86	4.50	5.15

Figure 32-16 Costs per Square Yard for Steel – 6-Inch Pavement Thickness

This table reflects continuously reinforced concrete pavement courses where cost per pound is the pay unit (based on cross-sectional area of steel expressed as a percentage of cross-sectional area of pavement course).

Cost per Pound	For 6-Inch Pavement Thickness			
	Cross-Sectional Area of Steel			
	0.5%	0.6%	0.7%	0.8%
1.00	11.03	13.23	15.44	17.64
1.05	11.58	13.89	16.21	18.52
1.10	12.13	14.55	16.98	19.41
1.15	12.68	15.22	17.75	20.29
1.20	13.23	15.88	18.52	21.17
1.25	13.78	16.54	19.30	22.05
1.30	14.33	17.20	20.07	22.93
1.35	14.89	17.86	20.84	23.82
1.40	15.44	18.52	21.61	24.70
1.45	15.99	19.19	22.38	25.58
1.50	16.54	19.85	23.16	26.46
1.55	17.09	20.51	23.93	27.35
1.60	17.64	21.17	24.70	28.23
1.65	18.19	21.83	25.47	29.11
1.70	18.74	22.49	26.24	29.99
1.75	19.30	23.16	27.01	30.87

Division 32

Figure 32-17   Costs per Square Yard for Steel – 7-Inch Pavement Thickness

This table reflects continuously reinforced concrete pavement courses where cost per pound is the pay unit (based on cross-sectional area of steel expressed as a percentage of cross-sectional area of pavement course).

For 7-Inch Pavement Thickness				
Cost per Pound	**Cross-Sectional Area of Steel**			
	0.5%	**0.6%**	**0.7%**	**0.8%**
1.00	12.86	15.44	18.01	20.58
1.05	13.51	16.21	18.91	21.61
1.10	14.15	16.98	19.81	22.64
1.15	14.79	17.75	20.71	23.67
1.20	15.44	18.52	21.61	24.70
1.25	16.08	19.30	22.51	25.73
1.30	16.72	20.07	23.41	26.76
1.35	17.37	20.84	24.31	27.79
1.40	18.01	21.61	25.21	28.82
1.45	18.65	22.38	26.11	29.84
1.50	19.30	23.16	27.01	30.87
1.55	19.94	23.93	27.92	31.90
1.60	20.58	24.70	28.82	32.93
1.65	21.23	25.47	29.72	33.96
1.70	21.87	26.24	30.62	34.99
1.75	22.51	27.01	31.52	36.02

Figure 32-18 Costs per Square Yard for Steel – 8-Inch Pavement Thickness

This table reflects continuously reinforced concrete pavement courses where cost per pound is the pay unit (based on cross-sectional area of steel expressed as a percentage of cross-sectional area of pavement course).

Cost per Pound	For 8-Inch Pavement Thickness			
	Cross-Sectional Area of Steel			
	0.5%	0.6%	0.7%	0.8%
1.00	14.70	17.64	20.58	23.52
1.05	15.44	18.52	21.61	24.70
1.10	16.17	19.41	22.64	25.88
1.15	16.91	20.29	23.67	27.05
1.20	17.64	21.17	24.70	28.23
1.25	18.38	22.05	25.73	29.40
1.30	19.11	22.93	26.76	30.58
1.35	19.85	23.82	27.79	31.76
1.40	20.58	24.70	28.82	32.93
1.45	21.32	25.58	29.84	34.11
1.50	22.05	26.46	30.87	35.28
1.55	22.79	27.35	31.90	36.46
1.60	23.52	28.23	32.93	37.64
1.65	24.26	29.11	33.96	38.81
1.70	24.99	29.99	34.99	39.99
1.75	25.73	30.87	36.02	41.17

Division 32

Figure 32-19   Costs per Square Yard for Sawed Joints – 24-Foot Pavement Width

In this table the pay unit reflects the cost per linear foot.

	For Pavements 24 Ft. Wide Having 1 Longitudinal Joint									
	Transverse Joint Spacing, Ft.									
Cost per L.F.	10	15	20	25	30	35	40	45	50	60
3.00	3.83	2.93	2.48	2.21	2.03	1.90	1.80	1.73	1.67	1.58
3.10	3.95	3.02	2.56	2.28	2.09	1.96	1.86	1.78	1.72	1.63
3.20	4.08	3.12	2.64	2.35	2.16	2.02	1.92	1.84	1.78	1.68
3.30	4.21	3.22	2.72	2.43	2.23	2.09	1.98	1.90	1.83	1.73
3.40	4.34	3.32	2.81	2.50	2.30	2.15	2.04	1.96	1.89	1.79
3.50	4.46	3.41	2.89	2.57	2.36	2.21	2.10	2.01	1.94	1.84
3.60	4.59	3.51	2.97	2.65	2.43	2.28	2.16	2.07	2.00	1.89
3.70	4.72	3.61	3.05	2.72	2.50	2.34	2.22	2.13	2.05	1.94
3.80	4.85	3.71	3.14	2.79	2.57	2.40	2.28	2.19	2.11	2.00
3.90	4.97	3.80	3.22	2.87	2.63	2.47	2.34	2.24	2.16	2.05
4.00	5.10	3.90	3.30	2.94	2.70	2.53	2.40	2.30	2.22	2.10
4.10	5.23	4.00	3.38	3.01	2.77	2.59	2.46	2.36	2.28	2.15
4.20	5.36	4.10	3.47	3.09	2.84	2.66	2.52	2.42	2.33	2.21
4.30	5.48	4.19	3.55	3.16	2.90	2.72	2.58	2.47	2.39	2.26
4.40	5.61	4.29	3.63	3.23	2.97	2.78	2.64	2.53	2.44	2.31
4.50	5.74	4.39	3.71	3.31	3.04	2.84	2.70	2.59	2.50	2.36
4.60	5.87	4.49	3.80	3.38	3.11	2.91	2.76	2.65	2.55	2.42
4.70	5.99	4.58	3.88	3.45	3.17	2.97	2.82	2.70	2.61	2.47
4.80	6.12	4.68	3.96	3.53	3.24	3.03	2.88	2.76	2.66	2.52
4.90	6.25	4.78	4.04	3.60	3.31	3.10	2.94	2.82	2.72	2.57
5.00	6.38	4.88	4.13	3.68	3.38	3.16	3.00	2.88	2.78	2.63

Example:

10,000 SY pavement 24' wide, joints at 30', $2.70/LF

10,000 SY x $2.70/SY = $27,700

Figure 32-20 Costs per Square Yard for Sawed Joints – 36-Foot Pavement Width

In this table the pay unit reflects the cost per linear foot.

Cost per L.F.	For Pavements 36 Ft. Wide Having 2 Longitudinal Joints — Transverse Joint Spacing, Ft.									
	10	15	20	25	30	35	40	45	50	60
3.00	4.20	3.30	2.85	2.58	2.40	2.27	2.18	2.10	2.04	1.95
3.10	4.34	3.41	2.95	2.67	2.48	2.35	2.25	2.17	2.11	2.02
3.20	4.48	3.52	3.04	2.75	2.56	2.42	2.32	2.24	2.18	2.08
3.30	4.62	3.63	3.14	2.84	2.64	2.50	2.39	2.31	2.24	2.15
3.40	4.76	3.74	3.23	2.92	2.72	2.57	2.47	2.38	2.31	2.21
3.50	4.90	3.85	3.33	3.01	2.80	2.65	2.54	2.45	2.38	2.28
3.60	5.04	3.96	3.42	3.10	2.88	2.73	2.61	2.52	2.45	2.34
3.70	5.18	4.07	3.52	3.18	2.96	2.80	2.68	2.59	2.52	2.41
3.80	5.32	4.18	3.61	3.27	3.04	2.88	2.76	2.66	2.58	2.47
3.90	5.46	4.29	3.71	3.35	3.12	2.95	2.83	2.73	2.65	2.54
4.00	5.60	4.40	3.80	3.44	3.20	3.03	2.90	2.80	2.72	2.60
4.10	5.74	4.51	3.90	3.53	3.28	3.10	2.97	2.87	2.79	2.67
4.20	5.88	4.62	3.99	3.61	3.36	3.18	3.05	2.94	2.86	2.73
4.30	6.02	4.73	4.09	3.70	3.44	3.26	3.12	3.01	2.92	2.80
4.40	6.16	4.84	4.18	3.78	3.52	3.33	3.19	3.08	2.99	2.86
4.50	6.30	4.95	4.28	3.87	3.60	3.41	3.26	3.15	3.06	2.93
4.60	6.44	5.06	4.37	3.96	3.68	3.48	3.34	3.22	3.13	2.99
4.70	6.58	5.17	4.47	4.04	3.76	3.56	3.41	3.29	3.20	3.06
4.80	6.72	5.28	4.56	4.13	3.84	3.63	3.48	3.36	3.26	3.12
4.90	6.86	5.39	4.66	4.21	3.92	3.71	3.55	3.43	3.33	3.19
5.00	7.00	5.50	4.75	4.30	4.00	3.79	3.63	3.50	3.40	3.25

Example

10,000 SY pavement 36' wide, joints at 30', $4.00/LF

10,000 SY × $3.20/SY = $32,000

Division 32

Figure 32-21 Costs per Square Yard for Sawed Joints – 48-Foot Pavement Width

In this table the pay unit reflects the cost per linear foot.

| | For Pavements 48 Ft. Wide Having 3 Longitudinal Joints | | | | | | | | | |
| | Transverse Joint Spacing, Ft. | | | | | | | | | |
Cost per L.F.	10	15	20	25	30	35	40	45	50	60
3.00	4.39	3.49	3.04	2.77	2.59	2.46	2.36	2.29	2.23	2.14
3.10	4.53	3.60	3.14	2.86	2.67	2.54	2.44	2.36	2.30	2.21
3.20	4.68	3.72	3.24	2.95	2.76	2.62	2.52	2.44	2.38	2.28
3.30	4.83	3.84	3.34	3.04	2.85	2.70	2.60	2.52	2.45	2.35
3.40	4.97	3.95	3.44	3.14	2.93	2.79	2.68	2.59	2.52	2.42
3.50	5.12	4.07	3.54	3.23	3.02	2.87	2.76	2.67	2.60	2.49
3.60	5.27	4.19	3.65	3.32	3.11	2.95	2.84	2.75	2.67	2.57
3.70	5.41	4.30	3.75	3.41	3.19	3.03	2.91	2.82	2.75	2.64
3.80	5.56	4.42	3.85	3.51	3.28	3.11	2.99	2.90	2.82	2.71
3.90	5.70	4.53	3.95	3.60	3.36	3.20	3.07	2.97	2.90	2.78
4.00	5.85	4.65	4.05	3.69	3.45	3.28	3.15	3.05	2.97	2.85
4.10	6.00	4.77	4.15	3.78	3.54	3.36	3.23	3.13	3.04	2.92
4.20	6.14	4.88	4.25	3.87	3.62	3.44	3.31	3.20	3.12	2.99
4.30	6.29	5.00	4.35	3.97	3.71	3.52	3.39	3.28	3.19	3.06
4.40	6.44	5.12	4.46	4.06	3.80	3.61	3.47	3.36	3.27	3.14
4.50	6.58	5.23	4.56	4.15	3.88	3.69	3.54	3.43	3.34	3.21
4.60	6.73	5.35	4.66	4.24	3.97	3.77	3.62	3.51	3.42	3.28
4.70	6.87	5.46	4.76	4.34	4.05	3.85	3.70	3.58	3.49	3.35
4.80	7.02	5.58	4.86	4.43	4.14	3.93	3.78	3.66	3.56	3.42
4.90	7.17	5.70	4.96	4.52	4.23	4.02	3.86	3.74	3.64	3.49
5.00	7.31	5.81	5.06	4.61	4.31	4.10	3.94	3.81	3.71	3.56

Example:

10,000 SY pavement 48' wide, joints at 30', $4.00/LF

10,000 SY x $3.45/SY = $34,500

Figure 32-22 Brick Paver Units for Mortared Paving

Brick Paver Unit Dimensions					Paver Units per S.F.	C.F. of Mortar Joints per 1000 Units	
W	x	L	x	H		3/8" Joint	1/2" Joint
3-5/8		8		2-1/4*	4.3	5.68	—
3-5/8		7-5/8		2-1/4	4.5	5.49	—
3-3/4		8		2-1/4	4.0	—	7.65
3-5/8		7-5/8		1-1/4	4.5	3.05	—
3-3/4		8		1-1/8	4.0	—	3.82

This table does not include provisions for waste. Allow at least 5% for brick and 10% to 25% for mortar waste.

*Running bond pattern only

(courtesy Brick Institute of America)

Figure 32-23 Mortar Quantities for 3/8" and 1/2" Setting Beds [a, b]

Mortar Type and Material	C.F. Mortar per 100 S.F.		Material Weight Lb./Ft.³	Material Quantity by Weight/100 S.F. (Lb.)[c]	
	3/8" Joint	1/2" Joint		3/8" Joint	1/2" Joint
Type N (1:1:6)	3.13	4.17			
Portland Cement			15.67	49.05	65.34
Hydrated Lime			6.67	20.88	27.81
Sand			80.00	250.40	333.60
Type S (1:1/2:4-1/2)	3.13	4.17			
Portland Cement			20.89	65.38	87.11
Hydrated Lime			4.44	13.90	18.51
Sand			80.00	250.00	333.60
Type M (1:1/4:3)	3.13	4.17			
Portland Cement			31.33	98.06	130.64
Hydrated Lime			3.33	10.42	13.89
Sand			80.00	250.40	333.60

[a] The quantities are only for setting bed.

[b] The table does not include provisions for waste. Allow 10% to 25% for waste.

[c] The quantities represent material weights required for 100 SF, depending on the mortar setting bed thickness. The provisions do not include waste.

(courtesy Brick Institute of America)

Figure 32-24   Brick Quantities for Use in Paving

Estimating Mortarless Paving Units			
Paver Face Dimensions (Actual Inches) W x L		Paver Face Area Sq. In.	Paver Units per S.F.
4	8	32.0	4.5
3-3/4	8	30.0	4.8
3-5/8	7-5/8	27.6	5.2
3-7/8	8-1/4	32.0	4.5
3-7/8	7-3/4	30.0	4.8
3-3/4	7-1/2	28.2	5.1
3-3/4	7-3/4	29.1	5.0
3-5/8	11-5/8	42.1	3.4
3-5/8	8	29.0	5.0
3-5/8	11-3/4	42.6	3.4
3-9.16	8	28.5	5.1
3-1/2	7-3/4	27.1	5.3
3-1/2	7-1/2	26.3	5.5
3-3/8	7-1/2	25.3	5.7
4	4	16.0	9.0
6	6	36.0	4.0
7-5/8	7-5/8	58.1	2.5
7-3/4	7-3/4	60.1	2.4
8	8	64.0	2.3
8	16	128.0	1.1
12	12	144.0	1.0
16	16	256.0	0.6
6	6 Hexagon	31.2	4.6
8	8 Hexagon	55.4	2.6
12	12 Hexagon	124.7	1.2

Note: The above table does not include waste. Allow at least 5% for waste and breakage.

(courtesy Brick Institute of America)

Figure 32-25 Landscape Systems and Graphics

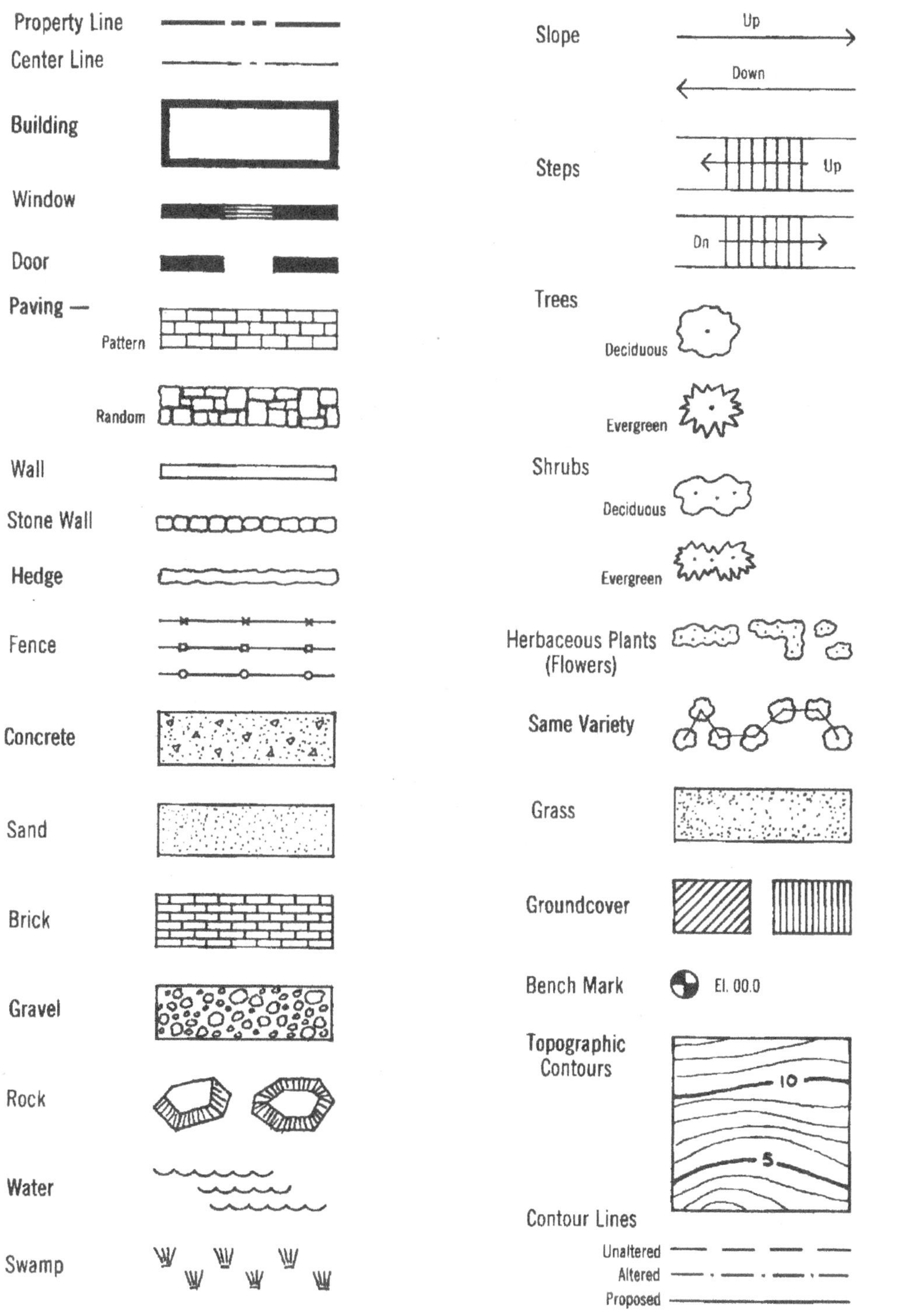

Property Line	
Center Line	
Building	
Window	
Door	
Paving — Pattern	
Random	
Wall	
Stone Wall	
Hedge	
Fence	
Concrete	
Sand	
Brick	
Gravel	
Rock	
Water	
Swamp	

Slope	Up / Down
Steps	Up / Dn
Trees Deciduous	
Evergreen	
Shrubs Deciduous	
Evergreen	
Herbaceous Plants (Flowers)	
Same Variety	
Grass	
Groundcover	
Bench Mark	El. 00.0
Topographic Contours	10 / 5
Contour Lines	
Unaltered	
Altered	
Proposed	

Figure 32-26 Tree Planting

This table can be entered either with the number of trees available to determine the area (acres or hectares) required, or with the area to determine the number of trees that can be planted.

Example:

2,000 trees to be planted at 10' x 10'
2,000 trees/436 trees/acre = 4.59 acres
or:
4 acres to be planted at 10' x 10'
436 trees/acre x 4 acres = 1,744 trees

Tree Spacing, Feet (Meters)	Number Trees per Acre (Hectares)		Tree Spacing, Feet (Meters)	Number Trees per Acre (Hectares)	
2 x 2 (.6 x .6)	10,890	(26,909)	8 x 10 (2.4 x 3.0)	544	(1,344)
3 x 3 (.91 x .91)	4,840	(11,959)	8 x 12 (2.4 x 3.7)	454	(1,122)
4 x 4 (1.2 x 1.2)	2,722	(6,726)	8 x 25 (2.4 x 7.6)	218	(539)
4 x 5 (1.2 x 1.5)	2,178	(5,382)	9 x 9 (2.7 x 2.7)	538	(1,329)
4 x 6 (1.2 x 1.8)	1,815	(4,485)	9 x 12 (2.7 x 3.7)	403	(996)
4 x 8 (1.2 x 2.4)	1,362	(3,365)	10 x 10 (3.0 x 3.0)	436	(1,077)
4 x 10 (1.2 x 3.0)	1,089	(2,691)	10 x 12 (3.0 x 3.7)	363	(897)
5 x 5 (1.5 x 1.5)	1,742	(4,304)	11 x 11 (3.4 x 3.4)	360	(890)
5 x 6 (1.5 x 1.8)	1,452	(3,588)	12 x 12 (3.7 x 3.7)	302	(746)
5 x 8 (1.5 x 2.4)	1,089	(2,691)	12 x 18 (3.7 x 5.5)	202	(499)
5 x 10 (1.5 x 3.0)	871	(2,152)	13 x 13 (4.0 x 4.0)	258	(638)
6 x 6 (1.8 x 1.8)	1,210	(2,990)	14 x 14 (4.3 x 4.3)	222	(549)
6 x 8 (1.8 x 2.4)	908	(2,244)	15 x 15 (4.6 x 4.6)	194	(479)
6 x 10 (1.8 x 3.0)	726	(1,794)	16 x 16 (4.9 x 4.6)	170	(420)
6 x 12 (1.8 x 3.7)	605	(1,495)	18 x 18 (5.5 x 5.5)	134	(331)
7 x 7 (2.1 x 2.1)	889	(2,197)	18 x 20 (5.5 x 6.0)	121	(299)
7 x 10 (2.1 x 3.0)	622	(1,537)	20 x 20 (6.0 x 6.0)	109	(269)
8 x 8 (2.4 x 2.4)	681	(1,683)	25 x 25 (7.6 x 7.6)	70	(173)

Figure 32-27 Tree Pits and Tree Balls – Cubic Feet per Tree for Estimating Excavation and Topsoil

		Diameters										
	Depth	1'	1-1/4'	1-1/2'	1-3/4'	2'	2-1/4'	2-1/2'	2-3/4'	3'	3-1/4'	3-1/2'
Tree Pit	1'	.94	1.47	2.13	2.88	3.77	4.78	5.89	7.13	8.48	9.96	11.5
Ball		.68	1.07	1.54	2.10	2.73	3.42	4.30	5.20	6.19	7.38	8.4
Tree Pit	1-1/4'	1.16	1.85	2.65	3.60	4.71	5.93	7.37	8.90	10.6	12.4	14.4
Ball		.85	1.35	1.93	2.63	3.42	4.29	5.36	6.49	7.7	9.2	10.5
Tree Pit	1-1/2'	1.40	2.22	3.08	4.32	5.65	7.16	8.83	10.7	12.7	15.0	17.3
Ball		1.02	1.62	2.32	3.15	4.10	5.19	6.39	7.8	9.2	10.9	12.7
Tree Pit	1-3/4'	1.65	2.58	3.70	5.04	6.60	8.20	10.3	12.5	14.8	17.4	20.2
Ball		1.20	1.88	3.72	3.68	4.78	6.03	7.5	9.1	10.7	12.7	14.7
Tree Pit	2'	1.87	2.95	4.26	5.76	7.54	9.55	11.8	14.3	17.0	19.9	23.0
Ball		1.38	2.15	3.10	4.20	5.49	6.92	8.5	10.4	12.3	14.5	16.8
Tree Pit	2-1/4'	2.10	3.32	4.78	6.48	8.48	10.7	13.2	16.0	19.1	22.4	26.0
Ball		1.55	2.43	3.48	4.73	6.19	7.5	9.6	11.7	13.8	16.4	18.9
Tree Pit	2-1/2'	2.34	3.69	5.30	7.20	9.42	11.9	14.7	17.8	21.2	24.9	28.9
Ball		1.70	2.70	3.87	5.25	6.89	8.7	10.7	13.0	15.4	18.2	21.0
Tree Pit	2-3/4'	2.57	4.06	5.83	7.92	10.3	13.1	16.2	19.6	23.3	27.4	31.8
Ball		1.87	2.98	4.25	5.77	7.6	9.6	11.8	14.3	17.0	20.0	23.1
Tree Pit	3'	2.81	4.43	6.37	8.64	11.3	14.3	17.7	21.4	25.4	29.9	34.6
Ball		2.05	3.24	4.65	6.30	8.3	10.5	12.9	15.6	18.6	21.8	25.2
Tree Pit	3-1/4'	3.00	4.80	6.90	9.36	12.2	15.5	19.2	23.2	27.5	32.4	37.5
Ball		2.21	3.50	5.03	6.83	8.9	11.4	14.0	16.9	20.1	23.6	27.3
Tree Pit	3-1/2'	3.28	5.17	7.41	10.1	13.2	16.7	20.7	25.0	29.6	34.9	40.4
Ball		2.39	3.77	5.42	7.4	9.6	12.2	15.0	18.2	21.7	25.4	29.4
Tree Pit	3-3/4'	3.50	5.53	7.96	10.8	14.2	17.9	22.2	26.7	31.7	37.1	43.3
Ball		2.56	4.04	3.80	7.9	10.3	13.1	16.1	19.5	23.2	27.2	31.5
Tree Pit	4'	3.74	5.90	8.50	11.5	15.1	19.1	23.7	28.5	33.8	39.9	46.2
Ball		2.73	4.31	6.19	8.4	11.0	14.0	17.2	20.8	24.8	29.0	33.6
Tree Pit	4-1/4'	3.97	6.28	9.01	12.3	16.0	20.3	25.1	30.3	36.0	42.3	49.1
Ball		2.90	4.58	6.58	8.9	11.8	14.8	18.3	22.1	26.3	30.8	35.7
Tree Pit	4-1/2'	4.20	6.64	9.55	13.0	17.0	21.5	26.6	32.0	38.1	44.8	52.0
Ball		3.07	4.85	6.97	9.5	12.4	15.7	19.4	23.4	27.8	32.6	37.8
Tree Pit	4-3/4'	4.43	7.00	10.0	13.7	17.9	22.7	28.0	33.8	40.2	47.3	54.9
Ball		3.24	5.12	7.4	10.0	13.0	16.5	20.4	24.7	29.1	34.2	40.0
Tree Pit	5'	4.68	7.38	10.6	14.4	18.8	23.9	29.8	35.6	42.3	49.9	57.7
Ball		3.41	5.38	7.8	10.5	13.7	17.4	21.5	26.0	30.9	36.1	42.1
Tree Pit	5-1/2'	5.14	8.12	11.7	15.8	20.7	26.3	32.4	39.2	46.7	54.8	63.5
Ball		3.76	5.93	8.5	11.6	15.1	19.1	23.7	28.6	34.0	40.0	46.3
Tree Pit	6'	5.41	8.86	12.7	17.3	22.6	28.7	35.4	42.8	50.9	57.8	69.3
Ball		4.10	6.46	9.3	12.6	16.5	20.9	25.9	31.2	37.1	43.6	50.5

This table lists the cubic feet of excavation necessary for holes for planting trees. The diameters of holes necessary are listed across the top. The depths are listed down the second column. For example, if a planting guide calls for a hole 2' deep, the pit will have 7.54 CF of earth to be excavated and, therefore, you will need 2.05 CF of topsoil for the planting.

Figure 32-28 Tree and Shrub Data Including Labor-Hour Requirements

Ball Size Diam. x Depth	Soil in Ball	Weight of Ball	Hole Diam. Req'd.	Hole Exca-vation	Amount of Soil Displ.	Topsoil Handled	Time Required in Man-Hours					
							Dig & Lace	Handle Ball	Dig Hole	Plant & Prune	Water & Guy	Total M.H.
Inches	C.F.	Lbs.	Feet	C.F.	C.F.	C.F.						
12 x 12	.7	56	2	4	3	11	.25	.17	.33	.25	.07	1.1
18 x 16	2	160	2.5	8	6	21	.50	.33	.47	.35	.08	1.7
24 x 18	4	320	3	13	9	38	1.00	.67	1.08	.82	.20	3.8
30 x 21	7.5	600	4	27	19.5	76	.82	.71	.79	1.22	.26	3.8
36 x 24	12.5	980	4.5	38	25.5	114	1.08	.95	1.11	1.32	.30	4.76
42 x 27	19	1520	5.5	64	45	185	1.90	1.27	1.87	1.43	.34	6.8
48 x 30	28	2040	6	85	57	254	2.41	1.60	2.06	1.55	.39	8.0
54 x 33	38.5	3060	7	127	88.5	370	2.86	1.90	2.39	1.76	.45	9.4
60 x 36	52	4160	7.5	159	107	474	3.26	2.17	2.73	2.00	.51	10.7
66 x 39	68	5440	8	196	128	596	3.61	2.41	3.07	2.26	.58	11.9
72 x 42	87	7160	9	267	180	785	3.90	2.60	3.71	2.78	.70	13.7

Figure 32-29 Plant Spacing Chart

This chart may be used when plants are to be placed equidistant from each other, staggering their position in each row.

Plant Spacing (Inches)	Row Spacing (Inches)	Plants per C.S.F.	Plants Spacing (Feet)	Row Spacing (Feet)	Plants per M.S.F.
			4	3.46	72
6	5.20	462	5	4.33	46
8	6.93	260	6	5.20	32
10	8.66	166	8	6.93	18
12	10.39	115	10	8.66	12
15	12.99	74	12	10.39	8
18	15.59	51.32	15	12.99	5.13
20	18.19	37.70	20	17.32	2.89
24	20.78	28.87	25	21.65	1.85
30	25.98	18.48	30	25.98	1.28
36	31.18	12.83	40	34.64	0.72

Figure 32-30 Conversion Chart for Loam, Baled Peat, and Mulch

Loam Requirements		Mulch Requirements (Bulk Peat, Crushed Stone, Shredded Bark, Wood Chips)	
Depth	Cubic yards per 10,000 S.F.	Depth	Cubic yards per 1,000 S.F.
2"	70	1"	3-1/2
4"	140	2"	7
6"	210	3"	10-1/2
		4"	14
	Cubic yards per 1,000 S.F.		Cubic yards per 100 S.F.
2"	7	1"	1/3
4"	14	2"	2/3
6"	21	3"	1
		4"	1-1/3
Bales (or Compressed) Peat			
Depth		**Bales Required per 100 S.F.**	
1"		one 4-cubic foot bale	
2"		two 4-cubic foot bales	
3"		two 6-cubic foot bales	
4"		one 4-cubic foot bales and two 6-cubic foot bales	

Note: Opened and spread, peat from bales will be about 2-1/2 times the volume of the unopened bale.

Figure 32-31 Installation Time in Labor-Hours for Pavement Recycling

Description	Labor-Hours	Unit
Asphalt Pavement Demolition		
Hydraulic Hammer	.035	S.Y.
Ripping Pavement, Load and Sweep	.007	S.Y.
Crush and Screen, Traveling Hammermill		
3" Deep	.004	S.Y.
6" Deep	.007	S.Y.
12" Deep	.012	S.Y.
Pulverizing, Crushing, and Blending		
into Base 4" Pavement		
Over 15,000 S.Y.	.027	S.Y.
5,000 to 15,000 S.Y.	.029	S.Y.
8" Pavement		
Over 15,000 S.Y.	.029	S.Y.
5,000 to 15,000 S.Y.	.032	S.Y.
Remove, Rejuvenate and spread, Mixer-Paver		
Profiling, Load and Sweep		
1" Deep	.002	S.Y.
3" Deep	.006	S.Y.
6" Deep	.011	S.Y.
12" Deep	.019	S.Y.

Division 32

Figure 32-32   Installation Time in Labor-Hours for Asphalt Pavement

Description	Labor-Hours	Unit
Subgrade, Grade and Roll		
Small Area	.024	S.Y.
Large Area	.011	S.Y.
Base Course		
Bank Run Gravel, Spread Compact		
6" Deep	.004	S.Y.
18" Deep	.013	S.Y.
Crushed Stone, Spread Compact		
6" Deep	.016	S.Y.
18" Deep	.029	S.Y.
Asphalt Concrete Base		
4" Thick	.053	S.Y.
8" Thick	.089	S.Y.
Stabilization Fabric, Polypropylene, 6 oz./S.Y.	.002	S.Y.
Asphalt Pavement, Wearing Course		
1-1/2" Thick	.026	S.Y.
3" Thick	.052	S.Y.

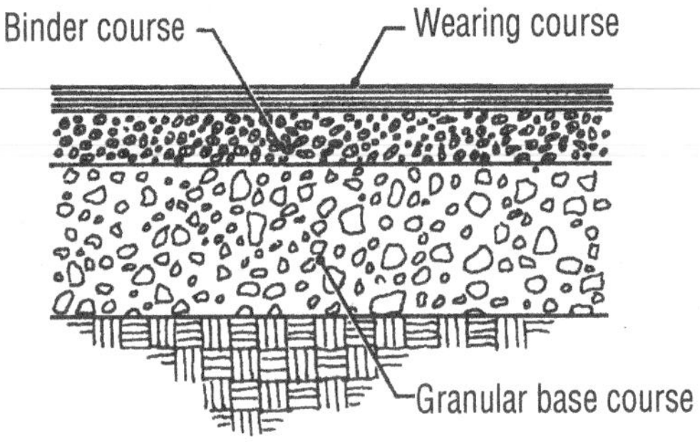

Asphalt Pavement 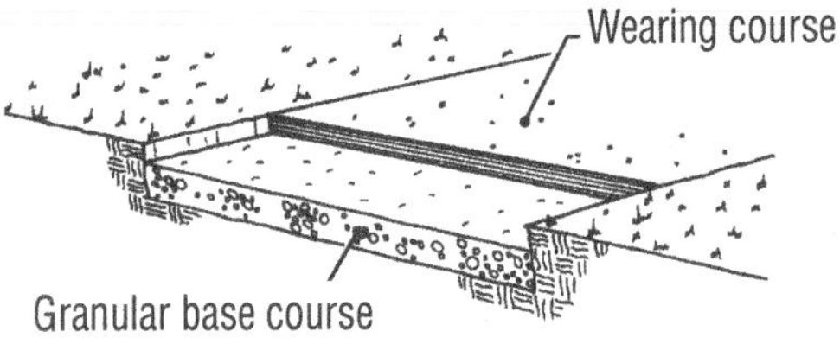   **Bituminous Sidewalk**

Figure 32-33 Installation Time in Labor-Hours for Brick, Stone, and Concrete Paving

Description	Labor-Hours	Unit
Brick Paving without Joints (4.5 Brick/S.F.)	.145	S.F.
Grouted, 3/8" Joints (3.9 Brick/S.F.)	.178	S.F.
Sidewalks		
Brick on 4" Sand Bed		
Laid on Edge (7.2/S.F.)	.229	S.F.
Flagging		
Bluestone, Irregular, 1" Thick	.198	S.F.
Snapped Random Rectangular 1" Thick	.174	S.F.
1-1/2" Thick	.188	S.F.
2" Thick	.193	S.F.
Slate		
Natural Cleft, Irregular, 3/4" Thick	.174	S.F.
Random Rectangular, Gauged, 1/2" Thick	.152	S.F.
Random Rectangular, Butt Joint, Gauged,		
1/4" Thick	.107	S.F.
Granite Blocks, 3-1/2" x 3-1/2" x 3-1/2"	.174	S.F.
4" to 12" Long, 3" to 5" Wide, 3" to 5" Thick	.163	S.F.
6" to 15" Long, 3" to 6" Wide, 3" to 5" Thick	.152	S.F.

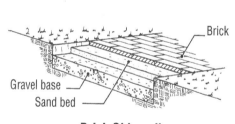

Brick Sidewalk

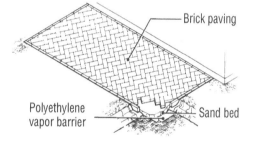

Brick Paving on Sand Bed

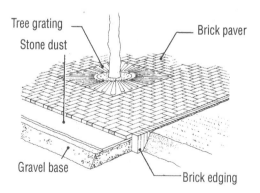

Plaza Brick Paving System

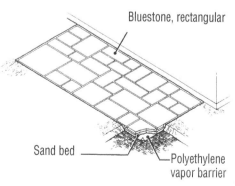

Stone Paving on Sand Bed

Division 32

(continued on next page)

Figure 32-33 Installation Time in Labor-Hours for Brick, Stone, and Concrete Paving *(continued from previous page)*

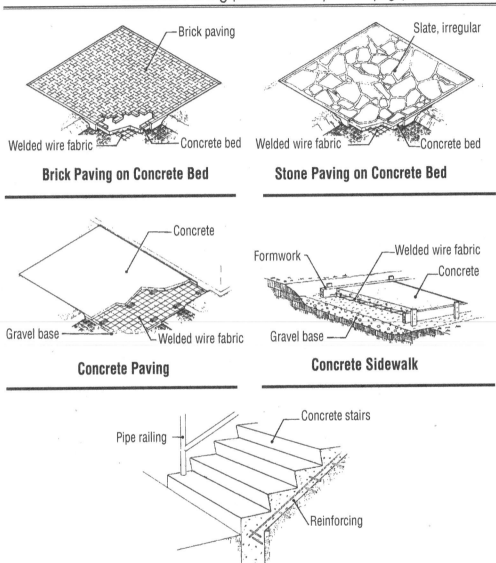

Brick Paving on Concrete Bed

Stone Paving on Concrete Bed

Concrete Paving

Concrete Sidewalk

Concrete Stairs

Figure 32-34 Installation Time in Labor-Hours for Fencing

Description	Labor-Hours	Unit
Fence, Chain Link, Industrial Plus 3 Strands Barbed Wire, 2" Line Post @ 10' O.C.		
1-5/8" Top Rail, 6' High	.096	L.F.
Corners, Add	.600	Ea.
Braces, Add	.300	Ea.
Gate, Add	.686	Ea.
Residential, 11 Gauge Wire, 1-5/8" Line Post @ 10' O.C. 1-3/8"		
Top Rail, 3' High	.048	L.F.
4' High	.060	L.F.
Gate, Add	.400	Ea.
Tennis Courts, 11 Gauge Wire, 1-3/4" Mesh, 2-1/2" Line Posts, 1-5/8" Top Rail,		
10' High	.155	L.F.
12' High	.185	L.F.
Corner Posts, 3" Diameter, Add	.800	Ea.
Fence, Security, 12' High	.960	L.F.
16' High	1.200	L.F.
Fence, Wood, Cedar Picket, 2 Rail, 3' High	.150	L.F.
Gate, 3'-6", Add	.533	Ea.
Cedar Picket, 3 Rail, 4' High	.160	L.F.
Gate, 3'-6", Add	.585	Ea.
Open Rail Rustic, 2 Rail, 3' High	.150	L.F.
Stockade, 6' High	.150	L.F.
Board, Shadow Box, 1" x 6" Treated Pine, 6' High	.150	L.F.

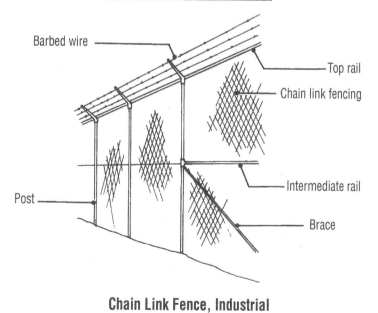

Chain Link Fence, Industrial

Barbed wire — Top rail — Chain link fencing — Intermediate rail — Post — Brace

Division 32

Figure 32-35   Installation Time in Labor-Hours for Curbs

Description	Labor-Hours	Unit
Curbs, Bituminous, Plain, 8" Wide, 6" High, 50 L.F./ton	.032	L.F.
8" Wide, 8" High, 44 L.F./ton	.036	L.F.
Bituminous Berm, 12" Wide, 3" to 6" High, 35 L.F./ton, before Pavement	.046	L.F.
12" Wide, 1-1/2" to 4" High, 60 L.F./ton, Laid with Pavement	.030	L.F.
Concrete, 6" x 18", Cast-in-Place, Straight	.096	L.F.
6" x 18" Radius	.106	L.F.
Precast, 6" x 18", Straight	.160	L.F.
6" x 18" Radius	.172	L.F.
Granite, Split Face, Straight, 5" x 16"	.112	L.F.
6" x 18"	.124	L.F.
Radius Curbing, 6" x 18", Over 10' Radius	.215	L.F.
Corners, 2' Radius	.700	Ea.
Edging, 4-1/2" x 12", Straight	.187	L.F.
Curb Inlets, (Guttermouth) Straight	1.366	Ea.
Monolithic Concrete Curb and Gutter		
Cast-in-Place with 6' High Curb and 6" Thick Gutter		
24" Wide, .055 C.Y. per L.F.	.128	L.F.
30" Wide, .066 C.Y. per L.F.	.141	L.F.

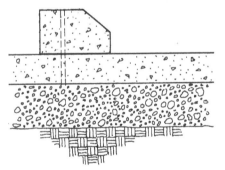

Precast Concrete Parking Bumper

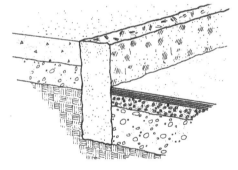

Granite Curb

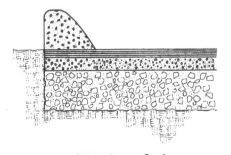

Bituminous Curb

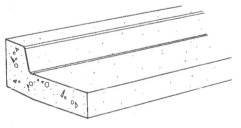

Cast-in-Place Concrete Curb and Gutter

Figure 32-36 Installation Time in Labor-Hours for Site Irrigation

Description	Labor-Hours	Unit
Sprinkler System, Golf Course, Fully Automatic	.600	9 holes
12' Radius Heads, 15' Spacing		
Minimum	.343	Head
Maximum	.600	Head
30' Radius Heads, Automatic		
Minimum	.857	Head
Maximum	1.040	Head
Sprinkler Heads		
Minimum	.267	Head
Maximum	.320	Head
Trenching, Chain Trencher, 12 H.P.		
4" Wide, 12" Deep	.010	L.F.
6" Wide, 24" Deep	.015	L.F.
Backfill and Compact		
4" Wide, 12" Deep	.010	L.F.
6" Wide, 12" Deep	.030	L.F.
Trenching and Backfilling, Chain Trencher, 40 H.P.		
6" Wide, 12" Deep	.007	L.F.
8" Wide, 36" Deep	.010	L.F.
Compaction		
6" Wide, 12" Deep	.003	L.F.
8" Wide, 36" Deep	.005	L.F.
Vibrating Plow		
8" Deep	.004	L.F.
12" Deep	.006	L.F.
Automatic Valves, Solenoid		
3/4" Diameter	.363	Ea.
2" Diameter	.500	Ea.
Automatic Controllers		
4 Station	1.500	Ea.
12 Station	2.000	Ea.

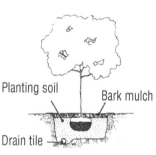

Planting soil   Bark mulch

Drain tile

Tree/Shrub Irrigation and Drainage

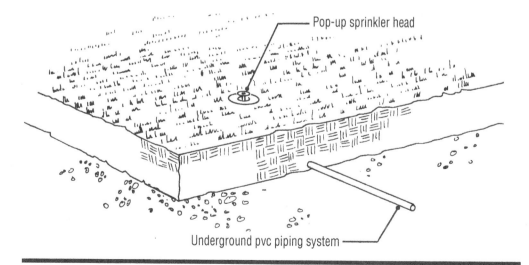

Pop-up sprinkler head

Underground pvc piping system

Notes

Division Thirty-Three
Utilities

Introduction

Division 33 addresses utility related items. Under the previous Master Format 95, these were located in Division 2.

Division 33 has several sections for utility systems, beginning with general site work maintenance costs, including pipe repair, pipe cleaning, and pipe relining. Trenchless excavation for utilities is also addressed, with data for microtunneling, horizontal boring, and directional drilling.

Also included in this division:

- Accessories: utility markers and utility boxes
- Water utilities: concrete pipe, ductile iron pipe, polyethylene pipe, polyvinyl chloride pipe, steel pipe, and copper pipe
- Data on tapping, fittings, valves, fire hydrants, water storage tanks, and water wells
- Sanitary sewerage piping: cast iron and plastics
- Sewerage accessories: wastewater pumping stations, manholes, and septic tank systems
- Storm drainage: corrugated metal pipe, HDPE plastic pipe, concrete pipe and PVC plastic pipe, culverts, catchbasins, and manholes. Also includes information on geotextiles and reservoir lines.
- Natural gas distribution: polyethylene and steel pipe, including fittings and valves
- HVAC chilled, hot water, and steam underground systems
- Electric utilities: underground electrical and telephone conduits
- Site lighting, electrical transmission lines and radio towers.

Note: Concrete pipe is located in the drainage section, but can be used for sewerage or drainage situations depending upon the joint seal.

It is important to understand that Division 31 – Earthwork will have to be used for data involving excavation, backfill, compaction, and bedding of the utility pipes.

Estimating Data

The following tables present estimating guidelines for items found in Division 33 – Utilities. Please note that these guidelines are intended as indicators of what may generally be expected, but that each project must be evaluated individually.

Table of Contents

Division 33

Estimating Tips

Drainage Utilities

Never assume that the water, sewer, and drainage lines will go in at the early stages of the project. Consider the site access needs before dividing the site in half with open trenches, loose pipe, and machinery obstructions. Always inspect the site to establish that the site drawings are complete. Check off all existing utilities on your drawings as you locate them. If you find any discrepancies, mark up the site plan for further research. Differing site conditions can be very costly if discovered later in the project.

New Materials

Use of new types of piping materials (particularly for restoration of pipe where removal/replacement may be undesirable) can reduce the overall project cost. Owners/design engineers should consider the installing contractor a valuable source of information on utility products and local conditions, which could lead to significant cost savings.

Checklist ✓

For an estimate to be reliable, all items must be accounted for. A complete estimate can also limit contingencies. The following checklist can be used to help ensure that all items are included.

☐ **Water Utilities**
 - ☐ Piping
 - ☐ Equipment
 - ☐ Tanks
 - ☐ Wells

☐ **Sanitary Sewer Utilities**
 - ☐ Piping
 - ☐ Equipment
 - ☐ Pumping stations
 - ☐ Force mains
 - ☐ Septic systems
 - ☐ Structures

☐ **Storm Drainage Utility**
 - ☐ Piping
 - ☐ Culverts
 - ☐ Drains
 - ☐ Structures

☐ **Trenchless Utility**
 - ☐ Microtunneling
 - ☐ Horizontal boring

☐ **Natural Gas Distribution**
 - ☐ Piping
 - ☐ Watering

☐ **Liquid Fuel Distribution**
 - ☐ Piping
 - ☐ Pumps

☐ **Fuel Storage Tanks**
 - ☐ Above ground
 - ☐ Underground
 - ☐ AV fuel
 - ☐ Compressed gas

☐ **Hydronic & Steam Energy Distribution**
 - ☐ Hydronic distribution
 - ☐ Steam distribution

☐ **Electrical Utilities**
 - ☐ Electrical utility transmission & distribution
 - ☐ Substations
 - ☐ Transformer
 - ☐ Overhead wiring
 - ☐ High voltage switchgear
 - ☐ Medium voltage switchgear
 - ☐ Grounding

☐ **Communications Utilities**
 - ☐ Structures
 - ☐ Distribution
 - ☐ Communications distribution
 - ☐ Copper cabling
 - ☐ Optical fiber cabling
 - ☐ Coaxial cabling
 - ☐ Grounding and bonding
 - ☐ Wireless communications cistribution
 - ☐ Laser
 - ☐ Microwave
 - ☐ Infrared
 - ☐ UHF/VHF

Division 33

Figure 33-1 Piping Designations

There are several systems currently in use to describe pipe and fittings. The following information will help clarify classification of piping systems used for water distribution.

Piping may be classified by schedule. Piping schedules include 5S, 10S, 10, 20, 30, Standard, 40, 60, Extra Strong, 80, 100, 120, 140, 160, and Double Extra Strong. These schedules depend on the pipe wall thickness. The wall thickness of a particular schedule may vary with pipe size.

Ductile iron pipe for water distribution is classified by American Water Works Association (AWWA) Pressure Classes such as Class 150, 200, 250, 300, and 350. These classes are actually the rated water working pressure of the pipe in pounds per square inch (psi). The pipe in these pressure classes is designed to withstand the rated water working pressure plus a surge allowance of 100 psi.

The American Water Works Association (AWWA) provides standards for various types of **plastic pipe**. C-900 is the specification for polyvinyl chloride (PVC) piping used for water distribution in sizes ranging from 4" through 12". C-901 is the specification for polyethylene (PE) pressure pipe, tubing, and fittings used for water distribution in sizes ranging from 1/2" through 3". C-905 is the specification for PVC piping sizes 14" through 48".

PVC pressure-rated pipe is identified using the standard dimensional ratio (SDR) method. This method is defined by the American Society for Testing and Materials (ASTM) Standard D 2241. This pipe is available in SDR numbers 64, 41, 32.5, 26, 21, 17, and 13.5. Pipe with an SDR of 64 will have the thinnest wall while pipe with an SDR of 13.5 will have the thickest wall. When the pressure rating (PR) of a pipe is given in psi, it is based on a line supplying water at 73° F.

The National Sanitation Foundation (NSF) seal of approval is applied to products that can be used with potable water. These products have been tested to ANSI/NSF Standard 14.

Valves and strainers are classified by American National Standards Institute (ANSI) classes 125, 150, 200, 250, 300, 400, 600, 900, 1500, and 2500. Within each class there is an operating pressure range depending on temperature. Design parameters should be compared to the appropriate material-dependent, pressure-temperature rating chart for accurate valve selection.

Figure 33-2 Service Electric Utilities

Service includes the excavation and structures for bringing power to the building, and the actual installation of the primary cables. The figure below illustrates typical methods by which power is carried to the project.

Service covers the distribution methods used to route power, control, and communications cables to a facility's property and between its buildings and structures. There are four basic options: (1) direct burial cables, (2) direct burial conduits; (3) underground in duct banks, and (4) overhead on poles.

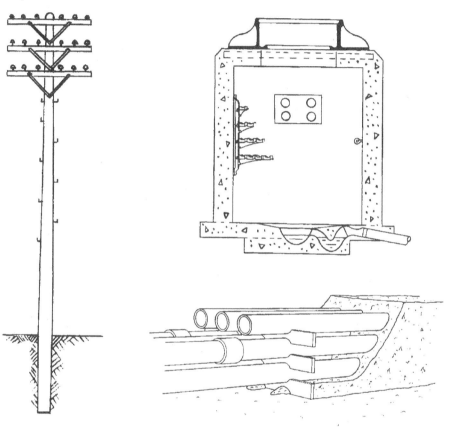

Electric and Telephone Site Work

Figure 33-3   Concrete for Conduit Encasement

The table below lists CY of concrete for 100 LF of trench. Conduit separation center to center should meet 7.5" (NEC).

Number of Conduits	1	2	3	4	6	8	9	Number of Conduits
Trench Dimension	11.5" x 11.5"	11.5" x 19"	11.5" x 27"	19" x 19"	19" x 27"	19" x 38"	27" x 27"	Trench Dimension
Conduit Diameter 2"	3.29	5.39	7.64	8.83	12.51	17.66	17.72	Conduit Diameter 2"
2.5"	3.23	5.29	7.49	8.62	12.19	17.23	17.25	2.5"
3.0"	3.15	5.13	7.24	8.29	11.71	16.59	16.52	3.0"
3.5"	3.08	4.97	7.02	7.99	11.26	15.98	15.84	3.5"
4.0"	2.99	4.80	6.76	7.65	10.74	15.30	15.07	4.0"
5"	2.78	4.37	6.11	6.78	9.44	13.57	13.12	5"
6.0"	2.52	3.84 	5.33 	5.74 	7.87 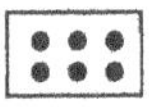	11.48 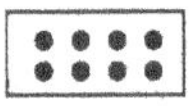	10.77 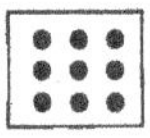	6.0"

Figure 33-4 Installation Time in Labor-Hours for Water
Distribution Systems*

Description	Labor-Hours	Unit
Water Distribution Piping, Not Including Excavation and Backfill		
Mains, Ductile Iron, 4" Diameter	0.2	L.F.
6" Diameter	0.25	L.F.
8" Diameter	0.3	L.F.
12" Diameter	0.38	L.F.
16" Diameter	0.55	L.F.
Polyvinyl Chloride, 4" Diameter	0.064	L.F.
6" Diameter	0.076	L.F.
8" Diameter	0.092	L.F.
12" Diameter	0.10	L.F.
Concrete,		L.F.
12" Diameter	0.292	L.F.
24" Diameter	0.438	L.F.
Fittings for Mains, Ductile Iron, Bend, MJ		
4" Diameter	2	Ea.
8" Diameter	3	Ea.
16" Diameter	5.5	Ea.
Wye, 4" Diameter	3	Ea.
8" Diameter	4.5	Ea.
16" Diameter	8.25	Ea.
Increaser, 4" x 6"	2.25	Ea.
6" x 16"	4	Ea.
Flange, 4" Diameter	1.600	Ea.
8" Diameter	3.080	Ea.
12" Diameter	4.000	Ea.
Polyvinyl Chloride, Bend, 4" Diameter	.240	Ea.
8" Diameter	.300	Ea.
12" Diameter	.800	Ea.
Tee, 4" Diameter	.267	Ea.
8" Diameter	.343	Ea.
12" Diameter	1.200	Ea.
Concrete, Bend, 12" Diameter	1.75	Ea.
16" Diameter	5.00	Ea.
Tee, 12" Diameter	2.65	Ea.
16" Diameter	7.5	Ea.
Service, Copper, Type K, 3/4" Diameter	0.04	L.F.
1" Diameter	0.05	L.F.
2" Diameter	0.07	L.F.
4" Diameter	0.168	L.F.
Polyvinyl Chloride, 1-1/2" Diameter	0.013	L.F.
2-1/2" Diameter	0.02	L.F.
Fittings for Service, Copper, Corp. Stop		
3/4" Diameter	.421	Ea.
2" Diameter	.727	Ea.
Wye, 3/4" Diameter	.667	Ea.
2" Diameter	1.140	Ea.
Curb Box, 3/4" Diameter Service	.667	Ea.
2" Diameter Service	1.000	L.F.
Valves for Mains, 4" Diameter	4	Ea.
8" Diameter	4	Ea.
12" Diameter	4	Ea.
Valves for Service, Curb Stop, 3/4" Diameter	.421	Ea.
1" Diameter	.500	Ea.
2" Diameter	.727	Ea.

*See Division 31 – Earthwork for trenching, bedding, and backfilling labor-hours.

Division 33

(continued on next page)

Figure 33-4 Installation Time in Labor-Hours for Water Distribution Systems *(continued from previous page)*

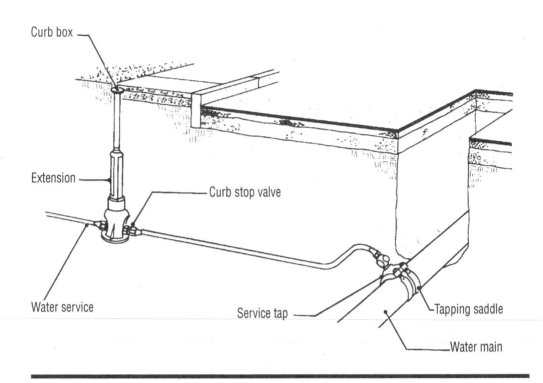

Figure 33-5   Installation Time in Labor-Hours for Sewage and Drainage
Collection Systems

Description	Labor-Hours	Unit
Catch Basins or Manholes, Not Including Excavation and Backfill, Frame and Cover		
Brick, 4' I.D., 6' Deep	22.857	Ea.
8' Deep	32.000	Ea.
10' Deep	40	Ea.
Concrete Block, 4' I.D., 6' Deep	16.000	Ea.
8' Deep	22.857	Ea.
10' Deep	28.675	Ea.
Precast Concrete, 4' I.D., 6' Deep	10	Ea.
8' Deep	15	Ea.
10' Deep	18.75	Ea.
Cast-in-place Concrete, 4' I.D., 6' Deep	32	Ea.
8' Deep	48	Ea.
10' Deep	60	Ea.
Frames and Covers, 18" Square, 160 lbs.	2.400	Ea.
270 lbs.	2.791	Ea.
24" Square, 220 lbs.	2.667	Ea.
400 lbs.	3.077	Ea.
26" D Shape, 600 lbs.	3.429	Ea.
Curb Inlet Frame & Grate, 24" x 36"	12	Ea.
Light Traffic, 18" Diameter, 100 lbs.	2.4	Ea.
24" Diameter, 300 lbs.	2.759	Ea.
36" Diameter, 900 lbs.	4.138	Ea.
Heavy Traffic, 24" Diameter, 400 lbs.	3.077	Ea.
36" Diameter, 1150 lbs.	8.000	Ea.
Raise Frame and Cover 2", for Resurfacing,		
20" to 26" Frame	2.183	Ea.
30" to 36" Frame	2.667	Ea.
Drainage and Sewage Piping, Not Including Excavation and Backfill		
Concrete, up to 8" Diameter	.2	L.F.
12" Diameter	.320	L.F.
15" Diameter	.320	L.F.
18" Diameter	.364	L.F.
21" Diameter	.400	L.F.
24" Diameter	.480	L.F.
27" Diameter	.609	L.F.
30" Diameter	.636	L.F.
36" Diameter	.778	L.F.
Currugated Metal, 8" Diameter	.135	L.F.
12" Diameter	.218	L.F.
18" Diameter	.234	L.F.
24" Diameter	.274	L.F.
30" Diameter	.431	L.F.
36" Diameter	.431	L.F.
Cast Iron, 6" Diameter	.286	L.F.
8" Diameter	.457	L.F.
12" Diameter	.561	L.F.
15" Diameter	.653	L.F.
Polyvinyl Chloride, 4" Diameter	.064	L.F.
8" Diameter	.072	L.F.
12" Diameter	.088	L.F.
15" Diameter	.117	L.F.

Division 33

(continued on next page)

Figure 33-5 Installation Time in Labor-Hours for Sewage and Drainage
Collection Systems *(continued from previous page)*

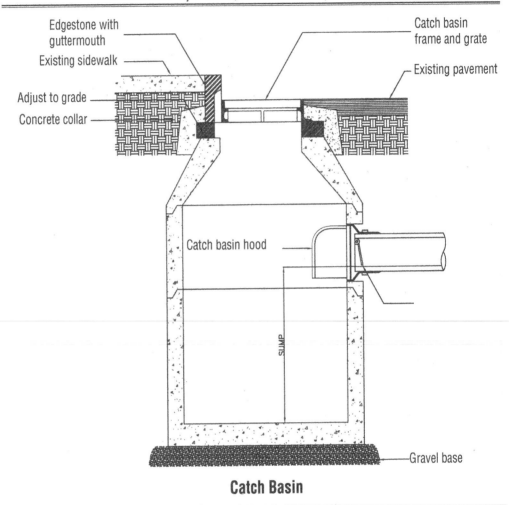

Catch Basin

Figure 33-6 Installation Time in Labor-Hours for Septic Systems

Description	Labor-Hours	Unit
Septic Tanks Not Including Excavation or		
Piping, Precast, 1000 Gallon	3.500	Ea.
1500 Gallon	4.0	Ea.
2000 Gallon	5.600	Ea.
5000 Gallon	16.000	Ea.
HDPE 1000 Gallon	4.667	Ea.
1500 Gallon	7.000	Ea.
Excavation, 3/4 C.Y. Backhoe	.110	C.Y.
Distribution Boxes, Concrete, 7 Outlets	1.000	Ea.
9 Outlets	2.000	Ea.
Leaching Field Chambers, 13' x 3'-7" x 1'-4"		
Standard	3.500	Ea.
Heavy Duty, 8' x 4' x 1'-6"	4.000	Ea.
13' x 3'-9" x 1'-6"	4.667	Ea.
20' x 4' x 1'-6"	11.200	Ea.
Leaching Pit, Precast Concrete, 3' Pit	3.500	Ea.
6' Pit	5.957	Ea.
Disposal Field		
Excavation, 4' Trench, 3/4 C.Y. Backhoe	.048	L.F.
Crushed Stone	.160	C.Y.
PVC, Perforated, 4" Diameter	0.153	L.F.
6" Diameter	0.160	L.F.

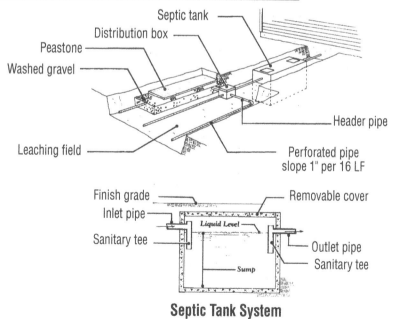

Septic tank
Distribution box
Peastone
Washed gravel
Leaching field
Header pipe
Perforated pipe
slope 1" per 16 LF

Finish grade
Inlet pipe
Sanitary tee
Liquid Level
Removable cover
Outlet pipe
Sanitary tee
Sump

Septic Tank System

Division 33

Figure 33-7   Installation Time in Labor-Hours for Open Site
Drainage Systems

Description	Labor-Hours	Unit
Paving, Asphalt, Ditches	.185	S.Y.
Concrete, Ditches	.360	S.Y.
Filter Stone Rubble	.258	C.Y.
Paving, Asphalt, Aprons	.320	S.Y.
Concrete, Aprons	.620	S.Y.
Drop Structure	8.000	Ea.
Flared Ends, 12" Diameter	.879	Ea.
24" Diameter	1.96	Ea.
Reinforced Plastic, 12" Diameter	.280	L.F.
Precast Box Culvert, 6' x 3'	.343	L.F.
8' x 8'	.480	L.F.
12' x 8'	.716	L.F.
Aluminum Arch Culvert, 17" x 13"	.240	L.F.
35" x 24"	.48	L.F.
57" x 38"	.747	L.F.
Multi-Plate Arch Steel	.014	Lb.

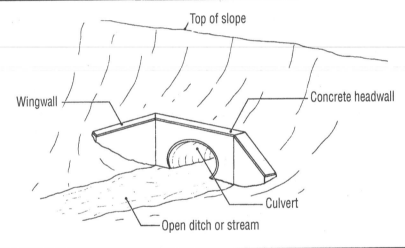

Figure 33-8 Installation Time in Labor-Hours for Trench Drains

Description	Labor-Hours	Unit
Trench Forms		
1 Use	.200	S.F.C.A.
4 Uses	.173	S.F.C.A.
Reinforcing	15.240	ton
Concrete		
Direct Chute	.320	C.Y.
Pumped	.492	C.Y.
With Crane and Bucket	.533	C.Y.
Trench Cover, Including Angle Frame		
To 18" Wide	.400	L.F.
Cover Frame Only		
For 1" Grating	.178	L.F.
For 2" Grating	.229	L.F.
Geotextile Fabric in Trench		
Ideal Conditions	.007	S.Y.
Adverse Conditions	.010	S.Y.
Drainage Stone		
3/4"	.092	C.Y.
Pea Stone	.092	C.Y.

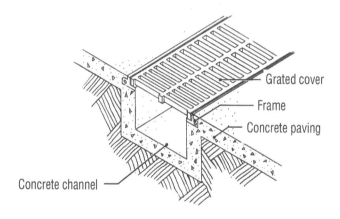

Grated cover
Frame
Concrete paving
Concrete channel

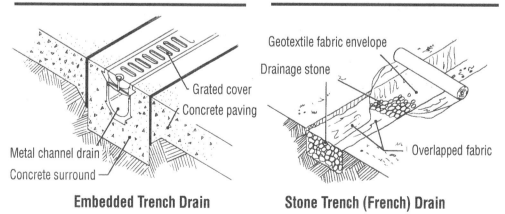

Grated cover
Concrete paving
Metal channel drain
Concrete surround

Embedded Trench Drain

Geotextile fabric envelope
Drainage stone
Overlapped fabric

Stone Trench (French) Drain

Figure 33-9 Installation Time in Labor-Hours for Radio Towers

Description	Labor-Hours
RADIO TOWERS Guyed, 50'high, 40 lb. sec., 70 MPH basic wind spd.	16
Wind load 90 MPH basic wind speed	16
190' high, 40 lb. section wind load 70 MPH basic wind speed	72.727
200' high, 70 lb. section, wind load 90 MPH basic wind speed	72.727
300' high, 70 lb. section, wind load 70 MPH basic wind speed	120
270' high, 90 lb. section, wind load 90 MPH basic wind speed	120
400' high, 100 lb. section, wind load 70 MPH basic wind speed	171
Self-supporting, 60 high, wind load 70 MPH basic wind speed	30
60' high, wind load 90 MPH basic wind speed	53.333
120' high, wind load 70 MPH basic wind speed	60
190' high, wind load 90 MPH basic wind speed	120
For states west of Rocky Mountains, add for shipping	

Figure 33-10 Installation Time for Ground Storage Tanks

Ground Storage Tanks	Crew Makeup	Daily Output	Labor-Hours	Unit
TANKS Not incl. pipe or pumps				
Vinyl coated fabric pillow tanks, freestanding				
5000 gallons	4 Building Laborers	4	8.000	Ea.
Supporting embankment not included				
25,000 gallons	6 Building Laborers	2	24.000	Ea.
50,000 gallons	8 Building Laborers	1.50	42.670	Ea.
100,000 gallons	9 Building Laborers	.90	80.000	Ea.
150,000 gallons		.50	144.000	Ea.
200,000 gallons		.40	180.000	Ea.
250,000 gallons		.30	240.000	Ea.

Division Thirty-Four
Transportation

Introduction

Division 34 addresses transportation related items, which were in Division 2 of the previous MasterFormat 95 classification system.

Included in this division are the following:

- Operation and maintenance of railroads
- Railroad related items including tracks, ties, and track accessories
- Railroad construction, including railroad sidings and railroad turnouts
- Traffic signal systems, including traffic control signal systems
- Airfield signaling and control systems, including airport lighting and airfield wind cones
- Roadway construction, including security vehicle barriers, vehicle guide rails, and vehicle delineators
- Aircraft loading bridges

LED signals use only about 10% the amount of electricity as incandescent signals. Energy Star's eligibility criteria require maximum wattages that range from 12 to 17 watts depending on the size, shape, and color of the module. All traffic signals manufactured on or after 2006 must meet these Energy Star requirements.

Estimating Data

The following tables present estimating guidelines for items found in Division 31 – Earthwork. Please note that these guidelines are intended as indicators of what may generally be expected, but that each project must be evaluated individually.

Table of Contents

Division 34

Estimating Tips

Rail Tracks

This subdivision includes items that may involve either repair of existing, or construction of new, railroad tracks. Additional preparation work, such as roadbed earthwork, would be found in Division 31.

Checklist ✓

For an estimate to be reliable, all items must be accounted for. A complete estimate can also limit contingencies. The following checklist can be used to help ensure that all items are included.

☐ **Rail Tracks**
- ☐ Rails
- ☐ Cross ties
- ☐ Signal and control equipment
- ☐ Track accessories

☐ **Roadways**
- ☐ Signals
- ☐ Controls
- ☐ Monitoring equipment
- ☐ Vehicle barriers
- ☐ Guide rails and guard rails
- ☐ Vehicle delineators
- ☐ Fare collection equipment

☐ **Airfields**
- ☐ Runway signals and lights
- ☐ Landing equipment
- ☐ Control tower equipment
- ☐ Weather observation equipment

☐ **Bridges**
- ☐ Bridge machinery
- ☐ Bridge specialties

Figure 34-1   Single Track RR Siding

The unit price costs for a single track RR siding include the components shown in the table below.

Description of Component	Qty. per L.F. of Track	Unit
Ballast, 1-1/2" crushed stone	.667	C.Y.
6" x 8" x 8'-6" Treated timber ties, 22" O.C.	.545	Ea.
Tie plates, 2 per tie	1.091	Ea.
Track rail	2.000	L.F.
Spikes, 6", 4 per tie	2.182	Ea.
Splice bars w/ bolts, lock washers & nuts, @ 33' O.C.	.061	Pair
Crew B-14 @ 57 L.F./Day	.018	Day

Figure 34-2 Single Track, Steel Ties, Concrete Bed

The unit price costs for a RR siding with steel ties and a concrete bed include the components shown in the table below.

Description of Component	Qty. per L.F. of Track	Unit
Concrete bed, 9' wide, 10" thick	.278	C.Y.
Ties, W6x16 x 6'-6" long, @ 30" O.C.	.400	Ea.
Tie plates, 4 per tie	1.600	Ea.
Track rail	2.000	L.F.
Tie plate bolts, 1", 8 per tie	3.200	Ea.
Splice bars w/bolts, lock washers & nuts, @ 33' O.C.	.061	Pair
Crew B-14 @ 22 L.F./Day	.045	Day

Figure 34-3 Aeronautical Lighting

In-runway Lighting

a. Runway Centerline Lighting System (RCLS). Runway centerline lights are installed on some precision approach runways to facilitate landing under adverse visibility conditions. They are located along the runway centerline and are spaced at 50-foot intervals. When viewed from the landing threshold, the runway centerline lights are white until the last 3,000 feet of the runway. The white lights begin to alternate with red for the next 2,000 feet, and for the last 1,000 feet of the runway, all centerline lights are red.

b. Touchdown Zone Lights (TDZL). Touchdown zone lights are installed on some precision approach runways to indicate the touchdown zone when landing under adverse visibility conditions. They consist of two rows of transverse light bars disposed symmetrically about the runway centerline. The system consists of steady-burning white lights which start 100 feet beyond the landing threshold and extend to 3,000 feet beyond the landing threshold or to the midpoint of the runway, whichever is less.

c. Taxiway Centerline Lead-Off Lights. Taxiway centerline lead-off lights provide visual guidance to persons exiting the runway. They are color-coded to warn pilots and vehicle drivers that they are within the runway environment or instrument landing system/microwave landing system (ILS/MLS) critical area, whichever is more restrictive. Alternate green and yellow lights are installed, beginning with green, from the runway centerline to one centerline light position beyond the runway holding position or ILS/MLS critical area holding position.

d. Taxiway Centerline Lead-On Lights. Taxiway centerline lead-on lights provide visual guidance to persons entering the runway. These "lead-on" lights are also color-coded with the same color pattern as lead-off lights to warn pilots and vehicle drivers that they are within the runway environment or instrument landing system/microwave landing system (ILS/MLS) critical area, whichever is more conservative. The fixtures used for lead-on lights are bidirectional, i.e., one side emits light for the lead-on function while the other side emits light for the lead-off function. Any fixture that emits yellow light for the lead-off function shall also emit yellow light for the lead-on function.

e. Land and Hold Short Lights. Land and hold short lights are used to indicate the hold short point on certain runways which are approved for Land and Hold Short Operations (LAHSO). Land and hold short lights consist of a row of pulsing white lights installed across the runway at the hold short point. Where installed, the lights will be on anytime LAHSO is in effect. These lights will be off when LAHSO is not in effect.

2-1-7 Control of Lighting Systems

a. Operation of approach light systems and runway lighting is controlled by the control tower (ATCT). At some locations the FSS may control the lights where there is no control tower in operation.

b. Pilots may request that lights be turned on or off. Runway edge lights, in-pavement lights and approach lights also have intensity controls which may be varied to meet the pilots request. Sequenced flashing lights (SFL) may be turned on and off. Some sequenced flashing light systems also have intensity control.

(continued on next page)

Figure 34-3 Aeronautical Lighting *(continued from previous page)*

TBL 2-1-1
Runways With Approach Lights

Lighting System	No. of Int. Steps	Status During Nonuse Period	Intensity Step Selected Per No. of Mike Clicks		
			3 Clicks	5 Clicks	7 Clicks
Approach Lights (Med. Int.)	2	Off	Low	Low	High
Approach Lights (Med. Int.)	3	Off	Low	Med	High
MIRL	3	Off or Low	◆	◆	◆
HIRL	5	Off or Low	◆	◆	◆
VASI	2	Off	☆	☆	☆

NOTES: ◆ *Predetermined intensity step.*
☆ *Low intensity for night use. High intensity for day use as determined by photocell control.*

TBL 2-1-2
Runways Without Approach Lights

Lighting System	No. of Int. Steps	Status During Nonuse Period	Intensity Step Selected Per No. of Mike Clicks		
			3 Clicks	5 Clicks	7 Clicks
MIRL	3	Off or Low	Low	Med.	High
HIRL	5	Off or Low	Step 1 or 2	Step 3	Step 5
LIRL	1	Off	On	On	On
VASI☆	2	Off	◆	◆	◆
REIL☆	1	Off	Off	On/Off	On
REIL☆	3	Off	Low	Med.	High

NOTES: ◆ *#32; Low intensity for night use. High intensity for day use as determined by photocell control.*
☆ *#32; The control of VASI and/or REIL may be independent of other lighting systems.*

a. With FAA approved systems, various combinations of medium intensity approach lights, runway lights, taxiway lights, VASI and/or REIL may be activated by radio control. On runways with both approach lighting and runway lighting (runway edge lights, taxiway lights, etc.) systems, the approach lighting system takes precedence for air-to-ground radio control over the runway lighting system which is set at a predetermined intensity step, based on expected visibility conditions. Runways without approach lighting may provide radio controlled intensity adjustments of runway edge lights. Other lighting systems, including VASI, REIL, and taxiway lights may be either controlled with the runway edge lights or controlled independently of the runway edge lights.

Division 34

(continued on next page)

Figure 34-3 Aeronautical Lighting (continued from previous page)

b. The control system consists of a 3-step control responsive to 7, 5, and/or 3 microphone clicks. This 3-step control will turn on lighting facilities capable of either 3-step, 2-step, or 1-step operation. The 3-step and 2-step lighting facilities can be altered in intensity, while the 1-step cannot. All lighting is illuminated for a period of 15 minutes from the most recent time of activation and may not be extinguished prior to end of the 15-minute period (except for 1-step and 2-step REILs which may be turned off when desired by keying the mike 5 or 3 times, respectively).

c. Suggested use is to always initially key the mike 7 times; this assures that all controlled lights are turned on to the maximum available intensity. If desired, adjustment can then be made, where the capability is provided, to a lower intensity (or the REIL turned off) by keying 5 and/or 3 times. Due to the close proximity of airports using the same frequency, radio controlled lighting receivers may be set at a low sensitivity, requiring the aircraft to be relatively close to activate the system. Consequently, even when lights are on, always key mike as directed when overflying an airport of intended landing or just prior to entering the final segment of an approach. This will assure the aircraft is close enough to activate the system and a full 15 minutes lighting duration is available. Approved lighting systems may be activated by keying the mike (within 5 seconds) as indicated in TBL 2-1-3.

TBL 2-1-3
Radio Control System

Key Mike	Function
7 times within 5 seconds	Highest intensity available
5 times within 5 seconds	Medium or lower intensity (Lower REIL or REIL-off)
3 times within 5 seconds	Lowest intensity available (Lower REIL or REIL-off)

d. For all public use airports with FAA standard systems, the Airport/Facility Directory contains the types of lighting, runway, and the frequency that is used to activate the system. Airports with IAPs include data on the approach chart identifying the light system, the runway on which they are installed, and the frequency that is used to activate the system.

NOTE-
Although the CTAF is used to activate the lights at many airports, other frequencies may also be used. The appropriate frequency for activating the lights on the airport is provided in the Airport/Facility Directory and the standard instrument approach procedures publications. It is not identified on the sectional charts.

e. Where the airport is not served by an IAP, it may have either the standard FAA approved control system or an independent type system of different specification installed by the airport sponsor. The Airport/Facility Directory contains descriptions of pilot controlled lighting systems for each airport having other than FAA approved systems, and explains the type lights, method of control, and operating frequency in clear text.

2-1-9. Airport/Heliport Beacons

a. Airport and heliport beacons have a vertical light distribution to make them most effective from one to ten degrees above the horizon; however, they can be seen well above and below this peak spread. The beacon may be an omnidirectional capacitor-discharge device, or it may rotate at a constant speed which produces the visual effect of flashes at regular intervals. Flashes may be one or two colors alternately. The total number of flashes are:

(continued on next page)

Figure 34-3 Aeronautical Lighting *(continued from previous page)*

1. 24 to 30 per minute for beacons marking airports, landmarks, and points on federal airways.

2. 30 to 45 per minute for beacons marking heliports.

b. The colors and color combinations of beacons are:

1. White and Green- Lighted land airport.

2. *Green alone- Lighted land airport.

3. White and Yellow- Lighted water airport.

4. *Yellow alone- Lighted water airport.

5. Green, Yellow, and White- Lighted heliport.

NOTE-
Green alone or yellow alone is used only in connection with a white-and-green or white-and-yellow beacon display, respectively.

c. Military airport beacons flash alternately white and green, but are differentiated from civil beacons by dual-peaked (two quick) white flashes between the green flashes.

d. In Class B, Class C, Class D, and Class E surface areas, operation of the airport beacon during the hours of daylight often indicates that the ground visibility is less than 3 miles and/or the ceiling is less than 1,000 feet. ATC clearance in accordance with 14 CFR Part 91 is required for landing, takeoff, and flight in the traffic pattern. Pilots should not rely solely on the operation of the airport beacon to indicate if weather conditions are IFR or VFR. At some locations with operating control towers, ATC personnel turn the beacon on or off when controls are in the tower. At many airports the airport beacon is turned on by a photoelectric cell or time clocks, and ATC personnel cannot control them. There is no regulatory requirement for daylight operation, and it is the pilot's responsibility to comply with proper preflight planning as required by 14 CFR Section 91.103.

2-1-10. Taxiway Lights

a. Taxiway Edge Lights. Taxiway edge lights are used to outline the edges of taxiways during periods of darkness or restricted visibility conditions. These fixtures emit blue light.

NOTE-
At most major airports these lights have variable intensity settings and may be adjusted at pilot request or when deemed necessary by the controller.

b. Taxiway Centerline Lights. Taxiway centerline lights are used to facilitate ground traffic under low visibility conditions. They are located along the taxiway centerline in a straight line on straight portions, on the centerline of curved portions, and along designated taxiing paths in portions of runways, ramp, and apron areas. Taxiway centerline lights are steady burning and emit green light.

c. Clearance Bar Lights. Clearance bar lights are installed at holding positions on taxiways in order to increase the conspicuity of the holding position in low visibility conditions. They may also be installed to indicate the location of an intersecting taxiway during periods of darkness. Clearance bars consist of three in-pavement steady-burning yellow lights.

d. Runway Guard Lights. Runway guard lights are installed at taxiway/runway intersections. They are primarily used to enhance the conspicuity of taxiway/runway intersections during low visibility

(continued on next page)

Figure 34-3 Aeronautical Lighting *(continued from previous page)*

conditions, but may be used in all weather conditions. Runway guard lights consist of either a pair of elevated flashing yellow lights installed on either side of the taxiway, or a row of in-pavement yellow lights installed across the entire taxiway, at the runway holding position marking.

NOTE-
Some airports may have a row of three or five in-pavement yellow lights installed at taxiway/ runway intersections. They should not be confused with clearance bar lights described in paragraph 2-1-9c, Clearance Bar Lights.

e. Stop Bar Lights. Stop bar lights, when installed, are used to confirm the ATC clearance to enter or cross the active runway in low visibility conditions (below 1,200 ft Runway Visual Range). A stop bar consists of a row of red, unidirectional, steady-burning in-pavement lights installed across the entire taxiway at the runway holding position, and elevated steady-burning red lights on each side. A controlled stop bar is operated in conjunction with the taxiway centerline lead-on lights which extend from the stop bar toward the runway. Following the ATC clearance to proceed, the stop bar is turned off and the lead-on lights are turned on. The stop bar and lead-on lights are automatically reset by a sensor or backup timer.

CAUTION-
Pilots should never cross a red illuminated stop bar, even if an ATC clearance has been given to proceed onto or across the runway.

NOTE-
If after crossing a stop bar, the taxiway centerline lead-on lights inadvertently extinguish, pilots should hold their position and contact ATC for further instructions.

(Courtesy FAA)

Figure 34-4 Transportation Abbreviations

AASHTO	American Association of State Highway and Transportation Officials
ANSI	American National Standards Institute
AREMA	American Railway Engineering and Maintenance of Way Association
ASCE	American Society of Civil Engineers
ASTM	American Society for Testing and Materials
FED-STD	Federal Standard
FHWA	Federal Highway Administration
FSS	Federal Specifications and Standards, General Services
IMSA	International Municipal Signal Association
ITE	Institute of Transportation Engineers
MUTCD	Manual on Uniform Traffic Control Devices
NCHRP	National Cooperative Highway Research Program
NOAA	National Oceanic and Atmospheric Administration
ROW	Right-of-Way
SI	International System of Units

(Courtesy of Department of Transportation)

Notes

Division Forty-Four
Pollution Control Equipment

Introduction

Division 44 – Pollution Control Equipment includes pollution control equipment that could be used for a variety of facilities including municipal, commercial, and industrial projects. In some cases project costs will have to be estimated using individual components. In other cases prepackaged items such as a packaged water treatment plant can be used.

Division 44 includes air pollution control, noise pollution control, water and wastewater treatment, and solid waste control. Under the previously used Master Format 95 most of these items were in Division 02 or Division 11.

Estimating Data

The following tables present guidelines for pollution control equipment.

Table of Contents

Division 44

Estimating Tips

Integrated Systems

Pollution control equipment addresses four basic types of pollution: air, noise, water, and solids. These systems may be interrelated, and care must be taken that the complete systems are estimated. For example, air pollution equipment may include dust and air-entrained particles that have to be collected. The vacuum systems could be noisy, requiring silencers to reduce noise pollution, and the collected solids must be disposed of to prevent solid pollution.

Water/Wastewater Treatment

Estimates for large water/wastewater treatment facilities require input from several trades or disciplines. They usually include inground structures, aboveground structures, mechanical equipment, and supporting electrical components. The overall cost of a facility is affected by the level of treatment required. Some wastewater treatment plants only need to provide conventional primary and secondary treatment for conventional pollutants. Others may be required to provide tertiary treatment for the removal of additional pollutants such as phosphorous or nitrogen. Some plants also need to control the discharge of heavy metals and organic chemicals.

Storm water often requires less extensive treatment than sanitary wastewater or industrial wastewater. Storm water does need to be managed and/or treated to control the discharge of pollutants, especially suspended solids, petroleum products, and nutrients.

Checklist ✓

For an estimate to be reliable, all items must be accounted for. A complete estimate can also limit contingencies. The following checklist can be used to help ensure that all items are included.

☐ *Air Pollution Control*
- ☐ Cyclonic separators
- ☐ Industrial dust collectors
- ☐ Air pollution filters
- ☐ Fugitive dust control
- ☐ Precipitators
- ☐ Scrubbers
- ☐ Thermal oxidizers
- ☐ Vacuum extraction systems

☐ *Noise Pollution Control*
- ☐ Noise abatement barriers
- ☐ Noise pollution silencers
- ☐ Frequency cancellers

☐ *Water Treatment Equipment*
- ☐ Packaged water treatment

- ☐ General water treatment equipment
- ☐ Water filtration equipment
- ☐ Water treatment chemical systems equipment
- ☐ Water treatment biological systems equipment
- ☐ Sludge treatment and handling equipment for water treatment system

☐ *Solid Waste Control*
- ☐ Compactors
- ☐ Baling equipment
- ☐ Fluffing equipment
- ☐ Liquid extraction equipment
- ☐ Containers
- ☐ Transfer trailers
- ☐ Transfer stations

Figure 44-1 Municipal Wastewater Characterization

Contaminants	Unit	Weak	Medium	Strong
Solids, total (TS)	mg/L	350	720	1200
Dissolved, total (TDS)	mg/L	250	500	850
Fixed	mg/L	145	300	525
Volatile	mg/L	105	200	325
Suspended solids (SS)	mg/L	100	220	350
Fixed	mg/L	20	55	75
Volatile	mg/L	80	165	275
Settleable solids	mg/L	5	10	20
BOD, 20 degree C	mg/L	110	220	400
Total organic carbon (TOC)	mg/L	80	160	290

Contaminants	Unit	Weak	Medium	Strong
Chemical oxygen demand (COD)	mg/L	250	500	1000
Nitrogen (total as N)	mg/L	20	40	85
Organic	mg/L	8	15	35
Free ammonia	mg/L	12	25	50
Nitrites	mg/L	0	0	0
Nitrates	mg/L	0	0	0
Phosphorus (total as P)	mg/L	4	8	15
Organics	mg/L	1	3	5
Inorganics	mg/L	3	5	10

Contaminants	Unit	Weak	Medium	Strong
Chlorides (See note a)	mg/L	30	50	100
Sulfate (See note a)	mg/L	20	30	50
Alkalinity (as $CaCO_3$)	mg/L	50	100	200
Grease	mg/L	50	100	150
Total coliform	mg/L	10^6 to 10^7	10^7 to 10^8	10^8 to 10^9
Volatile organic compounds (VOCs)	ug/L	< 100	100 to 400	> 400

Note a: Values should be increased by amount present in domestic water supply.

Figure 44-2 Municipal Wastewater Discharge

Residential Source	Unit	Flow (L/unit/d)
High-rise apartment	Person	50
Low-rise apartment	Person	65
Individual Residence:		
Typical home	Person	70
Better home	Person	80
Luxury home	Person	95
Older home	Person	45
Summer cottage	Person	40
Trailer park	Person	40

Commercial Source	Unit	Flow (gal/unit/d)
Airport	Passenger	3
Bar	Customer	3
Department store	Toilet Room	550
Hotel	Guest	45
Laundry	Machine	550
Office	Employee	15
Restaurant	Customer	9
Shopping center	Employee	10
Trailer park	Person	40

Institutional Source	Unit	Flow (gal/unit/d)
Hospital, medical	Bed	150
Hospital, mental	Bed	120
Prison	Inmate	120
Rest home	Resident	90
School, day:		
With cafeteria, gym, showers	Student	25
With cafeteria only	Student	15
Without cafeteria and gym	Student	10
School, boarding	Student	75

Notes

Appendix

List of Appendices

The Appendix is divided into ten sections, A – J. Each contains tables, illustrations, and/or listings of information that may be useful in creating an estimate. The specific contents of each Appendix section are listed in the table of contents below.

Historical Cost Indexes

The table below lists both the RSMeans Historical Cost Index based on Jan. 1, 1993 = 100 as well as the computed value of an index based on Jan. 1, 2009 costs. Since the Jan. 1, 2009 figure is estimated, space is left to write in the actual index figures as they become available through either the quarterly "RSMeans Construction Cost Indexes" or as printed in the "Engineering News-Record." To compute the actual index based on Jan. 1, 2009 = 100, divide the Historical Cost Index for a particular year by the actual Jan. 1, 2009 Construction Cost Index. Space has been left to advance the index figures as the year progresses.

Year	Historical Cost Index Jan. 1, 1993 = 100		Current Index Based on Jan. 1, 2009 = 100		Year	Historical Cost Index Jan. 1, 1993 = 100	Current Index Based on Jan. 1, 2009 = 100		Year	Historical Cost Index Jan. 1, 1993 = 100	Current Index Based on Jan. 1, 2009 = 100	
	Est.	Actual	Est.	Actual		Actual	Est.	Actual		Actual	Est.	Actual
Oct 2009					July 1994	104.4	60.3		July 1976	46.9	27.1	
July 2009					1993	101.7	54.7		1975	44.8	24.1	
April 2009					1992	99.4	53.5		1974	41.4	22.3	
Jan 2009	185.9		100.0	100.0	1991	96.8	52.1		1973	37.7	20.3	
July 2008		180.4	97.0		1990	94.3	50.7		1972	34.8	18.7	
2007		169.4	91.1		1989	92.1	49.6		1971	32.1	17.3	
2006		162.0	87.1		1988	89.9	48.3		1970	28.7	15.4	
2005		151.6	81.5		1987	87.7	47.2		1969	26.9	14.5	
2004		143.7	77.3		1986	84.2	45.3		1968	24.9	13.4	
2003		132.0	71.0		1985	82.6	44.4		1967	23.5	12.6	
2002		128.7	69.2		1984	82.0	44.1		1966	22.7	12.2	
2001		125.1	67.3		1983	80.2	43.1		1965	21.7	11.7	
2000		120.9	65.0		1982	76.1	41.0		1964	21.2	11.4	
1999		117.6	63.3		1981	70.0	37.6		1963	20.7	11.1	
1998		115.1	61.9		1980	62.9	33.8		1962	20.2	10.9	
1997		112.8	60.7		1979	57.8	31.1		1961	19.8	10.7	
1996		110.2	59.3		1978	53.5	28.8		1960	19.7	10.6	
1995		107.6	57.9		1977	49.5	26.6		1959	19.3	10.4	

Adjustments to Costs

The Historical Cost Index can be used to convert National Average building costs at a particular time to the approximate building costs for some other time.

Time Adjustment using the Historical Cost Indexes:

$$\frac{\text{Index for Year A}}{\text{Index for Year B}} \times \text{Cost in Year B} = \text{Cost in Year A}$$

Example:

Estimate and compare construction costs for different years in the same city.

To estimate the National Average construction cost of a building in 1970, knowing that it cost $900,000 in 2009:

INDEX in 1970 = 28.7
INDEX in 2009 = 185.9

$$\frac{\text{INDEX } 1970}{\text{INDEX } 2009} \times \text{Cost } 2009 = \text{Cost } 1970$$

$$\frac{28.7}{185.9} \times \$900,000 = .154 \times \$900,000 = \$138,600$$

The construction cost of the building in 1970 is $138,600.

Note: The City Cost Indexes for Canada can be used to convert U.S. National averages to local costs in Canadian dollars.

How to Use the City Cost Indexes

Means City Cost Indexes (CCI) are an extremely useful tool to use when you want to compare costs from city to city and region to region.

Keep in mind that a City Cost Index number is a *percentage ratio* of a specific city's cost to the national average cost of the same item at a stated time period.

In other words, these index figures represent relative construction *factors* (or, if you prefer, multipliers) for Material and Installation costs, as well as the weighted average for Total In Place costs for each CSI MasterFormat division. Installation costs include both labor and equipment rental costs.

The 30 City Average Index is the average of 30 major U.S. cities and serves as a National Average, base 100.0.

Please note that the index does not account for differences in productivity, building codes, and local customs.

How to Use

Compare costs from city to city.

In using the Means Indexes, remember that an index number is not a fixed number but a *ratio*.

Therefore, when making cost comparisons between cities, do not subtract one city's index number from the index number of another city and read the result as a percentage difference. Instead, divide one city's index number by that of the other city. The resulting number may then be used as a multiplier to calculate cost differences from city to city.

The formula used to find cost differences between cities for the purpose of comparison is as follows:

$$\frac{\text{City A Index}}{\text{City B Index}} \times \text{City B Cost (Known)} = \text{City A Cost (Unknown)}$$

In addition, you can use the RSMeans CCI to calculate and compare costs division by division between cities using the same basic formula. (Just be sure that you're comparing similar divisions.)

CSI DIV. NO.	JULY 1, 2008 ITEM	PHOENIX, AZ			LOS ANGELES, CA			SAN DIEGO, CA			SAN FRANCISCO, CA			DENVER, CO			WASHINGTON, D.C.		
		MAT.	INST.	TOTAL	MAT.	INST.	TOTAL	MAT.	INST.	TOTAL	MAT.	INST.	TOTAL	MAT.	INST.	TOTAL	MAT.	INST.	TOTAL
015433	CONTRACTOR EQUIPMENT	0.0	98.6	98.6	0.0	100.4	100.4	0.0	99.7	99.7	0.0	108.0	108.0	0.0	99.0	99.0	0.0	103.1	103.1
3120, 3130	Earth Moving & Earthwork	76.3	103.7	99.5	88.2	108.8	105.6	103.7	103.5	103.5	156.6	110.3	117.5	74.2	103.1	99.0	110.2	92.7	95.4
3160, 3170	Load-Bearing Elements & Tunneling	102.3	91.0	96.0	104.4	98.5	101.1	104.4	99.8	101.8	104.4	110.4	107.7	104.4	91.4	97.1	99.0	111.2	105.8
3210	Bases, Ballasts & Paving	106.3	110.7	106.7	114.6	109.6	114.0	112.7	99.8	111.3	129.4	110.3	127.4	94.9	103.9	103.9	118.0	85.8	114.7
3310, 3330, 3340, 3350	Utility Services & Drainage	101.3	67.2	98.7	94.3	124.4	96.7	101.3	117.2	102.6	96.6	139.3	100.0	96.6	78.5	95.2	110.7	87.2	108.9
3230	Site Improvements	98.7	109.3	102.4	100.7	114.9	105.7	103.0	115.2	107.3	103.0	128.4	111.9	109.3	94.8	104.2	102.5	88.6	97.6
3290	Planting	76.9	96.3	85.8	105.3	106.8	106.0	105.3	102.7	104.1	170.0	109.3	142.2	105.3	102.8	104.1	105.3	92.9	99.6
0241, 31 - 34	SITE & INFRASTRUCTURE, DEMOLITION	86.7	102.9	98.4	98.5	108.6	105.7	105.6	103.5	104.1	144.3	110.4	119.9	89.0	102.8	98.9	110.3	92.9	97.8
0310	Concrete Forming & Accessories	104.8	69.2	74.0	104.2	126.7	123.6	108.1	111.0	110.6	108.2	144.2	139.3	99.9	82.1	84.5	101.3	81.9	84.5
0320	Concrete Reinforcing	97.4	79.7	90.2	104.4	118.6	110.2	99.3	118.2	107.0	111.8	119.9	115.1	106.9	79.7	95.8	113.5	91.1	104.3
0330	Cast-in-Place Concrete	104.9	77.9	94.6	108.8	122.0	113.9	107.0	105.8	106.6	136.3	122.7	131.1	96.3	84.2	91.7	127.6	88.5	112.7
0340	Precast Concrete	122.2	75.9	116.0	112.7	100.5	111.1	117.8	100.9	115.6	124.1	105.1	121.5	110.5	86.1	107.3	120.0	114.0	119.2
03	CONCRETE	108.1	74.3	92.7	108.7	122.4	115.0	108.5	109.9	109.1	126.4	130.8	128.4	102.5	82.6	93.5	121.1	87.3	105.8
0405	Basic Masonry Materials & Methods	102.1	73.5	90.0	107.1	121.5	113.2	105.2	114.6	109.1	125.9	132.5	128.7	101.7	80.4	92.7	118.7	83.4	103.8
0420	Unit Masonry	96.9	64.3	77.2	93.9	122.3	110.0	97.6	111.8	106.2	162.6	136.2	146.6	104.2	78.6	88.7	99.5	81.6	88.7
0440	Stone Assemblies	110.7	64.5	80.3	98.4	120.9	113.2	103.5	111.4	108.0	144.1	136.1	138.9	82.0	79.4	80.3	90.6	82.2	85.0
04	MASONRY	97.8	64.9	78.1	95.5	121.8	111.2	98.7	111.5	106.4	157.2	136.0	144.5	103.3	79.0	88.7	101.7	81.9	89.8
0510	Structural Metal Framing	103.5	75.1	95.9	114.3	103.1	111.3	110.7	101.5	108.3	114.4	112.4	113.8	98.1	86.4	95.0	101.7	107.9	103.4
0520, 0530	Metal Joists & Decking	90.6	76.9	86.3	99.8	102.9	100.8	100.7	103.3	101.5	101.2	109.4	103.8	107.6	85.9	100.7	96.0	110.3	100.5
0550	Metal Fabrications	100.0	78.0	96.7	100.0	109.2	101.4	100.0	109.0	101.4	100.0	115.6	102.4	100.0	86.1	97.9	100.0	103.1	100.5
05	METALS	95.1	76.6	89.7	104.0	103.2	103.8	103.5	103.1	103.4	104.9	110.5	106.5	104.2	86.1	98.9	98.0	109.3	101.3
0610	Rough Carpentry	96.7	70.0	78.1	103.7	126.2	119.4	104.3	107.6	106.6	114.9	146.3	136.8	97.4	84.2	88.2	99.2	81.3	86.7
0620	Finish Carpentry	97.6	68.8	94.7	69.4	127.3	75.2	86.0	107.6	88.2	97.6	147.3	126.6	107.0	83.9	104.7	102.7	80.6	100.4
06	WOOD, PLASTICS & COMPOSITES	97.1	70.0	82.2	86.8	122.6	108.5	95.3	107.6	102.0	106.4	146.4	128.3	102.1	84.2	92.3	100.9	81.2	90.1
0710	Dampproofing & Waterproofing	120.4	68.8	77.6	119.8	127.3	126.1	122.7	107.6	110.2	126.9	147.3	143.8	93.6	83.9	85.5	93.6	80.6	82.8
0720, 0780	Thermal, Fire & Smoke Protection	106.9	52.8	91.6	106.9	121.2	109.7	113.2	116.7	109.9	117.0	132.3	121.3	101.9	66.0	91.7	107.4	83.3	100.6
0740, 0750	Roofing & Siding	109.9	77.4	98.3	101.5	123.2	109.3	113.6	107.6	109.2	118.8	136.7	114.3	94.6	79.2	89.1	103.1	82.2	95.6
0760	Flashing & Sheet Metal	100.0	81.8	86.5	100.0	116.3	112.1	100.0	102.0	101.5	100.0	153.6	139.7	100.0	83.1	87.4	100.0	93.0	94.8
07	THERMAL & MOISTURE PROTECTION	107.3	68.1	92.1	104.2	118.8	109.9	108.4	105.6	107.3	117.8	138.3	120.9	98.9	76.6	90.2	105.3	84.7	97.3
0810, 0830	Doors & Frames	99.3	73.3	92.8	96.8	118.6	102.3	96.4	107.9	99.3	106.9	130.5	112.8	99.2	85.1	95.6	107.8	91.7	103.8
0840, 0880	Glazing & Curtain Walls	91.0	68.5	82.2	98.9	124.2	108.8	99.1	117.2	106.1	110.5	140.9	122.3	105.5	86.3	98.0	104.7	106.7	105.5
08	OPENINGS	97.3	71.5	90.9	98.3	121.4	104.0	99.9	109.8	102.3	108.6	135.8	115.3	99.4	85.1	95.9	107.0	92.7	103.4
0920	Plaster & Gypsum Board	87.3	68.8	75.2	104.4	127.3	119.3	117.3	107.6	111.0	110.7	147.3	134.5	95.8	83.9	88.1	111.0	80.6	91.2
0930, 0966	Tile & Terrazzo	106.5	64.0	85.4	109.6	121.2	115.4	106.3	116.7	111.5	115.5	132.3	123.9	95.9	81.2	88.6	90.5	89.7	90.1
0950, 0980	Ceilings & Acoustic Treatment	108.7	68.8	84.0	126.1	121.2	126.9	107.2	107.6	107.4	127.8	147.3	139.8	106.3	83.9	92.4	107.4	80.6	90.8
0960	Flooring	103.2	58.0	91.0	102.3	113.6	105.3	105.4	107.6	106.3	117.6	126.2	119.9	109.2	88.8	103.7	127.7	100.2	120.3
0970, 0990	Wall Finishes & Painting/Coating	94.0	57.2	71.9	97.5	111.7	111.7	103.8	111.7	106.5	108.1	156.9	137.4	106.0	76.5	88.3	125.2	87.4	102.5
09	FINISHES	102.4	65.3	82.8	109.6	122.9	116.6	107.6	110.1	108.9	117.8	142.9	131.1	103.1	83.0	92.4	111.0	85.3	97.5
COVERS	DIVS 10 - 14, 25, 28, 41, 43, 44	100.0	82.4	96.5	100.0	114.3	102.8	100.0	111.7	102.3	100.0	130.7	106.1	100.0	88.6	97.7	100.0	98.9	99.8
2210, 2230, 2240, 2320	Piping, Pumps, Plumbing Equipment	100.0	76.4	89.9	100.0	117.1	107.3	100.0	115.0	106.4	100.0	163.7	127.2	100.0	86.1	94.1	100.0	93.1	97.0
2113	Fire Suppression Sprinkler Systems	100.0	79.4	86.3	100.0	110.0	106.7	100.0	109.8	106.5	100.0	171.7	147.7	100.0	90.3	93.5	100.0	96.8	97.9
2350	Central Heating Equipment	101.1	75.4	93.0	102.1	116.7	106.7	102.1	114.5	106.0	102.5	159.1	120.2	100.3	83.1	95.0	101.1	92.6	98.4
2330, 2340, 2360, 2370, 2380	Air Conditioning & Ventilation	100.0	73.5	98.1	100.0	111.4	101.4	100.0	113.8	105.8	100.0	158.0	126.0	100.0	82.9	94.7	100.0	89.6	97.6
21, 22, 23	FIRE SUPPRESSION, PLUMBING & HVAC	100.1	76.9	90.9	100.3	115.5	106.3	100.3	113.3	105.6	100.3	164.9	126.0	100.0	86.7	94.7	100.1	93.8	97.6
26, 27, 3370	ELECTRICAL, COMMUNICATIONS & UTIL.	100.3	64.5	82.9	100.4	114.5	107.3	97.7	104.1	100.8	104.9	162.2	132.8	106.7	85.4	96.3	103.4	101.5	102.5
MF2004	WEIGHTED AVERAGE	99.9	74.2	88.0	102.2	116.4	108.3	102.4	108.7	105.1	111.2	140.8	123.7	101.8	85.8	95.0	104.6	92.9	99.7

CSI DIV. NO.	JULY 1, 2008 — ITEM	ATLANTA, GA MAT.	INST.	TOTAL	CHICAGO, IL MAT.	INST.	TOTAL	INDIANAPOLIS, IN MAT.	INST.	TOTAL	NEW ORLEANS, LA MAT.	INST.	TOTAL	BOSTON, MA MAT.	INST.	TOTAL	BALTIMORE, MD MAT.	INST.	TOTAL
015433	CONTRACTOR EQUIPMENT	0.0	94.1	94.1	0.0	93.2	93.2	0.0	92.9	92.9	0.0	90.2	90.2	0.0	106.2	106.2	0.0	102.7	102.7
3120, 3130	Earth Moving & Earthwork	106.7	95.2	97.0	101.6	92.6	94.0	66.9	98.8	93.8	152.2	88.9	98.8	73.5	107.4	102.1	82.4	94.8	92.9
3160, 3170	Load-Bearing Elements & Tunneling	100.1	89.4	94.1	99.0	104.6	102.2	98.0	87.9	92.4	99.0	85.1	91.2	104.4	106.6	105.6	99.0	116.4	108.7
3210	Bases, Ballasts & Paving	105.5	94.8	104.4	100.5	103.5	100.8	90.5	94.7	91.0	98.2	99.8	98.4	106.4	98.8	105.6	95.6	100.7	96.1
3310, 3330, 3340, 3350	Utility Services & Drainage	99.0	75.3	97.1	82.5	152.1	88.0	82.5	84.8	82.7	108.4	61.5	104.7	91.9	141.9	95.9	106.0	77.5	103.8
3230	Site Improvements	97.7	85.5	93.4	104.7	118.3	109.4	96.3	103.9	98.4	94.8	80.4	89.7	100.4	114.0	105.2	104.1	107.9	105.4
3290	Planting	101.2	97.3	99.4	97.2	95.7	96.5	105.3	98.7	102.3	101.2	87.2	94.8	81.0	108.2	93.4	117.4	90.3	105.0
0241, 31 - 34	SITE & INFRASTRUCTURE, DEMOLITION	104.1	95.1	97.7	98.1	93.5	94.8	83.0	98.5	94.1	122.9	88.7	98.3	85.3	107.5	101.3	96.0	94.9	95.2
0310	Concrete Forming & Accessories	97.6	76.1	79.0	100.5	153.2	146.1	95.5	83.7	85.3	104.8	66.4	71.7	99.3	137.7	132.5	101.7	77.9	81.1
0320	Concrete Reinforcing	98.4	91.7	95.6	106.3	157.6	127.2	94.5	80.9	88.9	100.9	64.9	86.2	99.3	146.1	118.4	100.9	86.7	95.1
0330	Cast-in-Place Concrete	112.0	74.5	97.7	105.9	145.2	120.9	91.4	85.9	89.3	92.5	71.9	84.6	100.0	144.7	117.1	95.2	79.8	89.3
0340	Precast Concrete	87.3	84.9	86.9	101.8	126.4	105.1	89.4	79.0	88.1	90.9	78.4	89.2	138.2	120.2	135.8	112.0	102.6	110.7
03	CONCRETE	101.6	78.6	91.1	104.5	149.8	125.1	91.7	83.8	88.1	94.4	68.7	82.7	110.3	140.6	124.1	101.3	81.3	92.2
0405	Basic Masonry Materials & Methods	107.6	78.3	95.2	106.0	146.6	123.2	93.3	87.3	90.8	94.1	63.5	81.2	100.1	145.4	119.2	95.9	78.3	88.5
0420	Unit Masonry	85.2	73.1	77.9	92.8	153.1	129.3	94.0	83.4	87.6	108.0	59.2	78.5	116.5	155.8	140.3	97.2	70.7	81.2
0440	Stone Assemblies	74.0	73.9	73.9	92.5	151.6	131.5	77.1	83.3	81.2	107.6	60.0	76.2	139.4	157.1	151.0	93.1	71.7	79.0
04	MASONRY	87.7	73.7	79.3	94.4	152.2	129.0	93.5	83.6	87.6	106.3	59.7	78.4	114.9	155.6	139.3	96.9	71.5	81.7
0510	Structural Metal Framing	89.0	82.9	87.4	89.1	130.8	100.2	99.9	80.5	94.7	132.4	75.9	117.4	92.7	130.1	102.6	99.9	96.7	99.1
0520, 0530	Metal Joists & Decking	94.8	84.2	91.4	89.2	129.5	102.0	103.6	80.5	96.3	106.4	77.1	97.1	98.3	123.7	106.3	95.8	99.1	96.8
0550	Metal Fabrications	100.0	76.2	96.4	100.0	130.7	102.8	100.0	83.7	97.5	100.0	71.5	95.7	100.0	143.5	106.6	100.0	85.0	97.7
05	METALS	93.6	83.5	90.7	90.1	130.7	102.0	102.2	80.7	95.9	113.3	76.7	102.6	96.8	126.2	105.4	97.3	97.9	97.5
0610	Rough Carpentry	100.6	77.5	84.5	102.1	152.2	137.0	95.1	83.3	86.9	106.6	69.2	80.6	102.0	136.6	126.0	101.7	80.0	86.6
0620	Finish Carpentry	94.0	77.0	92.3	108.4	154.1	113.0	112.0	83.0	109.1	103.4	68.4	99.9	106.3	137.2	109.4	94.0	79.4	92.5
06	WOOD, PLASTICS & COMPOSITES	97.3	77.4	86.4	105.2	152.3	131.1	103.6	83.3	92.4	105.0	69.2	85.3	104.1	136.6	121.9	97.9	79.9	88.0
0710	Dampproofing & Waterproofing	90.1	77.0	79.3	111.8	154.1	146.9	92.8	83.0	84.7	115.0	68.4	76.4	95.2	137.2	130.0	90.9	79.4	81.4
0720, 0780	Thermal, Fire & Smoke Protection	95.8	76.9	90.5	101.0	143.7	113.1	95.4	75.7	89.9	100.8	51.1	86.8	98.1	165.7	117.2	95.8	81.2	91.7
0740, 0750	Roofing & Siding	90.4	76.0	85.3	100.5	144.7	116.3	91.4	92.2	91.7	108.0	67.1	93.4	89.4	136.2	106.1	95.4	80.0	92.8
0760	Flashing & Sheet Metal	100.0	85.8	89.5	100.0	133.0	124.4	100.0	84.4	88.4	100.0	73.5	80.4	100.0	130.1	122.3	100.0	80.7	85.7
07	THERMAL & MOISTURE PROTECTION	94.3	78.2	88.0	100.7	142.7	117.1	94.4	83.9	90.3	103.7	62.8	87.8	94.8	148.7	115.8	97.6	80.2	90.8
0810, 0830	Doors & Frames	98.8	76.0	93.0	98.1	155.9	112.7	105.3	83.0	99.7	103.8	69.9	95.3	102.1	140.6	111.8	93.9	81.5	90.8
0840, 0880	Glazing & Curtain Walls	95.8	77.4	86.7	93.2	154.5	117.0	101.9	84.8	95.2	102.4	62.4	86.8	108.7	145.8	123.2	93.3	104.5	97.7
08	OPENINGS	99.1	75.4	93.2	101.6	155.3	114.8	103.6	83.4	98.6	103.4	68.1	94.7	100.9	141.0	110.8	93.7	85.8	91.8
0920	Plaster & Gypsum Board	118.8	77.0	91.7	92.0	154.1	132.3	94.0	83.0	86.8	104.0	68.4	80.9	108.0	137.2	127.0	104.6	79.4	88.2
0930, 0966	Tile & Terrazzo	95.2	70.8	83.1	91.8	153.0	122.2	88.0	88.1	88.0	100.4	63.3	82.0	98.6	151.2	124.7	99.3	72.0	85.8
0950, 0980	Ceilings & Acoustic Treatment	109.8	77.0	89.5	101.5	134.0	113.1	95.3	83.0	87.7	96.9	68.4	79.3	98.2	137.2	122.3	95.4	79.4	85.5
0960	Flooring	89.4	71.8	84.6	92.0	149.5	107.5	92.7	89.6	91.9	107.7	68.7	97.2	97.3	167.4	116.2	88.9	78.2	86.7
0970, 0990	Wall Finishes & Painting/Coating	102.8	86.2	92.9	86.5	148.0	123.4	95.1	90.8	92.5	108.1	63.4	81.3	94.4	158.2	132.7	94.6	81.0	86.4
09	FINISHES	100.7	76.2	87.7	93.8	152.7	124.9	92.5	85.6	88.9	102.9	66.8	83.8	99.1	145.7	123.7	96.0	77.8	86.4
COVERS	DIVS 10 - 14, 25, 28, 41, 43, 44	100.0	85.7	97.2	100.0	126.8	105.3	100.0	91.7	98.4	100.0	79.8	96.0	100.0	121.9	104.3	100.0	86.0	97.2
2210, 2230, 2240, 2320	Piping, Pumps, Plumbing Equipment	100.0	77.4	90.4	100.0	136.3	115.5	100.0	85.9	94.0	100.0	65.5	85.3	100.0	132.6	113.9	100.0	86.9	94.4
2113	Fire Suppression Sprinkler Systems	100.0	80.3	86.9	100.0	124.8	116.5	100.0	85.4	90.3	100.0	69.0	79.4	100.0	128.1	118.7	100.0	94.4	96.3
2350	Central Heating Equipment	100.0	77.3	92.9	99.6	136.9	111.2	99.2	84.9	94.7	100.0	63.5	86.6	101.4	131.6	110.8	99.2	87.2	95.5
2330, 2340, 2360, 2370, 2380	Air Conditioning & Ventilation	100.0	75.3	98.2	100.0	140.3	102.9	100.0	85.1	94.9	100.0	62.9	97.3	100.0	133.2	102.4	100.0	82.4	98.7
21, 22, 23	FIRE SUPPRESSION, PLUMBING & HVAC	100.0	78.0	91.2	99.9	133.8	113.4	99.9	85.7	94.2	100.0	66.1	86.5	100.2	131.5	112.6	99.9	88.5	95.4
26, 27, 3370	ELECTRICAL, COMMUNICATIONS & UTIL.	98.7	76.7	88.0	97.5	135.8	116.1	98.3	93.0	95.7	96.3	68.6	82.8	96.3	137.1	116.2	98.2	91.7	95.0
MF2004	WEIGHTED AVERAGE	98.2	79.5	90.3	97.9	137.6	114.7	98.0	86.9	93.3	102.9	69.8	88.9	100.5	135.6	115.4	98.2	85.9	93.0

CSI DIV. NO.	JULY 1, 2008 — ITEM	DETROIT, MI MAT.	INST.	TOTAL	MINNEAPOLIS, MN MAT.	INST.	TOTAL	KANSAS CITY, MO MAT.	INST.	TOTAL	ST. LOUIS, MO MAT.	INST.	TOTAL	BUFFALO, NY MAT.	INST.	TOTAL	NEW YORK, NY MAT.	INST.	TOTAL
015433	**CONTRACTOR EQUIPMENT**	0.0	98.3	98.3	0.0	104.0	104.0	0.0	103.0	103.0	0.0	106.8	106.8	0.0	96.8	96.8	0.0	113.3	113.3
3120, 3130	Earth Moving & Earthwork	85.4	97.9	96.0	92.1	106.8	104.5	87.7	97.5	96.0	99.1	96.6	97.0	86.1	97.2	95.4	154.5	124.1	128.8
3160, 3170	Load-Bearing Elements & Tunneling	101.2	95.6	98.1	102.3	110.7	107.0	93.6	110.5	103.1	93.6	119.6	108.1	101.2	94.5	97.5	102.3	109.5	106.3
3210	Bases, Ballasts & Paving	88.6	98.8	89.7	104.9	106.3	105.0	90.5	77.9	89.2	88.5	99.7	89.7	109.4	102.9	108.7	115.8	114.5	115.7
3310, 3330, 3340, 3350	Utility Services & Drainage	99.0	116.9	100.4	113.1	132.1	114.6	103.7	108.2	104.0	87.2	111.8	89.2	101.3	110.4	102.0	99.0	174.0	104.9
3230	Site Improvements	108.3	114.5	110.5	93.5	116.0	101.4	99.1	103.7	100.7	94.6	97.5	95.6	93.2	98.6	95.1	99.0	117.6	105.6
3290	Planting	97.2	98.1	97.6	64.8	106.4	83.8	117.4	104.7	111.6	89.1	97.8	93.1	97.2	93.6	95.6	121.5	133.4	126.9
0241, 31 - 34	**SITE & INFRASTRUCTURE, DEMOLITION**	90.9	98.0	96.0	91.9	107.0	102.7	96.7	98.1	97.7	93.2	97.1	96.0	95.5	97.0	96.6	130.7	124.6	126.3
0310	Concrete Forming & Accessories	99.1	119.6	116.8	99.4	135.9	131.0	103.3	107.7	107.1	98.8	105.3	104.4	97.1	113.7	111.4	107.0	184.8	174.2
0320	Concrete Reinforcing	95.4	128.7	109.0	102.2	122.7	110.6	95.8	114.7	103.5	96.1	109.4	101.5	98.4	102.3	100.0	106.7	190.7	140.9
0330	Cast-in-Place Concrete	91.3	118.0	101.5	111.9	121.2	115.4	93.2	107.9	99.8	94.2	112.3	101.1	110.7	116.6	112.9	108.8	168.8	131.7
0340	Precast Concrete	90.9	98.7	91.9	94.5	131.5	99.4	91.6	114.6	94.7	85.8	123.9	90.8	110.5	93.3	108.3	122.2	135.3	123.9
03	**CONCRETE**	92.5	119.5	104.7	104.4	128.4	115.3	93.9	109.3	100.9	92.5	109.3	100.2	107.4	111.8	109.4	112.0	177.7	141.9
0405	Basic Masonry Materials & Methods	93.9	117.6	103.9	107.1	126.3	115.2	94.6	107.0	99.8	94.1	108.9	100.4	106.5	114.9	110.1	107.4	164.6	131.5
0420	Unit Masonry	91.6	119.7	106.6	99.8	132.7	119.7	99.8	107.7	104.5	100.6	110.6	106.6	109.5	118.8	115.1	108.0	172.3	146.9
0440	Stone Assemblies	87.1	119.6	108.5	100.0	131.6	122.7	86.1	107.6	100.3	92.3	110.0	103.9	125.1	118.3	120.6	109.3	171.2	150.1
04	**MASONRY**	91.8	119.5	108.4	100.9	131.8	119.4	98.8	107.6	104.1	99.5	110.3	106.0	109.3	118.4	114.8	107.8	171.5	145.9
0510	Structural Metal Framing	81.8	105.2	88.0	78.2	131.0	92.3	103.5	113.0	106.0	99.9	118.7	104.9	96.3	97.1	96.5	103.5	141.6	113.6
0520, 0530	Metal Joists & Decking	92.3	103.2	95.7	96.9	131.0	100.7	102.3	113.6	105.8	98.3	120.5	105.3	92.3	95.0	93.1	111.8	139.7	120.6
0550	Metal Fabrications	100.0	120.2	103.1	100.0	155.3	108.4	100.0	114.2	102.2	100.0	109.9	101.5	100.0	91.9	98.8	100.0	160.9	109.3
05	**METALS**	89.9	104.6	94.2	91.8	131.8	103.5	102.4	113.4	105.6	98.9	119.5	104.9	94.1	95.5	94.5	108.4	140.9	117.9
0610	Rough Carpentry	91.9	120.0	111.5	104.3	135.8	126.2	104.3	107.6	106.6	93.8	102.5	99.9	95.6	113.4	108.0	116.0	188.1	166.2
0620	Finish Carpentry	102.7	119.6	104.3	126.5	137.5	127.6	102.7	107.7	103.2	111.3	102.6	110.4	101.2	113.6	102.4	104.8	191.1	135.5
06	**WOOD, PLASTICS & COMPOSITES**	97.2	119.6	109.7	115.2	135.9	126.6	103.5	107.6	105.8	102.4	102.5	102.5	98.4	113.4	106.6	110.5	188.2	153.2
0710	Dampproofing & Waterproofing	88.8	119.6	114.3	95.4	137.5	130.3	95.2	107.7	105.6	93.3	102.6	101.0	93.1	113.6	110.1	95.7	191.1	174.8
0720, 0780	Thermal, Fire & Smoke Protection	92.2	128.5	112.5	100.7	150.8	114.9	95.9	105.6	98.7	99.3	110.9	102.6	99.6	108.6	102.1	111.9	181.7	131.6
0740, 0750	Roofing & Siding	109.4	118.4	112.6	100.5	134.9	116.0	95.9	107.1	99.9	106.3	108.8	107.2	98.9	113.1	104.0	108.9	151.9	124.3
0760	Flashing & Sheet Metal	100.0	115.1	111.2	100.0	105.9	104.4	100.0	113.8	102.8	100.0	102.2	101.6	100.0	94.0	95.6	100.0	167.9	150.3
07	**THERMAL & MOISTURE PROTECTION**	99.3	121.3	107.8	102.5	133.8	114.7	96.3	106.2	100.1	101.9	107.8	104.2	99.1	109.0	102.9	109.3	167.8	132.1
0810, 0830	Doors & Frames	97.9	120.2	103.5	100.6	150.0	113.0	97.6	111.4	101.1	99.0	106.1	100.8	95.3	102.4	97.1	95.1	177.8	115.9
0840, 0880	Glazing & Curtain Walls	97.0	106.9	100.8	96.3	122.2	101.8	96.3	109.6	101.5	98.9	126.9	109.8	98.6	91.2	95.8	108.7	162.6	129.7
08	**OPENINGS**	96.2	117.4	101.4	98.0	142.0	108.9	97.5	110.3	100.7	97.7	109.7	100.7	95.2	102.2	96.9	98.0	177.2	117.5
0920	Plaster & Gypsum Board	85.4	119.6	107.6	98.9	137.5	124.0	93.7	107.7	102.8	103.2	102.6	102.8	99.0	113.6	108.5	128.3	191.1	169.1
0930, 0966	Tile & Terrazzo	96.7	124.5	110.4	105.6	132.6	119.0	106.6	104.1	105.4	95.7	107.3	101.5	89.2	119.0	104.0	123.2	168.4	145.6
0950, 0980	Ceilings & Acoustic Treatment	89.7	119.6	108.2	99.7	137.5	123.1	90.8	107.7	101.3	91.1	110.9	98.2	98.5	113.6	107.8	109.1	191.1	159.8
0960	Flooring	92.4	126.2	101.6	105.3	126.6	111.0	98.9	99.9	99.2	94.0	102.4	96.3	89.1	117.2	96.7	93.3	169.6	113.9
0970, 0990	Wall Finishes & Painting/Coating	94.4	118.3	102.7	98.9	128.0	116.4	99.5	120.1	111.9	98.3	109.2	104.8	86.1	112.6	102.0	90.3	156.7	130.2
09	**FINISHES**	92.0	120.1	106.8	102.7	134.3	119.4	98.3	107.5	103.2	95.3	104.3	100.1	92.5	115.0	104.4	109.2	180.1	146.7
COVERS	DVS 10 - 14, 25, 28, 41, 43, 44	100.0	109.0	101.8	100.0	106.0	101.2	100.0	99.9	99.9	100.0	103.9	100.8	100.0	107.5	101.5	100.0	139.6	107.8
2210, 2230, 2240, 2320	Piping, Pumps, Plumbing Equipment	100.0	113.9	105.9	100.0	119.2	108.2	100.0	103.7	101.6	100.0	109.4	104.0	100.0	99.1	99.6	100.0	162.1	126.5
2113	Fire Suppression Sprinkler Systems	100.0	110.0	105.0	100.0	109.3	106.2	100.0	103.4	102.3	100.0	105.9	103.9	100.0	90.6	93.7	100.0	168.5	145.6
2350	Central Heating Equipment	99.2	112.1	103.3	98.5	120.4	105.4	100.0	108.2	102.5	99.2	110.2	102.7	99.2	97.4	98.7	101.8	170.3	123.2
2330, 2340, 2360, 2370, 2380	Air Conditioning & Ventilation	100.0	113.9	101.0	100.0	123.1	101.7	100.0	105.3	100.4	100.0	109.7	100.7	100.0	102.4	100.2	100.0	173.2	125.8
21, 22, 23	**FIRE SUPPRESSION, PLUMBING & HVAC**	99.9	112.8	105.0	99.8	117.1	106.7	100.0	104.1	101.6	99.9	108.7	103.4	99.9	97.1	98.8	100.2	164.6	140.4
26, 27, 3370	**ELECTRICAL, COMMUNICATIONS & UTIL.**	96.5	115.9	106.0	101.9	115.3	108.5	101.2	105.2	103.1	103.6	111.7	107.6	99.8	98.5	99.1	109.4	173.1	140.1
MF2004	**WEIGHTED AVERAGE**	95.3	113.9	103.2	99.3	123.9	109.7	99.2	106.3	102.2	98.6	108.6	102.8	99.0	104.5	101.4	106.1	164.3	130.7

CSI DIV NO.	JULY 1, 2008 ITEM	CINCINNATI, OH MAT.	INST.	TOTAL	CLEVELAND, OH MAT.	INST.	TOTAL	COLUMBUS, OH MAT.	INST.	TOTAL	PHILADELPHIA, PA MAT.	INST.	TOTAL	PITTSBURGH, PA MAT.	INST.	TOTAL	MEMPHIS, TN MAT.	INST.	TOTAL
015433	CONTRACTOR EQUIPMENT	0.0	101.0	101.0	0.0	96.9	96.9	0.0	96.0	96.0	0.0	96.5	96.5	0.0	111.9	111.9	0.0	101.7	101.7
3120, 3130	Earth Moving & Earthwork	56.1	105.9	98.2	84.0	105.0	101.7	79.3	100.8	97.5	111.3	96.6	98.9	121.6	106.4	108.7	95.3	92.8	93.2
3160, 3170	Load-Bearing Elements & Tunneling	99.0	94.5	96.5	100.1	85.9	92.2	99.0	89.9	93.9	99.0	99.1	99.1	99.0	112.9	106.8	93.6	104.7	99.8
3210	Bases, Ballasts & Paving	90.1	109.7	92.1	94.4	99.2	94.9	90.6	96.2	91.2	102.4	103.7	102.5	96.2	107.2	97.3	88.8	87.8	86.9
3310, 3330, 3340, 3350	Utility Services & Drainage	87.2	91.5	87.6	91.9	108.0	93.2	89.6	90.0	89.6	94.3	133.6	97.4	94.3	95.6	94.4	101.3	59.9	98.1
3230	Site Improvements	97.7	94.1	96.5	109.5	102.6	107.1	110.7	102.0	107.6	90.4	101.8	94.4	94.3	90.4	90.4	95.2	79.4	89.7
3290	Planting	81.0	107.5	93.1	89.1	103.2	95.5	101.2	100.3	100.8	93.1	98.4	95.5	105.3	108.8	106.9	76.9	92.6	84.1
0241, 31 - 34	SITE & INFRASTRUCTURE, DEMOLITION	73.9	105.8	96.8	89.0	104.5	100.2	88.5	100.5	97.1	103.0	97.0	98.7	108.6	106.6	107.1	90.4	92.8	92.1
0310	Concrete Forming & Accessories	96.2	85.4	86.8	101.1	103.3	103.0	98.3	83.6	85.6	100.8	135.8	131.0	99.7	93.0	93.9	94.8	60.5	65.2
0320	Concrete Reinforcing	89.7	84.8	87.7	96.6	93.8	96.6	92.8	86.2	90.1	101.3	140.6	117.3	96.1	104.8	99.7	91.0	65.6	80.6
0330	Cast-in-Place Concrete	81.6	87.3	83.8	91.4	109.9	98.5	89.5	93.0	90.8	109.1	127.5	116.1	96.2	91.5	94.4	86.6	66.5	78.9
0340	Precast Concrete	95.3	88.8	94.4	94.5	81.2	92.8	96.7	81.3	94.7	113.4	123.9	114.8	101.1	122.5	103.9	83.6	97.4	85.5
03	CONCRETE	87.8	86.1	87.0	94.3	102.9	98.2	92.6	87.1	90.1	108.2	133.1	119.6	97.7	95.9	96.9	87.1	65.3	77.2
0405	Basic Masonry Materials & Methods	85.8	93.4	89.0	95.8	106.6	100.4	92.8	95.2	93.8	104.0	127.1	113.8	95.5	93.4	94.6	89.3	68.5	80.5
0420	Unit Masonry	74.1	92.4	85.1	93.8	107.5	102.1	92.7	92.9	92.8	95.0	135.2	119.3	85.1	92.8	89.8	87.9	67.5	75.6
0440	Stone Assemblies	72.0	92.2	85.3	102.5	106.8	105.4	103.5	93.2	96.7	111.5	135.5	127.3	91.2	93.6	92.8	71.8	68.8	69.8
04	MASONRY	75.7	92.4	85.7	94.4	107.2	102.1	93.1	93.2	93.1	96.6	134.8	119.5	86.6	93.2	90.5	87.8	68.0	76.0
0510	Structural Metal Framing	89.0	88.4	88.8	87.2	86.3	87.0	87.2	84.1	86.4	103.5	128.6	110.2	89.0	117.4	96.6	114.3	91.5	108.2
0520, 0530	Metal Joists & Decking	101.1	88.5	97.1	95.5	85.1	92.2	93.0	83.1	89.9	102.0	126.1	109.6	95.3	119.4	102.9	101.5	92.0	98.5
0550	Metal Fabrications	100.0	80.7	97.1	100.0	89.7	98.4	100.0	81.5	97.2	100.0	151.3	107.8	100.0	115.4	102.4	100.0	73.8	96.0
05	METALS	97.5	88.1	94.8	93.5	85.7	91.3	91.9	83.4	89.4	102.3	127.7	109.7	93.9	118.6	101.1	105.1	91.0	101.0
0610	Rough Carpentry	93.6	82.9	86.1	94.4	100.9	98.9	95.1	81.2	85.5	107.3	137.2	128.1	105.0	93.3	96.8	95.2	61.2	71.5
0620	Finish Carpentry	94.0	82.5	92.8	97.6	100.5	97.9	101.2	80.5	99.1	92.5	138.5	97.1	86.7	92.8	87.3	88.2	60.1	85.4
06	WOOD, PLASTICS & COMPOSITES	98.7	82.8	87.8	96.0	100.8	98.6	98.1	81.2	88.8	100.0	137.2	120.5	96.0	93.2	94.5	91.7	61.1	74.9
0710	Dampproofing & Waterproofing	90.7	82.5	83.9	102.9	100.5	100.9	95.4	80.5	83.0	97.5	138.5	131.5	99.9	92.8	94.0	97.0	60.1	66.4
0720, 0780	Thermal, Fire & Smoke Protection	97.3	96.3	97.0	100.0	112.5	103.5	96.4	90.7	94.8	97.1	143.2	110.1	100.1	102.0	100.6	90.2	62.8	82.4
0740, 0750	Roofing & Siding	100.9	90.9	97.3	108.7	109.6	109.0	105.4	94.8	101.6	102.4	123.0	109.8	106.8	90.2	100.8	87.3	67.8	80.4
0760	Flashing & Sheet Metal	100.0	99.5	99.7	100.0	117.2	112.7	100.0	102.0	101.5	100.0	133.3	124.7	100.0	98.7	99.0	100.0	71.0	78.5
07	THERMAL & MOISTURE PROTECTION	98.8	93.5	96.8	103.2	110.2	106.0	100.2	93.8	97.7	99.5	133.9	112.9	102.6	96.6	100.3	90.1	67.5	81.3
0810, 0830	Doors & Frames	98.1	80.5	93.7	99.3	93.8	97.9	101.1	79.6	95.7	99.6	146.8	111.5	95.4	104.5	97.7	96.5	67.0	89.1
0840, 0880	Glazing & Curtain Walls	98.9	90.4	95.6	95.8	109.0	100.9	96.3	87.7	92.9	98.6	128.6	110.3	93.3	105.9	98.2	105.2	55.4	85.8
08	OPENINGS	97.8	82.9	94.1	97.8	98.2	97.9	100.2	81.4	95.6	97.7	141.5	108.5	91.8	102.6	94.5	98.4	63.3	89.7
0920	Plaster & Gypsum Board	92.6	82.5	86.0	89.9	100.5	96.8	88.0	80.5	83.1	103.8	138.5	126.3	99.1	92.8	95.0	91.4	60.1	71.1
0930, 0966	Tile & Terrazzo	92.4	93.9	93.1	107.6	109.9	108.7	98.0	91.9	95.0	122.2	130.8	126.5	103.6	92.4	98.1	75.9	51.2	63.6
0950, 0980	Ceilings & Acoustic Treatment	99.2	82.5	88.9	90.6	100.5	96.8	91.1	80.5	84.5	96.4	138.5	122.4	85.5	92.8	90.0	98.7	60.1	74.8
0960	Flooring	106.6	93.6	103.1	102.4	103.3	102.6	93.6	91.5	93.0	84.3	130.6	98.5	97.3	99.5	97.9	94.2	40.5	79.6
0970, 0990	Wall Finishes & Painting/Coating	106.9	88.9	96.1	105.7	112.3	109.7	97.3	94.2	95.4	97.3	144.5	125.6	97.0	105.7	102.2	101.3	57.0	74.7
09	FINISHES	99.4	86.8	92.7	99.5	104.2	102.0	93.6	85.6	89.4	100.4	137.3	119.9	96.3	94.8	95.5	90.8	55.8	72.3
COVERS	DIVS 10 - 14, 25, 28, 41, 43, 44	100.0	91.9	98.4	100.0	103.5	100.7	100.0	93.2	98.6	100.0	120.3	104.0	100.0	98.5	99.7	100.0	79.0	95.9
2210, 2230, 2240, 2320	Piping, Pumps, Plumbing Equipment	100.0	87.4	94.6	100.0	106.1	102.6	100.0	90.3	95.8	100.0	126.9	111.5	100.0	93.5	97.2	100.0	64.4	84.8
2113	Fire Suppression Sprinkler Systems	100.0	84.5	89.7	100.0	103.1	102.1	100.0	89.5	93.0	100.0	128.4	118.9	100.0	94.7	96.5	100.0	72.8	81.9
2350	Central Heating Equipment	100.3	90.2	97.2	99.6	106.4	101.7	97.8	87.4	94.6	100.3	130.4	109.7	99.6	94.9	98.1	98.9	67.4	89.0
2330, 2340, 2360, 2370, 2380	Air Conditioning & Ventilation	100.0	88.1	99.1	100.0	107.2	100.5	100.0	89.7	99.3	100.0	129.5	102.1	100.0	91.6	99.4	100.0	65.8	97.6
21, 22, 23	FIRE SUPPRESSION, PLUMBING & HVAC	100.0	87.0	94.9	99.9	105.5	102.1	99.7	89.8	95.8	100.0	127.6	111.0	99.9	93.8	97.5	99.9	66.6	86.6
26, 27, 3370	ELECTRICAL, COMMUNICATIONS & UTIL.	94.7	80.0	87.5	97.3	104.2	100.7	98.1	85.1	91.8	98.7	137.5	117.6	96.7	98.0	97.3	103.2	68.7	86.4
MF2004	WEIGHTED AVERAGE	95.3	88.4	92.4	97.2	102.7	99.5	96.3	88.8	93.1	101.0	129.5	113.0	96.9	98.9	97.7	97.5	70.5	86.1

CSI DIV. NO.	JULY 1, 2008 — ITEM	NASHVILLE, TN			DALLAS, TX			HOUSTON, TX			SAN ANTONIO, TX			SEATTLE, WA			MILWAUKEE, WI		
		MAT.	INST.	TOTAL	MAT.	INST.	TOTAL	MAT.	INST.	TOTAL	MAT.	INST.	TOTAL	MAT.	INST.	TOTAL	MAT.	INST.	TOTAL
015433	CONTRACTOR EQUIPMENT	0.0	104.9	104.9	0.0	98.4	98.4	0.0	98.7	98.7	0.0	91.0	91.0	0.0	103.1	103.1	0.0	89.5	89.5
3120, 3130	Earth Moving & Earthwork	98.1	99.8	99.5	138.7	87.9	95.8	152.2	85.4	95.8	77.9	91.3	89.2	102.1	111.6	110.1	86.1	96.6	95.0
3160, 3170	Load-Bearing Elements & Tunneling	95.8	103.8	100.3	95.8	104.4	100.6	95.8	111.6	104.6	100.1	87.1	92.9	106.6	89.7	97.1	103.3	83.7	92.4
3210	Bases, Ballasts & Paving	95.7	100.6	96.2	91.8	94.5	92.1	99.4	86.9	98.1	81.9	100.7	83.8	106.9	105.9	106.8	93.6	95.2	93.8
3310, 3330, 3340, 3350	Utility Services & Drainage	87.2	63.2	86.4	120.1	57.2	115.2	115.4	64.3	111.4	124.8	60.8	119.8	115.4	102.7	114.4	103.7	113.0	104.4
3230	Site Improvements	94.8	81.4	90.1	98.0	80.9	92.0	97.0	85.8	93.0	98.2	84.1	93.2	115.2	93.4	107.5	100.9	94.9	98.8
3290	Planting	101.2	99.1	100.2	113.4	89.2	102.3	97.2	84.8	91.5	101.2	88.6	95.5	89.1	110.4	98.9	93.1	94.7	93.8
0241, 31 - 34	SITE & INFRASTRUCTURE, DEMOLITION	96.9	99.6	98.8	119.7	88.1	97.0	123.0	85.7	96.2	90.2	91.0	90.8	102.3	111.0	108.5	91.9	96.4	95.1
0310	Concrete Forming & Accessories	94.0	64.0	68.1	94.8	59.3	64.2	94.1	65.9	69.8	92.7	55.6	60.6	103.4	102.7	102.8	99.6	113.8	111.8
0320	Concrete Reinforcing	91.0	64.5	80.2	91.6	52.9	78.8	98.0	58.2	81.8	104.4	49.3	81.9	111.7	93.3	104.2	100.6	102.7	101.5
0330	Cast-in-Place Concrete	90.5	67.4	81.7	92.3	55.4	78.2	97.2	68.9	86.4	74.7	67.5	71.9	106.1	108.7	107.1	97.1	108.5	101.5
0340	Precast Concrete	86.1	95.7	86.5	75.6	89.4	77.4	79.3	96.7	81.4	75.6	71.9	75.1	82.9	88.8	83.7	94.5	92.2	94.2
03	CONCRETE	89.3	66.9	79.1	88.7	58.4	75.0	92.2	67.2	80.8	82.0	59.5	71.7	100.7	102.5	101.5	97.2	109.0	102.6
0405	Basic Masonry Materials & Methods	91.8	63.7	79.9	95.0	64.4	82.1	98.5	64.9	84.3	83.2	65.1	75.6	109.2	99.0	104.9	97.8	114.1	104.7
0420	Unit Masonry	82.2	62.3	70.2	102.9	58.1	75.8	96.0	60.6	74.6	92.6	60.1	73.0	136.4	101.1	115.0	103.8	117.7	112.2
0440	Stone Assemblies	75.9	63.0	67.4	100.0	58.5	76.3	90.2	60.6	70.7	91.2	60.6	71.0	147.6	100.6	116.6	112.8	117.6	115.9
04	MASONRY	83.4	62.6	71.0	102.2	58.6	76.1	96.1	60.8	75.0	91.5	60.6	73.0	133.2	100.8	113.8	103.2	117.5	111.8
0510	Structural Metal Framing	114.3	88.9	107.5	107.1	81.2	100.2	103.5	86.3	98.9	96.3	69.2	89.1	103.5	90.6	100.1	96.3	97.5	96.6
0520, 0530	Metal Joists & Decking	101.5	90.6	98.1	100.8	84.2	95.5	109.4	89.6	103.1	101.3	70.3	91.5	116.4	90.2	108.1	104.7	95.2	101.7
0550	Metal Fabrications	100.0	73.7	96.0	100.0	56.5	93.4	100.0	61.7	94.2	100.0	54.3	93.0	100.0	88.2	98.2	100.0	106.0	100.9
05	METALS	105.1	89.4	100.5	102.5	82.2	96.6	106.9	87.5	101.2	99.7	69.4	90.9	111.2	90.3	105.1	101.8	96.2	100.2
0610	Rough Carpentry	93.2	64.9	73.5	92.7	60.3	70.1	92.6	66.7	74.6	97.6	54.1	67.3	99.9	102.1	101.4	103.4	113.6	110.5
0620	Finish Carpentry	101.2	64.0	97.5	104.8	59.2	100.2	101.2	65.9	97.7	94.0	52.9	89.9	89.6	102.1	90.9	121.4	114.3	120.7
06	WOOD, PLASTICS & COMPOSITES	97.1	64.9	79.4	98.7	60.2	77.6	96.8	66.7	80.3	95.8	54.0	72.9	94.8	102.1	98.8	112.3	113.6	113.0
0710	Dampproofing & Waterproofing	87.5	64.0	68.0	92.5	59.2	64.9	92.5	65.9	70.5	89.6	52.9	59.2	115.3	102.1	104.4	105.2	114.3	112.8
0720, 0780	Thermal, Fire & Smoke Protection	93.7	65.8	85.8	97.2	61.5	87.2	101.2	64.6	90.8	90.6	61.0	82.2	101.1	93.3	98.9	95.4	119.9	102.3
0740, 0750	Roofing & Siding	91.4	62.8	81.2	86.1	64.7	78.4	90.4	71.3	83.5	90.5	68.3	82.6	111.0	102.1	107.0	100.2	108.6	103.2
0760	Flashing & Sheet Metal	100.0	63.9	73.3	100.0	67.5	75.9	100.0	71.3	78.8	100.0	71.4	78.8	100.0	99.2	99.4	100.0	98.8	99.1
07	THERMAL & MOISTURE PROTECTION	93.4	64.6	82.2	93.2	63.4	81.6	96.8	67.4	85.4	91.4	64.9	81.1	105.0	97.3	102.0	97.8	112.1	103.4
0810, 0830	Doors & Frames	96.8	68.8	89.8	101.9	57.2	90.6	103.2	63.1	93.1	103.5	52.0	90.5	105.5	94.4	102.7	105.2	110.4	106.5
0840, 0880	Glazing & Curtain Walls	101.9	57.6	84.7	107.3	40.0	81.1	104.7	63.0	88.5	99.5	59.3	83.9	99.6	104.4	101.5	97.9	110.8	102.9
08	OPENINGS	97.6	65.6	89.7	103.4	54.0	91.2	104.5	63.6	94.4	103.9	53.6	91.4	106.0	97.9	104.0	105.6	111.3	107.0
0920	Plaster & Gypsum Board	98.7	64.0	76.2	90.5	59.2	70.1	94.8	65.9	76.0	90.9	52.9	66.2	107.3	102.1	103.9	96.5	114.3	108.1
0930, 0966	Tile & Terrazzo	97.3	66.8	82.2	102.4	57.2	79.9	98.6	62.7	80.8	85.7	61.9	73.9	100.1	104.8	102.1	101.2	117.3	109.2
0950, 0980	Ceilings & Acoustic Treatment	96.1	64.0	76.3	94.2	59.2	72.5	105.8	65.9	83.5	87.9	52.9	66.3	108.2	102.1	104.4	86.8	114.3	103.8
0960	Flooring	101.6	71.6	93.5	100.6	53.2	87.8	98.5	58.7	87.7	91.6	66.5	84.8	114.7	102.9	111.5	107.4	120.5	110.9
0970, 0990	Wall Finishes & Painting/Coating	108.4	66.4	83.2	103.5	53.1	73.2	101.4	61.2	77.3	92.7	50.0	67.1	109.3	89.6	97.5	95.6	113.1	106.1
09	FINISHES	99.3	65.7	81.5	98.4	57.5	76.7	99.9	64.0	80.9	89.2	55.9	71.6	108.1	101.4	104.5	98.6	115.5	107.5
COVERS	DIVS 10 - 14, 25, 28, 41, 43, 44	100.0	79.9	96.0	100.0	79.2	95.9	100.0	86.0	97.2	100.0	78.0	95.7	100.0	101.4	100.3	100.0	105.1	101.0
2210, 2230, 2240, 2220	Piping, Pumps, Plumbing Equipment	100.0	77.8	90.5	100.0	63.5	84.4	100.0	72.0	88.1	100.0	66.6	85.7	100.0	107.3	103.1	100.0	100.5	100.4
2113	Fire Suppression Sprinkler Systems	100.0	83.8	89.2	100.0	65.9	77.3	100.0	73.1	82.1	100.0	69.3	79.6	100.0	106.7	104.5	100.0	100.5	100.4
2350	Central Heating Equipment	98.9	75.8	91.7	98.9	61.2	87.1	99.6	68.8	90.0	98.9	65.3	88.4	101.4	104.8	102.5	99.2	105.8	101.3
2330, 2340, 2360, 2370, 2380	Air Conditioning & Ventilation	100.0	73.5	98.1	100.0	61.5	97.2	100.0	68.7	97.8	100.0	64.6	97.5	100.0	106.5	100.5	100.0	107.1	100.5
21, 22, 23	FIRE SUPPRESSION, PLUMBING & HVAC	99.9	78.9	91.5	99.9	63.8	85.5	99.9	71.9	88.8	99.9	67.0	86.8	100.2	107.0	102.9	99.9	101.4	100.5
26, 27, 3370	ELECTRICAL, COMMUNICATIONS & UTIL.	103.0	65.4	84.7	97.3	69.2	83.6	97.2	68.5	83.2	99.8	64.9	82.8	103.1	99.4	101.3	99.6	104.3	101.9
MF2004	WEIGHTED AVERAGE	98.5	74.0	88.1	99.4	66.5	85.5	100.7	71.4	88.3	96.2	65.7	83.3	105.4	101.9	103.9	100.3	106.2	102.8

Preliminary Building Size Determination

The following tables are helpful when estimating conceptually.

With basic information such as type of occupancy and number of people intended to occupy the structure, and/or rough dimensions, it is possible to calculate other needed information.

Example:
A potential client wants to know how big a building he will need and how much it will cost. All he knows at this time is that there will be approximately 75 people working in the building. To begin a conceptual estimate, some size information is needed.

Step 1: *Determine the net area.* Consult the "Occupancy Determinations—Net Areas" table. The table states that for each occupant, 100 S.F. (net) of office space is required. (Note that in this case all codes have the same requirements. Check your area for the code used.)

Step 2: *Determine the gross area.* (Net area does not include common areas, stairwells, toilet area, mechanical areas, etc.) To obtain the gross area, consult the "Floor Area Requirements" table. The *gross to net ratio* for offices is 135%. Therefore, 100 S.F. (net) x 1.35 (gross/net), or 135 square feet per person, is needed for this office building. Since we expect an occupancy of 75 persons, the total office space will be 10,125 square feet.

(Note: In some cases, such as in apartment buildings, schools, hospitals, etc., the basic information will be in unit form, such as "I want an apartment building with 50 units," or "a hospital with 200 beds," etc. In these cases, use the "Unit Gross Area" table provided. To use this table, it is necessary to know a little about the quality of the structure. If the quality is to be "above average," then the units will be larger; if they are to be of "economy" construction, then you can assume the units will be smaller. If, in the case of an apartment building, the client wants 50 units of average quality, to obtain the overall size, multiply the number of units by the median gross square footage—in this case 50 units x 860 square feet per unit or 43,000 square feet *gross area.*)

Step 3: *Determine the size.* A 10,000+ square foot building lends itself nicely to a 100' x 100' economical square building. If the land available does not allow for this design, adjust the dimensions to fit. You now have the basic building size and shape to start the estimate.

Occupancy Determinations — Net Areas

	Description	S.F. Required Per Person*			
		BBC	BOCA	SBC	UBC
Assembly Areas	Fixed Seats	6	**	6	7
	Movable Seats	15		15	15
	Concentrated		7		
	Unconcentrated		15		
	Standing Space		3		
Educational	Unclassified	40			
	Classrooms		20	40	20
	Shop Areas		50	100	50
Institutional	Unclassified	150		125	
	In-Patient Areas		240		
	Sleeping Areas		120		
Mercantile	Basement	30	30	30	20
	Ground Floor	30	30	30	30
	Upper Floors	60	60	60	50
Office		100	100	100	100

*BBC = Basic Building Code

BOCA = Building Officials & Code Administrators

SBC = Southern Building Code

UBC = Uniform Building Code

** The occupancy load for assembly area with fixed seats shall be determined by the number of fixed seats installed.

Unit Gross Area Requirements

The figures in the table below indicate typical ranges in square feet as a function of the "occupant" unit. This table is best used in the preliminary design stages to help determine the probable size requirement for the total project. The "1/4" means that of all buildings of this type built, 1/4 will be smaller in unit size and 3/4 will be larger; "median" means that 50% of the units will be smaller and 50% will be larger; the "3/4" indicates that 3/4 of the building units will be smaller and 1/4 will be larger.

Building Type	Unit	Gross Area in S.F.		
		1/4	Median	3/4
Apartments	Unit	660	860	1,100
Auditorium & Play Theaters	Seat	18	25	38
Bowling Alleys	Lane		940	
Churches & Synagogues	Seat	20	28	39
Dormitories	Bed	200	230	275
Fraternity & Sorority Houses	Bed	220	315	370
Garages, Parking	Car	325	355	385
Hospitals	Bed	685	850	1,075
Hotels	Rental Unit	475	600	710
Housing for the Elderly	Unit	515	635	755
Housing, Public	Unit	700	875	1,030
Ice Skating Rinks	Total	27,000	30,000	36,000
Motels	Rental Unit	360	465	620
Nursing Homes	Bed	290	350	450
Restaurants	Seat	23	29	39
Schools, Elementary	Pupil	65	77	90
Junior High & Middle		85	110	129
Senior High		102	130	145
Vocational		110	135	195
Shooting Ranges	Point		450	
Theaters & Movies	Seat		15	

Floor Area Ratios: Commonly Used Gross-to-Net and Net-to-Gross Area Ratios Expressed in % for Various Building Types

Building Type	Gross-to-Net Ratio	Net-to-Gross Ratio	Building Type	Gross-to-Net Ratio	Net-to-Gross Ratio
Apartment	156	64	School Buildings (campus type)		
Bank	140	72			
Church	142	70	Administrative	150	67
Courthouse	162	61	Auditorium	142	70
Department Store	123	81	Biology	161	62
Garage	118	85	Chemistry	170	59
Hospital	183	55	Classroom	152	66
Hotel	158	63	Dining Hall	138	72
Laboratory	171	58	Dormitory	154	65
Library	132	76	Engineering	164	61
Office	135	75	Fraternity	160	63
Restaurant	141	70	Gymnasium	142	70
Warehouse	108	93	Science	167	60
			Service	120	83
			Student Union	172	59

The gross area of a building is the total floor area based on outside dimensions. The net area of a building is the usable floor area for the function intended and excludes such items as stairways, corridors, and mechanical rooms. In the case of a commercial building, it might be considered as the "leasable area."

Standard Weights and Measures

Linear Measure		Square Measure	
1000 mils =	1 inch	144 square inches =	1 square foot
12 inches =	1 foot	9 square feet =	1 square yard
3 feet =	1 yard		
2 yards =	{ 1 fathom { 6 feet	30-1/4 square yds. =	{ 1 square rod { 272-1/4 square feet
5-1/2 yards =	{ 1 rod { 16-1/2 feet	160 square rods =	{ 1 acre { 43,560 square feet
40 rods =	{ 1 furlong { 660 feet	640 acres =	{ 1 square mile { 27,878,400 { square feet
8 furlongs =	{ 1 mile { 5280 feet	A circular mil is the area of a circle 1 mil, or 0.001 inch in diameter.	
1.15156 miles =	{ 1 nautical mile, { or knot { 6080.26 feet	1 square inch =	1,273,239 circular mils
3 nautical miles =	{ 1 league { 18,240.78 feet	A circular inch is the area of a circle = 1 inch in diameter	0.7854 square inches
		1 square inch =	1.2732 circular inches

Dry Measure		Weight—Avoirdupois or Commercial	
2 pints =	1 quart	437.5 grains =	1 ounce
8 quarts =	1 peck	16 ounces =	1 pound
4 pecks =	{ 1 bushel { 2150.42 cubic in. { 1.2445 cubic feet	112 pounds =	1 hundredweight
		20 hundredweight =	{ 1 gross, or long ton { 2240 pounds
		2000 pounds =	1 net, or short ton
		2204.6 pounds =	1 metric ton
		1 lb. of water (39.1°F) = = = =	27.681217 cu. in. 0.016019 cu. ft. 0.119832 U.S. gallon 0.453617 liter

Decimals of a Foot for Each 1/32 of an Inch

Inch	0	1	2	3	4	5	6	7	8	9	10	11
0	0	.0833	.1667	.2500	.3333	.4167	.5000	.5833	.6667	.7500	.8333	.9167
1/32	.0026	.0859	.1693	.2526	.3359	.4193	.5026	.5859	.6693	.7526	.8359	.9193
1/16	.0052	.0885	.1719	.2552	.3385	.4219	.5052	.5885	.6719	.7552	.8385	.9219
3/32	.0078	.0911	.1745	.2578	.3411	.4245	.5078	.5911	.6745	.7578	.8411	.9245
1/8	.0104	.0938	.1771	.2604	.3438	.4271	.5104	.5938	.6771	.7604	.8438	.9271
5/32	.0130	.0964	.1797	.2630	.3464	.4297	.5130	.5964	.6797	.7630	.8464	.9297
3/16	.0156	.0990	.1823	.2656	.3490	.4323	.5156	.5990	.6823	.7656	.8490	.9323
7/32	.0182	.1016	.1849	.2682	.3516	.4349	.5182	.6016	.6849	.7682	.8516	.9349
1/4	.0208	.1042	.1875	.2708	.3542	.4375	.5208	.6042	.6875	.7708	.8542	.9375
9/32	.0234	.1068	.1901	.2734	.3568	.4401	.5234	.6068	.6901	.7734	.8568	.9401
5/16	.0260	.1094	.1927	.2760	.3594	.4427	.5260	.6094	.6927	.7760	.8594	.9427
11/32	.0286	.1120	.1953	.2786	.3620	.4453	.5286	.6120	.6953	.7786	.8620	.9453
3/8	.0313	.1146	.1979	.2812	.3646	.4479	.5313	.6146	.6979	.7813	.8646	.9479
13/32	.0339	.1172	.2005	.2839	.3672	.4505	.5339	.6172	.7005	.7839	.8672	.9505
7/16	.0365	.1198	.2031	.2865	.3698	.4531	.5365	.6198	.7031	.7865	.8698	.9531
15/32	.0391	.1224	.2057	.2891	.3724	.4557	.5391	.6224	.7057	.7891	.8724	.9557
1/2	.0417	.1250	.2083	.2917	.3750	.4583	.5417	.6250	.7083	.7917	.8750	.9583
17/32	.0443	.1276	.2109	.2943	.3776	.4609	.5443	.6276	.7109	.7943	.8776	.9609
9/16	.0469	.1302	.2135	.2969	.3802	.4635	.5469	.6302	.7135	.7969	.8802	.9635
19/32	.0495	.1328	.2161	.2995	.3828	.4661	.5495	.6328	.7161	.7995	.8828	.9661
5/8	.0521	.1354	.2188	.3021	.3854	.4688	.5521	.6354	.7188	.8021	.8854	.9688
21/32	.0547	.1380	.2214	.3047	.3880	.4714	.5547	.6380	.7214	.8047	.8880	.9714
11/16	.0573	.1406	.2240	.3073	.3906	.4740	.5573	.6406	.7240	.8073	.8906	.9740
23/32	.0599	.1432	.2266	.3099	.3932	.4766	.5599	.6432	.7266	.8099	.8932	.9766
3/4	.0625	.1458	.2292	.3125	.3958	.4792	.5625	.6458	.7292	.8125	.8958	.9792
25/32	.0651	.1484	.2318	.3151	.3984	.4818	.5651	.6484	.7318	.8151	.8984	.9818
13/16	.0677	.1510	.2344	.3177	.4010	.4844	.5677	.6510	.7344	.8177	.9010	.9844
27/32	.0703	.1536	.2370	.3203	.4036	.4870	.5703	.6536	.7370	.8203	.9036	.9870
7/8	.0729	.1563	.2396	.3229	.4063	.4896	.5729	.6563	.7396	.8229	.9063	.9896
29/32	.0755	.1589	.2422	.3255	.4089	.4922	.5755	.6589	.7422	.8255	.9089	.9922
15/16	.0781	.1615	.2448	.3281	.4115	.4948	.5781	.6615	.7448	.8281	.9115	.9948
31/32	.0807	.1641	.2474	.3307	.4141	.4974	.5807	.6641	.7474	.8307	.9141	.9974

Water: Various Pressure and Flow Units

Comparison of Heads of Water in Feet with Pressures in Various Units
One foot of water at 39.1° F = 62.425 pounds per square foot (psf)
= 0.4335 pounds per square inch
= 0.0295 atmosphere
= 0.8826 inches of mercury at 30° F
= 773.3 feet of air at 32° F and atmospheric pressure
One foot of water at 62° F = 62.355 pounds per square foot
= 0.43302 pounds per square inch
One pound of water on the square inch at 62° F = 2.3094 feet of water
One ounce of water on the square inch at 62° F = 1.732 inches of water
1 atmosphere at sea level (32° F) = 14.697 lbs. per sq. in.
= 29.921 in. of mercury
1 inch of mercury (32° F) = 0.49119 lbs. per sq. in.
Flowing Water
cfs = cubic feet per second, or second feet
gpm = gallons per minute
1 cfs = 60 cu. ft. per min.
= 86,400 cu. ft. per 24 hrs.
= 448.83 U.S. gals. per min.
= 646,317 U.S. gals. per 24 hrs.
= 1.9835 acre-foot per 24 hrs. (usually taken as 2)
= 1 acre-inch per hour (approximate)
= .028317 cu. meter per second
1 U.S. gpm = 1440 U.S. gals per 24 hrs.
= 0.00442 acre-foot per 24 hrs.
= 0.0891 miner's inches, Ariz., Calif.
1 million U.S. gal. per day = 1.5472 cfs
= 3.07 acre-feet
= 2.629 cu. meters per min.

Volume and Capacity

Units and Equivalents	
1 cu. ft. of water at 39.1°F	= 62.425 lbs.
1 United States gallon	= 231 cu. in.
1 imperial gallon	= 277.274 cu. in.
1 cubic foot of water	= 1728 cu. in.
	= 7.480519 U.S. gallons
	= 6.232103 imperial gallons
1 cubic yard	= 27 cu. ft. = 46,656 cu. in.
1 quart	= 2 pints
1 gallon	= 4 quarts
1 U.S. gallon	= 231 cu. in.
	= 0.133681 cu. ft.
	= 0.83311 imperial gallons
	= 8.345 lbs.
1 barrel	= 31.5 gallons = 4.21 cu. ft.
1 U.S. bushel	= 1.2445 cu. ft.
1 fluid ounce	= 1.8047 cu. in.
1 acre-foot	= 43,560 cu. ft.
	= 1,613.3 cu. yds.
1 acre-inch	= 3,630 cu. ft.
1 million U.S. gallons	= 133,681 cu. ft.
	= 3.0689 acre-ft.
1 ft. depth on 1 sq. mi.	= 27,878,400 cu. ft.
	= 640 acre-ft.
1 cord	= 128 cu. ft.

Mechanical-Electrical Equivalents

Power

1 horsepower (hp)	= 550 foot-pounds (ft.-lbs.) per second (sec.)
	= 33,000 ft.-lbs. per minute (min.)
	= 1,980,000 ft.-lbs. per hour (hr.)
	= .275 ft.-tons per sec.
	= 16.5 ft.-tons per min.
	= 990 ft.-tons per hr.
1 horsepower-second (hp-sec.)	= 550 ft.-lbs.
	= .275 ft.-tons.
1 horsepower-minute (hp-min.)	= 33,000 ft.-lbs.
	= 16.5 ft.-tons.
1 horsepower-hour (hp-hr.)	= 1,980,000 ft.-lbs.
	= 990 ft.-tons
1 horsepower (hp)	= 746 watts (w)
	= .746 kilowatts (kw)

Energy

1 horsepower-hour	= 2544 BTU
	= .746 kw-hr.
1 kilowatt-hour	= 3413 BTU

Pressure

1 lb. per sq. in.	= 2.0360" of mercury at 32°F
	= 27.71" of water at 32°F
	= 2.3091 ft. of water at 60°F
	= 144 lbs. per sq. ft.
1 in. of mercury	= .491 lbs. per sq. in.
1 in. of water	= 5.2 lbs. per sq. ft. = .0361 PSI

Average Weights for Various Materials

Substance	Weight Lbs. per C.F.	Substance	Weight Lbs. per C.F.
Ashlar Masonry		Excavations in Water	
Granite, syenite, gneiss	165	Sand or gravel	60
Limestone, marble	160	Sand or gravel and clay	65
Sandstone, bluestone	140	Clay	80
Mortar Rubble Masonry		River mud	90
Granite, syenite, gneiss	155	Soil	70
Limestone, marble	150	Stone riprap	65
Sandstone, bluestone	130	Minerals	
Dry Rubble Masonry		Asbestos	153
Granite, syenite, gneiss	130	Barytes	281
Limestone, marble	125	Basalt	184
Sandstone, bluestone	110	Bauxite	159
Brick Masonry		Borax	109
Pressed brick	140	Chalk	137
Common brick	120	Clay, marl	137
Soft brick	100	Dolomite	181
Concrete Masonry		Feldspar, orthoclase	159
Cement, stone, sand	144	Gneiss, serpentine	159
Cement, slag, etc.	130	Granite, syenite	175
Cement, cinder, etc.	100	Greenstone, trap	187
Expanded slag aggregate	100	Gypsum, alabaster	159
Haydite (burned clay agg.)	90	Hornblende	187
Vermiculite/perlite, load bearing	70–105	Limestone, marble	165
Vermiculite and perlite, non-load bearing	25–50	Magnesite	187
Concrete Masonry Reinforced		Phosphate rock, apatite	200
Stone aggregate	150	Porphyry	172
Slag aggregate	138	Pumice, natural	40
Lightweight aggregates	30–106	Quartz, flint	165
Various Building Materials		Sandstone, bluestone	147
Ashes, cinders	40–45	Shale, slate	175
Cement, Portland, loose	90	Soapstone, talc	169
Cement, Portland, set	183	Stone, Quarried, Piled	
Lime, gypsum, loose	53–64	Basalt, granite, gneiss	96
Mortar, set	103	Limestone, marble, quartz	95
Slags, bank slag	67–72	Sandstone	82
Slags, bank screenings	98–117	Shale	92
Slags, machine slag	96	Greenstone, hornblende	107
Slags, slag sand	49–55	Bituminous Substances	
Earth, Etc., Excavated		Asphaltum	81
Clay, dry	63	Coal, anthracite	97
Clay, damp, plastic	110	Coal, bituminous	84
Clay and gravel, dry	100	Coal, lignite	78
Earth, dry, loose	76	Coal, peat, turf, dry	47
Earth, dry, packed	95	Coal, charcoal, pine	23
Earth, moist, loose	78	Coal, charcoal, oak	33
Earth, moist, packed	96	Coal, coke	75
Earth, mud, flowing	108	Graphite	131
Earth, mud, packed	115	Paraffine	56
Riprap, limestone	80–85	Petroleum	54
Riprap, sandstone	90	Petroleum, refined	50
Riprap, shale	105	Petroleum, benzine	46
Sand, gravel, dry, loose	90–105	Petroleum, gasoline	42
Sand, gravel, dry, packed	100–120	Pitch	69
Sand, gravel, wet	118–120	Tar, bituminous	75

(continued on next page)

Average Weights for Various Materials *(continued from previous page)*

Substance	Weight Lbs. per C.F.	Substance	Weight Lbs. per C.F.
Coal and Coke, Piled		Rubber goods	94
Coal, anthracite	47–58	Salt, granulated, piled	48
Coal, bituminous, lignite	40–54	Saltpeter	67
Coal, peat, turf	20–26	Starch	96
Coal, charcoal	10–14	Sulphur	125
Coal, coke	23–32	Wool	82
Metals, Alloys, Ores		Timber, U.S. Seasoned	
Aluminum, cast, hammered	165	Moisture Content by Weight:	
Brass, cast, rolled	534	Seasoned timber 15 to 20%	
Bronze, 7.9 to 14% Sn	509	Green timber up to 50%	
Bronze, aluminum	481	Ash, white, red	40
Copper, cast, rolled	556	Cedar, white, red	22
Copper ore, pyrites	262	Chestnut	41
Gold, cast, hammered	1205	Cypress	30
Iron, cast, pig	450	Fir, Douglas spruce	32
Iron, wrought	485	Fir, eastern	25
Iron, spiegel-eisen	468	Elm, white	45
Iron, ferro-silicon	437	Hemlock	29
Iron ore, hematite	325	Hickory	49
Iron ore, hematite in bank	160–180	Locust	46
Iron ore, hematite loose	130–160	Maple, hard	43
Iron ore, limonite	237	Maple, white	33
Iron ore, magnetite	315	Oak, chestnut	54
Iron slag	172	Oak, live	59
Lead	710	Oak, red, black	41
Lead ore, galena	465	Oak, white	46
Magnesium, alloys	112	Pine, Oregon	32
Manganese	475	Pine, red	30
Manganese ore, pyrolusite	259	Pine, white	26
Mercury	849	Pine, yellow, long-leaf	44
Monel Metal	565	Pine, yellow, short-leaf	38
Nickel	565	Poplar	30
Platinum, cast, hammered	1330	Redwood, California	26
Silver, cast, hammered	565	Spruce, white, black	27
Steel, rolled	490	Walnut, black	38
Tin, cast, hammered	459	Walnut, white	26
Tin ore, cassiterite	418	Various Liquids	
Zinc, cast, rolled	440	Alcohol, 100%	49
Zinc ore, blende	253	Acids, muriatic 40%	75
Various Solids		Acids, nitric 91%	94
Cereals, oats, bulk	32	Acids, sulphuric 87%	112
Cereals, barley, bulk	39	Lye, soda 66%	106
Cereals, corn, rye, bulk	48	Oils, vegetable	58
Cereals, wheat, bulk	48	Oils, mineral, lubricants	57
Hay and straw, bales	20	Water, 4°C. max. density	62.428
Cotton, flax, hemp	93	Water, 100°C.	59.830
Fats	58	Water, ice	56
Flour, loose	28	Water, snow, fresh fallen	8
Flour, pressed	47	Water, sea water	64
Glass, common	156	Gases	
Glass, plate or crown	161	Air, 0°C. 760 mm	.08071
Glass, crystal	184	Ammonia	.0478
Leather	59	Carbon dioxide	.1234
Paper	58	Carbon monoxide	.0781
Potatoes, piled	42	Gas, illuminating	.028–.036
Rubber, caoutchouc	59	Gas, natural	.038–.039

(continued on next page)

Average Weights for Various Materials *(continued from previous page)*

Material	Weight Lbs. per S.F.	Material	Weight Lbs. per S.F.
Gases (cont.)		Partitions	
Hydrogen	.00559	Clay Tile	
Nitrogen	.0784	3 in.	17
Oxygen	.0892	4 in.	18
Ceilings		6 in.	28
Channel suspended system	1	8 in.	34
Lathing and plastering	See Partitions	10 in.	40
Acoustical fiber tile	1	Gypsum Block	
		2 in.	9-1/2
Floors		3 in.	10-1/2
Steel Deck	See Mfg.	4 in.	12-1/2
		5 in.	14
Concrete–Reinforced 1 in.		6 in.	18-1/2
Stone	12-1/2	Wood Studs 2 x 4	
Slag	11-1/2	12–16 in. o.c.	2
Lightweight	6 to 10	Steel partitions	4
		Plaster 1 inch	
Concrete–Plain 1 in.		Cement	10
Stone	12	Gypsum	5
Slag	11	Lathing	
Lightweight	3 to 9	Metal	1/2
Fills 1 in.		Gypsum Board 1/2 in.	2
Gypsum	6		
Sand	8	Walls	
Cinders	4	Brick	
		4 in.	40
Finishes		8 in.	80
Terrazzo 1 in.	13	12 in.	120
Ceramic or Quarry Tile 3/4 in.	10	Hollow Concrete Block	
Linoleum 1/4 in.	1	(Heavy Aggregate)	
Mastic 3/4 in.	9	4 in.	30
Hardwood 7/8 in.	4	6 in.	43
Softwood 3/4 in.	2-1/2	8 in.	55
		12-1/2 in.	80
Roofs		Hollow Concrete Block	
Copper or tin	1	(Light Aggregate)	
3-ply ready roofing	1	4 in.	21
3-ply felt and gravel	5-1/2	6 in.	30
5-ply felt and gravel	6	8 in.	38
		12 in.	55
Shingles		Clay Tile	
Wood	2	(Load Bearing)	
Asphalt	3	4 in.	25
Clay tile	9 to 14	6 in.	30
Slate 1/4 in.	10	8 in.	33
		12 in.	45
Sheathing		Stone 4 in.	55
Wood 3/4 in.	3	Glass Block 4 in.	18
Gypsum 1 in.	4	Windows, Glass, Frame & Sash	8
		Curtain Walls	See Mfg.
Insulation 1 in.		Structural Glass 1 in.	15
Loose	1/2	Corrugated Cement	
Poured in place	2	Asbestos 1/4 in.	3
Rigid	1-1/2		

Minutes to Decimals

Minutes	Hours	Minutes	Hours
1	.017	31	.517
2	.033	32	.533
3	.050	33	.550
4	.067	34	.567
5	.083	35	.583
6	.100	36	.600
7	.117	37	.617
8	.133	38	.633
9	.150	39	.650
10	.167	40	.667
11	.183	41	.683
12	.200	42	.700
13	.217	43	.717
14	.233	44	.733
15	.250	45	.750
16	.267	46	.767
17	.283	47	.783
18	.300	48	.800
19	.317	49	.817
20	.333	50	.833
21	.350	51	.850
22	.367	52	.867
23	.383	53	.883
24	.400	54	.900
25	.417	55	.917
26	.433	56	.933
27	.450	57	.950
28	.467	58	.967
29	.483	59	.983
30	.500	60	1.000

Basic Metric Units and Prefixes

Basic Metric Units		Prefixes for Metric Units		
Quantity	Unit	Multiple and Submultiple	Prefix	Symbol
length	meter (m)	$1,000,000,000,000 = 10^{12}$	tera	T
mass	kilogram (kg)	$1,000,000,000 = 10^{9}$	giga	G
time	second (s)	$1,000,000 = 10^{6}$	mega	M
electric current	ampere (A)	$1,000 = 10^{3}$	kilo	k
temperature (thermodynamic)	kelvin (K)	$100 = 10^{2}$	hecto	h
amount of substance	mole (mol)	$10 = 10$	deka	da
luminous intensity	candela (cd)	$0.1 = 10^{-1}$	deci	d
		$0.01 = 10^{-2}$	centi	c
		$0.001 = 10^{-3}$	milli	m
		$0.000\ 001 = 10^{-6}$	micro	μ
		$0.000\ 000\ 001 = 10^{-9}$	nano	n
		$0.000\ 000\ 000\ 001 = 10^{-12}$	pico	p
		$0.000\ 000\ 000\ 000\ 001 = 10^{-15}$	femto	f
		$0.000\ 000\ 000\ 000\ 000\ 001 = 10^{-18}$	atto	a

Metric Equivalents

Length	
CM. = 0.3937 in.	In. = 2.5400 cm.
Meter = 3.2808 ft.	Ft. = 0.3048 m.
Meter = 1.0936 yd.	Yd. = 0.9144 m.
Km. = 0.6214 mile	Mile = 1.6093 km.

Area	
Sq. cm. = 0.1550 sq. in.	Sq. in. = 6.4516 sq. cm.
Sq. m. = 10.7639 sq. ft.	Sq. ft. = 0.0929 sq. m.
Sq. m. = 1.1960 sq. yd.	Sq. yd. = 0.8361 sq. m.
Hectare = 2.4710 acres	Acre = 0.4047 hectare
Sq. km. = 0.3861 sq. mile	Sq. mile = 2.5900 sq. km.

Volume	
Cu. cm. = 0.0610 cu. in.	Cu. in. = 16.3872 cu. cm.
Cu. m. = 35.3145 cu. ft.	Cu. ft. = 0.0283 cu. cm.
Cu. m. = 1.3079 cu. yd.	Cu. yd. = 0.7646 cu. m.

Capacity	
Liter = 61.0250 cu. in.	Cu. in. = 0.0164 liters
Liter = 0.0353 cu. ft.	Cu. ft. = 28.3162 liters
Liter = 0.2642 gal. (U.S.)	Gal. = 3.7853 liters
Liter = 0.0284 bu. (U.S.)	Bu. = 35.2383 liters

Liter = { 1000.027 cu. cm. / 1.0567 qt. (liquid) or 0.9081 qt. (dry) / 2.2046 lbs. of pure water at 4°C + 1 kg.

Weight	
Gram = 15.4324 grains	Grain = 0.0648 g.
Gram = 0.0353 oz.	Oz. = 28.3495 g.
Kg. = 2.2046 lbs.	Lb. = 0.4536 kg.
Kg. = 0.0011 ton (short)	Ton (short) = 907.1848 kg.
Ton (met.) = 1.1023 ton (short)	Ton (short) = 0.9072 ton (met.)
Ton (met.) = 0.9842 ton (large)	Ton (large) = 1.0160 ton (met.)

Pressure

1 kg. per sq. cm. = 14.223 lbs. per sq. in.
1 lb. per sq. in. = 0.0703 kg. per sq. cm.
1 kg. per sq. m. = 0.2048 lbs. per sq. ft.
1 lb. per sq. ft. = 4.8824 kg. per sq. m.
1 kg. per sq. cm. = 0.9678 normal atmosphere

1 normal atmosphere = { 1.0332 kg. per sq. cm. / 1.0133 bars / 14.696 lbs. per sq. in.

Metric Conversion Table

Conversion Formulas: Inches × 2.54 = Centimeters
Feet × .3048 = Meters
Pounds × .4536 = Kilograms

	Inches to Centimeters	Feet to Meters	Pounds to Kilograms		Inches to Centimeters	Feet to Meters	Pounds to Kilograms
1	2.54	.3048	.4536	51	129.54	15.5448	23.1336
2	5.08	.6096	.9072	52	132.08	15.8496	23.5872
3	7.62	.9144	1.3608	53	134.62	16.1544	24.0408
4	10.16	1.2192	1.8144	54	137.16	16.4592	24.4944
5	12.7	1.524	2.268	55	139.7	16.764	24.948
6	15.24	1.8288	2.7216	56	142.24	17.0688	25.4016
7	17.78	2.1336	3.1752	57	144.78	17.3736	25.8552
8	20.32	2.4384	3.6288	58	147.32	17.6784	26.3088
9	22.86	2.7432	4.0824	59	149.86	17.9832	26.7624
10	25.4	3.048	4.536	60	152.4	18.288	27.216
11	27.94	3.3528	4.9896	61	154.94	18.5928	27.6696
12	30.48	3.6576	5.4432	62	157.48	18.8976	28.1232
13	33.02	3.9624	5.8968	63	160.02	19.2024	28.5768
14	35.56	4.2672	6.3504	64	162.56	19.5072	29.0304
15	38.1	4.572	6.804	65	165.1	19.812	29.488
16	40.64	4.8768	7.2576	66	167.64	20.1168	29.9376
17	43.18	5.1816	7.7112	67	170.18	20.4216	30.3912
18	45.72	5.4864	8.1648	68	172.72	20.7264	30.8448
19	48.26	5.7912	8.6184	69	175.26	21.0312	31.2984
20	50.8	6.096	9.072	70	177.8	21.336	31.752
21	53.34	6.4008	9.5256	71	180.34	21.6408	32.2056
22	55.88	6.7056	9.9792	72	182.88	21.9456	32.6592
23	58.42	7.0104	10.4328	73	185.42	22.2504	33.1128
24	60.96	7.3152	10.8864	74	187.96	22.5552	33.5664
25	63.5	7.62	11.34	75	190.5	22.86	34.02
26	66.04	7.9248	11.7936	76	193.04	23.1648	34.4736
27	68.58	8.2296	12.2472	77	195.58	23.4696	34.9272
28	71.12	8.5344	12.7008	78	198.12	23.7744	35.3808
29	73.66	8.8392	13.1544	79	200.66	24.0792	35.8344
30	76.2	9.144	13.608	80	203.2	24.384	36.288
31	78.74	9.4488	14.0616	81	205.74	24.6888	36.7416
32	81.28	9.7536	14.5152	82	208.28	24.9936	37.1952
33	83.82	10.0584	14.9680	83	210.82	25.2984	37.6488
34	86.36	10.3632	15.4224	84	213.36	25.6032	38.1024
35	88.9	10.668	15.876	85	215.9	25.908	38.556
36	91.44	10.9728	16.3296	86	218.44	26.2128	39.0096
37	93.98	11.2776	16.7832	87	220.98	26.5176	39.4632
38	96.52	11.5824	17.2368	88	223.52	26.8224	39.9168
39	99.06	11.8872	17.6904	89	226.06	27.1272	40.3704
40	101.6	12.192	18.144	90	228.6	27.432	40.824
41	104.14	12.4968	18.5976	91	231.14	27.7368	41.2776
42	106.68	12.8016	19.0512	92	233.68	28.0416	41.7312
43	109.22	13.1064	19.5048	93	236.22	28.3464	42.1848
44	111.76	13.4112	19.9584	94	238.76	28.6512	42.6384
45	114.3	13.716	20.412	95	241.3	28.956	43.092
46	116.84	14.0208	20.8656	96	243.84	29.2608	43.5456
47	119.38	14.3256	21.3192	97	246.38	29.5656	43.9992
48	121.92	14.6304	21.7728	98	248.92	29.8704	44.4528
49	124.46	14.9352	22.2264	99	252.46	30.1752	44.9064
50	127	15.24	22.68	100	254	30.48	45.36

Standard and Metric Linear and Area Conversion Tables

Linear Conversions							
Inches	Feet	Yards	Rods	Miles	Centi-meters	Meters	Kilo-meters
1	0.083	0.028	0.005	—	2.540	0.0254	—
12	1	0.333	0.061	0.0002	30.480	0.305	0.0003
36	3	1	0.182	0.0006	91.440	0.914	0.0009
0.3937	0.033	0.011	—	—	1	0.01	—
39.37	3.281	1.094	0.199	0.0006	100	1	0.001
					Furlongs		
198	16.5	5.5	1	0.003	0.025	5.029	0.005
	5,280	1,760	320	1	8	1,609,347	1.609
	660	220	40	0.125	1	201.168	0.201
	3,280.83	1,093.61	198.838	0.621	4.971	1,000	1

Area Conversions							
Square Inches	Square Feet	Square Yards	Acres	Square Centi-meters	Square Meters	Hectares	Square Kilo-meters
1	0.007	—	—	6.452	0.0006	—	—
144	1	0.111	0.00002	929.034	0.093	—	—
1,296	9	1	0.0002	8,361.31	0.836	—	—
0.155	0.001	—	—	1	0.0001	—	—
1,549.997	10.764	1.196	0.0002	10,000	1	0.0001	—
				Square Miles			
	43,560	4,840	1	0.002	4,046.87	0.405	0.004
	27,878,400	3,097,600	640	1	2,589,998	258.999	2.590
	107,638.7	11,959.9	2.471	0.004	10,000	1	0.01
	10,763,867	1,195,985	247.104	0.386	1,000,000	100	1

General Conversion Factors

Multiply	By	To Obtain
acres	43,560	square feet
acres	4047	square meters
acres	1.562×10^{-3}	square miles
acres	5645.8	square varas
acres	4840	square yards
amperes	1/10	abamperes
amperes	3×10^9	statamperes
atmospheres	76.0	cms. of mercury
atmospheres	29.92	inches of mercury
atmospheres	33.90	feet of water
atmospheres	10.333	kgs. per sq. meter
atmospheres	14.70	pounds per sq. inch
atmospheres	1.058	tons per sq. foot
British thermal units	0.2520	kilogram-calories
British thermal units	777.5	foot-pounds
British thermal units	3.927×10^{-4}	horse-power-hours
British thermal units	1054	joules
British thermal units	107.5	kilogram-meters
British thermal units	2.928×10^{-4}	kilowatt-hours
B.t.u. per min.	12.96	foot-pounds per sec.
B.t.u. per min.	0.02356	horse-power
B.t.u. per min.	0.01757	kilowatts
B.t.u. per min.	17.57	watts
B.t.u. per sq. ft. per min.	0.1220	watts per sq. inch
bushels	1.244	cubic feet
bushels	2150	cubic inches
bushels	0.03524	cubic meters
bushels	4	pecks
bushels	64	pints (dry)
bushels	32	quarts (dry)
centimeters	0.3937	inches
centimeters	0.01	meters
centimeters	393.7	mils
centimeters	10	millimeters
centimeter-grams	980.7	centimeter-dynes
centimeter-grams	10^{-5}	meter-kilograms
centimeter-grams	7.233×10^{-5}	pound-feet
centimeters of mercury	0.01316	atmospheres
centimeters of mercury	0.4461	feet of water
centimeters of mercury	136.0	kgs. per sq. meter
centimeters of mercury	27.85	pounds per sq. foot
centimeters of mercury	0.1934	pounds per sq. inch
centimeters per second	1.969	feet per minute
centimeters per second	0.03281	feet per second
centimeters per second	0.036	kilometers per hour
centimeters per second	0.6	meters per minute
centimeters per second	0.02237	miles per hour
centimeters per second	3.728×10^{-4}	miles per minute
cubic centimeters	3.531×10^{-5}	cubic feet
cubic centimeters	6.102×10^{-2}	cubic inches
cubic centimeters	10^{-6}	cubic meters
cubic centimeters	1.308×10^{-6}	cubic yards
cubic centimeters	2.642×10^{-4}	gallons
cubic centimeters	10^{-3}	liters

(continued on next page)

General Conversion Factors *(continued from previous page)*

Multiply	By	To Obtain
cubic centimeters	2.113×10^{-3}	pints (liquid)
cubic centimeters	1.057×10^{-3}	quarts (liquid)
cubic feet	62.43	pounds of water
cubic feet	2.832×10^{4}	cubic cms.
cubic feet	1728	cubic inches
cubic feet	0.02832	cubic meters
cubic feet	0.03704	cubic yards
cubic feet	7.481	gallons
cubic feet	28.32	liters
cubic feet	59.84	pints (liquid)
cubic feet	29.92	quarts (liquid)
cubic feet per minute	472.0	cubic cms. per sec.
cubic feet per minute	0.1247	gallons per sec.
cubic feet per minute	0.4720	liters per second
cubic feet per minute	62.4	lbs. of water per min.
cubic inches	16.39	cubic centimeters
cubic inches	5.787×10^{-4}	cubic feet
cubic inches	1.639×10^{-5}	cubic meters
cubic inches	2.143×10^{-5}	cubic yards
cubic inches	4.329×10^{-3}	gallons
cubic inches	1.639×10^{-2}	liters
cubic inches	0.03463	pints (liquid)
cubic inches	0.01732	quarts (liquid)
cubic yards	7.646×10^{5}	cubic centimeters
cubic yards	27	cubic feet
cubic yards	46,656	cubic inches
cubic yards	0.7646	cubic meters
cubic yards	202.0	gallons
cubic yards	764.6	liters
cubic yards	1616	pints (liquid)
cubic yards	807.9	quarts (liquid)
cubic yards per minute	0.45	cubic feet per sec.
cubic yards per minute	3.367	gallons per second
cubic yards per minute	12.74	liters per second
degrees (angle)	60	minutes
degrees (angle)	0.01745	radians
degrees (angle)	3600	seconds
dynes	1.020×10^{-3}	grams
dynes	7.233×10^{-5}	poundals
dynes	2.248×10^{-6}	pounds
ergs	9.486×10^{-11}	British thermal units
ergs	1	dyne-centimeters
ergs	7.376×10^{-8}	foot-pounds
ergs	1.020×10^{-3}	gram-centimeters
ergs	10^{-7}	joules
ergs	2.390×10^{-11}	kilogram-calories
ergs	1.020×10^{-8}	kilogram-meters
feet	30.48	centimeters
feet	12	inches
feet	0.3048	meters
feet	.36	varas
feet	1/3	yards
feet of water	0.02950	atmospheres
feet of water	0.8826	inches of mercury

(continued on next page)

General Conversion Factors *(continued from previous page)*

Multiply	By	To Obtain
feet of water	304.8	kgs. per sq. meter
feet of water	62.43	pounds per sq. ft.
feet of water	0.4335	pounds per sq. inch
feet per second	30.48	centimeters per second
feet per sec. per sec.	0.3048	meters per sec. per sec.
foot-pounds	1.286×10^{-3}	British thermal units
foot-pounds	1.356×10^{7}	ergs
foot-pounds	5.050×10^{-7}	horse-power hours
foot-pounds	1.356	joules
foot-pounds	3.241×10^{-4}	kilogram-calories
foot-pounds	0.1383	kilogram-meters
foot-pounds	3.766×10^{-7}	kilowatt-hours
foot-pounds per min.	1.286×10^{-3}	B.t. units per minute
foot-pounds per min.	0.01667	foot-pounds per sec.
foot-pounds per min.	3.030×10^{-5}	horse-power
foot-pounds per min.	3.241×10^{-4}	kg.-calories per min.
foot-pounds per min.	2.260×10^{-5}	kilowatts
foot-pounds per sec.	7.717×10^{-2}	B.t. units per minute
foot-pounds per sec.	1.818×10^{-3}	horse-power
foot-pounds per sec.	1.945×10^{-2}	kg.-calories per min.
foot-pounds per sec.	1.356×10^{-3}	kilowatts
gallons	8.345	pounds of water
gallons	3785	cubic centimeters
gallons	0.1337	cubic feet
gallons	231	cubic inches
gallons	3.785×10^{-3}	cubic meters
gallons	4.951×10^{-3}	cubic yards
gallons	3.785	liters
gallons	8	pints (liquid)
gallons	4	quarts (liquid)
gallons per minute	2.228×10^{-3}	cubic ft. per second
gallons per minute	0.06308	liters per second
grains (troy)	1	grains (av.)
grains (troy)	0.06480	grams
grains (troy)	0.04167	pennyweights (troy)
grams	980.7	dynes
grams	15.43	grains (troy)
grams	10^{-3}	kilograms
grams	10^{3}	milligrams
grams	0.03527	ounces
grams	0.03215	ounces (troy)
grams	0.07093	poundals
grams	2.205×10^{-3}	pounds
horse-power	42.44	B.t. units per min.
horse-power	33,000	foot-pounds per min.
horse-power	550	foot-pounds per sec.
horse-power	1.014	horse-power (metric)
horse-power	10.70	kg.-calories per min.
horse-power	0.7457	kilowatts
horse-power	745.7	watts
horse-power (boiler)	33,520	B.t.u. per hour
horse-power (boiler)	9.804	kilowatts
horse-power-hours	2547	British thermal units
horse-power-hours	1.98×10^{6}	foot-pounds

(continued on next page)

General Conversion Factors *(continued from previous page)*

Multiply	By	To Obtain
horse-power-hours	2.684×10^6	joules
horse-power-hours	641.7	kilogram-calories
horse-power-hours	2.737×10^5	kilogram-meters
horse-power-hours	0.7457	kilowatt-hours
inches	2.540	centimeters
inches	10^3	mils
inches	.03	varas
inches of mercury	0.03342	atmospheres
inches of mercury	1.133	feet of water
inches of mercury	345.3	kgs. per sq. meter
inches of mercury	70.73	pounds per sq. ft.
inches of mercury	0.4912	pounds per sq. in.
inches of water	0.002458	atmospheres
inches of water	0.07355	inches of mercury
inches of water	25.40	kgs. per sq. meter
inches of water	0.5781	ounces per sq. in.
inches of water	5.204	pounds per sq. ft.
inches of water	0.03613	pounds per sq. in.
kilograms	980,665	dynes
kilograms	10^3	grams
kilograms	70.93	poundals
kilograms	2.2046	pounds
kilograms	1.102×10^{-3}	tons (short)
kilogram-calories	3.968	British thermal units
kilogram-calories	3086	foot-pounds
kilogram-calories	1.558×10^{-3}	horse-power-hours
kilogram-calories	4183	joules
kilogram-calories	426.6	kilogram-meters
kilogram-calories	1.162×10^{-3}	kilowatt-hours
kg.-calories per min.	51.43	foot-pounds per sec.
kg.-calories per min.	0.09351	horse-power
kg.-calories per min.	0.06972	kilowatts
kilometers	10^5	centimeters
kilometers	3281	feet
kilometers	10^3	meters
kilometers	0.6214	miles
kilometers	1093.6	yards
kilowatts	56.92	B.t. units per min.
kilowatts	4.425×10^4	foot-pounds per min.
kilowatts	737.6	foot-pounds per sec.
kilowatts	1.341	horse-power
kilowatts	14.34	kg.-calories per min.
kilowatts	10^3	watts
kilowatt-hours	3415	British thermal units
kilowatt-hours	2.655×10^6	foot-pounds
kilowatt-hours	1.341	horse-power-hours
kilowatt-hours	3.6×10^6	joules
kilowatt-hours	860.5	kilogram-calories
kilowatt-hours	3.671×10^5	kilogram-meters
$\log^{10} N$	2.303	$\log_\epsilon N$ or $\ln N$
$\log^\epsilon N$ or $\ln N$	0.4343	$\log_{10} N$
meters	100	centimeters
meters	3.2808	feet
meters	39.37	inches

(continued on next page)

General Conversion Factors *(continued from previous page)*

Multiply	By	To Obtain
meters	10^{-3}	kilometers
meters	10^3	millimeters
meters	1.0936	yards
miles	1.609×10^5	centimeters
miles	5280	feet
miles	1.6093	kilometers
miles	1760	yards
miles	1900.8	varas
miles per hour	44.70	centimeters per sec.
miles per hour	88	feet per minute
miles per hour	1.467	feet per second
miles per hour	1.6093	kilometers per hour
miles per hour	0.8684	knots per hour
miles per hour	26.82	meters per minute
miles per hour per sec.	44.70	cms. per sec. per sec.
miles per hour per sec.	1.467	ft. per sec. per sec.
miles per hour per sec.	1.6093	kms. per hr. per sec.
miles per hour per sec.	0.4470	M. per sec. per sec.
months	30.42	days
months	730	hours
months	43,800	minutes
months	2.628×10^6	seconds
ounces	8	drams
ounces	437.5	grains
ounces	28.35	grams
ounces	.0625	pounds
ounces per square inch.	0.0625	pounds per sq. inch
pints (dry)	33.60	cubic inches
pints (liquid)	28.87	cubic inches
pounds	444,823	dynes
pounds	7000	grains
pounds	453.6	grams
pounds	16	ounces
pounds	32.17	poundals
pounds of water	0.01602	cubic feet
pounds of water	27.68	cubic inches
pounds of water	0.1198	gallons
pounds of water per min.	2.669×10^{-4}	cubic feet per sec.
pounds per cubic foot	0.01602	grams per cubic cm.
pounds per cubic foot	16.02	kgs. per cubic meter
pounds per cubic foot	5.787×10^{-4}	pounds per cubic in.
pounds per cubic foot	5.456×10^{-9}	pounds per mil foot
pounds per square foot	0.01602	feet of water
pounds per square foot	4.882	kgs. per sq. meter
pounds per square foot	6.944×10^{-3}	pounds per sq. inch
pounds per square inch	0.06804	atmospheres
pounds per square inch	2.307	feet of water
pounds per square inch	2.036	inches of mercury
pounds per square inch	703.1	kgs. per sq. meter
pounds per square inch	144	pounds per sq. foot
quarts	32	fluid ounces
quarts (dry)	67.20	cubic inches
quarts (liquid)	57.75	cubic inches
rods	16.5	feet

(continued on next page)

General Conversion Factors *(continued from previous page)*

Multiply	By	To Obtain
square centimeters	1.973×10^5	circular mils
square centimeters	1.076×10^{-3}	square feet
square centimeters	0.1550	square inches
square centimeters	10^{-6}	square meters
square centimeters	100	square millimeters
square feet	2.296×10^{-5}	acres
square feet	929.0	square centimeters
square feet	144	square inches
square feet	0.09290	square meters
square feet	3.587×10^{-8}	square miles
square feet	.1296	square varas
square feet	1/9	square yards
square inches	1.273×10^6	circular mils
square inches	6.452	square centimeters
square inches	6.944×10^{-3}	square feet
square inches	10^6	square mils
square inches	645.2	square millimeters
square miles	640	acres
square miles	27.88×10^6	square feet
square miles	2.590	square kilometers
square miles	3,613,040.45	square varas
square miles	3.098×10^6	square yards
square yards	2.066×10^{-4}	acres
square yards	9	square feet
square yards	0.8361	square meters
square yards	3.228×10^{-7}	square miles
square yards	1.1664	square varas
temp. (degs. C.) + 17.8	1.8	temp. (degs. Fahr.)
temp. (degs. F.) − 32	5/9	temp. (degs. Cent.)
tons (long)	2240	pounds
tons (short)	2000	pounds
yards	.9144	meters

Approximate Comparison of Drawing Scales

Metric			Approximate English Equivalent			Usage
1:1500	(means 1 cm	= 15 m)	1 in.	=	100 ft.	Location plans
1:1000	(means 1 cm	= 10 m)	1 in.	=	80 ft.	Site plans
1:500	(means 1 cm	= 5 m)	1 in.	=	40 ft.	Plot plans
1:200	(means 0.5 cm	= 1 m)	1/16 in.	=	1 ft.	Small scale plans
1:100	(means 1 cm	= 1 m)	1/8 in.	=	1 ft.	Normal scale plans
1:50	(means 2 cm	= 1 m)	1/4 in.	=	1 ft.	Large scale plans
1:20	(means 5 cm	= 1 m)	1/2 in.	=	1 ft.	Sections and details
1:10	(means 10 cm	= 1 m)	1 in.	=	1 ft.	Details
1:8	(means 12.5 cm	= 1 m)	1-1/2 in.	=	1 ft.	Details
1:1	(means 1 m	= 1 m)	Full size			Large details

Trigonomentric Functions

sine A	=	$\dfrac{\text{opposite}}{\text{hypotenuse}}$	=	$\dfrac{a}{c}$
cosine A	=	$\dfrac{\text{adjacent}}{\text{hypotenuse}}$	=	$\dfrac{b}{c}$
tangent A	=	$\dfrac{\text{opposite}}{\text{adjacent}}$	=	$\dfrac{a}{b}$
cotangent A	=	$\dfrac{\text{adjacent}}{\text{opposite}}$	=	$\dfrac{b}{a}$
secant A	=	$\dfrac{\text{hypotenuse}}{\text{adjacent}}$	=	$\dfrac{c}{b}$
cosecant A	=	$\dfrac{\text{hypotenuse}}{\text{opposite}}$	=	$\dfrac{c}{a}$

By using these relationships and the following Table of Natural Trigonometric Functions, unknown angles and sides of right triangles may be found.

Table of Natural Trigonometric Functions

Degrees	Sin	Cos	Tan	Cot	Sec	Csc	
0°00′	.0000	1.0000	.0000	—	1.000	—	90°00′
10	029	000	029	343.8	000	343.8	50
20	058	000	058	171.9	000	171.9	40
30	.0087	1.0000	.0087	114.6	1.000	114.6	30
40	116	9999	116	85.94	000	85.95	20
50	145	999	145	68.75	000	68.76	10
1°00′	.0175	.9998	.0175	57.29	1.000	57.30	89°00′
10	204	998	204	49.10	000	49.11	50
20	233	997	233	42.96	000	42.98	40
30	.0262	.9997	.0262	38.19	1.000	38.20	30
40	291	996	291	34.37	000	34.38	20
50	320	995	320	31.24	001	31.26	10
2°00′	.0349	.9994	.0349	28.64	1.001	28.65	88°00′
10	378	993	378	26.43	001	26.45	50
20	407	992	407	24.54	001	24.56	40
30	.0436	.9990	.0437	22.90	1.001	22.93	30
40	465	989	466	21.47	001	21.49	20
50	494	988	495	20.21	001	20.23	10
3°00′	.0523	.9986	.0524	19.08	1.001	19.11	87°00′
10	552	985	553	18.07	002	18.10	50
20	581	983	582	17.17	002	17.20	40
30	.0610	.9981	.0612	16.35	1.002	16.38	30
40	640	980	641	15.60	002	15.64	20
50	669	978	670	14.92	002	14.96	10
4°00′	.0698	.9976	.0699	14.30	1.002	14.34	86°00′
10	727	974	729	13.73	003	13.76	50
20	756	971	758	13.20	003	13.23	40
30	.0785	.9969	.0787	12.71	1.003	12.75	30
40	814	967	816	12.25	003	12.29	20
50	843	964	846	11.83	004	11.87	10
5°00′	.0872	.9962	.0875	11.43	1.004	11.47	85°00′
10	901	959	904	11.06	004	11.10	50
20	929	957	934	10.71	004	10.76	40
30	.0958	.9954	.0963	10.39	1.005	10.43	30
40	987	951	992	10.08	005	10.13	20
50	.1016	948	.1022	9.788	005	9.839	10
6°00′	.1045	.9945	.1051	9.514	1.006	9.567	84°00′
10	074	942	080	9.255	006	9.309	50
20	103	939	110	9.010	006	9.065	40
30	.1132	.9936	.1139	8.777	1.006	8.834	30
40	161	932	169	8.556	007	8.614	20
50	190	929	198	8.345	007	8.405	10
7°00′	.1219	.9925	.1228	8.144	1.008	8.206	83°00′
10	248	922	257	7.953	008	8.016	50
20	276	918	287	7.770	008	7.834	40
30	.1305	.9914	1317	7.596	1.009	7.661	30
40	334	911	346	7.429	009	7.496	20
50	363	907	376	7.269	009	7.337	10
8°00′	.1392	.9903	.1405	7.115	1.010	7.185	82°00′
10	421	899	435	6.968	010	7.040	50
20	449	894	465	6.827	011	6.900	40
30	.1478	.9890	.1495	6.691	1.011	6.765	30
40	507	886	524	6.561	012	6.636	20
50	536	881	554	6.435	012	6.512	10
9°00′	.1564	.9877	.1584	6.314	1.012	6.392	81°00′
	Cos	Sin	Cot	Tan	Csc	Sec	Degrees

(continued on next page)

Table of Natural Trigonometric Functions (continued from previous page)

Degrees	Sin	Cos	Tan	Cot	Sec	Csc	
9°00'	.1564	.9877	.1584	6.314	1.012	6.392	81°00'
10	593	872	614	197	013	227	50
20	622	868	644	084	013	166	40
30	.1650	.9863	.1673	5.976	1.014	6.059	30
40	679	858	703	871	014	5.955	20
50	708	853	733	769	015	855	10
10°00'	.1736	.9848	.1763	5.671	1.015	5.759	80°00'
10	765	843	793	576	016	665	50
20	794	838	823	485	016	575	40
30	.1822	.9833	.1853	5.396	1.017	5.487	30
40	851	827	883	309	018	403	20
50	880	822	914	226	018	320	10
11°00'	.1908	.9816	.1944	5.145	1.019	5.241	79°00'
10	937	811	974	066	019	164	50
20	965	805	.2004	4.989	020	089	40
30	.1994	.9799	.2035	4.915	1.020	5.016	30
40	.2022	793	065	843	021	4.945	20
50	051	787	095	773	022	876	10
12°00'	.2079	.9781	.2126	4.705	1.022	4.810	78°00'
10	108	775	156	638	023	745	50
20	136	769	186	574	024	682	40
30	.2164	.9763	.2217	4.511	1.024	4.620	30
40	193	757	247	449	025	560	20
50	221	750	278	390	026	502	10
13°00'	.2250	.9744	.2309	4.331	1.026	4.445	77°00'
10	278	737	339	275	027	390	50
20	306	730	370	219	028	336	40
30	.2334	.9724	.2401	4.165	1.028	4.284	30
40	363	717	432	113	029	232	20
50	391	710	462	061	030	182	10
14°00'	.2419	.9703	.2493	4.011	1.031	4.134	76°00'
10	447	696	524	3.962	031	086	50
20	476	689	555	914	032	039	40
30	.2504	.9681	.2586	3.868	1.033	3.994	30
40	532	674	617	821	034	950	20
50	560	667	648	776	034	906	10
15°00'	.2588	.9659	.2679	3.732	2.732	3.864	75°00'
10	616	652	711	689	036	822	50
20	644	644	742	647	037	782	40
30	.2672	.9636	.2773	3.606	1.038	3.742	30
40	700	628	805	566	039	703	20
50	728	621	836	526	039	665	10
16°00'	.2756	.9613	.2867	3.487	1.040	3.628	74°00'
10	784	605	899	450	041	592	50
20	812	596	931	412	042	556	40
30	.2840	.9588	.2962	3.376	1.043	3.521	30
40	868	580	994	340	044	487	20
50	896	572	.3026	305	045	453	10
17°00'	.2924	.9563	.3057	3.271	1.046	3.420	73°00'
10	952	555	089	237	047	388	50
20	979	546	121	204	048	356	40
30	.3007	.9537	.3153	3.172	1.049	3.326	30
40	035	528	185	140	049	295	20
50	062	520	217	108	050	265	10
18°00'	.3090	.9511	.3249	3.078	1.051	3.236	72°00'
	Cos	Sin	Cot	Tan	Csc	Sec	Degrees

(continued on next page)

Table of Natural Trigonometric Functions (continued from previous page)

Degrees	Sin	Cos	Tan	Cot	Sec	Csc	
18°00'	.3090	.9511	.3249	3.078	1.051	3.236	72°00'
10	118	502	281	047	052	207	50
20	145	492	314	018	053	179	40
30	.3173	.9843	.3346	2.989	1.054	3.152	30
40	201	474	378	960	056	124	20
50	228	465	411	932	057	098	10
19°00'	.3256	.9455	.3443	2.904	1.058	3.072	71°00'
10	283	446	476	877	059	046	50
20	311	436	508	850	060	021	40
30	.3388	.9426	.3541	2.824	1.061	2.996	30
40	365	417	574	798	062	971	20
50	393	407	607	773	063	947	10
20°00'	.3420	.9397	.3640	2.747	1.064	2.924	70°00'
10	448	387	673	723	065	901	50
20	475	377	706	699	066	878	40
30	.3502	.9367	.3739	2.675	1.068	2.855	30
40	529	356	772	651	069	833	20
50	557	346	805	628	070	812	10
21°00'	.3584	.9336	.3839	2.605	1.071	2.790	69°00'
10	611	325	872	583	072	769	50
20	638	315	906	560	074	749	40
30	.3665	.9304	.3939	2.539	1.075	2.729	30
40	692	293	973	517	076	709	20
50	719	283	.4006	496	077	689	10
22°00'	.3746	.9272	.4040	2.475	1.079	2.669	68°00'
10	773	261	074	455	080	650	50
20	800	250	108	434	081	632	40
30	.3827	.9239	.4142	2.414	1.082	2.613	30
40	854	228	176	394	084	595	20
50	881	216	210	375	085	577	10
23°00'	.3907	.9205	.4245	2.356	1.086	2.559	67°00'
10	934	194	279	337	088	542	50
20	961	182	314	318	089	525	40
30	.3987	.9171	.4348	2.300	1.090	2.508	30
40	.4014	159	383	282	092	491	20
50	041	147	417	264	093	475	10
24°00'	.4067	.9135	.4452	2.246	1.095	2.459	66°00'
10	094	124	487	229	096	443	50
20	120	112	522	211	097	427	40
30	.4147	.9100	.4557	2.194	1.099	2.411	30
40	173	088	592	177	100	396	20
50	200	075	628	161	102	381	10
25°00'	.4226	.9063	.4663	2.145	1.103	2.366	65°00'
10	253	051	699	128	105	352	50
20	279	038	734	112	106	337	40
30	.4305	.9026	.4770	2.097	1.108	2.323	30
40	331	013	806	081	109	309	20
50	358	001	841	066	111	295	10
26°00'	.4384	.8988	.4877	2.050	1.113	2.281	64°00'
10	410	975	913	035	114	268	50
20	436	962	950	020	116	254	40
30	.4462	.8949	.4986	2.006	1.117	2.241	30
40	488	936	.5022	1.991	119	228	20
50	514	923	059	977	121	215	10
27°00'	.4540	.8910	.5095	1.963	1.122	2.203	63°00'
	Cos	Sin	Cot	Tan	Csc	Sec	Degrees

(continued on next page)

Table of Natural Trigonometric Functions (continued from previous page)

Degrees	Sin	Cos	Tan	Cot	Sec	Csc	
27°00'	.4540	.8910	.5095	1.963	1.122	2.203	63°00'
10	566	897	132	949	124	190	50
20	592	884	169	935	126	178	40
30	.4617	.8870	.5206	1.921	1.127	2.166	30
40	643	857	243	907	129	154	20
50	669	843	280	894	131	142	10
28°00'	.4695	.8829	.5317	1.881	1.133	2.130	62°00'
10	720	816	354	868	143	118	50
20	746	802	392	855	136	107	40
30	.4772	.8788	.5430	1.842	1.138	2.096	30
40	797	774	467	829	140	085	20
50	823	760	505	816	142	074	10
29°00'	.4848	.8746	.5543	1.804	1.143	2.063	61°00'
10	874	732	581	792	145	052	50
20	899	718	619	780	147	041	40
30	.4924	.8704	.5658	1.767	1.149	2.031	30
40	950	689	696	756	151	020	20
50	975	675	735	744	153	010	10
30°00'	.500	.8660	.5774	1.732	1.155	2.000	60°00'
10	025	646	812	720	157	1.990	50
20	050	631	851	709	159	980	40
30	.5075	.8616	.5890	1.698	1.161	1.970	30
40	100	601	930	686	163	961	20
50	125	587	969	675	165	951	10
31°00'	.5150	.8572	.6009	1.664	1.167	1.942	59°00'
10	175	557	048	653	169	932	50
20	200	542	088	643	171	923	40
30	.5225	.8526	.6128	1.632	1.173	1.914	30
40	250	511	168	621	175	905	20
50	275	496	208	611	177	896	10
32°00'	.5299	.8480	.6249	1.600	1.179	1.887	58°00'
10	324	465	289	590	181	878	50
20	348	450	330	580	184	870	40
30	.5373	.8434	.6371	1.570	1.186	1.861	30
40	398	418	412	560	188	853	20
50	422	403	453	550	190	844	10
33°00'	.5446	.8387	.6494	1.540	1.192	1.836	57°00'
10	471	371	536	530	195	828	50
20	495	355	577	530	197	820	40
30	.5519	.8339	.6619	1.511	1.199	1.812	30
40	544	323	661	501	202	804	20
50	568	307	703	1.492	204	796	10
34°00'	.5592	.8290	.6745	1.483	1.206	1.788	56°00'
10	616	274	787	473	209	781	50
20	640	258	830	464	211	773	40
30	.5664	.8241	.6873	1.455	1.213	1.766	30
40	688	225	916	446	216	758	20
50	712	208	959	437	218	751	10
35°00'	.5736	.8192	.7002	1.428	1.221	1.743	55°00'
10	760	175	046	419	223	736	50
20	783	158	089	411	226	729	40
30	.5807	.8141	.7133	1.402	1.228	1.722	30
40	831	124	177	393	231	715	20
50	854	107	221	385	233	708	10
36°00'	.5878	.8090	.7265	1.376	1.236	1.701	54°00'
	Cos	Sin	Cot	Tan	Csc	Sec	Degrees

(continued on next page)

Table of Natural Trigonometric Functions (continued from previous page)

Degrees	Sin	Cos	Tan	Cot	Sec	Csc	
36°00'	.5878	.8090	.7265	1.376	1.236	1.701	54°00'
10	901	073	310	368	239	695	50
20	925	056	355	360	241	688	40
30	.5948	.8039	.7400	1.351	1.244	1.681	30
40	972	021	445	343	247	675	20
50	995	004	490	335	249	668	10
37°00'	.6018	.7986	.7536	1.327	1.252	1.662	53°00'
10	041	969	581	319	255	655	50
20	065	951	627	311	258	649	40
30	.6088	.7934	.7673	1.303	1.260	1.643	30
40	111	916	720	295	263	636	20
50	134	898	766	288	266	630	10
28°00'	.6157	.7880	.7813	1.280	1.269	1.624	52°00'
10	180	862	860	272	272	618	50
20	202	844	907	265	275	612	40
30	.6225	.7826	.7954	1.257	1.278	1.606	30
40	248	808	.8002	250	281	601	20
50	271	790	050	242	284	595	10
39°00'	.6293	.7771	.8098	1.235	1.287	1.589	51°00'
10	316	753	146	228	290	583	50
20	338	735	195	220	293	578	40
30	.6361	.7716	.8243	1.213	1.296	1.572	30
40	383	698	292	206	299	567	20
50	406	679	342	199	302	561	10
40°00'	.6428	.7660	.8391	1.192	1.305	1.556	50°00'
10	450	642	441	185	309	550	50
20	472	623	491	178	312	545	40
30	.6494	.7604	.8541	1.171	1.315	1.540	30
40	517	585	591	164	318	535	20
50	539	566	642	157	322	529	10
41°00'	.6561	.7547	.8693	1.150	1.325	1.524	49°00'
10	583	528	744	144	328	519	50
20	604	509	796	137	332	514	40
30	.6626	.7490	.8847	1.130	1.335	1.509	30
40	648	470	899	124	339	504	20
50	670	451	952	117	342	499	10
42°00'	.6691	.7431	.9004	1.111	1.346	1.494	48°00'
10	713	412	057	104	349	490	50
20	734	392	110	098	353	485	40
30	.6756	.7373	.9163	1.091	1.356	1.480	30
40	777	353	217	085	360	476	20
50	799	333	271	079	364	471	10
43°00'	.6820	.7314	.9325	1.072	1.367	1.466	47°00'
10	841	294	380	066	371	462	50
20	862	274	435	060	375	457	40
30	.6884	.7254	.9490	1.054	1.379	1.453	30
40	905	234	545	048	382	448	20
50	926	214	601	042	386	444	10
44°00'	.6947	.7193	.9657	1.036	1.390	1.440	46°00'
10	967	173	713	030	394	435	50
20	988	153	770	024	398	431	40
30	.7009	.7133	.9827	1.018	1.402	1.427	30
40	030	112	884	012	406	423	20
50	050	092	942	006	410	418	10
45°00'	.7071	.7071	1.0000	1.000	1.414	1.414	45°00'
	Cos	Sin	Cot	Tan	Csc	Sec	Degrees

Measure of Angles

De-grees	Rise in Inches per Ft.	Rise in Inches per Ft.	Degrees and Minutes	Percent Rise in Ft. per 100 Ft.	Degrees and Minutes	Percent Rise in Ft. per 100 Ft.	Degrees and Minutes
1	.210	1/4	1° 11'	1	34.4'	36	19° 48'
2	.419	1/2	2° 23'	2	1° 8.7'	37	20° 18'
3	.629	3/4	3° 35'	3	1° 43.1'	38	20° 48'
4	.839	1	4° 46'	4	2° 17.5'	39	21° 18'
5	1.050	1-1/4	5° 56'	5	2° 51.8'	40	21° 48'
6	1.261	1-1/2	7° 7'	6	3° 26.0'	41	22° 18'
7	1.473	1-3/4	8° 18'	7	4° 0.3'	42	22° 47'
8	1.686	2	9° 28'	8	4° 34.4'	43	23° 16'
9	1.901	2-1/4	10° 37'	9	5° 8.6'	44	23° 45'
10	2.116	2-1/2	11° 46'	10	5° 42.6'	45	24° 14'
11	2.333	2-3/4	12° 54'	11	6° 16.6'	46	24° 42'
12	2.551	3	14° 2'	12	6° 50.6'	47	25° 10'
13	2.770	3-1/4	15° 9'	13	7° 24.4'	48	25° 38'
14	2.992	3-1/2	16° 15'	14	7° 58.2'	49	26° 6'
15	3.215	3-3/4	17° 21'	15	8° 31.9'	50	26° 34'
16	3.441	4	18° 26'	16	9° 5.4'	51	27° 1'
17	3.669	4-1/4	19° 30'	17	9° 38.9'	52	27° 28'
18	3.900	4-1/2	20° 33'	18	10° 12.2'	53	27° 55'
19	4.132	4-3/4	21° 36'	19	10° 45.5'	54	28° 22'
20	4.368	5	22° 37'	20	11° 18.6'	55	28° 49'
21	4.606	5-1/4	23° 38'	21	11° 51.6'	56	29° 15'
22	4.843	5-1/2	24° 37'	22	12° 24.5'	57	29° 41'
23	5.094	5-3/4	25° 36'	23	12° 57.2'	58	30° 7'
24	5.313	6	26° 34'	24	13° 29.8'	59	30° 32'
25	5.596	6-1/4	27° 31'	25	14° 2.2'	60	30° 58'
26	5.853	6-1/2	28° 27'	26	14° 34.5'	61	31° 23'
27	6.114	6-3/4	29° 22'	27	15° 6.6'	62	31° 48'
28	6.381	7	30° 16'	28	15° 38.5'	63	32° 13'
29	6.652	7-1/4	31° 8'	29	16° 10.3'	64	32° 37'
30	6.928	7-1/2	32°	30	16° 42.0'	65	33° 1'
31	7.210	7-3/4	32° 51'	31	17° 13.4'	66	33° 25'
32	7.498	8	33° 41'	32	17° 44.7'	67	33° 49'
33	7.793	8-1/4	34° 30'	33	18° 15.8'	68	34° 13'
34	8.094	8-1/2	35° 19'	34	18° 46.7'	69	34° 36'
35	8.403	8-3/4	36° 5'	35	19° 17.0'	70	35° 0'

Area Calculations

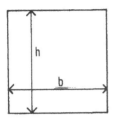

Square   Area = h x b

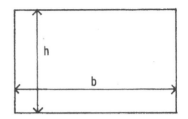

Rectangle   Area = h x b

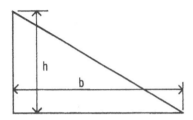

Triangle   Area = $\dfrac{h \times b}{2}$

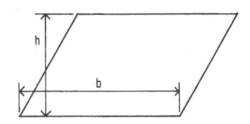

Parallelogram   Area = h x b

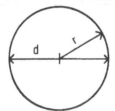

Circle   Area = $\dfrac{\pi d^2}{4}$ = πr^2

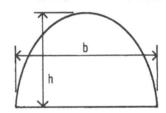

Parabola   Area = $\dfrac{2hb}{3}$

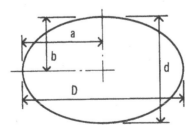

Ellipse   Area = 0.7854Dd

(continued on next page)

Area Calculations *(continued from previous page)*

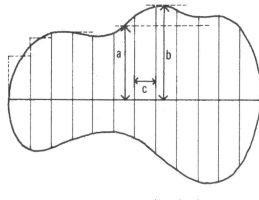

For each segment:

$$\text{Area} = \frac{(a + b)c}{2}$$

Irregular shape

Trapezoid area $= \dfrac{(a + b)h}{2}$

Trapezium area $= \dfrac{bd - ad - be + ac}{2}$

Volume Calculations

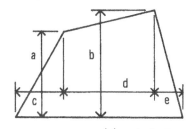

Sphere
Volume $= 0.5236\ D^3$
Surface area $= 4\pi r^2$ $(\pi = 3.1416)$

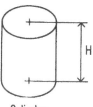

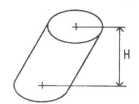

Cylinder   Slant top cylinder   Oblique cylinder

Volume = area of the base x height (H)

(continued on next page)

Volume Calculations *(continued from previous page)*

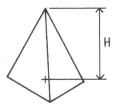

Pyramid

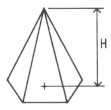

5 Sided pyramid

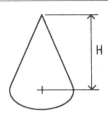

Cone

Volume = area of base x 1/3 height (H)

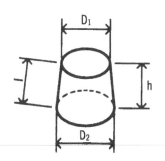

Frustum of a cone

Volume =

$$\frac{D_1 + D_2 + (\text{area of top} + \text{area of base}) \times h \times 0.7854}{3}$$

$$*\text{Surface area} = \frac{(D_1 + D_2)1}{2}$$

*Excludes top & base areas

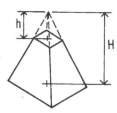

Truncated pyramid

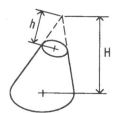

Truncated oblique cone

h = height of cut-off
H = height of whole

Volume = volume of the whole solid less volume of the portion cut off

How to Calculate Board Foot Measure

Board Feet

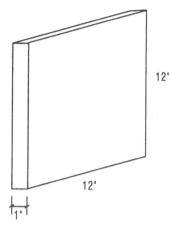

1 Board Foot = 1 foot wide × 1 foot high × 1 inch thick

A piece of lumber 2 × 6 × 12" long also = 1 B.F.

$$\frac{L \times W \times H}{12} = B.F.$$

Where L = Feet
W = Inches
H = Inches

Example:
For a 2 × 8 − 16 feet long

$$B.F. = \frac{16 \times 2 \times 8}{12} = 21.33 \ B.F.$$

How to Calculate Rafter Length

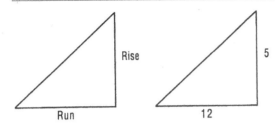

Ratio $\dfrac{Rise}{Run} = \dfrac{5}{12}$ or .42

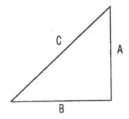

A = total rise
B = distance from ridge to outside of fascia
C = rafter length

If B = 12'- 7" (12.58'), Then A = $\dfrac{5}{12}$ × B

$$A = .42 \times 12.58 = 5.28'$$

Then C = $\sqrt{A^2 + B^2}$

$$C = \sqrt{12.58^2 + 5.28^2} = \sqrt{186.2} = 13.65'$$

To Calculate Roof Area

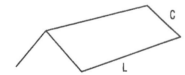

$$A = 2 (C \times L)$$

If our roof has L = 32.5'
Then   A = 2 (32.5 × 13.65)
A = 887.3 S.F.

Circumferences and Areas of Circles

Diameter In.	Circum. In	Area Sq. In.	Diameter In.	Circum. In.	Area Sq. In.
1/64	.04909	.00019	1	3.1416	.7854
1/32	.09818	.00077	1/16	3.3379	.8866
3/64	.14726	.00173	1/8	3.5343	.9940
1/16	.19635	.00307	3/16	3.7306	1.1075
5/64	.24544	.00479	1/4	3.9270	1.2272
3/32	.29452	.00690	5/16	4.1233	1.3530
7/64	.34361	.00940	3/8	4.3197	1.4849
			7/16	4.5160	1.6230
1/8	.39270	.01227			
9/64	.44179	.01553	1/2	4.7124	1.7671
5/32	.49087	.01917	9/16	4.9087	1.9175
11/64	.53996	.02320	5/8	5.1051	2.0739
3/16	.58905	.02761	11/16	5.3014	2.2365
13/64	.63814	.03241	3/4	5.4978	2.4053
7/32	.68722	.03758	13/16	5.6941	2.5802
15/64	.73631	.04314	7/8	5.8905	2.7612
			15/16	6.0868	2.9483
1/4	.78540	.04909			
17/64	.83449	.05542	2	6.2832	3.1416
9/32	.88357	.06213	1/16	6.4795	3.3410
19/64	.93266	.06922	1/8	6.6759	3.5466
5/16	.98175	.07670	3/16	6.8722	3.7583
21/64	1.03084	.08456	1/4	7.0686	3.9761
11/32	1.0799	.09281	5/16	7.2649	4.2000
23/64	1.1290	.10143	3/8	7.4613	4.4301
			7/16	7.6576	4.6664
3/8	1.1781	.11045			
25/64	1.2272	.11984	1/2	7.8540	4.9087
13/32	1.2763	.12962	9/16	8.0503	5.1572
27/64	1.3254	.13978	5/8	8.2467	5.4119
7/16	1.3744	.15033	11/16	8.4430	5.6727
29/64	1.4235	.16126	3/4	8.6394	5.9396
15/32	1.4726	.17257	13/16	6.8357	6.2126
31/64	1.5217	.18427	7/8	9.0321	5.4918
			15/16	9.2284	6.7771
1/2	1.5708	.19635			
17/32	1.6690	.22166	3	9.4248	7.0686
9/16	1.7671	.24850	1/16	9.6211	7.3662
19/32	1.8653	.27688	1/8	9.8175	7.6699
5/8	1.9635	.30680	3/16	10.014	7.9798
21/32	2.0617	.33824	1/4	10.210	8.2958
11/16	2.1598	.37122	5/16	10.407	8.6179
23/32	2.2580	.40574	3/8	10.603	8.9462
			7/16	10.799	9.2806
3/4	2.3562	.44179			
25/32	2.4544	.47937	1/2	10.996	9.6211
13/16	2.5525	.51849	9/16	11.192	9.9678
27/32	2.6507	.55914	5/8	11.388	10.321
7/8	2.7489	.60132	11/16	11.585	10.680
29/32	2.8471	.64504	3/4	11.781	11.045
15/16	2.9452	.69029	13/16	11.977	11.416
31/32	3.0434	.73708	7/8	12.174	11.793
			15/16	12.370	12.177

(continued on next page)

Circumferences and Areas of Circles *(continued from previous page)*

Diameter In.	Circum. In	Area Sq. In.	Diameter In.	Circum. In.	Area Sq. In.
4	12.566	12.566	**8**	25.133	50.265
1/16	12.763	12.962	1/8	25.525	51.849
1/8	12.959	13.364	1/4	25.918	53.456
3/16	13.155	13.772	3/8	26.311	55.088
1/4	13.352	14.186	1/2	26.704	56.745
5/16	13.548	14.607	5/8	27.096	58.426
3/8	13.744	15.033	3/4	27.489	60.132
7/16	13.941	15.466	7/8	27.882	61.862
1/2	14.137	15.904	**9**	28.274	63.617
9/16	14.334	15.349	1/8	28.667	65.397
5/8	14.530	16.800	1/4	29.060	67.201
11/16	14.726	17.257	3/8	29.452	69.029
3/4	14.923	17.721	1/2	29.845	70.882
13/16	15.119	18.190	5/8	30.238	72.760
7/8	15.315	18.665	3/4	30.631	74.662
15/16	15.512	19.147	7/8	31.023	76.589
5	15.708	19.635	**10**	31.416	78.540
1/16	15.904	20.129	1/8	31.809	80.516
1/8	16.101	20.629	1/4	32.201	82.516
3/16	16.297	21.135	3/8	32.594	84.541
1/4	16.493	21.648	1/2	32.987	86.590
5/16	16.690	22.166	5/8	33.379	88.664
3/8	16.886	22.691	3/4	33.772	90.763
7/16	17.082	23.221	7/8	34.165	92.886
1/2	17.279	23.758	**11**	34.558	95.033
9/16	17.475	24.301	1/8	34.950	97.205
5/8	17.671	24.850	1/4	35.343	99.402
11/16	16.868	25.406	3/8	35.736	101.62
3/4	18.064	25,867	1/2	36.128	103.87
13/16	18.261	26.535	5/8	36.521	106.14
7/8	18.457	27.109	3/4	36.914	108.43
15/16	18.653	27.688	7/8	37.306	110.75
6	18.850	28.274	**12**	37.699	113.10
1/8	19.242	29.465	1/8	38.092	115.47
1/4	19.635	30.680	1/4	38.485	117.86
3/8	20.028	31.919	3/8	38.877	120.28
1/2	20.420	33.183	1/2	39.270	122.72
5/8	20.813	34.472	5/8	59.663	125.19
3/4	21.206	35.785	3/4	40.055	127.68
7/8	21.598	37.122	7/8	40.448	130.19
7	21.991	38.485	**13**	40.841	132.73
1/8	22.384	39.871	1/8	41.233	135.30
1/4	22.776	41.282	1/4	41.626	137.89
3/8	23.169	42.718	3/8	42.019	140.50
1/2	23.562	44.179	1/2	42.412	143.14
5/8	23.955	45.664	5/8	42.804	145.80
3/4	24.347	47.173	3/4	43.197	148.49
7/8	24.740	48.707	7/8	43.590	151.20

(continued on next page)

Circumferences and Areas of Circles *(continued from previous page)*

Diameter In.	Circum. In	Area Sq. In.	Diameter In.	Circum. In.	Area Sq. In.
14	43.982	153.94	17-1/4	54.192	233.71
1/8	44.375	156.70	3/8	54.585	237.10
1/4	44.768	159.48	1/2	54.978	240.53
3/8	45.160	162.30	5/8	55.371	243.98
1/2	45.553	165.13	3/4	55.763	247.45
5/8	45.946	167.99	7/8	56.156	250.95
3/4	46.338	170.87	18	56.549	254.47
7/8	46.731	173.78	1/8	56.941	258.02
15	47.124	176.71	1/4	57.334	261.59
1/8	47.517	179.67	3/8	57.727	265.18
1/4	47.909	182.65	1/2	58.119	268.80
3/8	48.302	185.66	5/8	58.512	272.45
1/2	48.695	188.69	3/4	58.905	276.12
5/8	49.087	191.75	7/8	59.298	279.81
3/4	49.480	194.83	19	59.690	283.53
7/8	49.873	197.93	1/8	60.083	287.27
16	50.265	201.06	1/4	60.476	291.04
1/8	50.658	204.22	3/8	60.868	294.83
1/4	51.051	207.39	1/2	61.261	298.65
3/8	51.444	210.60	5/8	61.654	302.49
1/2	51.836	213.82	3/4	62.046	306.35
5/8	52.229	217.08	7/8	62.439	310.24
3/4	52.622	220.35	20	62.832	314.16
7/8	53.014	223.65	1/8	63.225	318.10
17	53.407	226.98	1/4	63.617	322.06
1/8	53.800	230.33	3/8	64.010	326.05

Weather Data and Design Conditions
(winter design @ 97.5% — summer design @ 2.5%)

City	Latitude (1) 0	1'	Winter Temperatures (1) Med. of Annual Extremes	99%	97-1/2%	Winter Degree Days (2)	Summer (Design Dry Bulb) Temperatures and Relative Humidity 1%	2-1/2%	5%
UNITED STATES									
Albuquerque, NM	35	0	6	12	16	4,400	96/61	94/61	92/61
Atlanta, GA	33	4	14	17	22	3,000	95/74	92/74	90/73
Baltimore, MD	39	2	12	14	17	4,600	94/75	92/75	89/74
Birmingham, AL	33	3	17	17	21	2,600	97/74	94/75	93/74
Bismark, ND	46	5	−31	−23	−19	8,800	95/68	91/68	88/67
Boise, ID	43	3	0	3	10	5,800	96/65	93/64	91/64
Boston, MA	42	2	−1	6	9	5,600	91/73	88/71	85/70
Burlington, VT	44	3	−18	−12	−7	8,200	88/72	85/70	83/69
Charleston, WV	38	2	1	7	11	4,400	92/74	90/73	88/72
Charlotte, NC	35	1	13	18	22	3,200	96/74	94/74	92/74
Casper, WY	42	5	−20	−11	−5	7,400	92/58	90/57	87/57
Chicago, IL	41	5	−5	−3	2	6,600	94/75	91/74	88/73
Cincinnati, OH	39	1	2	1	6	4,400	94/73	92/72	90/72
Cleveland, OH	41	2	−2	1	5	6,400	91/73	89/72	86/71
Columbia, SC	34	0	16	20	24	2,400	98/76	96/75	94/75
Dallas, TX	32	5	14	18	22	2,400	101/75	99/75	97/75
Denver, CO	39	5	−9	−5	1	6,200	92/59	90/59	89/59
Des Moines, IA	41	3	−13	−10	−5	6,600	95/75	92/74	89/73
Detroit, MI	42	2	0	3	6	6,200	92/73	88/72	85/71
Great Falls, MT	47	3	−29	−21	−15	7,800	91/60	88/60	85/59
Hartford, CT	41	5	−4	3	7	6,200	90/74	88/73	85/72
Houston, TX	29	5	24	28	33	1,400	96/77	94/77	92/77
Indianapolis, IN	39	4	−2	−2	2	5,600	93/74	91/74	88/73
Jackson, MS	32	2	17	21	25	2,200	98/76	96/76	94/76
Kansas City, MO	39	1	−2	2	6	4,800	100/75	97/74	94/74
Las Vegas, NV	36	1	18	25	28	2,800	108/66	106/65	104/65
Lexington, KY	38	0	0	3	8	4,600	94/73	92/72	90/72
Little Rock, AR	34	4	13	15	20	3,200	99/76	96/77	94/77
Los Angeles, CA	34	0	38	41	43	2,000	94/70	90/70	87/69
Memphis, TN	35	0	11	13	18	3,200	98/77	96/76	94/76
Miami, FL	25	5	39	44	47	200	92/77	90/77	89/77
Milwaukee, WI	43	0	−11	−8	−4	7,600	90/74	87/73	84/71
Minneapolis, MN	44	5	−19	−16	−12	8,400	92/75	89/73	86/71
New Orleans, LA	30	0	29	29	33	1,400	93/78	91/78	90/77
New York, NY	40	5	6	11	15	5,000	94/74	91/73	88/72
Norfolk, VA	36	5	18	20	22	3,400	94/77	91/76	89/76
Oklahoma City, OK	35	2	4	9	13	3,200	100/74	97/74	95/73
Omaha, NE	41	2	−12	−8	−3	6,600	97/76	94/75	91/74
Philadelphia, PA	39	5	7	10	14	4,400	93/75	90/74	87/72
Phoenix, AZ	33	3	25	31	34	1,800	108/71	106/71	104/71

(continued on next page)

Weather Data and Design Conditions
(winter design @ 97.5% — summer design @ 2.5%)

(continued from prevous page)

City	Latitude (1) 0	Latitude (1) 1'	Winter Temperatures (1) Med. of Annual Extremes	Winter Temperatures (1) 99%	Winter Temperatures (1) 97-1/2%	Winter Degree Days (2)	Summer (Design Dry Bulb) Temperatures and Relative Humidity 1%	Summer (Design Dry Bulb) Temperatures and Relative Humidity 2-1/2%	Summer (Design Dry Bulb) Temperatures and Relative Humidity 5%
UNITED STATES									
Pittsburgh, PA	40	3	1	3	7	6,000	90/72	88/71	85/70
Portland, ME	43	4	−14	−6	−1	7,600	88/72	85/71	81/69
Portland, OR	45	4	17	17	23	4,600	89/68	85/67	81/65
Portsmouth, NH	43	1	−8	−2	2	7,200	88/73	86/71	83/70
Providence, RI	41	4	0	5	9	6,000	89/73	86/72	83/70
Rochester, NY	43	1	−5	1	5	6,800	91/73	88/71	85/70
Salt Lake City, UT	40	5	−2	3	8	6,000	97/62	94/62	92/61
San Francisco, CA	37	5	38	38	40	3,000	80/63	77/62	83/61
Seattle, WA	47	4	22	22	27	5,200	81/68	79/66	76/65
Sioux Falls, SD	43	4	−21	−15	−11	7,800	95/73	92/72	89/71
St. Louis, MO	38	4	1	3	8	5,000	96/75	94/75	92/74
Tampa, FL	28	0	32	36	40	680	92/77	91/77	90/76
Trenton, NJ	40	1	7	11	14	5,000	92/75	90/74	87/73
Washington, DC	38	5	12	14	17	4,200	94/75	92/74	90/74
Wichita, KS	37	4	−1	3	7	4,600	102/72	99/73	96/73
Wilmington, DE	39	4	6	10	14	5,000	93/74	93/74	20/73
ALASKA									
Anchorage	61	1	−29	−23	−18	10,800	73/59	70/58	67/56
Fairbanks	64	5	−59	−51	−47	14,280	82/62	78/60	75/59
CANADA									
Edmonton, Alta.	53	3	−30	−29	−25	11,000	86/66	83/65	80/63
Halifax, N.S.	44	4	−4	1	5	8,000	83/66	80/65	77/64
Montreal, Que.	45	3	−20	−16	−10	9,000	88/73	86/72	84/71
Saskatoon, Sask.	52	1	−35	−35	−31	11,000	90/68	86/66	83/65
St. Johns, Nwf.	47	4	1	3	7	8,600	79/66	77/65	75/64
Saint John, N.B.	45	2	−15	−12	−8	8,200	81/67	79/65	77/64
Toronto, Ont.	43	4	−10	−5	−1	7,000	90/73	87/72	85/71
Vancouver, B.C.	49	1	13	15	19	6,000	80/67	78/66	76/65
Winnipeg, Man.	49	5	−31	−30	−27	10,800	90/73	87/71	84/70

Ground Snow Loads, p$_g$, for the United States (psf)

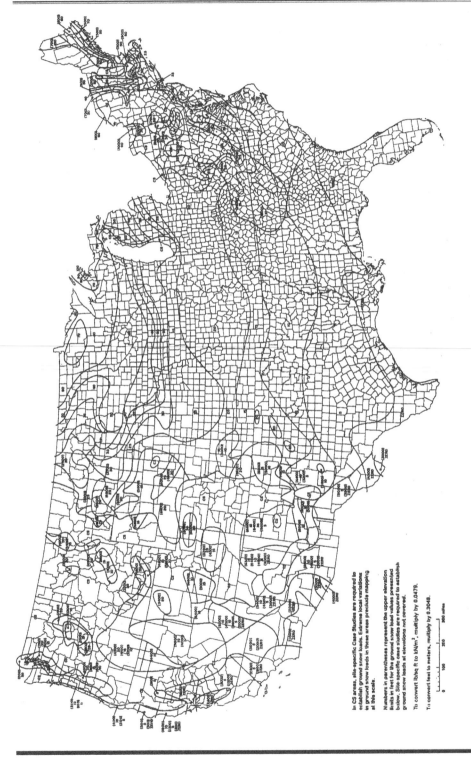

In CS areas, site-specific Case Studies are required to establish ground snow loads. Extreme local variations in ground snow loads in these areas preclude mapping at this scale.

Numbers in parentheses represent the upper elevation limits in feet for the ground snow load values presented below. Site-specific case studies are required to establish ground snow loads at elevations not covered.

To convert lb/sq ft to kN/m², multiply by 0.0479.

To convert feet to meters, multiply by 0.3048.

0   100   250   500 miles

Frost Penetration

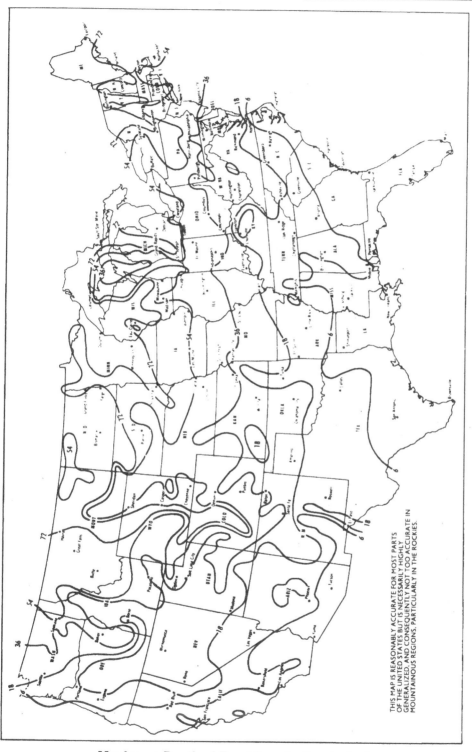

THIS MAP IS REASONABLY ACCURATE FOR MOST PARTS OF THE UNITED STATES BUT IS NECESSARILY HIGHLY GENERALIZED, AND CONSEQUENTLY NOT TOO ACCURATE IN MOUNTAINOUS REGIONS, PARTICULARLY IN THE ROCKIES.

Maximum Depth of Frost Penetration in Inches

Wind Conversion Table

Wind Speed Units:

1 mile per hour = 0.868391 knot = 1.609344 km/hr.
= 1.4667 ft./sec = 88 ft./min.
= 0.44704 m/sec. = 0.34754° at lat./day

Miles per Hour	Knots	Meters per Second	Feet per Second	Kilo-meters per Hour	Feet per Minute	Miles per Hour	Knots	Meters per Second	Feet per Second	Kilo-meters per Hour	Feet per Minute
1	0.9	0.4	1.5	1.6	88	51	44.3	22.8	74.8	82.1	4488
2	1.7	0.9	2.9	3.2	176	52	45.2	23.2	76.3	83.7	4576
3	2.6	1.3	4.4	4.8	264	53	46.0	23.7	77.7	85.3	4664
4	3.5	1.8	5.9	6.4	352	54	46.9	24.1	79.2	86.9	4752
5	4.3	2.2	7.3	8.0	440	55	47.8	24.6	80.7	88.5	4840
6	5.2	2.7	8.8	9.7	528	56	48.6	25.0	82.1	90.1	4928
7	6.1	3.1	10.3	11.3	616	57	49.5	25.5	83.6	91.7	5016
8	6.9	3.6	11.7	12.9	704	58	50.4	25.9	85.1	93.3	5104
9	7.8	4.0	13.2	14.5	792	59	51.2	26.4	86.5	95.0	5192
10	8.7	4.5	14.7	16.1	880	60	52.1	26.8	88.0	96.6	5280
11	9.6	4.9	16.1	17.7	968	61	53.0	27.3	89.5	98.2	5368
12	10.4	5.4	17.6	19.3	1056	62	53.8	27.7	90.0	99.8	5456
13	11.3	5.8	19.1	20.9	1144	63	54.7	28.2	92.4	101.4	5544
14	12.2	6.3	20.5	22.5	1232	64	55.6	28.6	93.9	103.0	5632
15	13.0	6.7	22.0	24.1	1320	65	56.4	29.1	95.3	104.6	5720
16	13.9	7.2	23.5	25.7	1408	66	57.3	29.5	96.8	106.2	5808
17	14.8	7.6	24.9	27.4	1496	67	58.2	30.0	98.3	107.8	5896
18	15.6	8.0	26.4	29.0	1584	68	59.1	30.4	99.7	109.4	5984
19	16.5	8.5	27.9	30.6	1672	69	59.9	30.8	101.2	111.0	6072
20	17.4	8.9	29.3	32.2	1760	70	60.8	31.3	102.7	112.7	6160
21	18.2	9.4	30.8	33.8	1848	71	61.7	31.7	104.1	114.3	6248
22	19.1	9.8	32.3	35.4	1936	72	62.5	32.2	105.6	115.9	6336
23	20.0	10.3	33.7	37.0	2024	73	63.4	32.6	107.1	117.5	6424
24	20.8	10.7	35.2	38.6	2112	74	64.3	33.1	108.5	119.1	6512
25	21.7	11.2	36.7	40.2	2200	75	65.1	33.5	110.0	120.7	6600
26	22.6	11.6	38.1	41.8	2288	76	66.0	34.0	111.5	122.3	6688
27	23.4	12.1	39.6	43.5	2376	77	66.9	34.4	112.9	123.9	6776
28	24.3	12.5	41.1	45.1	2464	78	67.7	34.9	114.4	125.5	6864
29	25.2	13.0	42.5	46.7	2552	79	68.6	35.3	115.9	127.1	6952
30	26.1	13.4	44.0	48.3	2640	80	69.5	35.8	117.3	128.7	7040
31	26.9	13.9	45.5	49.9	2728	81	70.3	36.2	118.8	130.4	7128
32	27.8	14.3	46.9	51.5	2816	82	71.2	36.7	120.3	132.0	7216
33	28.7	14.8	48.4	53.1	29.04	83	72.1	37.1	121.7	133.6	7304
34	29.5	15.2	49.9	54.7	2992	84	72.9	37.6	123.2	135.2	7392
35	30.4	15.6	51.3	56.3	3080	85	73.8	38.0	124.7	136.8	7480
36	31.3	16.1	52.8	57.9	3168	86	74.7	38.4	126.1	138.4	7568
37	32.1	16.5	54.3	59.5	3256	87	75.5	38.9	127.6	140.0	7656
38	33.0	17.0	55.7	61.2	3344	88	76.4	39.3	129.1	141.6	7744
39	33.9	17.4	57.2	62.8	3432	89	77.3	39.8	130.5	143.2	7832
40	34.7	17.9	58.7	64.4	3520	90	78.2	40.2	132.0	144.8	7920
41	35.6	18.3	60.1	66.0	3608	91	79.0	40.7	133.5	146.5	8008
42	36.5	18.8	61.6	67.6	3696	92	79.9	41.1	134.9	148.1	8096
43	37.3	19.2	63.1	69.2	3784	93	80.8	41.6	136.4	149.7	8184
44	38.2	19.7	64.5	70.8	3872	94	81.6	42.0	137.9	151.3	8272
45	39.1	20.1	66.0	72.4	3960	95	82.5	42.5	139.3	152.9	8360
46	39.9	20.6	67.5	74.0	4048	96	83.4	42.9	140.8	154.4	8448
47	40.8	21.0	68.9	75.6	4136	97	84.2	43.4	142.3	156.1	8536
48	41.7	21.5	70.4	77.2	4224	98	85.1	43.8	143.7	157.7	8624
49	42.6	21.9	71.9	78.9	4312	99	86.0	44.3	145.2	159.3	8712
50	43.4	22.4	73.3	80.5	4400	100	86.8	44.7	146.7	160.9	8800

Wind Chill Factors

Wind Speed M.P.H	Actual Thermometer Reading (°F)										
	50	40	30	20	10	0	–10	–20	–30	–40	–50
	Wind Chill Temperature (°F)										
0	50	40	30	20	10	0	–10	–20	–30	–40	–50
5	48	37	27	16	6	–5	–15	–26	–36	–47	–57
10	40	29	16	–4	–9	–21	–33	–46	–58	–70	–83
15	36	22	9	–5	–18	–36	–46	–58	–70	–85	–99
20	32	18	4	–10	–25	–39	–53	–67	–82	–96	–110
25	30	16	0	–15	–29	–44	–59	–74	–88	–104	–113
30	28	13	–2	–18	–33	–48	–63	–79	–94	–109	–123
35	27	11	–4	–20	–35	–49	–67	–82	–98	–113	–129
40	26	10	–6	–21	–37	–53	–69	–85	–100	–116	–132
	Very Cold				Bitter Cold			Extreme Cold			

Miscellaneous Temperature and Weather Data

Temperature

Freezing point of water = 32° Fahrenheit
                      = 0° Celsius

Boiling point of water (at normal air pressure) = 212° Fahrenheit
                                                = 100° Celsius

1 degree Fahrenheit = 0.5556 degree (Celsius)
1 degree Celsius = 1.8 degrees Fahrenheit

Ice and Snow

1 cubic foot of ice at 32°F weighs 57.50 pounds; 1 pound of ice at 32°F has a volume of 0.0174 cubic foot = 30.067 cubic inches.

1 cubic foot of fresh snow, according to humidity of atmosphere, weighs 5 pounds to 12 pounds. 1 cubic foot of snow moistened and compacted by rain weighs 15 pounds to 50 pounds.

Calculating the Time Value of Money

F = Future amount; the amount of money you will end up with or want to end up with at the end of the time period specified.

P = Present value; the amount of money you have or need to have at the beginning of the time period specified.

n = Number of time periods being calculated (coordinate with i).

i = Interest rate per time period. (Note: if n, the number of time periods, is in days, weeks, months or years, i must also be in that time period.)

Formulas for Compound Interest

A) $F = P(1 + i)^n$

Use: To find the value of funds at the end of the number of time periods (n) if the interest rate is i per time period. A time period is the time the interest rate is compounded. For example, if you wish to know how much you will have in an account at the end of five years and the bank compounds their rate monthly at 5% interest, then F is what you are looking for, n would equal 5 years x 4 quarters per year = 20, and i = 5% per year divided by 4 quarters per year = 1.25.

B) $P = F \left[\dfrac{1}{(1 + i)^n} \right] = F(1 + i)^{-n}$

Use: To find the amount needed today (present value) to end up with the desired amount F (future value) at the end of time period n, at the interest rate of i per time period.

C) Annuity Compound Amount

$$F = A \left[\frac{(1 + i)^n - 1}{i} \right]$$

Use: To find the future value of an annuity if amount A is added (or paid) to an account at regular intervals for the time period of n intervals, at interest rate of i per interval. This formula can also be used to determine what the total of all payments (principle and interest) of a loan will be.

D) Sinking Fund

$$A = F \left[\frac{i}{(1 + i)^n - 1} \right]$$

Use: To find the amount A needed to add to an account at regular intervals, to end up with a future amount F after a time period of n intervals at an interest rate of i per interval.

E) Capital Recovery (Loan Repayment)

$$A = P \left[\frac{i (1 + i)^n}{(1 + i)^n - 1} \right] = P \left[\frac{i}{1 - (1 + i)^{-n}} \right]$$

Use: To find what payments are necessary per time interval to pay back a loan R over the number of time intervals n at an interest rate of i per time interval. For example, if you wanted to find your *monthly* payment for a loan of $100,000 spread over 30 years at a fixed rate of 10% per year, then P = $100,000, n = 30 years x 12 months per year = 360, and i = 10%/year/12 months/year = 0.833.

F) Present Value of a Loan

$$P = A \left[\frac{((1 + i)^n) - 1}{(i(1 + i)^n)} \right] = A \left[\frac{1 - (1 + i)^{-n}}{i} \right]$$

Use: To find the present value of payments, if you are making payments A over a number of time intervals n at the interest rate i per time interval

LEED for New Construction and Major Renovation 2009
Project Scorecard

Project Name:
Project Address:

	Yes	?	No			
				Sustainable Sites	**26**	**Points**
	Y			Prereq 1 **Construction Activity Pollution Prevention**	Required
				Credit 1 **Site Selection**	1
				Credit 2 **Development Density & Community Connectivity**	5
				Credit 3 **Brownfield Redevelopment**	1
				Credit 4.1 **Alternative Transportation,** Public Transportation Access	6
				Credit 4.2 **Alternative Transportation,** Bicycle Storage & Changing Rooms	1
				Credit 4.3 **Alternative Transportation,** Low-Emitting & Fuel-Efficient Vehicles	3
				Credit 4.4 **Alternative Transportation,** Parking Capacity	2
				Credit 5.1 **Site Development,** Protect or Restore Habitat	1
				Credit 5.2 **Site Development,** Maximize Open Space	1
				Credit 6.1 **Stormwater Design,** Quantity Control	1
				Credit 6.2 **Stormwater Design,** Quality Control	1
				Credit 7.1 **Heat Island Effect,** Non-Roof	1
				Credit 7.2 **Heat Island Effect,** Roof	1
				Credit 8 **Light Pollution Reduction**	1

	Yes	?	No			
				Water Efficiency	**10**	**Points**
	Y			Prereq 1 **Water Use Reduction,** 20% Reduction	Required
				Credit 1.1 **Water Efficient Landscaping,** Reduce by 50%	2
				Credit 1.2 **Water Efficient Landscaping,** No Potable Use or No Irrigation	2
				Credit 2 **Innovative Wastewater Technologies**	2
				Credit 3 **Water Use Reduction**	2 to 4
				30% Reduction	2
				35% Reduction	3
				40% Reduction	4

	Yes	?	No			
				Energy & Atmosphere	**35**	**Points**
	Y			Prereq 1 **Fundamental Commissioning of the Building Energy Systems**	Required
	Y			Prereq 2 **Minimum Energy Performance:** 10% New Bldgs or 5% Existing Bldg Renovations	Required
	Y			Prereq 3 **Fundamental Refrigerant Management**	Required
				Credit 1 **Optimize Energy Performance**	1 to 19
				12% New Buildings or 8% Existing Building Renovations	1
				14% New Buildings or 10% Existing Building Renovations	2
				16% New Buildings or 12% Existing Building Renovations	3
				18% New Buildings or 14% Existing Building Renovations	4
				20% New Buildings or 16% Existing Building Renovations	5
				22% New Buildings or 18% Existing Building Renovations	6
				24% New Buildings or 20% Existing Building Renovations	7
				26% New Buildings or 22% Existing Building Renovations	8
				28% New Buildings or 24% Existing Building Renovations	9
				30% New Buildings or 26% Existing Building Renovations	10
				32% New Buildings or 28% Existing Building Renovations	11
				34% New Buildings or 30% Existing Building Renovations	12
				36% New Buildings or 32% Existing Building Renovations	13
				38% New Buildings or 34% Existing Building Renovations	14
				40% New Buildings or 36% Existing Building Renovations	15
				42% New Buildings or 38% Existing Building Renovations	16
				44% New Buildings or 40% Existing Building Renovations	17
				46% New Buildings or 42% Existing Building Renovations	18
				48% New Buildings or 44% Existing Building Renovations	19
				Credit 2 **On-Site Renewable Energy**	1 to 7
				1% Renewable Energy	1
				3% Renewable Energy	2
				5% Renewable Energy	3
				7% Renewable Energy	4
				9% Renewable Energy	5
				11% Renewable Energy	6
				13% Renewable Energy	7
				Credit 3 **Enhanced Commissioning**	2
				Credit 4 **Enhanced Refrigerant Management**	2
				Credit 5 **Measurement & Verification**	3
				Credit 6 **Green Power**	2

(continued on next page)

(continued from previous page)

LEED for New Construction and Major Renovation 2009
Project Scorecard

Project Name:
Project Address:

Yes	?	No	**Materials & Resources**	14 Points

Y			Prereq 1 **Storage & Collection of Recyclables**	Required
			Credit 1 **Building Reuse**	1 to 3
			Credit 1.1 ____ Maintain 55% of Existing Walls, Floors & Roof	1
			Credit 1.2 ____ Maintain 75% of Existing Walls, Floors & Roof	2
			Credit 1.3 ____ Maintain 95% of Existing Walls, Floors & Roof	3
			Credit 1.4 **Building Reuse**, Maintain 50% of Interior Non-Structural Elements	1
			Credit 2.1 **Construction Waste Management**, Divert 50% from Disposal	1
			Credit 2.2 **Construction Waste Management**, Divert 75% from Disposal	1
			Credit 3.1 **Materials Reuse**, 5%	1
			Credit 3.2 **Materials Reuse**, 10%	1
			Credit 4.1 **Recycled Content**, 10% (post-consumer + ½ pre-consumer)	1
			Credit 4.2 **Recycled Content**, 20% (post-consumer + ½ pre-consumer)	1
			Credit 5.1 **Regional Materials**, 10% Extracted, Processed & Manufactured Regionally	1
			Credit 5.2 **Regional Materials**, 20% Extracted, Processed & Manufactured Regionally	1
			Credit 6 **Rapidly Renewable Materials**	1
			Credit 7 **Certified Wood**	1

Yes	?	No	**Indoor Environmental Quality**	15 Points

Y			Prereq 1 **Minimum IAQ Performance**	Required
Y			Prereq 2 **Environmental Tobacco Smoke (ETS) Control**	Required
			Credit 1 **Outdoor Air Delivery Monitoring**	1
			Credit 2 **Increased Ventilation**	1
			Credit 3.1 **Construction IAQ Management Plan**, During Construction	1
			Credit 3.2 **Construction IAQ Management Plan**, Before Occupancy	1
			Credit 4.1 **Low-Emitting Materials**, Adhesives & Sealants	1
			Credit 4.2 **Low-Emitting Materials**, Paints & Coatings	1
			Credit 4.3 **Low-Emitting Materials**, Flooring Systems	1
			Credit 4.4 **Low-Emitting Materials**, Composite Wood & Agrifiber Products	1
			Credit 5 **Indoor Chemical & Pollutant Source Control**	1
			Credit 6.1 **Controllability of Systems**, Lighting	1
			Credit 6.2 **Controllability of Systems**, Thermal Comfort	1
			Credit 7.1 **Thermal Comfort**, Design	1
			Credit 7.2 **Thermal Comfort**, Verification	1
			Credit 8.1 **Daylight & Views**, Daylight 75% of Spaces	1
			Credit 8.2 **Daylight & Views**, Views for 90% of Spaces	1

Yes	?	No	**Innovation & Design Process**	6 Points

			Credit 1.1 **Innovation in Design**: Provide Specific Title	1
			Credit 1.2 **Innovation in Design**: Provide Specific Title	1
			Credit 1.3 **Innovation in Design**: Provide Specific Title	1
			Credit 1.4 **Innovation in Design**: Provide Specific Title	1
			Credit 1.5 **Innovation in Design**: Provide Specific Title	1
			Credit 2 **LEED® Accredited Professional**	1

Yes	?	No	**Regional Priority Credits**	4 Points

			Credit 1.1 **Regional Priority Credit**: Region Defined	1
			Credit 1.2 **Regional Priority Credit**: Region Defined	1
			Credit 1.3 **Regional Priority Credit**: Region Defined	1
			Credit 1.4 **Regional Priority Credit**: Region Defined	1

Yes	?	No	**Project Totals (Certification Estimates)**	110 Points

Not Certified   **Certified:** 40-49 points **Silver:** 50-59 points **Gold:** 60-79 points **Platinum:** 80+ points

Professional Associations

AA   Aluminum Association
1525 Wilson Blvd. Suite 600
Arlington, VA 22209
703-358-2967
www.aluminum.org

AACE   American Association of Cost Engineers
International
209 Prairie Avenue, Suite 100
Morgantown, WV 26501
800-858-2678
www.aacei.org

ABC   Associated Builders & Contractors, Inc.
4250 N. Fairfax Dr., 9th Floor
Arlington, VA 22203
703-812-2000
www.abc.org

ACEC   American Council of Engineering Companies
1015 15th Street, N.W.
Washington, DC 20005
202-347-7474
www.acec.org

ACI   American Concrete Institute
38800 Country Club Drive
Farmington Hills, MI 48331
248-848-3700
www.concrete.org

ACLA   Associated Landscape Contractors of America
950 Herndon Parkway, Suite 450
Herndon, VA 20170
703-736-9666
www.landcarenetwork.com

AED   Associated Equipment Distributors
615 W. 22nd Street
Oak Brook, IL 60523
603-574-0650
www.aednet.org

AEM   Association of Equipment Manufacturers
6737 W. Washington St., Suite 2400
Milwaukee, WI 53214
414-272-0943
www.aem.org

AGC   Associated General Contractors of America
            2300 Wilson Blvd., Suite 400
            Alexandria, VA 22201
            703-548-3118
            www.agc.org

AI   Asphalt Institute
            2696 Research Park Drive
            Lexington, KY 40511
            859-288-4960
            www.asphaltinstitute.org

AIA   American Institute of Architects
            1735 New York Avenue, N.W.
            Washington, DC 20006
            800-242-3837
            www.aia.org

AIC   American Institute of Constructors
            P.O. Box 26334
            Alexandria, VA 22314
            703-683-4999
            www.aicnet.org

AISI   American Iron and Steel Institute
            1140 Connecticut Avenue, Suite 705
            Washington, DC 20036
            202-452-7100
            www.steel.org

AISC   American Institute of Steel Construction
            One E. Wacker Drive, Suite 700
            Chicago, IL 60601
            312-670-2400
            www.aisc.org

AITC   American Institute of Timber Construction
            7012 S. Revere Parkway, Suite 140
            Centennial, CO 80112
            303-792-9559
            www.aitc-glulam.org

AMCA   Air Movement and Control Association
            30 West University Drive
            Arlington Heights, IL 60004
            847-394-0150
            www.amca.org

ANLA   American Nursery & Landscape Association
1000 Vermont Avenue, N.W., Suite 300
Washington, DC 20005
202-789-2900
www.anla.org

ANSI   American National Standards Institute
25 West 43rd Street, 4th floor
New York, NY 10036
212-642-4900
www.ansi.org

APA   The Engineered Wood Association
7001 S. 19th Street
Tacoma, WA 98466
253-565-6600
www.apawood.org

ARI   Air Conditioning, Heating and Refrigeration Institute
2111 Wilson Blvd., Suite 500
Arlington, VA 22201
703-524-8800
www.ari.org

ASA   American Subcontractors Association
1004 Duke Street
Alexandria, VA 22314
703-684-3450
www.asaonline.com

ASCE   American Society of Civil Engineers
1801 Alexander Bell Drive
Reston, VA 20191
800-548-2723
www.asce.org

ASHRAE   American Society of Heating,
Refrigerating and Air Conditioning Engineers
1791 Tullie Circle, N.E.
Atlanta, GA 30329
404-636-8400
www.ashrae.com

ASLA   American Society of Landscape Architects
636 Eye Street, N.W.
Washington, DC 20001
202-898-2444
www.asla.org

ASME   American Society of Mechanical Engineers
             Three Park Avenue
             New York, NY 10016
             800-843-2763
             www.asme.org

ASPE   American Society of Professional Estimators
             2525 Perimeter Place Dr., Suite 103
             Nashville, TN 37214
             615-316-9200
             www.aspenational.com

ASTM   American Society for Testing and Materials
             100 Barr Harbor Drive
             W. Conshohocken, PA 19428
             610-832-9500
             www.astm.org

AWS   American Welding Society
             550 N.W. LeJeune Road
             Miami, FL 33126
             800-443-9353
             www.aws.org

AWPA   American Wood Protection Association
             P.O. Box 361784
             Birmingham, AL 35236
             205-733-4077
             www.awpa.com

AWI   Architectural Woodwork Institute
             46179 Westlake Dr., Suite 120
             Potomac Falls, VA 20165
             571-323-3636
             www.awinet.org

AWC   American Wood Council
             1111 Nineteenth Street, S.W., Suite 800
             Washington, DC 20036
             202-463-2766
             www.awc.org

AWWA   American Water Works Association
             6666 West Quincy Avenue
             Denver, CO 80235
             303-794-7711
             www.awwa.org

BHMA   Builders Hardware Manufacturers Association
              355 Lexington Avenue, 15th floor
              New York, NY 10017
              212-297-2122
              www.buildershardware.com

BIA   Brick Industry Association
              1850 Centennial Park Dr., Suite 301
              Reston, VA 20191
              703-620-0010
              www.gobrick.org

CDA   Copper Development Association
              260 Madison Avenue
              New York, NY 10016
              212-251-7200
              www.copper.org

CFMA   Construction Financial Management Association
              29 Emmons Drive, Suite F-50
              Princeton, NJ 08540
              609-452-8000
              www.cfma.org

CMAA   Construction Management Association of America
              7926 Jones Branch Drive, Suite 800
              McLean, VA 22102
              703-356-2622
              www.cmaanet.org

CRSI   Concrete Reinforcing Steel Institute
              933 North Plum Grove Road
              Schaumburg, IL 60173
              847-517-1200
              www.crsi.org

CSI   Construction Specifications Institute
              99 Canal Center Plaza, Suite 300
              Alexandria, VA 22314
              800-689-2900
              www.csinet.org

DHI   Door & Hardware Institute
              14150 Newbrook Drive, Suite 200
              Chantilly, VA 20151
              703-222-2010
              www.dhi.org

FMR   FM Global
1301 Atwood Ave.
Johnston, RI 02919
401-275-3029
www.fmglobal.com

FPS   Forest Products Society
2801 Marshall Court
Madison, WI 53705-2295
608-231-1361
www.forestprod.org

GA   Gypsum Association
6525 Belcrest Rd., Suite 480
Hyattsville, MD 20782
301-277-8686
www.gypsum.org

GANA   The Glass Association of North America
2945 S.W. Wanamaker Drive, Suite A
Topeka, KS 66614
785-271-0208
www.glasswebsite.com

GSA   General Services Administration
Specifications and Consumer Information
Attn: Office of Public Affairs
1800 F. Street, N.W.
Washington, DC 20405
www.gsa.gov

IEEE   Institute of Electrical and Electronic Engineers
3 Park Avenue, 17th floor
New York, NY 10016
212-419-7900
www.ieee.org

IMI   International Masonry Institute
The James Brice House
42 East Street
Annapolis, MD 21401
800-803-0295
www.imiweb.org

MCAA   Mechanical Contractors Association of America, Inc.
1385 Piccard Drive
Rockville, MD 20850
301-869-5800
www.mcaa.org

NAAMM National Association of Architectural Metal Manufacturers
800 Roosevelt Rd.
Bldg. C, Ste. 312
Glen Ellyn, FL 60137
630-942-6591
www.naamm.org

NAHB National Association of Home Builders
1201 15th Street, N.W.
Washington, DC 20005
800-368-5242
www.nahb.org

NECA National Electrical Contractors Association
3 Bethesda Metro Center, Suite 1100
Bethesda, MD 20814
301-657-3110
www.necanet.org

NEMA National Electrical Manufacturers' Association
1300 N. 17th Street, Suite 1752
Rosslyn, VA 22209
703-841-3200
www.nema.org

NFPA National Fire Protection Association
1 Batterymarch Park
Quincy, MA 02169
617-770-3000
www.nfpa.org

NIBS National Institute of Building Science
1090 Vermont Avenue, N.W., Suite 700
Washington, DC 20005
202-289-7800
www.nibs.org

NPCA National Precast Concrete Association
10333 North Meridian Street, Suite 272
Indianapolis, IN 46290
800-366-7731
www.precast.org

NRCA National Roofing Contractors Association
10255 W. Higgins Road, Suite 600
Rosemont, IL 60018
847-299-9070
www.nrca.net

NSPE   National Society of Professional Engineers
              1420 King Street
              Alexandria, VA 22314
              703-684-2800
              www.nspe.org

NSWMA   National Solid Wastes Management Association
              Environmental Industry Associations
              4301 Connecticut Avenue, N.W., Suite 300
              Washington, DC 20008
              800-424-2869
              www.nswma.org

NUCA   National Utility Contractors Association
              4301 North Fairfax Drive, Suite 360
              Arlington, VA 22203
              703-358-9300
              www.nuca.com

PCA   Portland Cement Association
              5420 Old Orchard Road
              Skokie, IL 60077
              847-966-6200
              www.cement.org

PCEAA   Professional Construction Estimators Association
              P.O. Box 680336
              Charlotte, NC 28216
              704-489-1494
              www.pcea.org

PHCC   Plumbing-Heating-Cooling Contractors Association
              180 S. Washington Street
              P.O. Box 6808
              Falls Church, VA 22046
              703-237-8100
              www.phccweb.org

PS   Product Standard
              US Department of Commerce
              Technical Standards Activities Group
              NIST, 100 Bureau Drive, MS 2100
              Gaithersburg, MD 20889
              301-975-4000
              www.nist.gov

SDI   Steel Door Institute
        30200 Detroit Road
        Westlake, OH 44145
        440-899-0010
        www.steeldoor.org

SIGMA   Sealed Insulated Glass Manufacturers Association
        401 N. Michigan Ave., Suite 2400
        Chicago, IL 60611
        312-644-6610
        www.arcat.com

SMACNA Sheet Metal and Air Conditioning Contractors' National
        Association
        4201 Lafayette Center Drive
        Chantilly, VA 20151
        703-803-2980
        www.smacna.org

SSMA   Steel Stud Manufacturers' Association
        800 Roosevelt Rd.
        Bldg. C, Ste. 312
        Glen Ellyn, IL 60137
        603-942-6592
        www.ssma.com

SSPC   Steel Structures Painting Council
        40 24th Street, 6th Floor
        Pittsburgh, PA 15224
        412-281-2331
        www.sspc.org

TCA   Tile Council of America, Inc.
        100 Clemson Research Boulevard
        Anderson, SC 29625
        864-646-8453
        www.tileusa.com

UL   Underwriters' Laboratories, Inc.
        2600 N.W. Lake Road
        Camas, WA 98607
        877-854-3577
        www.ul.com

USFS   USDA Forest Service
Forest Products Laboratory
One Gifford Pinchot Drive
Madison, WI 53726
608-231-9200
www.fpl.fs.fed.us

WCLIB   West Coast Lumber Inspection Bureau
P.O. Box 23145
Tigard, OR 97281
503-639-0651
www.wclib.org

WWPA   Western Wood Products Association
522 S.W. Fifth Ave., Suite 500
Portland, OR 97204-2122
503-224-3930
www.wwpa.org

Abbreviations

(ANSI) = abbreviation by the American National Standards Institute

#	Pound; Number	AMM	Ammeter (ANSI)
%	Percent	AMP HR	Ampere hour (ANSI)
@	At	AMP	Ampere (ANSI)
~	Approximately	Amp.	Ampere
<	Less than	AMPL	Amplifier (ANSI)
>	Greater than	AMT	Amount (ANSI)
"	Delta	ANLG	Analog (ANSI)
Σ	Sum	ANN	Annunciator (ANSI)
Ø	Phase	Anod.	Anodized
3/C	Conductors, number of (ANSI)	ANSI	American National Standards Institute
A hr.	Ampere-hour	ANT	Antenna (ANSI)
A	Areas Square Feet/Ampere	AP	Access panel (ANSI); Acid proof (ANSI)
A.C.	Air conditioning; Asbestos cement	APPAR	Apparatus (ANSI)
A.G.A.	American Gas Association	Approx.	Approximate
A.H.	Ampere hours	APPX	Appendix (ANSI)
A.H.U.	Air handling unit	APT	Apartment (ANSI)
A.I.A.	American Institute of Architects	APU	Auxiliary power unit (ANSI)
A.S.B.C.	American Standard Building Code	ARCH	Architect (ANSI)
A.S.H.R.A.E.	American Society of Heating, Refrigeration & Air Conditioning Engineers	ARM SHT	Armature shunt (ANSI)
		Asb.	Asbestos
		Asbe.	Asbestos worker
A.S.M.E.	American Society of Mechanical Engineers	ASPH	Asphalt (ANSI)
A.S.T.M.	American Society for Testing and Materials	ASPRTR	Aspirator (ANSI)
		ASSN	Association (ANSI)
A.W.G.	American Wire Gauge	ASSY	Assembly (ANSI)
AB	Air blast (ANSI); Anchor bolt (ANSI)	ASW	Auxiliary switch (ANSI)
ABCD	Air blast circuit breaker (ANSI)	ASYM	Asymmetrical (ANSI)
ABRSV	Abrasive (ANSI)	AT	Air tight (ANSI)
ABS	Absolute acrylonitrile butadiene (ANSI); Acrylonitrile butadiene styrene; Asbestos bonded steel	ATCH	Attachment (ANSI)
		ATM	Atmosphere (ANSI)
		Attchmt.	Attachment
ABSW	Air break switch (ANSI)	AVE	Avenue (ANSI)
ABT	Air blast transformer (ANSI)	Avg.	Average
ABV	Above (ANSI)	AW	Acid waste (ANSI)
AC	Alternating current (ANSI)	AWG	American wire gage (ANSI)
ACB	Air circuit breaker (ANSI)	AZ	Azimuth (ANSI)
ACET	Acetylene (ANSI)	B & B	Balled & burlapped; Grade B and better
ACI	American Concrete Institute	B&B	Bell and bell (ANSI)
ACTR	Actuator (ANSI)	B&F	Bell and flange (ANSI)
ACV	Alarm check valve (ANSI)	B&S	Bell and spigot (ANSI)
AD	Area drain (ANSI)	B. & S.	Bell and spigot
Addit.	Additional	B. & W.	Black and white
ADH	Adhesive (ANSI)	b.c.c.	Body-centered cubic
ADJ	Adjustable (ANSI)	B.E.	Bevel end
ADPTR	Adapter (ANSI)	B.F.	Board feet
ADS	Automatic door seal (ANSI)	B.I.	Black iron
AF	Audio frequency (ANSI)	B.P.M.	Blows per minute
Af	Audio frequency	BAF	Baffle (ANSI)
AFC	Automatic frequency control (ANSI)	BAL	Balance (ANSI)
Agg.	Aggregate	BARO	Barometer (ANSI)
AGGR	Aggregate (ANSI)	BAT	Battery (ANSI)
AHR	Anchor (ANSI)	BB	Bulletin board (ANSI)
AIC	Ampere interrupting capacity	Bbl.	Barrel
AL	Aluminum (ANSI)	BBRG	Ball bearing (ANSI)
Allow.	Allowance	BC	Bolt circle (ANSI)
ALM	Alarm (ANSI)	BCT	Bushing current transformer (ANSI)
alt.	Altitude	BD	Board (ANSI)
Alum.	Aluminum	BDCT	Bus (ANSI)
AM	Amplitude modulation (ANSI)	BETW	Between (ANSI)
AMB	Ambient (ANSI)	BEV	Bevel (ANSI)

Abbreviations (continued)

BF	Bottom face		C.P.M.	Critical path method
BFW	Boiler feed water (ANSI)		C.Pr.	Hundred pair
Bg. Cem.	Bag of cement		C.S.F.	Hundred square feet
BHP	Boiler horse power; Brake horse power		C.T.	Current transformer
BI	Black iron (ANSI)		C.W.	Cool white; Cold water
Bit.; Bitum.	Bituminous		C.W.X.	Cool white deluxe
BITUM	Bituminous (ANSI)		C.Y./Hr.	Cubic yard per hour
Bk.	Backed		C/C	Center to center
Bkrs.	Breakers		CAB	Cabinet (ANSI)
BL	Base line (ANSI)		Cab.	Cabinet
BLDG	Building (ANSI)		Cair.	Air tool laborer
Bldg.	Building		CAL	Calibrate (ANSI)
BLK	Block (ANSI)		Calc.	Calculated
Blk.	Block		CAP	Capacity (ANSI)
BLKG	Blocking (ANSI)		Cap.	Capacity
BLO	Blower (ANSI)		CARP	Carpet (ANSI)
BLR	Boiler (ANSI)		CAT	Catalog (ANSI)
BLST	Ballast (ANSI)		CAV	Cavity (ANSI)
BLW	Below (ANSI)		CB	Catch basin (ANSI)
BLWDN	Blowdown (ANSI)		CBORE	Counterbore (ANSI)
BM	Beam (ANSI); Bench mark (ANSI)		CC	Closing coil (ANSI)
Bm.	Beam		CCPD	Coupling (ASNI)
BO	Blowoff (ANSI)		CCTV	Closed circuit television (ANSI)
Boil.	Boilermaker		CCW	Counter clockwise (ANSI)
BOT	Bottom (ANSI)		cd	Candela
BP	Base plate (ANSI); Boiling point (ANSI)		CD	Grade of plywood face & back
BPD	Bushing potential device (ANSI)		cd/sf	Candela per square foot
BR	Bedroom; Bend radius (ANSI)		CDX	Plywood, grade C&D, exterior blue
BRG	Bearing (ANSI)		Cefi.	Cement finisher
Brg.	Bearing		CEM	Cement (ANSI)
BRK	Brick (ANSI)		Cem.	Cement
Brk.	Brick		CER	Ceramic (ANSI)
BRKR	Breaker (ANSI)		CF	Hundred feet
BRKT	Bracket (ANSI)		CFLG	Counter flashing (ANSI)
Brng.	Bearing		CFM	Cubic feet per minute
BRS	Brass (ANSI)		CHAN	Channel (ANSI)
Brs.	Brass		CHEM	Chemical (ANSI)
BRZ	Bronze (ANSI)		CHHWR	Chilled and heating water return (ANSI)
Brz.	Bronze		CHHWS	Chilled and heating water supply (ANSI)
BS	Both sides (ANSI)		CHK	Check (ANSI)
BSHG	Bushing (ANSI)		CHW	Chilled water
BSMT	Basement (ANSI)		CI	Cast iron (ANSI)
Bsn.	Basin		CIP	Cast iron pipe (ANSI)
BTLF	Butterfly (ANSI)		CIR	Circle (ANSI)
Btr.	Better		CIRC	Circular (ANSI)
BTU	British thermal unit		Circ.	Circuit
BTUH	BTU per hour		CJ	Construction joint (ANSI)
BW	Both ways (ANSI)		CKT	Circuit (ANSI)
BWV	Back water valve (ANSI)		CL	Center line (ANSI)
BX	Interlocked armored cable		Clab.	Common laborer
c	Conductivity		CLF	Current limited use
C	Hundred; Centigrade		CLG	Ceiling (ANSI)
C.B.	Circuit breaker		CLJ	Control joint (ANSI)
C.C.A.	Chromate copper arsenate		CLO	Closet (ANSI)
C.C.F.	Hundred cubic feet		CLOS	Closure (ANSI)
C.F.	Cubic feet		CLP	Clamp (ANSI)
c.g.	Center of gravity		CLP	Cross linked polyethylene
C.I.	Cast iron		CLR	Clear (ANSI)
C.I.P.	Cast in place		cm	Centimeter
C.L.	Carload lot		CMP	Corrugated metal pipe
C.L.F.	Hundred linear feet		CMPST	Composite (ANSI)
C.M.U.	Concrete masonry unit		CND	Conduit (ANSI)

Abbreviations (continued)

CNTFGL	Centrifugal (ANSI)	d	Penny (nail size)
CNTR	Counter (ANSI)	d.f.u.	Drainage fixture units
CO	Cleanout (ANSI)	D.H.	Double hung
CO²	Carbon dioxide	D.L.	Deal load; Diesel
COEF	Coefficient (ANSI)	D.P.S.T.	Double pole, single throw
COL	Column (ANSI)	D.S.	Double strength
Col.	Column	D.S.A.	Double strength A grade
COM	Common (ANSI)	D.S.B.	Double strength B grade
COMB	Combination (ANSI)	DAT	Datum (ANSI)
Comb.	Combination	Db	Decibel
COMM	Communication (ANSI)	DB	Dry bulb (ANSI)
COMPL.	Complete (ANSI)	DBC	Diameter bolt circle (ANSI)
Compr.	Compressor	DBL ACT	Double acting (ANSI)
COMPT	Compartment (ANSI)	Dbl.	Double
CONC	Concrete (ANSI)	DC	Direct current
Conc.	Concrete	DCL	Door closure (ANSI)
COND	Condenser (ANSI)	Demob.	Demobilization
CONN	Connection (ANSI)	DEPT	Department (ANSI)
CONSTR	Construction (ANSI)	DET	Detail (ANSI): Detector (ANSI)
CONT	Continuous/continuation (ANSI)	DF	Drinking fountain (ANSI)
Cont.	Continuous/continuation	DHW	Domestic hot water
CONTR	Contract/contractor (ANSI)	Diag.	Diagonal
COORD	Coordinate (ANSI)	Diam.	Diameter
Corr.	Corrugated	DIFF	Difference (ANSI)
Cos.	Cosine	DIM	Dimension (ANSI)
Cot	Cotangent	DIR CONN	Direct connection (ANSI)
COV	Cover (ANSI)	Dis.; Disch.	Discharge
Cov.	Cover	DISP	Dispenser (ANSI)
CPA	Control point adjustment	DIST	Distance (ANSI)
CPLG	Coupling (ANSI)	DISTR	Distribute/distribution (ANSI)
Cplg.	Coupling	Distrib.	Distribution
CPRS	Compressible (ANSI)	DIV	Division (ANSI)
CPRSR	Compressor chlorinated polyvinyl (ANSI)	Dk.	Deck
CPVC	Chlorinated polyvinyl chloride	DL	Deal load (ANSI)
CRC	Cold rolled channel	DM	Demand meter (ANSI)
CRCMF	Circumference (ANSI)	DMH	Drop manhole (ANSI)
Creos.	Creosote	DMPR	Damper (ANSI)
CRN	Crown (ANSI)	DN	Down (ANSI)
Crpt.	Carpet & linoleum	DO	Ditto (ANSI)
CRS	Cold rolled steel	Do.	Ditto
CRT	Cathode-ray tube	DP	Dew point (ANSI)
CS	Carbon steel; Cast stone (ANSI)	Dp.	Depth
CSB	Concrete splash block (ANSI)	DPDT	Double pole, double throw (ANSI)
Csc	Cosecant	DPST	Double pole, single throw (ANSI)
CSG	Casing (ANSI)	DR	Drain (ANSI); Drive (ANSI)
CSI	Construction Specifications Institute	Dr.	Driver
CT	Current transformer (ANSI)	Drink.	Drinking
CTS	Copper tube size	DS	Downspout (ANSI)
Cu	Cubic	DST	Door stop (ANSI)
Cu. Ft.	Cubic foot	DT	Drain tile (ANSI); Dust tight (ANSI)
CUB	Cubicle (ANSI)	Dty.	Duty
CUR	Current (ANSI)	DVTL	Dovetail (ANSI)
cw	Continuous wave	DW	Dishwasher (ANSI); Distilled water (ANSI)
CW	Clockwise (ANSI); Cold water (ANSI); Cool white (ANSI)	DWG	Drawing (ANSI)
CWP	Circulating water pump (ANSI)	DWL	Dowel (ANSI)
Cwt	100 pounds	DWR	Drawer (ANSI)
CY	Cubic yards (27 cubic feet); Cycle (ANSI)	DWV	Drain waste vent
CYL	Cylinder (ANSI)	DX	Deluxe white; Direct expansion; Duplex (ANSI)
Cyl.	Cylinder	dyn	Dyne
CYLL	Cylinder lock (ANSI)	e	Eccentricity
D	Deep; depth; discharge		

Abbreviations (continued)

E	Equipment only; East (ANSI)		F.E.	Front end
E.D.R.	Equivalent direct radiation		F.G.	Flat grain
E.W.	Each way		F.H.A.	Federal Housing Authority
Ea.	Each		F.M.	Frequency modulation
EB	Encased burial		F.R.	Fire rating
ECC	Eccentric (ANSI)		Fab.	Fabricated
ECH	Exchanger (ANSI)		FAC	Factor (ANSI)
ECON	Economizer (ANSI)		FB	Flat bar (ANSI)
Econ.	Economy		FBGS	Fiberglass
EDP	Electronic data processing		FC	Footcandle (ANSI)
EDR	Equivalent direct radiation (ANSI)		FCU	Fan coil unit (ANSI)
EF	Each face (ANSI)		FD	Floor drain (ANSI)
EFL	Effluent (ANSI)		FDC	Fire department connection (ANSI)
EL	Elevation (ANSI)		FDN	Foundation (ANSI)
ELB	Elbow (ANSI)		FDR	Feeder (ANSI)
ELEC	Electrical (ANSI)		FDW	Feedwater (ANSI)
Elec.	Electrician; Electrical		FEM	Female (ANSI)
ELEV	Elevator (ANSI)		FEP	Fluorinated ethylene propylene (Teflon)
Elev.	Elevator; Elevating		FEXT	Fire extinguisher (ANSI)
EMER	Emergency (ANSI)		FG	Frog (ANSI)
EMF	Electromotive force (ANSI)		FHC	Fire hose cabinet (ANSI)
EMT	Electrical metal tubing (ANSI); Electrical metallic conduit		FHR	Fire hose rack (ANSI)
			FHY	Fire hydrant (ANSI)
ENAM	Enamel (ANSI)		Fig.	Figure
ENCL	Enclosure (ANSI)		Fin.	Finished
ENG	Engine (ANSI)		Fixt.	Fixture
Eng.	Engine		FL	Flashing (ANSI); Footlambert (ANSI)
ENGR	Engineer (ANSI)		Fl. Oz.	Fluid ounces
ENGY	Energy (ANSI)		FLDG	Folding (ANSI)
ENTR	Entrance (ANSI)		FLG	Flange (ANSI)
EPDM	Ethylene propylene diene monomer		FLL	Flow line (ANSI)
Eq.	Equation		Flr.	Floor
EQL SP	Equally spaced (ANSI)		FLRD	Flared (ANSI)
Equip.	Equipment		FLSW	Flow switch (ANSI)
EQUIV	Equivalent (ANSI)		FLTR	Filter (ANSI)
ERECT	Erection (ANSI)		FLUOR	Fluorescent (ANSI)
ERW	Electric resistance welded		FM	Frequency modulation (ANSI)
EST	Estimate (ANSI)		Fmg.	Framing
Est.	Estimated		Fndtn.	Foundation
esu	Electrostatic units		FO	Fuel oil (ANSI)
EVAP	Evaporate (ANSI)		Fount.	Fountain
EW	Each way (ANSI)		FP	Freezing point (ANSI)
EWC	Electrical water cooler (ANSI)		FPM	Feet per minute
EWT	Entering water temperature		FPRF	Fireproofing (ANSI)
EXC	Excavate (ANSI)		FPT	Female pipe thread
Excav.	Excavation		FR	Frame (ANSI)
EXCH	Exchanger (ANSI)		Fr.	Frame
EXCTR	Exciter (ANSI)		FREQ	Frequency (ANSI)
EXH	Exhaust (ANSI)		FRK	Foil reinforced kraft
EXP JT	Expansion joint (ANSI)		FRP	Fiberglass reinforced plastic
EXP	Expansion (ANSI)		FS	Far side (ANSI); Forged steel; Full size (ANSI)
Exp.	Expansion			
EXST	Existing (ANSI)		FSC	Cast body; Cast switch box
EXT	Exterior (ANSI)		FSTNR	Fastener (ANSI)
Ext.	Exterior		Ft. Lb.	Foot pound
Extru.	Extrusion		Ft.	Foot; Feet
f.	Fiber stress		FTF	Flared tube fitting (ANSI)
F.	Fahrenheit; Female; Fill		FTG	Fitting (ANSI); Footing (ANSI)
f.c.	Compressive stress in concrete extreme compressive stress		Ftg.	Footing
			Ftng.	Fitting
F.C.	Footcandles		FU	Fuse (ANSI)
f.c.c.	Face-centered cubic		FUR	Furnace (ANSI)

Abbreviations (continued)

FURN	Furniture (ANSI)	HH	Hand hole (ANSI)
Furn.	Furniture	HIC	High interrupting capacity
FUT	Future (ANSI)	HM	Hollow metal (ANSI)
FVNR	Full voltage non-reversing	HNDRL	Hand rail (ANSI)
FXM	Female by male	HNDWL	Handwheel (ANSI)
FXTR	Fixture (ANSI)	HO	High output
Fy.	Minimum yield stress of steel	HORIZ	Horizontal (ANSI)
G	Gauss	Horiz.	Horizontal
g	Gram	HP	High pressure (ANSI); Horsepower
G.F.I.	Ground fault interrupter	HPF	High power factor
GA	Gage (ANSI)	HPS	High pressure steam (ANSI)
Ga.	Gauge	HPT	High point (ANSI)
GAL	Gallon (ANSI)	HR	Hour (ANSI)
Gal.	Gallon	Hr.	Hour
Gal./Min.	Gallon per minute	Hrs./Day	Hours per day
GALV	Galvanized (ANSI)	HSC	High short circuit
Galv.	Galvanized	HT	Heat (ANSI)
GALVI	Galvanized iron (ANSI)	Ht.	Height
GALVS	Galvanized steel (ANSI)	HTG	Heating (ANSI)
GDR	Guard Rail (ANSI)	Htg.	Heating
GEN	Generator (ANSI)	HTR	Heater (ANSI)
Gen.	General	Htrs.	Heaters
GENL	General (ANSI)	HVAC	Heating, ventilating & air conditioning
GENL CONTR	General contractor (ANSI)	HVY	Heavy (ANSI)
GL	Glass (ANSI)	Hvy.	Heavy
GLV	Globe valve (ANSI)	HW	Hot water (ANSI)
GLZ	Glazing (ANSI)	HWC	Hot water, circulating (ANSI)
GND	Ground (ANSI)	HWH	Hot water heater (ANSI)
GOVT	Government (ANSI)	HWY	Highway (ANSI)
GPD	Gallons per day	Hyd.; Hydr.	Hydraulic
GPH	Gallons per hour	HYDR	Hydraulic (ANSI)
GPM	Gallons per minute	Hz.	Hertz (cycles)
GPS	Gallons per second (ANSI)	I.C.	Interrupting capacity
GR	Grade; Grade/grading (ANSI)	I.D.	Inside dimension
Gran.	Granular	I.F.	Inside frosted
GRL	Grille (ANSI)	I.M.C.	Intermediate metal conduit
Grnd.	Ground	I.P.	Iron pipe
GRTG	Grating (ANSI)	I.P.S.	Iron pipe size
GSKT	Gasket (ANSI)	I.P.T.	Iron pipe threaded
GT	Grease trap (ANSI)	I.W.	Indirect waste
GTV	Gate valve (ANSI)	ID	Identification; Inside diameter (ANSI)
GUT	Gutter	IE	Indirect waste (ANSI)
GVL	Gravel (ANSI)	IGN	Ignition (ANSI)
GYP	Gypsum (ANSI)	ILLUM	Illumination (ANSI)
H	High (ANSI); High strength bar joist; Henry	IMH	Inlet manhole (ANSI)
		In.	Inch
H.C.	High capacity	Incan.	Incandescent
H.D.	Heavy duty; High density	INCAND	Incandescent (ANSI)
HB	Hose bibb (ANSI)	INCIN	Incinerator (ANSI)
HD	Head (ANSI); Heavy duty (ANSI)	Incl.	Included; Including
HDO	High density overlaid	INDL	Industrial (ANSI)
HDR	Header (ANSI)	INL	Inlet (ANSI)
Hdr.	Header	Inst.	Installation
Hdwe.	Hardware	INSTL	Installation (ANSI)
Help.	Helper average	INSTR	Instrument (ANSI)
HEPA	High efficiency particulate air filter	INSUL	Insulation (ANSI)
HEX	Hexagonal (ANSI)	Insul.	Insulation
HF	High frequency (ANSI)	Int.	Interior
Hg	Mercury	INTERCOM	Intercommunication (ANSI)
HGR	Hanger (ANSI)	INTLK	Interlock (ANSI)
HGT	Height (ANSI)	INTR	Interior
		IP	Iron pipe (ANSI)

Abbreviations (continued)

IPS	International pipe standard (ANSI); Iron pipe size (ANSI)
IPT	Iron pipe thread (ANSI)
ISO	Isometric (ANSI)
J.	Joule
J.I.C.	Joint Industrial Council
JB	Junction box (ANSI)
JC	Janitor's closet (ANSI)
JCT	Junction (ANSI)
JR	Junior (ANSI)
JT	Joint (ANSI)
K	Kelvin (ANSI); Heavy wall copper tubing; Thousand; Thousand pounds
K.L.	Effective length factor
KAH	Thousand amp. hours
KB	Knee brace (ANSI)
KCP	Keene's cement plaster (ANSI)
KD	Kiln dried (ANSI); Knock down (ANSI)
KDAT	Kiln dried after treatment
kG	Kilogauss
kg	Kilogram
kgf	Kilogram form
kHz	Kilohertz
Kip	1000 pounds
KJ	Kiljoule
KLF	Kips per linear foot
km	Kilometer
KO	Knockout (ANSI)
KPL	Kick plate (ANSI)
KSF	Kips per square foot
KSI	Kips per square inch
kV	Kilovolt
kVA	Kilovolt ampere
KVAR	Kilovar (reactance)
kW	Kilowatt
kWh	Kilowatt-hour
KWY	Keyway (ANSI)
L	Labor only; Length; Long; Medium wall copper tubing
L	Left (ANSI)
L. & E.	Labor and equipment
L. & H.	Light and heat
L.B.	Load bearing; L conduit body
L.C.L.	Less than carload lot
L.F.	Linear foot
L.H.	Long span standard strength
L.L.	Live load
L.L.D.	Lamp lumen depreciation
L.O.A.	Length over all
L.P.F.s	Low power factor
L.P.s	Liquefied petroleum; Low pressure
L.S.	Lump sum
L.T.L.	Less than truckload lot
L.V.	Low voltage
Lab.	Labor
LAD	Ladder (ANSI)
LAM	Lamination (ANSI)
lat	Latitude
Lath.	Lather
LATL	Lateral (ANSI)
LAU	Laundry (ANSI)
LAV	Lavatory (ANSI)

Lav.	Labatory
LB	Pound (ANSI)
lb.	Pound
lb./hr.	Pounds per hour
lb./L.F.	Pounds per linear foot
lbf/sq.in.	Pound-force per square inch
LBL	Label (ANSI)
LBR	Lumber (ANSI)
LCMU	Lightweight (ANSI)
Ld.	Load
LDG	Landing (ANSI)
LDR	Leader (ANSI)
LG	Length (ANSI)
Lg.	Long; Length; Large
LH	Left hand (ANSI)
LIN	Linear (ANSI)
LIQ	Liquid (ANSI)
LKR	Locker (ANSI)
LKROT	Locked rotor (ANSI)
LL	Live load (ANSI)
lm	Lumen
lm/sf	Lumen per square foot
lm/W	Lumen per watt
LNTL	Lintel (ANSI)
log	Logarithm
LONG	Longitudinal (ANSI)
LP	Low pressure (ANSI)
LPW	Lumens per watt (ANSI)
LR	Long radius
LS	Loudspeaker (ANSI)
LT	Light (ANSI)
Lt. Ga.	Light gauge
Lt. Wt.	Lightweight
Lt.	Light
LTG	Lighting (ANSI)
LTL	Lateral (ANSI)
LV	Low voltage (ANSI)
LVL	Level (ANSI)
LVR	Lever (ANSI)
LVR	Louver (ANSI)
LWC	Lightweight concrete (ANSI)
M	Thousand; Material; Male; Light wall copper tubing
M.C.F.	Thousand cubic feet
M.C.F.M.	Thousand cubic feet per minute
M.C.M.	Thousand circular mils
M.F.B.M.	Thousand feet board measure
M.L.F.	Thousand linear feet
M.S.F.	Thousand square feet
M.S.Y.	Thousand square yards
M.V.A.	Million volt amperes
mA	Milliampere
MA	Milliampere (ANSI)
MACH	Machine (ANSI)
Mach.	Machine
Mag. Str.	Magnetic starter
MAINT	Maintenance (ANSI)
Maint.	Maintenance
Mat.; Mat'l.	Material
MAX	Maximum (ANSI)
Max.	Maximum
MBF	Thousand board feet

Abbreviations (continued)

MBH	Thousand BTUs per hour		MV	Megavolt
MBR	Member (ANSI)		MVAR	Million volt amperes reactance
MC	Medicine cabinet (ANSI); Metal clad cable; Momentary contact (ANSI)		MVBL	Movable (ANSI)
			MW	Megawatt
MCP	Motor circuit protector		MWP	Maximum working pressure (ANSI)
MD	Medium duty		MXM	Male by male
MDO	Medium density overlaid		MYD	Thousand yards
MECH	Mechanical (ANSI)		N	Natural; North
Med.	Medium		N	North (ANSI)
MEMB	Membrane (ANSI)		N.B.C.	National Building Code
MET	Metal (ANSI)		N.E.M.A.	National Electrical Manufacturers Association
METB	Metal base (ANSI)			
MEZZ	Mezzanine (ANSI)		N.O.C.	Not otherwise classified
MF	Thousand feet		N.P.T.	National pipe thread
MFG	Manufacturing (ANSI)		nA	Nonoampere
Mfg.	Manufacturing		NA	Not applicable (ANSI); Not available
Mfrs.	Manufacturers		NAT	Natural (ANSI)
mg	Milligram		NC	National coarse (thread) (ANSI); Normally closed (ANSI)
MG	Motor generated (ANSI); Motor generator (ANSI)			
			NEG	Negative (ANSI)
MGD	Million gallons per day		NEHB	Bolted circuit breaker to 600V
MGPH	Thousand gallons per hour		NEUT	Neutral (ANSI)
MH	Manhole; Metal halide;		NF	National fine (thread) (ANSI); Near face (ANSI)
MHT	Mean high tide (ANSI)			
MHz	Megahertz		NFSD	Non-fused (ANSI)
MI	Malleable iron (ANSI); Mineral insulated; Mile (ANSI)		NIC	Not in contract (ANSI)
			NIP	Nipple (ANSI)
Mi.	Mile		NKL	Nickel (ANSI)
Mill.	Millwright		NLB	Non-load bearing
MIN	Minimum (ANSI)		nm	Nanometer
Min.	Minimum		NM	Non-metallic cable
MIR	Mirror (ANSI)		NO	Normally open; Number (ANSI)
MISC	Miscellaneous (ANSI)		No.	Number
Misc.	Miscellaneous		NOM	Nominal (ANSI)
MK	Mark (ANSI)		Nose.	Nosing
MKR	Marker (ANSI)		NOZ	Nozzle (ANSI)
ml	Milliliter		NPL	Nameplate (ANSI)
MLDG	Molding (ANSI)		NPT	National taper pipe (thread) (ANSI)
MLT	Mean low tide (ANSI)		NQQB	Bolted circuit breaker to 240V
mm	Millimeter		NRC	Noise reduction coefficient
MO	Masonry opening (ANSI)		NRCP	Non-reinforced concrete pipe (ANSI)
Mo.	Month		NRS	Non-rising stem
Mobil.	Mobilization		ns	Nanosecond
MOD	Module (ANSI)		NS	Near side (ANSI)
Mog.	Mogul base		NTS	Not to scale (ANSI)
I.	Moment of inertia		nW	Nanowatt
MON	Monitor (ANSI); Monument (ANSI)		O & P	Overhead and profit
MOT	Motor (ANSI)		O.S.&Y	Outside screw and yoke
MP	Medium pressure (ANSI)		OA	Overall (ANSI)
MPH	Miles per hour		OB	Opposing blade
MPT	Male pipe thread		OC	On center
MRT	Mile round trip		OCB	Oil circuit burner (ANSI)
ms	Millisecond		OCR	Oil circuit recloser (ANSI)
Mstz.	Mosaic & terrazzo worker		OD	Outside diameter; Outside dimension
MTD	Mounted (ANSI)		ODS	Overhead distribution system
Mtd.	Mounted		OF	Outside face (ANSI)
MTG	Mounting (ANSI)		OFCI	Owner (ANSI)
Mthe.	Mosaic & terrazzo helper		Oper.	Operator
MTL	Material (ANSI)		OPNG	Opening (ANSI)
Mtng.	Mounting		Opng.	Opening
MULT	Multiple (ANSI)		OPP	Opposite (ANSI)
Mult.	Multi; multiply			

Abbreviations (continued)

ORIG	Original (ANSI)	PR	Pair (ANSI)
Orna.	Ornamental	Pr.	Pair
OVFL	Overflow (ANSI)	PRC	Point of reverse curve (ANSI)
OVHD	Overhead (ANSI)	PRCST	Precast (ANSI)
Ovhd.	Overhead	PREFAB	Prefabricated (ANSI)
OWG	Oil, water or gas	Prefab.	Prefabricated
OXY	Oxygen (ANSI)	Prefin.	Prefinished
OZ	Ounce (ANSI)	PRELIM	Preliminary (ANSI)
Oz.	Ounce	PREP	Preparation (ANSI)
P & T	Pressure and temperature	PRI	Primary (ANSI)
P	Pole (ANSI)	PROJ	Project (ANSI)
p.	Page	PROP	Property (ANSI)
P.	Pole; Applied load; Projection	Prop.	Propelled
P.A.P.R.	Powered air purifying respirator	PRV	Pressure reducing valve (ANSI)
P.C.	Portland cement; Power connector	PS	Point of switch (ANSI)
P.C.F.	Pounds per cubic foot	PSF; psf	Pounds per square foot (ANSI)
P.E.	Plain end; Porcelain enamel; Polyethylene;	PSI; psi	Pounds per square inch (ANSI)
	Professional engineer	PSIG	Pounds per square inch gauge
P.I.	Pressure injected	PSP	Plastic sewer pipe
P.T.	Potential transformer	PST	Point of spiral tangent (ANSI)
PA	Public address (ANSI)	PT	Pipe tap (ANSI); Point of tangency
Pape.	Paperhanger		(ANSI); Potential transformer
PAR	Weatherproof reflector	PTD	Painted (ANSI)
PAT	Patent (ANSI)	Ptd.	Painted
PB	Pull box (ANSI); Pushbutton (ANSI)	PTN	Partition (ANSI)
PC	Piece (ANSI); Plug cock (ANSI); Point of	Ptns.	Partitions
	curve (ANSI)	Pu	Ultimate load
Pc.	Piece	PVC	Polyvinyl chloride (ANSI)
PCC	Point of compound curve (ANSI)	Pvmt.	Pavement
PCF	Pounds per cubic foot (ANSI)	PWR	Power (ANSI)
PCM	Phase contrast microscopy	Pwr.	Power
PED	Pedestal (ANSI)	Q	Quantity heat flow
PERF	Perforated (ANSI)	Q.C.	Quick coupling
Perf.	Perforated	QTR	Quarter (ANSI)
PERM	Permanent (ANSI)	QUAL	Quality (ANSI)
PERP	Perpendicular (ANSI)	Quan; Qty	Quantity
PF	Point of frog (ANSI); Power factor (ANSI)	r	Radius of gyration
PG	Pressure gage (ANSI)	R	Resistance; Riser (ANSI)
PH	Phase (ANSI)	R.H.W.	Rubber, heat & water resistant; residential
Ph.	Phase		hot water
PHOTO	Photograph (ANSI)	R.S.	Rapid start
PI	Point of intersection (ANSI)	RAD	Radius (ANSI)
Pile.	Pile driver	RADN	Radiation (ANSI)
PIV	Post indicator valve (ANSI)	RBR	Rubber (ANSI)
PKG	Package (ANSI)	RC	Reinforced concrete (ANSI); Remote
Pkg.	Package		control (ANSI)
Pkwy	Parkway (ANSI)	RCP	Reinforced concrete pipe (ANSI)
PL	Plate (ANSI); Property line (ANSI)	RCVR	Receiver (ANSI)
Pl.	Plate	RD	Road (ANSI); Roof drain (ANSI)
PLAS	Plaster (ANSI)	REC	Recessed (ANSI)
Plum.	Plumber	RECD	Received (ANSI)
Ply.	Plywood	RECIRC	Recirculate (ANSI)
PLYWD	Plywood (ANSI)	RECPT	Receptacle (ANSI)
PNEU	Pneumatic (ANSI)	RECT	Rectangle (ANSI)
PNL	Panel (ANSI)	Rect.	Rectangle
PNT	Paint (ANSI)	REF	Reference (ANSI)
POL	Polished (ANSI)	REFL	Reflector (ANSI)
PORT	Portable (ANSI)	Reg.	Regular
POS	Positive (ANSI)	REINF	Reinforced/Reinforcing/Reinforcements
pp.	Pages		(ANSI)
PPl; PPL	Polypropylene	Reinf.	Reinforced
PPM	Parts per million	REM	Removable (ANSI)

Abbreviations (continued)

REPRO	Reproduce (ANSI)		SECT	Section (ANSI)
Req'd.	Required		SEG	Segment (ANSI)
REQD	Required (ANSI)		SF Shlf.	Square foot of shelf
Resi	Residential		SFR	Square foot of radiation
RESIL	Resilient (ANSI)		SGL	Single (ANSI)
RF	Radio frequency (ANSI); Raised face (ANSI)		SHLDR	Shoulder (ANSI)
			SHTHG	Sheathing (ANSI)
RFG	Roofing (ANSI)		SIG	Signal (ANSI)
RFGT	Refrigerant (ANSI)		SIM	Similar (ANSI)
Rgh.	Rough		Sin.	Sine
RGLTR	Regulator (ANSI)		SJ	Slip joint (ANSI)
RGS	Rigid galvanized steel		SK	Sink (ANSI)
RH	Relative humidity (ANSI); Right hand (ANSI)		SL	Sliding/sliding (ANSI); Slimline
			Sldr.	Solder
RLF	Relief (ANSI)		SLP	Slope (ANSI)
RLG	Railing (ANSI)		SLV	Sleeve (ANSI)
RM	Room (ANSI)		SMLS	Seamless (ANSI)
rms	Root mean square		SOLV	Solenoid valve (ANSI)
RND	Round (ANSI)		SP	Single pole (ANSI); Soil pipe (ANSI); Space/spacing (ANSI); Specific (ANSI); Standpipe (ANSI); Static pressure (ANSI)
Rnd.	Round			
RO	Rough opening (ANSI)			
ROW	Right of way		SP GR	Specific gravity (ANSI)
RPM	Revolutions per minute		SP HT	Specific heat (ANSI)
RPM	Revolutions per minute (ANSI)		SP VOL	Specific volume (ANSI)
RPS	Revolutions per second (ANSI)		SPCL	Special (ANSI)
RR	Direct burial feeder conduit; Railroad (ANSI)		SPDT	Single pole, double throw
			SPEC	Specification (ANSI)
RT	Round trip		SPKR	Speaker (ANSI)
RV	Relief valve (ANSI)		SPLY	Supply (ANSI)
RVS	Reverse (ANSI)		SPRT	Support (ANSI)
RVT	Riveted (ANSI)		SPT	Standard pipe thread
RWC	Rainwater conductor (ANSI)		SQ FT	Square foot (ANSI)
RWY	Runway (ANSI)		SQ IN	Square inch (ANSI)
S	South (ANSI)		SQ KM	Square kilometer (ANSI)
S.C.R.	Modular brick		SQ	Square (ANSI)
S.D.	Sound deadening		Sq. Hd.	Square head
S.D.R.	Standard dimension ratio		Sq. In.	Square inch
S.E.	Surfaced edge		Sq.	Square; 100 square feet
S.E.R.; S.E.U.	Service entrance cable		SSD	Sub-soil drain (ANSI)
S.F.	Square foot		SST	Stainless steel (ANSI)
S.F.C.A.	Square foot contact area		ST PR	Static pressure (ANSI)
S.F.G.	Square foot of ground		ST	Street (ANSI)
S.F. Hor.	Square foot horizontal		St., Stl.	Steel
S.L.	Saran lined		STA	Station (ANSI)
S.N.	Solid neutral		STD	Standard (ANSI)
S.P.	Self-propelled		Std.	Standard
S.S.	Single strength; Stainless steel		STIF	Stiffener (ANSI)
S.S.B.	Single strength B grade		STIR	Stirrup (ANSI)
S.T.C.	Sound transmission coefficient		STL	Steel (ANSI)
S.Y.	Square yard		STOR	Storage (ANSI)
S4S	Surface 4 sides		STP	Standard temperature & pressure
SALV	Salvage (ANSI)		Stpi.	Steamfitter/Pipefitter
SAN	Sanitary (ANSI)		STR	Straight (ANSI)
SAT	Saturation (ANSI)		Str.	Strength; Starter; Straight
SB	Soot blower (ANSI); Splash block (ANSI)		Strd.	Stranded
Scaf.	Scaffold		Struct.	Structural
Sch.; Sched.	Schedule		STWP	Steam working pressure (ANSI)
SCHED	Schedule (ANSI)		Sty.	Story
SCRN	Screen (ANSI)		Subj.	Subject
SCT	Structural clay tile (ANSI)		Subs.	Subcontractors
SDL	Saddle (ANSI)		SUCT	Suction (ANSI)
SEC	Second (ANSI)		S.	Suction; Single entrance

Abbreviations (continued)

SUPPL	Supplement (ANSI)		TS	Tensile strength (ANSI)
SURF	Surface (ANSI)		TV	Television (ANSI)
Surf.	Surface		TYP	Typical (ANSI)
SURV	Survey (ANSI)		U.H.F.	Ultra high frequency
SUSP	Suspended (ANSI)		U.L.	Underwriters Laboratory
SV	Safety valve (ANSI)		UCI	Uniform Construction Index
SVCE	Service (ANSI)		UF	Underground feeder
Sw.	Switch		UGND	Underground (ANSI)
SWBD	Switchboard (ANSI)		UH	Unit heater (ANSI)
Swbd.	Switchboard		UHF	Ultra high frequency (ANSI)
SWD	Sidewater depth (ANSI)		ULT	Ultimate (ANSI)
SWG	Swage (ANSI)		UN	Union (ANSI)
SWGR	Switchgear (ANSI)		UNFIN	Unfinished (ANSI)
SWP	Safe working pressure (ANSI)		Unfin.	Unfinished
SYM	Symbol (ANSI)		UNIF	Uniform (ANSI)
SYMM	Symmetrical (ANSI)		UNIV	Universal (ANSI)
Syn.	Synthetic		UR	Urinal (ANSI)
SYNTH	Synthetic (ANSI)		URD	Underground residential distribution
SYS	System (ANSI)		UTIL	Utility (ANSI)
Sys.	System		UV	Ultraviolet (ANSI)
T&B	Top and bottom (ANSI)		V	Valve (ANSI); Volt
T&G	Tongue and groove (ANSI)		V.A.	Volt amperes
T. & G.	Tongue and groove; Tar & gravel		V.A.C.	Vinyl composition tile
t.	Thickness		V.G.	Vertical grain
T.E.M.	Transmission electron microscopy		V.H.F.	Very high frequency
T.L.	Truckload		V.L.F.	Vertical linear foot
T.S.	Trigger start		VAC	Vacuum (ANSI)
T.W.	Thermoplastic water resistant wire		VAP PRF	Vapor proof (ANSI)
T	Temperature; Ton		VAV	Variable air volume
TACH	Tachometer (ANSI)		VB	Valve box (ANSI)
TAN	Tangent (ANSI)		VCT	Vitrified clay tile (ANSI)
TBE	Threaded both ends (ANSI)		VEH	Vehicle (ANSI)
TC	Thermocouple (ANSI)		VEL	Velocity (ANSI)
TCAC	Temperature (ANSI)		VENT	Ventilator (ANSI)
TD	Temperature differential (ANSI)		Vent.	Ventilating
TDH	Total dynamic head (ANSI)		VERT	Vertical (ANSI)
TE	Totally enclosed (ANSI)		Vert.	Vertical
TECH	Technical (ANSI)		VEST	Vestibule (ANSI)
TEL	Telephone (ANSI)		VHF	Very high frequency (ANSI)
TEMP	Temperature (ANSI); Temporary (ANSI)		VHO	Very high output
TER	Terrazzo (ANSI)		VIB	Vibration (ANSI)
TERM	Terminal (ANSI)		Vib.	Vibrating
TFE	Tetrafluoroethylene (Teflon)		VISC	Viscosity (ANSI)
Th; Thk	Thick		VIT	Vitreous (ANSI)
THD	Thread(ed) (ANSI)		VOL	Volume (ANSI)
THERM	Thermal (ANSI)		Vol.	Volume
THHN	Nylon jacketed wire		W	Waste (ANSI); Watt (ANSI); West (ANSI); Wide; Wire
Thn.	Thin			
Thrded	Threaded		W. Mile	Wire mile
THRU	Through (ANSI)		W.R.	Water resistant
THW; THWN	Insulated strand wire		W.S.P.	Water, steam, petroleum
TMPD	Tempered (ANSI)		w/	With
TNL	Tunnel (ANSI)		W/	With (ANSI)
TOE	Threaded one end (ANSI)		W/O	Without (ANSI)
TOL	Tolerance (ANSI)		WB	Wet bulb (ANSI)
TOT	Total (ANSI)		WC	Water column; water closet
Tot.	Total		WD	Width; Wood
Tr.	Trade		WF	Wash fountain (ANSI); Wide flange
TRANS	Transparent (ANSI)		WFR	Wafer (ANSI)
Transf.	Transformer		WG	Water gauge
Trlr.	Trailer		WGL	Wire glass (ANSI)
TRNKBKL	Turnbuckle (ANSI)		WH	Wall hydrant; Watt-hour (ANSI)

Abbreviations (continued)

WHSE	Warehouse (ANSI)
WI	Wrought iron; Water line; Wind load (ANSI)
WLD	Welded (ANSI)
Wldg.	Welding
WP	Working point (ANSI)
WPR	Working pressure (ANSI)
WS	Weather stripping; Wetted surface (ANSI)
WT, wt	Weight
WTR	Water (ANSI)
WWF	Welded wire fabric
X ARM	Cross arm (ANSI)
X SECT	Cross section (ANSI)
XFMR	Transformer
XFR	Transfer (ANSI)
XHD	Extra heavy duty
XHHW; XLPE	Cross-linked polyethylene wire insulation
Y	Wye
yd	Yard
yr	Year
ZFMR	Transformer (ANSI)

Index

Index

Notes

Notes

Notes

Notes

Notes

Notes

Notes